Living and Fossil

BRYOZOA

PROCEEDINGS OF THE INTERNATIONAL BRYOZOOLOGY ASSOCIATION CONFERENCE

Living and Fossil

BRYOZOA

RECENT ADVANCES IN RESEARCH

G. P. Larwood, Editor

Department of Geological Sciences
University of Durham
Durham, England

1973

ACADEMIC PRESS • London and New York

A subsidiary of Harcourt Brace Jovanovich, publishers

ACADEMIC PRESS INC. (LONDON) LTD.
24/28 Oval Road,
London NW1

United States Edition published by
ACADEMIC PRESS INC.
111 Fifth Avenue
New York, New York 10003

Library of Congress Catalog Card Number: 72–7709

ISBN: 0 12 436950 2

PRINTED IN GREAT BRITAIN BY ROBERT MACLEHOSE AND CO. LTD
THE UNIVERSITY PRESS, GLASGOW

List of Contributors

ABBOTT, M., *Marine Biological Laboratory, Woods Hole, Massachusetts, U.S.A.*
ANDERSON, F., *Department of Geology, University of Illinois, Urbana, Illinois, U.S.A.*
ANDROSOVA, H. I., *Zoological Institute, Academy of Science, Leningrad, U.S.S.R.*
ANNOSCIA, E., *Via Mincio, Milan, Italy*
ANSTEY, R. L., *Department of Geology, National Science Building, Michigan State University, East Lansing, Michigan, U.S.A.*
ASTROVA, G. G., *Paleontological Institute, Academy of Science, Moscow, U.S.S.R.*
BANTA, W., *Department of Biology, The American University, Washington, D.C., U.S.A.*
BIGEY, F. P., *Université de Paris, Laboratoire de Paléontologie des Invertébrés, Quai Saint Bernard, Paris, France*
BLAKE, D. B., *Department of Geology, University of Illinois, Urbana, Illinois, U.S.A.*
BOARDMAN, R. S., *Division of Invertebrate Paleontology, Smithsonian Institution, U.S. National Museum, Washington, D.C., U.S.A.*
BROOD, K., *Geologiska Institutet, Stockholms Universitet, Kungstensgatan, Stockholm, Sweden*
BULMAN, O. M. B., *Sedgwick Museum, Cambridge, England*
BUSHNELL, J. H., *Department of Biology, University of Colorado, Boulder, Colorado, U.S.A.*
CHEETHAM, A., *Division of Invertebrate Paleontology, Smithsonian Institution, U.S. National Museum, Washington, D.C., U.S.A.*
COOK, P. L., *Zoology Department, British Museum (Natural History), Cromwell Road, London, England*
CUFFEY, R., *Department of Geology and Geophysics, Pennsylvania State University, State College, Pennsylvania, U.S.A.*
CUTLER, J. F., *Department of Geology, University of Kentucky, Lexington, Kentucky, U.S.A.*
DUNAEVA, N. N., *Geological Institute Ukranian Academy of Science, Kiev, U.S.S.R.*
EMSCHERMANN, P., *Anatomisches Institut der Albert-Ludwigs University, Albertstrasse, Germany*
FIERRO, G., *Istituto di Geologia dell'Università, Via Balbi, Genova, Italy*
FLOR, F. D., *Hamburg, Germany*
FORESTER, R. M., *Department of Geology, University of Illinois, Urbana, Illinois, U.S.A.*
GAUTIER, R. G., *Department of Paleobiology, Smithsonian Institution, Washington, U.S.A.*
GHIURCA, V., *Universitates "Babes Bolya", Catedra de Geologie Str. Kogilniceanu, Roumania*

GORDON, D. P., *Department of Biology, Dalhousie University, Halifax, Nova Scotia, Canada*
HÅKANSSON, E., *Institut for Historisk Geologi og Paleontologi, København K, Denmark*
HARMELIN, J. G., *Station Marine d'Endoume et Centre d'Oceanographie, Marseille, France*
HAYWARD, P. J., *Zoology Department, University College, Singleton Park, Swansea, Wales*
HINDS, R. W., *Department of Geology, Slippery Rock State College, Slippery Rock, Pennsylvania, U.S.A.*
HOROWITZ, A., *Department of Geology, Indiana University, Bloomington, Indiana, U.S.A.*
HUMPHRIES, E. M., *Department of Zoology, University of Florida, Gainesville, Florida, U.S.A.*
ILLIES, G., *Technische Hochschule Karlsruhe, Institut für Geologie, Kaiserstrasse, Karlsruhe, Germany*
JAROSHINSKAYA, A. M., *Ob-Irtish Expedition, Tomsk, By-street Nachanovicha, U.S.S.R.*
JEBRAM, D., *Zoologisches Institut der Universität Kiel, Germany*
KING, P. E., *Zoology Department, University College, Singleton Park, Swansea, Wales*
LABRACHERIE, M., *Laboratoire de Géologie, Université de Bordeaux, Talence, Gironde, France*
LAGAAIJ, R., *Doormanlaan, Wassenaar, The Netherlands*
LARWOOD, G. P., *Department of Geological Sciences, University of Durham, England*
LUTAUD, G., *Laboratoire d'Anatomie comparées, Faculté des Sciences, Quai Saint Bernard, Paris, France*
MCKINNEY, F. K., *Department of Geology and Geography, Appalachian State University, Boone, North Carolina, U.S.A.*
MATURO, F., *Department of Zoology, University of Florida, Gainesville, Florida, U.S.A.*
MENON, N. R., *College of Fisheries, University of Agricultural Sciences, Mangalore, S. Kanara, India*
MOORE, P. G., *Wellcome Marine Laboratory, University of Leeds, Robin Hood's Bay, North Yorkshire, England*
MOROZOVA, I. P., *Paleontological Institute, Academy of Science, Moscow, U.S.S.R.*
NAKAMURA, R. M., *Department of Tuberculosis, National Institute of Health, Tokyo, Japan*
ODA, S., *Biological Laboratory, St. Paul's University, Ikebukure, Tokyo, Japan*
PERRY, R. G., *Department of Geology, Indiana University, Bloomington, Indiana, U.S.A.*
PINTER, P., *San Francisco State College, Department of Marine Biology, San Francisco, California, U.S.A.*
POHOWSKY, R. A., *Vlaskamp, The Hague, Netherlands*
RAO, K. S., *Marine Organisms Scheme, Department of Zoology, Andhra University, Waltair, India*

Ross, J. P., *Department of Biology, Western Washington, State College, Bellingham, Washington, U.S.A.*
Ryland, J. S., *Zoology Department, University College, Singleton Park, Swansea, Wales*
Sandberg, P. A., *Department of Geology, University of Illinois, Urbana, Illinois, U.S.A.*
Schopf, R. J. M., *Department of Geophysical Sciences, University of Chicago, Chicago, Illinois 60637, U.S.A.*
Smith, L. W., *Department of Zoology, University College of Swansea, Swansea, Wales*
Soule, D. F., *Allan Hancock Foundation, The University of Southern California, University Park, Los Angeles, California, U.S.A.*
Soule, J. D., *Allan Hancock Foundation, The University of Southern California, University Park, Los Angeles, California, U.S.A.*
Stebbing, R. D., *Institute for Marine Environmental Research, Plymouth, England*
Tavener-Smith, R., *Department of Geology, The Queen's University, Belfast, Northern Ireland*
Viskova, L. A., *Paleontological Institute, Academy of Science, Moscow, U.S.S.R.*
Voigt, E., *Geologisch-Paläontologisches Institut, Hamburg, Germany*
Wass, R. E., *Department of Geology and Geophysics, University of Sydney, New South Wales, Australia*
Wiebach, F., *Plön, Schlossgebeit, Holstein, Germany*
Wyer, D., *Department of Zoology, University College, Singleton Park, Swansea, Wales*
Zimmer, E. L., *Santa Carolina Marine Biological Laboratory, Allan Hancock Foundation, University of Southern California, Avalon, California, U.S.A.*

Preface

The second conference of the International Bryozoology Association was held in the University of Durham, England, 6–16 September 1971. Professor Nils Spjeldnaes (University of Aarhus) presided, the Conference Chairman was Dr G. P. Larwood (University of Durham) and Miss P. L. Cook [British Museum (Natural History) London] was Conference Secretary. The Association records its sincere thanks to the University of Durham and particularly to the Department of Geological Sciences for the facilities provided.

At the Conference a wide range of topics relating to both living and fossil Bryozoa was considered in an extensive series of papers and seminar discussions. A number of field visits were made to freshwater lake localities and to parts of the northeast coastline where especially low tides allowed investigation of low-water communities. Palaeontological excursions were held to various Carboniferous and Permian rock sequences characteristic of the area.

The scope and variety of the sixty-five papers presented at the Conference, and those comprising the present volume, reflect the almost explosive burst of research activity in bryozoology. This stems from a community of interest of both neontologists and palaeontologists. The dialogue which has developed recently between zoologists and those primarily concerned with the prolific fossil remains of Bryozoa has proved most fruitful.

Fifty-eight papers are published here reflecting the wide variety of interest in the Bryozoa. These interests centre on morphology and structure, on ecology and palaeoecology, and on evolution and phylogeny.

Morphological studies are currently concerned with taxonomic problems, but more particularly with the functional interpretation of morphology and with budding patterns of colony development. In this respect the Conference was fortunate to hear Professor O. M. B. Bulman (University of Cambridge) who was invited to talk on graptolite morphology and colonial budding patterns. This contribution appears here as Chapter 5 and is a valuable résumé of graptolite characteristics of interest to bryozoologists in that the understanding of colonial development raises parallel problems in the two groups.

Our understanding of skeletal structure in the Bryozoa—as with many groups—recently has been greatly enhanced by scanning electron-microscopy studies. Many members of the Conference referred to such studies and some provided illustrated demonstrations on this topic. Professor Alwyn Williams (University of Belfast) was invited to talk on "Ultrastructure of the Growing Tip and the Process of Calcification in some Bryozoa". In talking on this topic, and in a subsequent stimulating seminar, Professor Williams covered much of the material discussed in his important paper, written jointly with Dr Tavener-

Smith, on secretion and structure in the skeleton of living and fossil Bryozoa. Much of this is concerned with ultra-structural scanning microscopy studies and had been accepted prior to the I.B.A. Conference for publication and subsequently appeared in the *Phil. Trans. Roy. Soc. Lond.* **264,** 97–159 (1972).

Two other invited papers which stimulated much interest and discussion at the Conference were those by Dr G. G. Astrova on "Polymorphism in the Trepostomata" and by Dr P. G. Moore on "Bryozoa as a Community Component on the Northeast Coast of Britain". Dr Moore's paper was a valuable prelude to an equally rewarding coastal excursion.

A principal field of interest at the Conference was ecology and palaeoecology. The President of the I.B.A., Professor Nils Spjeldnaes, gave the opening lecture of the Conference. His paper on the "Palaeoecology of the Bryozoan Limestone in the Danian of Denmark" was presented jointly with Dr E. Thomsen. Numerous papers presented the results of recent and current surveys and research on seasonal variation in growth and settlement of bryozoan colonies and studies of their distribution and relation to other elements of the ecosystem. Chapters 3 to 6, 9 to 18, 33, 49 and 51 are all concerned to a greater or lesser extent with ecological and palaeoecological considerations. Professor Voigt's elegant studies of beautifully preserved material from late Cretaceous hard-grounds and Professor Bushnell's comprehensive global survey of freshwater ectoproct zoogeography are particularly notable.

Many workers are most active in the field of morphological and structural studies in the Bryozoa. It is here that the effectiveness of the interplay of zoological techniques and studies with those of palaeontology is most evident. Papers considering structural or ultrastructural features of both skeletal or soft tissues are Chapters 19, 20, 22 to 26, 28, 30 to 32, 34 and 46. Among these Dr Tavener-Smith's Chapter (34) on "Skeletal Organization in Bryozoa", Dr Boardman's Chapter (21) comparing structures and relationships in Recent Cyclostomata and Palaeozoic Trepostomes, Dr Emschermann's Chapter (24) on cuticular pores and spines and Dr Geneviève Lutaud's elegant account (30) of the "Great Tentacle Sheath Nerve in *Electra pilosa*" have wide significance in bryozoology. Forester *et al.* (Chapter 8) make an important contribution on isotope studies of bryozoan skeletal tissues.

Chapters 27 and 29 deal particularly with growth and budding patterns in restricted groups of Bryozoa. Many other papers touch on this topic also. Morphological studies of particular or general features are recorded in Chapters 35, 36, 39, 44, 47 and 50. Among these Dr Pohowsky's Chapter (44) on the earliest known Cheilostome from the Jurassic and Dr Cheetham's Chapter (38) on "Cheilostome Polymorphism" are particularly notable. Dr Cheetham's use of principal components analysis in elucidating systematics and evolution in *Poricellaria* and *Vincularia* is of especial significance and the method will undoubtedly be used more widely in bryozoan studies.

A numbers of authors present more conventional systematic studies in Chapters 40, 42, 43, 45 and 52. Chapters 7, 36, 37, 41 and 48 are valuable reviews of some fossil and recent bryozoan faunas.

Classification, genetics and phylogeny of the Bryozoa are the subjects of a group of papers toward the end of the volume. Dr Maturo (Chapter 55) presents a rare study of observed offspring variation in *Parasmittina* and Dr Schopf (Chapter 56) reviews population genetics generally for the phylum. Dr Jebram (Chapter 54) gives an intricate paper on growth direction and phylogeny and Dr Zimmer (Chapter 57) considers wider matters of lophophorate affinities. Dr Cuffey in Chapter (53) attempts a revised higher taxa classification for the Bryozoa and reviews jointly with Dr McKinney, in the final Chapter of the volume, information services required by bryozoologists.

Editing a series of papers of this variety by authors from 14 different countries presents many problems. It is evident there are still varied usages of some bryozoological terms even within contributions from the same country. Semantic difficulties may well account for some of these variations in understanding and it is regretted that it has not been possible to include, for reasons of space restrictions, résumés of papers in languages other than English.

For the same reasons of limited space it is not possible to print here abstracts of other papers read at the Conference but not published in this volume. These included, in the order of presentation at the Conference, papers by Dr D. Jebram (Germany) on "Advances in the Cultivation of Bryozoa", by Dr P. A. Sandberg (U.S.A.) on "The Nature and Development of Skeletal Ultrastructure in Cheilostome Bryozoan Skeletons", by Dr O. L. Karklins (U.S.A.) on "Tentative Taxonomic Groupings in the Cryptostome Bifoliate Bryozoans", by Dr F. J. S. Maturo (U.S.A.) on "Larval Behaviour and Metamorphosis in the Cupuladriidae (Bryozoa, Anasca)", by Dr J. Utgaard (U.S.A.) on "An Undescribed Constellariid Genus with strong Crystoporate Affinities", by Dr G. Hillmer (Germany) on "Lower Cretaceous Bryozoa of Germany", by Mr B. J. Taylor (G.B.) on "A Cheilostomatous Bryozoan (*Pyripora* sp.) from the Lower Cretaceous of S.E. Alexander Island, Antarctica", by Dr R. W. Hinds (U.S.A.) on "Body Wall Structure and Development in Post-Palaeozoic Cyclostome Bryozoa", by Dr R. A. Pohowsky (U.S.A.) on "The Palaeontology of Shell Penetrating Ctenostomata—a Prospectus", and a paper by Dr G. Braga and M. Munari (Italy) on "Biometric Studies of Two Populations of species of *Conescharellina* from the Upper Eocene of Venetia, Italy" [this work is to be published in full in Italian in *Studi Trent. Sc. Nat. di Trento* (*Italia*)].

From the papers published here, and from the additional list of papers also read at the Conference, it is evident there is a wide and varied interest in the phylum Bryozoa. There is a concentration of study on ecology, functional morphology and on problems of skeletal micro- and ultrastructure. This latter field of interest raises a new set of characters and reveals undetected relationships which need to be integrated with structural and systematics studies already established. The exploration of new techniques and their application in no way eclipses either the fact or the importance of continued systematic taxonomic researches on all groups of living and fossil Bryozoa. The accumulation of observed data is essential and continues to be a central interest of bryozoologists. It is certain that this volume serves both to consolidate understanding of the phylum and to stimulate further research on the Bryozoa by both zoologists and palaeontologists.

February, 1973 G. P. LARWOOD

Contents

		PAGE
LIST OF CONTRIBUTORS	.. ..	v
PREFACE	.. ..	ix
1	Polymorphism and Its Development in the Trepostomatous Bryozoa G. G. ASTROVA	1
2	Graptolite Periderm Structure and Budding Patterns: a Résumé .. O. M. B. BULMAN	11
3	Bryozoa as a Community Component on the Northeast Coast of Britain .. .. P. G. MOORE	21
4	Seasonal Diversity and Density in Bryozoan Populations of Block Island Sound (New York, U.S.A.) .. .. MARIE B. ABBOTT	37
5	Bryozoan Ecology in Relation to Sediment Texture in the "Golfo dell'Asinara" (Sardinia, Italy) .. .. E. ANNOSCIA and G. FIERRO	53
6	Settlement and Early Colony Development in some Cheilostomata PATRICIA L. COOK	65
7	Carboniferous Bryozoa of the Donetz Basin .. .. N. N. DUNAEVA	73
8	Isotopic Variability of Cheilostome Bryozoan Skeletons .. .. RICHARD M. FORESTER, PHILIP A. SANDBERG and THOMAS F. ANDERSON	79
9	Morphological Variations and Ecology of the Recent Cyclostome Bryozoan "*Idmonea*" *atlantica* from the Mediterranean .. .. J. G. HARMELIN	95

10 Preliminary Observations on Settlement and Growth in Populations of *Alcyonidium hirsutum* (Fleming) 107
P. J. HAYWARD

11 Seasonal Settlement of Bryozoans in Rehoboth Bay, Delaware, U.S.A. 115
E. M. HUMPHRIES

12 Functional Morphology and Habitat of Bryozoa in the Eocene of the Northern Aquitaine Basin, France 129
MONIQUE LABRACHERIE

13 Shallow-water Bryozoa from Deep-sea Sands of the Principe Channel, Gulf of Guinea 139
R. LAGAAIJ

14 Vertical and Horizontal Distribution of Fouling Bryozoans in Cochin Backwaters, Southwest Coast of India 153
N. R. MENON

15 The Analysis of Spatial Distribution Patterns 165
J. S. RYLAND

16 Observations on Colony Overgrowth and Spatial Competition .. 173
A. R. D. STEBBING

17 Environmental Conditions of Bryozoan Ecology of the Hardground Biotope of the Maastrichtian Tuff-Chalk, near Maastricht (Netherlands) 185
EHRHARD VOIGT

18 Relationships Between Some British Littoral and Sublittoral Bryozoans and Pycnogonids 199
D. W. WYER and P. E. KING

19 The Significance of Areolae in Cheilostome Bryozoa 209
WILLIAM C. BANTA

20 Acanthopore Ultrastructure in the Palaeozoic Bryozoan Family Rhabdomesidae 221
D. BRYAN BLAKE

21 Body Walls and Attachment Organs in Some Recent Cyclostomes and Palaeozoic Trepostomes 231
RICHARD S. BOARDMAN

22 Palaeozoic Cyclostomata (A preliminary report) 247
KRISTER BROOD

23 Nature of "Acanthopores" and Related Structures in the Ordovician Bryozoan *Constellaria* 257
JOHN F. CUTLER

24 Cuticular Pores and Spines, Ion Regulatory Organs in Entoprocta .. 261
P. EMSCHERMANN

25 Growth in Bryozoans of the Order Fenestrata 271
T. GARY GAUTIER

26 A Fine-structure Study of Brown Bodies in the Gymnolaemate *Cryptosula pallasiana* (Moll) 275
D. P. GORDON

27 Mode of Growth of the Cupuladriidae (Bryozoa, Cheilostomata) .. 287
ECKART HÅKANSSON

28 Intrazooecial Structures in Some Tubuliporinid Cyclostome Bryozoa 299
ROBERT W. HINDS

29 Different Budding Patterns in the Genus *Stomatopora* (Bryozoa, Cyclostomata) 307
GISELA ILLIES

30 The Great Tentacle Sheath Nerve as the Path of an Innervation of the Frontal Wall Structures in the Cheilostome *Electra pilosa* (Linné) 317
GENEVIÈVE LUTAUD

31 New and Little-known Structures of Fenestelloidea 327
I. P. MOROZOVA

32 Ultrastructure of the Tentacles of *Flustrellidra hispida* (Fabricius) .. 335
L. W. SMITH

33 Histological and Histochemical Studies on the Bryozoan-substrate Interface 343
JOHN D. SOULE

34 Some Aspects of Skeletal Organization in Bryozoa 349
R. TAVENER-SMITH

35 Frontal Wall Calcification in a Recent Cribrimorph Genus from Australia 361
ROBIN E. WASS

36 Bryozoa Cheilostomata (Anasca) of the Antarctic and Subantarctic 369
E. I. ANDROSOVA

37 Devonian Bryozoa from the Southeastern Armorican Massif, Western France 375
FRANÇOISE BIGEY

38 Study of Cheilostome Polymorphism Using Principal Components Analysis 385
ALAN H. CHEETHAM

39 Variation and Habitat in Some Erect Fossil Bryozoa (Preliminary Report) 411
FERDINAND D. FLOR

40 Types of Miocene Bryozoa described in Roumania 417
VIRGIL GHIURCA

41 Ordovician Bryozoa of the Altai-Sayan Region 421
A. M. JAROSHINSKAYA

42 Restudy of Some Nicholson's Types of Palaeozoic Bryozoans .. 429
T. G. PERRY, A. S. HOROWITZ and R. L. ANSTEY

43 The *Hippothoa hyalina* (L.) Complex with a New Species from the Pacific Coast of California 437
PENNY A. PINTER

44 A Jurassic Cheilostome from England 447
ROBERT A. POHOWSKY

45 New Species of *Pyripora* d'Orbigny from the Cretaceous and the Miocene 463
G. P. LARWOOD

46 Calcification in the Colonial Stenolaemate, *Heteropora* 475
JUNE R. P. ROSS

47 Morphogenesis of Giant Avicularia and Ovicells in Some Pacific Smittinidae 485
DOROTHY F. SOULE

48 On the Morphology and Systematics of Some Late Cretaceous Cyclostomatous Bryozoa 497
L. A. VISKOVA

49 The Freshwater Ectoprocta: A Zoogeographical Discussion .. 503
JOHN H. BUSHNELL

50 The Occurrence of Double Polypides in Freshwater Bryozoa .. 523
S. ODA and R. M. NAKAMURA

51 Studies on Freshwater Bryozoa—III. The Bryozoa of the Narmada River System 529
K. S. RAO

52 Preliminary Notes on a Revision of the Genus *Hyalinella* 539
F. WIEBACH

53 An Improved Classification, Based upon Numerical-Taxonomic Analyses, for the Higher Taxa of Entoproct and Ectoproct Bryozoans 549
ROGER J. CUFFEY

54 The Importance of Different Growth Directions in the Phylactolaemata and Gymnolaemata for Reconstructing the Phylogeny of the Bryozoa 565
D. JEBRAM

55 Offspring Variation from Known Maternal Stocks of *Parasmittina nitida* (Verrill) 577
FRANK J. S. MATURO, JR.

56 Population Genetics of Ectoprocts: Status as of January, 1972 .. 585
THOMAS J. M. SCHOPF

57 Morphological and Developmental Affinities of the Lophophorates 593
RUSSEL L. ZIMMER

58 Information Services Desired by Bryozoologists 601
F. KENNETH McKINNEY and ROGER J. CUFFEY

AUTHOR INDEX 607

SUBJECT INDEX 615

1 | Polymorphism and Its Development in the Trepostomatous Bryozoa

G. G. ASTROVA

Paleontological Institute, Academy of Science, Moscow, USSR

Abstract: In Palaeozoic Trepostomatous Bryozoa polymorphism is expressed in differential development of zooecia (cells) in macules and between macules, in the development of mesozooecia (mesopores), exilapores and acanthopores, distinguished by their different morphology, size and distribution in the colony.

In Trepostomata it is supposed that the absence of highly specialized individuals adapted for brooding embryos reflects a primitive type of sexual reproduction. The large zooecia in macules, characteristic for the mature colonies of all members of the order, are considered as fertile individuals.

The isolation of two types of specialized individuals—mesozooecia and exilapores, never found together, distinguishes two directions of development in the order Trepostomata. These features may be considered as distinguishing suborders.

Bryozoa, as a unique type of multicellular attached organism, are characterized by a high degree of adaptation to colonial life. Each bryozoan colony, consisting of many individuals, created by agamic reproduction from a single individual, represents a highly integrated individuum. The individual forms of the colony are genetically identical and joined together by a common metabolism and a coordinating system which provides for the existence of the colony as a whole.

Among many colonial multicellular organisms their chief adaptation to colonial life consists of developing different organs, common for the whole colony, which maintain the basic life processes of the colony. Among the Bryozoa the basic device for adaptation to colonial life is polymorphism, that is—the differential development of the individuals of a colony to accomplish different functions for the maintenance of life in the colony as a whole.

The autozooids which are the chief individuals of recent Bryozoa feed and generate sexual products independently; they bud off different types of heterozooids which have different degrees of rudimentary polypides and accomplish different functions. The gonozooids brood the embryos, avicularia, vibracula and different types of spine-shaped individuals protect the surface of the colony and the kenozooids strengthen the colony. Kenozooids are completely without polypides, have only layers of ectoderm and mesoderm in the cystid and are filled up by live undifferentiated tissue; also they often have some part in the budding process.

Polymorphism which is a discriminating feature of the Bryozoa, leads to a great systematic differentiation of the whole group (Kluge, 1962). Within the class Gymnolaemata, among the recent marine Bryozoa, the orders Ctenostomata, Cyclostomata and Cheilostomata differ greatly by the character and manifestation of their polymorphism and by the types of specialized individuals.

The diverse functional character of the individuals forming each colony is connected in a definite way, not only with differences in the structure of their soft tissues, but also in their hard skeletal structures making possible the interpretation of fossil Bryozoa. Studies of the Trepostomata show that they have many features which are very similar to those of recent Bryozoa, indicating that in the earliest Bryozoa there was already a high degree of coloniality.

The amassing of vast data and the large number of specialized microscopic studies of Trepostomata of different ages, conducted in many countries during nearly a century after the works of Dybowski in Russia (Dybowski, 1877) and of Ulrich and Nicholson in North America (Ulrich, 1879, 1882; Nicholson, 1879, 1881) make it possible to sum up to some extent polymorphism in this ancient group and the direction of its development. Such considerations may be used to achieve systematic differentiation in the order Trepostomata.

In addition to autozooecia, which are prismatic tubes with different shaped openings, the Trepostomata also possess smaller tubular structures—the mesopores or mesozooecia (Ross, 1967) and acanthopores. Much more rarely there

---

Plate I

Fig. 1. *Dekayia maculata* James. Hypotype. PIN, N 2586/9; Upper Ordovician (Cincinnatian); U.S.A. Longitudinal section ×17. Macrozooecia in maculae raised above the surface of a colony.

Fig. 2. *Diplotrypa pseudopetropolitana* Astrova. Holotype. PIN, N 1244/40. Middle Ordovician, U.S.S.R., Vajgatch Island. Longitudinal section ×20. Budding of mesopores in different parts of a colony.

Fig. 3. *Hallopora elegantula* (Hall). Hypotype. PIN, N 1706/189. Lower Silurian, U.S.S.R., Podolia. Longitudinal section ×23. Budding of mesopores in the exozone of a colony.

Fig. 4. *Leptotrypella gurievensis* Astrova. Holotype. PIN, N 2218/459. Middle Devonian, U.S.S.R., Salair. Longitudinal section ×45. Exilapores in the exozone of a colony.

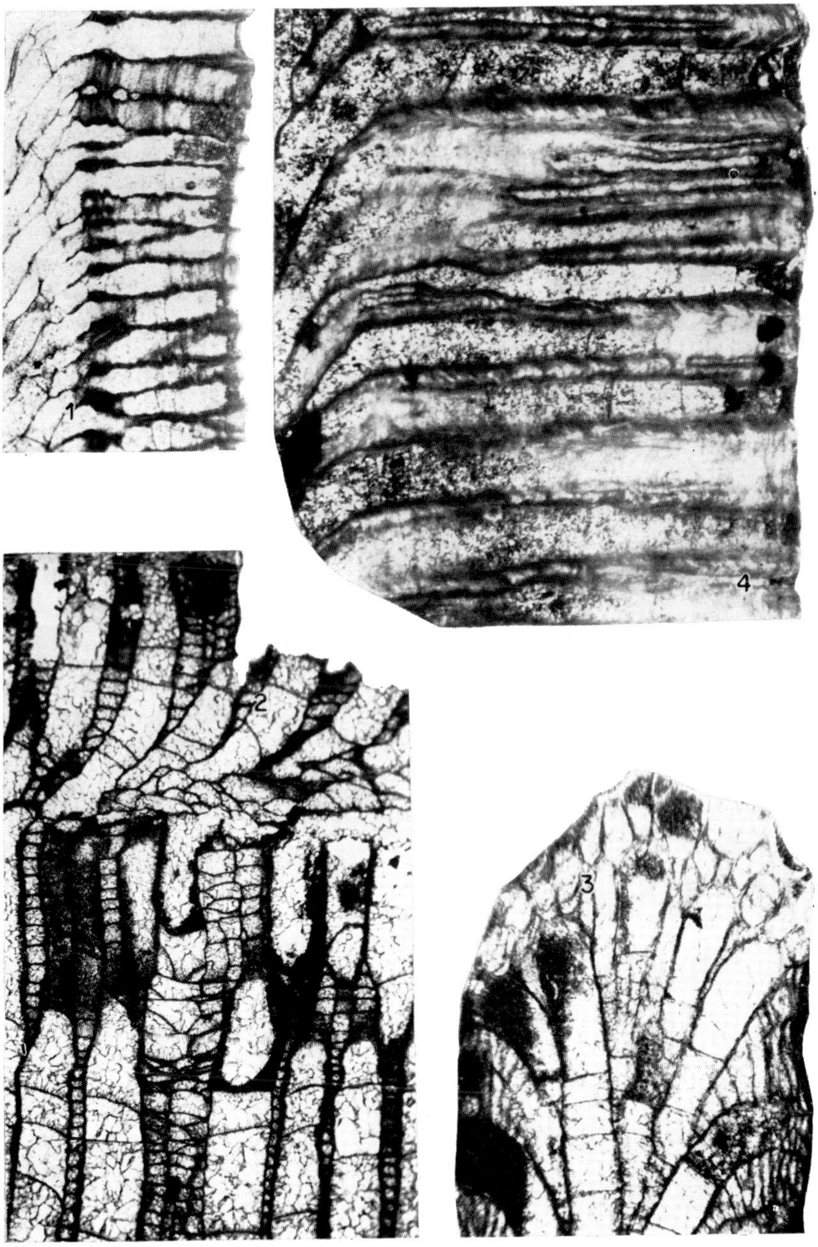
1
4
2
3

may also be small short tubes, which Russian authors call underdeveloped cells or aborted zooecia and which recently have been termed exilapores (Dunaeva and Morozova, 1967).

The analysis of large data on the trepostomatous Bryozoa, ranging from the Early Ordovician to the Triassic, allows us to trace the origin and development of all these heteozooecia. All the representatives of the Trepostomata were also characterized by differential development among the autozooecia.

Already at early stages of the study of Trepostomata different authors described exceptionally large autozooecia forming macules, which are sometimes raised above the surface of the colony as tubercles or monticules. In addition to macrozooecia other types of zooecia often develop in the macules (Plate I, Fig. 1). The inner structure of the macrozooecia within the macules is not in any way different from that of the common smaller autozooecia and the peculiarities of their appearance and distribution in the colonies allows us to establish the fact that they occurred at mature astogenetic stages only.

The regular development of the macrozooecia among all representatives of the order Trepostomata leads us to consider their significance in the life of the colony. The absence of any other skeletal remains that can be looked upon as gonozooecia, which exist among other Bryozoa, suggests a possible connection with the process of reproduction (Borg, 1965). Gonozooecia are diverse among Recent Bryozoa and often develop rather complex devices for the brooding of fecundated eggs. But even among recent marine Bryozoa (in a few groups of species of Cheilostomata and Ctenostomata) a primitive type of sexual reproduction is preserved in which the fecundated eggs are ejected from the fertilized zooid straight into the water (Braiko, 1970). Among some recent Ctenostomata, which reproduce in this way, the cystids of the fertilized zooids are somewhat enlarged (Waters, 1912).

Small eggs (cysts), preserved in the zooecia of some Carboniferous Trepostomata, which were found and described by Dunaeva (1968), confirm it seems that the whole order was characterized by this most primitive type of sexual reproduction, when the development of the eggs went on in an outer medium. Specialization of different individuals with respect to brooding of the embryos reflects higher levels of evolution of the Bryozoa than was achieved by Trepostomata. The isolation of the macrozooecia within the macules is therefore considered a manifestation of polymorphism, a trait of the whole order Trepostomata.

The much smaller skeletal structures of the mesopores and exilapores characterize separate groups of Trepostomata.

The mesopores, rather short tubular structures with the same wall-microstructure as the autozooecia, are characterized by abundant regularly distributed diaphragms. The thickening of the diaphragms in a distal direction and increased carbonate precipitation in the region of the walls leads to the complete overgrowth of the openings of the mesopores in some groups.

The mesopores are budded out of the autozooecia chiefly in the exozone or in the most peripheral parts of the endozone (Plate I, Figs 2, 3). Mesopores develop around the autozooecia in varying numbers and they emerge all over the lateral and upper surfaces of the colony sometimes forming accumulations.

The structure of the walls of the mesopores and of the diaphragms connected with these reveals the character and position of the epithelium which secreted them. The group of families with mesopores developing from the Early Ordovician and reaching a maximum from the middle Ordovician up to the end of the Silurian is certainly a special branch of the order Trepostomata, which has been designated (Astrova, 1965) as the suborder Halloporoidea which is thus characterized by a special type of polymorphism.

Abundant deposits of carbonate tissue, in the cavities of the mesopores, in their walls and often in the region of their terminal ends, which were thus sealed, suggests the mesopores may be the specialized individuals of kenozooid type as with more recent Bryozoa. It is obvious that the cystids of the mesopores deprived of polypides, filled up by mesenchymatous cells and carrying apparently a strengthening function, preserved the capacity for asexual reproduction.

In another group of the order Trepostomata, which originated also in the Early Ordovician, but reached a maximum in the Devonian-Carboniferous and which in a large part existed to the end of the Palaeozoic (suborder Amplexoporoidea Astrova, 1965), there developed another type of specialized individuals —exilapores.

These short tubular structures, very rare among the more ancient representatives of the group but more abundant in its later Palaeozoic branches, are distinguished in addition to their smaller size by the absence of diaphragms or by their only very slight development. (Plate I, Fig. 4; Plate II, Fig. 1). The exilapores, budding from the autozooids, developed as a rule in the peripheral parts of the colonies. Among a few Late Palaeozoic Amplexoporoidea with numerous exilapores (Crustoporidae, Dyscritellidae) there existed fan-shaped budding of whole groups of exilapores out of one exilapore, which had primary-budded out of an autozooecium. The exilapores probably did not have the same strengthening function in the colonies as the mesopores. It is easier to compare them to nannozoids, known among some recent Cyclostomata.

The greater part of the Trepostomata also have acanthopores. These structures

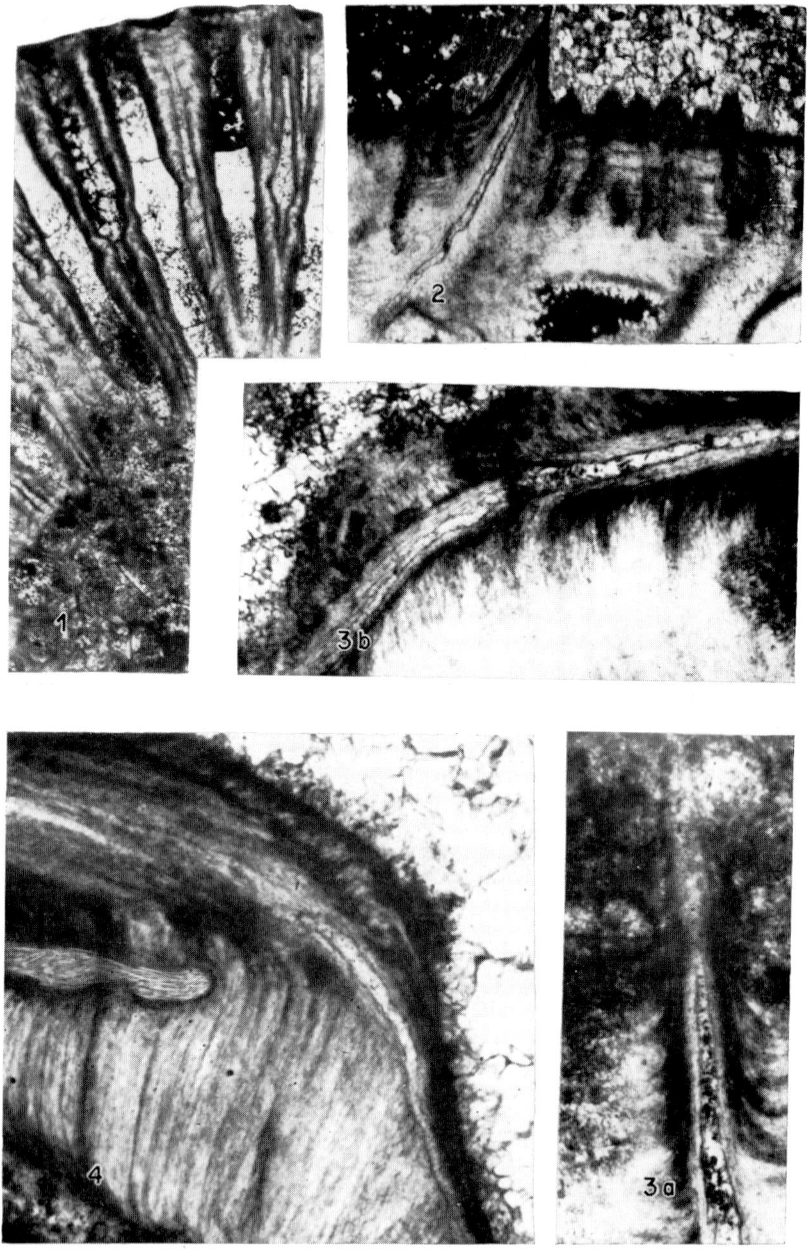
1
2
3b
4
3a

occur on the surface of the colony and are singular spine-shaped structures with highly thickened walls and closed up terminal ends (Plate II, Fig. 2). With the exception of some microacanthopores among the family Stenoporidae, and some small indistinct acanthopores of the family Amplexoporidae, the acanthopores are very diverse in size, form and place in the colonies for they certainly are specialized individuals.

Despite the small size of the acanthopores, the existence inside them of a free narrow cavity, generally filled up by secondary calcite, is confirmed by the frequent presence, as it is in the autozooecia, of mineral grains from the sediment which covered the colony after it perished (Plate II, Fig. 3).

Thin diaphragms sometimes seen inside the acanthopores show very convincingly the former presence here of the coelom, covered inside by living layers of epithelium. In colonies of some Late Palaeozoic genera (*Rhombotrypella, Permopora*) short pores (capillaries) pierced the walls of the autozooecia surrounding especially large acanthopores, and opened into their cavities thus connecting them apparently both with the nearest autozooecia and the surrounding space (Astrova, 1971).

The interpretation of acanthopores as calcareous rods (Tavener-Smith, 1969; Armstrong, 1970) does not take the above considerations into account.

The acanthopores are budded from the autozooecia in the initial part of the exozone or from the most peripheral parts of the endozone, where the calcareous walls of the autozooecia were still thin. Each acanthopore always evidently preserved its connection with the neighbouring autozooecium, first branching out of it at a particular angle, but later, lengthening with growth and taking a more or less parallel position (Plate II, Fig. 4).

Plate II

Fig. 1. *Leptotrypella gurievensis* Astrova. Holotype. PIN, N 2218/459. Middle Devonian, U.S.S.R., Salair. Transverse section ×45. Budding of exilapores.

Fig. 2. *Rhombotrypella pentagonalis* (Schulga-Nesterenko). Hypotype. PIN, N 136/6201. Upper Carboniferous U.S.S.R., Moscow district. Longitudinal section ×90 Different types of acanthopores

Fig. 3. *Dyscritella* sp. PIN, N 613c/732. Lower Permian, U.S.S.R., Middle Ural. Longitudinal sections ×200. (a) The central cavity of a macroacanthopore, filled up by microgranular calcite and pelitomorphic carbonate. (b) The same in the high part of the broken acanthopore.

Fig. 4. *Dyscritella parallela* Morozova. Holotype. PIN, N 2210/122. Upper Permian, U.S.S.R., Primorie. Longitudinal section ×170. Macroacanthopore, bent away from the cavity of a zooecium.

The very thin primary wall of the acanthopore, formed evidently very soon after the budding, can generally be seen clearly in longitudinal and cross sections. As the acanthopores rising from the surface of the colony get longer, their primary walls are more and more embraced by the plates of the calcareous walls of neighbouring autozooecia, which grow out obliquely upward and in the region of joining the primary wall of the acanthopore rise nearly vertically. Surrounding the acanthopore on all sides, the intensively growing calcareous secretions evidently rose above the surface and generally united, covering over completely the opening of the acanthopore from above, converting the uppermost part of the acanthopore into a closed spine-like structure.

The morphology of such small and numerous individuals of Trepostomatous colonies, as well as their position, allows them to be interpreted as a special type of kenozooid very probably protecting the surface of the colony.

The function of acanthopores was probably similar to that of the chitinous or calcareous tube-like kenozooids of many modern Bryozoa of the order Cyclostomata or the large spines of some families of Cheilostomata. The larger number of acanthopores around the macrozooecia in the macules, often seen among many Trepostomata, confirms also their protective function.

Acanthopores occur in most of the Trepostomata in both the Halloporoid and in the Amplexoporoid branches, developing in the first instance together with the mesopores and in the second with exilapores.

Our knowledge of the character and development of polymorphism among the Trepostomata is as yet far from being complete, but even the data that we possess permits us to judge the differential development of various heterozooecia. Each Trepostome family in a general way is characterized by its specific features and by a special combination in the colony inherent to their specialized individuals.

In the order Halloporoidea all the most ancient families (Halloporidae, Monticuliporidae, Mesotrypidae and Heterotrypidae), which reached a maximum in the second half of the Ordovician, are chiefly characterized by numerous mesopores with thick diaphragms and free openings. In the family Trematoporidae, which was especially numerous during the Silurian, the mesopores are characterized by closed openings and evidently in consequence have a smaller number of diaphragms. The reduction in the number of diaphragms is regularly observed in the development of some of the late representatives of the family and also characterizes the whole family Araxoporidae (D_1–P_2) of less ancient origin, which is characterized also by slight development of mesopores. The ancient Halloporoidea families are also characterized either by the absence of acanthopores in the colonies or by the development of very small and short acanthopores. The more long-living families Heterotrypidae and Trematoporidae have abundant and often large acanthopores, especially their later representatives. Among some Heterotrypidae there also may appear differentiated acanthopores (the genera *Dekayia, Heterotrypa*). This last direction of development, beginning in the Ordovician, is developed greatly from the end of the Early Devonian and especially during the Permian in the family Araxoporidae.

In the suborder Amplexoporoidea its most ancient family, the Amplexoporidae which flourished in the Ordovician, is distinguished by a very feeble development of exilapores, which are almost absent in many genera. In the later families, which reach a maximum from the Devonian to the beginning of the Permian, there appear single genera with relatively large accumulations of exilapores in the colonies. The late Palaeozoic families Crustoporidae and Dyscritellidae are characterized by an exceptional development of exilapores.

In the Amplexoporidae, acanthopores are very small or are absent. Among the Atactotoechidae, besides the prevalent small acanthopores, in some genera they become larger and longer. In the families Eridotrypellidae, Stenoporidae, Dyscritellidae in each colony acanthopores of various sizes and structures develop at the same time.

Thus the analysis of morphologically different skeletal structures, preserved in the colonies of the trepostomatous Bryozoa, permits us to think that the development of polymorphism is the most important biologic device of this ancient group.

The morphology of the skeletal structure of individuals forming the colony, as well as the data concerning the changes that occur in it at different times allow us to define the more general and the narrower directions of the development of polymorphism within the order. So these last become the most important criteria for distinguishing taxa of different ranks in the order Trepostomata.

The development of accumulations of macrozooecia, evidently originating during periods of sexual reproduction, characterizes the whole order. The development of mesopores and exilapores defines two large groups which may be treated as suborders. Combined in a definite manner, mesopores and acanthopores or exilapores and acanthopores, always differing both morphologically and by their distribution in the colony, characterize successive phases of evolution of taxa of family level.

Genera, forming the families, are generally characterized by smaller, chiefly quantitative features which show the peculiarities of the autozooecia and heterozooecia.

REFERENCES

ARMSTRONG, J. (1970). Zoarial microstructures of two Permian species of the Bryozoan Genus *Stenopora. Palaeontology* **13** (4), 581–587.

ASTROVA, G. G. (1965). Morphology, history of development and system of Ordovician and Silurian Bryozoa. *Trudi Paleont. In-ta AN SSSR* **106,** 1–432.

ASTROVA, G. G. (1971). On the morphology and the functional significance of the acanthopores among the Trepostomata (Bryozoa). *Paleont. J.* **N3,** 72–79.

BORG, F. (1965). A comparative study on fossil and recent Bryozoa of the suborders Cyclostomata and Trepostomata. *Arkiv. Zool.* **17,** H.I., 1–91.

BRAIKO, V. D. (1970). Biological reproduction of the Black Sea Bryozoa. Sea biology, *vip. 18, Biol. obrast. A.N. Ukr. Inst. biol. iugn. morei,* 147–167.

DUNAEVA, N. N. (1968). Sexual Reproduction of some Trepostomatous Bryozoa. *Proceedings of the First Intern. Confer. on Bryozoa. Milano,* pp. 61–63.

DUNAEVA, N. N. and MOROZOVA, I. P. (1967). Evolution and Systematic Position of certain Late Palaeozoic Trepostomata. *Paleont. J.* **4,** 64–72. (Translation of paper in *Paleont. Zh.,* 1967. **4,** 86–94.)

DYBOWSKI, W. (1877). Die Chaetetiden der Ostbaltischen Silur-Formation. *Verhandl. Russ. Keiserl. Miner. Ges. St. Petersburg* **2,** 1–134.

KLUGE, G. A. (1962). Bryozoa of the Northern Seas of the USSR. Guides to the fauna of the USSR. *Zool. Inst. AN SSSR* **8,** 1–582.

NICHOLSON, H. A. (1879). "On the structure and affinities of the 'Tabulate Corals' of the Paleozoic period with critical descriptions of illustrative species," pp. 253–327, Edinburgh.

NICHOLSON, H. A. (1881). "On the structure and affinities of the genus *Monticulipora* and its subgenera with critical descriptions of illustrative species," pp. 1–240, Edinburgh.

ROSS, J. P. (1967). Evolution of Ectoproct Genus *Prasopora* in Trentonian (Middle Ordovician) in Northern and Central United States. *J. Paleont.* **2,** 403–416.

TAVENER-SMITH, R. (1969). Wall structure and acanthopores in the bryozoan *Leioclema asperum. Lethaia* **12,** 89–97.

ULRICH, E. O. (1879). Descriptions of new Genera and species of fossils from the Lower Silurian about Cincinnati. *Soc. Nat. Hist.* **2,** 1–30.

ULRICH, E. O. (1882). American Paleozoic Bryozoa. *J. Cincinnati Soc. Nat. Hist.* **5,** 121–175; 232–257.

WATERS, A. W. (1912). A Structure in *Adeonella* (*Laminopora*) *contorta* and some other Bryozoa . . . *Ann. Mag. nat. Hist.* **8,** *ix,* 489–500.

2 | Graptolite Periderm Structure and Budding Patterns: a Résumé

O. M. B. BULMAN

Sedgwick Museum, Cambridge, England

Abstract: The graptolite periderm is composed of scleroproteic material forming an inner layer of fusellar tissue and an outer laminated cortical layer. The presence of cortical tissue combined with a stolon system constitutes a significant difference between the graptolites and the pterobranchs. A brief survey is given of the distinctive regularity of these organisms as exemplified in growth patterns, changes in thecal shape, and the development of "pseudobranches" or cladia.

INTRODUCTION

The Graptolithina are a class of colonial organisms which first appeared in rocks of Middle Cambrian age, flourished during the Ordovician, Silurian and Devonian, and became extinct in the lower part of the Upper Carboniferous. They have been increasingly studied over the past 150 years, chiefly because of the stratigraphical value of one order, but only during the last 30 years have their affinities become clarified. Well-preserved material can be freed from the rock matrix by solution with appropriate acids and rendered transparent with Schultze's solution, or embedded and sectioned with a microtome. It is naturally from such material, only sporadically available, that our principal morphological information has been derived.

They are now considered a class of the phylum Hemichordata (Bulman, 1970; see also Kozlowski, 1947, 1966). This phylum includes also the living Pterobranchia, which are represented at intervals in rocks from Lower Ordovician to Tertiary age. Six orders of the class are at present recognized and all secrete a scleroproteic skeleton (Florkin, 1965).

Five of these orders are sessile and often encrusting, possess a well-developed stolon system (Fig. 1a, b) and are polymorphic, comprising autothecae, bithecae and stolothecae. The former may have housed female and male zooids respectively; the latter enclosed the stolon system, but contained no distinctive zooids; rather, these tubes appear to have been secreted by the immature autothecal

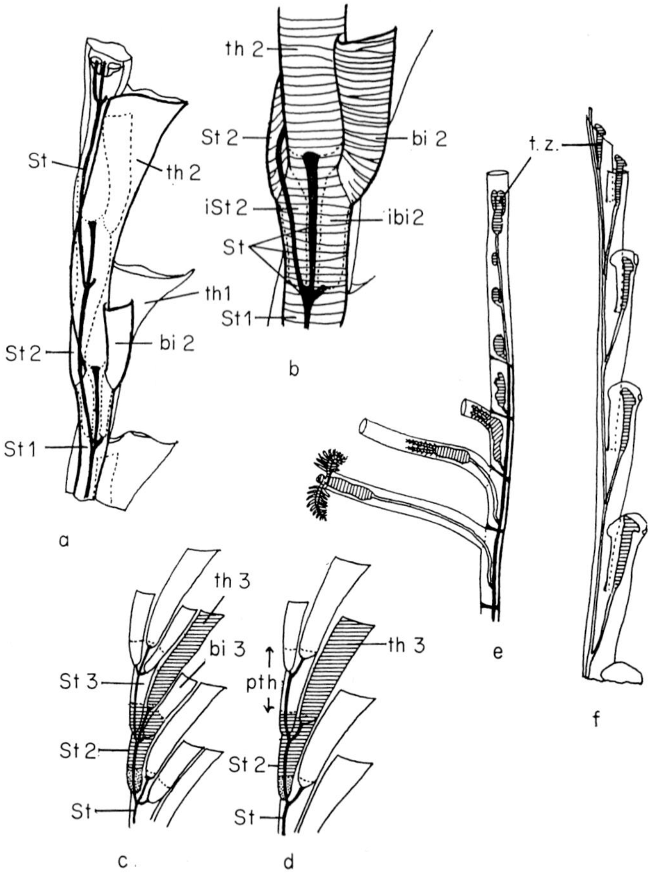

Fig. 1. Structural features of dendroid and graptoloid graptolites.

(a) stipe of dendroid graptolite (×25 approx.) showing stolon system (solid black) with triad budding of stolothecae (St), autothecae (th) and bithecae (bi).

(b) portion of same enlarged (×50 approx.) to show fusellar growth-bands and internal portions of stolotheca (iSt2) and bitheca (ibi2).

(c), (d) diagrammatic figures illustrating dendroid (c) branch structure compared with more simple graptoloid structure (d). St, stolons; pth, prothecal part of graptoloid theca corresponding to stolothecae (St2, St3) of dendroid.

(e), (f) diagrammatic comparison of pterobranch (*Rhabdopleura*) and graptoloid (*Monograptus*). Sclerotized portion of stolon (pectocaulus) solid black, unsclerotized portion (gymnocaulus) unshaded; t.z., terminal zooid.

zooids and to constitute the proximal portions of successive autothecae (Fig. 1b).

The sixth order, Graptoloidea, is in many respects the most specialized and is the most useful stratigraphically, being planktonic or epi-planktonic and hence widely distributed. The colony comprises autothecae only (possibly hermaphrodite), with no bithecae, stolothecae or sclerotized stolons. They probably possessed a gymnocaulus (as in the distal portions of the living *Rhabdopleura*) and the prothecal segment of each autotheca (Fig. 1d) corresponds morphologically to the stolotheca of other orders. Moreover, it has recently been shown that the "intermediate" dendroid graptolite *Kiaerograptus kiaeri* lacks proximal bithecae, but retains them distally, implying the presence of stolons (sclerotized or unsclerotized) throughout the rhabdosome.

THECAL WALL STRUCTURE

The graptolite periderm is composed of two layers, an inner *fusellar* layer, and an outer *cortical* layer (Fig. 2).

The fusellar layer consists of alternating left and right fusellae (segments), closely comparable to those of the skeleton of *Rhabdopleura* and especially those of the adnate portions of the zooidal tubes. Assuming the existence of a preoral lobe in the graptozooid, it was perhaps secreted by this portion of the autothecal and bithecal zooids.

The cortical layer is a laminated, encrusting layer covering the fusellar layer, and resembling the interstitial tissue in the coenoecium of *Cephalodiscus*.

In *Cephalodiscus*, however, the coenoecium represents an "association" rather than a true colony; zooids can leave their tubes (e.g. nocturnally) and creep about on the surface secreting this binding substance while "external". In *Rhabdopleura*, the zooids are captive and cannot leave their tubes on account of their proximal attachment to the gymnocaulus and pectocaulus; consequently, in *Rhabdopleura* the coenoecium possesses no cortical tissue. In the graptolites, the zooids were also captive (being attached proximally to the stolon), but here the rhabdosome *does* possess cortical tissue. This layer must consequently have been secreted by some external living tissue; the origin and relations of this are entirely conjectural but it presumably constituted a sheath more or less enveloping the colony. The postulated existence of this tissue constitutes one of the most significant differences between the graptolites and the pterobranchs.

This cortical tissue is always more abundant at the proximal end of a rhabdosome—the first-formed portion—and is absent at the growing ends of the branches (Fig. 2c, d). Other structures, often far removed from the thecae, which are also composed of cortical tissue include the dissepiments or connecting bars

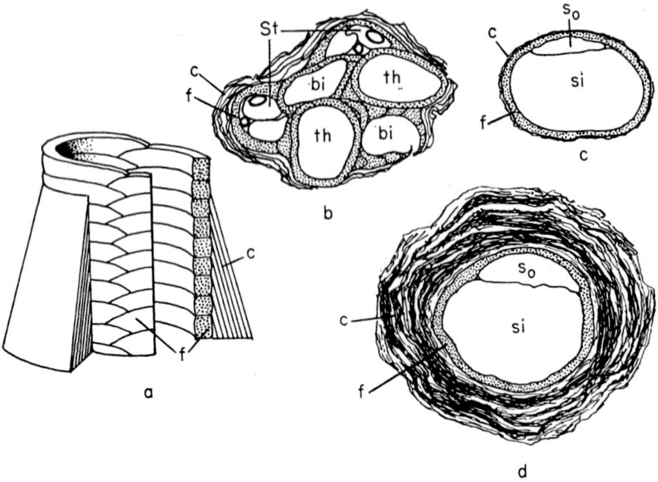

Fig. 2. Periderm structure of dendroid graptolites.

(a) Diagram showing fusellar tissue (f) laid down in alternating half-rings surrounded by cortical tissue (c).

(b) transverse section of dendroid *Koremagraptus* with laminated cortical tissue (c) surrounding fusellar tissue (f) in complete rings for autothecae (th) and "split tubes" for bithecae (bi) and stolothecae (St); where growth-bands are oblique to plane of section (e.g. bitheca at lower right) several such bands are cut by the plane of section; fusellar tissue stippled. ×85 approx.

(c), (d) sections through apex of sicula (si) and initial bud (s_0) of *Dendrograptus*. (c) early growth stage with only a film of cortical tissue; (d) corresponding level in a mature colony with thick cortical deposits forming a strong stem at base of rhabdosome. Fusellar tissue stippled. ×120 approx.

between branches in the conical dendroid *Dictyonema*, and various web-structures and "floats" in the Graptoloidea.

In recent years, several attempts have been made to examine the ultrastructure of these layers under the electron microscope. Under high-power Stereoscan (up to ×130,000 has been attained) the cortical tissue reveals a spongy, granular texture; superficially, these spherical granules are less compacted, but show greater coalescence in deeper, earlier-formed layers, which suggests (Rickards *et al.*, 1971) some penetration of living tissue into the cortical substance. Fusellar tissue exhibited a more "lumpy", granular texture, with an unexplained lineation at right angles to the growth lines, but scanning pictures of this at higher magnifications than ×30,000 have not yet been obtained. It is unknown to what extent this granular texture reflects the effects of post-mortem changes. Thin sections under the transmission microscope by various authors have so far (1971) given rather conflicting results but see Addendum.

THE SICULA

All graptolite colonies originate in a cell called the sicula, representing the skeleton of the only sexually-produced zooid in the colony, which comprises an apical prosicular portion and a distal metasicular segment. In the Graptoloidea (Fig. 3a, b) the prosicula contains a spiral thread, which is believed to be internal, and a number of external, longitudinal ridges; but it was evidently secreted in a single piece and shows no growth stages other than a progressive development of the full complement of longitudinal ridges. In contrast to this, the metasicula results from accretionary growth and shows normal fusellar growth-bands (Fig. 3c–e). The difference between these two portions of the sicula is so great

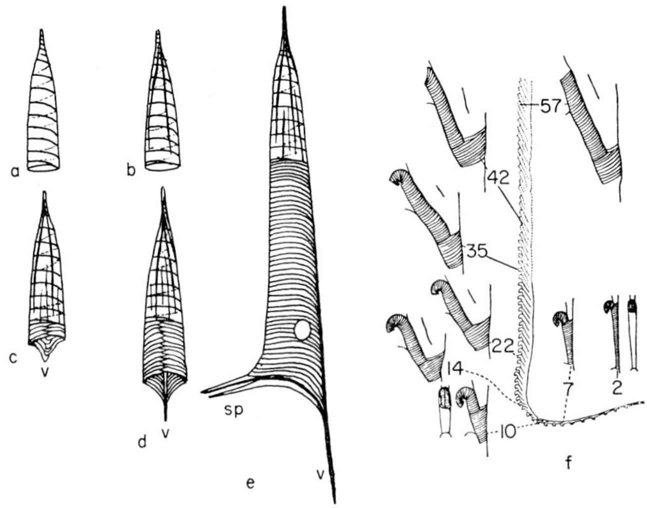

Fig. 3. (a–e) Growth stages in formation of sicula of graptoloid (*Orthograptus*). ×25 approx.

(a), (b) successive stages of prosicula with spiral line (a) and longitudinal ridges (b). (c), (d) early stages in development of metasicula with fusellar tissue. (e) mature sicula with resorption foramen for initial bud. v, virgella spine; sp, apertural spines.

(f) *Monograptus argenteus* to show steady and gradual change in shape of autothecae throughout the rhabdosome; $\times 1\frac{1}{3}$, thecal enlargements ×6 approx.

that Kozlowski (1966) postulated complete metamorphosis of the prosicular zooid before the beginning of metasicular growth. There is also a perceptible difference in the microstructure of the periderm, that of the metasicula comparing with fusellar tissue elsewhere in the rhabdosome while the prosicula is composed of an irregular aggregate of minute branching and anastomosing rods.

B

From this sicula, all types of graptolite rhabdosome are derived by linear budding from a single initial bud, generally passing to the exterior through a resorption foramen, but in the latest graptoloid sub-order originating as an apertural notch during growth of the metasicula.

REGULARITY OF GROWTH OF COLONIES

The general form of the rhabdosome in many orders of graptolites is irregular, though there may be uniformity (within the order) in the type of stolonal budding; and the development of stolonal triads, with alternately left- and right-handed bithecae (Fig. 1) is diagnostic of the Dendroidea. It is, however, in the Graptoloidea that rhabdosomal regularity is outstanding, expressed here in: early astogeny (with a considerable range of developmental types); branching patterns; precisely controlled thecal sequences in a rhabdosome; and the production of cladia. All these are discussed by Bulman (1970), and a brief account of the last three is given below.

THECAL CHANGES AND SEQUENCES

Thecae of a particular kind (e.g. autothecae, bithecae) are virtually uniform throughout the colony in most orders except the Graptoloidea. Here the autothecae generally show at least a progressive increase in size up to a certain specific maximum, and may show very considerable change in shape within a single colony. This variation in shape during growth of a rhabdosome is regular and is gradual, never abrupt (Fig. 3f); the appearance of different thecal types bears no relation to branching nor to any nodal points along the rhabdosome.

Phyletically, a new thecal type is most commonly introduced at the proximal end, gradually spreading distally in successive descendants. This was first adequately demonstrated by Sudbury (1958) for "triangulate" monograptids and *Rastrites* from the Lower Silurian (Fig. 4a–d), and involves two processes later called penetrance and expressivity by Urbanek (1960). Well-formed thecae of the new type penetrate further and further into the rhabdosome of successive descendants, while at the same time all thecae of the rhabdosome are slightly but increasingly affected towards the new type (expressivity). Sudbury also deduced that there must also, but more rarely, be instances of the distal introduction of new thecal types, spreading proximally, and this was later conclusively demonstrated by Urbanek (1960, 1966) in morphological and phyletic studies on the Upper Silurian *Lobograptus* and *Cucullograptus* (Fig. 4e–i). These processes he explained on a hypothesis of increasing stimulator and decreasing inhibitor hormones rising above or falling below a certain threshold level.

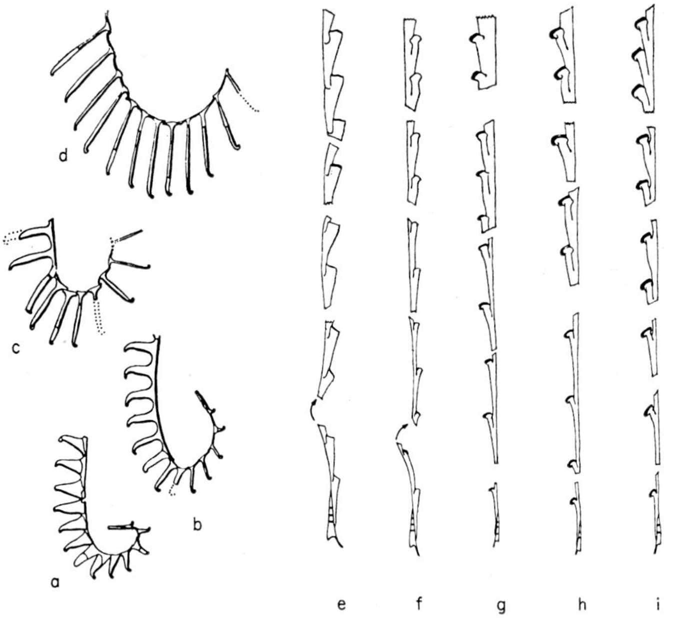

Fig. 4. Introduction of new thecal types in monograptid lineages. (a–d) proximal introduction and distal spread of isolate thecae; (e–i) distal introduction and proximal spread of bilaterally lobate type.

(a) *Monograptus triangulatus separatus;* (b) *M. t. triangulatus;* (c) *M. t. extremus;* (d) *Rastrites longispinus.* (sicula and extreme proximal end missing in Figs. (c) and (d)). ×3 approx.

(e) *Cucullograptus (Lobograptus) progenitor;* (f) *C. (L.) simplex;* (g) *C. (L.) scanicus parascanicus;* (h) *C. (L.) s. amphirostris;* (i) *C. (L.) s. scanicus.* ×13 approx.

BRANCHING

The branching of graptolite rhabdosomes may also show remarkable regularity and conformity to particular patterns—sufficiently so for early workers to base the majority of their genera on rhabdosome form. Regularity is seen in a few dendroids, especially in the transitional family Anisograptidae, but it is more characteristic of the Graptoloidea, where there is a wide range from multiramous dichograptids to the single uniserial stipe of the monograptids.

Branching in the Dendroidea is due to the replacement of a bithecal bud by a second stolothecal bud in the normal stolonal triad; in the Graptoloidea, it is due to the formation of two buds in place of the normal single bud—i.e. the replacement of a normal theca by a *dicalycal* theca. Regularity in branching in the Graptoloidea is consequently due to the regular siting of these dicalycal thecae, and hence to the almost synchronous development of dicalycal thecae at all growing points of the rhabdosome. Examples of such regularity are shown in Fig. 5 (*Goniograptus*, *Brachiograptus* and several species of the anisograptid *Clonograptus*).

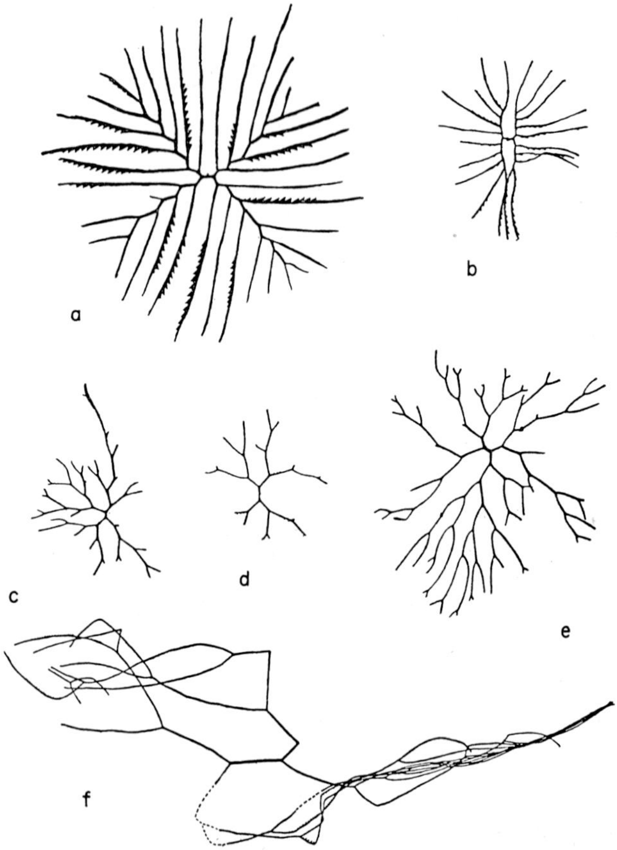

Fig. 5. Regularity in branching patterns.
(a) *Goniograptus thureaui*, $\times\frac{4}{5}$: (b) *Brachiograptus etaformis*, $\times\frac{4}{5}$; (c–f) various species of *Clonograptus* showing different branching frequencies; (c) *C. tenellus callavei*, $\times\frac{2}{5}$; (d) *C. t. tenellus*, $\times\frac{2}{5}$; (e) *C. rigidus*, $\times\frac{2}{5}$; (f) *C. ramulosus*, $\times\frac{1}{5}$.

CLADIA FORMATION

This final and most remarkable instance of regularity occurs among the latest and most highly specialized family of the Graptoloidea, the Monograptidae.

Basically the rhabdosome here has been reduced to a single undivided uniserial stipe, and cladia are "false branches" or "pseudobranches" produced from the apertural region of a particular theca and/or from the apertural region of the sicula, at a time when the normal thecal succession has long passed this particular point of rhabdosome development. The siting of the parent theca may be almost constant for a given species, at $n \pm 1$. Thus in *Cyrtograptus rigidus* (Thorsteinsson, 1955), the single thecal cladium occurs between th8 and th10, usually at th9, and its development only begins when growth of the main stipe has reached th13 $\pm$ 1. The first theca of the cladium is completed by the time the main stipe has reached th16 $\pm$ 1; it is of the same type as that theca, and thereafter thecal change is integrated with change in the main stipe (Fig. 6d), so that thecae of the same shape are being added at all growing points on the compound rhabdosome. This is claimed to be equally true of many species with multiple cladia. The morphogenetic field covering the rhabdosome appears as the

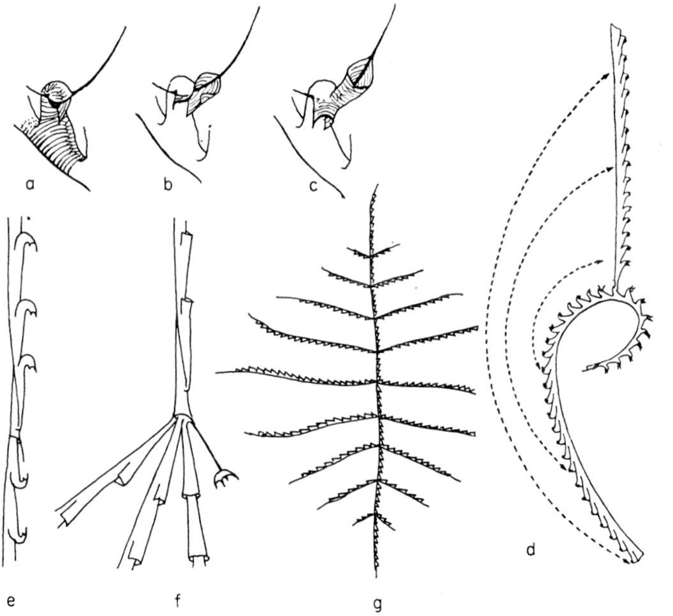

Fig. 6. Cladia formation in various monograptids.

(a, b, c) successive stages in development of initial theca of cladium from apertural region of a hooked theca on main stipe of *Cyrtograptus rigidus*; adult rhabdosome showing development of single cladium and thecal relations of cladium and main stipe, $\times 1\frac{1}{4}$ approx.

(e) *Diversograptus runcinatus* illustrating development of a single sicular cladium.

(f) *Linograptus posthumus* with three sicular cladia (and buoyancy structure on virgella spine).

(g) *Abiesgraptus* with three sicular cladia and paired thecal cladia.

astogenetic counterpart of the growth-gradient systems so well displayed in arthropod or vertebrate ontogeny.

Monograptid genera can be arranged in ascending order of cladial complexity; thus in *Monograptus* the rhabdosome consists of the main stipe only; *Cyrtograptus* develops one or more thecal cladia (which may themselves bear second- or even third-order cladia); *Diversograptus* (Fig. 6e) and *Linograptus* (Fig. 6f) possess one or several sicular cladia respectively; and in *Abiesgraptus* (Fig. 6g) there may be three sicular cladia, with paired thecal cladia spaced along the main stipe and on its opposing sicular cladium.

REFERENCES

BULMAN, O. M. B. (1970). Graptolithina. *In* "Treatise on Invertebrate Paleontology", 2nd edition (TEICHERT, C., ed.), pp. xxxii–163. Geol. Soc. Am. & Univ. Kansas, Lawrence, Kansas.

FLORKIN, M. (1965). Paléoprotéins. *Acad. royale Belg.* (Sci.), *Bull.* (5) **51,** 156–169.

KOZLOWSKI, R. (1947). Les affinités des graptolithes. *Biol. Rev.* **22,** 93–108.

KOZLOWSKI, R. (1966). On the structure and relationships of graptolites. *J. Paleont.* **40,** 489–501.

RICKARDS, R. B., HYDE, P. J. W. and KRINSLEY, D. H. (1971). Periderm ultrastructure of a species of *Monograptus* (Phylum Hemichordata). *Proc. R. Soc. Lond.* B **178,** 347–356.

SUDBURY, M. (1958). Triangulate monograptids from the *Monograptus gregarius* zone of the Rheidol gorge. *Phil. Trans. R. Soc. Lond.* B **241,** 485–555.

THORSTEINSSON, R. (1955). The mode of cladial generation in *Cyrtograptus*. *Geol. Mag.* **92,** 37–49.

URBANEK, A. (1960). An attempt at biological interpretation of evolutionary changes in graptolite colonies. *Acta Palaeont. Polonica* **5,** 127–234.

URBANEK, A. (1966). On the morphology and evolution of the Cucullograptinae (Monograptidae, Graptolithina). *Acta Palaeont. Polonica* **11,** 291–544.

ADDENDUM

Since the above was written, an important paper by K. M. Towe and A. Urbanek has appeared in *Nature*, 1972, **237,** 443–445, Collagen-like structures in Ordovician graptolite periderm.

3 | Bryozoa as a Community Component on the Northeast Coast of Britain

P. G. MOORE*

Wellcome Marine Laboratory, University of Leeds, Robin Hood's Bay, North Yorkshire, England

Abstract: Existing knowledge of bryozoan/algal associations and of Bryozoa on the Northeast coast of Britain is summarized. A widescale survey of the fauna associated with holdfasts of kelp (*Laminaria hyperborea* (Gunn) Fosl.) along the Northeast coast is described. The influence of physical environmental factors, in particular that of seawater turbidity, is discussed in relation to the bryozoan component of this fauna. The clear waters off the coast of Berwickshire and North Northumberland are characterized by *Membraniporella nitida*, *Escharina spinifera*, *Escharoides coccineus* and *Cellepora* (=*Celleporaria*) *pumicosa*. Other species, notably *Electra pilosa*, *Callopora lineata*, *Celleporella hyalina* and *Alcyonidium* sp. are apparently indifferent to the higher turbidity level of the coastal waters off South Northumberland, Durham and Yorkshire. The interrelationship between water movement and sedimentation is restated and discussed in relation to the ecology of kelp holdfast Bryozoa. Various biological interactions within the sessile element of the holdfast fauna, and between the Bryozoa and their algal substratum are described. Consideration is also given to the role of bryozoans as physical settlement substrata in their own right, in particular in relation to the primary settlement of *Mytilus* spat. Predation on holdfast-dwelling bryozoans locally is discussed and a list of bryozoan species which have been recorded locally from laminarian holdfasts is presented.

INTRODUCTION

The Bryozoa form an interesting component of the total kelp holdfast fauna under consideration. It is with the physical and biological interactions of this component that this paper is concerned.

The ecology of the fauna associated with marine algae in general is poorly known. French workers first noted associations between Bryozoa and algae around Roscoff (Joliet, 1877; Prenant and Teissier, 1924; Prenant, 1927, 1932). In Britain, Colman (1940) discussed aspects of the ecology of some commoner

* Present address: University Marine Biological Station Millport, Isle of Cumbrae, Scotland.

bryozoan species in his classic paper on intertidal algal faunas. More extensive studies of bryozoan/algal associations were made in the region of Woods Hole, Mass., U.S.A. by Rogick and Croasdale (1949), and in the Baltic by Bock (1950, 1954). To date, the paper by Ryland (1962) provides the most comprehensive British treatment of the topic. More recently Hagerman (1966) has described the Bryozoa associated with *Fucus serratus* L. in Swedish waters. The unpublished work of Scarratt (1961) dealing with the fauna of *Laminaria* holdfasts along the South and West coasts of Britain also contains useful ecological observations on the Bryozoa.

Unfortunately contemporary published works relating to Bryozoa on the

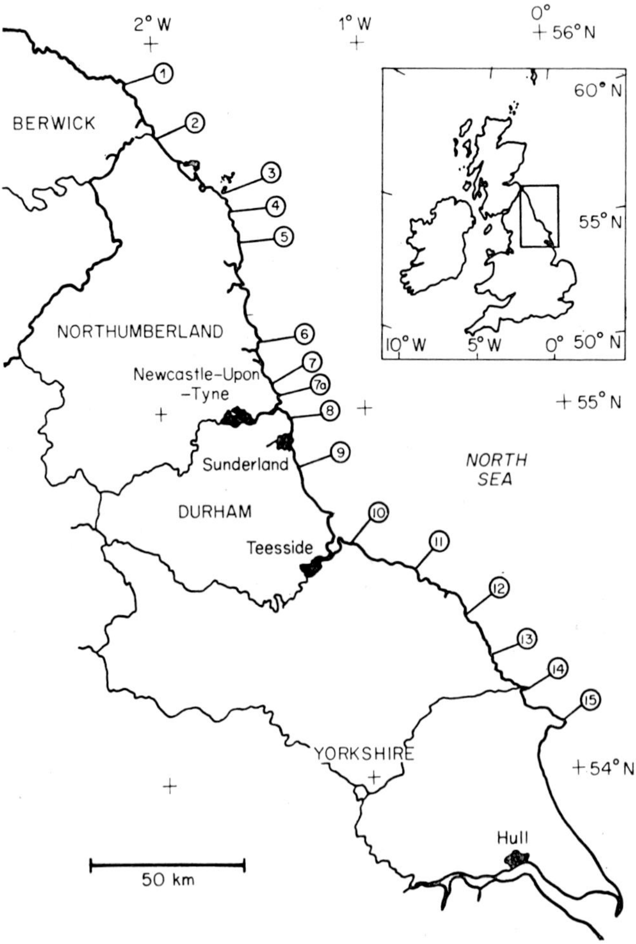

Fig. 1. The Northeast coast of Britain showing the position of sampling sites.

Northeast coast of Britain are scarce. The most comprehensive coverage is by Roper (1913) in a paper on the Polyzoa of Northumberland. Thompson *et al.* (1966) have also listed the bryozoans recorded during a descriptive survey of the Inner Farne Island off the coast of Northumberland.

METHODS

The area surveyed was the wave exposed Northeast coastline from St Abbs, Berwickshire in the North to Flamborough Head, Yorkshire in the South. Seventy-two mature holdfasts of *Laminaria hyperboarea* (Gunn.) Foslie were taken from fifteen approximately equidistant sites (Fig. 1) in July 1969. As near

Table 1. Faunal composition of survey samples

Total=72 samples	387 species	*ca* 90,000 individuals
Group	No. species	No. individuals
Protozoa	not included	
Coelenterata	6	
Kinorhyncha	1	
Sipunculoidea	1	
Platyhelminthes	not included	
Nematoda	61	20,700
Rhynchocoela	2+	
Polychaeta	50	3700
Oligochaeta	rare, not included	
Copepoda	132	17,500
Ostracoda	14	
Cirripedia	2	10,400
Isopoda	6	
Amphipoda	20	6000
Tanaidacea	1	
Mysidacea	1	
Decapoda	6	
Halacarida	9	
Pycnogonida	6	
Polyplacophora	3	
Prosobranchia	19 }	2625
Opisthobranchia	12 }	
Bivalvia	11	25,000
Bryozoa	15	
Kamptozoa	1	
Echinodermata	5	
Ascidiacea	4	
Pisces	1	

as possible instantaneous sampling was achieved over the period of one spring tide. Five holdfasts were taken from most sites (depth range 1 to 5 m) by Scuba divers. Each holdfast was put into a polythene bag and secured underwater.

Details of the laboratory extraction method for motile forms have been given elsewhere (Moore, 1971). Subsequent to this extraction, the individual samples were stored in labelled polythene bags in a deep freeze for later examination of

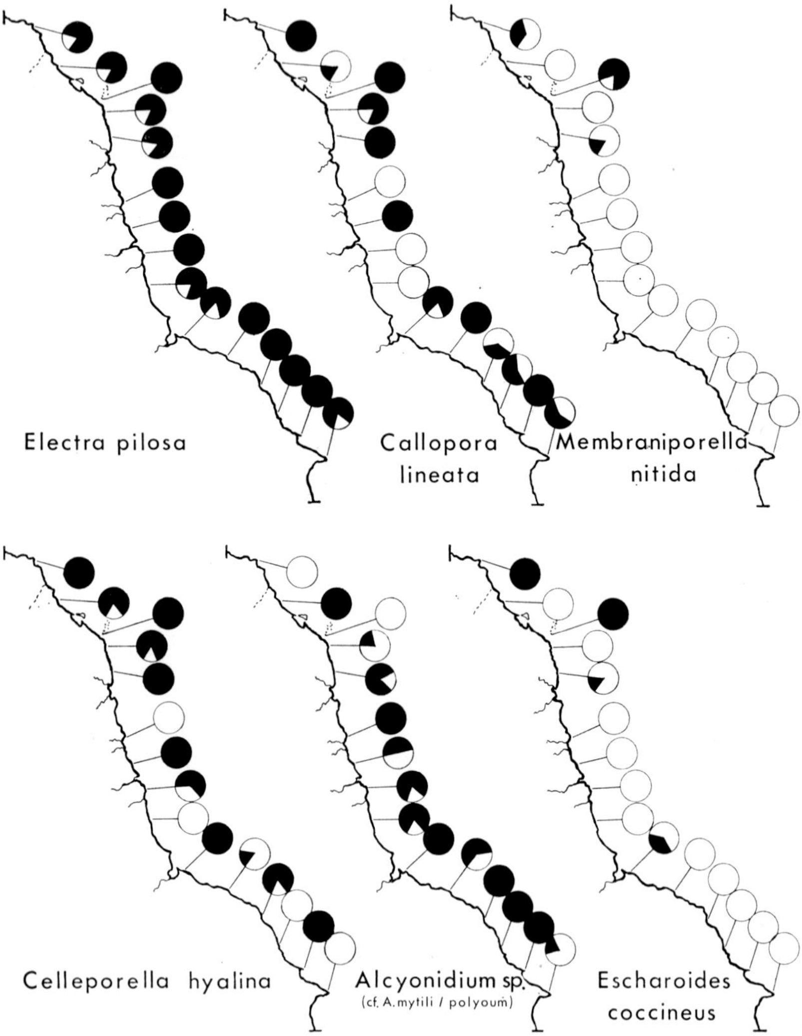

Fig. 2. The distribution of kelp holdfast Bryozoa along the Northeast coast. Black segments at each site represent the percentage frequency of occurrence of each species.

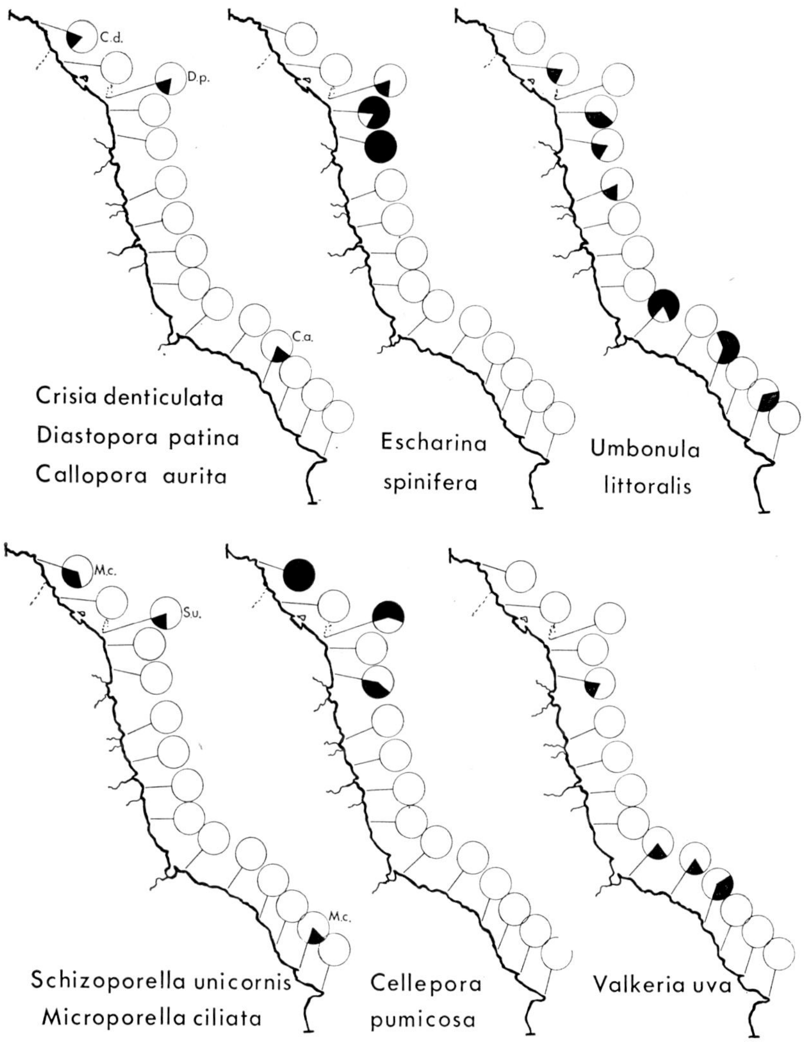

Fig. 3. The distribution of kelp holdfast Bryozoa along the Northeast coast—continued. Symbols as Fig. 2.

the sessile element. Dissection and microscopic examination of the entire surface of the hapteron was undertaken for this purpose. Quantification of colonial Bryozoa, Hydrozoa and Ascidiacea was not attempted. Counts of numbers of individuals would have been prohibitive of time, and colony numbers would have been artificial when dealing with dissected pieces of alga. A qualitative "presence–absence" approach was thus adopted for these groups.

RESULTS

The total fauna of the samples was established (with the major exceptions of the Protozoa, Turbellaria and Porifera) and amounted to some 90,000 animals of 387 species from most invertebrate phyla (Table 1). Ordering of these data has been accomplished using a number of multivariate statistical techniques and a preliminary assessment of these analyses has been given by Moore (1972).

The Bryozoa recorded comprised the following fifteen species, *Crisia denticulata* (Lamarck), *Diastopora* (=*Berenicea*) *patina* (Lamarck), *Electra pilosa* (L.), *Callopora lineata* (L.), *Callopora aurita* (Hincks), *Membraniporella nitida* (Johnston), *Celleporella hyalina* (L.), *Schizoporella unicornis* (Johnston), *Escharina spinifera* (Johnston), *Microporella ciliata* (Pallas), *Escharoides coccineus* (Abildgaard), *Umbonula littoralis* (Hastings), *Cellepora* (=*Celleporaria*) *pumicosa* L., *Alcyonidium* sp. and *Valkeria* (= *Walkeria*) *uva* (L.).

Records of other species found outside this survey are presented at the end of this paper. Figures 2 and 3 show the pattern of distribution of these fifteen species.

KELP HOLDFAST BRYOZOA AND PHYSICAL FACTORS ON THE NORTHEAST COAST

1. *Water Movement*

The importance of water movement to the distribution of marine biota, in its two components of wave action and tidal current surge, is widely known (Lewis 1964, 1968). The high faunal diversity of the laminarian holdfast habitat on wave exposed shores is partly a reflection of the protection and shelter it offers from otherwise inclement conditions. The dominant bryozoan life form in this habitat is the encrusting type which appears to characterize littoral and sublittoral hard bottom habitats especially in areas subject to wave turbulence. The less frequent bushy forms, in the main the jointed stemmed *Crisia* species, might be under-represented in the present report due to their fragility. Only colonies which were observed to be actually attached to a piece of alga were recorded. Loose crisiid fragments were more common but were not included. Water currents, when divorced from the destructive action of waves are generally productive of high diversity, especially of sessile filter-feeding animals. This tendency is best seen in tidal rapids areas, e.g. Lough Inc, Ireland (see summary by Kitching and Ebling, 1967) and certain Scottish sea lochs (Lewis, 1957, 1964, 1968; P. G. Moore, unpublished), which support rich assemblages of Porifera, Hydrozoa, Bryozoa and Ascidiacea.

On the Northeast coast no such rapids conditions prevail. Known rheophilic species (Ursin, 1960) such as the ophiuroids *Ophiothrix fragilis* (Abildgaard) and *Ophiopholis aculeata* (L.) were plentiful amongst holdfasts from St Abbs (site 1) and Inner Farne (site 3). The latter site experiences particularly strong currents caused by the tide being funnelled through the narrow passages between the complex of islands, often producing standing waves at the junctions between current systems (see also Thompson *et al.*, 1966). High bryozoan diversity was found at these sites (Figs 1 and 2) but such diversity was found also at Craster (site 5) which was sheltered to a degree by offshore reefs.

2. Turbidity

In a preliminary assessment of the total survey data the seawater turbidity regime over the transect was considered as being of overriding significance to the holdfast fauna (Moore, 1972). Essentially the survey area can be divided into two parts—a northern clear water area (Berwickshire and North Northumberland) and a southern turbid water area (South Northumberland, Durham and Yorkshire); the difference being caused primarily by differential coast erosion.

The highest bryozoan diversity was associated with the northern clear water area. The species characteristic of this area were *Membraniporella nitida, Escharina spinifera, Escharoides coccineus* and *Cellepora pumicosa.* The latter two species contribute, along with encrusting "lithothamnion", to the delicate red and pink colouration of the hapteron surface often shown by holdfasts from this region. Other species were ubiquitous with respect to the turbidity factor, for instance *Electra pilosa, Callopora lineata, Celleporella hyalina, Alcyonidium* sp. and *Umbonula littoralis. Alcyonidium* seems to flourish in the turbid water and was the dominant bryozoan on kelp holdfasts from this area, accounting for their comparatively drab colouration. *Celleporella* on the other hand, although tolerating the turbid conditions, grew best in the clear water area.

In their discussion of the ecology of the sedentary fauna of sublittoral boulders in the Lough Ine rapids, Lilly *et al.* (1953) recorded the preference of *Cellepora pumicosa* and *Membraniporella nitida** for sheltered bays; of *Escharoides coccineus* and *Diastopora patina* for the main stream or sheltered bays, and of *Celleporella hyalina** and *Umbonula littoralis* for the main stream. Presumably exposed coast holdfasts offer a sufficiently sheltered microclimate for the former species in the clear water area in the Northeast. These authors also showed that *Cellepora pumicosa* was less common on the tops of boulders in sheltered bays where sediment deposition was greater than at other stations.

The ramifications of the inverse relationship between water movement and sedimentation have been well summarized by Lilly *et al.* (1953). Thus suspended sediment interferes with feeding and respiratory mechanisms, coarser particles in suspension abrade exposed surfaces and material settling out can smother

* Dr Ryland has expressed doubts as to the validity of these identifications.

sessile animals. Loosanoff and Tommers (1948) have demonstrated a marked reduction in the rate of feeding of oysters (*Crassostrea virginica*) when experimentally subjected to suspended silt, but no work of a similar nature appears to have been undertaken with bryozoans.

Abrasion by coarse sediment (along with grazing effects) probably accounts for the poor fauna of sublittoral rock surfaces locally, except on vertical faces to the lee of large boulders or else beneath rock overhangs.

The deleterious smothering effect of settling sediment is well known. Round *et al.* (1961) have shown in Lough Ine that the hydroid *Sertularia operculata* (L.) which favours a strong current, dies when transferred to sheltered conditions through the accumulation of sediment. Bryozoa are particularly susceptible to sediment deposition (Maturo, 1959; Ryland, 1970). The only forms which can tolerate even moderate rates of deposition are those warm water species with lunulitiform colonies (Cook, 1963; Ryland, 1970).

The causation of the observed turbidity correlation is probably a complex of such direct physical effects, compounded with indirect biological consequences, the analysis of which requires an experimental investigation.

BRYOZOA–SUBSTRATUM RELATIONSHIPS

1. *Shore Zonation*

The selection of algal substrata by free swimming larvae accounts for the association between bryozoans and particular species of algae (Ryland, 1959a, b; Crisp and Williams, 1960). Ryland (1962) has interpreted the adaptive value of this phenomenon in terms of the maintenance of species at a suitable tide level on the shore. Thus the low alga *Fucus serratus* L. is highly preferentially selected by larvae of *Flustrellidra hispida* (Fabricius) and *Alcyonidium hirsutum* (Fleming), resulting in the maximum abundance of these bryozoans on this alga in the field (Ryland, 1962; Hagerman, 1966). The infrequent occurrence of these species on *Laminaria hyperborea* holdfasts along the Northeast coast is a further confirmation of the generality of this selective process.

The sublittoral bryozoan species are in general less discriminating. There is a high degree of resemblance between the bryozoan fauna of sublittoral rock surfaces and kelp holdfasts. Species such as *Callopora lineata*, *Scrupocellaria scruposa*, *S. reptans*, *Escharoides coccineus*, *Umbonula littoralis*, together with *Electra pilosa* being equally represented on either substratum (Ryland, 1962). The specific constitution of the laminarian holdfast bryozoan fauna on the Northeast coast has proved to be essentially similar to that recorded by Ryland (1962) from Welsh surveys.

2. Epiphyte Zonation—the Problem of Membranipora membranacea (*L.*)

This species is well known for its abundance on *Laminaria* fronds, to which habitat its flexible zooids and directional growth towards the growth zone of the alga, are perfect adaptations. However *Membranipora* appears largely to be *confined* to the frond and has not been recorded from holdfasts in the present study (Scarratt, 1961 recorded *M. membranacea* as "frequently found on the stipes and fronds of *Laminaria*, less frequently found on the holdfasts", his holdfast records pertained to three Scottish sites only).

Is the larva capable of discriminating between different regions of the kelp plant? It is possible that surface texture could be important in such a preference: the frond is very smooth and mucilaginous, whereas the stipe (in *L. hyperborea* at any rate) is very rough. However hapteron surfaces are also fairly smooth. Another possible explanation involves detection on the part of the larva of the age of the algal tissue (the blade is annual but the stipe and holdfast are perennial). Such has been suggested as controlling the directional growth of *Membranipora* towards the youngest part of the frond. *Alcyonidium polyoum* larvae have been shown by Ryland (1959) to prefer the tips of *Fucus* thalli, in a choice between tips, "sub-tips" and pieces from the centre and base of the frond. However if this were the case one would at least expect to find *Membranipora* on the youngest hapteron branches. The question is intriguing especially since other species, e.g. *Electra pilosa* and *Celleporella hyalina* are to be found on both blades and holdfasts.

It might be added in passing that another characteristic frond-dwelling sessile organism, the hydroid *Obelia geniculata* (L.) shows an analogous distribution pattern. The common holdfast hydroids locally are *Clytia johnstoni* (Alder) and *Obelia flabellata* Hincks. Sloane *et al.* (1957) recorded *O. geniculata* from all parts (fronds, stipes, frills and holdfasts) of the alga *Saccorhiza polyschides* Batt. in Lough Ine, but noted its preference for fronds. The advantage of any frond-selection mechanism could well reside in an increased food availability brought about by the greater circulation of water above the bufferred kelp forest.

3. Intra-holdfast Zonation

The faunal paucity shown by the newest whorl of holdfast branches has been noted on East (P. G. Moore, unpublished) and West coasts of Britain (Scarratt, 1961 unpublished). Ghelardi (1960, unpublished) has suggested that bacterial inhibitors might be secreted by the giant kelp *Macrocystis pyrifera* (L.) Agardh. Scarratt suggested that some such mechanism might be responsible for the poverty of organisms attached to the actively growing region of the holdfast. Thus, larvae might be deterred directly by the secretion of such chemicals or else indirectly due to the elimination of the primarily attractive bacterial or diatomaceous epiflora. The effect on the holdfast bryozoan fauna is striking, only few small colonies of *Electra pilosa* and *Celleporella hyalina* were ever recorded on this new growth during these investigations. The apparent contradiction of these observations with the generally attractive properties of young algal tissue noted above is probably attributable to the largely unknown role and response specificity of external algal metabolites in the sea.

Besides a possibly significant gradient of tissue age, the distribution of bryozoans within individual holdfasts is likely to be significantly influenced by the intimate structure of the holdfast in terms of its effect on sedimentation. Thus sediment tolerant forms will be most likely to penetrate deeper into the holdfast interstices and sediment susceptible forms will be likely to prefer external attachment sites.

INTERSPECIFIC FAUNAL RELATIONSHIPS

1. Bryozoa as Substrata

Bryozoa may also serve as substrata in their own right especially in relation to settling bivalve spat. Bayne (1964) described primary and seondary settlement phases in young mussels *Mytilus edulis* L. Early plantigrades settle offshore on filamentous red algae and, after growing until their thigmotactic requirements cease to be met by this primary substratum, migrate secondarily onto the definitive adult bed. Littoral populations of *Mytilus* on the Northeast coast are characterized by extremely erratic spatial and temporal fluctuations in numbers (Seed, 1969*a*, *b*). Young mussels can appear on the adult bed at all times of the year. The dominant offshore alga is the brown kelp *Laminaria hyperborea* which supports a prolific red algal epiflora. At first sight, the commoner "leafy" epiphytes such as *Rhodymenia palmata* (L.), *Delesseria sanguinea* Lamour and *Membranoptera alata* Kylin would appear to be unsuitable primary settlement substrata for young mussels. The basal regions of these algae however support epiphytic populations of the long-spined form of *Electra pilosa* (see also Rogick and Croasdale, 1949 p. 31) which are particularly attractive to *Mytilus* spat (<3 mm in length). It is not unusual to find such regions so crowded with spat as practically to conceal the bryozoan. According to Hagerman (1966) hydroids and non-encrusting bryozoans epiphytic on *Fucus serratus* are settled on in quantity by *Mytilus* spat, and Stebbing (1971) has noted the attraction of the bryozoan *Flustra foliacea* (L.) for spat and juvenile mussels. A similar relationship between *Pecten* spat and bryozoan substrata has also been described by Eggleston (1962). The spiny nature of the substratum is obviously of pre-eminent importance in this connection and bryozoans probably play an important role in the provision of primary settlement surfaces for bivalve plantigrades.

Observation suggests that the physical irregularities provided by the surface of most bryozoan colonies are also important in providing leverage points which are exploited during locomotion by many vermiform animals within the holdfast community e.g. nematodes, errant polychaetes and nemertines.

2. Interspecific Competition

The sedentary fauna of *Laminaria* holdfasts is made up of a taxonomically heterogeneous assemblage which it is logical to consider in two parts, viz. those forms which are attached to the surface of the alga (hydroids, compound ascidians, sponges, bryozoans, certain polychaetes, barnacles and saddle oysters), and those which occupy the holdfast interstices (tubicolous amphipods, most of the sedentary tube-dwelling polychaetes and the remaining bivalves).

The surface attached forms would seem likely *a priori* to compete interspecifically for space. In a discussion of the importance of intraspecific competition to sedentary marine animals, Knight-Jones and Moyse (1961) described various mechanisms which have evolved to mitigate hardship, notably the spacing behaviour of some settling larvae and the ability of some young encrusting forms to fuse with one another and thus co-operate rather than compete. Colonial ascidians and bryozoans may possess the latter property and frequently form blanket colonies. In the present investigation, such colonies of *Alcyonidium* were often found growing over empty shells of the barnacle *Verruca* and the saddle oyster *Anomia*, but it is not possible to say whether the *Alcyonidium* growth smothered these animals or covered them subsequent to their death. Live *Verruca* have however been observed almost occluded by *Alcyonidium* so that only the cirri of the barnacle were visible. Smaller organisms, e.g. spirorbid polychaetes might be more easily overgrown, although these particular animals are extremely scarce on East coast kelp holdfasts.

Recent observations by Stebbing (1971) have shown that epizooites on the bryozoan *Flustra foliacea* (L.) exhibit a distinctive microzonation pattern, thereby reducing direct interspecific competition. Such mechanisms could well apply to the holdfast epifauna, in response perhaps to the intimate sediment regime.

3. Feeding Relationships

Representatives of all the important groups of bryozoan predators noted by Ryland (1970), i.e. sea-urchins, nudibranchs and pycnogonids, are to be found within kelp holdfasts.

The common sublittoral urchin *Echinus esculentus* L. is omnivorous and readily devours epiphytic bryozoans as well as their algal substrata. The smaller species *Psammechinus miliaris* (Gmelin) is fairly common in local holdfasts and was cited by Ryland (1970) as scraping *Electra pilosa* off mussel shells. Bryozoans probably contribute to the food of these species but the bulk of specialized bryozoan grazing is due to the nudibranchs.

The most spectacular of these predators, in the Northeast as elsewhere, is undoubtedly *Polycera quadrilineata* (Müller). It creates havoc with the large mats of *Membranipora membranacea* on kelp fronds, clearing its path of the bryozoan with its wide radula. *Polycera* is however largely confined to kelp fronds due to its preference for *M. membranacea*. Although it has been recorded as accepting *Electra pilosa* (Miller, 1961), it has only occasionally been taken in holdfasts. The commonest holdfast dwelling nudibranch along this coast is the small dorid *Onchidoris muricata* (Müller), occasional individuals of the related, predominantly intertidal, species *Acanthodoris pilosa* (Müller) being of small significance. Other holdfast nudibranchs prey on hydroids e.g. *Doto coronata* (Gmelin) and *Trinchesia viridis* (Forbes), or sponges e.g. *Jorunna tomentosa* (Cuvier).

According to Miller (1961) *Onchidoris muricata* feeds on a variety of bryozoan species but most commonly in algal habitats on *Membranipora membranacea, Electra pilosa* and *Celleporella hyalina*. During the course of the present data analyses a high statistical association (χ^2 with 1 d.f.$=13{\cdot}2$: correlation coefficient$=+0{\cdot}46$) was found between the occurrence of *Onchidoris* and the bryozoan *Alcyonidium* sp. (cf. *mytili/polyoum*). Whether in this instance this reflects a feeding relationship is not clear. Known food species such as *Electra* and *Celleporella* which were ubiquitously distributed would not be expected to show a high statistical association. Swennen (1961) has however added *Alcyonidium polyoum* to the list of bryozoans preyed upon by *Onchidoris muricata* (summarized by Thompson, 1964), and the above observation at least fulfils his first criterion for assessing the importance of a particular substratum as food for a benthic carnivore, i.e. that the carnivore should customarily be found in the field on or near the suspected food organism.

Pycnogonids, although consistently present in kelp holdfasts along the Northeast coast were rather few in number and probably make little impact on the Bryozoa.

BRYOZOA RECORDED FROM *Laminaria* HOLDFASTS IN NORTHEAST BRITAIN

This list is appended for the purpose of convenient comparison with other bryozoan supporting habitats and with kelp holdfast Bryozoa from different geographical areas. It is in no way intended to anticipate the complete treatment of the Bryozoa in the series "Marine Fauna of the Cullercoats District" which is in a state of advanced preparation by Dr D. Eggleston.

Abbreviations used below:

RER = Roper (1913)
PGM = author's personal observation
DML = Dove Marine Laboratory, unpublished record
WML = Wellcome Marine Laboratory, unpublished record.

CYCLOSTOMATA

Crisidia cornuta (L.) rather rare (RER: DML). WML
Crisia denticulata (Lamarck) rare (PGM). WML
C. ramosa Harmer WML
Berenicea patina (Lamarck) occasional, as *Diastopora* (RER: PGM: DML)

CHEILOSTOMATA

Scruparia chelata (L.) WML as *Eucratea*
Eucratea loricata (L.) WML as *Gemellaria*
Electra pilosa (L.) common (RER: PGM: DML: WML)
Conopeum reticulum (L.) WML as *Membranipora lacroixii* Audouin
Callopora lineata (L.) common (RER: PGM). DML. WML
C. aurita (Hincks) rare (PGM)
Membraniporella nitida (Johnston) rare (RER), Frequent (PGM). DML. WML
Flustra foliacea (L.) WML
Scrupocellaria scruposa (L.) common (RER). WML
S. reptans (L.) WML
Cribrilaria radiata (Moll) WML as *Cribrilina*
Celleporella hyalina (L.) common (RER: PGM). DML. WML
Schizoporella unicornis (Johnston) rare (PGM)
Schizomavella linearis (Hassall) WML. as *Schizoporella*
Escharina spinifera (Johnston) locally common (PGM). RER. DML
Microporella ciliata (Pallas) not common (PGM: RER)
Escharoides coccineus (Abildgaard) locally common (PGM), frequent (RER)
Umbonula littoralis Hastings infrequent (PGM). RER. DML. WML
Celleporaria pumicosa (Pallas) locally common (PGM: RER. DML) as *Cellepora*
Celleporina hassallii (Johnston) RER. as *Siniopelta*
**Smittina affinis* (Hincks) rare (PGM)

* This is only the third British record of *S. affinis* (Dr J. S. Ryland, personal communication identified by him).

CTENOSTOMATA

Alcyonidium hirsutum (Fleming) rare (PGM). WML

**A. polyoum* (Hassall) *A. mytili* Dalyell	common (PGM. DML)
†*A. albidum* Alder	rare (PGM)
Walkeria uva (L.)	occasional (PGM) as *Valkeria*

ACKNOWLEDGEMENTS

Dr J. R. Lewis is thanked for supervision. Facilities provided by Professors J. M. Dodd and R. McNeill Alexander are acknowledged. Dr J. S. Ryland is thanked for his interest and opinion on material quoted above. I am grateful to Professor R. B. Clark for allowing me to quote unpublished records of holdfast Bryozoa from the Dove Marine Laboratory and to Dr J. R. Lewis for allowing the inclusion of unpublished records from our own laboratory. Invaluable diving assistance was rendered by members of the Department of Botany, University of Durham and the East Yorkshire branch of the British Sub-Aqua Club. Messrs A. E. Simpson and J. E. Smithson provided photographic assistance.

The study was financed by N.E.R.C. and forms part of the Northeast Pollution Group's contribution to I.B.P. (section PM).

REFERENCES

Bayne, B. L. (1964). Primary and secondary settlement in *Mytilus edulis* L. (Mollusca). *J. Anim. Ecol.* **33,** 513–523.

Bock, K. J. (1950). Über die Bryozoen und Kamptozoen der Kieler Bucht. *Kieler Meeresforsch.* **7,** 161–166.

Bock, K. J. (1954). Einige Zahlen zur Bewuchsdichte von Epizoen auf Laminarien aus der östlichen Kieler Bucht. *Veröff. Inst. Meeresf. Bremerhaven* **3,** 42–45.

Colman, J. (1940). On the fauna inhabiting intertidal seaweeds. *J. mar. biol. Ass. U.K.* **24,** 129–183.

Cook, P. L. (1963). Observations on live lunulitiform zoaria of Polyzoa. *Cah. Biol mar.* **4,** 407–413.

Crisp, D. J. and Williams, G. B. (1960). Effect of extracts from fucoids in promoting settlement of epiphytic Polyzoa. *Nature, Lond.* **188,** 1206–1207.

Eggleston, D. (1962). Spat of the Scallop (*Pecten macimus* (L.)) from off Port Erin, Isle of Man. *Rep. Mar. biol. Sta. Port Erin,* **74,** 29–32.

Ghelardi, R. J. (1960). Structure and dynamics of the animal community found in *Macrocystis pyrifera* holdfasts. Unpublished Ph.D. thesis; University of California, La Jolla.

* It is as yet uncertain whether *A. polyoum* and *A. mytili* should be regarded as distinct species. Dr Ryland has examined some of this material and considered some specimens to be cf. *polyoum* and others to be cf. *mytili*. No attempt at separation has therefore been made.

† My provisional identification was *A. ?mamillatum*. Dr Ryland referred the material to *A. albidum* whilst remaining sceptical of the reality of the distinction between these species.

HAGERMAN, L. (1966). The macro- and microfauna associated with *Fucus serratus* L. with some ecological remarks. *Ophelia* **3,** 1–43.

JOLIET, L. (1877). Contributions à l'histoire naturelle des bryozoaires des côtes de France. *Arch. Zool. exp. Gen.* **6,** 193–304.

KITCHING, J. A. and EBLING, F. J. (1967). Ecological studies at Lough Ine. *In* "Advances in Ecological Research", Vol. 4, pp. 197–291. Academic Press, London and New York.

KNIGHT-JONES, E. W. and MOYSE, J. (1961). Intraspecific competition in sedentary marine animals. *Symp. Soc. exp. Biol.* **15,** 72–95.

LEWIS, J. R. (1957). An introduction to the intertidal ecology of the rocky shores of an Hebridean Island. *Oikos* **8,** 130–160.

LEWIS, J. R. (1964). "The ecology of rocky shores", 323 pp. English Univ. Press, London.

LEWIS, J. R. (1968). Water movements and their role in rocky shore ecology. *2nd European Symposium on Marine Biology, Sarsia* **34,** 13–36.

LILLY, S. J., SLOANE, J. F., BASSINDALE, R., EBLING, F. J. and KITCHING, J. A. (1953). The Ecology of the Loch Ine rapids with special reference to water currents. IV. The Sedentary Fauna of subilittoral boulders. *J. Anim Ecol.* **22,** 1, 87–122.

LOOSANOFF, V. L. and TOMMERS, F. D. (1948). Effect of suspended silt and other substances on rate of feeding of oysters. *Science, N.Y.* **107,** 69–70.

MATURO, F. J. S. (1959). Seasonal distribution and settling rates of estuarine Bryozoa. *Ecology* **40** (1), 116–127.

MILLER, M. C. (1961). Distribution and food of the nudibranchiate Mollusca of the south of the Isle of Man. *J. Anim. Ecol.* **30,** 95–116.

MOORE, P. G. (1971). The nematode fauna associated with holdfasts of kelp (*Laminaria hyperborea*) in North East Britain. *J. mar. biol. Ass. U.K.* **51** (3), 589–604.

MOORE, P. G. (1972). The kelp fauna of the North East coast—a preliminary assessment of pollution. *Rep. Challenger Soc.*

PRENANT, M. (1927). Notes éthologiques sur la faune marine sessile des environs de Roscoff, II. Spongiaires, Tuniciers, Anthozoaires, associations de la faune fixée. *Trav. Sta. Biol. Roscoff* **6,** 1–64.

PRENANT, M. (1932). Études de bionomie intercotidale. La baie et la pointe de Quiberon. *Trav. Sta. Biol. Roscoff* **10,** 37–103.

PRENANT, M. and TEISSIER, G. (1924). Notes éthologiques sur la faune marine sessile des environs de Roscoff. Cirripèdes, Bryozoaires, Hydraires. *Trav. Sta. Biol. Roscoff* **2,** 3–49.

ROGICK, M. D. and CROASDALE, H. (1949). Studies on marine Bryozoa. III. Woods Hole region Bryozoa associated with algae. *Biol. Bull., Woods Hole* **96,** 32–69.

ROPER, R. E. (1913). The marine Polyzoa of Northumberland. *Rep. Dove Mar. Lab.* n.s. II., 36–57.

ROUND, F. E., SLOANE, J. F., EBLING, F. J. and KITCHING, J. A. (1961). The ecology of Lough Ine. X. The hydroid *Sertularia operculata* (L.) and its associated flora and fauna: effects of transference to sheltered water. *J. Ecol.* **49,** 617–629.

RYLAND, J. S. (1959*a*). The settlement of Polyzoa larvae on algae. *Proc. XVth Int. Congr. Zool.* 236–239.

RYLAND, J. S. (1959*b*). Experiments on the selection of algal substrates by polyzoan larvae. *J. exp. Biol.* **36,** 613–631.

RYLAND, J. S. (1962). The association between Polyzoa and algal substrata. *J. Anim. Ecol.* **31,** 331–338.

RYLAND, J. S. (1970). "Bryozoans," 175 pp. Hutchinson University Library, London.

SCARRATT, D. J. (1961). The fauna of *Laminaria* holdfasts. Unpublished Ph.D. thesis: University of Wales, Aberystwyth.

SEED, R. (1969*a*). The ecology of *Mytilus edulis* L. (Lamellibranchiata) on exposed rocky shores I. Breeding and settlement. *Oecologia (Berl.)* **3,** 277–316.

SEED, R. (1969*b*). The ecology of *Mytilus edulis* L. (Lamellibranchiata) on exposed rocky shores II. Growth and mortality. *Oecologia (Berl.)* **3,** 317–350.

SLOANE, J. F., EBLING, F. J., KITCHING, J. A. and LILLY, S. J. (1957). The ecology of the Lough Ine rapids with special reference to water currents V. The sedentary fauna of the laminarian algae in the Louth Ine area. *J. Anim. Ecol.* **26,** 197–211.

STEBBING, A. R. D. (1971). The epizoic fauna of *Flustra foliacea* (Bryozoa). *J. mar. biol. Ass. U.K.* **51** (2), 283–301.

SWENNEN, C. (1961). Data on distribution, reproduction and ecology of the nudibranchiate molluscs occurring in the Netherlands. *Neth. J. Sea Res.* **1,** 191–240.

THOMPSON, T. E. (1964). Grazing and the life cycles of British nudibranchs. *In* "Grazing in Terrestrial and Marine Environments." (D. J. CRISP, ed.), pp. 275–297, *4th Symp. Br. Ecol. Soc.* Blackwell, Oxford.

THOMPSON, T. E., SMITH, S. T., JENKINS, M., BENSON-EVANS, K., FISK, D., MORGAN, G., DELHANTY, J. E. and WADE, A. E. (1966). Contributions to the biology of the Inner Farne. *Trans. nat. Hist. Soc. Northumberland* **15,** No. 5 (n.s.), 197–225.

URSIN, E. (1960). A quantitative investigation of the echinoderm fauna of the central North Sea. *Medd. Komm. Havundersøg., Kbn* N.S. **2** (23), 1–204.

4 | Seasonal Diversity and Density in Bryozoan Populations of Block Island Sound (New York, U.S.A.)

MARIE B. ABBOTT

The University of Connecticut, Storrs, Connecticut, U.S.A.
and
Systematics-Ecology Program, Marine Biological Laboratory, Woods Hole, Massachusetts, U.S.A.

Abstract: Relatively high diversity in a bryozoan fauna of 29 species is attributed to the stable oceanic character of the water masses of Block Island Sound. Throughout an annual temperature range of 1·8°–18·2°C nine encrusting cheilostome species function vegetatively, and two of them, *Cribrilina punctata* and *Hippoporina* cf. *H. verrilli*, also reproduce sexually.

Somewhat lower winter diversity (H=2·62–2·97 cf. 3·05–3·60 in summer) reflects a smaller total number rather than a less equitable distribution of species, owing to the reduction in the number of cellulariform and stolonate species which either require or prefer the soft substrates more abundant in summer.

Density, which averages 235 colonies/0·1 m^2, but varies extremely, appears dependent more on the character and availability of hard and soft substrates than directly on temperature.

The known ranges of two species, *Chorizopora brongniartii* and *Porella reduplicata*, are extended northward and southward, respectively.

This is a first report on a long-range intensive study, in progress since 1969, of the bryozoan component of the epifaunal community at a permanent oceanographic station, hereafter called Station A, located in northwestern Block Island Sound, New York, U.S.A. (Fig. 1).

MATERIALS AND METHODS

The central study collection is a suite of 42 quantitative biological samples taken at Station A at regular intervals between 26 June 1969 and 17 November 1970. The sampling dates, together with depth, salinity, and surface and bottom temperatures for each, are given in Table 1. They encompass the seasonal temperature changes of about one and a half years. Since the station was re-located by Decca navigation each time it was occupied, and the maximum deviation

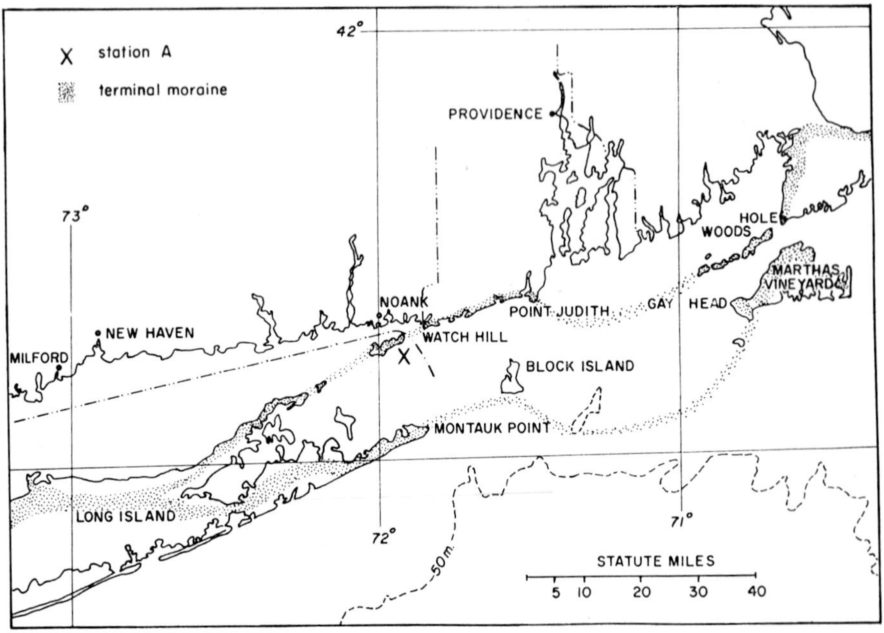

Fig. 1. Locality index map, showing location of oceanographic Station A in relation to the terminal moraines of southern New England, including their submerged sections.

Table 1. Oceanographic data for biological samples taken at Station A, Block Island Sound (New York, U.S.A.): June 1969–November 1970

DATE	DEPTH (m)	TEMPERATURE, °C Bottom		TEMPERATURE, °C Surface	SALINITY ‰
26 June 1969	26·8	16·4		16·8	31·29
10 July 1969	27·7	16·2		17·8	31·46
11 Aug. 1969	36·9	17·1		19·5 High	31·69
18 Aug. 1969*	36·0	16·7		19·0	31·71
25 Sept. 1969*	38·4	17·5 High		18·0	31·46
02 Oct. 1969	28·6	17·4		18·2	31·64
09 Oct. 1969*	35·4	17·2		17·8	31·66
17 Oct. 1969	35·4	16·7		17·1	31·59
24 Oct. 1969	36·6	15·0		14·7	31·97
30 Oct. 1969	28·6	12·4	=	12·4	32·00
08 Nov. 1969	28·6	13·3		13·4	31·64
13 Nov. 1969	34·1	12·5		12·7	31·59
18 Nov. 1969	30·5	11·7		11·9	31·57

Table 1. (*cont'd*)

DATE	DEPTH (m)	TEMPERATURE, °C Bottom		TEMPERATURE, °C Surface	SALINITY ‰
13 Dec. 1969	36·3	8·0	=	8·0	31·22
17 Dec. 1969	32·6	7·6		7·6	31·23
29 Dec. 1969	48·8	5·1		5·0	31·60
08 Jan. 1970*	33·5	3·7		3·9	30·12
16 Jan. 1970	37·8	2·8	=	2·8	31·17
26 Jan. 1970*	29·6	1·8 Low		1·7 Low	31·23
30 Jan. 1970	29·6	2·2		2·1	31·48
14 Feb. 1970	26·2	2·4		2·2	31·59
20 Feb. 1970	31·1	2·1		2·0	30·95
05 Mar. 1970*	29·9	2·5		2·4	30·80
12 Mar. 1970	29·9	2·7		2·3	31·10
19 Mar. 1970	25·9	2·9	=	2·9	31·27
01 Apr. 1970	29·6	3·6		3·8	30·72
08 Apr. 1970	32·6	4·1		4·4	30·36
17 Apr. 1970	37·2	4·9		5·7	30·96
23 Apr. 1970	37·2	6·1		6·8	31·07
30 Apr. 1970	36·9	6·1		6·8	31·07
14 May 1970	38·1	6·2		8·0	31·03
21 May 1970	45·7	7·8		9·2	30·22
28 May 1970	33·5	9·4		10·7	30·46
24 June 1970	36·0	12·3		14·4	30·74
23 July 1970	44·2	15·1		16·7	30·99
05 Aug. 1970	40·0	15·8		19·1 } High	31·59
20 Aug. 1970	32·3	18·2 High		19·1 } High	31·36
17 Sept. 1970	38·1	17·2		17·5	31·35
01 Oct. 1970	31·7	16·7		17·5	31·57
15 Oct. 1970	34·4	16·1		17·0	31·50
29 Oct. 1970	32·9	14·9		15·1	31·18
17 Nov. 1970	41·4	12·7	=	12·7	31·38

Note: Depths by ship's fathometer, temperatures by bathythermograph, salinities by salinometer.
* Discussed quantitatively in this paper.

at sampling time represents a horizontal distance of probably 300 m at most, the bryozoan and other populations sampled may reasonably be considered completely sympatric and within a given sample synchronous as well.

Only substrate units (shell and pebble) greater than $\frac{1}{4}$ in. (6 mm) in size and their attached algae and epifauna were retained after shipboard screening. These were sorted into barren and bryozoan-bearing units. Some bryozoan colonies on substrate units smaller than $\frac{1}{4}$ in. (6 mm) were undoubtedly lost and the quantitative distribution data are therefore not fully representative of the bryozoan content of the 0·1 m^2 area encompassed by the grab. Nevertheless they are at least consistent within this and related studies based on the same collections, and the method itself can be readily duplicated.

Six population samples, taken at and near the times of the highest and lowest bottom water temperatures, were selected for a comparison of summer and winter density and diversity in this bryozoan community.

CHARACTER OF THE BRYOZOAN FAUNA

The sample suite contains several thousand living colonies representing at least 29 species in 23 genera (Table 2). It was not possible to assign firm names to some of the species present; however, all of the species were distinguishable from each other by clear-cut qualitative gaps usually requiring no further quantitative confirmation. The one exception is the genus *Amphiblestrum*, in which young or non-ovicelled colonies of the two species apparently cannot be distinguished from each other. For density and diversity counts the two species are therefore treated together under *A. flemingii* (s.l.).

Table 2. List of bryozoan species taken alive, Station A, Block Island Sound (New York, U.S.A.): June 1969–November 1970

NAME	ABUNDANCE	
CYCLOSTOMATA Busk 1852		
Order Articulata Busk 1859		
1. *Crisia eburnea* (L., 1758)		
Order Rectangulata Waters 1887		
2. *Disporella* cf. *D. hispida* (Fleming, 1828)	RR	
GYMNOLAEMATA Allman 1856		
Order Ctenostomata Busk 1852		
Suborder Carnosa Gray 1841		
3. *Alcyonidium* cf. *A. polyoum* (Hassall, 1841)	RR	
Order Cheilostomata Busk 1852		
Suborder Anasca Levinson 1909		
Superfamily Inovicelloidea Jullien 1888		
4. *Aetea anguina* (L., 1758)		
Superfamily Malacostegoidea Levinsen 1902		
Family Membraniporidae Busk 1854		
5. *Membranipora tenuis* Desor 1848*		P
Family Electridae Lagaaij 1952		
6. *Electra monostachys* (Busk, 1854)*		P
7. *Pyripora catenularia* (Jameson, 1814)*	RR	
8. *Electra* sp. (? n. sp.)*	RR	

Table 2. (*cont'd*)

NAME	ABUNDANCE	
Family Calloporidae Norman 1903		
9. *Callopora craticula* (Alder, 1857)		P
10. *C. aurita* (Hincks, 1877)	RR	
11. *Amphiblestrum flemingii* (Busk, 1854)		P
12. *A. osburni* Powell, 1968		
13. *Tegella unicornis* (Fleming, 1828)	RR	
Superfamily Cellularioidea Smitt 1867		
14. *Bicellariella ciliata* (L., 1758)		
15. *Bugula* cf. *B. turrita* (Desor, 1848)		
Suborder Cribrimorpha Lang 1916		
16. *Cribrilina punctata* (Hassall, 1841)		P
Suborder Ascophora Levinsen 1909		
17. *Hippothoa divaricata* Lamouroux, 1821		
18. *H. hyalina* (L., 1767)	R	
19. *Chorizopora brongniartii* (Audouin, 1826)		
Family Microporellidae Hincks 1880		
20. *Microporella ciliata* (Pallas, 1766)	R	
Family Schizoporellidae Jullien 1903		
21. *Schizoporella unicornis* (s.l.) (Johnston, 1847)	RR	
Family Hippoporinidae Osburn 1952		
22. *Hippoporina porosa* (Verrill, 1879)	D	
23. *H. americana* (Verrill, 1875)	D	
24. *H.* cf. *H. verrilli* Maturo and Schopf, 1968	D	
Family Smittinidae Levinson 1909		
25. *Porella reduplicata* (Osburn, 1933)		P
26. *Smittina* sp. (n. sp. ?)	R	
27. *Parasmittina nitida* (Verrill, 1875)		
28. *Escharella* cf. *E. immersa* (Fleming, 1828)		
Family Celleporidae Busk 1852		
29. *Cellepora avicularis* Hincks, 1860		

D = dominant: present at all temperatures; together constitute 32–65% of all samples.
P = persistent: present at all temperatures but not necessarily in each sample; together constitute 6–53% of all samples.
R = rare: <25 colonies in several thousand scanned.
RR = very rare: <5 colonies in several thousand scanned.
* planktotrophic (cyphonautes) larva.

Perhaps the most immediately striking characteristic of the bryozoan fauna at Station A is the high (29) species diversity in an area certainly less than 1 km². The nearest parallel bryozoan fauna (38 spp.) is probably that described from one day's dredging at 50 fathoms in the Bay of Fundy off Digby, Nova Scotia (Powell, 1968). Comparison with that fauna and with the 88 Woods Hole area and 80 Mount Desert Island species (Osburn, 1912, 1933) and 42 Long Island

Sound species (Hutchins, 1945), each from a very much larger geographic and bathymetric range, indicates that this Block Island Sound fauna is not an impoverished one.

The 29 species so far recorded from this station may be classed as dominant, persistent (D and P, respectively, in Table 2) or accessory and possibly opportunistic. The three species of the encrusting ascophoran genus *Hippoporina* Neviani (*H. porosa*, *H. americana*, and *H.* cf. *H. verrilli*) are not only present but dominant in all population samples, although in varying proportions, throughout the full annual temperature cycle of 1·8–18·2°C. Together they constitute 32–65% of the bryozoan component of the epifauna at all seasons in all 20 samples quantitatively processed to date. Six additional, also encrusting, cheilostomes (*Cribrilina punctata*, *Amphiblestrum flemingii* [s.l.], *Callopora craticula*, *Porella reduplicata*, *Electra monostachys*, and *Membranipora tenuis*) are present at all temperatures but do not necessarily appear in each population sample. Together they constitute a smaller (16–53%) but highly persistent fraction of the bryozoan fauna.

The occurrence and seasonal distribution pattern of the remaining 19 accessory species—of which at least two are undescribed new species—appear much more irregular and unpredictable. Of these, seven (RR in Table 2), including the only ctenostome and one of the only two cyclostomes, are extremely rare, with less than five and usually only one or two colonies of each recorded. Fewer than 25 colonies have so far been listed for each of three other (R in Table 2) somewhat less rare species.

The bryozoan fauna presents a distinctly boreal aspect. One of the persistent species, *Porella reduplicata*, is a northern one whose known range is extended southward in this study; two others, *Cribrilina punctata* and *Callopora craticula*, have been widely regarded as boreal and appear to reach their maximum development well north of Cape Cod (Osburn, 1933). On the other hand, one of the not uncommon accessory species, *Chorizopora brongniartii*, has not previously been reported north of Hudson Canyon (Maturo, 1968). A further zoogeographic anomaly lies in the absence or comparative rarity of species abundant (Osburn, 1912; Rogick and Croasdale, 1944; Hutchins, 1945; Stickney and Stringer, 1957; Abbott, 1971) in warmer and shallower waters of and south of Cape Cod, e.g.: *Callopora aurita*, *Schizoporella unicornis* (s.l.), *Bowerbankia gracilis*, *Parasmittina nitida*, *Microporella ciliata*, *Hippothoa hyalina*.

DIVERSITY AND DENSITY: SUMMER *VS* WINTER

The average density (number of bryozoan colonies/0·1 m²) at Station A, based on all data in hand to date, is about 235, but varies extremely widely between 15 (on 5 March 1970) and 561 (on 9 October 1969). The number of species/0·1 m² averages about 14 and ranges between 8 (also on 5 March 1970) and 22 (on 17 November 1970, replicate sample No. 3).

To assess the possible role of temperature in producing the great ranges in density and species diversity, and to detect any concomitant changes in dominance diversity (the distribution pattern of the individual colonies among the species present), six population samples—three each at or near the annual minimum and maximum bottom water temperatures—were analysed in detail. For each sample the following observations or calculations were made: (1) The total number of substrate units and the number and percentage of total of bryozoan bearing units. (2) The diversity of the bryozoan community expressed as a Shannon–Wiener index (H). (3) For each species present: (a) the number and percentage of total of living colonies; (b) the presence or absence of very young colonies, arbitrarily defined here as an ancestrula plus less than four generations of zooids; (c) for ovicelled species: the presence or absence of ovicells; if present, their stage of development—whether incipient, fully formed, or already inhabited; (d) for species present on both hard (shell and pebble) and soft (algae and hydroids) substrates, the number and percentage of total colonies occupying the latter. The data for the summer samples are shown in Table 3, those for winter in Table 4.

Both the highest and the lowest densities occur at or near the highest and lowest bottom water temperatures and at the end of relatively long periods (see Fig. 2) of more or less stable near-maximum or -minimum temperatures. The percentage of substrate units carrying bryozoa varies between relatively narrow limits (21–34%), regardless of bryozoan population density, temperature or absolute number of substrate units present. This ratio may have no biological significance, in that it may represent only those pieces of shell and pebble lying on the plane of the sediment–water interface, and therefore available to passing larvae, out of the entire hemi-cylinder of sediment and in- and epi-fauna gouged out by the sampler.

The three winter populations sampled exhibit markedly lower species diversity than the three summer populations (12 *vs* 22). Only one of the dominant-persistent species (*Porella reduplicata*) is not represented in these winter samples, and that species has been recorded at similar temperatures in other samples. The principal loss in species number lies in the markedly fewer accessory species, most especially those species, mainly cellulariform or stolonate, that appear able to settle facultatively on either hard or soft substrates (*Bugula* cf. *B. turrita*, *Bicellariella ciliata*, *Aetea anguina*, *Hippothoa hyalina*, *H. divaricata*) and in the absence of the single species (*Cellepora avicularis*) which is virtually never seen on other than soft substrates. It is possible that prolonged low winter

Table 3. Diversity and density: bryozoan populations at Station A, Block Island Sound (New York, U.S.A.): at or near summer maximum temperatures

SPECIES	16·7°—18 August 1969[1] Colonies/0·1 m^2, % *total*			17·5°—25 September 1969 (High) Colonies/0·1 m^2, % *total*			17·4°—9 October 1969 Colonies/0·1 m^2, % *total*		
Hippoporina porosa	5·5	*3·82*	y, ●	27	*5·32*	y, √	54	*9·62*	y, ●
H. americana	22	*15·28*	y, ●	66	*13·02*	y, ●	80	*14·26*	y, ●
H. cf. *H. verrilli*	18	*12·50*	y, ●	25	*4·93*	y, ●	75	*13·37*	y, ●
Cribrilina punctata	22	*15·28*	y, ●	95	*18·74*	y, ●	71	*12·66*	y, ●
Amphiblestrum flemingii (s.l.)[2]	11·5	*7·99*	y, ●	48	*9·47*	y, ●	63	*11·23*	y, ●
Callopora craticula	2	*1·39*	y, ○	36	*7·10*	y, ○	33	*5·88*	y, ○
Porella reduplicata	0·5	*0·35*		5	*0·99*	y	6	*1·07*	y
Electra monostachys	6·5	*4·51*	y, *	30	*5·92*	y, *	17	*3·03*	*
Membranipora tenuis	1	*0·69*	*	6	*1·18*	y, *	4	*0·71*	*
Aetea anguina	1·5 (s: 0·5)	*1·04*	y	10 (s: 6)	*1·97*	y	20 (s: 7)	*3·56*	y
Electra sp. (n. sp. ?)		—		1	*0·20*	*		—	
Callopora aurita		—		1	*0·20*	○		—	
Bugula cf. *B. turrita*	38·5 (s: 0·5)	*26·74*	y	26 (s: 5)	*5·13*	y	43 (s: 5)	*7·66*	y
Bicellariella ciliata	12·5 (s: 55)	*8·68*	y	39 (s: 34)	*7·69*	y	52 (s: 30)	*9·27*	y
Hippothoa hyalina		—			[D]		1 (s: 1)	0·*18*	y

H. divaricata	1	*0·69*	y	8 (s: 8)	*1·58*	y	3 (s: 2)	*0·53*	y	
Chorizopora brongniartii		—		16	*3·16*	y, ○	15	*2·67*	y, ○	
Parasmittina nitida	0·5	*0·35*		3	*0·59*			[D]		
Escharella cf. *E. immersa*	1	*0·69*	y	2	*0·39*		1	*0·18*		
Smittina sp. (n. sp.)		—			—		1	*0·18*	y	
Cellepora avicularis		—		63 (s: 62)	*12·43*	y	21 (s: 20)	*3·74*	y	
Alcyonidium cf. *A. polyoum*		—			[D]		1	*0·18*		
DENSITY										
Living colonies/0·1 m²	144 (s: 6·5	*4·51*)		507 (s: 115	*22·44*)		561 (s: 65	*11·59*)		
DIVERSITY										
Living species	15			19			19			
Total species in 3 summer samples										22
Diversity index (H)	3·0515			3·6036			3·5561			
SUBSTRATE UNITS										
With bryozoa	175	*21·19*		281	*33·86*		281	*28·02*		
Without bryozoa	651			549			722			
Total	826			830			1003			

[1] Based on 0·2 m² sample. [2] Includes *A. osburni*.

Note: Explanation of symbols used: [D] = only dead colonies present; * = larvae planktotrophic; y = very young colonies (ancestrula plus <4 generations) present.

√ developing ovicells present ○ fully developed ovicells present ● ovicells inhabited by embryoes present.

(s:—) colonies on soft substrate (algae or hydroids).

Table 4. Diversity and density: bryozoan populations at Station A, Block Island Sound (New York, U.S.A.): at or near winter minimum temperatures

SPECIES	3·7°—8 January 1970			1·8°—26 January 1970 (Low)			2·5°—5 March 1970		
	Colonies/0·1 m², % *total*			Colonies/0·1 m², % *total*			Colonies/0·1 m², % *total*		
Hippoporina porosa	9	*9·57*	y, √	1	*1·96*	√	3	*20·00*	√
H. americana		*9·57*	√		*21·57*	y, ○	1	*6·67*	
H. cf. *H. verrilli*	37	*39·36*	y, ●	8	*15·69*	y, ●	3	*20·00*	√
Cribrilina punctata	15	*15·96*	y, ●	11	*21·57*	y, ○	3	*20·00*	y
Amphiblestrum flemingii (s.l.)[1]	10	*10·64*	y, ○	6	*11·76*	y	2	*13·33*	
Callopora craticula	8	*8·51*	y, ○	3	*5·88*	y, ○	1	*6·67*	○
Electra monostachys	2	*2·13*	*	5	*9·80*	*		—	
Membranipora tenuis	1	*1·06*	y, *	2	*2·92*	y, *	1	*6·67*	y, *
Pyripora catenularia		—		2	*3·92*	*		—	
Hippothoa divaricata		—		2	*3·92*	y		—	
Parasmittina nitida	1	*1·06*	y		—		1	*6·67*	
Escharella cf. *E. immersa*	2	*2·13*	y		[D]			—	
DENSITY									
Living colonies/0·1 m²	94			51			15		
DIVERSITY									
Living species	10			10			8		
Total species in 3 winter samples									12
Diversity index (H)	2·6224			2·9671			2·8226		
SUBSTRATE UNITS									
With bryozoa	56	*24·45*		58	*25·00*		9	*29·03*	
Without bryozoa	173			174			22		
Total	229			232			31		

[1] Includes *A. osburni*.
Note: For explanation of symbols used, see Table 3.

temperatures may repress either sexual reproduction and larval dispersal in these species or the proliferation of any required substrate, or both. In this connection, the restriction of the allegedly warm or temperature species (*Callopora aurita*, *Chorizopora brongniartii*) to the highest summer temperatures and the appearance of the presumably subarctic *Pyripora catenularia* at the 1·8°C minimum temperature may be significant.

As measured by the six Shannon–Wiener indices, the diversity of the bryozoan community is relatively high (2·62 or more) in both summer and winter, but appears to be significantly greater (3·05 or more) in summer. Since this diversity index is sensitive to both the total number and the proportional representation of the species present, the contribution of both components of the index must be segregated. Buzas and Gibson (1969, p. 73) suggest that the ratio of the natural log *e* raised to the (H) power to the number of species may be a measure of species equitability, the ratio being 1 when the species are perfectly evenly distributed. For the winter populations at Station A this ratio averages 0·10, for the summer 0·07. This, together with inspection of the data tables, may indicate that the higher summer diversity is more the expression of a greater number than of a more equitable distribution of species.

The association of very young colonies with colonies with fully developed ovicells inhabited by embryos constitutes at least presumptive evidence that gametogenesis and larval growth, release, and settlement are occurring in the populations observed. Such an association was seen at both winter and summer temperature extremes in only two species, *Hippoporina* cf. *H. verrilli* and *Cribrilina punctata*. The citation (Friedman and Sanders, 1970) of *Cribrilina punctata* as an organism very sensitive to temperature and one not extending south of Cape Cod is therefore not supported in this study.

Very young colonies and colonies with fully developed but uninhabited ovicells in the same population (e.g. *Callopora craticula* in both summer and winter, *Chorizopora brongniartii* in summer, *Amphiblestrum flemingii* (s.l.) and *Hippoporina americana* in winter) suggests that sexual reproduction may be possible in these species at the given temperatures. In the absence of data on zooid growth rate and its relation to remperature in the species under study, the significance of incipient ovicells, with or without associated very young colonies, is difficult to assess.

THE ABIOTIC ENVIRONMENT

The hydrography and topography of the continental shelf of southeastern New England are dominated by the fact that this region is one of submerged terminal and ground moraine. Station A itself occupies the steep, southeastward facing

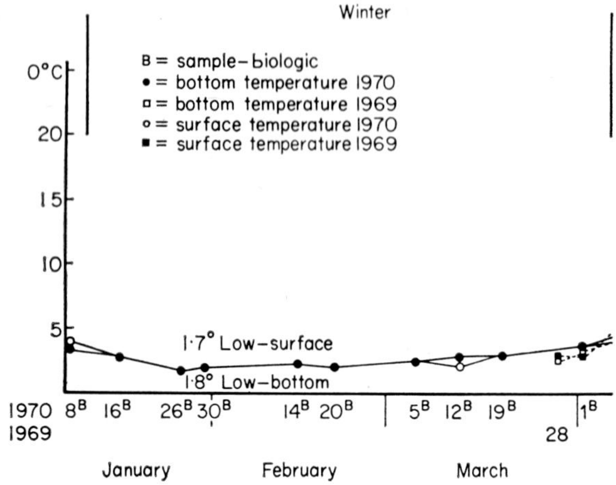

Fig. 2 (i)

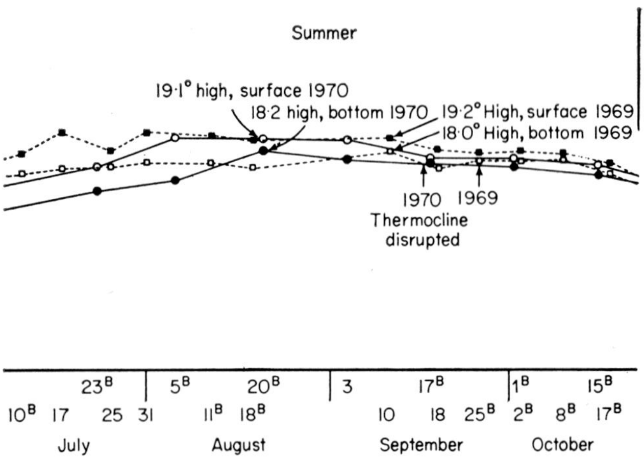

Fig. 2 (ii)

slope of a northern branch of terminal moraine (Fig. 1) which forms a series of islands and shoals restricting circulation in and out of the inner shallow bays and sounds despite the existence of strong tidal currents which can frequently reach velocities of 4 or 5 knots. South of this barrier, however, waters from the open outer continental shelf flow freely into the Sound, bringing in larvae, stabilizing salinity and removing both the upper and lower extremes of temperatures that obtain in the shallow protected inner waters (Verrill, 1874; Riley, 1952).

The temperature regime at the station at both surface and bottom during the

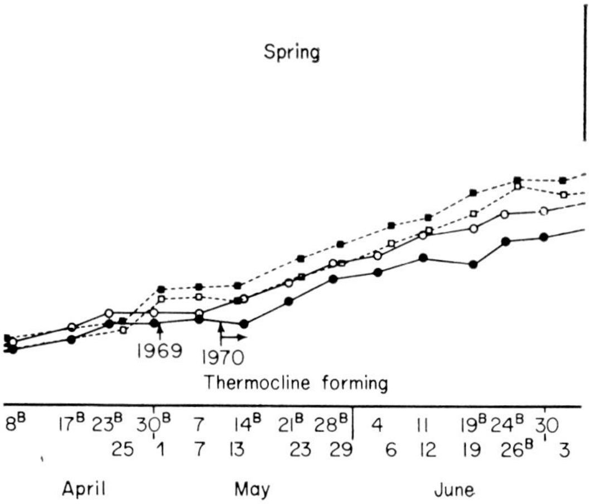

Fig. 2 (iii)

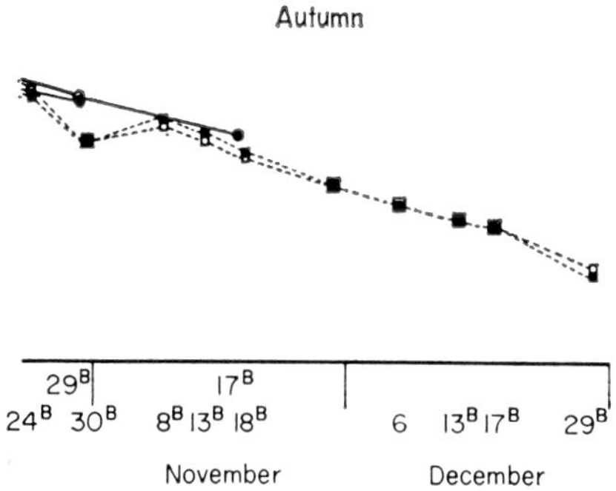

Fig. 2 (iv)

Fig. 2(i–iv). Charts showing surface and bottom temperature regime at Station A, Block Island Sound (New York, U.S.A.) between March 1969 and November 1970. Data in part from Table 1.

18 month period from late March 1969 to mid-November 1970 is plotted in Fig. 2. During this interval bottom temperatures ranged between 1·8°C in late January and 18·2°C in mid-August 1970, a range throughout which the nine dominant-persistent species can survive and reproduce vegetatively and at least two of them can reproduce sexually. Water temperatures are isothermal except during the four-month period between mid-May and mid-September, when a shallow thermocline forms. The thorough mixing of waters for eight months of the year may bring in the occasional rare accessory species.

Sample depths range between 27 and 49 m (Table 1), but only between 30 and

38 m in the six analysed in this study. Salinity fluctuates narrowly between 30·12 and 32·36‰. Both depth and salinity can therefore probably be dismissed as ecological limiting factors, at least for this phase of the study.

The nature and availability of substrate on a microenvironmental scale may be the most important factor in determining the overall high but extremely variable bryozoan population density. The potential substrate units (pebble and shell fragments largely slumped from above) are mixed in varying proportions with blue-gray mud, loose silt, and sand. All are partly sorted and constantly redistributed by the scouring tidal currents. The ratio of substrate units more than $\frac{1}{4}$ in. (6 mm) in diameter to the smaller units taken in the 0·1 m² of a given sample is probably largely a matter of chance. Since densities below a certain, as yet undetermined, critical value undoubtedly affect diversity, the validity of the 5 March 1970 data may be open to question. As mentioned previously, the presence or abundance of soft substrate, especially of macroalgae, on the larger and therefore more stable hard substrate units significantly increase diversity and possibly density as well in the bryozoan populations.

CONCLUSIONS

Diversity in marine benthic communities has been reviewed recently (Sanders, 1968) with regard to soft bottom infauna, but rather little is known of the diversity and bioenergetics of epifaunal communities. This study demonstrates that despite their small size and highly varying biomass bryozoan colonies can comprise a significant element in epifaunal communities in both abundance and diversity of species.

This study has also yielded data on the reproductive ecology and latitudinal and bathymetric distribution of some 29 species in the little known southern New England bryozoan fauna. The high overall diversity of this fauna may be attributed to the relatively stable oceanic character of the water masses of Block Island Sound.

The nine dominant-persistent species owe their success to their ability to function at the site, regardless of the availability or abundance of soft substrate, over the entire annual temperature range. Data presented in this paper suggest that this ability may be due to one of two mechanisms: (1) continuous sexual reproduction and dispersal of larvae at all times and temperatures, as has been demonstrated for two species, or (2) cyclic repopulation by alternating and overlapping yet distinct and possibly temperature-sensitive phases of sexual reproduction and vegetative growth and regeneration.

ACKNOWLEDGEMENTS

The biologic samples were collected under U.S. Navy Contract No. N 00140–69–C–0031 to the Marine Research Laboratory of the University of Connecticut through the courtesy of Dr A. J. Nalwalk, chief scientist. This paper is Contribution No. 84 of the Marine Research Laboratory.

REFERENCES

ABBOTT, M. B. (1971). Systematics and Ecology of Populations of "*Hippoporina*" Neviani from Block Island Sound, New York. PhD Thesis (unpublished), University of Connecticut, 166 pp.

BUZAS, M. A. and GIBSON, T. G. (1969). Species diversity: Benthonic foraminifera in western North Atlantic. *Science, N.Y.* **163,** 72–75.

FRIEDMAN, G. M. and SANDERS, J. E. (1970). Coincidence of high sea level with cold climate and low sea level with warm climate: Evidence from carbonate rocks. *Bull. Geol. Soc. Am.* **81,** 2457–2458.

HINCKS, T. (1880). "A History of the British Marine Polyzoa." 601 pp. J. van Voorst, London.

HUTCHINS, L. W. (1945). An annotated check list of the saltwater bryozoa of Long Island Sound. *Trans. Conn. Acad. Arts and Sci.* **36,** 533–551.

MATURO, F. J. S., Jr. (1968). The distributional pattern of the bryozoa of the east coast of the United States exclusive of New England. *Atti Soc. It. Nat. Museo Cir. St. Nat.* (Milano) **108,** 261–284.

OSBURN, R. C. (1912). The bryozoa of the Woods Hole region. *Bull. Bur. Fish.* **30,** 205–266.

OSBURN, R. C. (1933). Bryozoa of the Mt. Desert Island region. *In* "Biol. Surv. Mt. Desert Is." (Wistar Inst., Philadelphia), pp. 291–385.

POWELL, N. A. (1968). Studies on bryozoa of the Bay of Fundy region. Part II: Bryozoa from fifty fathoms, Bay of Fundy. *Cah. Biol. Mar.* **9,** 247–259.

RILEY, G. A. (1952). Hydrography of Long Island and Block Island Sounds. *Bull. Bingham Oceanog. Inst.* **13,** 5–39.

ROGICK, M. D. and CROASDALE, H. (1949). Studies on marine bryozoa. Part III: Woods Hole region bryozoa associated with algae. *Biol. Bull.* **96,** 32–69.

SANDERS, H. L. (1968). Marine benthic diversity: A comparative study. *Am. Nat.* **102,** 243–282.

STICKNEY, A. P. and STRINGER, L. D. (1957). A study of the invertebrate bottom fauna of Greenwich Bay, Rhode Island. *Ecology* **38,** 111–122.

VERRILL. A. E. (1874). Report upon the invertebrate animals of Vineyard Sound and adjacent waters. *In* Report, U.S. Fish Comm. 1871–1872, Pt. VIII: 296–747.

5 | Bryozoan Ecology in Relation to Sediment Texture in the "Golfo dell'Asinara" (Sardinia, Italy)

E. ANNOSCIA

Via Mincio, Milano, Italy

and

G. FIERRO

Istituto di Geologia dell'Università, Via Balbi, Genova, Italy

Abstract: The results of a preliminary study of bryozoan ecology, using Stach's method, in relation to textural analyses of the sediments from the "Golfo dell'Asinara" and "Bocche di Bonifacio" are reported here.

The samples studied were of at least four distinct environmental types.

The samples collected in the straits between the islands of Corsica and Sardinia ("Bocche di Bonifacio") were the more definite reflecting strong bottom current action with a "negative skewness" and the presence of the "pralines facies" with *Venus casina*.

INTRODUCTION

An Oceanographic Cruise was organized by the "Istituto Idrografico della Marina Italiana" and "Consiglio Nazionale delle Ricerche" in the "Golfo dell'Asinara" (Sardinia) and "Bocche di Bonifacio" (Corsica) during the season of 1964.

One of us (G. Fierro) dredged 120 samples of the bottom sediments for sedimentological studies.

Trotti (1967) published the oceanographic data collected during the Cruise and Bregant (1969) published the results of the determination of the phosphate and nitrate concentrations. The morphological observations, lithological map of the sea bottom and study of the sediment transport with the aid of the heavy minerals, were carried out and published by Fierro (1965, 1969, 1970) (see also Fig. 3).

The preliminary results of the sedimentological study having made evident different zones of more or less strong hydrodynamic energy, the study of the zoarial habit of the Bryozoans was deemed interesting in order to differentiate ecological zones, if any. A study in this direction was carried out by G. Podestà (1969) for his thesis of Master of Geological Sciences with the aid of the writers, who continued the investigation further.

The long time needed to study all the samples collected and the results of this preliminary investigation prompted us to publish this brief note.

A tentative scheme of at least four environmental types (named here "A", "B", "C", "D") resulted from the preliminary comparison of the ecological study of the Bryozoans with the sedimentological data in the area examined (Fig. 2).

An interesting positive correlation was observed between the "B" environment, defined by the analysis of the Bryozoa, and the "negative Skewness" zone of the sedimentological study.

This investigation confirms the results obtained by Bellan *et al.* (1961) in the southwest waters of Corsica based on the study of the biocenosis distribution.

REMARKS ON THE OBSERVED MATERIAL

Only a few samples (about 10%) were examined from the point of view of the Bryozoan zoarial habit. The preliminary choice of these samples was possible following the results of the sedimentological investigations in order to evaluate the possible extension of such a comparative study (whose value was pointed out by Lagaaij and Gautier (1965) in the Rhone delta) to a particular geographic zone such as the strait of "Bocche di Bonifacio" and surrounding areas.

An analysis of the samples studied follows. Much of the data is summarized in Fig. 1.

The weight of each sample being 100 g, two fractions were separated, particle size 0·149 to 0·425 mm (fraction A) and 0·425 to 0·833 (fraction B), see Lagaaij and Gautier (1965).

1. *Sample* O_2

Collected close to the island of Razzoli, Lat 41° 20′·1 N, Long 09° 20′·3 E in 74 m. Weight of fraction A 26·666 g, and fraction B 8·378 g.

Zoarial habits of species present. Membraniporiform: *Onychocella* sp., *Reptadeonella violacea*, *Cribrilaria radiata*, *Lichenopora* sp. and *Diastopora* sp. Cellariiform: *Cellaria* spp. 1, 2, 3, *Scrupocellaria* sp., *Caberea boryi*, *Tubucellaria* sp., *Crisia* sp. Reteporiform: *Sertella cellulosa* Catenicelliform: *Filicrisia* sp. Vinculariiform: *Entalophora* sp. and *Idmidronea* sp.

Ecology. The abundance of cellariiform zoaria (rich in number and in species) are typical of the moderate-deposition environments where the effect of the wave action is overcome by the articulation of the long narrow stems (envir. B).

2. Sample O_6

Collected at Punta Falcone, Lat 41° 17′·3 N, Long 09° 14′·7 E in 74 m. Weight of fraction A 0·342 g, fraction B 0·112 g.

Zoarial habits of species present. Membraniporiform: *Cribrilaria radiata*, (?) *Hippoporina* sp., *Schizomavella rudis*, *Schizoporella* sp., *Micropora* sp., *Batopora* sp., *Diplosolen* sp., *Stomatopora* sp. Cellariiform: *Cellaria* spp. 1, 2, 3, *Scrupocellaria* sp., *Caberea boryi*, *Crisia* sp. Reteporiform: *Sertella* cf. *cellulosa*. Vinculariiform: *Tessaradoma* sp., *Entalophora* sp., *Idmidronea* sp. Adeoniform: *Dimorphocella* sp.

Ecology. Environment "B" as for sample O_2 inferred from the abundance of cellariiform and membraniporiform zoaria.

3. Sample O_{11}

Collected close to S. Teresa di Gallura, Lat 41° 15′·5 N, Long 09° 11′·3 E in 57 m. Weight of fraction A was 16·134 g, and fraction B 13·163 g.

Zoarial habits of species present. Membraniporiform: *Hippopodinella kirchenpaueri*, *Escharella ventricosa*, *Calpensia* sp., *Micropora* sp., *Schizoporella* sp., *Microporella* sp., *Cribrilaria radiata*, undetermined *Cheilostomata* belonging to 7 different species, *Diastopora* sp. Cellariiform: *Cellaria* sp., *Scrupocellaria* sp., *Caberea boryi*, *Crisia* sp. Reteporiform: *Sertella* cf. *cellulosa*. Catenicelliform: *Filicrisia* sp. Vinculariiform: *Tessaradoma* sp., *Entalophora* sp., *Idmidronea* sp.

Ecology. Environment "B" as for samples O_2 and O_6 inferred from the abundance of cellariiform and membraniporiform zoaria.

4. Sample O_{36}

Collected NNE of Castel Sardo, Lat 41° 02′·1 N, Long 08° 43′·4 E in 98·7 m. Weight of fraction B was 26·902 g, and fraction A 37·432 g.

Zoarial habit of species present. Membraniporiform: *Cribrilaria radiata*, *Lichenopora* sp., *Microporella* sp. Cellariiform: *Cellaria* sp., *Scrupocellaria* sp., *Caberea boryi*, *Crisia* sp. Reteporiform: *Sertella* cf. *cellulosa*. Vinculariiform: *Tessaradoma* sp., *Entalophora* sp., *Idmidronea* sp.

Ecology. Neritic environment (env. "C") where the wave action and currents are scarcely active inferred from the presence and abundance of the cellariiform and vinculariiform zoaria as for sample O_{67}, below.

5. Sample O_{45}

Collected at Punta Trabucato, Island of Asinara, Lat 41° 01′·6 N, Long 08° 20′·0 E in 46·6 m. Weight of fraction A was 35·396 g, and fraction B 32·335 g.

Zoarial habits of species present. Membraniporiform: *Escharella ventricosa*,

SAMPLE	WATER DEPTH (m)	LOCATION	LITHOLOGY	No. of the recovered specimen	No. of the species	Membraniporiform No. of the specimen	Membraniporiform No. of the species	Membraniporiform %	Ce pori No. of the specimen	N
O 2	74	Island of Razzoli	Poorly sorted sandy calcirudite	9244	16	42	5	31·2	—	-
O 6	74	Punta Falcone	Poorly sorted calcarenitic calcirudite	856	20	11	9	45	—	-
O 11	57	S. Teresa di Gallura	Poorly sorted sandy calcirudite	1142	24	94	15	62·5	—	-
O 36	98·7	NNE of Castel Sardo	Poorly sorted calciruditic calcarenite	790	11	3	3	27·27	—	-
O 45	46·6	P. Trabucato, Island of Asinara	Poorly sorted sandy calcarenite	55	14	37	9	64·5	—	-
O 67	49	S. of Capo di Feno (Corsica)	Moderately sorted calcarenite	186	10	42	1	10	—	-
O 92	45·7	S of Isola Rossa	Poorly sorted coarse sand with many pebbles	73	20	47	15	75	3	
O 97	86	Island of Santa Maria	Poorly sorted sandy calcarenite	128	7	—	—	—	—	-
O 109	40	Mouth of the Coghinas River	Moderately sorted and very pebbly sand	61	12	16	6	50	—	-
O 115	40	Castel Sardo	Poorly sorted coarsely pebbly sand	21	4	—	—	—	—	-

Fig. 1. Data of the samples ex

Crassimarginatella sp., Cheilostomata undetermined belonging to 4 species, and Cyclostomata belonging to 3 species. Cellariiform: *Cellaria* sp., *Scrupocellaria* sp., *Crisia* sp. Vinculariiform: *Tessaradoma* sp., *Entalophora* sp.

Ecology. Environment "B" as for samples O_2, O_6, O_{11}, inferred from the abundance of membraniporiform and cellariiform zoaria.

Sample O_{67}

Collected South of Capo di Feno (Corsica), Lat 41° 17′·6 N, Long 09° 03′·6 E in 49 m. Weight of fraction A was 12·036 g, and fraction B 0·785 g.

Zoarial habits of species present. Membraniporiform: *Membranipora* sp. Cellariiform: *Cellaria* sp., *Scrupocellaria* sp., *Caberea boryi*, *Crisia* sp. Reteporiform:

BRYOZOAN ZOARIAL HABITS

Cellarii-form		Retepori-form			Cateni-celliform			Vincularii-form			Adeoni form			
No. of the species	%	No. of the specimen	No. of the species	%	No. of the specimen	No. of the species	%	No. of the specimen	No. of the species	%	No. of the specimen	No. of the species	%	ENVIRONMENTAL INFERENCES
7	43·9	28	1	6·2	11	1	6·2	238	2	12·5	—	—	—	Environmental type "B"
6	30	16	1	5	—	—	—	141	3	15	3	1	5	Environmental type "B"
4	16·8	31	1	4·1	13	1	4·1	67	3	12·5	—	—	—	Environmental type "B"
4	36·36	1	1	9·09	—	—	—	497	3	27·27	—	—	—	Environmental type "C" connected to West with Mediterranean Sea
3	21·5	—	—	—	—	—	—	5	2	14	—	—	—	Environmental type "B"
4	40	13	1	10	1	1	10	18	3	30	—	—	—	Environmental type "C" connected to West with Mediterranean Sea
2	10	4	1	5	—	—	—	2	1	5	—	—	—	Environmental type "B"
4	57·15	4	1	14·28	—	—	—	17	2	28·57	—	—	—	Environmental type "D" connected to East with Tyrrenian Sea
4	33·34	8	1	8·33	—	—	—	1	1	8·33	—	—	—	Environmental type "B"
3	75	3	1	25	—	—	—	—	—	—	—	—	—	Environmental type "A" coastal

eference to Bryozoan zoarial habit.

Sertella cf. *cellulosa*. Catenicelliform: *Filicrisia* sp. Vinculariiform: *Tessaradoma* sp., *Entalophora* sp., *Idmidronea* sp.
Ecology. Neritic environment (env. "C") where the wave action and currents are scarcely active inferred from the presence and abundance of cellariiform and vinculariiform zoaria as for sample O_{36}.

7. *Sample* O_{92}

Collected South of Isola Rossa, Lat 40° 59′·2 N, Long 08° 49′·8 E in 45·7 m. Weight of fraction A was 24·007 g, and fraction B 21·375 g.
Zoarial habits of species present. Membraniporiform: *Ramphostomella* sp., small fragments of undetermined Cheilostomata belonging to 11 different species and

fragments of Cyclostomata belonging to 3 different species. The material was poorly preserved. Celleporiform: *Holoporella* sp. Cellariiform: *Cellaria* sp. and *Scrupocellaria* sp. Reteporiform: *Sertella* cf. *cellulosa*. Vinculariiform: *Idmidronea* sp.
Ecology. Environment "B" inferred from the abundance of membraniporiform and cellariiform zoaria, even if in poor states of preservation.

8. Sample O_{97}

Collected close to the Island of Santa Maria, Lat 41° 18′·8 N, Long 09° 25′·3 E in 86 m. Weight of fraction A was 26·666 g, and fraction B 36·115 g.
Zoarial habits of species present. Cellariiform: *Cellaria* sp., *Scrupocellaria* sp., *Caberea boryi*, *Crisia* sp., Reteporiform: *Sertella* cf. *cellulosa*. Vinculariiform: *Entalophora* sp., *Idmidronea* sp.
Ecology. Neritic environment, freely connected with the deep Tyrrenian Sea, where the wave action and currents are scarcely active, inferred from the predominance of the cellariiform and vinculariiform zoaria. (Env. "D").

9. Sample O_{109}

Collected close to S. Pietro al Mare (Coghinas River mouth), Lat 40° 57′·7 N, Long 08° 47′·1 E in 40 m. Weight of fraction A was 38·245 g, and fraction B 34·995 g.
Zoarial habits of species present. Membraniporiform: *Schizoporella* sp., undetermined Cheilostomata belonging to at least 5 different species. Cellariiform: *Cellaria* sp., *Scrupocellaria* sp., *Tubucellaria* sp., *Crisia* sp. Reteporiform: *Sertella* cf. *cellulosa*. Vinculariiform: *Entalophora* sp.
Ecology. Environment "B" as samples O_2, O_6, O_{11}, O_{45}, O_{92}, inferred from the predominance of the membraniporiform and cellariiform zoaria.

Sample O_{115}

Collected close to Castel Sardo, Lat 40° 53′·3 N, and Long 08° 42′·2 E in 40 m. Weight of fraction A was 39·978 g and fraction B 6·466 g.
Zoarial habits of species present. Cellariiform: *Cellaria* sp., *Scrupocellaria* sp., *Tubucellaria* sp. Reteporiform: *Sertella* cf. *cellulosa*.
Ecology. Coastal detrital bottoms where the wave action and current are still noticeable (environment "A").

ECOLOGICAL INFERENCES FROM A COMPARATIVE STUDY OF THE BRYOZOA AND TEXTURAL ANALYSES OF THE SEDIMENTS

From this preliminary investigation of only a few samples, four environmental types were tentatively distinguished (Fig. 2).

"*A*". Immediately coastal zone, where wave action and currents are still noticeable, these factors being overcome by the rigidity and fenestration of the reteporiform zoaria and by the articulation of the long and slender stems of the cellariiform zoaria, which are the only zoarial habits observed. The sediments were of coarsely pebbly and poorly sorted sands. (Sample O_{115} Golfo dell'Asinara.)

"*B*". Bottoms with strong current action. The sediments were of calcirudite and more or less coarse and poorly sorted calcarenite with pebbles. This was the more widely sampled zone where membraniporiform and some cellariiform zoaria are dominant. (Samples O_2, O_6, O_{11}, O_{45}, O_{92}, O_{109} Bocche di Bonifacio and Golfo dell'Asinara.)

"*C*". Neritic environment, where the wave action and currents are scarcely active, connected to the Mediterranean Sea to the west. The sediments were of moderately-to-poorly sorted calcarenite sediments and cellariiform and vinculariiform zoaria were dominant. (Samples O_{67} and O_{36}, SW Corsica and Golfo dell'Asinara.)

"*D*". Neritic environment, like "C" but freely connected with the deep Tyrrenian Sea to the east. The sediments were of poorly sorted sandy calcarenite and cellariiform, reteporiform and vinculariiform zoaria were found. (Sample O_{97} Island of Santa Maria, NE Sardinia.)

Sample O_{115} presented a peculiar assemblage of only two zoarial habits (reteporiform and cellariiform). This sample is distinguished from all other samples as representative of a restricted and distinct environment, the "A" (coastal) environment.

The samples O_2, O_6, O_{11} were similar and scattered in a well defined area (the strait) where a negative "Skewness" was observed. Here strong bottom currents are present in a narrow lane. A similar feature was pointed out by Bellan *et al.* (1965) on the Corsica side of the strait and along the SW coasts of Corsica. The samples of the coastal detrital environment type, generally corresponding to the sand, calcarenite and calcirudite zones (sedimentological zone 2 and 3 in Fig. 3) are typified by "pralines" associated with *Venus casina*. Such an association was found in the "Bocche di Bonifacio" and in some isolated banks and this location was considered to be influenced by open waters (Pérès and Picard, 1958).

The samples which do not exhibit such an association corresponded to those from coastal detrital zones influenced by neritic waters with higher thin terrigenous sediment which hinders the growth of *Melobesia* forming the "pralines".

Examining the distribution of the "pralines facies" with *Venus casina* in our samples, our conclusions about the presence of a lane (environment "B") with

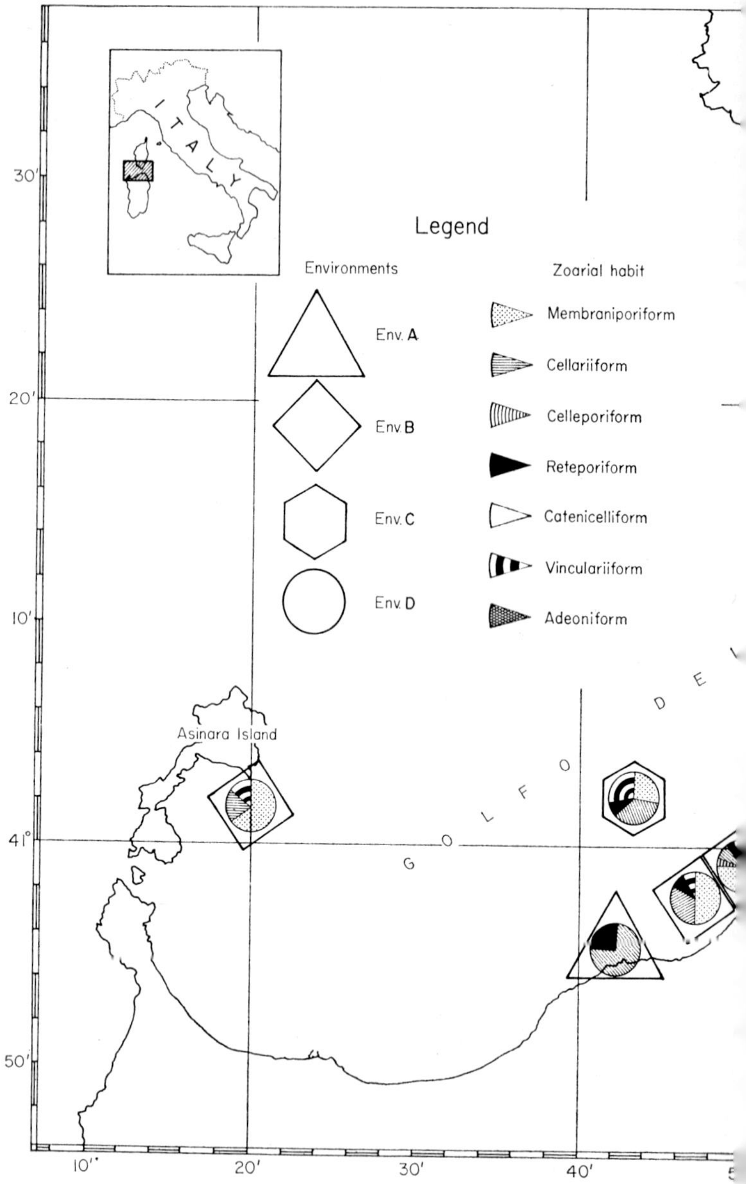

Fig. 2. Map of the association of Bryozoan Zoarial Habit and tentat

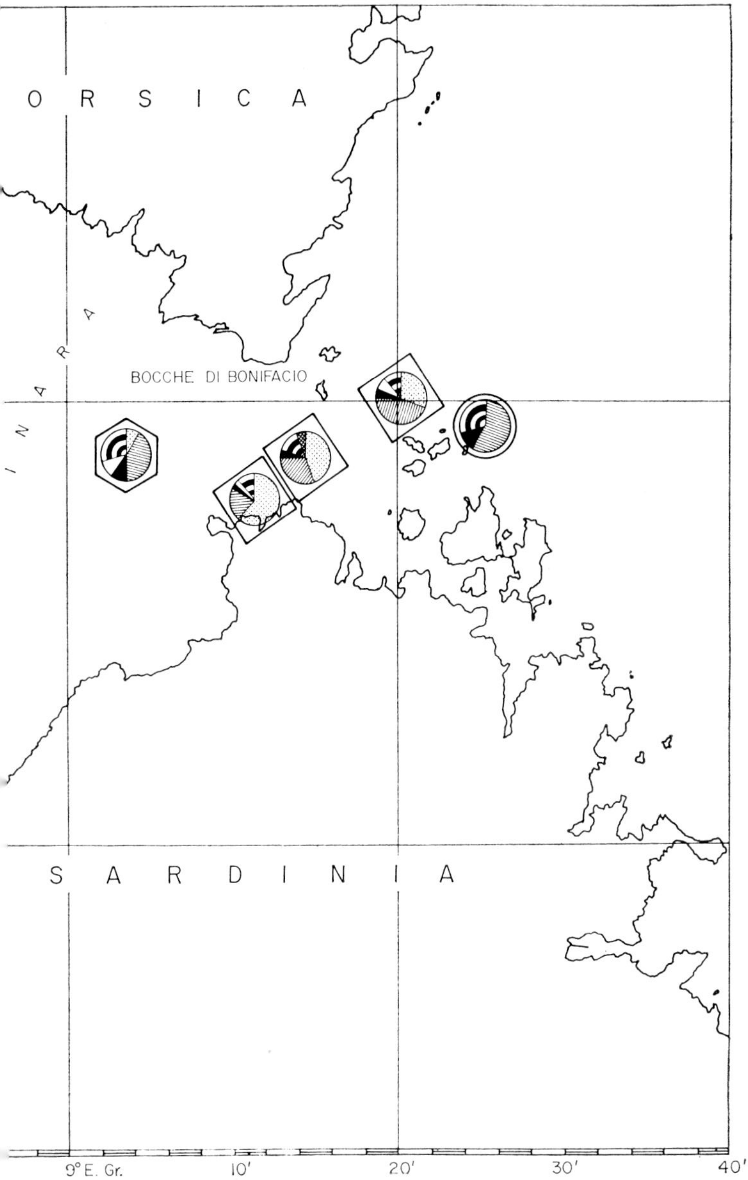

ed environment. (The legend is comprised in the figure.)

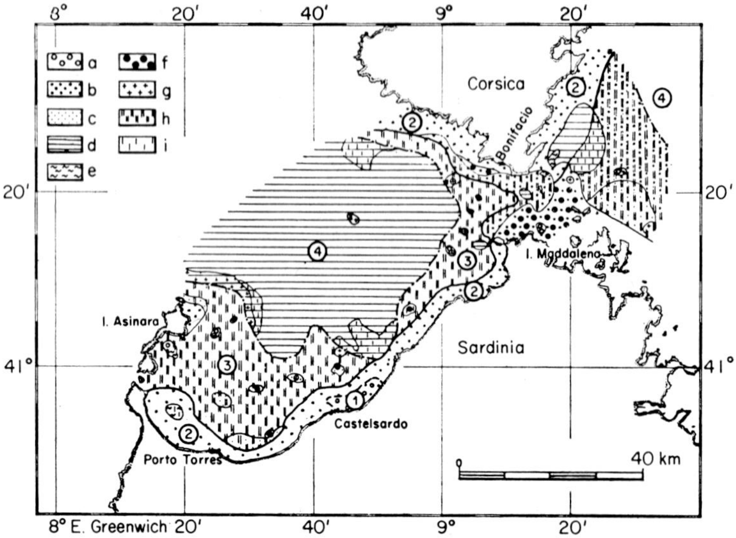

Fig. 3. Lithological map of the bottom sediments. (From Fierro, 1969).

LITHOLOGY

Detrital mobile sediments

a—gravel (diameter <2 mm)
b—sand (diameter 2–0·05 mm) >30 p. 100
c—sand as above 5–30 p. 100
d—mud (diameter $<0{\cdot}05$ mm) calcareous
e—mud as above but clayey

Zoogenal mobile sediments

f—with calciruditic composition (diameter >2 mm) >30 p. 100
g—as above, 5–10 p. 100
h—with calcarenitic composition (diameter 2–0·05 mm) >30 p. 100
i—as above, 5–10 p. 100

SEDIMENTOLOGICAL ZONES

1—Gravel Zone
2—Sand Zone
3—Calcarenite and calcirudite Zone
4—Calcareous and clayey mud Zone

strong currents (strait between Corsica and Sardinia) were well supported. In effect:

(1) The "pralines facies" with *Venus casina* was found in the lane of "B" environment and in the lane along the SW coasts of Corsica by Bellan *et al.* (1965).

(2) The "pralines facies" was also found close to the NW coasts of Sardinia (Asinara).

(3) No "pralines facies" was found in the fluviomarine fan of the Coghinas River.

The differences among samples O_{97}, O_{36} and O_{67} are small.

The last two samples showed high percentages of cellariiform and vinculariiform zoaria (with some reteporiform and membraniporiform). We include these two samples in the "C" environmental type, differentiating it from the type "D" one (sample O_{97}) because of the presence in this sample of only three zoarial habits (cellariiform, vinculariiform and reteporiform) with more restricted significance. The location of the samples on two opposite sides of the lane of "B" environment in two different basins of deposition suggested this distinction.

The investigation of all other samples should allow us to define better the tentatively established environmental zones and complete the ecological panorama inferred from the study of the Bryozoan zoarial habit. The possible transport of Bryozoan debris also will be investigated.

ACKNOWLEDGEMENT

We wish here to thank Dr G. Podestá for his preliminary investigations on the Bryczoans, Dr R. Lagaaij for his kind suggestions on the beginning of this study, and Mr F. Settepassi for the determination of the Mollusca.

REFERENCES

ANNOSCIA, E. (1968). I Briozoi. Introduzione allo studio con particolare riguardo ai Briozoi Italiani e Mediterranei. *Palaeontographia Italica* 1–400.

ANNOSCIA, E. (Ed.) (1968). Proceedings of the First International Conference on Bryozoa. *Atti Soc. It. Sc. Nat.* **108,** 1–377.

BELLAN, G., MOLINIER, R. and PICARD, J. (1961). Distribution et particularités des peuplements benthiques de l'étage circalittoral des parages de Bonifacio (Corse). *Rapp. Pr. Verb. CIESMM* **VI.**

BREGANT, D. (1969). Distribuzione dei sali nutritivi nell'area delle Bocche di Bonifacio e del Golfo dell'Asinara. Crociera Bannock 1964. *Boll. Pesca Piscic. Idrobiol.* **XXII** 2, 113–120.

CALVET, L. (1902). Bryozaires marine des Côtes de Corse. *Inst. Zool. Montpellier et Stat. Zool. de Cette.* ser. 2 **12,** 1–52.

CANO', M. and STOCCHINO, C. (1966). Sulla circolazione delle correnti nel Golfo dell'Asinara e nelle Bocche di Bonifacio. Misure di corrente eseguite a bordo della nave "Bannock". *Atti XVI Con. Ass. Geof. Ital.* 523–535.

FIERRO, G. (1965). Observations morphologiques et sédimentologiques sur les Bouches de Bonifacio et le Golfe de l'Asinara. *Cah. Océanogr.* **XVII,** 565–571.

FIERRO, G. (1966). Profils echographiques dans les Bouches de Bonifacio et dans le Golfe de l'Asinara. *Rapp. Pr. Verb. CIESMM* **XVIII** (3), 931.

Fierro, G. (1969). Répartitions des sédiments dans les Bouches de Bonifacio et dans le Golfe de l'Asinara. *Rapp. Pr. Verb. CIESMM* **XIX** (3), 20–21.

Fierro, G. (1970). I minerali pesanti nei sedimenti marini del Golfo dell'Asinara e delle Bocche di Bonifacio. *Atti Soc. It. Sc. Nat.* **110** (2), 155–197.

Gautier, Y. V. (1962). Recherches écologiques sur les Bryozoaires Cheilostomes en Méditerranée occidentale. *Rec. Trav. Stat. Mar. Endoume* (**38**) **24,** 1–434.

Lagaaij, R. and Gautier, Y. V. (1965). Bryozoan assemblages from marine sediments of the Rhone delta, France. *Micropaleontology* **11** (1), 39–58.

Pérès, J. M. and Picard, J. (1958). Manuel de bionomie benthique de la Mer Méditerranée. *Rec. Trav. Stat. Mar. Endoume*, (**23**) **14**.

Podestá, G. (1969). Interpretazione statistica dell'analisi meccanica ed ecologica di sedimenti marini recenti. Thesis for the Master's degree in Geological Sciences, University of Genoa, pp. 1–120.

Stach, L. W. (1936). Correlation of zoarial form with habitat *J. Geol.* **XLVI** (1), 60–65.

Trotti, L. (1967). Crociera Bocche di Bonifacio 1964. *Dati oceanografici Ist. Sper. Talass. Trieste*, no. 438.

6 | Settlement and Early Colony Development in some Cheilostomata

PATRICIA L. COOK

British Museum (Natural History) London, England

Abstract: Observations on three cheilostome species from West Africa illustrate diversity in larval form and behaviour and in the nature of the ancestrular complex.

The three species observed belong to the Anasca (*S. levinseni*), the umbonuloid Ascophora (*M. unguiculatum cookae*) and the cryptocystidean Ascophora (*S. duboisii*) respectively. Although their methods of brooding embryos and the nature of their early development differ considerably, these differences are not correlated systematically with the brooding and development known for other, related species. There are, however, indications that the form of the ancestrular complex and nature of the primary budding may be reflected in the subsequent astogeny. Many more observations are needed before such possible correlations as type of brooding and larvae, type of larva and extent of larval life, form of ancestrular complex and rate of formation of feeding zooids, and early astogeny and colony structure will be available for assessment.

1. Smittipora levinseni (Canu and Bassler)

Specimens were collected in 35–40 m, on a muddy sea-bottom, south of Tema, Ghana, exclusively encrusting the large foliaceous foraminiferan, *Jullienella foetida* Schlumberger. Embryos in *Smittipora levinseni* are brooded in internal ovisacs and up to 75% of zooids in a colony contain a large, orange embryo. Brooding zooids have larger opesiae than autozooids, but both have feeding lophophores with 24–26 tentacles. In some brooding zooids degeneration of the calcification of the proximal cryptocyst had occurred, similar to that seen in *Steganoporella buskii* Harmer during the growth of the embryo (see Cook, 1964, p. 50). The larval cilia become active 48 h before extrusion, when the lophophores are retracted, the operculum opened and the parietal muscles slightly contracted. In *Steganoporella buskii* and *S. magnilabris*, which have no feeding lophophore when the embryos are fully developed, the same contraction of parietal muscles is seen on extrusion of the larva. Extrusion takes 1–2 min and the larvae (average diameter 250 μm) which are without pigment spots, are immediately active and strongly positively phototropic. After 3 h, negative reaction begins and most larvae settle 5 h after extrusion. About 12% of approximately 80 larvae observed continued swimming for a further 24 h, but settled successfully. *Jullienella* was overwhelmingly the preferred substratum. About 40% of larvae metamorphosed successfully, forming a small cuticle-covered mass. Within 2 h calcification of peripheral vertical walls was seen beneath the cuticle, and a simple ancestrula, with fully developed operculum and parietal muscles, and a rudimentary lophophore was formed in 48 h. Protrusion of the lophophore, which has 20 tentacles, and feeding, were seen 60 h after settlement, although the cryptocyst is not fully developed until 12 h later. Primary distal and distal-lateral buds are developed as cuticular swellings at this stage. One-hundred and forty hours after settlement these buds had become almost fully developed zooids.

The development of *Onychocella alula* Hastings is similar to that of *S. levinseni*.

2. *Metrarabdotos unguiculatum cookae* Cheetham

Very large colonies, over 1000 cm² in area were collected from under rocks and large stones embedded in sand in 2 m at low tide at Paradise Beach, east of Tema. Embryos are brooded in gonoecia which have an enlarged distal chamber and wide, slit-like orifices. Both gonoecia and autozooids have a lophophore with 18 tentacles, and an opercular flap, which is hardly differentiated from the frontal membrane, although it has well developed muscles. Normally, the opercula are in the open position, with the introverted tentacles within the tentacle sheath protruding through them. The everted base of the lophophore is thickened and blocks the aperture of the extended peristome. The tentacle sheaths are apparently fully retracted and the opercula closed only under unfavourable conditions. The tentacle sheath and lophophore of the gonoecium is elongated from side to side and the proximal base of the lophophore is enlarged and swollen. It completely fills the wide orifice, moving inward and downward as the tentacles are everted. The tentacle sheaths are retracted as the larvae are extruded; this process takes from 20 to 140 min. The larva constantly twists and is considerably constricted during extrusion. Although the cilia are active, released larvae adhere to the colonial cuticle from 5 to 15 min before moving. The larvae are very large (average diameter 500 μm), reddish purple in colour, and have no pigment spots. They move slowly across any substratum, but do not seem capable of swimming upward, and if released into water unsupported, fall steadily downward, although the cilia are active. Larvae are positively phototropic from 2 to 3 h after release. Settlement occurs after 12 to 18 h and metamorphosis results in a large rounded mass of pigmented cells covered by transparent cuticle. Of 40 larvae seen, about 33% metamorphosed successfully. Two to three hours after settlement the primary mass becomes differentiated into 4 smaller masses beneath the expanding cuticular cover. Twenty-four hours later vertical peripheral calcified walls have formed round one mass (see Fig. 1c). This is the primary ancestrular zooid of a tetrad and the re-

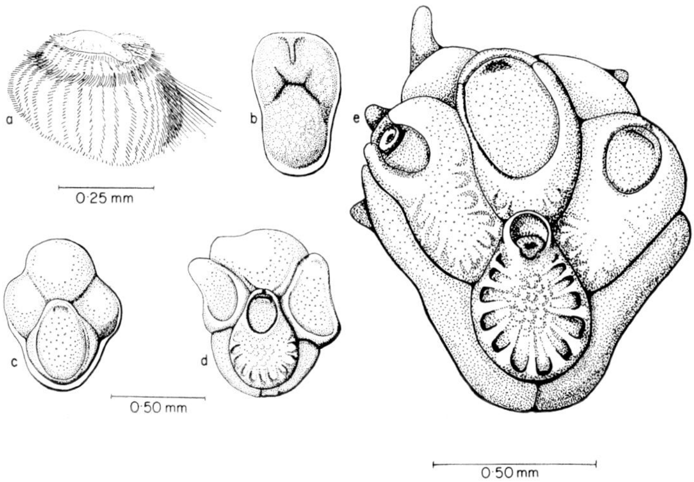

Fig. 1. *Metrarabdotos unguiculatum cookae* Cheetham.
(a) Larva. (b) 2 h after settlement. (c) 28 h after settlement. Calcification of peripheral walls of primary member of tetrad complete. (d) 48 h after settlement. Development of frontal shield of primary member nearly complete, proximal kenozooids formed. (e) 60 h after development. Distal intercalary buds formed.

maining 3 masses of cells may now be regarded as distal in position to this primary zooid. They are not directly budded from it, however, and comprise the remaining members of the ancestrular tetrad. At the same time paired, elongated masses of pigmented cells, covered by cuticle, are left proximally to the primary ancestrular zooid; these grow and extend distally and laterally beside the outer lateral walls of the two lateral distal members of the tetrad. Their walls do not calcify and they form paired proximal kenozooids. No other zooids are formed proximally and the growth of the colony is thus in one direction only. In preserved specimens the cuticle of the proximal kenozooids collapses against the calcified walls of the adjacent members of the tetrad. Cheetham (1968) postulated from preserved specimens that the astogeny of Recent encrusting *Metrarabdotos* was derived from the erect growth form of fossil ancestors.

Forty-eight hours after settlement, the lateral walls of the secondary members of the tetrad are well formed and calcification of the frontal shields above the frontal membranes has commenced. The calcification of the shield of the primary member has reached the position of the future secondary orifice, above the operculum, which is clearly differentiated from the frontal membrane. Ninety-six hours after settlement an intercalary bud on either side of the central distal zooid of the tetrad is well developed, and each of the secondary members has produced a small distal bud. This pattern of distal and distal-lateral budding is repeated in subsequent generations. One-hundred and twenty hours after settlement the primary member of the tetrad has formed a complete calcified secondary orifice, which is at first oval, the "sinus" being formed by rapid subsequent thickening and upward extension of the frontal calcification, forming a peristome (see Fig. 2a). At this stage the central part of the frontal shield is

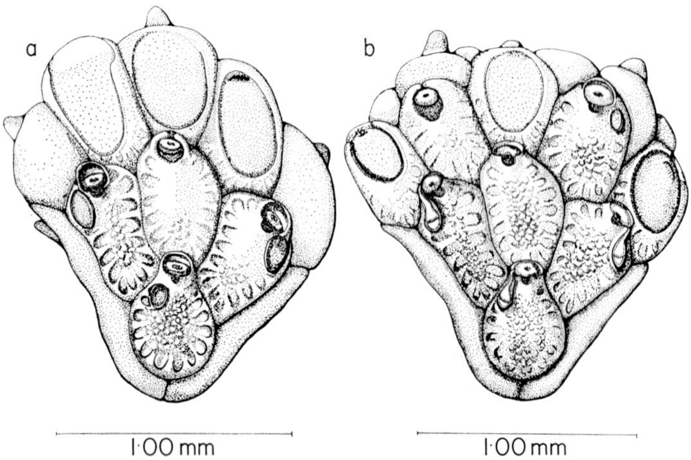

Fig. 2. *Metrarabdotos unguiculatum cookae* Cheetham.
(a) 120 h after settlement. Calcification of tetrad almost complete, avicularia beginning to develop, intercalary and distal zooids continue fan-shaped growth of colony.
(b) 140 h after settlement. Avicularia formed in some tetrad members, frontal calcification of intercalary zooids almost complete.

tuberculate, and the marginal pores become hidden by intervening calcified ribs. In some zooids, occasionally including the primary ancestrular zooid, an area lateral and proximal to the secondary orifice develops small vertical calcified walls. Twenty-four hours later this has become a fully formed avicularium with mandible. The members of the tetrad have lophophores with 16–17 tentacles;

in a few cases feeding was observed before the calcification of the secondary orifice was complete.

3. Stylopoma duboisii (Audouin)

Specimens were collected from calcareous accretions in 10 m on the Vernon Bank, south of Tema. The embryos are brooded in very large, hyperstomial ovicells, and the lophophores of both autozooids and fertile zooids have 18–20 tentacles. Larvae are extruded in 2–10 min, becoming very constricted as they pass through the small ovicellular orifice. Larvae are yellowish-pink (average diameter 250 μm), with 4 large, bright red pigment spots, and a variable number of smaller spots; they are very active and strongly positively phototropic. Negative reaction commences 30 min after extrusion, and about half the 30 larvae observed settled immediately, the remainder settled after 2 h swimming. Successful metamorphosis occurred in about 33% of the larvae. After metamorphosis a mass of pigmented cells remains in the centre of a transparent cuticular sac, which expands rapidly forming a triangular capsule. Four hours after settlement the pigmented mass is tripartite, the divisions marked by three vertical walls of calcification, which commence at the centre of the mass and extend outwards. The cuticular capsule continues to expand and three primary ancestrular zooids are formed by the calcification of curved distal walls (see Fig. 3e). Forty-three hours after settlement the frontal walls show thin calcification beneath the

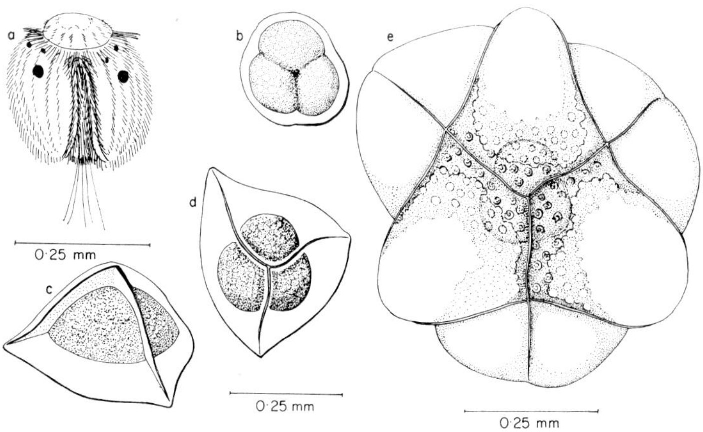

Fig. 3. *Stylopoma duboisii* (Audouin).

(a) Larva. (b) 2 h after settlement. (c) 12 h after settlement. Growth of vertical calcified walls has commenced. (d) Vertical view of same stage as (c). (e) 36 h after settlement. Peripheral calcified walls of triad complete, with beginning of frontal wall calcification.

cuticle (and presumably hypostegal coelom). The position of the pseudopores, and the rudiment of the lophophore can be seen. Fifty-four hours after settlement the outline of the primary calcified orifice is delineated, the lophophore and other viscera are well-developed, and the compensation sac can be seen growing beneath the calcified frontal wall. At the same time, further extension and branching of the vertical calcified walls, and expansion of the cuticle has produced the next generation of zooids (see Fig. 4a). The ancestrular triad and first

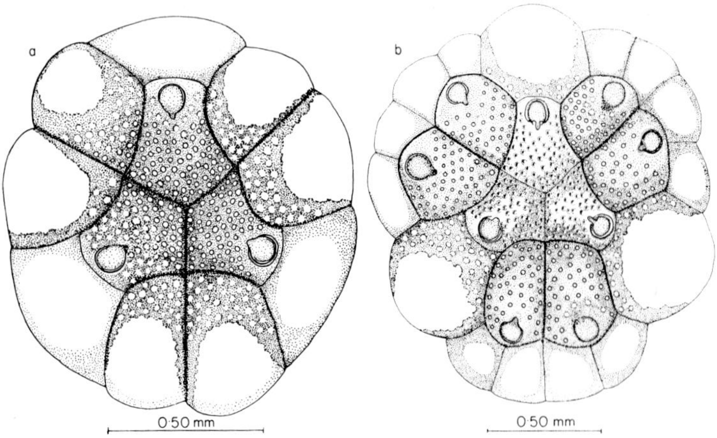

Fig. 4. *Stylopoma duboisii* (Audouin).
(a) 54 h after settlement. Primary calcified orifice of triad zooids almost complete.
(b) 70 h after settlement. Calcification of triad and 3 pairs of buds almost complete. Note that establishment of second circle of zooids includes primary and secondary distal zooid buds.

subsequent generation of zooids have lophophores with 17 tentacles, and 80 h after settlement the colonies, which consisted of more than 9 zooids, were actively feeding.

An increasing number of ancestrular complexes is becoming known. Twinned ancestrulae are known in *Membranipora* and a few other genera. The differentiation of a primary ancestrular zooid from a post-larval mass of tissue is known in some species of *Steganoporella* (see Cook, 1964). Although the development of *Metrarabdotos* differs from this, there is a similarity in that the secondary members of the tetrad are not directly budded from the primary member. The development of *Smittipora* and *Onychocella*, which otherwise have several characters in common with *Steganoporella*, shows nothing of this type of early astogeny. In

Stylopoma duboisii, an ancestrular triad is differentiated simultaneously, and it should be noted that at the growing edge of the primary layer of zooids in this multi-laminar species, series of vertical walls continue to be calcified within an undifferentiated cuticle-covered common bud.

REFERENCES

CHEETHAM, A. H. (1968). Morphology and systematics of the Bryozoan genus *Metrarabdotos. Smithson. misc. Collns* **153** (1), i–vii, 1–121.

COOK, P. L. (1964). Polyzoa from West Africa. 1. Notes on the Steganoporellidae, Thalamoporellidae and Onychocellidae (Anasca, Coilostega). *Annls. Inst. océanogr.* ("*Calypso*" 6), *Monaco* **41,** 43–78.

7 | Carboniferous Bryozoa of the Donetz Basin

N. N. DUNAEVA

Geological Institute, Ukrainian Academy of Science, Kiev, U.S.S.R.

Abstract: The remains of fossil bryozoans occur irregularly in several strata of the Ukrainian Carboniferous especially in the Carboniferous of the Donetz basin. The Carboniferous bryozoan fauna of the Donetz basin is characterized by considerable diversity of genera and species. Up to 80 species assigned to 26 genera are reported from the Lower Carboniferous, about 60 species and 24 genera are found in the Middle Carboniferous, and about 60 species and 16 genera in the Upper Carboniferous.

The vertical distribution of the species and even the genera may be used for stratigraphical division of the Carboniferous sequence of the Donetz basin. Additionally, the bryozoans are of value for facies analysis.

The great variability of bryozoan colonies indicates the necessity of studying bryozoan species on a sufficient number of specimens. It is obvious that most of the Palaeozoic species erected upon a single or few colonies will have to be revised in future when additional palaeontological material becomes available.

The remains of fossil Bryozoa are widespread in the Carboniferous of the Great Donbass, especially in the outcrops of the Donetz basin which, due to its completeness, diversity of facies and abundance of fossils is one of the most remarkable Carboniferous sequences in the world. The fragments of bryozoan colonies occur irregularly in the Donetz basin throughout the whole Carboniferous sequence. In the lower parts of the Carboniferous (namely the massive homogeneous limestones in the Tournaisian and Lower Visean stages) bryozoans are extremely rare, except in the lowermost zone $C_{\mathrm{I}}^{t}a$ where the trepostomes *Volnovachia distincta* and *Aisenvergia cylindrica* occur in abundance. Various bryozoans commonly occur in several predominantly carbonate strata in the overlying thick clastic unit of the Upper Visean, Namurian and Middle Carboniferous and are fairly abundant in bioclastic and silty limestones interbedded with sandstones and shales in the Upper Carboniferous.

The Carboniferous bryozoan fauna of the Donetz basin is characterized by a considerable diversity of genera and species. In the Lower Carboniferous up to 80 species assigned to 26 genera have been recognized but are at present only partly described. In the Middle Carboniferous 24 genera and up to 60 species are listed; in the Upper Carboniferous 16 genera and 64 species have been recorded.

Several species of Bryozoa occur in certain strata in profusion and are represented in collections by a great number (several hundreds) of specimens; the others are very rare (only single or several specimens). As a rule the most abundant are the long-ranging species, for example *Fenestella donaica*, which occurs from the Upper Viséan to the lower part of the Upper Carboniferous, or *Nikiforopora concentrica* which occurs in profusion in the Namurian, but is known also in the Middle Carboniferous; *Polypora martis* which is very typical for the whole Upper Carboniferous.

In spite of these long ranging species the vertical distribution of bryozoan fauna as a whole has considerable stratigraphical value. Even the genera of the bryozoan fauna can be regarded as valuable for stratigraphic division of the Carboniferous in the Donetz basin. Fenestelloids, predominating throughout the whole Carboniferous sequence, are represented in the Lower and Middle Carboniferous mostly by the species of *Fenestella*. In the Upper Carboniferous the most abundantly represented fenestrate bryozoans belong to the genus *Polypora*.

Trepostomes are most abundant and diversified in the Namurian; *Nikiforopora, Stenophragmidium, Anisotrypa, Dyscritella, Crustopora, Tabulipora* are common. In the Middle Carboniferous abundant *Stenodiscus* is extremely typical; *Tabulipora, Crustopora, Anisotrypa* occur rarely. In the Upper Carboniferous the numerous trepostomes all belong to a single genus, *Rhombotrypella*. Rhabdomesonata are infrequent in the Lower Carboniferous (some representatives of *Clausotrypa* are found in the Lower Viséan—zone C_I^ve). In the Upper Namurian *Nikiforovella donbassica*, which is common in the Lower Bashkirian (zone C_I^ne) appears. In the Middle and Upper Carboniferous Rhabdomesonata are abundant and diversified. *Rhabdomeson, Ascopora, Streblotrypa, Nikiforovella* occur sometimes in profusion.

In addition to the stratigraphic significance of the Carboniferous bryozoans of the Donetz basin these fossils may be used as good facies indicators. As mentioned above the bryozoans occur in the Donetz basin predominantly in carbonate rocks, namely in silty, sandy and bioclastic limestones, more rarely in carbonate shales and mudstones. They are abundant in the peculiar cherty marls, occurring at the top of the Lower Viséan (zone C_I^ve); single species of fenestrate bryozoans occur sparsely in the massive crystalline limestones of the Tournaisian and Lower Visean.

A distinct correlation is noted between the lithofacies and associated bryozoan fauna: in the finegrained cherty marls thin slender colonies of *Diploporaria, Penniretepora, Nematopora, Clausotrypa* and *Hexites* predominate. They would suggest a relatively quiet water environment during the deposition of the sediments. The same may be suggested for the mudstones and shales containing abundant remains of the fragile colonies of *Fenestella*. But in the latter case the occurrence of bryozoans belonging only to a single genus appears to suggest abnormal salinity or gas conditions during the sedimentation.

The most diversified and richest bryozoan fauna occurs in the silty and bioclastic limestones interbedded with the sandstones, shales and coals in the clastic rock sequence of the Upper Viséan, Middle and Upper Carboniferous. It contains representatives of the various genera of Cryptostomata (e.g. *Fenestella Polypora, Septopora, Penniretepora, Sulcoretepora, Goniocladia* etc.) and Trepostomata (e.g. *Tabulipora, Nikiforopora, Anisotrypa, Stenophragmidium, Dyscritella*) as well as rare Cystoporata (*Fistulipora, Cyclotrypa, Eridopora*), characterized by various growth forms—encrusting, ramose, flat, ribbon or fan-shaped. This abundance and diversity of bryozoans, as well as their numerous growth forms, indicate most favourable shallow water environments during sedimentation and the proper conditions for fossilization due to the rapid deposition of terrigenous muds. The different state of preservation of the bryozoan colonies, which are broken into small pieces in the bioclastic limestones and finely preserved in the silty limestones, suggest deposition under rough and under slight current or wave action respectively.

A sparse occurrence of bryozoans in the massive fossiliferous limestones of the Tournaisian and Lower Viséan appears to coincide with vigorous development of competing groups, such as Tabulata, inhabiting the bottom of the shallow sea during the early stages of the Carboniferous in the Donetz basin. Unfavourable conditions for the preservation of fragile colonies in rough water environments are also not excluded.

Study of a great number of specimens belonging to some species of Trepostomata (for example *Volnovachia distincta, Aisenvergia cylindrica, Nikiforopora concentrica, Stenodiscus beralicus, Rhombotrypella rectangulata, Megacanthopora maculosa*) and Cryptostomata (e.g. *Fenestella karakubensis, F. frutex, F. donaica, Polypora martis*) has shown the extremely high variability of these bryozoans.

It is well known that the bryozoans as colonial forms have all types of variability. The variations of taxonomic features in bryozoan species may be connected either with the environmental conditions and are reflected in the growth form of colonies (ecological variability), or may depend upon the individual stage of development (astogenetic variability); even a single colony might display a very high variability due to the particular conditions of growth, especially in the encrusting colonies closely depending on the substratum. Likewise, some variations in tangential views may be technical artefacts.

All this indicates the necessity of studying great numbers of bryozoan specimens; unfortunately this has not always been done. It is obvious that most Palaeozoic bryozoan species which have been erected upon a single or few colonies will have to be revised in future when additional palaeontological material is available.

The investigation of numerous representatives of some Carboniferous bryozoans from the Donetz basin has allowed identification of several species, particularly among the fenestrate bryozoans, which have been previously regarded as different species due to slight variations in their meshwork.

This observation indicates the wide stratigraphical as well as geographical distribution of some bryozoan species. In fact there exists a very great resemblance between species from the Carboniferous of the USSR and other countries though they have been described by Russian and foreign palaeontologists under different names. By direct comparison of thin-sections it is difficult to note differences between the American species, *Megacanthopora fallacis* Moore, and numerous representatives of this genus found in the Viséan of the Great Donbass. Likewise the representatives of *Nipponostenopora* known in the Upper Carboniferous of Japan and now detected in the Namurian of the Great Donbass are very similar.

The existence of many synonyms among the Carboniferous bryozoan species resulting from inaccessibility to reference type thin-sections indicates much revisional work will have to be carried out in the future.

Table 1. Carboniferous sequence of the Donetz basin

	Stage	Zone	Bryozoan species
Upper Carboniferous	Gjelian	C_3^d	*Fistulipora labiatiformis, Rhombotrypella rectangulata, Rh. bachmutica, Rh. grandis, Fenestella praeornata, F. aisenvergi, Polypora subornamentata, P. subfenestelloides, P. timanica, Penniretepora beresovensis, P. donaica, Rhabdomeson manocyclum, Ascopora magniseptata, Goniocladia subpulchra, Nematopora ivanovi, Clausotrypa sp.*
		C_3^c	
	Kasimowian	C_3^b	*Fistulipora petaloida, Rhombotrypella rectangulata, Rhombocladia kasimovensis, Rh. punctata, Fenestella veneris, F. ex gr. donaica, F. massiva, F. tschernyschevi, F. spinosa, Polypora praeornamentata, P. martis, P. abundans, P. magnocellata, P. longifenestrata, Rhombopora mariae, Rh. verae, Rhabdomesom bispinatum, Sulcoretepora tenuata, S. festonata.*
		C_3^a	
Middle Carboniferous	Moscowian	$C_2^m e$	*Fistulipora mariae, Fenestella praespecifica, Penniretepora inconstans, Ascopora oblonga, Sulcoretepora pyriformis.*
		$C_2^m d$	*Megacanthopora maculosa, Stenodiscus beralicus, Crustopora?, Rhombocladia, Fenestella, Penniretepora, Septopora, Polypora, etc.*
		$C_2^m c$	
		$C_2^m b$	*Stenodiscus beralicus, Nikiforopora concentrica, Nikiforovella, Streblotrypa, Sulcoretepora, etc.*
		$C_2^m a$	*Nikiforopora concentrica, Fenestella ex gr. donaica, F. ex gr. virgosa, Nikiforovella ex gr. donbassica, Crustopora ramosa, etc.*
	Bashkirian	$C_2^b e$	
		$C_2^b d$	*Nikiforopora concentrica, Fenestella ex gr. donaica, Penniretepora, Rhombopora, Streblotrypa, etc.*
		$C_2^b b+c$	*Fenestella ex gr. donaica, Penniretepora sp., Nikiforopora voltchensis, Polypora sp.*
		$C_2^b a_2$	*Fistulipora steshevensis, Nikiforopora concentrica, Stenodiscus beralicus.*
		$C_2^b a_1 = C_1^n e_2$	
	Namurian	$C_1^n e_1$	*Fenestella frutex, Penniretepora cruciformis, Polypora cesteriensis, Nikiforovella donbassica.*
Lower Carboniferous		$C_1^n d$	*Eridopora macrostoma, Fistulipora glarea, Anisotrypa beshevensis, Nikiforopora concentrica, Stenophragmidium parasiticum, Dyscritella tenuata, D. incrustans, Fenestella donaica, F. kalmiussensis, F. polyporellata.*
		$C_1^n c$	*Eridopora macrostoma, Crustopora tubulosa, Cr. tuberculata, Anisotrypa obscura, Tabulipora quadrata, Fenestella beshevensis, F. polyporata, F. donaica.*
		$C_1^n b$	
		$C_1^n a$	*Crustopora ramosa, Septopora cesteriensis, Fenestella cf. taxata, F. cf. erkovae.*
	Visean	$C_1^v g$	*Fistulipora steshevensis, Cyclotrypa grabovensis, Tabulipora diaphragmata, Nikiforopora rhombicellata, Fenestella donaica.*
		$C_1^v f$	*Fenestella kmaensis.*
		$C_1^v e$	
		$C_1^v d$	
		$C_1^v b+c$	
		$C_1^v a$	
	Tournaisian	$C_1^t d$	*Fenestella karakubensis, F. stylensis.*
		$C_1^t c$	
		$C_1^t b$	
		$C_1^t a$	*Aisenvergia cylindrica, Volnovachia distincta, Leptotrypa singularis, F. stylensis.*

8 | Isotopic Variability of Cheilostome Bryozoan Skeletons

RICHARD M. FORESTER, PHILIP A. SANDBERG and THOMAS F. ANDERSON

Department of Geology, University of Illinois, Urbana, Illinois, U.S.A.

Abstract: Stable carbon and oxygen isotope data have been determined for material of selected modern cheilostome bryozoan species from scattered geographical localities with recorded bottom temperatures. The results of this study indicate that oxygen in cheilostome skeletal carbonate is deposited in isotopic equilibrium with surrounding seawater. Cheilostomes thus appear useful as paleotemperature indicators. Analyses of forms with high organic content in the skeleton tend to give varying, unreliable values which are strongly influenced by the method of sample preparation; such forms should be avoided in isotopically based temperature studies. Comparisons of stable carbon and oxygen isotope ratios of different age parts of the colony may become a useful tool in determining changes in environmental variables, such as temperature or food source, that have occurred during the growth of a colony, whether ancient or modern.

INTRODUCTION

Urey *et al.* (1951) demonstrated that the stable oxygen ($^{18}O/^{16}O$) isotope ratios of skeletal carbonate precipitated in equilibrium with sea water could serve as temperature indicators of ancient seas in which the organisms secreted their skeletons. This isotopic ratio was later shown to be related to the isotopic composition of the water, as well as temperature (Epstein and Mayeda, 1953). This means that the isotopic composition of the water must be known with reasonable accuracy ($\pm 0{\cdot}3‰$) before accurate temperature values can be obtained. Since these initial works the isotopic equilibrium relationships between water and calcite (Clayton, 1961), high magnesium calcite (Tarutani *et al.*, 1969) and aragonite (Rubinson and Clayton, 1969) have been determined or refined. Other studies have shown what processes are responsible for variations in the oxygen and carbon isotope distribution in nature (e.g. Craig, 1953; Epstein and Mayeda, 1953), as well as the possibility of using stable carbon ($^{13}C/^{12}C$) ratios as paleosalinity indicators (Clayton and Degens, 1959). Further studies have examined the isotopic changes that occur when carbonate material undergoes diagenesis (e.g. Gross, 1964, Gross and Tracey, 1966; Hodgson, 1966).

Most of the phyla other than bryozoans occurring in the fossil record have been investigated by earlier workers to determine whether or not skeletal carbonate is secreted in isotopic equilibrium with the water in which the organism lived. In certain calcareous algae (Keith and Weber, 1965), some

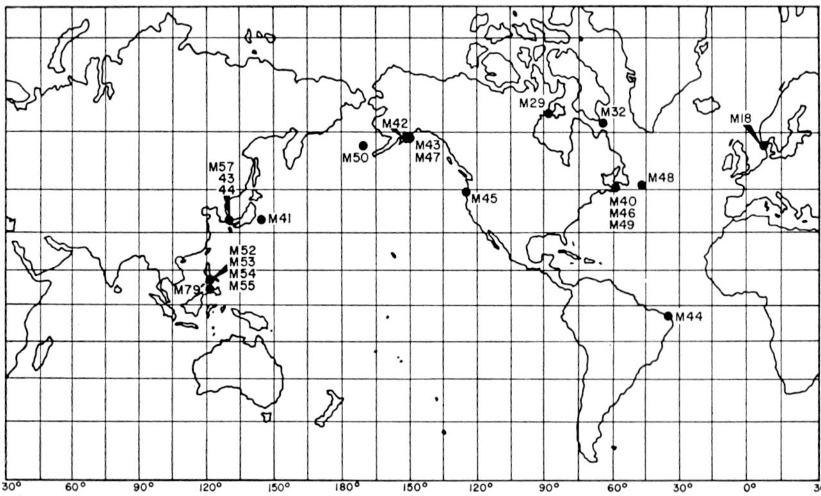

Fig. 1. Collection localities for isotopically analysed Bryozoans. M number refers to an identification slide with a fragment of the bryozoan colony for reference purposes. Samples without M numbers are cyclostomes.

foraminifera (Emiliani and Epstein, 1953; Emiliani, 1955, 1970), Brachiopoda (Lowenstam, 1961), and Mollusca (Epstein *et al.*, 1951; Lloyd, 1964) the skeletal carbonate is in isotopic equilibrium with sea water. Such skeletal material can be used for paleotemperature and paleosalinity determinations, providing that it can be shown that the original isotopic composition of the carbonate has not been altered. Some groups of calcareous marine organisms exhibit a "vital" effect (Urey *et al.*, 1951) in their isotopic composition, i.e. a preferential selection of isotopes by the organism with no systematic temperature control of the skeletal isotopic ratios. It has been demonstrated that coelenterates (Keith and Weber, 1965) and echinoderms, (Weber and Raup, 1966) exhibit a vital effect and thus cannot be used for quantitative paleoenvironmental determinations. Furthermore, caution must be exercised with fauna from marginal marine environments since the oxygen isotopic composition of the water can vary significantly due to precipitation, run-off from the land, storms and evaporation, thus making oxygen isotope temperature determinations difficult, if not impossible (Lloyd, 1964).

Similarly the carbon isotopes reflect the isotopic composition of dissolved HCO_3^- in the environment, which is related to photosynthetic fractionation of ^{13}C and ^{12}C. The $^{13}C/^{12}C$ composition of the skeletons will be progressively lighter as the influence of any freshwater sources on the water in which the organisms live increases (Clayton and Degens, 1959). Terrestrial and fresh water plants preferentially incorporate $^{12}CO_2$, due to selective diffusion of atmospheric $^{12}CO_2$ through cell membranes. This process produces organic matter relatively depleted in ^{13}C. The decomposition of this organic matter produces a HCO_3^- depleted in ^{13}C. In contrast marine plants use dissolved CO_2 and HCO_3^- for photosynthesis without the gaseous diffusion step, producing organic matter somewhat enriched in ^{13}C. The resulting HCO_3^- and eventual skeletal carbonates are relatively enriched in ^{13}C compared to HCO_3^- and carbonates in fresh water. The mixing of fresh water and sea water bicarbonates results in HCO_3^- and skeletal carbonate of intermediate carbon composition in brackish waters. Since the carbon isotopic composition of $CaCO_3$ precipitated in marine, brackish or freshwater bodies reflects that of the HCO_3^- in those waters, those isotopes may be useful in paleosalinity determinations.

Methods of preparation of modern samples for isotopic analysis can influence the nature of the analytical results (Emiliani, 1966). Special preparation techniques are often required to remove or stabilize the organic portions of the sample. Otherwise isotopic exchange may occur between the organic material and the skeletal carbonate, thereby altering the original isotopic composition of the carbonate. Organic material can also produce substances such as C_2H_5OH, CS, NO_2 and propane, which have molecular masses in the same range as the important isotopic species of carbon dioxide (mass 44–46) and thus produce molecular mass ratios which do not reflect the isotopic composition of the CO_2 (Urey *et al.*, 1951). Various methods have been used to treat modern skeletal carbonate (see below under specimen preparation) Lowenstam and Epstein, 1957; Epstein *et al.*, 1953; Naidin *et al.*, 1956.

EXPERIMENTAL PROCEDURES

1. *Specimen Preparation*

Each specimen selected for isotope analysis was carefully hand cleaned, ultrasonorated, or both, to remove foreign particles. A small portion of each species sample was catalogued in an identification slide collection maintained in our department. Fifteen to thirty milligrams of each sample were prepared in one or more of the following ways (Fig. 2):

Cloroxing. Samples were soaked in a mixture of one part commercial Clorox bleach and four parts distilled water (approx. 1% soln NaOCl) for either 30 or 60 min, then washed thoroughly in several distilled water baths for a total of 30–40 min.

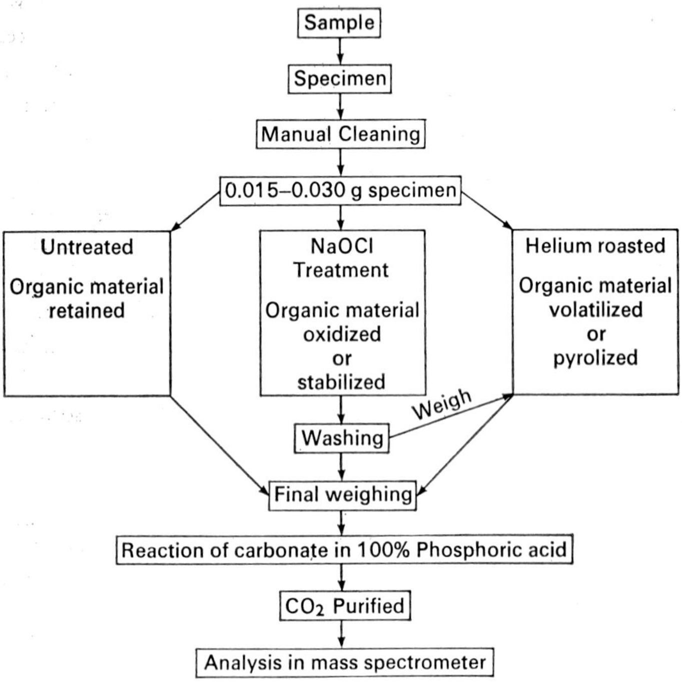

Fig 2 Flow chart of preparation techniques used in the study.

Helium roasting. Samples were roasted for 20 min at 420–430°C, then cooled rapidly to room temperature and removed from the furnace.

Untreated. These samples received no further treatment.

After a final weighing the samples were transferred to reaction vessels and reacted in 100% phosphoric acid (McCrea, 1950).

2. *Laboratory Procedures*

The raw values obtained from the mass spectrometer (Nuclide Corp.) (Nier, 1947; McKinney *et al.*, 1950) are reported in δ notation (parts per thousand ‰

deviation of the sample gas isotopic ratio from that of the standard gas). The working standard was CO_2 evolved from an optical grade Iceland Spar. δ is calculated according to the following equation:

$$\delta_{\text{sample}} = \left[\frac{R_{\text{sample}}}{R_{\text{standard}}} - 1\right] 10^3,$$

where R is the ratio of mass 46 to 44+45 in the case of $\delta^{18}O$ and the ratio of mass 45 to 44 in the case of $\delta^{13}C$. In addition to the corrections derived from the recorder traces, the isotope analyses were corrected for valve leakage, tail contribution, background and the other corrections discussed by Craig (1957). The resultant values have been converted to PDB-I standard values for reporting here.

3. Reproducibility of Isotopic Values

The reproducibility for a single carbonate sample is about $\pm 0{\cdot}1‰$ for both carbon and oxygen, based on the identical preparation of two carbonate samples from the same part of a single colony (*Celleporina surcularis* IIA and IIB, Table 1), and other non-bryozoan carbonate samples prepared under identical conditions.

RESULTS OF PREPARATION TECHNIQUES

1. Discussion of Preparation Techniques

Three species of cheilostomes (Table I) were selected to test the effects of three preparation techniques on the isotopic composition of the skeletons. The approximate percentages of organic material (dry weight) for the three species were: *Flustra foliacea*, (Linné) 40%; *Microporina articulata*, (Fabricius) 10%; *Celleporina surcularis*, (Packard) 4%.

Six samples from the same section of the colony were taken from each species and prepared in the following manner: untreated (I), cloroxed for 30 min (IIA) or 60 min (IIB), cloroxed for either 30 or 60 min, then helium roasted (IIIA and IIIB respectively) and helium roasted only (IV). Samples I, IIA, and IIB gave similar results for the same species, as did samples IIIA, IIIB and IV (Table 1). However, the oxygen isotopic temperature values obtained from the latter three samples are too high, particularly in the case of *Flustra foliacea* (Linné). From these data it is concluded that helium roasting alters the original isotopic composition of the skeleton, probably by isotopic exchange between the carbonate and the organic material during the roasting process. In all cases the δ values for both oxygen and carbon have been reduced, that is, depleted in the heavy isotopes. This can be explained if the organic material present is

Table 1. Results from the comparison of preparation techniques (see Fig. 2)

Genus species	Code number	Location	Depth	$\delta^{13}C$ values	$\delta^{18}O$ values	δw	Calc. T°C	Rec. T°C
Flustra foliacea (Linné)	I(u)[1] M18	N 57°48′ E 11°00′ (Danish coast, Skagerrak)	5 m[2]	−1·35	+0·02	−1·00	12·8	2–15
	IIA(Cl, 30 min)			−1·46	+0·46	−1·00	11·1	2–15
	IIB(Cl, 60 min)			−1·45	+0·35	−1·00	11·5	2–15
	IIIA(Cl, 30 min, H)			−2·63	−3·21	−1·00	37·9	2–15
	IIIB(Cl, 60 min, H)			−2·74	−3·45	−1·00	39·2	2–15
	IV(H)			−2·59	−3·37	−1·00	38·7	2–15
Celleporina surcularis (Packard)	I(u) M32	N 62°74′ W 65°26′5″ (Hudson Strait)	46 m	+1·33	+2·61	−1·00	3·4	
	IIA(Cl, 30 min)			+1·38	+2·58	−1·00	3·5	
	IIB(Cl, 60 min)			+1·34	+2·53	−1·00	3·7	
	IIIA(Cl, 30 min, H)			+1·19	+1·41	−1·00	7·5	*[3]
	IIIB(Cl, 60 min H)			+0·96	+1·47	−1·00	7·3	
	IV(H)			+1·04	+1·33	−1·00	7·8	
Microporina articulata (Fabricius)	I(u) M29	N 65°07′ W 88°45′ (Hudson Bay)	45 m	−1·01	+1·06	−1·00	8·8	
	IIA(Cl, 30 min)			−0·92	+0·97	−1·00	9·1	
	IIB(Cl, 60 min)			−0·24	+1·10	−1·00	8·6	*
	IIIA(Cl, 30 min, H)			−0·94	−0·35	−1·00	23·8	
	IIIB(Cl, 60 min, H)			−0·32	−0·08	−1·00	22·6	
	IV(H)			−0·52	−0·37	−1·00	23·9	

[1] u = untreated; Cl = cloroxed; H = Helium roasted. [2] collected on beach. [3] * = no data available.

isotopically lighter than the skeletal carbonate. It has been shown that the carbon isotope composition of organic material is depleted in ^{13}C relative to skeletal carbonate, (Craig, 1953; Parker, 1964). Possibly the oxygen isotopes in organic matter are also depleted in the heavy isotope (^{18}O) since the isotopic composition of the skeletal carbonate for helium roasted *Flustra foliacea* (Linné) was depleted in both ^{18}O and ^{13}C.

The temperature calculated from the analyses of samples I, IIA and IIB for each species are similar. However, sample I of the *Flustra foliacea* series gave a value of 2°C higher than samples IIA and IIB, but still within the temperature range for the collecting area (2–15°C). The high value for that sample may be due to isotopic variability within the colony or to isotopic exchange between the organic material and the skeletal carbonate during the reaction period in phosphoric acid. If the latter is the case then it would indicate that specimens with a high organic content cannot be left untreated, but must be cloroxed to ensure that the original isotopic composition of the skeleton is maintained. Although the cloroxing process does not remove all of the organic material, it may stabilize the remaining organic material and thus prevent isotope exchange with the skeletal material or the CO_2 evolved.

2. *Oxygen—18 Ratios in Cheilostome Bryozoans*

Once the method of preparation was established, samples collected by the U.S.S. Albatross (U.S. Bureau Fish, 1900, 1902, 1903, 1906, 1907, 1910) from a broad spectrum of geographic localities (Fig. 1) were selected to determine how accurately the oxygen isotope ratios of cheilostome skeletal carbonate reflected the local average temperature of the surrounding water. The bottom and surface water temperatures were available for each Albatross station at the time when the specimens were collected, but annual temperature ranges are not. Each specimen selected for analysis displayed no obvious abrasion and where possible, was a complete colony, sometimes containing zooid remains in the zooecia. Specimens possessing these criteria were presumed to have lived at the location where they were collected. In addition the specimens selected had no or only easily removable external contamination, e.g. encrusting organisms. Whenever possible several different species were selected from each station to check for possible taxonomic control of isotopic ratios, masking a temperature control.

The relationship between temperature and the oxygen isotope composition of carbonates and of sea water is expressed in the following equation from Craig (1965):

$$T°\text{C} = 16{\cdot}9 - 4{\cdot}2(\delta s - \delta w) + 0{\cdot}13(\delta s - \delta w)^2$$

where

$T°C$ = the temperature of the water in which the carbonate is precipitated

δs = the corrected stable oxygen isotope composition of the sample measured against a known standard

δw = the corrected oxygen isotopic composition of the CO_2 gas of the known standard equilibrated with water from the area where the carbonate sample was precipitated.

Because the isotopic composition of sea water is variable, obtaining a valid temperature is impossible without a reasonably accurate estimation of δw. Factors causing the variation in the isotopic composition of the water were discussed by Epstein and Mayeda (1953), Emiliani (1955) and Dansgaard (1961). Epstein and Mayeda (1953) found that an average for all modern ocean water measured in their study was $-0{\cdot}04‰$, relative to PDB-I. They also estimated a variation of the isotopic composition of sea water by no more than $\pm 1{\cdot}0‰$ for non-glacial times, representing a temperature uncertainty of 5°C. However, modern polar water samples tended to be isotopically lighter than the average, usually ranging from $0{\cdot}0‰$ to $-2{\cdot}0‰$, presumably due to mixing of isotopically light continental run off and glacial melt waters with sea water. Waters from temperate and equatorial zones tend to be isotopically heavier than the average value, usually ranging from $0{\cdot}0‰$ to $+2{\cdot}0‰$. This range is partially explained by the fractionation of ^{18}O and ^{16}O, with $H_2{}^{16}O$ preferentially going to the vapor state. Local conditions of run off or high evaporation (Lloyd, 1964) may produce exceptions to the regional ranges.

Using the ^{18}O values (Table 2) and an equilibrated CO_2-water value (δw) of $-0{\cdot}04‰$, it was found that in most cases the calculated bryozoan isotopic temperature was several degrees Celsius higher or lower than the measured temperature from the sampling station. Since most of the samples used in the present study were obtained from shelf waters (20–75 fms) in subpolar to equatorial regions, it is reasonable to assume that the isotopic composition of the water was different at each collecting station. Generally the isotopic temperatures from warm water areas were too low, and the temperatures from cold water areas too high. This discrepancy can be explained by a variation in the isotopic composition of the water: in the case of cold water areas the value of $-0{\cdot}04‰$ is probably too heavy and in warm water areas it is probably too light, thereby causing large errors in the calculated temperatures. Using the average equilibrated CO_2–water values (δw) for the regional area where the samples were collected, the calculated temperature values vary by no more than 3°C from the measured bottom temperature at the time when the specimens were collected. Considering that the majority of the samples were collected during the summer months it is assumed that the bottom temperature values probably represent a rough average temperature for the time of maximum skeletal growth. If this assumption is correct, then the calculated temperature from the oxygen isotopes represents a close approximation of the average summer temperatures, or the average yearly temperature for those deeper water or equatorial samples where little or no seasonal variation occurs.

Because the calculated temperatures approximate the recorded single-value temperatures within the limits of $\pm 3°C$ the skeletal isotopic composition of cheilostomes is interpreted as being in isotopic equilibrium with sea water. Thus cheilostomes can serve as palaeoenvironmental indicators as long as their original skeletal composition is not altered.

Urey *et al.* (1951) believed that the skeletal structure most likely to retain its original isotopic composition would be one that was essentially solid carbonate, so that any isotopic exchange would occur only on the exterior portions of the skeleton leaving the interior relatively unaltered for long periods of geological time. This was the reasoning behind selection of a belemnoid for the first isotopic studies. Bryozoans do not fit these criteria since they construct small porous thin-walled colonies which will probably not maintain their original isotopic composition for more than a few million years. Many species of cheilostomes contain aragonite, and the presence of this mineral might indicate little or no alteration of the skeletal carbonate. It would appear at this time that the most useful application of stable isotopes in cheilostome Bryozoa will be with Pleistocene and modern forms.

3. *Isotopic Variability of* Myriozoum coarctatum (*M. Sars*)

The final phase of this research consisted of systematically examining a single bryozoan colony to determine if stable carbon and oxygen isotopes can be used to facilitate the ecological study of modern forms. An accretionary mode of growth produces skeletal increments in the older parts of the colony which are synchronous with new growth at colony tip or edge. This "averaging" effect must be taken into consideration in growth studies. Samples for this study were taken from an erect branching form (*Myriozoum coarctatum* (M. Sars) (Fig. 3).

The isotopic temperature values indicate that the basal portion (number 47) grew in colder water temperatures than any of the other specimens analysed from this colony (Table 3). Since the temperature obtained from isotopic values is an average temperature, the lower average temperature for the basal portion of the colony may indicate carbonate deposition during colder water seasons or over longer periods of the year than the other parts of the colony. Examination of the basal portion of the colony shows considerable secondary calcification,

Table 2. Comparison of isotopic values with recorded temperatures

Genus species	Code number Treatment	Location Albatross Station number	Depth	$\delta^{13}C$ values	$\delta^{18}O$ values	δw	Calc. T°C	Rec. T°C
Celleporina surcularis (same specimen as in Table 1 but different part of colony)	M32(NC, U)	N 62°74′ W 65°26′5″	46 m	+1·02	+2·36	−1·00	4·2	*
Myriozoum subgracile (D'Orbigny, 1852)	M40(NC, Cl)	N 45°14′30″ W 59°06′45″ D2494	50 fathoms	+1·45	+2·77	−1·30	2·0	+0·28
Celleporina ? sp.	M41(C, Cl)	N 35°03′15″ E 139°37′42″ D5093	302 fathoms	+0·33	+2·70	−0·02	6·4	6·6
Micropora sp.	M42(NC, U)	N 58°07′00″ W 151°36′00″ D2856	68 fathoms	+1·39	+1·73	−1·00	6·4	6·6
Micropora sp.	M43(C, Cl)	N 58°05′00″ W 150°46′00″ D2857	51 fathoms	+0·38	+1·65	−1·00	6·7	6·8
Holoporella sp.	M44(C, Cl)	N 6°59′30″ W 34°47′00″ D2758	20 fathoms	+2·70	−0·56	+1·65	26·8	26·1
Veleroa ? sp.	M45(NC, Cl)	N 44°00′00″ W 124°57′00″ D3052	48 fathoms	+2·86	+2·73	+0·7	8·9	9·4
Cysticella saccata (Busk)	M46(NC, U)	N 45°14′30″ W 59°06′45″ D2494	50 fathoms	+1·69	+2·72	−1·3	2·1	+0·28
Myriozoum coarctatum (M. Sars)	M47(NC, Cl)	N 58°07′00″ W 151°36′00″ D2856	68 fathoms	+1·18	+1·96	−1·00	5·6	6·6

Cystisella saccata (Busk)	M48(NC, U)	N 46°23′00″ W 52°45′00″ D2459	88 fathoms	+1·81	+3·05	−1·3	1·1	−1·4
Porella compressa (Hincks)	M49(NC, Cl)	N 45°14′30″ W 59°06′45″ D2494	50 fathoms	+1·44	+2·53	−1·3	2·7	+0·28
Porella compressa (Hincks)	M50(NC, Cl)	N 57°39′00″ W 170°02′00″ D3523	39 fathoms	+1·29	+3·08	−1·00	1·9	3·3
Triphyllozoon sp. 1	M52(NC, Cl)	N 12°38′15″ E 122°12′30″ D5179	37 fathoms	+1·18	−1·54	+0·13	24·3	24·3
Adeonella gibbera (Canu and Bassler)	M53(NC, Cl)	N 12°38′15″ E 122°12′30″ D5179	37 fathoms	+1·55	−1·29	+0·13	23·1	24·3
Adeonella gibbera (Canu and Bassler)	M54(NC, Cl)	N 12°38′15″ E 122°12′30″ D5179	37 fathoms	+1·33	−1·16	+0·13	22·5	24·3
Conescharellina grandiporosa (Canu and Bassler)	M55(NC, U)	N 12°38′15″ E 122°12′30″ D5179	37 fathoms	+1·36	−1·44	+0·13	23·8	24·3
Triphyllozoon sp. ?	M57(NC, U)	N 34°16′ E 130°16′ D4880	59 fathoms	+1·30	+1·01	−0·02	12·7	14·4
Cupuladria sp.	M79(NC, U)	N 6°44′45″ E 121°48′ D5134	25 fathoms	+2·74	+1·10	+0·13	22·3	24·5
Crisia sp.	43(NC, U)	N 34°16′ E 130°16′ D4880	59 fathoms	+0·16	+0·58	−0·02	14·4	14·4
Idmonea sp.	44(NC, U)	N 34°16′ E 130°16′	59 fathoms	+0·71	+0·54	−0·02	14·6	14·4

NC = not crushed C = crushed U = untreated Cl = cloroxed (1 hr.). * = no data available.

Table 3. Isotopic variability of *Myriozoum coarctatum* (M. Sars)
Sta. D2856, N 58°07′00″, W 151°36′00″, Depth 68 fathoms

Code number	45	46	47	48	49A	49B
$\delta^{13}C$	+1·63	+1·46	+1·11	+1·23	+1·39	+1·33
$\delta^{18}O$	+2·02	+2·16	+2·55	+2·07	+2·14	+2·14
δw	−1·00	−1·00	−1·00	−1·00	−1·00	−1·00
Calculated T°C	5·4	4·9	3·6	5·2	5·0	5·0
Recorded T°C	6·6	6·6	6·6	6·6	6·6	6·6

typically forming a solid base covering, occluding the zooecia. Perhaps this additional calcification takes place over longer periods of the year than normal colony growth, giving a lower isotopic temperature than the rest of the colony. The warmer temperatures calculated from the younger portions of the colony indicate that they were probably depositing carbonate in late spring through early fall. If periodic bottom temperatures (monthly, weekly) were available for the various collection sites it should be possible to determine the season or perhaps the month when the colony was growing at the greatest rate.

Specimens 49A and 49B (Fig. 3B) gave identical ^{18}O results indicating that they were deposited at the same average temperature. Since the total length of 49A and 49B was only 5 mm, this probably indicates that the constricted portion between the two specimens is just an irregularity of growth in one season.

From the preliminary results discussed above it appears that stable isotopes may prove to be a useful tool for the study of growth rates or times of maximum skeletal deposition.

4. *Cyclostome Oxygen and Carbon Ratios*

Carbon and oxygen values obtained from two cyclostome species (Table 2, numbers 13 and 44) differ from a cheilostome (*Triphyllozoon* sp.) from the same station. Because the calculated temperature is closer to the recorded bottom temperature than *Triphyllozoon* sp. it would appear that cyclostomes are also in isotopic equilibrium with sea water. However, the higher temperature values obtained may suggest that the skeletal carbonate is precipitated at higher average temperatures than the corresponding cheilostome. The variability of the carbon isotopic values may reflect a different food source, perhaps a more selective feeding than cheilostomes, or possibly a different average HCO_3^- source, due to a difference in the length of the maximum growth period. Presumably it is the latter since they appear to be in isotopic equilibrium with sea water based on the oxygen isotopes.

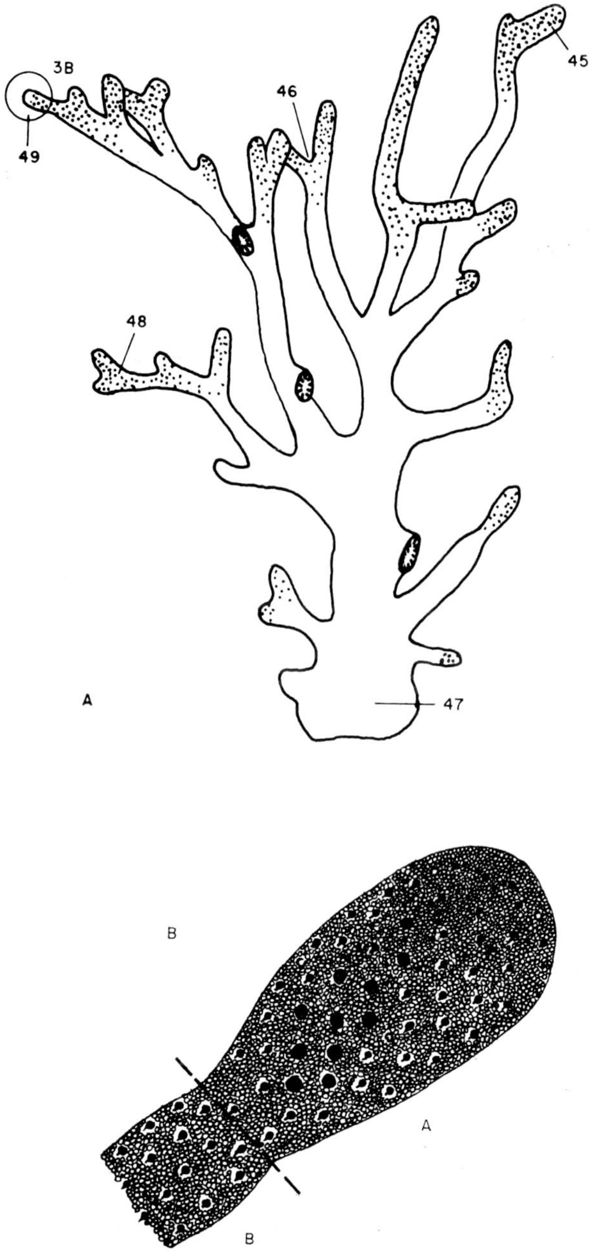

Fig. 3(A). Diagrammatic representation of *Myriozoum coarctatum* (Sars) colony with sample sites indicated by the numbers. Magnification approx. ×2. (B) Detail of upper left tip of the same colony showing constriction between sample 49A and 49B. Magnification approx. ×14.

5. Carbon—13 Ratios in Cheilostome Bryozoans

Little or no information can be obtained from the carbon isotope values, because they all come from isolated, widely scattered collection sites. Significant information can be obtained only if the carbon isotope values represent a large number of points from a single area (Keith and Parker, 1965; Lloyd, 1964; Clayton and Degens, 1959). However, most of the values are greater than 1·0‰ (ranging from −1·46‰ to +2·86‰) and thus represent the upper part of the range of carbon isotopes in marine limestone (Degens, 1967, p. 195). This suggests that most of the specimens collected (Table 2) grew in normal marine conditions, without significant influence of fresh water sources, and adds credibility to the oxygen isotope values.

CONCLUSIONS

From this study it can be concluded that:

1. Cheilostome bryozoans secrete their carbonate skeletons in isotopic equilibrium with sea water. This implies that they can be used for quantitative paleoenvironmental studies and as modern ecological indicators.

2. The best preparation technique for modern skeletal carbonate is cloroxing. However, in many cases, untreated skeletal material also reproduces the correct temperature values.

3. The application of stable isotopes to the study of modern and submodern cheilostomes seems feasible at this time. Parameters such as temperature of skeletal deposition, seasons of skeletal growth and growth rates can be obtained using stable isotopes.

4. Variations in oxygen isotope composition of seawater at differing latitudes and distances from shore must be taken into account if reliable temperatures are to be obtained. Because of difficulties in estimating the extent of these variations for ancient seas, there is some doubt as to the accuracy of absolute temperatures from the Pleistocene epoch and perhaps from earlier epochs.

ACKNOWLEDGEMENTS

This study was supported by the National Science Foundation grant GB 27346 to Philip Sandberg. The Mass Spectrometer and specimen preparation line was funded by the National Science Foundation grant GA 1680 to Thomas F. Anderson. The writers are indebted to Neil Powell of the Canadian National Museum and Alan Cheetham of the U.S. National Museum for providing all but one of the bryozoan samples run (*Flustra foliacea* collected by P. A. Sandberg). The writers wish to thank Elly Brouwers for drawing Fig. 3 and Leslie Lewis for doing the drafting.

REFERENCES

CLAYTON, R. N. (1961). Oxygen isotope fractionation between calcite and water. *J. Chem. Phys.* **34,** 724–726.

CLAYTON, R. N. and DEGENS, E. T. (1959). Use of carbonate isotope analysis of carbonates for differentiating fresh and marine sediments. *Bull. Am. Ass. Petrol. Geol.* **43,** 890–897.

CRAIG, H. (1953). The geochemistry of the stable carbon isotopes. *Geochim. Cosmochim. Acta* **3,** 53–93.

CRAIG, H. (1965). The measurement of oxygen isotope paleotemperatures: In *Proc., Spoleto Conf. on Stable Isotopes in Oceanographic Studies and Paleotemperatures,* July 26–30. Vol. 3, 23 pp.

DANSGAARD, W. (1961). The isotopic composition of natural waters. *Medd. om Gronland* **165,** 1–120.

DEGENS, E. T. (1967). Stable Isotope Distribution in Carbonates. In "Carbonate Rocks", (CHILINGAR, R., ed.), Vol. 9B, pp. 193–208.

EMILIANI, C. (1955). Pleistocene temperatures. *J. Geol.* **63,** 538–578.

EMILIANI, C. (1966). Paleotemperature analysis of Caribbean cores P6304–8 and P6304–9 and a generalized temperature curve for the past 425,000 years. *J. Geol.* **74,** 109–126.

EMILIANI, C. (1970). Pleistocene temperatures. *Science, N.Y.* **168,** 822–825.

EMILIANI, C. and EPSTEIN, S. (1953). Temperature variations of the lower Pleistocene of lower California. *J. Geol.* **61,** 171–181.

EPSTEIN, S. and MAYEDA, T. (1953). Variation of oxygen-18 content of waters from natural sources. *Geochim. Cosmochim. Acta* **4,** 213–224.

EPSTEIN, S., BUCHSBAUM, R., LOWENSTAM, H. A. and UREY, H. C. (1951). Carbonate-water isotopic temperature scale. *Geol. Soc. Am. Bull.* **62,** 417–426.

EPSTEIN, S., BUCHSBAUM, R., LOWENSTAM, H. A. and UREY, H. C. (1953). Revised carbonate-water isotopic temperature scale. *Geol. Soc. Am. Bull.* **64,** 1315–1326.

GROSS, M. G. (1964). Variations in the O–18/O–16 and C–13/C–12 ratios of diagenetically altered limestone in the Bermuda Islands. *J. Geol.* **72,** 170–194.

GROSS, M. G. and TRACEY, J. I. jr. (1966). Oxygen and carbon isotopic composition of limestone and dolomites, *Bikini and Eniwetok Atolls. Sci.* **151,** 1082–1084.

HODGSON, W. A. (1966). Carbon and oxygen isotope ratios in diagenetic carbonates from marine sediments. *Geochim. Cosmochim. Acta* **30,** 1223–1233.

KEITH, M. L. and PARKER, R. H. (1965). Local variation of C^{13} and O^{18} content of mollusk shells and the relatively minor temperature effect in marginal marine environments. *Marine Geol.* **3,** 115–129.

KEITH, M. L. and WEBER, J. N. (1965). Systematic relationships between carbon and oxygen isotopes in carbonate deposited by modern corals and algae. *Science, N.Y.* **150,** 498–501.

LLOYD, R. M. (1964). Variations in the oxygen and carbon isotope ratios of Florida Bay mollusks and their environmental significance. *J. Geol.* **72,** 84–111.

LOWENSTAM, H. A. (1961). Mineralogy, O–18/O–16 ratios, and strontium and magnesium content of recent and fossil brachiopods and their bearing on the history of the oceans. *J. Geol.* **69,** 241–260.

LOWENSTAM, H. A. and EPSTEIN, S. (1957). On the origin of sedimentary aragonite needles of the Great Bahama Bank. *J. Geol.* **65,** 364–375.

McCrea, J. M. (1950). On the isotopic chemistry of carbonates and a paleotemperature scale. *J. Chem. Phys.* **18,** 849–857.

McKinney, C. R., McCrea, J. M., Epstein, S., Allen, H. A. and Urey, H. C. (1950). Improvements in mass spectrometers for the measurement of small differences in isotope abundance ratios. *Rev. Sci. Instr.* **21,** 724–730.

Naidin, D. P., Theis, R. V. and Chupakhin, M. S. (1956). Determinations of climatic conditions of some regions of the U.S.S.R. in the upper Cretaceous Period by the method of isotopic paleothermometry. *Geokhimiya* **8,** 23–34.

Nier, A. O. (1947). A mass spectrometer for isotope and gas analysis. *Rev. Sci. Instr.* **18,** 398.

Parker, P. L. (1964). The biochemistry of the stable isotopes of carbon in a marine bay. *Geochim. Cosmochim. Acta* **28,** 1155–1164.

Rubinson, M. and Clayton, R. N. (1969). Carbon-13 fractionation between calcite and argonite. *Geochim. Cosmochim. Acta* **33,** 997–1002.

Tarutani, T., Clayton, R. N. and Mayeda, T. (1969). The effects of polymorphism and magnesium substitution on oxygen isotope fractionation between calcium carbonate and water. *Geochim. Cosmochim. Acta* **33,** 987–996.

Urey, H. C., Lowenstam, H. A., Epstein, S. and McKinney, C. R. (1951). Measurements of paleotemperatures and temperatures of the Upper Cretaceous of England, Denmark and the southeastern United States. *Geol. Soc. Am. Bull.* **62,** 399–416.

U.S. Bureau of Fisheries (1900). Dredging and hydrographic records of the U.S. fisheries steamer *Albatross*. Reports: 387–562; (1902). Reports: 397–432; (1903). Reports: 123–138; (1906). Document 604; (1907). Document 621; (1910). Document 741.

Weber, J. N. and Raup, D. M. (1966). Fractionation of the stable isotopes of carbon and oxygen in marine calcareous organisms—the Echinoides. Part I. *Geochim. Cosmochim. Acta* **30,** 681–736.

9 | Morphological Variations and Ecology of the Recent Cyclostome Bryozoan "*Idmonea*" *atlantica* from the Mediterranean

J. G. HARMELIN

Station Marine d'Endoume et Centre d'Oceanographie,
Faculté des Sciences de Marseille,
Rue de la Batterie-des-Lions, Marseille, France

Abstract: The Cyclostome "*Idmonea*" *atlantica*, very abundant in the sciaphilous communities of the rocky grounds of Marseilles, presents a strongly marked zoarial polymorphism. This variability has been estimated by statistical analyses of 9 measurable characters. It appeared that these characters vary with the coupled ecological factors, light and water motion, which present a gradient within the distribution area of *I. atlantica*, i.e. from the precoralligenous biocoenosis to that of the semiobscure caves. Four characters seem to be directly influenced by the environment. The others depend more or less on the first ones. The zoarial appearance results from these variable characters and a progressive variation can be found between the poorly branched slender form (semiobscure caves) and the regularly branched stout form (precoralligenous biocoenosis and filiform artificial substrates). The morphological reactivity is sufficiently subtle to appear within a same biotope, being influenced by microclimates and by interspecific competition.

"*Idmonea*" *atlantica* Forbes *in* Johnston (1847) is, among the Cyclostomes, one of the most quoted taxa. But, in fact, the morphologic boundaries of this species are poorly defined; so it is not surprising to meet this name in faunal lists from every part of the world, nor to find in the literature, especially from Mediterranean authors (Meneghini, Heller, Waters), so many species and varieties created for differences which are, in reality, intraspecific morphoses.

The purpose of this study is to demonstrate the range of variation of a species in a single geographical area, to identify the various acting factors and so to get a better definition of this species.

HABITAT IN MARSEILLES

I. atlantica, a very widespread species in Marseilles, is often strikingly abundant. Its vertical distribution extends from a few metres beneath the surface, in shadowed situations, to more than 200 metres on canyon slopes. It can be found on soft bottoms, settling small substrates or other Invertebrates. But its favourite habitat is located on the rocky grounds of the Circalittoral zone (Fig. 6) where it begins appearing at the deeper limit of the Photophilic Algae of the Infralittoral zone (Pérès and Picard, 1964; Pérès, 1967). It can be collected first in the relatively well illuminated Precoralligenous biocoenosis (*Eunicella cavolini* and *Halimeda tuna—Udotea petiolata* aspects). It reaches a maximal abundance when the decreasing light is suitable for the Coralligenous biocoenosis (*Paramuricea clavata* aspect) or for the semiobscure cave biocoenosis (*Corallium rubrum* and sciaphilous sponges aspects). In the darkest parts of the caves, only rare and weak specimens are to be found. It must be noticed that, if this species is present everywhere in these sciaphilous communities, there are some limited areas where its abundance is particularly important.

The thickness of the epibiotic layer, epifauna and epiflora, is an important common character of these circalittoral communities. These organisms, by their successive epibioses, form a very irregular microrelief and are subject to a strong interspecific competition. This microrelief modifies the surrounding physical data: light, water motion, sedimentation, form and nature of the settling points. The result is a mosaic of small ecological niches.

When favourable life conditions are realized, *I. atlantica* is not exacting with regard to the nature and the form of the supports. The substrates can be living or inert, flexible or rigid, flat or filiform. According to the space offered to the colonies, three types of settling points can be distinguished: on a flat surface, in a nook, on an edge or a filiform substrate.

METHODS OF STUDY

The statistical analysis of the specimens, to be correctly interpreted, needed very carefully conducted samplings. To this end, diving has been used, allowing individual collection of these fragile specimens. It also enables the collector to reach them in the smallest nooks, to perceive the ecological surroundings and, above all, to carry out the sampling in a homogeneous area or in a particular ecological situation.

Nine series of measurements (Fig. 1) have been made on chloride cleaned material.

(1) Width of branches, *l.Br*: measured on the basal side, half-way up between two bifurcations.

(2) Number of series between two bifurcations, *n.S*: counted up on the internal side of the branch.

(3) Interserial space, *e.S*.

(4) Number of tubes in series, *n.T*: apical and basal parts of the zoarium are neglected.
(5) Width of tubes, *l.T*: measured in the axis of the series, on the innermost tube, and half-way up of it.
(6) Bifurcation angle, α : angle determined by the theoretical axis joining a proximal bifurcation point to the two succeeding distal one, measured on *camera-lucida* drawings.
(7) Gonozoid height, *h.Gz*.
(8) Gonozoid width, *l.Gz*: greatest width.
(9) Width of the ooeciostome tube, *l.Ooe*: measured by the front view.

A measurement form was established for every specimen and completed with a *camera-lucida* drawing of the basal side outlines.

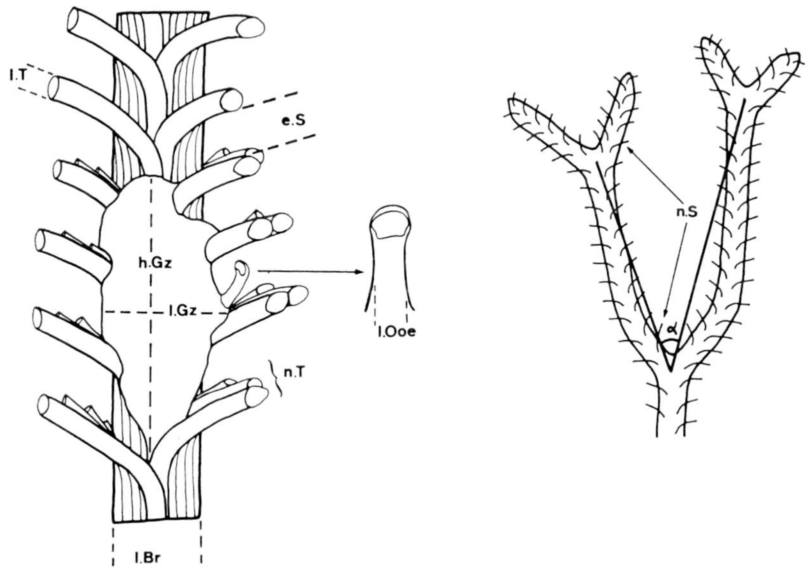

Fig. 1. Key to measurements taken.

These sample data were statistically analysed using the method as recommended by Cheetham (1966) and by Maturo and Schopf (1968). The mean ($\overline{X}$), variance, standard deviation (σ), coefficient of variation ($Cv=100\sigma/\overline{X}$) were calculated.

SOURCES OF MATERIALS

The main part of the specimens analysed here came from natural communities, but some of them had grown on artificial filiform substrates. These experimental supports consisted of small pieces of nylon string tightened on PVC frames put against cliffs rich in *I. atlantica*.

This study is based upon 50 specimens from 11 samples:

Precoralligenous Bioc.: 71·20 (Sarraniers, –30 m), 70·28 (Planier, –18 m).

Coralligenous Bioc.: 70·30 (Grand-Conglue, –37 m), 71·2–B and 71·2–C (Grand-Conglue, –30 m).

Semi-Obscure Caves: 71·18–A (Plane, –18 m), 71·4 and 71·8–A (Jarre, –15 m).

Precoralligenous "strings": IT 1·12 and IT 13·1 (Impériaux de Terre, –18 m, age: 6 to 7 months).

Coralligenous "strings": IL 20·2 (Impériaux du Large, –67 m, age: 8 months).

NUMERICAL RESULTS

The primary results are grouped according to the biotopes to which the various samples belong. The following table (opposite) gives for each measurement the number of specimens measured (nb. spec.), the number of data (nb. data), the mean ($\overline{X}$), the extreme values (v.e.), and the average coefficient of variation (Cv my.) in the four biotopes. These biotopes are arranged according to an increasing gradient of the light and water motion factors, from the caves to the "strings".

VARIABILITY AMONG BIOTOPES

The numerical table shows that the means fit together in a noticeable manner with an increasing or a decreasing order in the classified sequence of the biotopes. So, the light-water motion increase produces an increase of the mean values of l.Br, n.T, l.T, α, l.Gz, h.Gz, l.Ooe (?), and inversely a decrease of e.S and n.S.

These different characters do not vary independently one from another and the value of some is logically derived from that of others which directly influenced by the environment. These basic measurements are: l.Br, n.S, e.S, l.T.

Angle: the value is an inverse function of n.S, as the divergence of the branches diminishes after a certain length.

Number of tubes in series: depends on the width of the branches and of the tubes.

Gonozoid width: is directly connected to the branch width. High values are due to gonozoids spread at a bifurcation level and involve important Cv.

Gonozoid height: seems to be linked to the l.Br value, perhaps through a h.Gz/l.Gz constant. It may be more directly influenced by the surrounding factors acting on the growth rapidity and on the sexual maturation. So its low

	nb. spec.	nb. don.	$\overline{X}$	v.e.	Cv. my.	nb. spec.	nb. don.	$\overline{X}$	v.e.	Cv. my.
			n.T					**n.S**		
Caves	17	158	3·17	2–5	21·9	17	128	9·61	3–37	72·5
Corall. bioc.	13	159	3·92	2–5	15·7	13	176	6·56	2–18	50·1
Precorall. bioc.	9	80	4·33	3–6	18·7	9	70	6·60	2–16	38·7
"Strings"	11	102	4·21	3–6	16·0	11	101	5·06	2–14	38·4
			I.T (μ)					**e.S** (μ)		
Caves	17	124	95·9	85–110	10·2	17	152	288·7	220–380	11·1
Corall. bioc.	13	132	101·64	85–110	5·9	13	137	286·2	220–345	9·3
Precorall. bioc.	9	77	104·25	95–120	6·2	9	75	281·2	220–315	11·4
"Strings"	11	88	106·4	95–125	6·3	11	92	278·9	190–350	11·5
			h.Gz (μ)					**l.Br** (μ)		
Caves	6	19	1191	850–1640	12·4	17	144	348·4	250–530	18·9
Corall. bioc.	7	38	1274	655–1700	12·6	13	148	466·2	280–630	12·3
Precorall. bioc.	4	14	1367	1250–1570	12·0	9	77	481·4	315–660	27·5
"Strings"	4	9	1063	570–1320	—	11	82	511·6	315–756	13·5
			l.Gz (μ)					**angles**		
Caves	6	19	624	470–980	14·0	17	94	46·7	14–78	31·0
Corall. bioc.	7	38	667	393–1180	19·2	13	97	51·3	19–91	58·5
Precorall. bioc.	4	14	693	400–980	103·3	9	53	57·9	13–95	29·8
"Strings"	4	10	700	535–1050	—	11	46	59·4	23–100	28·7
			I.Ooe (μ)							
Ca and C and PC	12	41	93	82–115	16·7					

value in the "string" specimens could be interpreted as a sign of a sexual precocity during a rather rapid growth.
Ooeciostome width: here the number of data was too small to authorize a biotope analysis. The primary results seem to indicate a slightly higher mean in the Precoralligenous samplings, compared with the Coralligenous and caves values. The ooeciostome probably varies as the autozooid tubes.

The zoarial shape is a resultant of these different characters. Consequently zoarial polymorphism is often strikingly pronounced depending on the ecological differences in the biotopes. In the caves, for instance, under poor illumination and diminished water motion, the zoaria are particularly slender, weakly branched, poorly calcified and sometimes sinuous. In the Precoralligenous biocoenosis, specimens are stout and well branched. The Coralligenous community offers a certain heterogeneity in the zoarial types, but the average values of the measurements are very near to the general means. This biotope represents the typical habitat of this species.

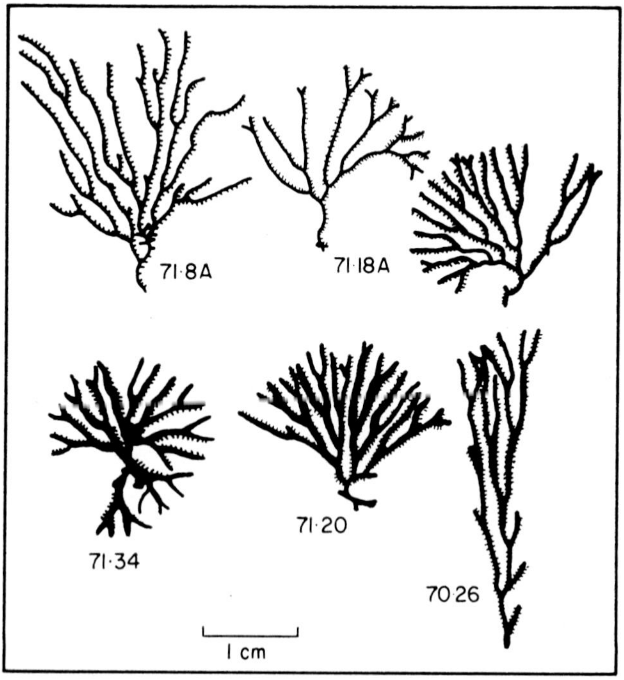

Fig. 2. Zoarial shapes according to the biotope: 71·8–A, obscure zone of a cave; 71·18–A, semiobscure zone of another cave; 71·20 and 71·26, Precoralligenous rocks; 71·34, deep coastal detritic sand (–70 m).

VARIABILITY IN THE SAME BIOTOPE

The above variability represents the average according to the general conditions of the water masses surrounding the various biotopes. But within these biotopes a local variability can be found which is the more pronounced where the epifaunal covering is more marked and where nooks are more numerous; it corresponds to an ecological heterogeneity at the colony level and to a strong interspecific competition. This is particularly clear in the Coralligenous grounds

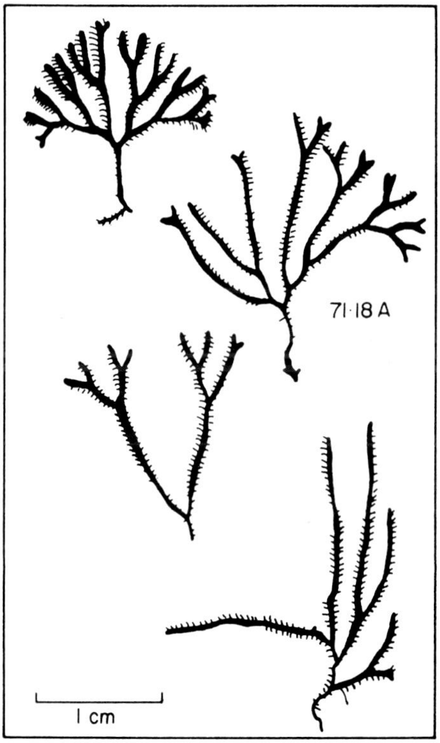

Fig. 3. Zoarial variability in a single biotope, gradual change from the stretched form of an obscure nook to the regularly branched one of the upper layer of the semi-obscure cave epifauna.

and in the sponges fauna of the semiobscure caves. In this last biotope the values of the coefficient of variation for the n.S measurements are particularly high: 66·7, 73·5, 77·4, and correspond to a great diversity in the zoarial shape. On the other hand, for the same measurement in the Precoralligenous biocoenosis, the Cv values are lower: 39·7 and 37·5.

A gradual change from the stretched forms to the fan-shaped zoaria can be found among the cave specimens (Fig. 3). An interpretative scheme (Fig. 4) of these different shapes can be proposed from *in situ* observations. The stretched forms are found in nooks or beneath a massive sessile organism. When the branches can reach the more homogeneous external surroundings, those which are free from obstacles and better supplied with food, the rhythm of the bifurcations increases and the zoarium becomes fan-shaped. It is possible to see, in

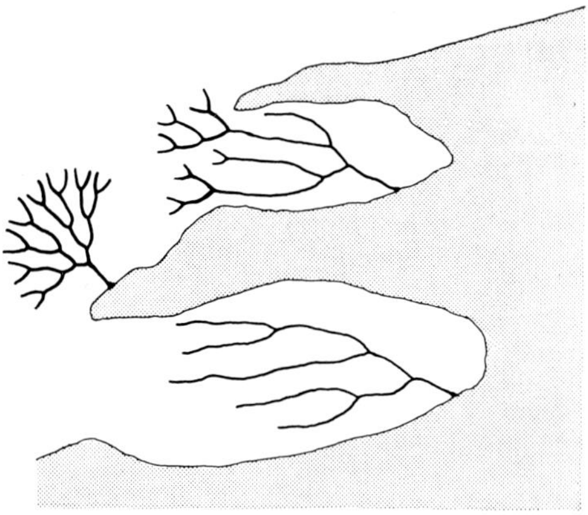

Fig. 4. Interpretative scheme of the shapes of the cave specimens.

specimens showing intermediate forms, an increase of the branch widths from the base to the top of the colony. In contrast with these heterogeneous forms are the specimens grown on experimental lines (Fig. 5). On these substrates, the colonies are able to grow freely in a homogeneous water volume, without competition from incrusting species. In these conditions, zoaria show a tendency to become circular with great regularity in the arrangement and the width of the branches.

TAXONOMIC REMARKS

The knowledge of the numerical data of the species and of their range of variation is an important aid for taxonomic works, especially those dealing with Cyclostomes. But only relatively constant biometric data can be specifically used as discriminating elements. So in order to know the viability of the measurements calculated for a small population from a given area, i.e. on the specimens of a sample, it is interesting to consider the dispersion of the means obtained from the eleven samples considered here. The following table gives, for every measurement, the mean and the coefficient of variation of the series of these eleven means.

	n.T	l.T	n.S	e.S	l.Br	h.Gz	l.Gz	l.Ooe	α
$\overline{X}_{\bar{x}}$	3·87	102	6·89	284	449	1248	660	93	53·5
$Cv_{\bar{x}}$	14·2	*4·43*	32·2	*5·32*	16·9	11·9	12·4	16·7	18·4

Fig. 5. Circular shaped specimen growing on an experimental line (8 months immersion).

It is clear that only two measurements, l.T and e.S, are sufficiently constant, in the various samples, to be considered as good characters. The ooeciostome diameter is probably also a good one, but its small size and its position make precise measurements difficult.

The examination of five or six species stored in museum collections under the name *atlantica* has convinced me that the relative size of the gonozoid is one of the most important character and that the relation h.Gz/l.Br could be a useful point.

I propose to class the discriminating characters of the "*Idmonea*" species from a same geographical zone as follow: 1°—Tube diameter and interserial space; 2°—size and form of the ooeciostome; 3°—size and form of the gonozoid; 4°—width of the branches; 5°—other characters.

When species from different regions are considered, biometrics have to be used with a greater carefulness; it would be interesting to follow within a species the variation of some parameters according to the temperature.

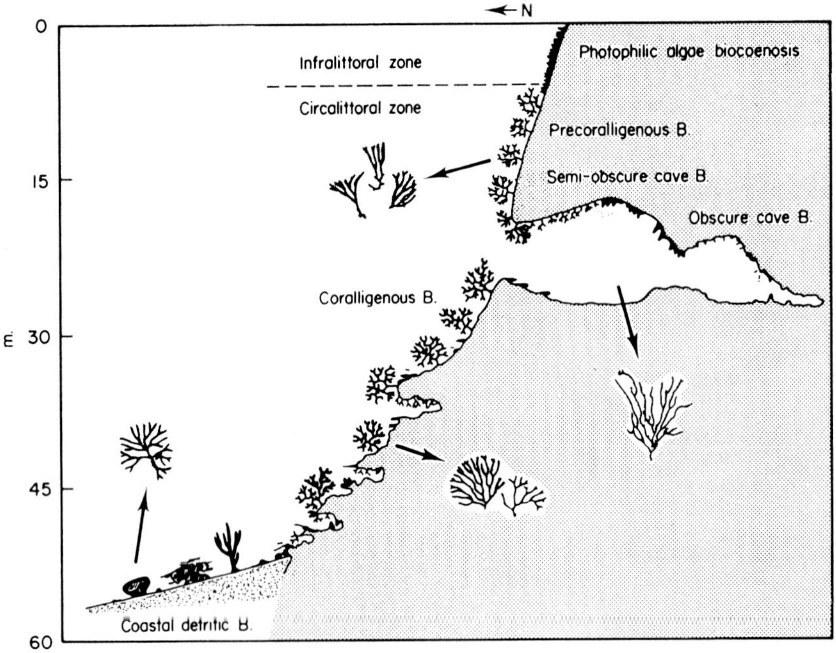

Fig. 6. Distribution of the zoarial shapes in the vertical zonation of the Marseilles' rocky grounds.

The specimens analysed here belong to the species, the distribution of which is limited to the North Atlantic. This species is present in the collections by the following specimens.

Musée Océanographique de Monaco:

Idmonea atlantica, st.060, 1907, Calvet coll., Cap Martin, Monaco.

U.S. National Museum, Washington:

Idmonea atlantica, MC.2, lot 6, Grand Manan I., Aug. 1910, with a slight doubt.

British Museum, Natural History, London:
Idmonea atlantica, 1899.7.1.4094, Busk coll., Porcupine, Atlantic.
Idmonea atlantica, 1879.4.25.75, dredge Secca, Italy.
Idmonea atlantica, 1959.1.6.12, Lagaaij coll., Gulf of Mexico.
Idmonea atlantica, 1911.10.1.119, Norman coll., Madeira.
Idmonea atlantica, 1879.5.28.9, Rev. R. Watson, Madeira.
Idmonea atlantica, 1899.7.1.4099, Busk coll., coasts of Norway and Finmark.

This species is clearly different, from the characters of its gonozoid, its ooeciostome and its measurements, from the sub-arctic species figured by Smitt (1866), Osburn (1912) and Kluge (1962) under the same name. Unfortunately it is hardly possible to judge which of these two species is that described by Johnston (1847).

REFERENCES

ANDROSOVA, E. I. (1968). Bryozoans of the orders Cyclostomata and Ctenostomata of the Antarctic and Sub-antarctic. *Issled. Faunï Morei, Moskva Leningrad* **6** (14) 35–84.

BORG, F. (1926). Studies on recent Cyclostomatous Bryozoa. *Zöol. Bidr. Uppsala* **10,** 181–507.

BORG, F. (1944). The Stenolaematous Bryozoa. "Further Zoological Results of the Swedish Antarctic Expedition 1901–1903," **3** (5) 1–276. Stockholm.

BUGE, E. (1948). Révision du genre *Idmidronea* (Canu et Bassler MSS) Canu 1919, (Bryozoa, Cyclostomata). I—Caractères généraux et discussion. *Bull. Mus. nat. Hist. nat.* **20** (2) 1, 115–119. II—Systématique et conclusion. *Bull. Mus. nat. Hist. nat.* **20** (2) 2, 183–189.

BUSK, G. (1875). Catalogue of the Cyclostomatous Polyzoa in the collection of the British Museum, Part III. 41 pp., 25 pls. London.

CALVET, L. (1902). Bryozoaires marins de la région de Cette. *Trav. Inst. zool. Univ. Montpellier* (2) **11,** 1–103.

CHEETHAM, A. H. (1966). Polyzoa from the Upper Bracklesham Beds (Eocene) of Sussex. *Br. Mus. nat. Hist., Geol., Bull.* **13** (1) 1–115.

DAVID, L. and MONGEREAU, N. (1961). Un exemple d'étude statistique en Paléontologie: *Cellaria fistulosa Auct.* (Bryozoa Cheilostomata) du Vindobonien de la région lyonnaise. *Bull. Bur. Rech. géol. min.* **3,** 29–41.

HARMELIN, J. G. (1969). Bryozoaires des grottes sous-marines obscures de la région marseillaise, Faunistique et Ecologie. *Téthys* **1** (3) 793–806.

HARMER, S. F. (1915). The Polyzoa of the Siboga Expedition, part 1. Entoprocta, Ctenostomata and Cyclostomata. *Siboga Expedite* **28a,** 1–180.

HELLER, C. (1867). Die Bryozoen des Adriatischen Meeres: *Verhandl. zool. bot. Gesellschaft. Wien* **17,** 77–136.

HINCKS, T. (1880). "A history of the British marine Polyzoa," Vol. 1, Text, I–CXLI, 1–601; Vol. 2, Plates, pls. 1–83. London.

JOHNSTON, G. (1847). "A history of British Zoophytes," 2nd Edn. Vol. 1, 1–488; Vol. 2, 74 pls. John van Voorst, London.

KLUGE, G. A. (1962). Les Bryozoaires des mers septentrionales de l'U.R.S.S. (in Russian). *Opredel. Faune S.S.S.R., Moscou* **76,** 584 pp.

MATURO, F. J. S. and SCHOPF, T. J. M. (1968). Ectoproct and Entoproct type material: reexamination of species . . . *Postilla* **120,** 1–95.

MONGEREAU, N. (1969a). Le genre *Idmonea*, Lamouroux, 1821 (Bryozoa, Cyclostomata) dans le tertiaire d'Europe. *Geobios* **2,** 205–264.

MONGEREAU, N. (1969b). Les Bryozoaires Cyclostomes branchus du Miocène rhodanien. *Docum. Lab. Géol. Fac. Sci. Lyon* **36.**

NEVIANI, A. (1939). I briozoi Adriatici del Museo Civico di Moria Naturale de Venizia. *R. Istit. Veneto Sc. Lett. ed. Art.* **30** (4) 3–131.

NORMAN, A. M. (1909). The Polyzoa of Madeira and neighbouring islands. *J. Linn. Soc., Zool.* **30,** 275–314.

OSBURN, R. C. (1912). Bryozoa from Labrador, Newfoundland and Nova Scotia. *Proc. U.S. nat. Mus.* **43,** 275-289.

SMITT, F. A. (1866). Kritisk förteckning öfver Skandinaviens Hafs-Bryozoer. *Öfvers. Kongl. Vetensk. Akad. Förh.* **23,** 395–533.

SMITT, F. A. (1872). Floridan Bryozoa, collected by Count L. F. de Pourtales. Part 1. *K. Vetensk. Akad. Handl.* **10** (11), 1–20.

PÉRÈS, J. M. (1967). The Mediterranean Benthos. *Oceanogr. mar. Biol. ann. Rev.* **5,** 449–533.

PÉRÈS, J. M. and PICARD, J. (1964). Nouveau manuel de Bionomie benthique de la mer Méditerranée. *Rec. Trav. Sta. mar. Endoume* **47** (31) 5–137.

WATERS, A. W. (1879). On the Bryozoa of the bay of Naples. *Ann. Mag. nat. Hist.* (5) **3,** 267–281.

WATERS, A. W. (1922). On the mediterranean *Tervia* and *Idmonea*. *Ann. Mag. nat. Hist.* (9) **10,** 1–16.

10 | Preliminary Observations on Settlement and Growth in Populations of *Alcyonidium hirsutum* (Fleming)

P. J. HAYWARD

Department of Zoology, University College, Swansea, Wales

Abstract: A preliminary report is presented of studies on populations of *Alcyonidium hirsutum* (Fleming) encrusting the alga *Fucus serratus* L. Two problems are considered: the spatial distribution of the ancestrulae on the algal fronds, and the growth of the resulting colonies.

Samples of *Fucus serratus* bearing ancestrulae and adult colonies of *Alcyonidium hirsutum* were collected from Mid Tide Level to Low Water Springs on the shore at Solva, on the Pembrokeshire coast, at approximately monthly intervals over a period of eighteen months, from February 1970. The spatial distribution of the settled larvae was analysed statistically and observed results related to subsequent patterns of growth and reproduction. The mean surface area of the growing colonies was measured for each monthly sample. The patterns of growth for the two seasons studied are shown to be closely comparable. The extent to which the algal thallus may influence the growth pattern of the bryozoan is considered briefly. The data presented here form part of a wider study of the ecology of *Alcyonidium hirsutum* (Fleming) currently being carried out at the University College, Swansea.

INTRODUCTION

The material on which this study is based is collected at monthly intervals from the shore at Solva, Pembrokeshire, where large populations of *A. hirsutum* occur encrusting the thallus of the alga *Fucus serratus* L. The shore studied is situated on the southern side of the mouth of a small drowned valley opening into the sheltered north-eastern corner of St Brides Bay. The valley has two branches, from each of which a stream flows, but, except in the stream beds, the flora and fauna are of a marine, not brackish character. Considerable protection from prevailing winds and currents, and from wave surge is effected by the steep sides of the valley, by the offshore island of Green Scar, and by two low stacks. Further shelter is provided by a reef of rock which runs from the eastern side of the estuary to halfway across its mouth. The landward side of the reef supports a rich growth of *F. serratus* on the lower shore.

The samples of weed collected at each spring tide are initially fixed and preserved in 5% sea water formalin, and later washed and transferred to a 1% aqueous solution of propylene phenoxetol, the preservative properties of which have been demonstrated by Owen and Steedman (1956, 1958). This procedure cleans the samples of the large quantities of mucilage exuded by the alga and facilitates comfortable examination.

Two problems are considered in the present paper: the spatial distribution of the ancestrulae, and the rate of growth of the colonies.

SETTLEMENT

At Solva fertile zooids of *A. hirsutum* appear in the colonies from late October and the larvae are liberated during February. Eggleston (1963), working on the Bryozoa around the Isle of Man, found the breeding season for this species to extend from September to April, although Ryland (personal communication) found that in the Menai Straits all larvae were liberated by December. The larvae typically settle in concavities on the surface of the algal frond; initial settlement is thus largely in the groove flanking the midrib of the frond, just as Wisely (1960) has described for the tubeworm *Spirorbis borealis* Daudin [*S. spirorbis* (L.)]. Settlement also occurs in pits and depressions on the wings of the frond and, to a lesser degree, on the flat of the frond. The question to be considered here is: whether the larvae actively space themselves out on settling, as has been shown to be the case in balanomorph cirripedes (Knight-Jones and Moyse, 1961) and spirorbid tubeworms (Wisely, 1960), or is their spatial distribution, within the limits of their displayed preference for grooves, quite random?

In tackling this problem it has proved necessary to divide the settled populations into several components and in the present investigation only the larger one, that which settles along the midrib groove, has been analysed statistically. Three samples were available for analysis: the natural spat settlement of February 1970 and February 1971, collected at Solva, and a sample of ancestrulae from larvae liberated and settled in the laboratory in February 1971. The larvae settle in localized series along the midrib groove, the disposition of each series probably indicating suitable areas for settlement. Thus, on a particular frond, the proximal part, closest to the holdfast, may be rendered unsuitable by a heavy growth of epiphytes or by dense coverage of adult colonies of *Alcyonidium*, or by the occurrence of large areas of advanced necrosis. Similarly, the distal areas of the frond may be unsuitable through the large quantities of mucilage produced by fruiting tips. The larvae thus tend to settle on the central areas of the frond, although the limits of suitability may vary according to the age and reproductive state of the alga.

From each sample the distances between successive ancestrulae were measured in a large number of these series. The measurements were taken from edge to edge of consecutive ancestrulae, beginning at the distal end of the series and proceeding basipetally along the midrib groove. The figures obtained were grouped into classes of 1 mm intervals, and a frequency histogram was prepared

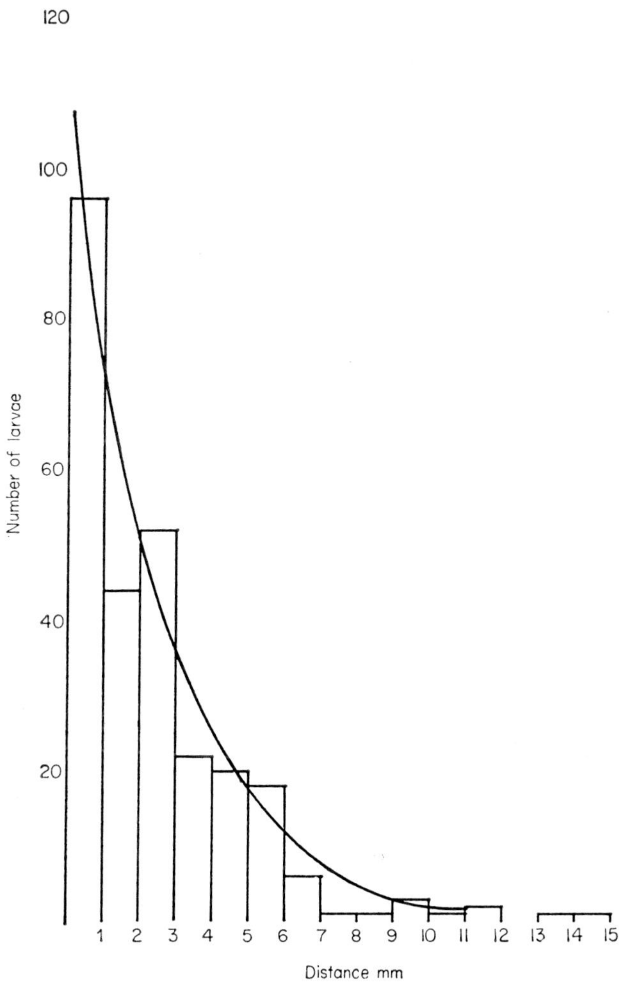

Fig. 1. Analysis of a settlement of *Alcyonidium* larvae with a frequency histogram for the observed values of z (mean 2·62), compared with the theoretical curve for random distribution. February 1971 settlement.

for each sample. Finally, the arithmetic mean interval $\bar{z}$ was calculated. The figures were analysed by the method outlined by Ryland (this volume, p. 165) which states that if settlement is random the numbers per standard unit of length will be distributed as a Poisson series, and the frequency distribution of the intervals, z, will be exponential. The theoretical exponential distribution is calculated from the equation $f(z)=N\lambda e^{-\lambda z}$, where λ is the mean population density, estimated in the sample by $1/\bar{z}$, for $z=0$, $z=1$, $z=2$ etc., and can be plotted as a smooth curve. The expected frequencies for each class may be calculated from $F(z)=1-e^{\lambda z}$, for $z=0$, $z=1$, $z=2$ etc., and the products obtained will give the frequencies from the series of equations $N(z_1-z_0)$, $N(z_2-z_1)$, where z_0, z_1, z_2, are the results of the first equation. These may be compared with the observed frequencies by means of x^2 tests.

The analysis showed the histogram of observed frequencies of z to correspond closely with the theoretical curve for random distribution, in all three samples. No significant difference was found between the observed and expected values. It appears, therefore, that spacing, *within the settled lengths of groove*, is quite random. This implies that not only are all parts within the settled lengths equally attractive to prospecting larvae, but that each larva settles quite independently of previously settled spat. That is to say, there is no evidence either for gregariousness, in the sense of an attraction to settled spat on the part of prospecting larvae, or for deliberate spacing out. As yet no statistical tests have been applied to the other components of the settled population, namely, the spat settled in depressions in the surface of the alae, and those settled on the flat of the frond. However, inspection suggests that the same state of affairs prevails; the larvae do not show any clear indication of deliberate spacing out on settlement, and in the necrotic depressions up to a dozen or more are frequently found in juxtaposition.

GROWTH

Samples for the study of growth rates were obtained at monthly intervals, or fortnightly during the period of accelerating growth. From each sample the areas of one hundred colonies were measured. The outline of each colony was drawn under a camera lucida attached to a Wild stereomicroscope on heavy gauge cartridge paper. The tracings were then cut out and weighed, and the areas calculated with reference to the weight of paper of known area. The specimens drawn were not selected entirely at random; for the first six months of growth only colonies with unobstructed growing edges were drawn, thereafter accurate measurement becomes more difficult as individual colonies tend to abut closely on to one another. Also, only those colonies which do not extend over the sides of the frond can be measured and thus the mean figures for the last few months may be a little lower than they actually are.

In the two generations studied (1970 and 1971), settlement in February was followed by a prolonged period of slow growth until early summer, when rapid expansion began. The measurements for February 1970 and 1971 are of colonies consisting of the ancestrula, with a functional polypide, and up to four undifferentiated zooid buds. The two values, 0·21 mm^2 and 0·23 mm^2 respectively, are closely similar. In 1970 the period of rapid expansion was initiated between May 4 (70 days after settlement) and June 3 (day 100). By June 21 (day 120) the colonies had largely coalesced and the rate of growth began to slacken. From day 150 the mean values vary little, and what variation there is can be largely attributed to sampling discrepancies. In contrast, the 1971 generation, arising from a noticeably sparser settlement than in 1970, from April 11 (day 60) onwards shows higher mean values than for comparable dates in the preceding generation. Up to July 23 (day 163), the last date for which figures are at present available, there is still no indication of coalescence, although it may be assumed that subsequent samples will show the same sharp decrease in growth rate as appeared in the 1970 generation.

Fertile zooids, or brood sacs, appear in the colonies from late October to early

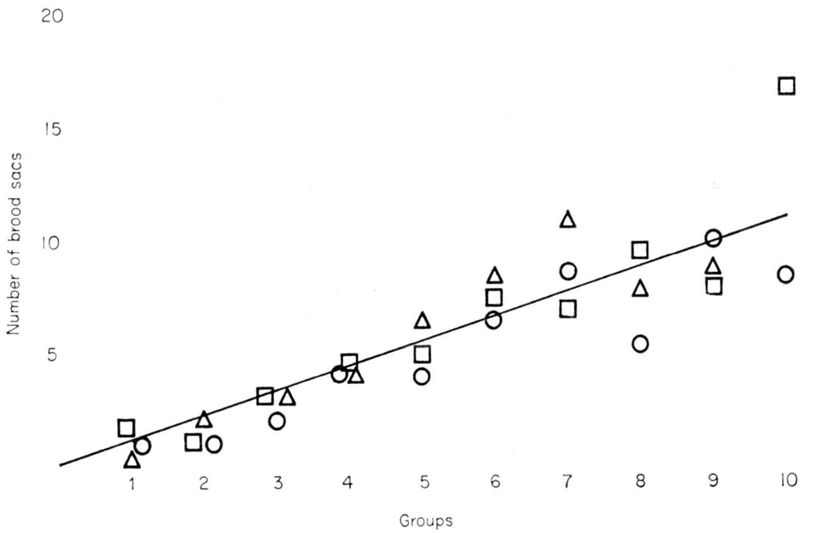

Fig. 2. Number of brood sacs plotted against colony size with colonies grouped at 10 sq mm intervals: grouped results for all three months. □=November, ○=December, △=January 1970. Data expressed as mean number of brood sacs per colony per group.

E

November and persist until the embryos are released in February. When measuring colony areas for November, December and January of the first generation the number of brood sacs present was noted for each colony measured and plotted against colony size. The results for all three months are combined in Fig. 2; for each month the colonies have been grouped into size classes of 10 mm² intervals and the mean number of brood sacs per colony plotted for each class. The regression line shown was calculated from the total data, ungrouped, to avoid weighting due to different numbers of observations falling into each interval. This figure demonstrates clearly a linear relationship between the two variables and establishes an important principle, namely that, even in very small colonies, the number of brood sacs is linearly proportional to the surface area of the colony.

DISCUSSION

Wisely (1960) showed that the larvae of *Spirorbis borealis* Daudin, when settling along the midrib groove of *Fucus serratus* L., tended to space themselves out such that none was closer than 0·5 mm to its nearest neighbour, and in the majority the distance, measured from centre to centre of two individuals, was between 1 and 2 mm. Knight-Jones and Moyse (1961) showed that spacing out also occurs in the settlement of barnacle spat. In contrast, the larvae of *Alcyonidium hirsutum* (Fleming), although choosing broad areas of preference, do not actively space out on settlement, but are randomly distributed. Wisely stated that the spacing out behaviour of *Spirorbis* allowed each individual sufficient room to grow to a mature size; this was also suggested by Knight-Jones and Moyse to be the reason for similar behaviour among balanomorph cirripedes. In both cases the mature size would be that at which the individual was able to reproduce. Unlike these two examples, *A. hirsutum* does not have a definitive adult size (i.e. surface area at maturity), and as was demonstrated above the linear relationship between colony size and fecundity indicates that the small colony is proportionally just as effective a reproductive unit as the larger. The conclusion to be derived from this is that spacing out on the part of the larvae confers no extra benefit to the population as a whole. This view is perhaps reinforced if one considers the record of growth plotted for two seasons' settlement. The density of settlement of that part of the population which was analysed is inversely proportional to the mean interval $\bar{z}$. Thus, the settlement of February 1970, with $\bar{z}=1{\cdot}76$, was far denser than that of February 1971, with $\bar{z}=2{\cdot}62$. The period of slow growth, perhaps related to the shortage of suspended nutriment available, or lower sea temperatures (in fact, a combination of both factors is probably involved) in the early part of the year, is followed by rapid expansion with the advent of summer. During the latter phase the 1971 generation grew at a slightly faster rate than its predecessor and maintained the rate for a longer period of time; these facts must be attributed to the sparser settlement of 1971 and the greater amount of space available for expansion. However, in both generations growth is slowed down and, in many colonies, halted as the population coalesces to form the typical, tight terrazo arrangement of mature colonies. It seems that when a growing edge reaches the limits of a frond it may curve round and continue growth on the other side of the frond, or form fleshy extensions independent of the algal substrate. But, when two growing edges come into contact they appear to inhibit each other, preventing further growth. The pattern which emerges is thus of rapid growth from early summer to fill all the space available to the population, the rate of growth and the point in time at which all colonies have coalesced being dependent on the original density of settlement. There is no reason to suppose that spacing out by the settling larvae would be of advantage to the species unless reproductive potential was geometrically related to colony size, which is clearly not the case. It may even be true that, with the observed high levels of predation by grazing molluscs and other invertebrates, a higher rate of survival of breeding colonies results from a large number of small colonies than from the reverse.

ACKNOWLEDGEMENTS

I wish to thank my supervisor, Dr J. S. Ryland, for the help and encouragement he has given me during the course of my work, which was carried out during the tenure of a Research grant from the Natural Environmental Research Council.

REFERENCES

EGGLESTON, D. (1963). "The Marine Polyzoa of the Isle of Man." Thesis, University of Liverpool.

KNIGHT-JONES, E. W. and Moyse, J. (1961). Intraspecific competition in sedentary marine animals. *Symp. Soc. Exp. Biol.* **XV**: 72–95.

OWEN, G. and STEEDMAN, H. F. (1956). Preservation of animal tissues, with a note on staining solutions. *Quart. Journal Micr. sci.* **97** (3), 319–321.

OWEN, G. and STEEDMAN, H. F. (1958). Preservation of Molluscs. *Proc. malac. soc.* Lond. **33** (3), 101–103.

WISELY, B. (1960). Observations on the settling behaviour of larvae of the tubeworm *Spirorbis borealis* (Daudin), (Polychaeta). *Austr. J. mar. freshw. res.* **11** (1), 55–72.

11 | Seasonal Settlement of Bryozoans in Rehoboth Bay, Delaware, U.S.A.*

E. M. HUMPHRIES†

Department of Biology, University of Delaware, Newark, Delaware, U.S.A.

Abstract: The seasonal and spatial distribution of ctenostomatous (*Bowerbankia gracilis*) and cheilostomatous (*Bugula fulva*, *B. stolonifera*, *B. rylandi*, *Microporella ciliata*, and *Conopeum tenuissimum*) bryozoans on lathing strips in Rehoboth Bay, Delaware is discussed. Seasons of attachment corresponded with seasonal temperature changes. Degree of settling was determined by (1) total number of colonies and their respective areas for the encrusting bryozoans, and (2) "size range" and density within this range for the remaining bryozoans. Results indicate a denser set on shaded surfaces, on strips located in an area of greater water movement, and on strips located in the Bay in contrast to those within an enclosed yacht basin. A new zoogeographical distribution record was set by the occurrence of *Bugula rylandi* and *Bugula fulva* in Rehoboth Bay, Delaware.

INTRODUCTION

The primary objectives of this study were (1) determination of the distribution, abundance and seasonal occurrence of bryozoans on test panels, (2) description of the relationships between the species, with reference to vertical distribution, density and size of the organisms on individual collecting strips and (3) association of the degree and the type of fouling with the environmental factors of wave action, sun exposure and turbidity. A secondary objective was to report the presence of reproductive structures.

Of the approximately 2000 species of fouling organisms, i.e. organisms growing on other than natural objects, bryozoans constitute only 7%; however, they play a significant role in the succession of a fouling community.

AREA OF INVESTIGATION

This study was conducted at a marina located in the northeast corner of

* Part of a thesis submitted to the Graduate Faculty of the University of Delaware in partial fulfilment of the requirements for the degree of Master of Science.

† Present address: Dept. Zoology, University of Florida, Gainesville, Florida, U.S.A. 32601.

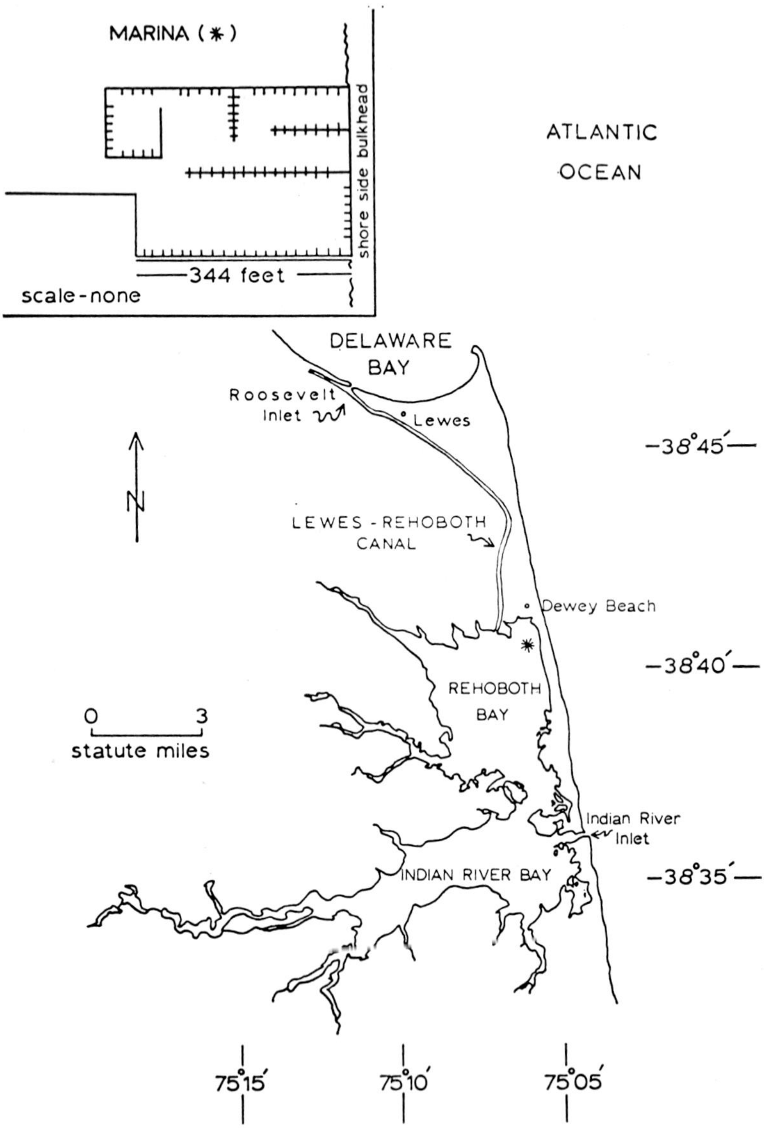

Fig. 1. Chart of the east coast of the state of Delaware indicating the Indian River Drainage Basin (Rehoboth Bay and Indian River Bay).

Approximate location of the Rehoboth Bay Marina, Inc., Dewey Beach, Delaware, marked by an asterisk. Dock plan of the Marina shown in insert diagram.

Rehoboth Bay, Delaware (Fig. 1). According to the Delaware State Board of Health this area of the Bay is polluted as determined by a median coliform MPN greater than 700/100 ml and greater than 10% of the samples with a coliform MPN in excess of 2300/100 ml.

The tidal conditions in Rehoboth Bay are influenced by Indian River Inlet to the South and Roosevelt Inlet via the Lewes-Rehoboth Canal to the North. The tide in Delaware Bay entering through Roosevelt Inlet has the greater effect upon the water movement within this Bay (Kaplovsky and Aulenbach, 1956). Tidal surveys (Table 1), using a tide stick, indicated tidal fluctuations of approximately one foot over a 24 h period. The effective tidal area, which is affected primarily by tidal range and wave action, the latter a function of wind direction and velocity, was 15 in.

Table 1. Hydrographic data from 24 h tidal studies conducted at the end of the dock of the Rehoboth Bay Marina

Date	Tidal range (in)	Effective tidal area (in)	Temp. (°C)	Salinity (‰)
August 1968 (neap tide)	12	15	26–32·5	29·6–30·7
September 1968 (full moon)	11	15	23–25·0	31·6–32·3
November 1968 (new moon)	11½	15	2–6·5	27·3–29·2

The water level measured from the dock surface at the marina at the time of collection showed a maximum amplitude of 91·4 cm and a minimum of 24·9 cm above the substratum over the year period. The average level was 67·1 cm.

The salinity recorded at the time of collection with a hydrometer ranged from 27·3‰ to 32·9‰ over the year collecting period (Fig. 2). Surface and bottom salinity samples had an average variability of only 0·19‰, which is indicative of the shallowness of the Bay. Over a 24 h period (Table 1) the salinity range varied from 0·75‰ to 1·9‰, with variations of only 0·02‰ to 0·60‰ between high and low tides.

Water temperatures recorded at the time of collection (Fig. 2) ranged from 0°C (February) to 27°C (August) for 1967–1968. There was a gradual rise in water temperature from March to August, with the temperature falling within the range of air temperatures recorded during this period. Over a 24 h period (Table 1) the water temperature range fluctuated 2° to 6·5°C with an increase in surface temperatures from 1100 to 1600 h irrespective of the tide.

Turbidimetric analysis of water from the Bay proper and from within the basin revealed no significant difference between these two areas.

MATERIALS AND METHODS

Six stations were established along the southern pier of the marina, one in the

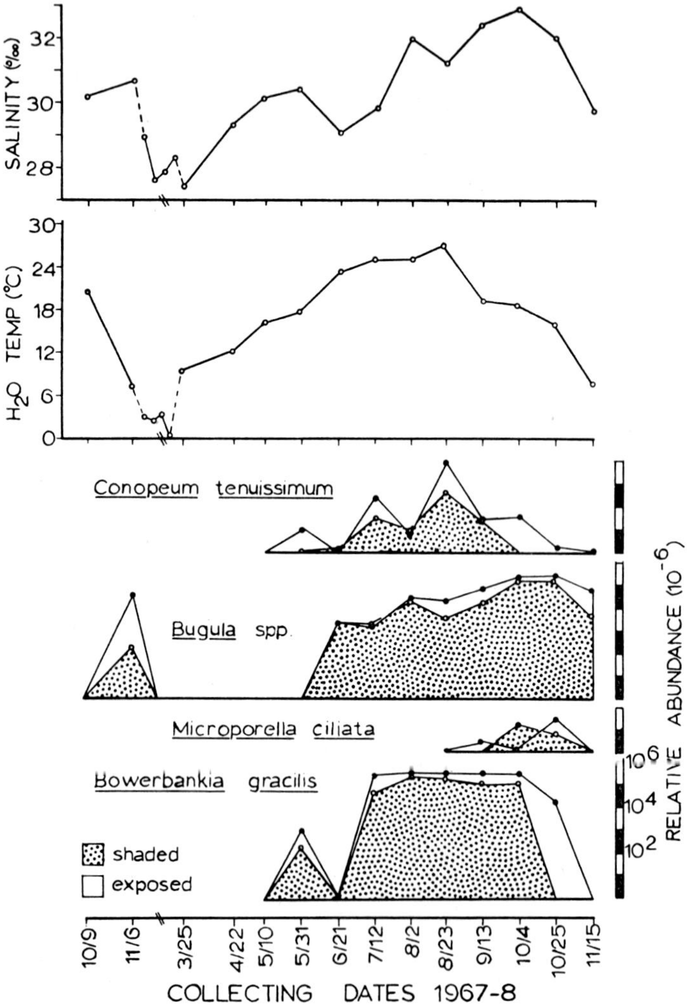

Fig. 2. Seasonal settling of bryozoans on collecting strips in Rehoboth, Delaware, 1967–1968.

Surface salinity and water temperature recorded at time of collection.

$$\text{Relative abundance} = \frac{\text{total area occupied by bryozoans on the 7 collecting strips}}{\text{total area available on the 7 collecting strips}}$$

yacht basin and five in the Bay proper (Table 2). Three South-facing collecting strips were set up to compare the effect of wave action on the settling of sessile organisms. One collecting strip was established under the dock in the shade as a comparison with that strip located at the same station but exposed to direct sunlight. Two collecting strips were oriented in a westward direction for a comparison of the set with that of two of the South-facing strips.

Table 2. Location of collecting strips on the southern dock of the Rehoboth Bay Marina

Stations	Collecting strip no.	Strip orientation	Horiz. dist. west end of dock
1	1 W	West	0
2	2 S	South	5′ 9″
6	6 W	West	29′ 6″
7	7 S	South	35′ 5″
12	12 N	North (under pier)	62′ 9″
	12 S	South	62′ 9″
slip	s–15	North	166′ 0″

At each station a Ponderosa pine lathing strip, $4\frac{1}{2}$ ft × $1\frac{5}{8}$ in. of uniform color but varying wood grain, was nailed to a piling or some structure of the dock perpendicular to the substrate though not necessarily touching it. All strips, except that suspended free underneath the dock, were held as tightly against the pilings as was possible without removing the pile-dwelling organisms present. The shaded side of the collecting strip was the side against the piling, or the side facing the bulkhead in the case of the suspended strip.

All strips were collected and replaced with new strips at the same vertical height at 4-week intervals initially (9 October 1967 to 22 April 1968); however due to the increased settling in the spring, a 3-week collecting period was initiated for the termination of the study (15 November 1968). The strips were brought back to the laboratory unpreserved for analysis.

All bryozoans were identified, counted, and measured. The number of individuals and the surface area of each colony were recorded for the encrusting bryozoans, *Microporella ciliata* (Pallas) and *Conopeum tenuissimum* (Canu). In the case of *Bugula* spp., an erect bryozoan, the number and height of the colonies were determined from a 1 cm section from three regions (upper, middle and lower third) within the species vertical distribution and/or from each discrete settling area. Quantitative density measurements for *Bowerbankia gracilis* Leidy were not feasible because of the compact network of stolons; however, area determinations were calculated.

As a means of comparing the vertical distribution of the bryozoans on the collecting strips, the surface of the dock was taken as the fixed reference line.

RESULTS

1. *Seasonal Distribution*

Conopeum tenuissimum and *Bowerbankia gracilis* were the first bryozoans to settle in the spring (Fig. 2), followed by *Bugula* spp. three weeks later. *Microporella ciliata*, in contrast, set in the early autumn after a drop in the summer water temperatures. The lengths of the settling periods varied from approximately nine weeks for *Microporella* sp. to approximately twenty-six weeks for *Bugula* spp.

The settling pattern for *Bowerbankia gracilis* is interesting as no individuals were recorded during early June after a May set. However, with its return in late June to September its relative abundance increased 365 times over the previous set, and fluctuated only 2% during this period. *Bugula* spp. showed a moderately heavy initial set which increased steadily, particularly on the exposed sides of the strips, to its peak in late October. The relative abundance of the encrusting bryozoans, *Microporella* sp. and *Conopeum* sp., fluctuated throughout their settling period.

2. Microporella ciliata

From a total area of 12·4 mm² occupied by this species (Table 3) 12·1 mm² occurred on the shaded side of the collecting strips, thus indicating a probable preference for this side. However, the strip located beneath the dock and presumably more shaded does not substantiate this hypothesis, as attachment

Table 3. Seasonal and vertical distribution of *Microporella ciliata*

Collection	Collecting strip no.	No. of colonies	Colony size mm²	Colony size no. zooids
9/13/68	1 W	1	1·2	7
10/4	6 W	1 e	6·0	38
10/25	1 W	1 e	0·3	4
	2 S	1	0·3	3
	6 W	1	1·0	5
	7 S	7	0·4–1·3 (T=4·4)	2–8 (T=30)
	12 N	2	0·9 and 1·1	—
	12 S	4	0·1–1·8 (T=2·9)	1–15 (T=26)
	s–15	1	1·5	10

e=colony on exposed side of collecting strip
T=total area occupied

occurred only on one side and the number of colonies was less than that reported from the shaded side of the exposed strip located at the same station. The size of the colonies varied from 0·1 mm² (1 zooid) to 6·0 mm² (38 zooids) with the majority of the colonies recorded from a 129 mm vertical range. The West-oriented strips were the only strips on which attachment occurred during the early settling period. No substantial difference in settling was observed between the strips located within the basin and the Bay proper.

3. Conopeum tenuissimum

During the heaviest set in August, 9 colonies (area=413·9 mm²) out of a total of 15 colonies (area=654·9 mm²) settled on the shaded side of the collecting strips (Table 4). This shade preference was demonstrated throughout the entire settling period of this species as evidenced by (1) 76% of the colonies occupying

Table 4. Seasonal and vertical distribution of *Conopeum tenuissimum*

Collection date	Collecting strip no.	No. of colonies	Colony size mm²	Colony size no. zooids
5/31/68	s–15	2	2·5 (T=5·0)	12 and 16
6/21	12 N	1	0·5	2
7/12	1 W	4	0·1–70·0 (T=125·6)	3–300 (T=648)
	2 S	1	7·0	45
		1 e	—	—
	6 W	1	9·0	—
	7 S	1	7·5	55
	12 N	4	0·4–8·1 (T=18·5)	4–70 (T=169)
8/2	6 W	1	3·1	18
	12 N	1 e	—	—
	12 S	2 e	2 and 2·2 (T=4·2)	—
8/23	s–15	1	130·0	—
	1 W	4	4·4–78·0 (T=95·6)	30–800 (T=920)
	2 S	3 e	1·8–22·0 (T=32·6)	9 and 35 (T>190)
	6 W	1 e	100·0	—
	7 S	1	101·5	700
		1 e	48·0	200
	12 N	3	3·8–75 (T=86·8)	35–500 (T=615)
9/13	6 W	1	6·0	70
	7 S	1	9·8	54
	12 N	1	9·6	50
10/4	7 S	1	19·5	100
10/25	7 S	1	0·5	2

e=colony on exposed side of collecting strip
T=total area occupied

the shaded side of the collecting strips; and (2) attachment on the strip beneath the dock throughout the majority of the settling period, whereas only one colony was recorded from the exposed strip located at the same station.

Colonies ranged in size from 0·1 mm² (3 zooids) to 130 mm² (greater than 700 zooids) with no apparent correlation between the colony size, and the side of the strip to which it was attached or its location in the Bay proper. However, where settling occurred on both sides of the same strip, the colonies were larger on the shaded side.

The second heaviest set, occurring in early July, had colonies occupying approximately the same vertical range as those in August, however, they were approximately half the size. Total vertical range for this species was 650 mm.

4. Bowerbankia gracilis

The seasonal settling pattern of this species is treated differently from that of the encrusting forms previously discussed due to the different growth form. The quantitative measurement for the degree of settling is "size range", which is the sum total of the vertical distances occupied by this species on the collecting strip.

Throughout the entire 21-week settling period (Fig. 3), the shaded side of the collecting strips showed the greatest attachment. The "size range" of attachment on the strip located in the shade underneath the dock was approximately equal on both sides, and inversely related to that on the shaded side of the exposed strip at the same station. No correlation was observed between the degree of settling, and aspect (South *vs* West-oriented strips) and horizontal location along the dock.

The vertical range on those strips located in the Bay averaged 720 mm (maximum range 820 mm), with the densest set occupying 26% to 83% of the vertical range reported for a particular collecting period.

5. Bugula *spp.*

The reproductive period of *Bugula* spp., determined by the presence of mature ovicells, was late July to September with the major period in late September. The major portion of the reproductive colonies during this period occurred within 190 mm of the 600 mm total vertical range of this genus. Exposure was probably a prime factor in the development of mature reproductive colonies, as the majority of these colonies occupied the shaded side of the collecting strips. The largest colony was reported from the collecting strip located within the basin.

No attempt was made to differentiate between the relative abundance of

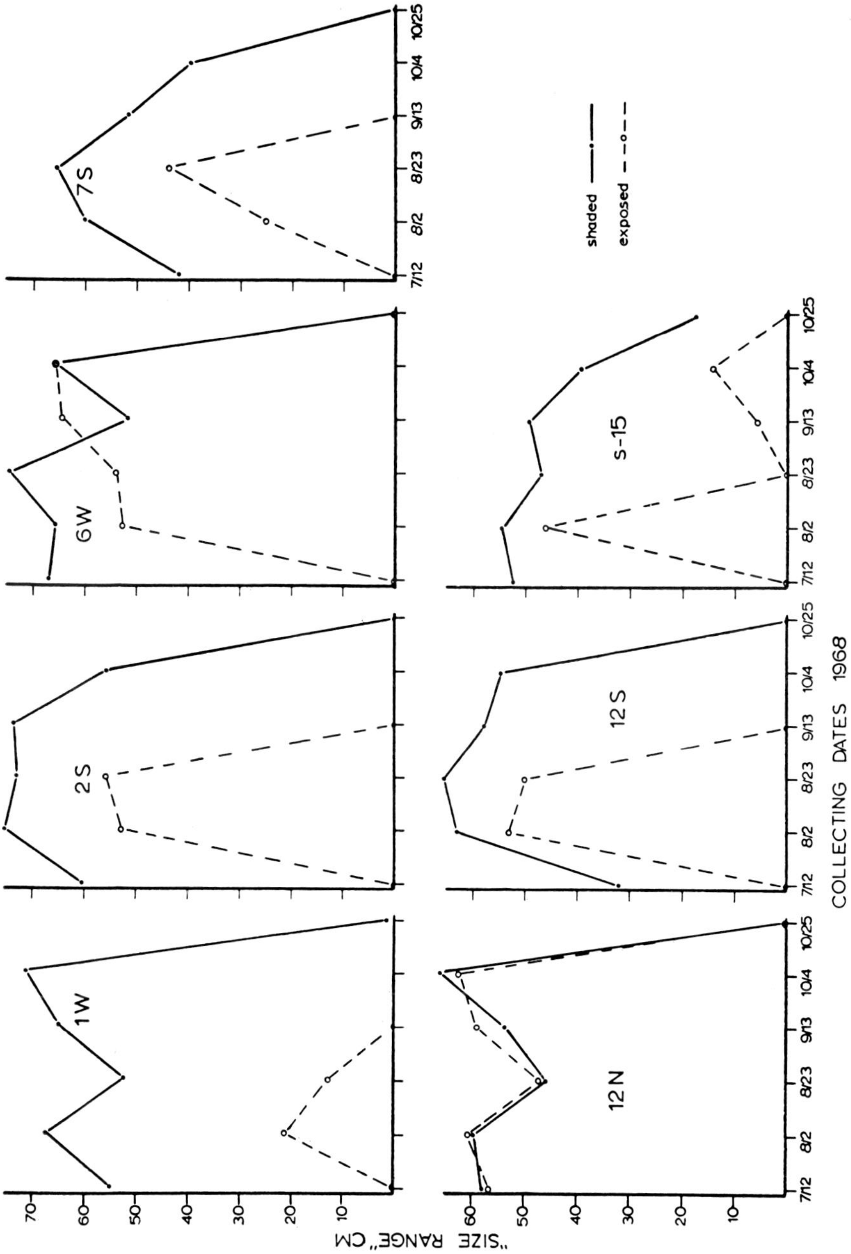

Fig. 3. "Size range" (sum of the vertical distances occupied by a species on a collecting strip) occupied by *Bowerbankia gracilis* on the shaded and exposed sides of the collecting strips.

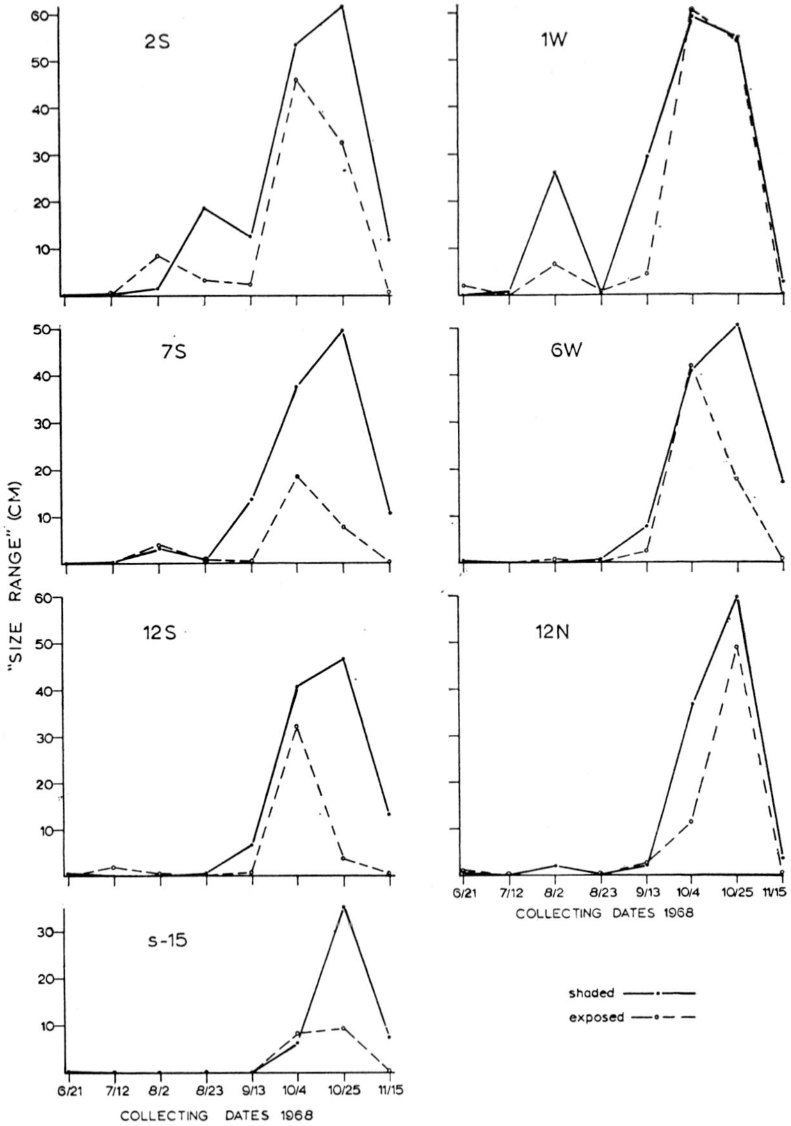

Fig. 4. "Size range" (sum of the vertical distances occupied by a species on a collecting strip) occupied by *Bugula* spp. on the shaded and exposed sides of the collecting strips.

Bugula stolonifera Ryland, *B. fulva* Ryland, and *B. rylandi* Maturo. It should be noted, however, that *Bugula rylandi* occurred infrequently in this study.

Bugula spp. settled on the average with a greater frequency, density, vertical range, and "size range" (Fig. 4) on the shaded side of the collecting strips. A comparison of South and West-oriented strips revealed no consistent pattern of attachment; however, an analysis of similarly oriented strips along the dock revealed an increase in density and "size range" during the majority of the settling period on those strips located farthest from shore.

DISCUSSION

1. *Seasonal Distribution*

The bryozoans in Rehoboth Bay, Delaware exemplified a type 3 breeding behavior according to the classification proposed by Paul in 1942 and modified by the Woods Hole Oceanographic Institution (1952). This type is particular to temperate regions and is characterized by attachment limited to a definite portion of the year only, with the duration of the breeding season dependent on the length of time during which the water temperature remains above a critical level for reproduction for that species.

Microporella ciliata settled on strips in Rehoboth Bay, Delaware from September to October, with the heaviest set during October; whereas further South at Beaufort Harbor, North Carolina larvae commenced settling earlier (late April) and showed the heaviest set in late July (Maturo, 1959). This species can apparently withstand higher temperatures (up to 29°C) and salinity (up to 35‰) than were reported during the period of its attachment in Rehoboth Bay (maximum temperature 20°C, and maximum salinity 32·5‰).

Conopeum tenuissimum settled on strips in Rehoboth Bay from late May to October, with the heaviest set during August. Information compiled on the seasonal distribution of this species along the east coast of the United States tends to indicate a type 3 attachment extending into November in the upper reaches of the Chesapeake Bay and northward; and a type 4 attachment, characterized by attachment during two separate periods of the year (spring and autumn), South of Hampton Roads to Florida. The absence of settling of this species during the summer months in its southern range apparently reflects its lack of tolerance of temperatures greater than 28°C. Calder and Brehmer (1967) reported this species in salinities of 19·5‰ in Hampton Roads, Virginia, which is 10‰ lower than that reported in this study.

Based on the settling data available for *Bowerbankia gracilis* along the Atlantic coast of the United States (Cory, 1967; Maturo, 1959) this species occurs in the upper portion of its temperature and salinity range in Rehoboth Bay.

Bugula spp. (*B. fulva* and *B. stolonifera*) attached to the collecting strips in Rehoboth Bay from June to mid-November, with the heaviest set in October. Farther South at Beaufort, North Carolina it has been reported from March to August (Maturo, 1966). *Bugula rylandi* was identified from strips during August; and this represented the first account of its occurrence from an area other than Sapelo Island, Georgia.

2. General Relationships

Obelia spp. (*O. bicuspidata*, *O. commissuralis*, *O. dichotoma*, and *O. longissima*), the distribution of which is not treated in this paper, settled primarily below the maximum vertical height of *Bugula* spp.; and the density and size of these colonies were less than that of *Bugula* spp. in the area in which both genera occurred. These latter observations may have been due to the subsequent settlement of *Obelia* spp., which would be contrary to the biotic successional pattern proposed by Scheer (1945), or to the slower growth of *Obelia* spp.

Bowerbankia gracilis appears to have no effect on the density of *Bugula* spp. as indicated by the lack of change in density with either the presence or absence of *Bowerbankia* sp. However, it does appear to have an inhibitory effect on the size of the *Bugula* spp. colonies.

Of the nineteen colonies of *Microporella ciliata*, 90% were reported among colonies of *Bugula* spp. This high degree of association, particularly evident in October, was undoubtedly due to the high percentage of settling of both species during this period.

Whereas only 10% of the colonies of *Microporella ciliata* occurred in the vicinity of *Bowerbankia gracilis*; 84% of the colonies of *Conopeum tenuissimum* were reported among the latter, with 76% of them occurring among only *Bowerbankia* sp. This association was particularly noticeable during early July and August, which were major attachment periods for both species. In contrast to *Microporella* sp., colonies of *Conopeum* sp. occurred in areas on the collecting strips unoccupied by either *Obelia* spp. or other bryozoans.

3. Factors Influencing Attachment

The reduced density of set on the strips located within the basin was probably due to the concentration of oil and petroleum, which resulted from a lack of circulation within the semienclosed basin and the absence of a significant tidal fluctuation in the Bay. This situation, however, appeared to have a beneficial effect on the growth of the encrusting bryozoans, as colonies of *Microporella* sp. and *Conopeum* sp. were larger in this area.

The greater abundance of the bryozoans on the shaded side of the collecting strips is undoubtedly a function of the larvae's photonegative reaction prior to attachment (Thorson, 1964; Ryland, 1960). McDougall (1943) observed that colonies of *Bugula neritina* grew more luxuriantly and faster in the presence of direct sunlight, however, this phenomenon was not observed for the species of *Bugula* or the other bryozoans reported from Rehoboth Bay. No conclusive information is available in the literature or from the present study concerning the degree of attachment of bryozoans on South and West-oriented strips relative to each other.

Water movement, a physical factor affecting settlement and growth, may be either beneficial or detrimental to epifuanal species. The increased "size range", density, and colony size of *Bugula* spp. on strips located at the end of the dock is probably a function of (1) the increased water movement, which theoretically carries a greater quantity of larvae per unit time to these strips and a continuous supply of food and oxygen, which prevents sediment accumulation, and which aids in the rapid removal of waste products; and (2) the increased wave action, which aids in the vertical dispersal of larvae on the collecting strips. It may be assumed that the water movement was of a moderate degree at the end of the dock for this genus, since too violent a water action would have prevented attachment of the larvae or torn previously attached young colonies loose. One or both of these cases were apparently operative for *Microporella* sp., which colonized strips primarily farther toward shore.

REFERENCES

Calder, D. R. and Brehmer, M. L. (1967). Seasonal occurrence of epifauna on test panels in Hampton Roads, Virginia. *Int. J. Oceanol. Limnol.* **1,** 149–164.

Cory, R. L. (1967). Epifauna of the Patuxent River estuary, Maryland, for 1963 and 1964. *Chesapeake Sci.* **8,** 71–89.

Kaplovsky, A. J. and Aulenbach, D. B. (1956). A comprehensive study of pollution and its effect on the waters within the Indian River Drainage Basin. *Delaware Water Pollution Commission Tech. Rep.* 207 pp.

McDougall, K. D. (1943). Sessile marine Invertebrates of Beaufort, North Carolina. *Ecol. Monogr.* **13,** 322–374.

Maturo, F. J. S., Jr. (1959). Seasonal distribution and settling rates of estuarine Bryozoa. *Ecology* **40,** 116–127.

Maturo, F. J. S., Jr. (1966). Bryozoa of the southeast coast of the United States: Bugulidae and Beaniidae (Cheilostomata: Anasca). *Bull. Mar. Sci.* **16,** 556–583.

Ryland, J. S. (1960). Experiments on the influence of light on the behavior of Polyzoan larvae. *J. exp. Biol.* **37,** 783–800.

Scheer, B. T. (1945). The development of marine fouling communities. *Biol. Bull.* **89,** 103–121.

THORSON, G. (1964). Light as an ecological factor in the dispersal and settlement of larvae of marine bottom Invertebrates. *Ophelia* **1,** 167–208.

WOODS HOLE OCEANOGRAPHIC INSTITUTION (1952). Marine fouling and its prevention. Contrib. No. 580. 388 pp. Prep. for Bur. Ships, Navy Dep. U.S. Naval Institute, Annapolis, Maryland.

12 Functional Morphology and Habitat of Bryozoa in the Eocene of the Northern Aquitaine Basin, France

MONIQUE LABRACHERIE

Laboratoire de Géologie, Université de Bordeaux I Talence, France

Abstract: Fossil Bryozoa, analysed in terms of structural forms in a continuous series from numerous wells, reflect certain changes in paleoenvironment and indicate some correlations between the sections studied.

The vertical distribution of zoarial assemblages in the northern Aquitanian Eocene shows that during the Eocene the sea covering the northern Aquitaine basin underwent shallowing. However, distinct depth fluctuations occurred during each stage; they may indicate the different transgressive and regressive phases.

INTRODUCTION

Studies of Recent Bryozoa have shown that some correlation exists between zoarial growth form and certain physical conditions of the environment.

For several years now I have tried to apply the data pertaining to Recent faunas for palaeoecological purposes. To this end the distribution of the Bryozoa has been analysed in terms of zoarial growth form in a continuous series of samples from the Eocene of a deep well at Ludon (Gironde) (Labracherie and Prud'homme, 1966). We were able to distinguish a number of shifts in growth form composition as one goes up in the section, which may well reflect certain changes in palaeoenvironment during the Eocene. Admittedly the interpretation of these changes remains rather hypothetical and incomplete, but on the other hand our Ludon study did show that our method had a certain attraction because of the simplicity by which it could provide information on changes in palaeobathymetry and substrate.

It occurred to me that observations of this kind on large series of samples in a sufficiently large number of vertical sections in a sedimentary basin would give a fairly good insight into the sedimentary history of the basin.

CHOICE OF AREA; MODE OF PRESENTATION OF RESULTS

I have tried to put this theory to the test in the Eocene of northern Aquitaine

(Fig. 1), which is ideally suited for four reasons: (1) This area constitutes a well-defined geographical unit; (2) A great number of wells have been drilled through the Eocene into the Cretaceous; (3) The stratigraphy is well-established by foraminiferal studies (Veillon, 1962); and (4) One has the additional advantage of being able to use the extensive records of lithological and mineralogical data (Klingebiel, 1967).

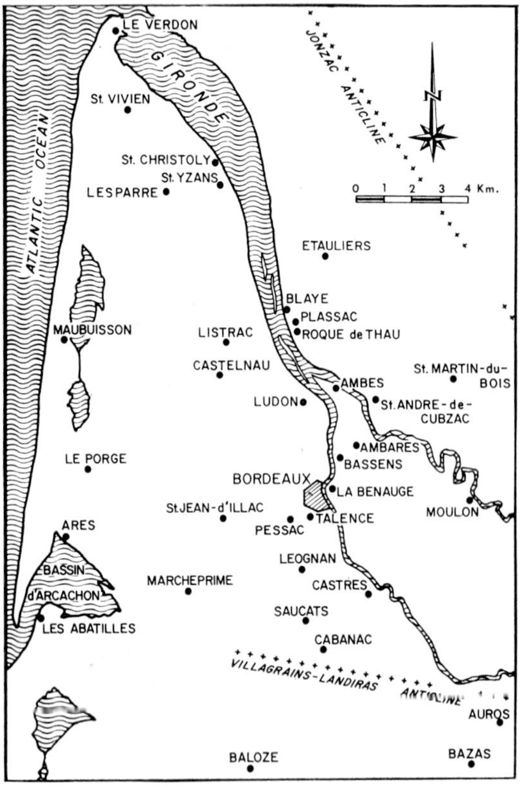

Fig. 1. Map of the study area showing the location of wells.

In order to facilitate detailed correlations I have improved the original form of graphical presentation by combining the percentage distribution graphs of the various zoarial types into one sawblade diagram for each section. I found composite frequency diagrams less satisfactory because they do not give an exact idea of the changes in relative abundance of each individual zoarial type. After long experimentation I have adopted the following way of plotting the percentages: those of the cellariiformes are placed centrally, with the percentages of the reteporiformes, vinculariiformes, conescharelliniformes and lunulitiformes on their right, and those of the adeoniformes and the various encrusting types on their left hand side (Fig. 2).

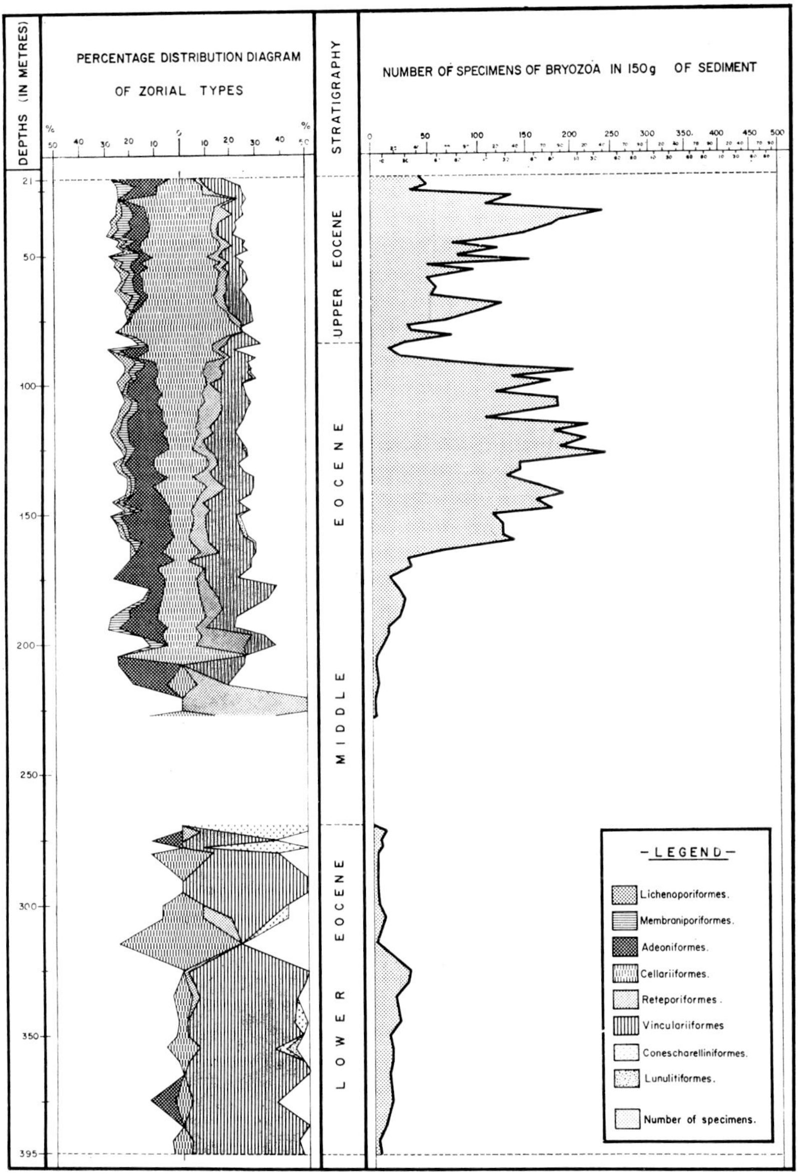

Fig. 2. Ludon No. 1 and 2.

REVIEW OF RECENT CRITERIA

This way of presentation not only keeps the distortion of individual graphs at a minimum, but also has the advantage of enabling one to take the results of studies on Recent bryozoan assemblages into account, such as:

(1) that the cellariiformes, although widely distributed in all depths, yet reach their highest percentages in relatively shallow waters; this zoarial type appears to be fairly independent of the substratum, water turbulence and the rate of sedimentation.

(2) that high percentages of vinculariiformes characterize relatively deep and calm waters. This would correspond to a shift to the right on the sawblade diagram.

(3) that the adeoniformes, eschariformes and membraniporiformes, which, like the cellariiformes, can occur at different depths, are representatives of the infralittoral zone (from 30 to 100 m for the adeoniformes and somewhat shallower for the others); that they reflect, on the other hand, the nature of the bottom and of the available substrates; and finally that they are, contrary to the previous groups, very sensitive to the rate of sedimentation.

Thus the sawblade diagrams constructed for each of the numerous Eocene well sections in the northern Aquitaine basin illustrate the percentage variations in the various zoarial form groups. Moreover, each sawblade diagram is supplemented by a graph for the number of specimens per unit weight.* As previously shown (Labracherie, 1970a), this quantity may be taken as grosso modo indicative of the approximate position of the shore line.

EVALUATION OF RESULTS

As a first step towards evaluation I shall now consider only the gross vertical changes. These should allow us to recognize throughout the area of our study a sequence of "zoarial complexes", from which the palaeobathymetry can be inferred. The overall trend throughout the Eocene appears to be fairly consistent for the various parts of the basin, viz.,

(1) In the Lower Eocene the vinculariiform group often exceeds 80% of the total count in the rich basal assemblages (Figs 3, 4). This suggests, on the basis of our present knowledge, a relatively deep, outer-shelf, environment of deposition. At a later stage, towards the top of the Lower Eocene, the appearance of cellariiformes, together with a decrease in the number of species and specimens, could indicate shallower depths. Most conspicuous is the complete disappearance of

* The complete set of sawblade diagrams of all wells studied forms part of my unpublished Ph.D. thesis, 1970, University of Bordeaux. A few of the diagrams have been selected here to illustrate the variations in zoarial form composition (Labracherie, 1970 *b*).

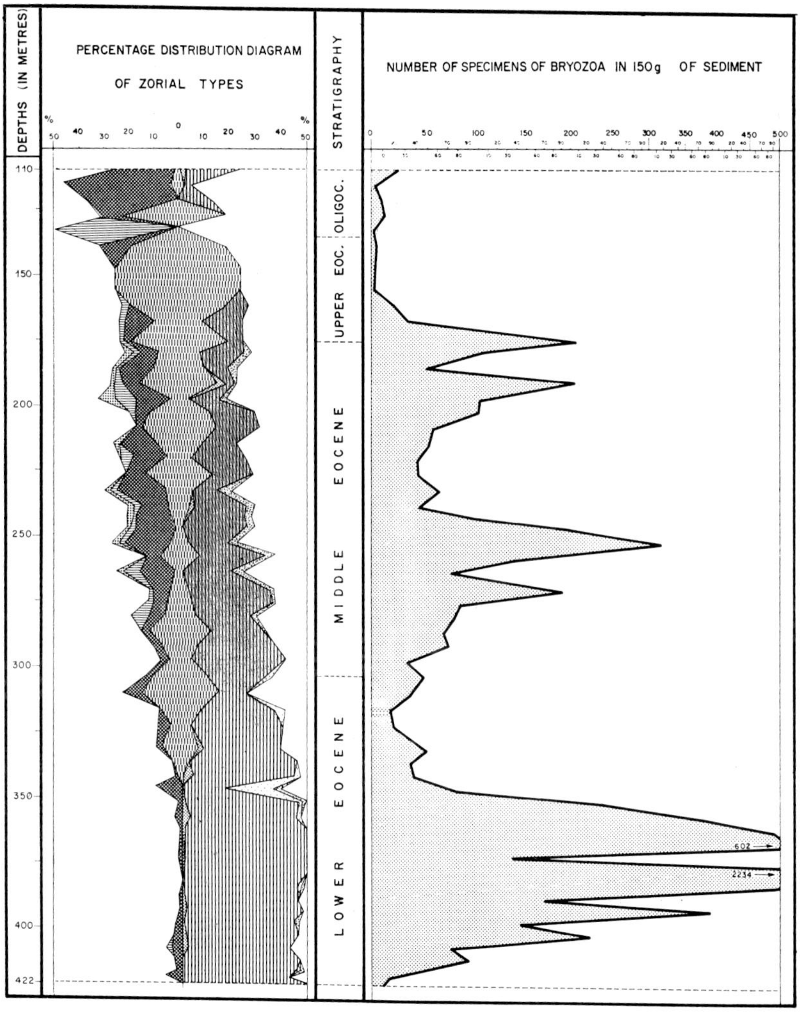

Fig. 3. Maubuisson No. 1.

the erect rigid forms which are associated with rock bottoms or indurated calcareous sediments. In general during this phase the influx of fine-grained terrigenous clastics and of colloidal clay has markedly increased.

(2) During the Middle Eocene the faunas, which had become greatly reduced at the end of the Early Eocene, gradually attains great abundance and diversity throughout the entire basin. The oldest coarse-grained quartz sands of the Middle Eocene still have very few species, which are predominantly cellariiform.

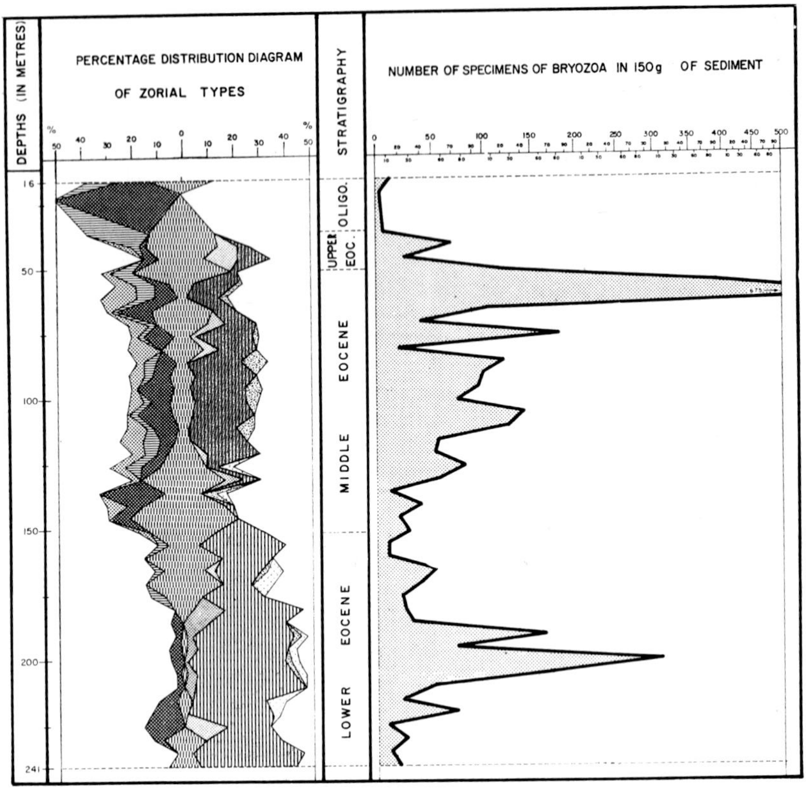

Fig. 4. Castelnau-Médoc No. 3.

In the course of the Middle Eocene, after this initial predominance of cellariiformes, and associated with the deposition of more calcareous sediments, rich faunas reappear in which the percentage of other zoarial types is always increased. This results in better-balanced assemblages. On the average the erect rigid forms, among which the adeoniformes and the vinculariiformes are the most important groups, persistently outnumber the erect flexible forms without, however, ever attaining the same high percentages they had in the Lower Eocene. The rich assemblages found in the middle part of the Middle Eocene must have populated a variety of favourable substrates. The zoarial growth form composition suggests a typically (inner to outer) sublittoral environment.

The youngest Middle Eocene beds generally yield assemblages which are less rich in species. The number of specimens remains sometimes quite high but may also become greatly reduced. The erect flexible forms always outnumber the erect rigid forms. Assemblages of this type indicate moderate depth in the infralittoral zone.

It would seem therefore that the Lutetian is characterized by infralittoral to circalittoral deposits. The deepest biofacies occurs in the Middle Lutetian but does not appear to have reached the depth maxima attained in the Lower Eocene. During the Late Middle Eocene a general shallowing becomes apparent. (3) Finally, in the Upper Eocene, the zoarial form composition remains fairly

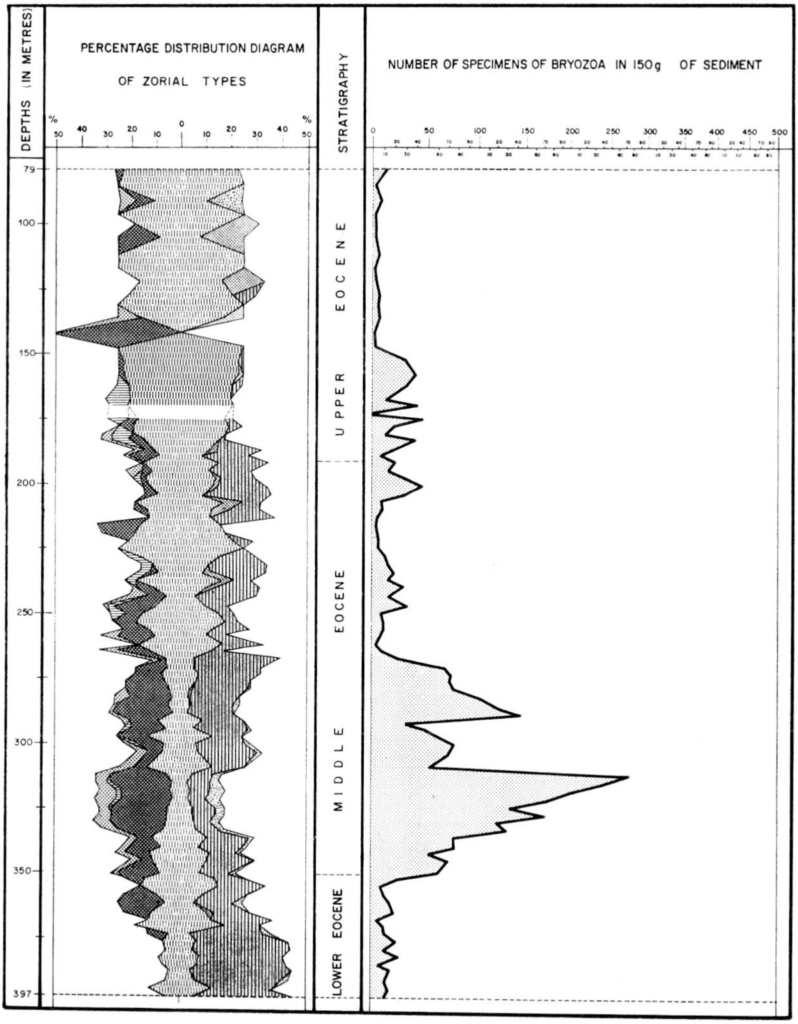

Fig. 5. Talence Faculté No. 2.

constant throughout. The cellariiformes regularly exceed 75%. Forms encrusting a solid substratum are present in most cases and sometimes attain high percentages, whereas the erect rigid forms are very rare and even mere chance finds. A purely cellariiform assemblage often characterizes the upper part of the marine Eocene before Bryozoa disappear completely in the Characeae facies

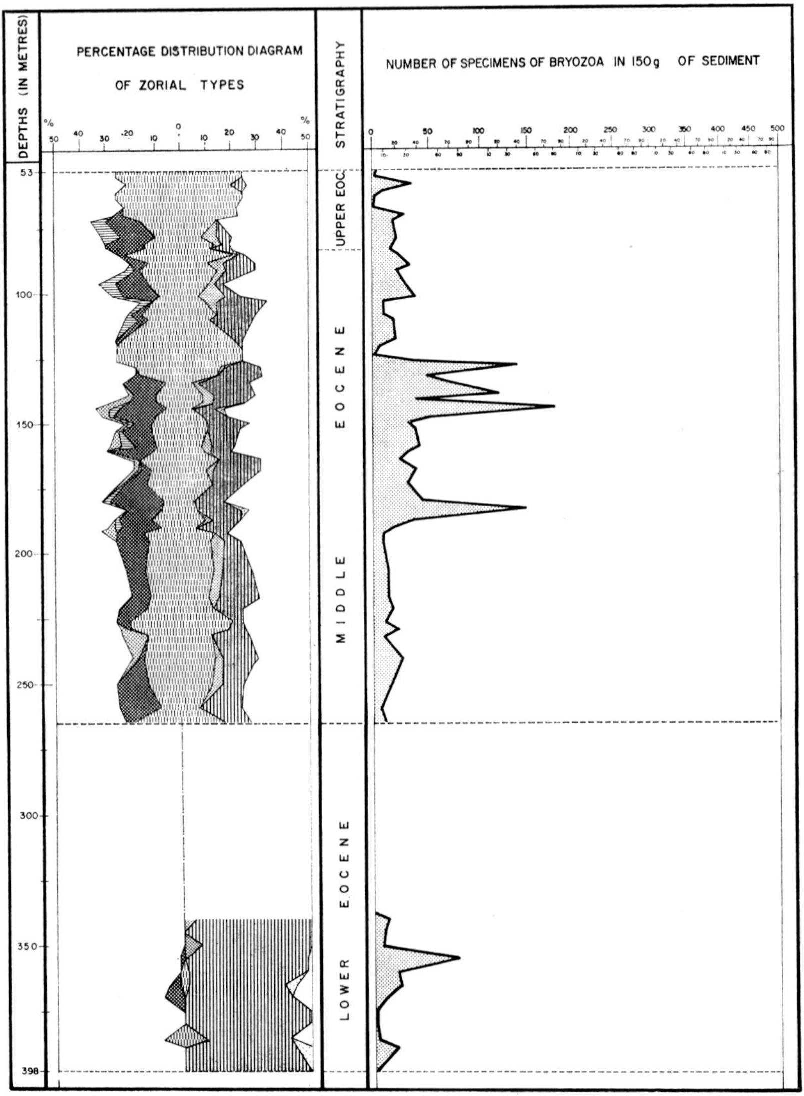

Fig. 6. Bassens Everitube and Bassens Pierrefitte No. 2.

towards the top of the Upper Eocene (Fig. 5, above 79 m and Fig. 6, above 53 m) and could well represent essentially inner infralittoral or even near-shore conditions.

Summarizing, it would seem that during the Eocene the sea covering the northern Aquitaine basin underwent considerable shallowing. Distinct depth fluctuations occurred, however, most markedly so in the Lower and Middle Eocene. They manifest themselves by a rather rapid deepening at the base followed by gradual shallowing which became most pronounced towards the close of the Eocene.

CONFIRMATORY EVIDENCE

The idea of biotopes becoming progressively shallower in the course of the Eocene is also suggested by the presence and distribution of a few extant species whose bathymetric requirements are relatively well-known. The extreme abundance of *Entalophora proboscidea* in the numerous Lower Eocene sections is particularly noteworthy. In Recent seas this species is only prolific in depths greater than 100 m. It disappears, however, in the course of the Early Eocene, only to reappear sporadically and in low percentages in the Middle Eocene, and disappears completely in the Upper Eocene.* This behaviour suggests again that in northern Aquitaine the sea was deepest during the Early Eocene. This would indeed be confirmed by the occurrence in the same assemblages of a number of monospecific cheilostome genera. As a matter of fact Gautier (1962) reports several vinculariiform genera living at several hundred metres in the Mediterranean which are monospecific.

Locally I have been able to prove the occurrence of characteristically euryhaline species, belonging to the genus *Electra*. They first appear and persist locally during the Late Eocene at which time the bryozoan fauna as a whole is so much reduced compared to that of the Early and Middle Eocene. The occurrence, although sporadic, of such brackish-water forms tallies with the above interpretation of shallow water, if not throughout the entire basin, then at least in a fairly wide area; the very area which during the Lutetian had been inhabited by abundant and diversified faunas, including numerous open-marine sublittoral elements.

An independent confirmation of the occurrence of the deepest biofacies in the Lower Eocene is derived from the conspicuous presence of planktonic Foraminifera. This group is lacking in the Middle and Upper Eocene of the northern Aquitaine basin, confirming that this deepest biofacies disappeared in the course of the Eocene.

* In northern Aquitaine it reappears once more with high percentages in the Miocene.

By using zoarial growth forms, and analysing changes in their relative abundance in the Eocene sedimentary sequence, I have been able to establish the bathymetric history of the north-aquitanian Eocene. It would seem to me that this approach may be applied more wisely because it opens the possibility of "ecozone correlation" between sections. Therefore it could, in biostratigraphic studies, supplement correlations based on individual marker fossils or marker fossil associations, which method, at least with Bryozoa, is bound to fail when one is trying to correlate beds of different facies.

The various terms of the arbitrary bathymetric scale which I have used above do not imply that the basin was equally deep everywhere at a given moment. Greater refinement in establishing the distribution of biotopes during each period could conceivably be achieved by tracing the spatial distribution of the various zoarial assemblages corresponding to the various bio- and ecozones.

ACKNOWLEDGEMENTS

My grateful thanks go to Dr R. Lagaaij for helpful advice and for his kind aid in translating the manuscript.

REFERENCES

Gautier, Y. V. (1962). Recherches écologiques sur les Bryozoaires Chilostomes en Méditerranée occidentale. *Thèse Marseille. Rec. Trav. stn. mar. Endoume* (38) **24,** 434.

Klingebiel, A. (1967). Etude sédimentologique du Paléogène nord-aquitain. Interprétation litho-stratigraphique et paléogéographique. *Thèse Doct. Sci. Nat. Bordeaux* (187). *Bull. Inst. géol. Bassin d'Aquitaine, Bordeaux* (2), 290.

Labracherie, M. (1970a). Valeur de l'abondance et de la diversité des Bryozoaires dans l'interprétation des sédiments fossiles. *Bull. Soc. géol. Fr. Paris* 7e sér. **12,** 134–136.

Labracherie, M. (1970b). Les Bryozoaires dans l'Eocène nord-aquitain. Signification biostratigraphique et paléoécologique. *Thèse Doct. Sci. Nat. Bordeaux* (316) (ronéotypée), 314.

Labracherie, M. and Prud'homme, J. (1966). Essai d'interprétation de paléomilieux grâce à la méthode de distribution des formes zoariales chez les Bryozoaires. *Bull. Soc. géol. Fr. Paris* 7e sér. **8,** 102–106.

Veillon, M. (1962). Les zones de Foraminifères du Paléogène nord-aquitain et leur valeur stratigraphique. *Colloque sur le Paléogène, Bordeaux 1962, Mém. Bur. rech. géol. min.*, Paris, (1964), (28) **1,** 227–241.

13 Shallow-water Bryozoa from Deep-sea Sands of the Principe Channel, Gulf of Guinea

R. LAGAAIJ

Shell Internationale Petroleum Maatschappij B.V., The Hague, Netherlands

Abstract: Numerous shallow-water non-deltaic Bryozoa occur in the deep-sea sands of the Principe Channel that extend for some 900 km from the nearest point of the continental shelf edge out into water nearly 4700 m deep.

These Bryozoa, obviously displaced, are identical to and must have derived from the fossil (Late Pleistocene) *Amphistegina* fauna, now occurring on submarine outcrops and in shallow cores on the Nigeria-Cameroon shelf between the 40 and 60 fathom lines. Their absolute age is estimated to be approximately 11,000 years BP. The mise-en-place of these deep-sea channel sands probably dates from the same period.

INTRODUCTION

The deep-sea sands filling the Principe Channel and extending beyond its mouth to a point approximately 900 km from the nearest part of the continental shelf edge into water nearly 4700 m deep (Houbolt, in press), contain numerous Bryozoa (Plates I and II). Samples were taken from seven short cores (Fig. 1) at depths varying from 0·31 to 1·40 m below the top of the core. They either originate from below a Recent to Holocene top layer with abundant *Globorotalia menardii* (cores 27, 29, 36) or contain a Pleistocene colder-water planktonic foraminiferal fauna (cores 29, $41^{\text{I and II}}$) (Brouwer, in press). In one case (core 29) both qualifications apply.

Three things come to mind immediately when contemplating this bryozoan fauna as a whole. The first is its pronounced shallow-water character: only one out of some 1200 identified specimens is a genuine deep-water form, *Euginoma vermiformis*; the remainder inhabit shallow-shelf depth. The second is the non-deltaic character of the fauna, which in its quantitative composition bears little similarity to the present-day bryozoan fauna on the Nigeria shelf. Conical species like *Cupuladria*, for instance, predominate in the Recent shelf fauna, but are barely represented in the channel sands. The third is a feeling of wonder that so many of these small and fragile forms could have travelled such a distance in a

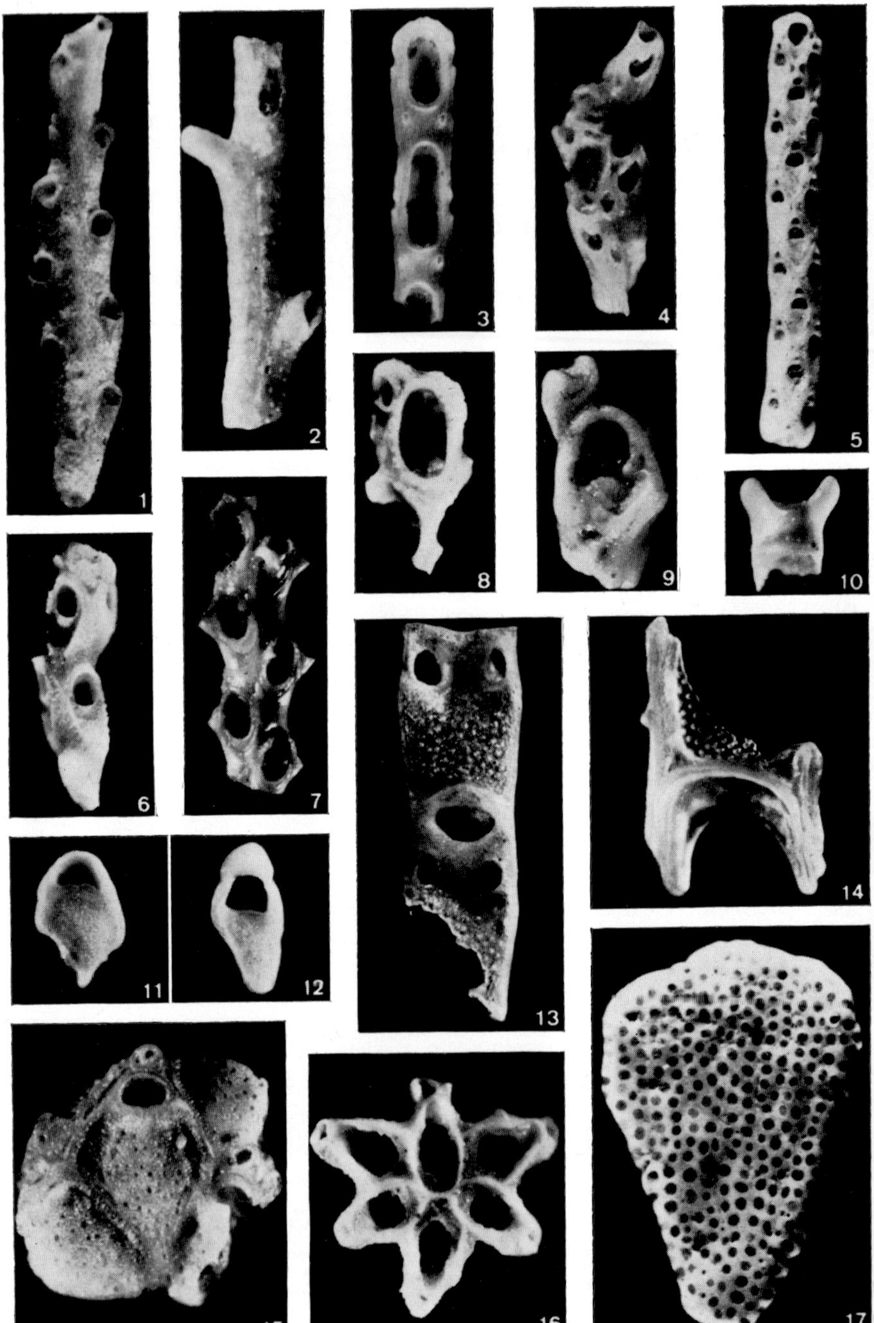
1
2
3
4
5
6
7
8
9
10
11
12
13
14
15
16
17

gravity flow without suffering critical damage. The percentage of unidentified bryozoan fragments varies from 30 to 60% without, however, consistently increasing in the downstream direction.

A fourth conclusion is not so immediately obvious, and was only reached after a systematic comparison of how the various channel species are distributed over the Nigeria shelf. From this exercise the following facts emerged. Of the 48 shallow-water species identified in the channel sands, 27 are commonly and 10 are rarely represented in the Recent Nigeria shelf fauna; 11 others are lacking altogether. On the other hand, 43 out of the 48 occur in the Late Pleistocene "*Amphistegina* fauna" (discussed below). This suggests that the displaced fossil shallow-water fauna in the deep-sea channel sands is identical to and must have derived from the "*Amphistegina* fauna".

THE AMPHISTEGINA FAUNA AND ITS AGE

The term "*Amphistegina* fauna" refers to the isolated concentrations of *Amphistegina gibbosa*. These occur together with a host of other shallow- and clear-water organisms such as other Foraminifera, calcareous Algae, corals and Bryozoa associated with the scattered outcrops of quartz sands and/or "proto"-glauconite mostly* between the 40 and 60 fathom lines (Fig. 3), and in a number

---

Plate I

Fig. 1. *Crisia* sp. Core 28.
Fig. 2. *Crisulipora* sp. Core 41, 0·55–0·61 m-surf.
Fig. 3. *Nellia oculata* Busk. Core 39.
Fig. 4. *Caberea boryi* (Audouin). Core 39.
Fig. 5. *Cellaria salicornioides* Auctt. var. *normani* Hastings. Core 29, × 20.
Fig. 6. *Euginoma vermiformis* Jullien. Core 32.
The only genuine deep-water form in the entire fauna.
Fig. 7. *Scrupocellaria maderensis* Busk. Core 28.
Fig. 8. *Copidozoum tenuirostre* (Hincks). Core 41, 0·55–0·61 m-surf.
Fig. 9. *Floridina antiqua* (Smitt). Core 39.
Fig. 10. *Membranipora tuberculata* (Bosc). Core 41, 0·55–0·61 m-surf.
Fig. 11. *Mollia patellaria* (Moll). Core 36.
Fig. 12. *Mollia patellaria* (Moll). Core 36.
Fig. 13. *Thalamoporella* sp. Core 28.
Fig. 14. *Steganoporella* sp. Core 28.
Fig. 15. *Micropora* sp. Core 28.
Fig. 16. *Cupuladria owenii* (Gray). Core 32.
Fig. 17. *Cupuladria canariensis* (Busk). Basal side of fragment. Core 41. 1·40–1·46 m-surf., × 20. (All specimens × 40, unless otherwise stated.)

* There are a few deeper occurrences however; for example at D. 137 in 109 fathoms.

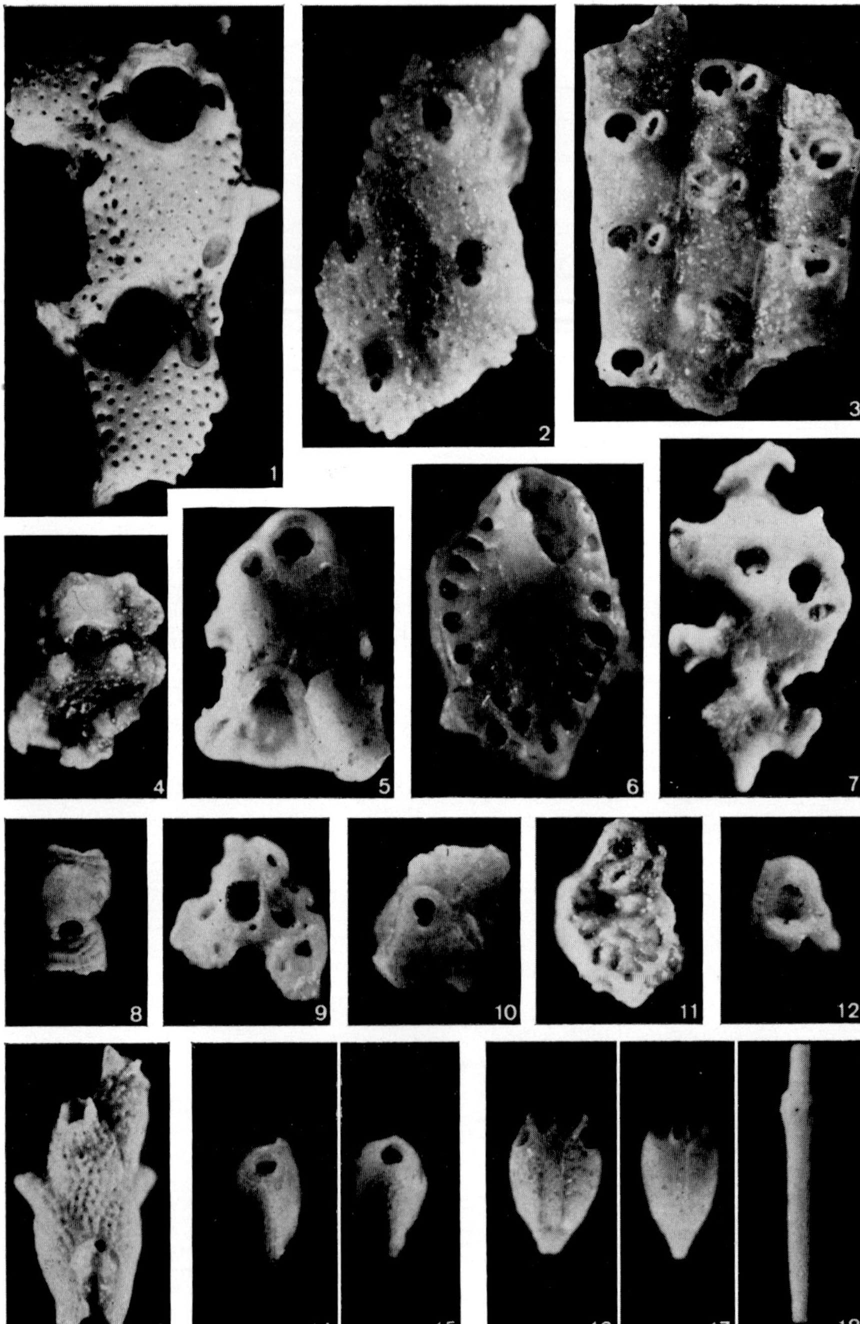
1
2
3
4
5
6
7
8
9
10
11
12
13
14
15
16
17
18

of shallow cores around the delta, also between 40 and 60 fathoms. The most extensive sand outcrops with *Amphistegina* occur west of the Avon Deep and east of the Calabar Canyon, where they escaped blanketing by Holocene to Recent sediments. It is most likely, however, that this fossil fauna extends around most of the delta in this depth interval beneath the cover of Holocene to Recent sediments, and that it occupies an equivalent position farther east and south, parallel to the coasts of Cameroon, Rio Muni and Gabon. Berthois *et al.* (1968), for instance, report abundant *Amphistegina gibbosa*, associated with corals, from two stations on the 100 m (=54 fathoms) isobath, north-northwest of Campo, Cameroon.

Allen (1964) introduced the term "Older Sands" for these sediments and proved, by radiocarbon dating on calcareous Algae and molluscan shells, a Late Pleistocene age (approx. 11,000–12,000 years BP) for his topographical features "outer terrace" and "outer rise", to which the *Amphistegina* fauna is apparently confined. Recently, Giresse (1969, p. 984) also recognized the fossil nature of an *Amphistegina*—Melobesiae—Bryozoa association occurring off Gabon at approximately 50 m depth and stressed the small size of the submarine outcrops.

It is interesting to note that fossil *Amphistegina* faunas, also dated as Late Pleistocene or Early Holocene, have been reported at comparable depths from

---

Plate II

Fig. 1. *Hippopetraliella africana* Cook. Core 28, × 33.
Fig. 2. *Schizomavella auriculata* (Hassall). Core 28.
Fig. 3. *Schizoporella* sp. Core 28.
Fig. 4. *Hippopodinella lata* (Busk). Core 39.
Fig. 5. *Escharina porosa* (Smitt). Core 39.
Fig. 6. *Metrarabdotos* (*Uniavicularium*) *unguiculatum* (Canu and Bassler) *cookae* Cheetham. Core 28.
Fig. 7. *Cleidochasma porcellanum* (Busk). Core 36.
Fig. 8. *Cribrilaria radiata* (Moll). Core 36.
Fig. 9. *Triporula stellata* (Smitt). Core 28.
Fig. 10. *Trypostega venusta* (Norman). Core 41, 0·55–0·61 m-surf.
Fig. 11. *Reptadeonella* sp. Core 28.
Fig. 12. *Escharina pesanseris* (Smitt). Core 36.
Fig. 13. *Margaretta* sp. Core 28, × 20.
Fig. 14. *Euteleia evelinae* Marcus. Core 36.
Fig. 15. *Euteleia evelinae* Marcus. Core 36.
Figs 16, 17 and 18. *Pasythea tulipifera* (Ellis and Solander). Triads and axial kaenozooecium. Cores 36 and 27 resp. (All specimens × 40, unless otherwise stated.)

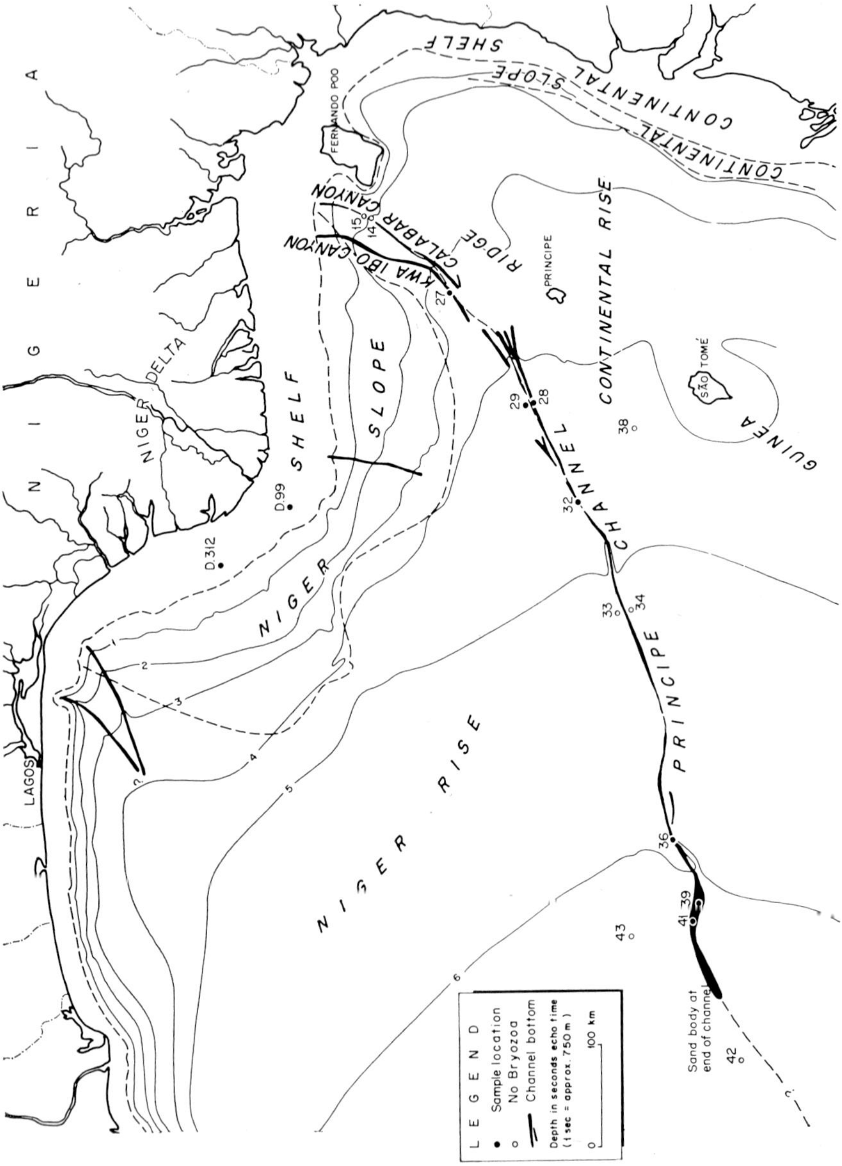
NIGERIA
NIGER DELTA
LAGOS
FERNANDO POO
CALABAR CANYON
KWA IBO CANYON
RIDGE
PRINCIPE
SÃO TOMÉ
CONTINENTAL SHELF
CONTINENTAL SLOPE
CONTINENTAL RISE
GUINEA
NIGER SHELF
SLOPE
NIGER RISE
PRINCIPE CHANNEL
D.99
D.312
Sand body at end of channel
LEGEND
Sample location
No Bryozoa
Channel bottom
Depth in seconds echo time (1 sec = approx. 750 m)
0 100 km

Fig. 1.

the middle and outer shelves throughout the entire Caribbean and Gulf of Mexico region (Seiglie, 1968), often associated with submerged reefs. They have also been reported to occur off Dakar, Sénégal (Masse, in press).

AGE AND ORIGINAL HABITAT OF THE DISPLACED FAUNA

We have already concluded that the displaced fauna in the channel sands must have derived from the Late Pleistocene *Amphistegina* fauna. *Amphistegina gibbosa* itself was found in six of the eight deep-sea core samples examined. The impressive number (40) of bryozoan species (Table 1) that the channel sands have in common with a single core sample with a rich *Amphistegina* fauna, D.312, taken off the Pennington River mouth in 55 fathoms depth at 1·14 m below the top of the core, also strongly supports this conclusion. Allen (1964, p. 313) reports a radiocarbon age of 10,750 ± 250 years BP for a shell sample from this core. No single Recent sample even remotely approaches the number of species in common with the channel sands. The most diversified Recent sample, D.99 (17 fathoms), with 40 different species, has only 18 species in common with the channel sands (Table 1).

The original depth habitat of the displaced bryozoan fauna can be inferred from its Recent component. Thirty-three of the 48 shallow-water species in the deep-sea channel sands occur off Nigeria both in the *in situ Amphistegina* fauna and in the Recent fauna. Their Recent depth distribution off Nigeria is almost exclusively shallower than 30 fathoms, and mostly falls in the 5 to 30 fathoms zone. In other words, at the time the ancient fauna now occurring between the 40 and 60 fathom lines were actually living there, the sea level rising from its Würm Glacial low of – 65 fathoms had reached a height some 30 fathoms below its present level. According to the various diagrams showing the post-Glacial rise in sea level that have been published in recent years, this happened some 11,000 years ago; an indirect dating which accords well with the radiocarbon dates mentioned earlier. The deep-sea channel sands with displaced fauna were probably also deposited in this period.

The development leading to the present distribution of Recent and fossil Bryozoa off Nigeria has been summarized in the two palaeogeographic reconstructions shown in Figs 2 and 3.

Still requiring some explanation is the frequent occurrence in the Late Pleistocene channel sands of *Pasythea tulipifera*, *Membranipora tuberculata* and *Euteleia evelinae*, species which have not been observed in the *in situ Amphistegina* fauna but occur off Nigeria at the present day. They appear conspicuously "clustered", i.e. all three attain high percentages in core samples 36 and 41^{II} (1·40–1·46 m—top of core). They also have a very shallow Recent depth distribution in

Table 1

Sample no. (for position of samples see Fig. 1)	41^{II}	41^{I}	39	36	32	29	28	27	D.99 Present in most diversified fauna in one single Recent sample*	D.312 Present in late Pleistocene Amphistegina fauna in one single core sample (1·14 m below top of core)**
	Displaced fauna in channel sands									
Lat. N.	0° 22·4′	0° 22·4′	0° 17·9′	0° 32·7′	1° 21·1′	1° 57·0′	1° 52·6′	2° 34·5′	4° 05·7′	4° 47·6′
Long. E.	1° 53·2′	1° 53·2′	2° 00·4′	2° 35·3′	5° 39·0′	6° 29·4′	6° 28·9′	7° 25·6′	5° 35·6′	5° 01·6′
Depth (m)	4658	4658	4617	4540	3509	3129	3240	2552	31·6	102·3
Interval in core (m–Surf.)	1·40–1·46	0·55–0·61	0·31–0·38	0·35–0·39	0·36–0·40	1·03–1·07	0·67–0·71	1·00–1·04	0	1·14
Weight of sample (g)	160·0	160·0	112·45	103·29	86·50	90·00	90·00	90·00	100	36
Weight of residue (g)	153·82	153·95	105·12	83·33	76·92	80·99	82·32	88·20		
Total number of specimens in the $>150\mu$ size fractions	392	207	218	375	91	265	338	29	24,894	3488
Species (% of total fauna)										
Cupuladria canariensis/biporosa	1·8	2·9	2·3	2·7		3·–	1·8		7·4	+
Metrarabdotos (Uniaviculārium) unguiculatum cookae	3·8	1·9	2·8	2·4	3·3	1·5	3·2	6·9		1·6
Cribrilaria radiata	+	+		+					+	+
Pasythea tulipifera	10·2	8·2	3·7	7·7	+	4·9	1·5	+		
Crisia spp.	15·8	18·4	17·9	10·9	2·2	21·9	18·3	24·1	13·6	4·8
Scrupocellaria spp.	9·2	12·6	8·3	5·6	2·2	11·7	12·7	20·7	2·2	26·–
Cellaria salicornioides var. *normani*	+	+	+	+		1·5	+	+	+	7·–
Crisulipora sp.	+	1·5	1·8	1·9		+	+		5·4	+
Membranipora tuberculata	4·8	1·9	3·7	15·2			2·4			
Mollia patellaria	+			1·6			+			+
Margaretta	1·–		+	+	+	1·1	+			+
Euteleia evelinae	1·8			1·9		+	+			
Biflustra	1·–	4·8		2·9	2·2	1·9	4·1		2·6	+
	[illegible]	[illegible]	[illegible]	[illegible]	4·4	3·8	7·1			3·4

Buffonellaria	+			+		+	+			1·7
Micropora sp.	+						+			+
Labioporella	1·6	1·–	1·8	1·1	2·2	+	3·6			2·4
Trypostega venusta	+	+	+						+	+
Reptadeonella	+					+	+			+
Hippopetraliella africana	1·–			+			+			1·4
Cupuladria owenii	+	+	+	+	2·2	2·3	+	+	1·5	+
Onychocella alula	+									+
Others	2·6	1·9	1·4	3·7	6·6	1·9	2·7			
Indet. Cheilostomata	35·–	28·5	37·6	24·–	54·9	31·7	17·4	31·–	2·3	16·1
Indet. Cyclostomata	2·8	3·4	7·3	5·1	6·6	4·9	11·2	6·9		6·7
Nellia oculata		1·–	+				+			
Escharina porosa		1·9	+			+			+	1·3
Copidozoum tenuirostre		+	+	+					+	+
"*Rosseliana*" sp.		+		+		+	+			1·8
Escharina pesanseris		+		+					+	+
Schizoporella sp.		1·–	+	+			+			+
Caberea boryi			+							+
Retevirgula			+	+			+			+
Hippopodinella lata			+			+				+
Floridina antiqua			+	+						+
genus indet.			+							+
Chaperia				+						
Aptonella violacea				+	2·2	+	+			+
Cleidochasma porcellanum				+	+		+		+	+
Euginoma vermiformis					+					
Cleidochasma oranense					+		+			+
Entalophora delicatula					+				2·7	3·5
Celleporaria aperta						+				+
Savignyella lafontii						+			+	
Exechonella antillea						+				+
Pachycleithonia cf. *nigra*						+				+
Cellaria elongatoides						+				+
Schizomavella auriculata							+		2·6	1·6
Cleidochasma contractum							+		+	
Halysisis diaphana							+			+
Triporula stellata							+		+	+
Total number of species	49 (columns 1–8 combined)								18*	40†

+ Stands for <1%, or for single specimen. * Out of the 40 species in this sample. † Out of some 65pecies in this sample.

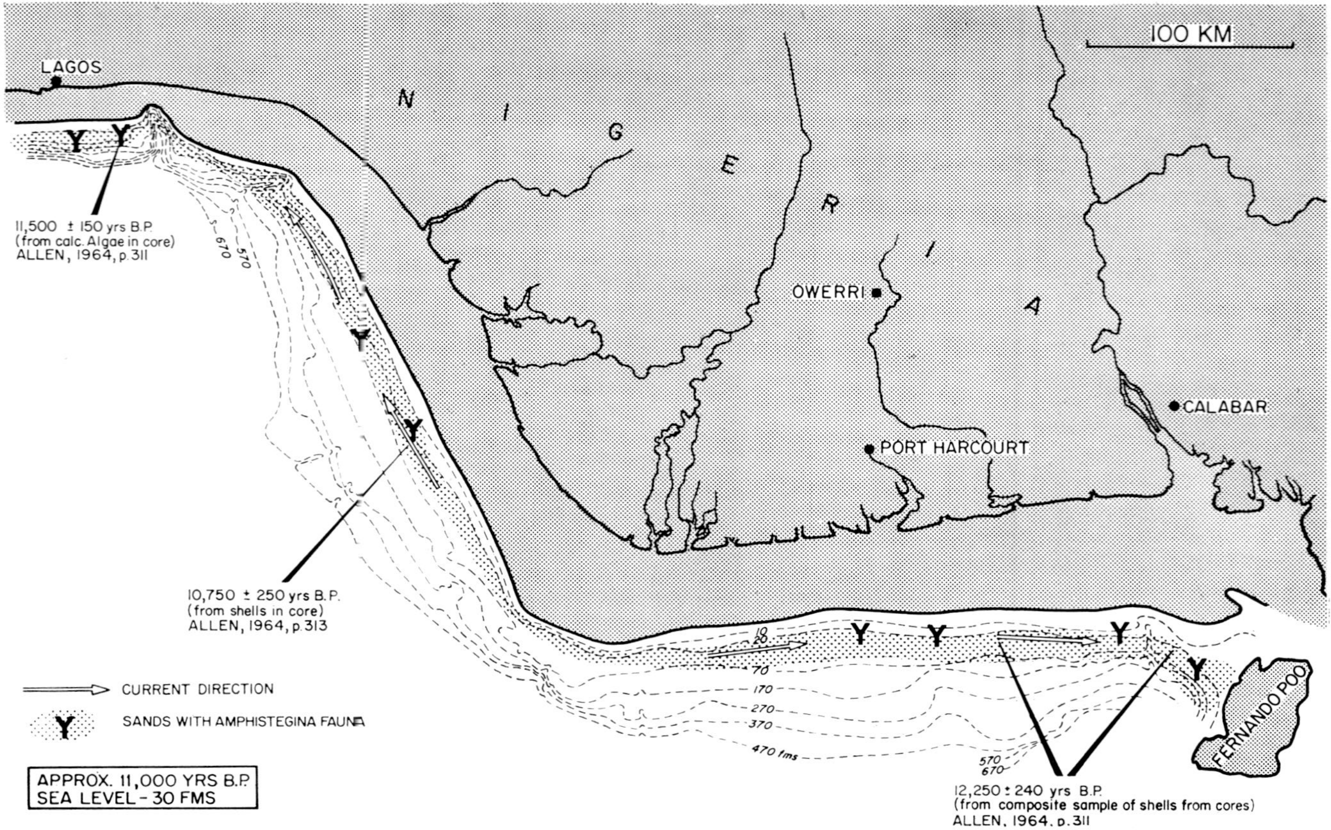

Fig. 2. Advancing Post-Würm Transgression. Narrow, sand-covered, mainly inner neritic shelf, inhabited by rich and diversified clear-water (non-deltaic) *Amphistegina* fauna, representing growth in 5–30 fathoms depth. Canyon heads all lie in this zone and

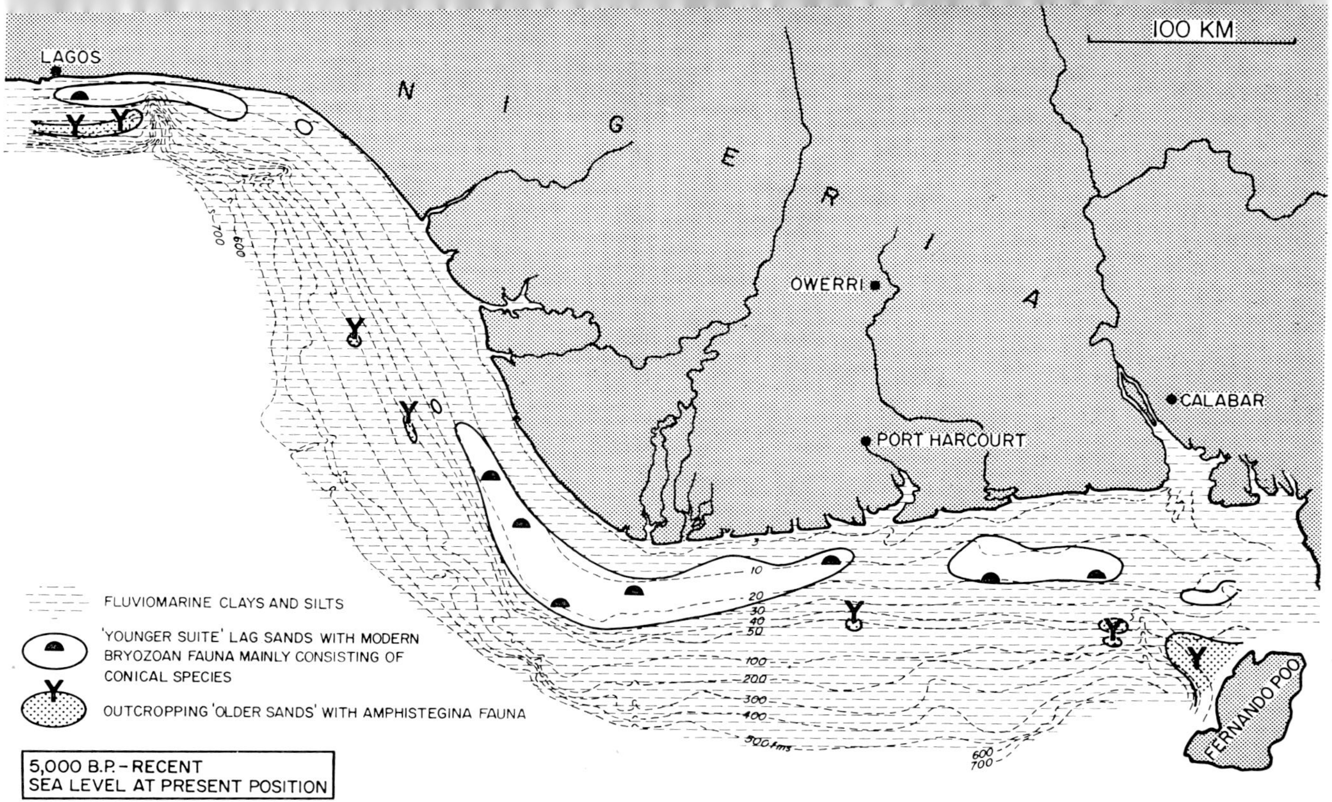

Fig. 3. Post-Würm Transgression Completed. Fluviomarine offlap has started: fluviomarine clays and silts, blanketing large parts of the shelf, leave only small outcrops with *Amphistegina* fauna at eastern and western extremities uncovered. A new "peri-deltaic" fauna, largely dominated by conical species, populates the "Younger Suite" of Allen (1964). Down-canyon movement of sand has come to a standstill.

common; namely, *Pasythea* not deeper than 8 fathoms (4 stations), *Euteleia* not exceeding 18 fathoms (28 stations) and *Membranipora* having a maximum depth of 20 fathoms (27 stations) in the Recent sediments off Nigeria. These three species in the channel sands presumably represent a Late Pleistocene very-shallow-water faunule not yet observed *in situ* at or below the present Nigerian shelf surface.

CONCLUSIONS

The displaced fauna in the deep-sea sands of the Principe Channel is identical to and must have derived from the *Amphistegina* fauna, dated at approximately 11,000 years BP, and represents a phase in the Late Pleistocene post-Glacial sea level rise. The *mise-en-place* of the channel sands containing this fauna probably dates from the same period. This is in line with recent views held by oceanographers and sedimentologists elsewhere; namely that during the lowered stand (of sea level) most submarine canyons actively diverted shallow-water sediments into the deep-sea and submarine basins (Dill, 1968).

ACKNOWLEDGEMENTS

The samples here reported on have been collected during the M.V. *Admiraal Nelson*'s deep-sea sedimentological survey in the Gulf of Guinea in behalf of the Koninklijke/Shell Exploratie en Produktie Laboratorium in 1967 and 1968.

I am grateful to my colleagues J. J. H. C. Houbolt and J. Brouwer for having brought the occurrence of Bryozoa in these deep-sea sediments to my attention, and for helpful information.

The photographs were taken by Mr W. Geluk. Permission by the Shell Research B.V. and Shell Internationale Petroleum Mij. B.V. to publish this paper is gratefully acknowledged.

REFERENCES

ALLEN, J. R. L. (1964). The Nigerian continental margin: bottom sediments, submarine morphology and geological evolution. *Marine Geol.* I, 289–332.

BERTHOIS, L. CROSNIER, A. and LE CALVEZ, Y. (1968). Contribution à l'étude sedimentologique du Plateau Continental dans la Baie de Biafra. *Cahiers O.R.S.T.O.M.*, sér. *Océanogr.* **VI** (3–4) 55–86.

BROUWER, J. (In press). Foraminiferal faunas from deep-sea sediments in the Gulf of Guinea.

DILL, R. F. (1968). Pleistocene sea levels and continental margin sedimentation. 18th Ann. AAPG and SEPM Gulf Coast Sections Mtg. (Jackson, Miss., 10/23–25/68) Pap.; *Abstr. Am. Ass. Petrol. Geol. Bull.* **LII,** 9, 1836.

GIRESSE, P. (1969). Carte sédimentologique des fonds sousmarins du delta de l'Ogooue. *Cah. Océanogr.* **XXI** (10) 965–994.

HOUBOLT, J. J. H. C. (In press). The deep-sea canyons in the Gulf of Guinea near Fernando Póo. *Geologie en Mijnbouw*, 1973.

MASSE, J. P. (In press). Sur la présence d'une thanatocoenose à Amphistegines d'âge quaternaire à la base du plateau continental de la région de la presqu'île du Cap-Vert (République du Sénégal). *Actes Congr. Panafric. Préhist. étude Quatern., Dakar*, 1967.

SEIGLIE, G. A. (1968). Relationships between the distribution of *Amphistegina* and the submerged Pleistocene reefs off Western Puerto Rico. *Tulane Stud. Geol.* **VI** (4) 139–147.

14 Vertical and Horizontal Distribution of Fouling Bryozoans in Cochin Backwaters, Southwest Coast of India

N. RAVINDRANATHA MENON

College of Fisheries, University of Agricultural Sciences, Mangalore, India

Abstract: The vertical and horizontal distribution of 15 species of fouling bryozoans occurring in the Cochin backwaters is discussed.

On the basis of the occurrence of bryozoans encrusting test panels suspended at different levels three categories may be recognized:

(1) Forms preferring comparatively higher levels, including *Nolella papuensis, Bowerbankia gracilis, Aeverrillia setigera* and *Bugula neritina*.

(2) Species preferring lower levels, including *Membranipora savartii, Electra crustulenta, E. bengalensis, Conopeum reticulum, Alderina arabianensis, Schizomavella linearis* var. *inarmata* and *Watersipora subovoidea*.

(3) Species distributed throughout the vertical volume column of water with greater settlement at intermediate levels, including *Victorella pavida, Bugula cucullata* and *Schizoporella cochinensis*.

An important aspect noticed from the nature of the horizontal distribution of fouling bryozoans in the backwaters was the successful colonization of the estuary by the brackish water forms during the low saline period connected with the monsoons; and the occurrence of marine forms during the hot season.

INTRODUCTION

Observations have been made on the horizontal and vertical distribution of fouling bryozoans of the Cochin backwaters. Important papers which deal with the study of fouling organisms in this area are those of Nair, 1965, 1967; Menon and Nair, 1967; Menon, 1970a, b; 1971. These studies have shown that marked changes in the faunal composition occur, the outstanding feature being total elimination of monsoon-settled forms during the non-monsoon period and colonization by other species which are totally absent during the monsoon period. Detailed examination is therefore necessary to determine the vertical and horizontal distribution of fouling bryozoans which form a major component of the fouling complex of this region.

ENVIRONMENT

A narrow channel of 450 m width, connects the Cochin backwater with the Arabian sea at Cochin (Lat 9°, 58′N. Long 76°, 17′E). Low lying belts of sand separate the backwaters from the Arabian sea on the west. Maximum rainfall is recorded in this region during the southwest monsoon, which bursts along the southwest coast of India during late May or in June. Some precipitation occurs during the northeast monsoon in October–December period. The rainfall during the two monsoons and the consequent drainage, influence the salinity of backwaters as well as the coastal waters in this region, and during July and August the surface water of Cochin backwater may almost be fresh (Menon and Nair, 1967). Except in the three dredged shipping channels the depth in the area investigated ranged from 1 to 5 m.

Based on the influence of the southwest monsoon the year is divided into three well defined periods. The pre-monsoon period (February to May) is comparatively dry with very little rainfall, and is characterized by fairly uniform high salinity and high temperature of the surface waters. The monsoon period (June to September) is characterized by very heavy rainfall and a consequent lowering of the surface water salinity, which may become almost fresh. The post-monsoon period (October to January) shows an upward fluctuating trend in surface salinity (for details on hydrographical conditions see Menon, 1971).

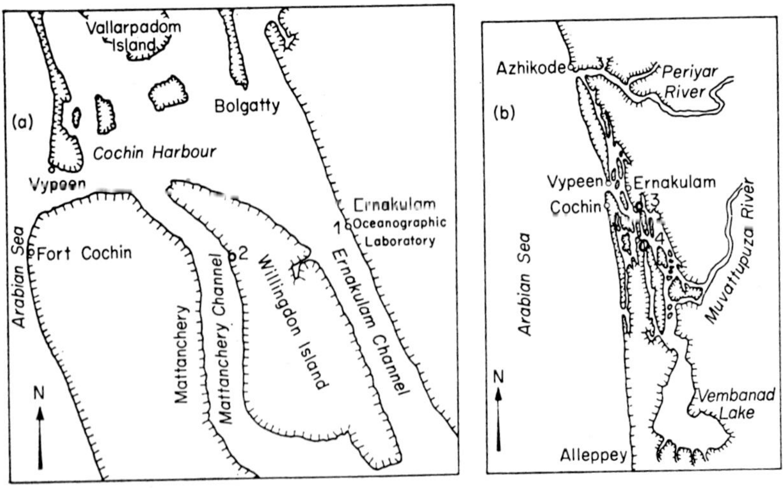

Fig. 1. (a) Cochin Harbour, India, stations 1 (Conch Pier) and 2 (Dry Dock); (b) Cochin Backwaters and Vembanad Lake, India, stations 3 (Thevara) and 4 (Aroor).

STATIONS INVESTIGATED

The four stations, Dry Dock, Conch Pier, Thevara, and Aroor represent four different ecological habitats in the Cochin backwaters. Dry Dock is situated 2 km inside the bar mouth, which connects the backwaters with the Arabian sea. The distance of the other stations from the bar mouth is 4 km, 10 km and 14 km respectively (Fig. 1, Maps A and B). The depth of the stations investigated varied from 3 to 4·5 m (see Menon, 1971).

MATERIAL AND METHODS

Glass panels (12 × 10 cm) were suspended and maintained at three levels, 0·5 m, 1·5 m and 2·5 m below low-tide mark at Conch Pier, Thevara and Aroor, and at 0·5, 1·0, 1·5, 2·0, 2·5 and 3·0 m at Dry Dock. The data gathered for the respective seasons derive from observations of both short-term and long-term panels (Menon, 1971). All the colonies recorded at different depths and at different localities are added and averaged to compile the data for the present paper. Thus the data presented here are the number of colonies presented at each level from glass panels exposed during different times of the years 1965–1966.

RESULTS

Fifteen species of bryozoans settled on the panels. The numerical details and general pattern of vertical and horizontal distribution are given in Tables 1 to 4 and Figs 2 to 4.

Table 1. Vertical distribution of fouling bryozoans in Cochin backwaters

Depth (m)	*Victorella pavida*	*Electra crustulenta*	*Electra bengalensis*	*Alderina arabianensis*	*Schizoporella cochinensis*	*Nolella papuensis*	*Bowerbankia gracilis*	*Aeverrillia setigera*	*Membranipora savartii*	*Conopeum reticulum*	*Bugula neritina*	*Bugula cucullata*	*Savignyella lafontii*	*Schizomavella linearis* var *inarmata*	*Watersipora subovoidea*
0·5	C	89	14	2	0	VC	C	R	4	5	51	45	0	0	0
1·0	VC	353	73	6	0	C	R	R	5	8	12	40	32	0	0
1·5	A	424	80	33	35	C	R	R	4	12	9	24	16	12	9
2·0	VC	924	118	73	52	C	R	R	12	16	9	37	0	16	20
2·5	A	806	139	60	27	C	N	N	25	4	9	40	0	0	40
3·0	N	105	87	167	32	R	N	N	0	0	0	40	0	0	0

A—Above 100 colonies VC—Between 50–100 colonies C—Between 25–50 colonies
R—Between 0–25 colonies N—Absent.

Table 2. Vertical distribution of *Victorella pavida* Kent in Cochin backwaters

Depth (m)	Monsoon (June–September) Dry dock	Conch pier	Thevara	Aroor	Post monsoon (October–January) Dry dock	Conch pier	Thevara	Aroor
0·5	VC	C	VC	N	N	C	C	C
1·0	A	—	—	—	VC	—	—	—
1·5	A	VC	A	A	C	A	VC	A
2·0	A	—	—	—	R	—	—	—
2·5	VC	VC	VC	A	C	VC	A	A
3·0	N	—	—	—	N	—	—	—

A—Above 100 colonies VC—Between 50–100 colonies C—Between 25–50 colonies
R—Between 0–25 colonies N—Nil.

Table 3. Vertical distribution of *Electra crustulenta* Pallas in Cochin backwaters

Depth (m)	Monsoon (June–September) Dry dock	Conch pier	Thevara	Aroor	Post Monsoon (October–January) Dry dock	Conch pier	Thevara	Aroor
0·5	481	98	18	99	0	5	1	10
1·0	671	—	—	—	35	—	—	—
1·5	1164	578	210	437	160	538	205	100
2·0	1417	—	—	—	432	—	—	—
2·5	729	1212	1152	1078	318	673	896	393
3·0	0	—	—	—	210	—	—	—

Table 4. Horizontal distribution of fouling Bryozoans in Cochin backwaters

Species	Dry dock 2 km away from bar mouth	Conch pier 4 km away from bar mouth	Thevara 10 km away from bar mouth	Aroor 14 km away from bar mouth
a. *Victorella pavida*	VC	A	A	A
b. *Nolella papuensis*	VC	N	N	N
c. *Bowerbankia gracilis*	VC	R	N	N
d. *Aeverrillia setigera*	R	N	N	N
e. *Electra crustulenta*	2231	1888	1380	1614
f. *Electra Bengalensis*	231	391	211	148
g. *Membranipora savartii*	13	—	—	—
h. *Conopeum reticulum*	13	—	—	—
i. *Alderina arabianensis*	269	96	41	51
j. *Bugula neritina*	48	18	0	0
k. *Bugula cucullata*	52	25	0	0
l. *Savignyella lafontii*	12	0	0	0
m. *Watersipora subovoidea*	29	14	0	0
n. *Schizoporella cochinensis*	73	41	0	0
o. *Schizomavella linearis* var *inarmata*	20	12	0	0

A—Above 100 colonies VC—Between 50–100 colonies R—Between 0–25 colonies
N—Nil.

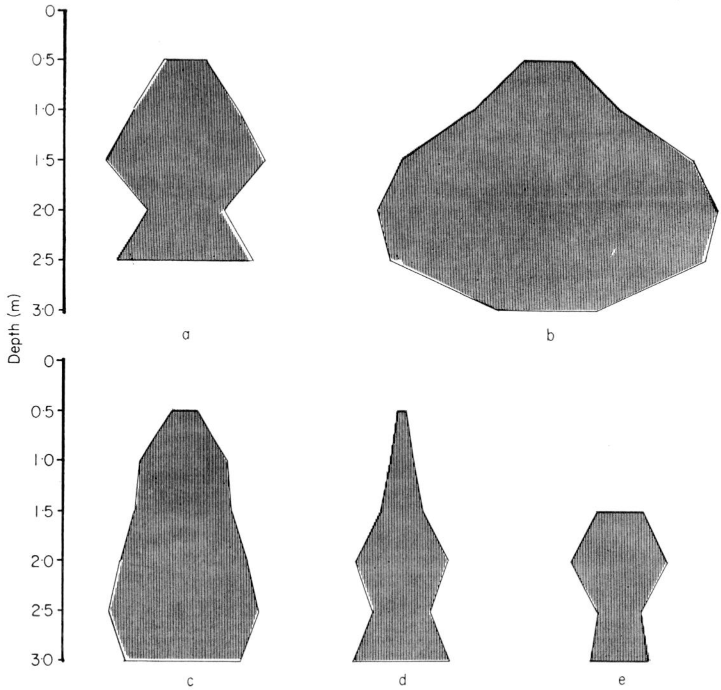

Fig. 2. Vertical Distribution of Fouling Bryozoans (for key to species see Table 4).

To have a generalized picture of vertical distribution, the forms recorded are grouped into three categories based on their incidence and comparative abundance at various levels.

(1) Those preferring the higher levels in the vertical column of water such as *Nolella papuensis, Bowerbankia gracilis, Aeverrillia setigera* and *Bugula neritina.*

(2) Species preferring the lower levels in the vertical column of water. *Membranipora savartii, Electra crustulenta, E. bengalensis, Conopeum reticulum, Alderina arabianensis, Schizomavella linearis* var. *inarmata* and *Watersipora subovoidea* may be included in this grouping.

(3) Species distributed throughout the vertical column of water with greater settlement at intermediate levels. *Victorella pavida, Bugula cucullata* and *Schizoporella cochinensis.*

Further *M. savartii, Conopeum reticulum, Schizomavella linearis* var *inarmata* and *Watersipora subovoidea* were conspicuous by their absence at the intertidal region.

In this connection it will be interesting to refer to a few records of the vertical distribution of identical or allied species. Nair (1961) noticed *Bowerbankia imbricata* in abundance on the intertidal panels in Western Norway. Mawatari

(1953) recorded more colonies of *Electra angulata* and allied species of *E. bengalensis* from 1·5 m. Kawahara and Iizima (1960) found them in greater numbers at 3·5 m, while Nair (1961) noticed increased numbers of *Electra pilosa* f. *pilosa* on the bottom panels. Mawatari (1951) found *Bugula neritina* occurring in depths ranging from 0·5 m to 5·0 m. Tokuoka (1939) found that *Watersipora cucullata* was more plentifully found near the bottom, even though Miyaku (1940) noted them mainly at 3·0 m or less and disappearing from the range of 4·0 m, to 8 m. Kawahara (1961) observed that this species abounds in shallow waters and noted them on all panels placed at various depths from 0·5 m to 5·0 m, though maximum numbers occurred at 3·0 m. McDougall (1943) observed a greater occurrence of *Schizoporella unicornis* within a limit of 30 cm above the bottom.

From Table 4 it is clearly evident that the horizontal distribution of these sedentary organisms was not uniform even though the area investigated was small. The station (Dry Dock) nearest to the open sea recorded the maximum number of species. Of the fifteen species only *V. pavida*, *E. crustulenta*, *E. benga-*

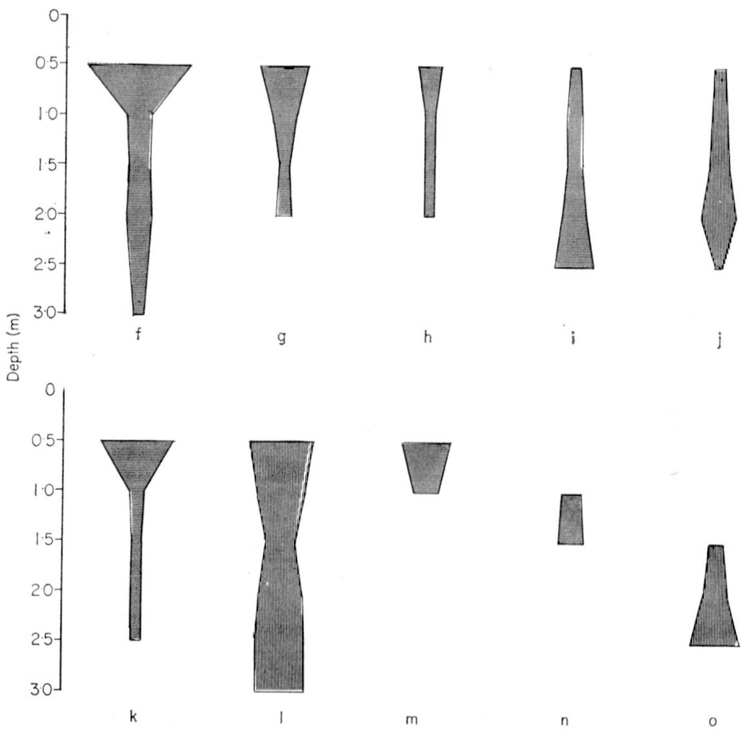

Fig. 3. Vertical Distribution of Fouling Bryozoans (for key to species see Table 4).

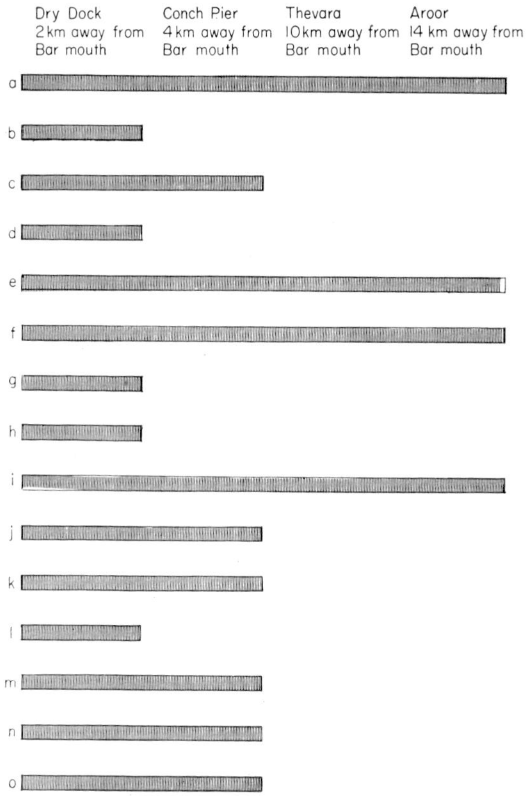

Fig. 4. Horizontal Distribution of Fouling Bryozoans (for key to species see Table 4).

lensis and *A. arabianensis* were represented at all the stations and forms like *Nolella papuensis*, *Bowerbankia imbricata*, *Aeverrillia setigera*, *Membranipora savartii* and *Savignyella lafontii* were restricted to Dry Dock.

DISCUSSION

The pattern of vertical distribution of species of sedentary organism is considerably influenced by the larval behaviour during settlement as well as by several ecological factors like illumination (Pyefinch and Downing, 1949) salinity, nature of substratum (Barnes and Powell, 1950) etc. Ryland (1962) commented that the selection of substratum by settling larvae plays an important role in maintaining a species at a suitable tidal level. In the present instance this factor cannot be given much importance as the material used was inert glass. The subsequent growth and establishment of a species on the substratum depends on many factors among which intra- and interspecific relationships of the organisms on the substrata are of utmost importance. Hence the pattern of vertical distribution shown on a series of panels, suspended at different levels, is the combined effects of larval and adult behaviour. Furthermore, aspects such as the succession of fouling communities, "biological extinction" of fouling species from the substrata, etc. are important aspects which have to be analysed thoroughly to arrive at definite conclusion on the incidence of fouling animals at different levels.

It is evident from Tables 2 and 3 that the vertical distribution of *V. pavida* and *E. crustulenta* showed no marked variation in the pattern of distribution during the two seasons. Even though these two species were the dominant fouling bryozoans of this locality for this period, both of them preferred more or less the same position in the vertical column of water. A noticeable decrease of *E. crustulenta* in the panels kept intertidally at the four stations during the post-monsoon period is rather interesting, as the surface water of the Cochin backwaters was reasonably mesohaline (Menon, 1971). It is clear that the larvae of *E. crustulenta* preferred bottom panels to top ones. This is indicative of the fact that during settlement the larvae were positively geotropic. In the case of forms settled during the high saline period, the ctenostomes always preferred the upper levels. The cheilostomes which shared this regime were *B. neritina* and *B. cucullata*. It is noteworthy that Lynch (1949) found that the shortening of the natatory period and the failure of the larvae to become geopositive may be due to changes in the environmental conditions, the most important of which are increased temperature and brackish conditions of the water. Lynch's experiments with the larvae of *Bugula neritina* have shown that by changing the environmental conditions in the laboratory the behaviour of the larvae can be reversed. Ryland (1960) found with the larvae of *Cryptosula pallasiana* that among the larvae pipetted out into a dish 65% became photonegative within 3 hours, whereas only 20% of the larvae liberated *in situ* became photonegative after 3 hours. It has also been noticed that water movements would cause the larvae to swim away from light.

From the available data it can be safely concluded that the larval behaviour during the phase of settlement reflects a very complex pattern and it is probable that each individual species has its own level of preference. Any definite conclusions on causative factors affecting behaviour under natural conditions can be arrived at only by a parallel study of settlement behaviour in the laboratory.

The successful colonization of the estuary by the brackish water forms during the monsoon and post-monsoon periods, and the settlement of marine forms during the hot pre-monsoon period is a characteristic feature of the fouling species occurring in this region (Menon, 1971). If the number of species settled could be taken as a criterion to assess the suitability of the environment for successful settlement and growth of this interesting group of fouling organisms, Dry Dock seems to offer the most congenial conditions in this area. This station being nearest to the sea experienced the maximum influence of marine conditions. This certainly accounted for the greater representation of marine bryozoans in this region. Similarly the absence of such typical marine forms as *Bugula neritina*, *B. cucullata* and *Watersipora subovoidea* at Thevara and Aroor could be explained as the result of the inability of the tidal currents to carry the larvae of these species to the sites. Even though twelve species of marine forms entered successfully and settled in the estuary the majority could colonize only the vicinity of the bar mouth, the number rapidly diminishing towards the upper reaches. The majority of bryozoan species possess lecithotropic larvae which settle shortly after liberation (Ryland, 1959). Such larvae might have failed to accomplish access to regions away from the open sea, which is the source of these larvae. The only possibility to settle at regions away from the open sea would be from a natural population of marine forms in the Cochin backwaters. Successful maintenance of a natural population of marine species is nearly impossible in this region. Experiments on salinity tolerance of two marine forms have shown that the members of populations growing during highly saline periods cannot tolerate salinities below 26‰ for more than 24 h (Menon, 1970b). Examination of dredged materials taken throughout the year from coastal waters has shown that sexually mature colonies are present almost throughout the year. Studies conducted on the reproductive biology of marine invertebrates of this region have shown that breeding is a continuous process with peaks (Pillay, 1969). So it is clear that the larvae of marine forms, though produced throughout the year, do not enter the backwaters during the period when the waters of this region are less saline. It is also probable that those which succeed in entering are killed by the very uncongenial environmental conditions prevailing in the region. It is certain therefore, that the settlement pattern of marine forms reflects the nature and extent of larval transport from the open sea into the backwaters. Settlement on a horizontal gradient shows that the intensity of occurrence is indirectly proportional to the distance of the locality from the open sea. This may be explained by the populations of marine fouling bryozoans being replenished every year during the hot pre-monsoon period by transport of larvae of marine species into this locality. The ephemeral planktonic existence does not help the larvae to colonize successfully the interior regions. Those that succeed must exist in a very dormant state during the monsoon period. The colonies which are not sloughed off or not smothered by the heavy settlement of barnacles (Nair, 1965) may come back to activity by regeneration and succeed in producing larvae. Observations on settlement have shown that the larvae of marine forms settle by the beginning of December.

The horizontal distribution of brackish water species is very different. The species which were found at all the stations during the monsoon and post-monsoon periods were *Victorella pavida* and *Electra crustulenta*. Studies on the salinity tolerance of these two species have shown that two distinct populations, characterized by different degrees of tolerance, inhabit this region (Menon, 1970b). The monsoon population of *V. pavida* showed a tolerance from fresh water to 10‰, whereas the post monsoon population showed tolerance limits of 16‰–22‰ (Menon and Nair, 1967). In the case of *E. crustulenta* the former population showed a tolerance from F.W. to 21‰ and the latter 16‰ to 32‰. Cook (1964) found that species like *E. monostachys* and *Conopeum reticulum* can tolerate wide ranges in salinity from brackish river estuaries to sea water. Commenting on the effect of salinity tolerance on the distribution of these two species, she reports that *C. reticulum* is more abundant in the higher, less saline parts of the river, and *E. monostachys* more abundant in the lower reaches where the higher salinities are more frequent. *E. bengalensis* never occurred on the panels during the low saline period. Tolerance to salinity changes thus determines the distribution of *V. pavida* and *E. crustulenta*. The salinity of the surface waters in a horizontal plane, during the monsoon period, shows that Aroor experiences nearly fresh water conditions for two months. During this period fewer colonies of *V. pavida* were recorded. During the post-monsoon phase, when this region is mesohaline, *V. pavida* is at a maximum.

Nair (1961) attributed the variations he noticed in the horizontal distribution of bryozoans along the fjords of Western Norway to such factors as high turbidity, pollution, low salinity and to the influence of wave action. Kawahara and Iizima (1960) were able to observe a correlation between the absence of *Watersipora cucullata* and a gradual increase in water pollution and temperature along with a decrease in dissolved oxygen towards the interior of Ago Bay. Only a few species of bryozoans succeed in colonizing estuaries. Fraser (1938) failed to record bryozoans in the Mersey estuary. According to Milne (1940), even though bryozoans appeared on seaward buoys, they were not generally exuberant in the Tamar estuary. The decrease in number of species was attributed to salinity changes which adversely affected the feeding mechanism of suspension feeders. It is probable that the combined operation of several extrinsic factors referred to, together with the ephemeral larval period, account for the observed variations in the regional distribution of bryozoans in Cochin backwaters.

ACKNOWLEDGEMENTS

I wish to express my gratitude to Prof. Dr N. B. Nair, Marine Biological Laboratory, University of Kerala, India for guidance and encouragement. I am grateful to Prof. Dr O. Kinne, Leading Director and Professor, Biologische Anstalt Helgoland, FRG for providing facilities. Help rendered by my wife, Mrs Indira R. Menon, in preparing the manuscript is acknowledged. I thank the Ministry of Education, New Delhi, India and DAAD, FRG for the award of fellowships and the authorities of the University of Agricultural Sciences, Bangalore, India for granting study leave.

REFERENCES

BARNES, H. and POWELL, H. T. (1950). The development, general morphology and subsequent elimination of barnacle populations, *Balanus crenatus* and *B. balnoides* after a heavy initial settlement. *J. Anim. Ecol.* **19** (2), 175–179.

COOK, P. L. (1964). The development of *Electra monostachys* (Busk) and *Conopeum reticulum* (Linnaeus), Polyzoa, Anasca. *Cahiers de Biologie Marine* **5,** 391–397.

FRASER, J. H. (1938). The fauna of fixed and floating structures in the Mersey estuary and Liverpool Bay. *Proc. Trans. Lpool biol. Soc.* **51,** 1–21.

KAWAHARA, T. (1961). Regional differences in the composition of fouling communities in Ago Bay. *Rep. Fac. Fish. Pref. Univ. Mie.* **4** (1), 65–80.

KAWAHARA, T. and IIZIMA, H. (1960). On the constitution of marine fouling communities at various depths in Ago Bay. *Rep. Fac. Fish. Pref. Univ. Mie.* **4** (3), 582–594.

LYNCH, W. F. (1949). Modifications on the responses of two species of *Bugula* larvae from Woods Hole to light and gravity: ecological aspects of the behaviour of *Bugula* larvae. *Biol. Bull. mar. biol. Lab., Woods Hole* **97,** 302–310.

McDOUGALL, K. D. (1943). Sessile marine invertebrates at Beaufort, North Carolina. *Ecol. Monogr.* **13,** 321–374.

MAWATARI, S. (1951). The natural history of a common fouling bryozoan *Bugula neritina* (Linnaeus). *Misc. Rept. Res. Inst. nat. Res. Tokyo* **19/21,** 47–54.

MAWATARI, S. (1953). On *Electra angulata* Levinsen, one of the fouling bryozoans in Japan. *Misc. Rept. Res. Inst. Nat. Res., Tokyo* **32,** 5–10.

MENON, N. R. and NAIR, N. B. (1967). Observations on the structure and ecology of *Victorella pavida* Kent (Bryozoa) from the Southwest Coast of India. *Int. Rev. ges. Hydrobiol.* **52** (2), 237–256.

MENON, N. R. (1970a). The growth rates of four species of intertidal bryozoans in Cochin backwaters. *Bull. natn. Inst. Sci. India.* (In press.)

MENON, N. R. (1970b). On the nature of tolerance of salinity in two euryhaline bryozoans *Victorella pavida* Kent and *Electra crustulenta* Pallas. *Bull. natn. Inst. Sci. India.* (In press.)

MENON, N. R. (1971). Ecology of fouling bryozoans in Cochin waters. *Mar. biol.* **8** (4), 280–307.

MILNE, A. (1940). The ecology of the Tamar estuary. IV. The distribution of the fauna on buoys. *J. mar. biol. Ass. U.K.* **24,** 69–87.

MIYAKU, S. (1940). On the development, growth, attachment and distribution of Balanus and Hydroides. *Misc. Rep. Naval. Tech. Inst. Japan* **273,** 54–80.

NAIR, N. B. (1961). Some observations on the distribution of bryozoans in the fjords of western Norway. *Sarsia* **3,** 37–45.

Nair, N. B. (1965). The seasonal settlement of marine fouling and wood boring crustaceans at Cochin Harbour, south west coast of India. "Proceedings of the International Symposium on Crustacea, *J. mar. biol. Ass. India*, Pt. IV, 1254–1268.

Nair, N. U. (1967). The settlement and growth of major fouling organisms in Cochin harbour. *Hydrobiologia* 503–512.

Pillay, K. K. (1969). Studies on the reproductive biology of some crustaceans from the south west coast of India. Ph.D.Thesis, University of Kerala.

Pyefinch, K. A. and Downing, F. S. (1949). Notes on the general biology of *Tubularia larynx* Ellis and Solander. *J. mar. biol. Ass. U.K.* **28,** 21–44.

Ryland, J. S. (1959). Experiments on the selection of algal substrates by Polyzoan larvae. *J. exp. Biol.* **36,** 613–631.

Ryland, J. S. (1960). Experiments on the influence of light on the behaviour of polyzoan larvae. *J. exp. Biol.* **37,** 783–800.

Ryland, J. S. (1962). The association between Polyzoa and algal substrata. *J. Anim. Ecol.* **31,** 331–338.

Tokuoka, Y. (1939). On the morphological, ecological and physiological studies of barnacles. *Misc. Rept. Naval Tech. Inst. Japan* **237,** 23–30.

15 | The Analysis of Spatial Distribution Patterns

J. S. RYLAND

Department of Zoology, University College of Swansea, Wales

Abstract: Bryozoa on the shore tend to occur in localized aggregations. This is true on the large scale, as in the dispersal of a species on individual algae within a zone or community, and on the small scale, as in the arrangement of colonies on a single frond. Aggregations on the large scale occur in many species having short-lived larvae, and have also been demonstrated in *Electra pilosa* (which settles from a cyphonautes) found on *Fucus serratus*. It is suggested that aggregation might favour cross-fertilization.

In small-scale aggregations interest centres on the possible occurrence of spacing out among colonies, in a manner comparable to that known in other invertebrate groups. For this kind of analysis quadrat sampling has proved unsuitable. The use of distance measurements in two niches (*Alcyonidium hirsutum* alongside the midrib of *Fucus serratus*, and *Crisia eburnea* on the fronds of *Flustra foliacea*) has surprisingly revealed a total absence of any tendency towards spacing out.

In a majority of marine bryozoa the colony forms an adherent crust over the surface of boulders, shells, algae or other substrata. Erect species, such as *Scrupocellaria reptans* (L.) and *Crisia denticulata* (Lamarck) on British shores, sometimes develop into a dense sward. Thus both encrusting and erect species may virtually cover rock faces or algal fronds, and it is readily apparent to a shore ecologist that such bryozoans tend to occur in dense and local aggregations of this kind. This evident patchiness poses two fundamental problems: the first concerns the mode of origin and subsequent delimitation of aggregations, and the second involves the competitive situation within them.

Until recently it may have been supposed that there would be little advantage to bryozoa from the development of single-species aggregations. However, the demonstration by Silén (1966) that spermatozoa may pass externally from zooid to zooid necessitates some reconsideration of this viewpoint. Clustered colonies may achieve interchange of spermatozoa, with the attendant advantages of cross-fertilization; whereas in large isolated colonies transference could occur only between zooids within the colony, which amounts genetically to self-fertilization.

How do aggregations arise? Do they simply reflect great differences in suitability between potentially available substrata? Are they strongly favoured by the limited dispersal generally achieved by non-cyphonautes larvae (Ryland, 1967, p. 357)? Or is there a behavioural mechanism which corresponds to the gregarious response (in which prospecting larvae are attracted to adults of their own species) found in acorn barnacles (Knight-Jones and Moyse, 1961)? Such questions have hardly been considered.

Competition is the endeavour of animals to obtain some particular element of the environment, the supply of which is insufficient for them all (Milne, 1961). For bryozoa food and space might constitute such elements. All bryozoa are filter feeders, probably depending mainly upon non-diatom phytoplankton. There is no evidence that competition for food occurs, even within densely populated habitats, although the general abundance of bryozoa certainly appears to be correlated with the availability of food. In non-turbulent waters the erect habit of growth may promote efficient filtration, but it also permits the colonies to continue growing above the encrusting layer. Populations become so dense in favoured localities that competition for space must be genuine and acute (also see Hayward, this volume p. 107 and Stebbing, this volume p. 173). Although some recent study has been devoted to settlement and growth phenomena which appear relevant to the subject of spatial competition (Ryland and Stebbing, 1971; Stebbing, 1971), the key question of spacing out at settlement—such as occurs in acorn barnacles and spirorbid tubeworms, thereby allowing room for growth (Knight-Jones and Moyse, 1961)—has remained unexamined.

Some of the above questions will be answered only by appropriate experiments, but much information can be obtained by analysing dispersal patterns in suitable habitats on the shore. Bryozoa are sessile organisms, and their communities can be studied by the methods developed by plant ecologists (Greig-Smith, 1964), but often applicable to the zoobenthos (Elliott, 1971).

Sampling units may be comparable natural entities, such as *Laminaria* holdfasts; but when the study area constitutes a continuum, arbitrary sampling units (usually quadrats) have to be employed. In all such cases the variate (variable quantity) will be discontinuous. Thus the number of colonies falling within any one of a series of quadrats can only be an integer; measurements of distance between colonies, on the other hand, can assume any value and are therefore continuous. When the counts or measurements (the latter appropriately grouped) have been made, the data can be assembled in the form of a frequency distribution table. Thus if x, the number of colonies per quadrat, say, has successive values of 0, 1, 2, 3, etc., f_0 is the frequency of class 0, f_1 the frequency of class 1, and so on. The number of quadrats sampled, n, which is often in the order of 150–200, equals the sum of the frequencies (Σf). The total number of colonies in the samples will be Σfx, and the sample mean, $\bar{x} = \Sigma fx/n$.

Should every point within the sample area be equally likely to support a bryozoan colony, and the chances that it actually does so be fairly low, the colonies would be randomly dispersed. The sample frequency distribution of $x=0, 1, 2$, etc., would then agree with a Poisson series having a mean m corresponding to the sample mean $\bar{x}$. Two points should be appreciated about a Poisson series: first, it represents a discontinuous distribution,* so that the frequencies indicated can be plotted only as a histogram; and, secondly, its variance equals its mean, so that the distribution depends on a single parameter λ (replaced in a sample by the statistic m). For low values of m the Poisson distribution is highly skewed, but for higher values of m the distribution is approximately normal (e.g. Elliott, 1971).

The values of $P(x)$, the probability that the number of colonies in any sample is 0, 1, 2, etc., can be found from tables of the Poisson distribution (Pearson and Hartley, 1966) or calculated from

$$P(x) = (e^{-m}m^x)/x!,$$

where e is the base of natural logarithms. The summed values of P equal 1, and are converted to expected frequencies by multiplying by the sample size n. The observed and expected frequencies may then be tabulated and compared by means of χ^2 (e.g. Greig-Smith, 1964 or Elliott, 1971). Close agreement of observed with predicted frequencies indicates a population dispersal that is random, while significant disagreement indicates some non-random arrangement. When disagreement is due to shortage of values near to the mean, and excess in low and high classes, the population is dispersed in clumps and the distribution described as "contagious"; conversely, when the disagreement is due to an excess of values near to the mean, the population is evenly dispersed and the distribution is "regular". Elliott (1971) gives examples.

One standard requirement of the sampling procedure used in this kind of analysis is that the unit or quadrat must be of constant size. If however it is required to study the dispersal of, say, *Electra pilosa* (L.) on *Fucus serratus* L., it is hardly possible to collect a random sample of *Fucus* fronds having even approximately the same area. This appears to present an entirely new statistical problem, but one which can be overcome using a computer. The details of the method, which could have considerable practical value, will be presented elsewhere,

* The word "distribution" is best confined to its statistical sense, avoiding the environmental meaning "pattern of dispersal".

but the findings are of immediate interest. It was established that the dispersal pattern of *E. pilosa* did not correspond with a Poisson series, but displayed a high degree of contagion. The result is particularly interesting for, unlike all the other species studied, *E. pilosa* settles from a planktonic cyphonautes larva. It is not possible, of course, to decide whether the aggregation in this case results from gregarious settlement or merely from the unequal favourability of the *Fucus* fronds; but it is clearly of possible biological significance in regard to the facilitation of cross-fertilization Ryland (1972).

On a smaller scale, the quadrat method has been employed to study the dispersal pattern of bryozoa on heavily encrusted fronds of *Laminaria digitata* Huds. (Lamx.). Here the primary objective was to find whether colonies were regularly spaced, thereby allowing themselves room to grow. The results soon confirmed one of the serious limitations of using quadrats to try to assess the degree of contagion or regularity. As plant ecologists have discovered (Greig-Smith, 1964; Pielou, 1969), the pattern revealed is highly dependent on the size of quadrat employed. Thus *Hippothoa hyalina* (L.) was randomly dispersed when the quadrat size was 1 cm² but highly aggregated when it was 4 cm². *Scrupocellaria reptans* showed a similar trend, with mild aggregation apparent when the quadrat size was 1 cm², but with marked aggregation (tenfold increase in the value of χ^2) when it was 4 cm². Aggregation on this scale presumably implies that not all parts of the frond are equally favourable, and, if larvae are spacing themselves out at settlement, any resultant regularity is being masked by the tendency to clump. It is concluded that quadrat sampling may not elucidate the true pattern, and alternative methods must be employed.

A rather unusual ecological situation is that in which settlement takes place in single file along a groove. Spirorbid tubeworms (Wisely, 1960) and spat of *Alcyonidium hirsutum* (Fleming) do this along each side of the midrib of *Fucus serratus* (Hayward, this volume, p. 107). As before, if settlement is random, i.e. independent, along the grooves, the numbers present per standard unit of length will be distributed as a Poisson series. Just as with quadrats, presumably, the pattern revealed may depend on the length of the sampling unit employed. Fortunately, however, it is possible to emply a different, and simpler, method.

It is well known that, if the disposition of events in a linear continuum is random, then the frequency distribution of w, the interval between the events, will be exponential. All that is necessary, therefore, is to measure the successive intervals between newly settled spat, calculate the mean interval $\bar{w}$, and divide the results into about 15 frequency classes of convenient length. Using this method Wisely (1960) found that *Spirorbis borealis* Daudin (= *S. spirorbis* (L.)) larvae rarely settled less than 1 mm apart, and concluded (quite correctly, though without statistical analysis) that spacing out had occurred at settlement. In fact the theoretical exponential distribution can readily be calculated as a smooth curve (Fig. 1) showing the expected frequencies of w as

$$f(w) = n\lambda e^{-\lambda w},$$

where λ is estimated as $1/\bar{w}$. The numbers expected in each frequency class can also be calculated readily, and the observed and expected frequencies compared by means of χ^2 (Ryland and Hayward, in preparation).

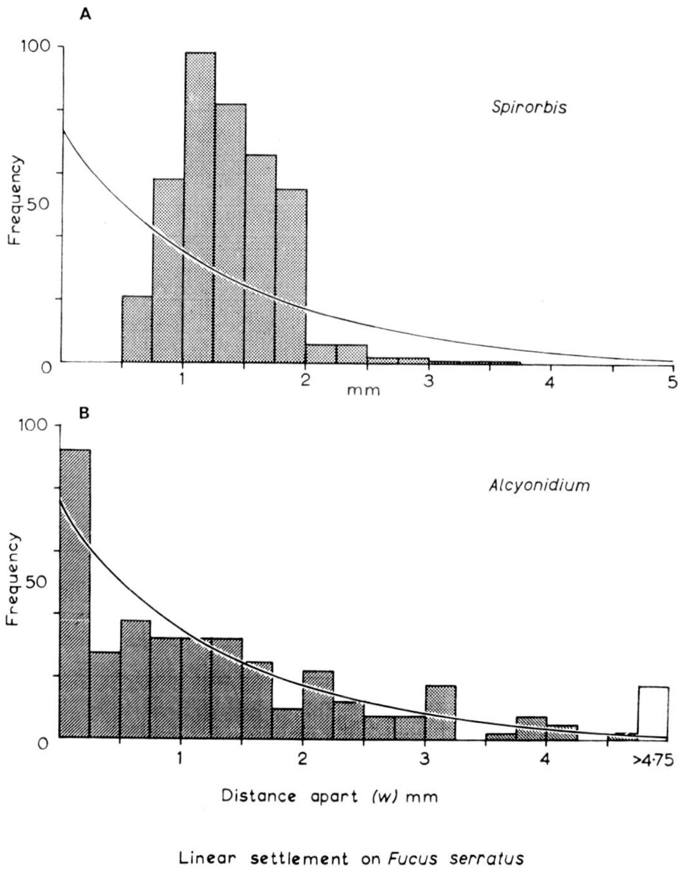

Fig. 1. Settlement alongside the midrib of *Fucus serratus*. The frequency histograms show distance apart of newly settled spat of (A) *Spirorbis spirorbis* and (B) *Alcyonidium hirsutum* compared with an exponential distribution (smooth curve). The mean distance apart is the same in both ($\bar{w} = 1{\cdot}36$ mm). Spacing out is clearly displayed by *Spirorbis* but not by *Alcyonidium*. Data from Wisely (1960) and Hayward (this volume).

This method of analysis has been applied to *Alcyonidium hirsutum* (Hayward, this volume, p. 107), and to Wisely's *Spirorbis* data, and the two sets of results compared. By a fortunate coincidence one of Wisely's samples has the same mean interval, $\bar{w}=1{\cdot}36$ mm, as one of Hayward's, which makes the comparison particularly striking. The differences are very obvious in Fig. 1, in which the two frequency distribution histograms have the theoretical curve superimposed. As the use of χ^2 confirms, the dispersal pattern of *Alcyonidium* is well described by the exponential distribution, and this leads to the clear though surprising conclusion that larvae settling alongside the midrib show no tendency at all to move away from each other. This is very different indeed from the situation revealed for *Spirorbis*.

Methods comparable to that just described, in that the variate is continuous, are also available for two dimensional dispersions (Pielou, 1969). These make use of the distance from randomly selected points to the nearest colony or (the "nearest neighbour" method) the distance from randomly selected colonies to the colony closest to them. This distance, r, represents the radius of a circle passing through the nearest neighbour, and the methods of analysis may utilize either r (Clark and Evans, 1954) or r^2 (Skellam, 1952) as variate. The latter is usually more convenient (Pielou, 1969), but for the present discussion a variate expressed in linear (unsquared) units is more appropriate. The frequency function of r, which gives the curve of expected values, is

$$f(r)=n2\lambda re^{-\lambda r^2}$$

where λ, the Poisson parameter, denotes the mean number of individuals per circle of unit radius, and is calculated as πD, where D is the density of colonies within the area studied. An independent estimate of density is required.

The situation that has been investigated concerns a population of *Crisia eburnea* L. occurring epizoically on the fronds of *Flustra foliacea* (L.) (see Stebbing, 1971, for description). A full consideration of the method and results must be deferred, but the general conclusion is relevant to the present paper. One hundred and sixty-six nearest-neighbour distances were measured on seven *Flustra* frond surfaces. Since all the *C. eburnea* colonies had been utilized, and the area of the frond could be simply determined, both density D and mean distance $\bar{r}$ were readily calculated.

On examination, the seven samples were seen to fall naturally into two groups according to the values of statistics D and $\bar{r}$. In the first group (64 measurements) $D_1=0{\cdot}0345$ colonies/mm² and $\bar{r}_1=2{\cdot}5$ mm. Analysis using χ^2 (and other methods) shows that the arrangement of the *Crisia* colonies is not significantly different from random (Fig. 2A). In the second group (102 measurements)

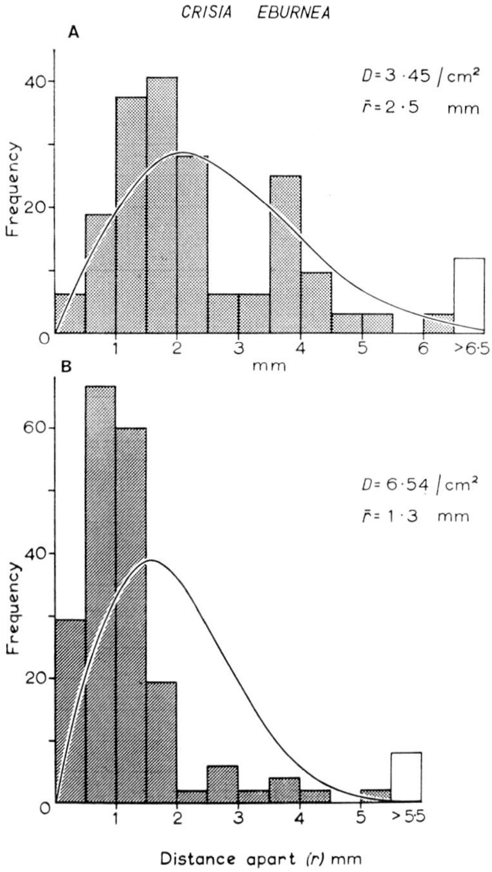

Fig. 2. Frequency histograms showing nearest-neighbour distance in two populations of *Crisia eburnea* growing on *Flustra foliacea* compared with the random distributions (smooth curves) for the same density. (A) is the less dense population ($D=3{\cdot}45/\text{cm}^2$) and the distribution does not depart significantly from random; (B) is denser ($D=6{\cdot}54/\text{cm}^2$) and the distribution is contagious: there is no evidence of spacing out.

$D_2=0{\cdot}0654$ colonies/mm² and $\bar{r}_2=1{\cdot}3$ mm. χ^2 here shows a significant departure from the theoretical distribution, and this is clearly due to the excess of low *r* values (Fig. 2B). Not only, therefore, were the *C. eburnea* colonies in the second, more dense, group aggregated, but the large numbers in the lowest classes ($r<1$ mm) shows beyond doubt that spacing out is not occurring. So *C. eburnea* provides a second instance of this rather surprising discovery.

REFERENCES

Clark, P. J. and Evans, F. C. (1954). Distance to nearest neighbour as a measure of spatial relationships in populations. *Ecology* **35,** 445–453.

Elliott, J. M. (1971). *Some methods for the statistical analysis of samples of benthic invertebrates.* Freshwater Biological Association Scient. Pub. No. 25.

Greig-Smith, P. (1964). "Quantitative Plant Ecology," 2nd Ed., Butterworths, London.

Hayward, P. J. (1972). Preliminary observations on settlement and growth in populations of *Alcyonidium hirsutum* (Fleming). "Living and Fossil Bryozoa" (G. Larwood, ed.), pp. 107–113. Academic Press, London and New York.

Knight-Jones, E. W. and Moyse, J. (1961). Intraspecific competition in sedentary marine animals. *Symp. Soc. exp. Biol.* **15,** 72–95.

Milne, A. (1961). Definition of competition among animals. *Symp. Soc. exp. Biol.* **15,** 40–61.

Pearson, E. S. and Hartley, H. O. (1966). "Biometrika Tables for Statisticians," 3rd Ed. Cambridge University Press, Cambridge.

Pielou, E. C. (1969). "An Introduction to Mathematical Ecology." John Wiley, New York.

Ryland, J. S. (1967). Polyzoa. *Oceanogr. mar. Biol., Ann. Rev.* **5,** 343–369.

Ryland, J. S. (1972). The analysis of pattern in communities of bryozoa. I. Discrete sampling methods. *J. exp. mar. Biol. Ecol.* **8,** 277–297.

Ryland, J. S. and Hayward, P. J. (In preparation). The analysis of pattern in communities of bryozoa. II. Distance sampling methods.

Ryland, J. S. and Stebbing, A. R. D. (1971). Settlement and orientated growth in epiphytic and epizoic bryozoans. *Proceedings of the 4th European Marine Biology Symposium.* 105–123. Cambridge University Press, Cambridge.

Silén, L. (1966). The fertilization problem in gymnolaematous bryozoa. *Ophelia* **3,** 113–140.

Skellam, J. G. (1952). Studies in statistical ecology. I. Spatial pattern. *Biometrika* **39,** 346–362.

Stebbing, A. R. D. (1971). The epizoic fauna of *Flustra foliacea* (Bryozoa). *J. mar. biol. Ass. U.K.* **51,** 283–300.

Stebbing, A. R. D. (1972). Observations on colony overgrowth and spatial competition. "Living and Fossil Bryozoa" (G. Larwood, ed.), pp. 173–183. Academic Press, London and New York.

Wisely, B. (1960). Observations on the settling behaviour of larvae of the tubeworm *Spirorbis borealis* Daudin (Polychaeta). *Aust. J. mar. freshwat. Res.* **11,** 55–72.

16 | Observations on Colony Overgrowth and Spatial Competition

A. R. D. STEBBING*

The Laboratory, Citadel Hill, Plymouth, Devon, England

Abstract: This paper describes the characteristics of two species of bryozoans which appear to prevent their being overgrown by other sessile animals. These are the peripheral calcareous margin of *Disporella hispida* and the long-spined zooids of *Electra pilosa*. Intra-specific competition and the fusion of colonial animals are discussed with some examples.

INTRODUCTION

With sessile animals like bryozoans, competition for space is often a dominant ecological factor. It occurs when the edges of separate sessile organisms grow towards one another and meet. Growth of both colonies may then stop or one colony may overgrow the other. The zooids overgrown usually die. Examples of overgrowth involving bryozoans have not been studied systematically, nor have any of their structures been shown to assist or prevent overgrowth. This paper describes some observations that have been made on bryozoans with the object of understanding their competitive interrelationships, when space is the limiting factor. Some other ways in which competition for space appears to be avoided, or reduced in intensity, in some epiphytic and epizoic bryozoans have already been described (Ryland and Stebbing, 1971; Stebbing, 1971a).

COLONY MARGIN OF TWO CYCLOSTOMES

Two cyclostomatous bryozoans, *Disporella hispida* and *Berenicea patina*, which occur on dead bivalve shells off Stoke Point near Plymouth, have a calcareous margin which is laid down in front of the growing edge and encircles each colony. Normally this margin is adherent to the substratum (Plate I, Fig. 1), but some times it is elevated at its outer edge, so that it is set at an angle to the substratum (Plate I, Figs 2, 3). The two forms are described by Hincks (1880).

Observations were made on the occurrence of these two types of margin to determine what function the less common type might serve. The observations

* Present address: Institute for Marine Environmental Research, Plymouth, Devon, England.

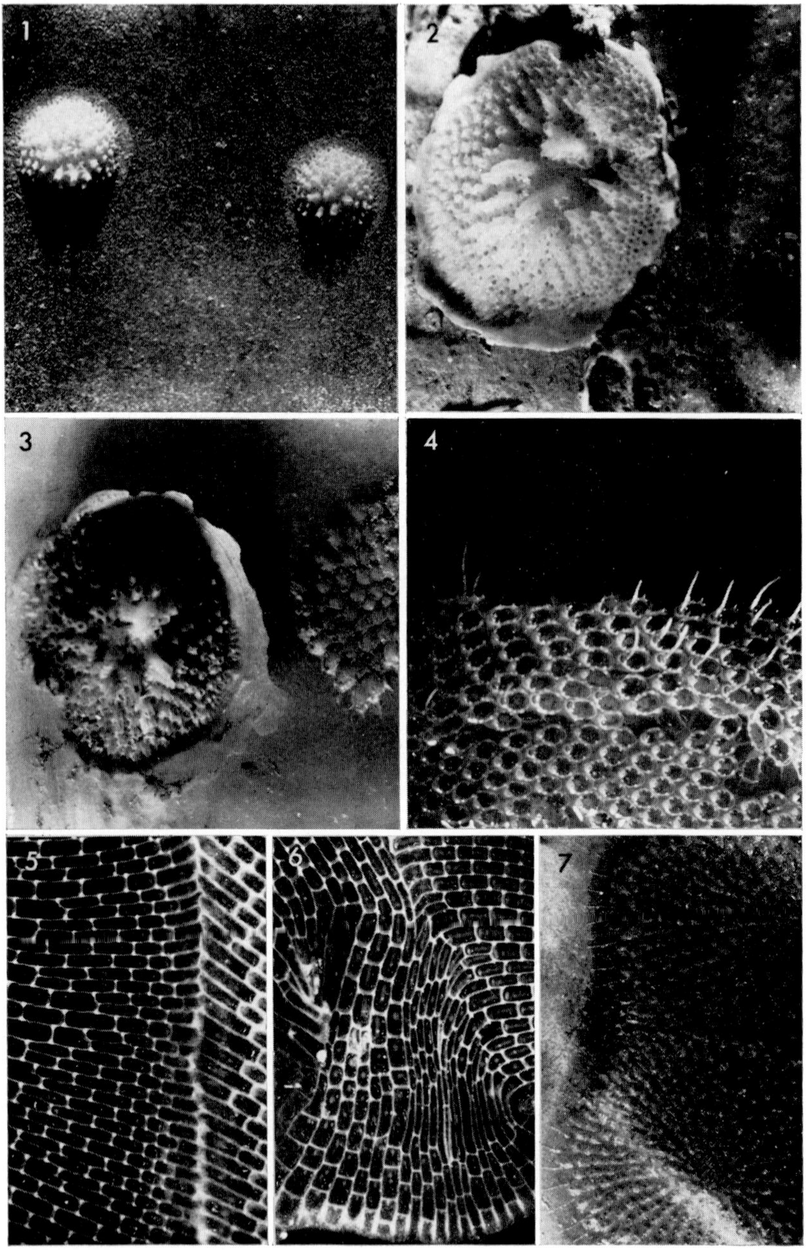
1
2
3
4
5
6
7

Table 1. *Disporella hispida.* The occurrence of an elevated margin in relation to the close proximity (<1 mm) of other sessile animals

	Isolated colonies	Colonies near other organisms
Colonies with adherent margin	55	1
Colonies with elevated margin	1	13

on *Disporella hispida* are summarized in Table 1. It may be seen that the adherent type of margin generally occurs in isolated colonies, whereas the elevated type of margin occurs most frequently when a colony of *D. hispida* is near another sessile organism. It appears, by its occurrence, that the ability to elevate the margin is a response to threatened overgrowth. Furthermore, it appears to be an effective means of doing so, for no living colony of *D. hispida* was found that had been overgrown. Far fewer colonies of *B. patina* were found, but the calcareous margin of this species too may be elevated from the substratum and this appears to prevent the colonies from being overgrown.

SPINES OF *Electra pilosa*

Zooids of *Electra pilosa* commonly occur in two forms (Hincks, 1880), both of which may be present in the same colony. The typical and more common type of zooid has a frontal wall encircled by 4–12 short spines, while the other has a greatly elongated vertical spine in place of one of the short ones at the proximal

Plate I

Fig. 1. *Disporella hispida.* Isolated colonies showing the adherent type of margin (× 5·5).
Figs 2, 3. *Disporella hispida.* Colonies showing the elevated type of margin, when overgrowth is threatened (× 5·5).
Fig. 4. *Electra pilosa.* Colony showing zooids with long spines and others with short spines (× 6·0).
Fig. 5. *Membranipora membranacea.* Two colonies that have met and stopped growing at the point where they touched. Note the shorter zooecia near the line along which the colonies met (× 6·75).
Fig. 6. Homosyndrome in *Membranipora membranacea.* At the top of the photograph are seen two separate colonies which are touching. At the bottom of the photograph the two colonies are seen to be fused producing a common growing edge (× 6·75).
Fig. 7. Homosyndrome in *Parasmittina trispinosa.* Two larvae have given rise to two colonies, which have grown towards one another and touched. At the growing edge along the line where they met, the colonies have begun to grow as one (× 4·5).

end of the frontal wall. It is usually 2–5 times the length of the zooecium. The occurrence of these two forms and the significance of the elongated spines does not appear to have been considered in detail, although Marcus (1926) observed one instance of intraspecific competition for space between living colonies of *E. pilosa* for three weeks and suggested that the spines may prevent overgrowth.

Large quantities of *Fucus serratus*, heavily encrusted by *Electra pilosa* and other epiphytes, were collected from the shore at Wembury near Plymouth on 25 June, 8 July and 7 October 1971. Observations were then made on the frequency of the two forms and the factors with which they appeared to be associated. The object of the first set of observations was to determine whether the physical characteristics of the substratum were associated with the occurrence of long-spined zooids. Initial observations seemed to indicate that these zooids occur more frequently on those parts of the colonies attached to convex areas of the fucoid fronds. These are the midrib and the edges of the fronds. The observations (Table 2) show that there is a correlation between the growth of colonies over the convexities of the fronds and the occurrence of long spines. It appears that

Table 2. *Electra pilosa*. The occurrence of long and short-spined zooids are given relative to the shape of the algal substratum in a sample of *Fucus serratus* collected at Wembury on 7 October 1971. (a) shows the occurrence of the two forms of zooid on those parts of 100 colonies that overly the midrib of *F. serratus* fronds. Generally the midrib projects on only one side of the fronds. Observations therefore were made only on colonies of *Electra pilosa* attached to the side on which it projects. (b) shows the occurrence of the two types of zooid at the points where 100 colonies had grown around the edge of the fronds, from one side to the other

	Numbers of colonies
(a) Long-spined zooids over midrib	88
only short-spined zooids over midrib	12
(b) long-spined zooids on edge of frond	70
only short-spined zooids on edge of frond	30

the long-spined zooids occurring on those parts of the colonies attached to convexities may afford some measure of physical protection from abrasion and buffetting by adjacent fronds during wave action.

The second set of observations was concerned with the extent to which some biological factors might be associated with the occurrence of long-spined zooids. Colonies of *E. pilosa* are frequently overgrown by their competitors for space, even though the long spines appear to be an impediment to overgrowing organisms. At Wembury the most common species encrusting the fronds of

Fucus serratus with *Electra pilosa* are *Flustrellidra hispida, Alcyonidium polyoum* and *A. hirsutum*. To what extent is contact with these competitors associated with the occurrence of long-spined zooids in colonies of *Electra pilosa*? The observations on this association are summarized in Table 3.

The figures show firstly that the numbers of *E. pilosa* colonies with long spines increased considerably over the period separating the samples. This may simply be a product of increasing age, because older and larger colonies do generally have more long-spined zooids. Secondly, the proportion of colonies

Table 3. *Electra pilosa*. The occurrence of long-spined zooids in colonies of *E. pilosa* when they were in contact with different species of competitors. The numbers of colonies of *E. pilosa* with at least 2 long-spined zooids and without any are given for each encounter with each of 3 species of competitors. The data are also expressed as percentages for each species. The data from the June 25 and July 8 samples have been presented as one for comparison with the sample taken on October 7. The data from these two samples did not appear significantly different from each other, and only together did they provide numbers comparable with those from the October sample

Competitor	Colonies with or without long-spined zooids	Numbers of colonies			
		June and July		October	
		Nos.	%	Nos.	%
Flustrellidra hispida	with	8	7·3	59	26·9
	without	102	92·7	160	73·1
Alcyonidium hirsutum	with	15	21·4	13	37·1
	without	55	78·6	22	62·9
Alcyonidium polyoum	with	113	68·5	154	77·4
	without	52	31·5	45	22·6

with long-spined zooids varied appreciably between the three species of competitors with which they came in contact. Colonies of *E. pilosa* in contact with those of *Alcyonidium polyoum* had long-spined zooids most frequently. They were least common when *Electra pilosa* encountered *Flustrellidra hispida*.

It is next important to attempt to determine the extent to which spines may prevent overgrowth by halting or diverting the growth of competitors. When two colonies of competing species grow towards one another and touch, one of three things may happen. One colony may overgrow the other or vice versa. The remaining possibility is that growth in that direction may stop. When this happens further growth is usually then diverted in another direction. Observations on over 1000 encounters between colonies of *Electra pilosa* and its competitors are summarized in Table 4. Although *E. pilosa* appears to grow much more quickly than the other three epiphytic bryozoans, it is clear from the observations that its colonies were consistently overgrown by each of its competitors in most of the instances when they grew together. However, there are variations between the observations for each species which should be discussed in the light of the figures on the frequency of long-spined zooids given in Table 3.

Table 4. *Electra pilosa*. The data show the summarized observations on numerous encounters between colonies of *E. pilosa* and their competitors for space. Records of overgrowth of *E. pilosa* by each of the 3 species represent the overgrowth horizontally by the overgrowing colony to a point at least 1 zooid inwards from the edge of the colony. In parentheses are the numbers of these records where the overgrowing colony had overgrown inward by at least 3 zooids from the edge of the colony. The percentages of the total number of colonies overgrown where overgrowth was at least 3 zooids inward are also included. As before, samples taken on June 25 and July 8 are considered as one

Competitor	Observation	Numbers of colonies			
		June and July		October	
		Nos.	%	Nos.	%
Flustrellidra hispida	overgrowth by *F. hispida*	140(28)	70·0(20·0)	157(38)	70·7(24·2)
	overgrowth by *E. pilosa*	54	27·0	30	13·5
	growth stops where they meet	6	3·0	35	15·8
Alcyonidium hirsutum	overgrowth by *A. hirsutum*	142(71)	95·3(50·0)	26(3)	74·3(11·5)
	overgrowth by *E. pilosa*	7	4·7	1	2·9
	growth stops where they meet	0	0·0	8	22·9
Alcyonidium polyoum	overgrowth by *A. polyoum*	130(61)	65·3(47·0)	116(36)	57·7(31·0)
	overgrowth by *E. pilosa*	0	0·0	1	0·5
	growth stops where they meet	69	34·7	84	41·8

Flustrellidra hispida consistently overgrew a large proportion of the *Electra pilosa* colonies, although it was occasionally overgrown itself by single rows of zooids meandering between the spines of *Flustrellidra hispida*. The proportion of colonies overgrown by *F. hispida* is the same in both samples, despite an increase in the number of *Electra pilosa* colonies with long-spined zooids (Table 3). *Alcyonidium hirsutum* also overgrew a large proportion of the *Electra pilosa* colonies with which it came in contact; although there was a decrease in the percentage of colonies overgrown by the time the October sample was taken (Table 4). Furthermore, in October a much lower proportion of the overgrowing colonies had extended far in from the edges of the *E. pilosa* colonies (Table 4, figures in parentheses). It appears that the increase in the numbers of *E. pilosa* colonies with long-spined zooids (Table 3) may explain these figures. The influence of long-spined zooids is most clearly seen in encounters between *E. pilosa* and *Alcyonidium polyoum*. The proportion of *Electra pilosa* colonies with long-spined zooids was highest when they encountered colonies of *Alcyonidium polyoum*, and here the proportion of *Electra pilosa* colonies overgrown was lowest. When the observations for *Flustrellidra hispida* and *Alcyonidium polyoum* are considered alone, it appears that there is an inverse relationship between the frequency of long-spined zooids and the number of occasions on which colonies of *Electra pilosa* are overgrown.

Finally, there is a correlation between the frequency of long-spined zooids in *E. pilosa* and the frequency with which growth apparently stops at the point where colonies met (Fig. 1). Growth of *Flustrellidra hispida* colonies rarely stopped where they met those of *Electra pilosa*, apparently because in these encounters *E. pilosa* colonies did not often have long-spined zooids. The reverse appears to be true when colonies of *Alcyonidium polyoum* met those of *Electra pilosa*. A high frequency of long-spined zooids seems to result in the frequent cessation of growth where colonies met.

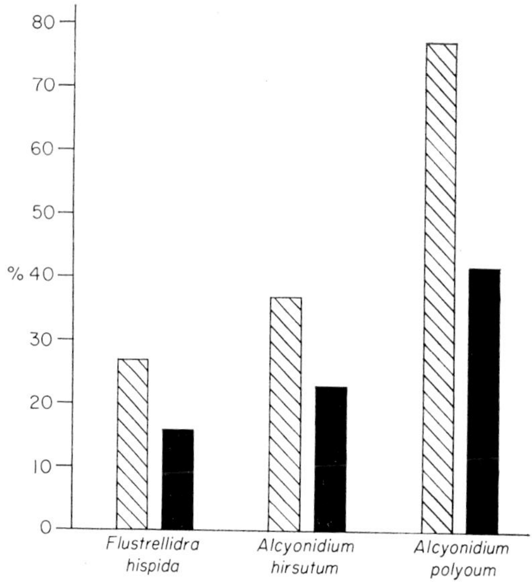

Fig. 1. Competition between *Electra pilosa* and the 3 other bryozoans epiphytic on *Fucus serratus* from Wembury. The cross-hatched columns represent the proportion of the colonies of *Electra pilosa*, touching each of the 3 competitors, which had long-spined zooids. The other columns represent the proportion of these encounters where growth of both colonies had apparently stopped at the point where they met. Data from Tables 3 and 4.

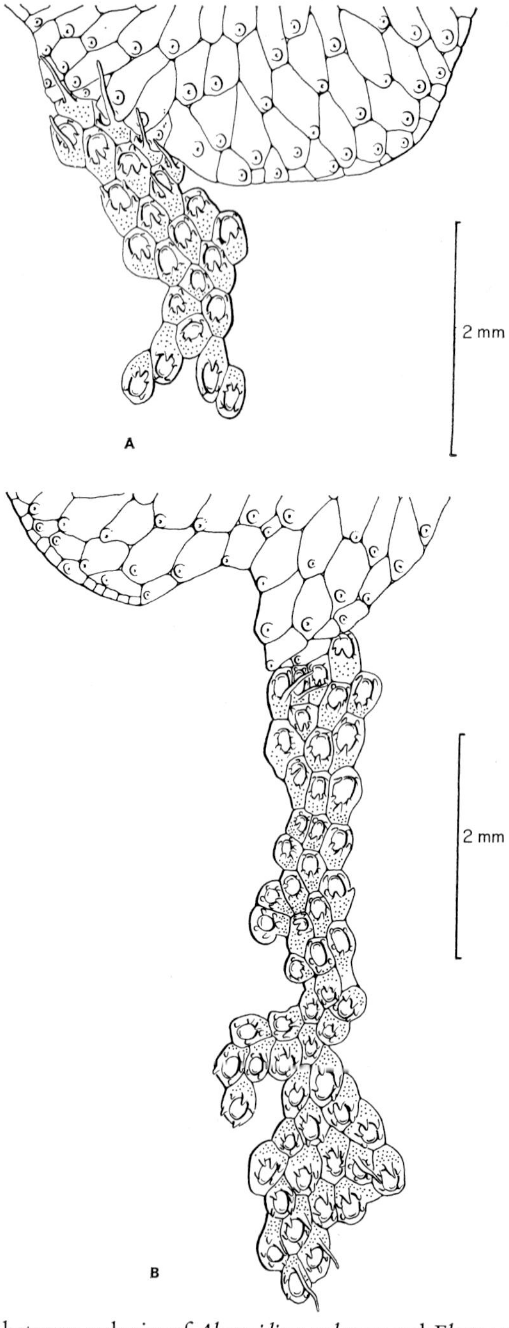

Fig. 2. Encounters between colonies of *Alcyonidium polyoum* and *Electra pilosa*. (A) shows how long-spined zooids occur typically at the point where overgrowth is occurring and frequently appear to stop it. (B) shows a similar encounter. Long-spined zooids of *E. pilosa* can be seen at the point where overgrowth appears imminent, but they also occur at the other extremity of the colony, where new zooids are being laid down.

No evidence has been given to show whether or not long-spined zooids of *E. pilosa* occur specifically on those parts threatened with overgrowth, where they would best prevent it. Often spines occurred sparsely over whole colonies, but when *E. pilosa* and *Alcyonidium polyoum* grew together and touched, spines were often restricted to that part of the colony that was about to be overgrown by *A. polyoum* (Fig. 2A). Sometimes colonies would be found like that illustrated in Fig. 2B. At one end the new growth of an *Electra pilosa* colony has come in contact with a colony of *Alcyonidium polyoum*. Long-spined zooids are found at the point threatened with overgrowth, but in addition to these there are also long spines at the other end of the colony, where other new zooids have been laid down. Sometimes, star-shaped colonies were found with growth occurring at the tip of each arm. When one arm of such a colony has encountered a colony of *A. polyoum*, the zooids constituting the tips of all the other arms may include some zooids with long spines. It appears that here is an example of interzooidal communication of a "message" of functional significance for the colony as a whole.

It has been suggested that the spines of *Electra pilosa* would appear to present a barrier to any overgrowing competitor. Although the spines were frequently responsible for stopping growth in one direction and diverting it in another, examples of each of the three species of competitors were observed growing over or between the spines. Occasionally the long spines would be found projecting upward through the middle of a colony of *Alcyonidium polyoum*, where a colony of *Electra pilosa* with long-spined zooids had been completely overgrown. However, the impression is that growth of bryozoans was usually orientated in the direction of least resistance to further growth.

In conclusion it may be said that, although it is difficult to draw definite conclusions from numerical data of this type, the long-spined zooids of *E. pilosa* appear to be of functional significance to the colonies. Firstly, they appear to occur on those parts which may require physical protection. Secondly, long-spined zooids seem to be associated with instances where there is interspecific competition for space and they prevent or slow overgrowth at those points where they occur densely.

INTRASPECIFIC COMPETITION

When colonies of the same species meet, they generally stop growing at the point where they touch. Growth is then diverted in other directions. This has been established in four epiphytic bryozoans on *Fucus serratus* (Stebbing, 1972) and appears to apply generally. Among the common littoral bryozoans, occasional exceptions may be found in *Electra pilosa* and *Membranipora membranacea*. However, they too usually stop growing where they meet (Plate, I, Fig. 5).

Some sessile animals have larvae which have been shown to space out from others of their own species at settlement, allowing the individuals space to grow to adult size without inhibiting each other's growth. The best known examples are the larvae of barnacles and *Spirorbis* spp. (see review by Knight-Jones and Moyse, 1961). Solitary species generally have a sigmoid growth curve and in time grow to an asymptotic size typical of their species. Individuals of each species require a relatively constant amount of space in which to reach their usual maximum size. Colonial animals are different. The zooids that make up the typical colony do not grow, for their size is determined by the dimensions of the zooecium which remains the same once it is laid down at the growing edge. However, the growth of the entire colony, in terms of numbers of zooids tends to be exponential (Stebbing, 1971b); and its eventual size seems to be limited by the space available for growth, rather than some intrinsic factor peculiar to the creature itself. One would therefore not expect the larvae of bryozoans, or other colonial organisms, to space out as do barnacles and tube worms. The absence of any tendency to space out has been demonstrated in two bryozoans, *Alcyonidium hirsutum* and *Crisia eburnea* (Ryland, this volume p. 165). It appears that the cessation of growth where colonies of the same species meet may have the same effect as the spacing out of larvae in solitary species. In both instances the result is a minimization of intraspecific competition for space, allowing individuals or colonies space to grow without interference.

It has already been pointed out that where colonies meet, one may overgrow the other or further growth in that direction may stop. There is a third possibility which has not been considered. When two colonies of the same species meet, they may occasionally fuse and grow as one colony, each in organic continuity with the other. Knight-Jones and Moyse (1961) introduced, in their review on intraspecific competition in sedentary animals, two important terms to describe different types of fusion in colonial animals. Autosyndrome refers to the fusion of different parts of one colony after they have grown apart. Homosyndrome refers to the fusion of separate colonies that have originated from different larvae of the same species. They note that autosyndrome has been observed in *Botryllus* and a number of bryozoans, while homosyndrome has been recorded in *Hydractinia*, some sponges and corals. Autosyndrome is implicit in the habits of growth of a number of bryozoans like *Aetea* spp. and some boring bryozoans. Since Knight-Jones and Moyse's review, Moyano (1967) has observed homosyndrome in a bryozoan. He found two colonies of *Membranipora hyadesi* that had grown together and fused to grow as one. However, he did not observe whether there were pore plates in the wall separating zooids of the two colonies, which would indicate that there was organic continuity between the zooids. If pore plates do make possible the passage of nutrients between the zooids of one colony, homosyndrome might result in the sharing of resources between the zooids of two colonies, perhaps of different genotypes.

Two probable examples of homosyndrome were observed during the present study. In *Membranipora membranacea*, two examples of two colonies fusing to grow as one were found (Plate I, Fig. 6). These are the only examples that have been found despite an extensive search. Examination of thick sections from both these examples showed that there were pore plates connecting zooids of the two colonies.

Parasmittina trispinosa is a common species of the fauna encrusting dead shells from Stoke Point. In 11 instances where two colonies had grown towards one another and touched, 10 had fused and the pair of colonies had begun to grow as one (Plate I, Fig. 7). It appears in this species that homosyndrome may be the rule, although it has not yet been confirmed that pore plates exist in the walls separating zooids of each colony.

ACKNOWLEDGEMENTS

I would like to thank Miss P. L. Cook for critical evaluation of the manuscript. I am also grateful to the Director and staff of the Marine Biological Association's Laboratory for providing facilities for my research, and to N.E.R.C. for a Post-doctoral Fellowship.

REFERENCES

HINCKS, T. (1880). "A history of the British marine Polyzoa." 2 Volumes. van Voorst, London.

KNIGHT-JONES, E. W. and MOYSE, J. (1961). Intraspecific competition in sedentary marine animals. *Symp. Soc. exp. Biol.* **15,** 72–95.

MARCUS, E. (1926). Beobachrungen und Versuche an lebendum Meeresbryozoen. *Zool. Jb.* (*Syst.*) **52,** 1–102.

MOYANO, H. I. (1967). Sobre la fusión de dos colonias de *Membranipora hyadesi* Jullien, 1888. *Notic. mens. Mus. nac. Hist. nat. Santiago* **11** (126), 1–4.

RYLAND, J. S. and STEBBING, A. R. D. (1971). Settlement and orientated growth in epiphytic and epizoic bryozoans. *Proc. IVth European Mar. Biol. Symp.* 105–123.

STEBBING, A. R. D. (1971a). The epizoic fauna of *Flustra foliacea* (Bryozoa). *J. mar. biol. Ass. U.K.* **51,** 281–300.

STEBBING, A. R. D. (1971b). Growth of *Flustra foliacea* (Bryozoa). *Mar. biol.* **9,** 267–272.

STEBBING, A. R. D. (1972). Competition for space between the epiphytes of *Fucus serratus* L. *J. mar. biol. Ass. U.K.* **53.** (In press.)

17 | Environmental Conditions of Bryozoan Ecology of the Hardground Biotope of the Maastrichtian Tuff-Chalk, near Maastricht (Netherlands)

EHRHARD VOIGT

Geologisch–Paläontologisches Institut, Hamburg, Germany

Abstract: The rich Bryozoan fauna of the Maastrichtian Tuff-Chalk occurs in the Bryozoan-beds overlying two hardgrounds and within the burrows penetrating these hard banks. The surface of the hard banks offered a favourable substratum for Bryozoa and a rich algal vegetation as well. An investigation of some thousand bryozoan bases suggests that many erect forms were attached to stalks or leaves of these algae. The walls of the burrows were also occupied by many encrusting forms, and small holes in the rock were ecological niches for a special bryozoan dwarf fauna. Analysis of the zoarial forms of the Bryozoa and their coexistence with *Lithothamnia* and reef corals indicates a depth of perhaps 20–30 m.

INTRODUCTION

One of the most famous localities of Cretaceous Bryozoa is the well-known Tuff-Chalk near Maastricht (Netherlands). This fauna was first described in the classic works of Goldfuss (1826) and von Hagenow (1851). As I have been investigating the Maastrichtian Bryozoa for some decades I am preparing a revision of that fauna.

The Upper Maastrichtian Tuff-Chalk represents a regressive carbonate facies consisting of a friable calcarenitic limestone, which was deposited in the sublittoral region of the Chalk sea in South Limburg and adjacent areas after a long period of sedimentation of a marly white Chalk.

It is well known that the distribution of the Bryozoa is controlled by the existence of several layers of hardground in the upper part of the Tuff-Chalk (Ubaghs 1865; Voigt 1959). The so-called Bryozoan Beds, consisting of immense masses of bryozoan fragments and other smaller fossils, are found mostly overlying these hardgrounds.

In general the Bryozoa are rare in the Tuff-Chalk itself, but they are extremely common in the roof or in the neighbourhood of the hard layers, which were hardened at an early stage of diagenesis, as proved by numerous burrowing and encrusting organisms settling upon the surface of the hardgrounds. In contrast to the loose, minute calcareous grains of the Tuff-Chalk sediment, which provided few possibilities for the fixation of larvae, the consolidated surface of the hardground presented a favourable habitat for the settlement both of Bryozoa and of a rich algal vegetation, which secondarily gave suitable possibilities for the attachment of bryozoan larvae. Consequently it seems reasonable to assume that the extraordinary concentration of bryozoan fragments in the Bryozoan Beds was conditioned by the hardground phenomenon itself (Voigt, 1959). This conclusion implies the general importance of the mode of fixation of the larvae upon hard substrata.

To gain a thorough understanding of the ecology of the Bryozoan Beds and their relationship to the hardgrounds, an attempt was made to study the settling of Bryozoa in the hardground-environment and in the tunnels burrowed in it and to research the mode of attachment on the different kinds of substratum. Apart from the investigation of the surface of the hardened Tuff-Chalk and of all other fossils which could serve as a substratum, a precise examination is required of all bases obtained of erect Bryozoa, from which conclusions as to the different kinds of attachment may be drawn. This method seems to be preferable to that using the total number of bryozoan fragments counted in a certain volume of the Bryozoan Beds, because the number of fragments, which depends on the size of the rigid erect zoaria and on the intensity of wave action or currents, gives insufficient information relating to the true former abundance of the different species. It may depend mainly on the total number of bryozoan fragments or on the distributional pattern of certain species, which lived very isolated in different ecological niches. This is evident from the rather different faunal associations found in samples taken during the last decade in the different localities near Maastricht.

These factors account for the extremely high number of bryozoan species collected in the Tuff-Chalk (about 300 species, many of which have not yet been described).

THE SUBSTRATA OF THE BRYOZOANS IN THE MAASTRICHTIAN TUFF-CHALK

1. The Hardgrounds

Two main hardgrounds exist in the upper Maastrichtian, and these have been known for a long time (Ubaghs, 1865; Voigt, 1959) The older one is situated at the base of the zone Md and the younger one is found some metres higher in this zone. A third hardground, at the top of the Upper Maastrichtian, is overlain by the Dano-Montian beds (which are also rich in Bryozoa) and is exposed in

Curfs quarry, near Berg in the Geul-Valley. It contains additionally within its burrows a rich bryozoan fauna resembling that of the Baltic Danian and the "Montian" of Ciply (Belgium). The surfaces of the two older hardgrounds are very often covered by numerous sedentary fossils, among which encrusting cheilostome Bryozoa are most dominant. The "membraniform" zoarial habit of many species of "*Membranipora*", *Aechmella*, *Onychocella* and certain Cribrimorphs, some of which may attain multilamellar growth and a wide lateral colony extension, as well as the relative scarcity of the bases of erect forms suggest that for a long time no sedimentation took place and strong current or wave action was not favourable for the growth of the rigid arborescent species. Based on the observation that most of the boring bivalves in the floor of the hardground are truncated and appear only as hemispherical holes in the surface of the hard layers, it becomes evident that the sea-bottom was exposed to violent submarine erosion before the settling of bryozoans was possible. Erosion is proved also by gullies cut into the hardgrounds and by the local concentration of hardground-pebbles. Evidence of the boring activity of both sponges and lithophages and of pre-existent thalassinoid burrows indicates that the eroded surface of the hardground was rather uneven before it was covered by encrusting Bryozoa, Foraminifera, calcarous sponges, brachiopods (*Crania*), Serpulids, bivalves (Spondylids) and numerous *Lithothamnia*. Hollows some cms in diameter beneath the encrusting zoaria may be assumed to represent former basal discs and holdfasts of seaweeds or horny *Alcyonaria*, which flourished in that biotope. The encrusting cyclostome genera, such as *Stomatopora*, *Proboscina*, *Filifascigera*, *Lichenopora* and some species of the "*Berenicea*"-habit are relatively rare. On the other hand fragments and even whole zoaria of erect rigid *Cyclostomata* are extremely abundant in the Bryozoan Beds and enriched in the burrows penetrating the hardgrounds. It must be assumed that in other places the erect bryozoans were widely distributed and settled on the hardground itself or on the algae growing there.

The burrows and galleries in the hardground, filled with loose bryozoan sediment, are assumed to be mined by Decapod crustacea (*Thalassinoides*) before the rock was indurated. In comparison with the bryozoan-detritus of the Bryozoan Beds covering the hardgrounds, the preservation of the Bryozoa, concentrated in these hollows, may locally be excellent (Plate I, Figs 1 and 8) resembling recent specimens by the preservation of spines (Plate I, Fig. 7) or long peristomes (Plate I, Fig. 6), fragile ovicells or whole arborescent or reticulate zoaria (Plate I, Fig. 8). The burrows were traps for the zoaria detached from their substrata by vagile animals or by wave or current action and swept into the galleries before they were broken and abraded on the sea floor. The walls

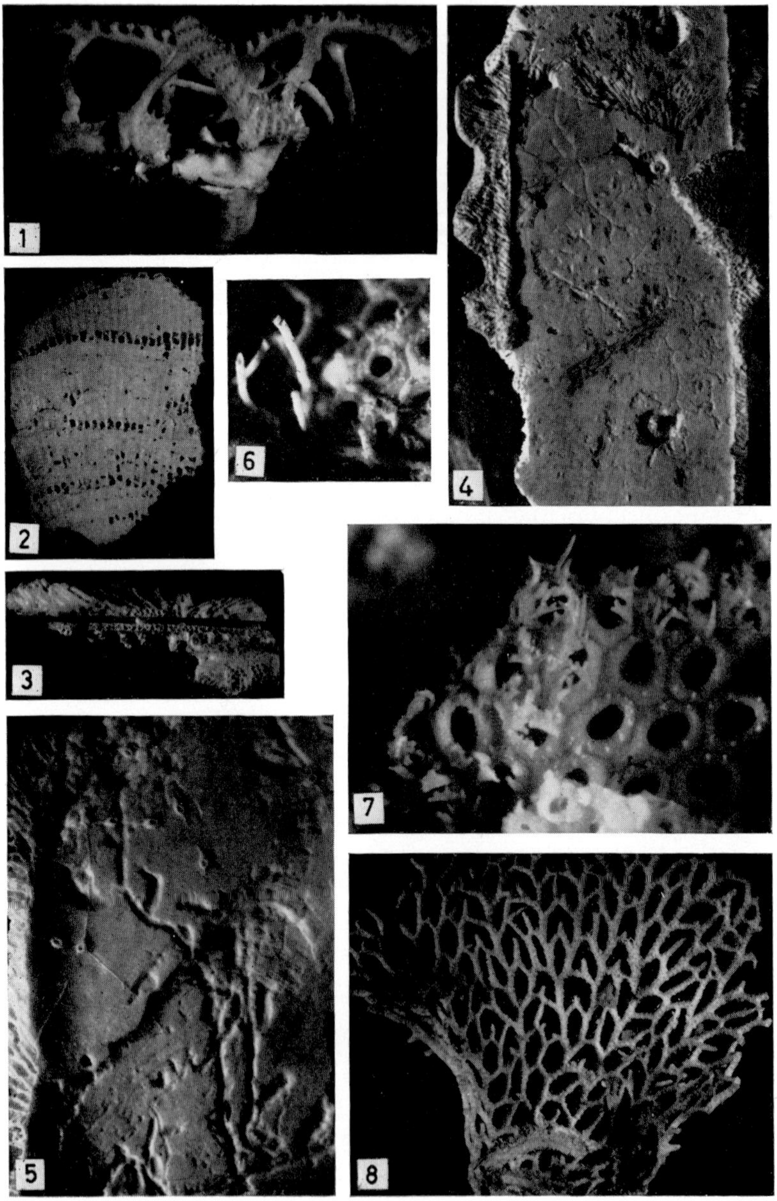
1
2
3
4
5
6
7
8

of the labyrinthic vertical and horizontal burrows are also often covered by encrusting unilamellar Bryozoa and other sessile benthonic fossils such as Serpulids, Cranias, encrusting Foraminifera or penetrated by burrowing organisms. Three conclusions of general interest about the behaviour of large encrusting zoaria may be postulated: firstly this biotope was rather dark, and may be completely without light and the Bryozoa living within these caverns must have preferred a dark environment, as known from the negative phototropic behaviour of the larvae of some recent forms. Secondly the water circulation within these thalassinoid tunnels, extending often many decimetres into the hardground, must be assumed, even after the death of the Crustacea, to have provided the sessile fauna with oxygen and planktonic nourishment. Thirdly the rate of sedimentation in these burrows was very low during the life-period of the sessile fauna attached to the wall.

Special interest was paid to a dwarf Bryozoan faunal assemblage, consisting of some still undescribed species, which readily escapes detection, in minute hollows of boring sponges and lithophages or narrow burrows. The Bryozoa

---

Plate I

Fig. 1. Base of *Idmonea dorsata* v. Hagenow, showing calcareous pillars and rootlets, ×4. Upper Maastrichtian, Quarry Curfs near Geulem-Berg (Netherlands).

Fig. 2. Unilamellar fragment of *Onychocella cyclostoma* (Goldfuss), showing growth-zones on the reverse side indicated by zooecia whose dorsal wall is thinner and mostly worn, *ca* ×1·6. Upper Maastrichtian, Albert-Canal near Néercanne (Belgium). Coll. Voigt.

Fig. 3. *Actinopora disticha* (v. Hagenow), showing median slit left by encased algal leaves ×5·7. Upper Maastrichtian, Quarry Curfs near Geulem-Berg (Netherlands). Coll. Voigt.

Fig. 4. Base of *Ditaxia anomalopora* (Goldfuss), split up, showing the hollows left by encased algal leaf with moulded incrusting Hydroids, ×4·3. Upper Maastrichtian, Albert-Canal near Néercanne (Belgium). Coll. Voigt.

Fig. 5. Base of *Ditaxia anomalopora* (Goldfuss), split up like Fig. 4, showing moulded Ctenostome *Stolonicella schindewolfi* Voigt with extremely small *Pyripora sp.* settling in the hollow left by the rotten alga, ×8·8. Upper Maastrichtian, Quarry Curfs near Geulem-Berg, (South Limburg Netherlands). Coll. Voigt.

Fig. 6. *Stomatopora* sp. with extremely long peristoms settling in microcaverns in the hardground, ×20. Upper Maastrichtian, St Pietersberg near Maastricht (Netherlands). Coll. Voigt.

Fig. 7. *Callopora* sp. with preserved spines, ×35. Upper Maastrichtian, Quarry Curfs near Geulem-Berg (South Limburg, Netherlands). Coll. Voigt.

Fig. 8. *Reticrisina maculata* (v. Hagenow), well preserved free reticulate zoarium from burrow in the hardground, ×2·7. Same locality as Fig. 7. Coll. Voigt.

settling on the walls of the burrows were protected against predators and mechanical destruction before they were buried by the sediment washed into the burrows. Occasionally the delicate projecting spines of the Membranimorphs and Cribrimorphs or the long projecting peristomes of minute *Stomatopora*-species have been preserved, because they were neither worn nor cemented by the matrix during later diagenesis.

FOSSILS AS SUBSTRATA OF BRYOZOA

Within the white Chalk and in most other Cretaceous fossiliferous sediments larger fossils commonly provide a substratum for Bryozoa ("secondary Hartboden" sensu Remane). In the Tuff-Chalk, however, most of the large echinoids (*Hemipneustes radiatus* (Lam.)) and the very common small oysters (*Pycnodonte*) the belemnites, the orbitoids and the tubes of *Ditrupa mosae* (Montf.), occurring in huge masses, lack or show only very few encrusting Bryozoa. As remarked by Schopf (1969, p. 237), Bryozoa make better use of rocks and shells as substrata in deeper areas than in shallower ones. On the other hand *Pycnodonte* and other shells often settled on arborescent rigid Bryozoa, and many bases of the erect forms are found attached to other Bryozoa or serpulid tubes as well.

ALGAE AND OTHER PLANT REMAINS AS SUBSTRATA FOR BRYOZOA

Although seaweeds have never been preserved in the Tuff-Chalk, their abundance has been proved indirectly by numerous discoveries of bryozoan bases or other sessile fossils, which have encrusted stalks or fronds of algae and other plants (Voigt, 1956; 1966). By no means all hollow zoaria always encased algae, as often erroneously stated in the literature. In many cases it is possible to discern between primarily hollow zoaria and the tubelike habit of those encrusting algae. Primarily tubular or unilamellar zoaria of Cheilostomata mostly represent a very regular architecture of their zoaria like true arborescent species. The

---

Plate II

Fig. 1. Bases of *Pustulopora* (=*Hammia*) *pustulosa* (Goldfuss) with cylindrical hollows corresponding to algal stalks or (perhaps) seagrass-rootlets, × 4·8. Coll. Voigt.

Fig. 2. Silicified axes of seagrass (*Thalassocharis bosqueti* Debey ex Miquel) with encrusting Bryozoa. In the left lower corner base of *Idmidronea macilenta* (v. Hagenow) showing the hollow left by the *Thalassocharis*-axis with its characteristic zigzag-pattern, × 4·8. Coll. Voigt.

Fig. 3. Silicified roots of seagrass (*Thalassocharis bosqueti* Debey ex Miquel) with encrusting Bryozoa, × 4·8. Coll. Voigt.
(All originals of Figs 1–3 from the Upper Maastrichtian of Kunrade 30 km ESE Maastricht (Netherlands).)

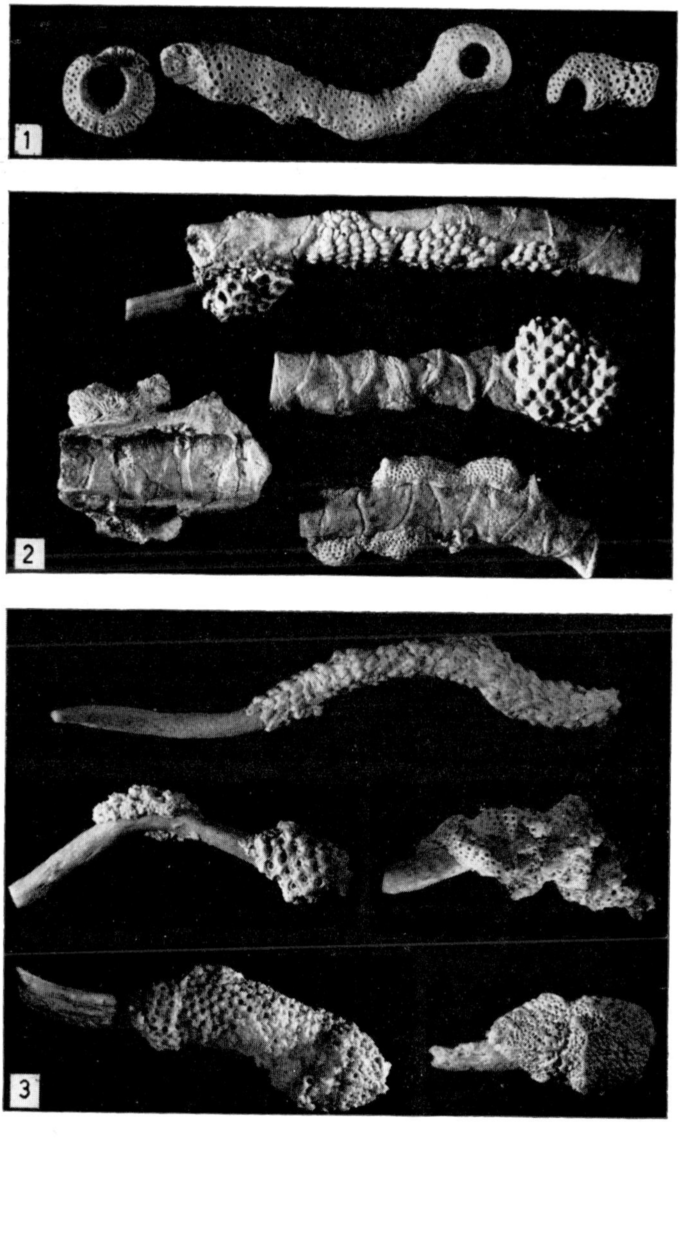
1
2
3

reverse side of the zooecia is usually slightly convex and the walls are equally calcified. Primary tubular Cyclostomata (such as *Cavaria pustulosa* v. Hagenow) often possess diaphragm-like calcareous floors traversing the internal channel. On the contrary, tubular encrustations of algal stalks show a more irregular growth of the zooecia and often the reverse side of the hollow zoarium is very smooth and glossy corresponding to the closely adherent zooecial dorsal face. But it must be admitted that in other cases no close contact exists between algal substrata and the adherent zoaria which seem to envelop their hosts without direct contact. Infallible criteria for bioimmuration of seaweeds are other smaller Bryozoa, Hydrozoa or Serpulids growing on the algae, which were immurated by encrusting Bryozoa and which appear on the inner concave side of the tubular zoaria after the disintegration of the algal remains. Many fossil species of erect rigid cylindrical or bilamellar Bryozoa exhibit basal parts, showing cylindrical or blade-like hollows (Pustulopora pustulosa (Goldfuss) Plate II, Fig. 1), which must be interpreted as encased algal stalks or fronds, which now have disappeared. (*Actinopora disticha* (v. Hagenow) Plate I, Fig. 3, *Ditaxia anomalopora* (Goldfuss) Plate I, Figs 4–5).

Numerous disciform colonies of *Actinopora* and *Lichenopora* from Maastricht showing a glossy basal face are recognized to have been fixed on flat algal fronds. Other encrusting Cyclostomata or sessile oysters have preserved by bioimmuration the delicate uniserial Ctenostomata with their stolons, e.g., *Stolonicella schindewolfi* Voigt (Plate I, Fig. 5) and *S. filosa* Voigt or the articulated zoarium of *Taeniocellaria setifera* Voigt (1966) with its serrated setae.

Within the minute hollows in bryozoan bases left by rotted algal fronds extremely small *Pyripora* sp. (length of zooecia 0·25 mm) have been found encrusting the internal surface of the substrata (Plate I, Fig. 5).

Actinopora disticha (v. Hagenow) and small *Lichenopora* sp. often have the cuticular pattern and leaf-veins of dicotyledonous leaves moulded on their basal faces (Voigt 1956, p. 109).

As a principal result of these investigations it must be concluded that many big erect bilamellar, cylindrical or even idmoneidal species were attached to algal stems. At the locality of Kunrade (30 km ESE Maastricht) silicified cylindrical rootlets and axis-stocks of the sea-grass *Thalassocharis bosqueti* (Debey ex Miquel) (Voigt and Domke, 1955), showing the characteristic zigzag-pattern of the axis, are common. Many of these seagrass-remains are encrusted by unilaminar or rigid erect zoaria, indicating the common settling of Bryozoa on these root-organs (Plate II, Figs 2, 3), while the fronds of *Thalassocharis* never have been observed at this locality. Certain species are found settling only on these sea-grass remains.

Table 1 shows the result of a statistical examination of about 2300 specimens belonging to the 32 representative species which were preserved with their bases. Although in about 500 zoaria the substratum was not more recognizable, the rest of 1800 bases revealed the nature of this substratum. The high number of algal bases in many erect rigid forms is quite remarkable and certain species must have preferred seaweed substrata. Table 2 shows the distribution of the different substrata of 1800 bases as percentages. It reveals that nearly 79% of the bases had algal substrata.

Numerous "*Vincularia*"-fragments ("*V.*" *canalifera* v. Hagenow and "*V.*" *procera* v. Hagenow) with pointed bases, have small pores of radicells at their proximal end suggesting cellariform jointed zoaria like the delicate joints of *Quadricellaria*, *Dimorphostylus tetrastichus* Voigt and *Fusicellaria hofkeri* Key. *Stamenocella cuvieri* (v. Hagenow), *Stamenocella scindulata* (v. Hagenow), *Onychocellaria rhombea* (v. Hagenow), *Virgocella esperi* (v. Hagenow) and the genus *Beisselinopsis* Voigt were also fixed by radicells. But it must be stressed that the number of flexible jointed species as observed to-day occurring dominantly in shallower waters was still less in the Upper Cretaceous than during Tertiary and Recent time.

Significant criteria indicating agitated water are calcareous anchoring and holding appendages such as pillars, rootlets or grapnel-like hooks observed in certain species of Cyclostomes. Stout vertical pillars were constantly observed in *Ceidmonea procera* (Hamm), *Filisparsa tubulifera* (v. Hagenow) and *Filisparsa* sp., *Idmonea dorsata* (v. Hagenow) (Plate I, Fig. 1) and other Idmoneids with horizontally expanded zoaria. *Crisisina carinata* (Roemer) developed basal pointed appendages with fine grapnel-like spines. All these anchoring elements are specific and it is obvious that not all species could produce holding organs like these forms. They never have been observed in the same species within the deeper and calm water of the White Chalk facies. The assemblage of abundant "membraniform", "eschariform", "vinculariform", "celleporiform", "reteporiform" and the more rare "lunulitiform" zoarial habits, which according to Stach (1936) represent different bathymetric ranges in regard to the extremely high total number of about 300 species suggest that this thanatocoenosis may represent different ecological conditions upon a well differentiated sea-bottom. This is proved by the occurrence of stout zoaria of *Inversaria tubiporacea* (Goldfuss), *Heteropora crassa* v. Hagenow, bulbous zoaria of *Ceriopora* or thick eschariform colonies of *Porina*, *Beisselina* or *Onychocella*, not far from populations including smaller species of *Entalophora*, *Idmonea*, *Crisina*, *Crisisina*, "*Vincularia*" etc. or the minute meshwork of *Filisparsa tubulifera* (v. Hagenow) and *Reticrisina maculata* (v. Hagenow) (Plate I, Fig. 8), indicating quieter

Tab

	Growth of Zoaria	LIST OF THE 32 REPRESENTATIVE BRYOZOA WITH PRESERVED BASES	St Pietersberg near Maastricht						
			1	2	3	4	5	6	7
			on Bryozoa	on other fossils	on seagrass-roots	on algal-stalks	on algal-leaves (narrow)	on algal-leaves (broad)	on Dicotyl.-leaves
CYCLOSTOMATA	Erect rigid species	*Plagioecia compressa* (Goldf.)				1			
		Pustulopora pustulosa (Goldf.)	1	1		1			
		Mecynoccia madreporacca (Goldf.)				1			
		Spiropora verticillata (Goldf.)							
		Lopholepis radians (v. Hag.)				1			
		Heterocrisina communis (d'Orb.)						1	
		Idmidronea macilenta (v. Hag.)	1					21	
		Reteporidea lichenoides (Goldf.)	14	2		7		1	
		Crisina cancellata (Goldf.)	6				1		
		Retecava clathrata (Goldf.)				5	1		
		Crisisina carinata (Roem.)	3			2			
		Ditaxia anomalopora (Goldf.)	2			2	1		
		Cavarinella ramosa (v. Hag.)	1			6			
		Grammascosoecia dichotoma (Goldf.)	9	2					
		Coelocochlea torquata (v. Hag.)				2			
		Petalopora fallax (Hamm.)							
		Osculipora truncata (Goldf.)		1		4	1	1	
		Truncatulipora repens (v. Hag.)	1	2					
	Adnate	*Truncatulipora filix* (v. Hag.)				3		24	
		Lichenopora reticulata (v. Hag.)		1		5		9	26
		Actinopora disticha (v. Hag.)		3			1	67	28
CHEILOSTOMATA	Erect rigid species	*Thyracella lesueuri* (v. Hag.)							
		Onychocella sexangularis (Goldf.)				2			
		Onychocella stigmatophora (Goldf.)	1						
		Onychocella lamarcki (v. Hag.)							
		Onychocella nysti (v. Hag.)				1			
		Onychocella dichotoma (Goldf.)							
		Onychocella propinqua (v. Hag.)				8			
		Coscinopleura lamourouxi (v. Hag.)	1			1			
		Escharifora müelleri (v. Hag.)				2			
		Beisselina striata (Goldf.)							
		Porina diff. sp.				8		1	
		Total	40	12	0	62	5	125	54

LOCALITIES OF THE MAASTRICHTIAN CHALK-TUFF

Curfs Quarry near Berg (Geul-Valley)							Albert-Canal near Néercanne								Schunk Quarry near Kunrade								
2	3	4	5	6	7	8	1	2	3	4	5	6	7	8	1	2	3	4	5	6	7	8	
on other fossils	on seagrass-roots	on algal-stalks	on algal-leaves (narrow)	on algal-leaves (broad)	on Dicotyl.-leaves	Substratum not recognizable	on Bryozoa	on other fossils	on seagrass-roots	on algal-stalks	on algal-leaves (narrow)	on algal-leaves (broad)	on Dicotyl.-leaves	Substratum not recognizable	on Bryozoa	on other fossils	on seagrass-roots	on algal-stalks	on algal-leaves (narrow)	on algal-leaves (broad)	on Dicotyl.-leaves	Substratum not recognizable	Number of controlled specimens
		1								2				2									9
2		14		30		2		1		14				3			10	131					212
						2	1			1				1									7
		3					1	1		1				3									9
							1			4	5	2		3									16
			1	1		5	1							1						2		4	19
				35		8	12			7	1	50		10									149
4		9				33	25	2		28	1	3		40		1	1	10		254		7	476
		1		1		2	3	1		2		1		7	3							8	43
														6									13
		10		1		9	4			4		3		12				10		1			67
1		2				7				1		1											20
		5				4	2			3				4				4					29
7		1		2		24	10	4		2		3		18	2			4					119
		11								8	1						2	28	1				53
			1															5		1			8
		1	1			1	1	1		4	5	16		8	1			18		8		4	80
			2	2		1	3			9	2	6		6			1	2		1		2	46
1				27		2						44		1						118			222
1				6	4	10	1					1		2				5		3		15	104
2		7	4	43	26	27	3	1			3	39								13	3	19	344
						4	2							4									11
		1				2				2		1		1				1					11
		2				2	1				1			4		1		1				4	19
										2				2									5
4		6				7	5	4			1			9	1	1		10				5	58
						4								1									5
		10				3				16							1	37		1		6	85
							2							5				2					15
1		5				6	2	1		12		3		10	2			41		1			89
		1				1	2			1				1									6
		8				9	9	3		24	6	7		24				26		3		4	137
23	0	98	9	148	30	175	91	19	0	147	26	180	0	188	9	3	15	335	1	406	3	78	2486

Table 2

Localities of the Upper Maastrichtian Chalk-Tuff near Maastricht	Recognizable substrata (%) Settling upon Bryozoa	Settling upon other Fossils	Settling upon Seagrass-roots	Settling upon Algal-stalks	Settling upon Algal-fronds (narrow)	Settling upon Algal-fronds (broad)	Settling upon Dicotyledonous leaves
St Pietersberg near Maastricht	2·2	0·5	0	3·4	0·3	6·9	3·0
Curfs Quarry near Berg (Geul-Valley)	3·5	1·3	0	5·2	0·5	8·4	1·7
Albert-Canal near Néercanne	5·1	1·1	0	8·5	1·4	10·3	0
Schunk Quarry near Kunrade	0·5	0·2	0·8	10·1	0·1	22·7	0·2
Total	11·3	3·1	0·8	27·2	2·3	48·3	4·9

depositional conditions below effective wave base, but influenced by currents which built up the Bryozoan Beds.

The abundant presence of *Lithothamnia* within the Bryozoan Beds, combined with the indirect proof of a rich algal vegetation and numerous species of reef corals and orbitoid Foraminifera living on algae, reflects the euphotic nature of the region for the whole sequence of the Tuff-Chalk, which was deposited in clear and warm water rich in planktonic food and oxygen and controlled by a depth of perhaps 20–30 m. This conclusion may contradict the experience that in recent seas most erect rigid species, which are very common within the Tuff-Chalk, are found to prefer deeper habitats below 35 m (Schopf, 1969, p. 237).

Interrupted or possibly seasonal growth has been observed especially on the reverse of unilaminar fronds of *Onychocella piriformis* (Goldfuss) or *O. cyclostoma* (Goldfuss). The sharp lines of growth-stop are only visible on the reverse of zooecia (Plate I, Fig. 2) where the wall is thinner and often destroyed, their interval varies between one and four or more zooecia.

ACKNOWLEDGEMENTS

The author is indebted to the Deutsche Forschungsgemeinschaft, Bad Godesberg, for a financial grant and to Dr J. Ross, Bellingham (U.S.A.) for reading the English text.

REFERENCES

Goldfuss, G. A. (1826). *Petrefacta Germaniae* **1** (1), 1–76 Düsseldorf.

Schopf, T. J. M. (1969). Paleoecology of Ectoprocts (Bryozoans). *J. Paleontology* **43** (2) 234–244.

Stach, L. W. (1936). Correlation of zoarial form with habitat. *J. Geol.* **44,** 60–65.

Ubaghs, J. G. (1865). Die Bryozoenschichten der Maastrichter Kreidebildung nebst einigen neuen Bryozoen-Arten aus der Maastrichter Tuffkreide. *Naturhistor. Verein Preuß. Rheinl. u. Westf. Bonn* **22,** 31–62.

Voigt, E. (1956). Der Nachweis des Phytals durch Epizoen als Kriterium der Tiefe vorzeitlicher Meere. *Geol. Rundschau, Stuttgart* **45,** 97–119.

Voigt, E. (1959). Die ökologische Bedeutung der Hartgründe ("Hardgrounds") in der oberen Kreide. *Paläont. Z. Stuttgart* **33,** 129–147.

Voigt, E. (1966). Die Erhaltung vergänglicher Organismen durch Abformung infolge Inkrustation durch sessile Tiere, unter besonderer Berücksichtigung einiger Bryozoen und Hydrozoen aus der Oberen Kreide.—*Neues Jahrb. Geol. Paläont. Abh., Stuttgart* **125** (Festb. Schindewolf), 401–422.

Voigt, E. and Domke, W. (1955). *Thalassocharis bosqueti* Debey ex Miquel, ein strukturell erhaltenes Seegras aus der holländischen Kreide. *Mitt. Geol. Staatsinst. Hamburg,* **24,** 87–102.

von Hagenow, F. (1851). Die Bryozoen der Maastrichter Kreidebildung. pp. 1–111. Kassel.

18 | Relationships Between Some British Littoral and Sublittoral Bryozoans and Pycnogonids

D. W. WYER and P. E. KING

Department of Zoology, University College of Swansea, Swansea

Abstract: The relationships between the bryozoans and pycnogonids are of three types. First, the bryozoan provides a firm substrate on which the pycnogonid occasionally walks. Second, the epiphytes and epizoites on the bryozoan provide food for the pycnogonid, e.g. the young red alga shoots on *Flustrellidra* are eaten by *Achelia longipes*, and *Laomedea angulata* growing on *Flustra foliacea* provides a food source for *Nymphon rubrum*. Third, the bryozoan itself is eaten by the pycnogonid, e.g. *Achelia echinata* feeds on *F. foliacea* and *Nymphon gracile* feeds on *Bowerbankia imbricata*.

The pycnogonida are a group of marine arthropods often referred to as sea spiders because of their superficial resemblance to the Araneae.

The body of pycnogonids consists basically of a cephalothorax and an abdomen or anal appendage. The latter is invariably greatly reduced and in living species consists of a single segment. The anterior part of the cephalothorax has a proboscis and usually, on the dorsal surface, a tubercle bearing the eyes. Most of the pycnogonids move by slowly crawling over the substratum but some of the species with an elongate body and limbs can swim. Those with limited powers of locomotion have the largest claws and usually the stoutest legs. The claws are responsible for clinging onto the substrate and their relative size as compared to the other leg segments varies between species. Pycnogonids occur from the intertidal regions to depths as great as 6800 m. They are sometimes abundant but more often they occur in small numbers. Species and individuals occur in large numbers and grow to a larger size in cold waters, particularly the Antarctic. The main source of food for them seems to be hydroids, bryozoans and sponges, although when these are not available most species will take other types of food. The larvae may feed on the same species as the adults, or on different ones. Some live parasitically on galls produced on the host species but this phenomenon has not yet been observed with the bryozoa. The feeding preferences and the range of diets for each species of pycnogonid are very poorly known; most of the knowledge consists solely of observations of associations between pycnogonids and other animals in dredge or trawl samples. Previous associations between bryozoans and pycnogonids recorded are between *Crisia* spp. and *Phoxichilidium femoratum* (Prell, 1910), an unnamed bryozoan and *Anoplodactylus petiolatus* (Helfer, 1909), an unnamed bryozoan and *Endeis* (*Phoxichilus*) *spinosus* (Dohrn 1881, *Serialaria* spp. and *Ascorhynchus arenicola* (Dohrn, 1881), and *Cellarinella foveolata* and *Austrodecus glaciale* (Fry, 1965).

On the shores of Great Britain representatives of five pycnogonid families occur and each family has at least one species which is associated in some way with Bryozoa though the exact nature has not been determined in all examples.

Nymphon gracile and *Nymphon rubrum*, which belong to the Nymphonidae, have well developed chelifores and palps and have been associated with Bryozoa, *Nymphon gracile* with *Bowerbankia imbricata* and *Nymphon rubrum* with *Flustra foliacea*.

Anoplodactylus pygmaeus, of the family Phoxichilidiidae, has well developed chelifores but the palps are lacking. This species has been found on *Bowerbankia imbricata* and *Scrupocellaria reptans*.

In the family *Ammotheidae* the chelifores are greatly reduced in the adult, but the palps are well developed. *Achelia echinata* is associated with *Flustra foliacea*, *Bugula turbinata*, *Electra pilosa* and *Scrupocellaria reptans* while *Achelia longipes* is associated with *Flustrellidra hispida* and *Electra pilosa*.

Both the Endeidae and Pycnogonidae lack both chelifores and palps and both have been associated with *Flustra foliacea*.

The associations between Bryozoa and Pycnogonida may be of three types. First, the bryozoan provides a substrate on which the pycnogonid may walk; second, the bryozoan is eaten by the pycnogonid; third, the epithytes, epizoites and accumulated detritus on the bryozoan provide food for the pycnogonid. It is the last two, the feeding associations, which will be dealt with here.

At Mumbles Head on the Gower Peninsula in South Wales, colonies of *Bowerbankia imbricata* are found growing on the underside of stones close to the low water mark. Specimens of *Nymphon gracile* have been observed feeding on the zooids of these colonies. The zooecium of one of the individuals of the colony is grasped by means of the well developed chelifores and the wall ruptured so that pieces of the zooid can be pulled free. These pieces of food are held in front of the proboscis tip and shredded by the chelae prior to ingestion via the mouth. The specimens of *N. gracile* observed seemed to be voracious feeders and so the damage done to *Bowerbankia* colonies would be extensive if these were the only source of food for *Nymphon gracile*. However, *Actinia equina* and *Dynamena pumila* also form a major part of their diet.

Some specimens of *Pycnogonum littorale* collected amongst *Flustra foliacea* at 25–30 m off Oxwich Point, Gower, have been observed to feed on the decaying edge of some damaged fronds. The pycnogonid clings to the frond by means of its first pair of ambulatory legs and inserts the distal half of the proboscis into the damaged area so that pieces of tissue can be removed.

The samples of *Flustra foliacea* from Oxwich Point also yielded abundant specimens of *Achelia echinata*, a gallon brevit of *Flustra* yielding up to 200 specimens at certain times of the year. These pycnogonids feed on the living zooids of the colony, usually those in the distal half of the frond, though seldom those at the extreme tip or edge.

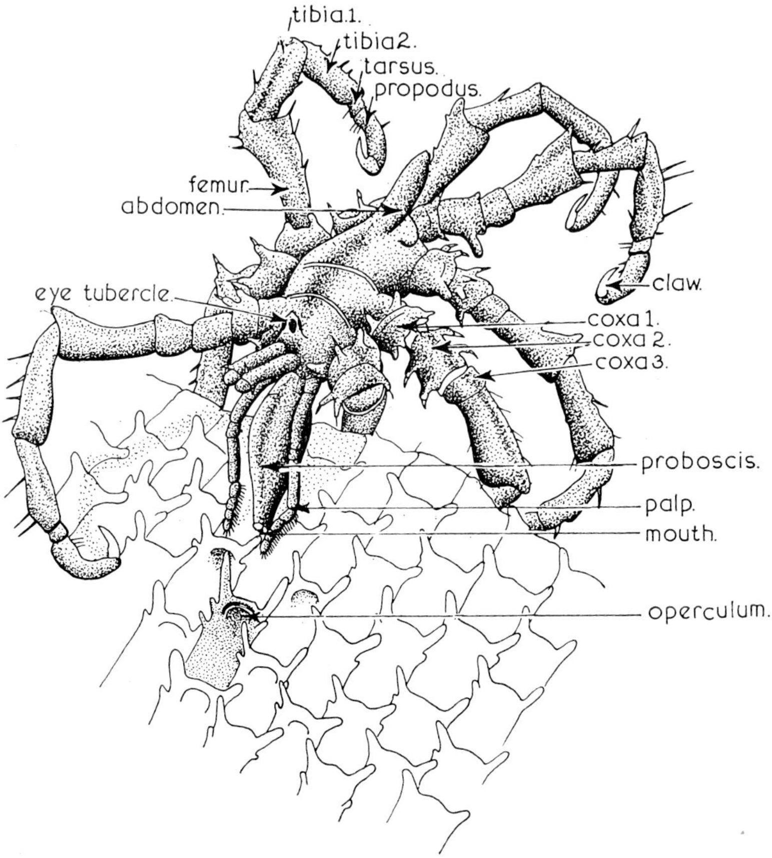

Fig. 1. Diagrammatic representation of *Achelia echinata* walking over *Flustra* showing the parts of a pycnogonid. Note the position of the proboscis and palps in relation to the spines and operculum of the proposed host zoecium (shaded).

The pycnogonid moves slowly over the surface of the *Flustra* colony, with its proboscis lowered towards the surface, and its palps held at the side of the proboscis in a characteristic "crooked" position, with the palp tip just in front of, and below, the proboscis tip (Fig. 1). The four distal segments of the palp are small and mobile and bear sensory bristles which are aggregated on one side. This mobile tip of each palp is bent into an S-shape so that the sensory bristles face the substrate on which the pycnogonid is walking.

The pycnogonid continues this movement until it encounters the large spines which occur at the opercular end of the *Flustra* autozooids. If during this time there are any sudden movements, as when an avicularium moves, the palps are moved rapidly sideways, thus raising them clear of any possible damage. If the spines curve away from the pycnogonid it may pause momentarily but soon passes on. If the spines curve towards it, which means that it is facing the side at which the operculum opens, it lowers the cephalon so that the proboscis is at an angle of about 60° to the autozooid and the proboscis tip is between the two large spines and thus in direct line with the operculum. If nothing further

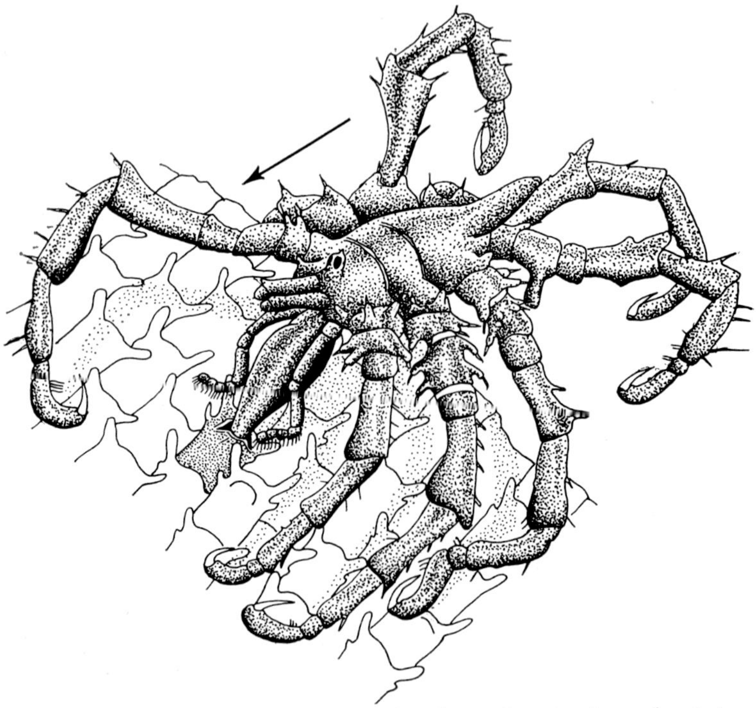

Fig. 2. Diagrammatic representation of *Achelia echinata* thrusting its proboscis into the opening operculum. The proboscis is held at 60° to the main axis of the body.

happens, if for instance the zooecium is empty, the pycnogonid will remain in this position for up to 30 min before moving off. When the operculum begins to open, the pycnogonid lunges forward so that the tip of the proboscis penetrates the orifice and enters the autozooid (Fig. 2). As the operculum is poorly calcified in this species and the proboscis is then completely filling the orifice, it is impossible for the proboscis to suffer damage once it has successfully gained entrance. The angle of attack of 60° is necessary since the operculum seldom opens to a vertical position. The entry into the autozooid must be made before the tentacles begin to emerge as they will block access to the orifice. Once access has been gained the pycnogonid moves forward so that the angle between the proboscis and zooid increases to the vertical. This results in the operculum being ruptured and the proboscis being pushed further into the autozooid, where it remains throughout the feeding (Fig. 3).

Small pieces of the autozooid are nipped off by the jaws which are situated at the tip of the finely tapering proboscis, and so able to function in a confined

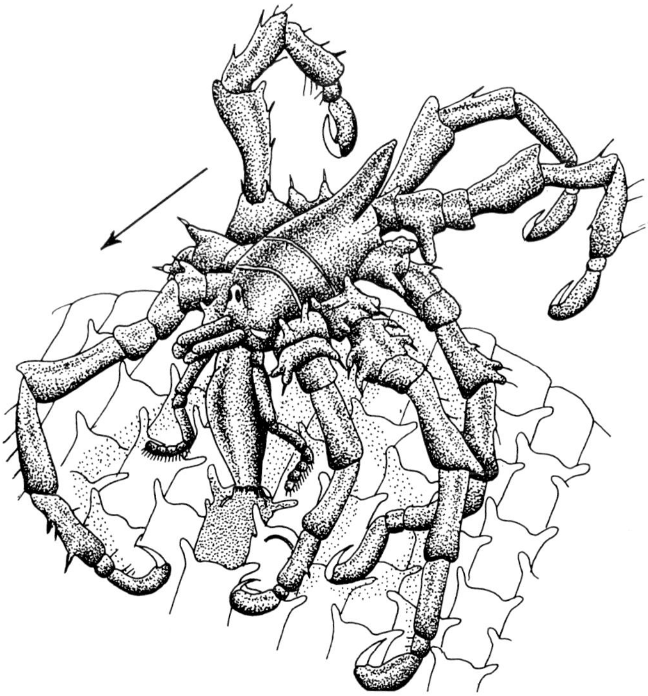

Fig. 3. Diagrammatic representation of the position taken up by the *Achelia echinata* when the body moves forwards so that the angle with the proboscis increases and the proboscis tip levers up the operculum.

space (Fig. 4). Feeding usually continues for about 10 min, after which the proboscis is withdrawn. The tip of the proboscis may then be cleaned by the grooming action of the ovigerous legs, the mobile distal segments of the ovigers being curled around the tip of the proboscis and moved distally so that they remove any remains of the food.

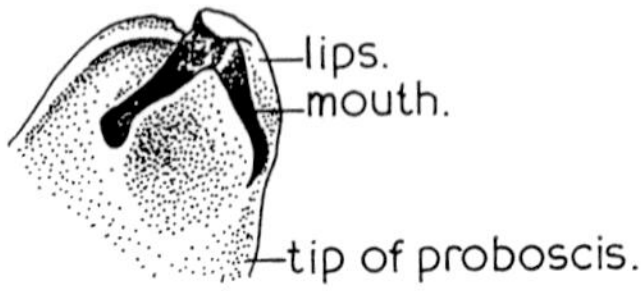

Fig. 4. Diagram of the tip of the proboscis of *Achelia echinata* showing the prominent pointed lips around the mouth which are capable of movement within the operculum of *Flustra* and can thus tear pieces off.

By no means all of the attacks are successful and there are a number of reasons for failure. If the pycnogonid aligns with a large spine on one autozooid and another large spine on a neighbouring autozooid it will not be in line with an operculum and so fail to penetrate. If the pycnogonid does not react quickly enough to the opening of the operculum this will close again and so prevent penetration by the proboscis. When an unsuccessful attempt has been made the pycnogonid usually moves to a different autozooid.

About a week after collection, the *Flustra* autozooids began to open less frequently and the reactions of the pycnogonid began to slow down. It was at this time that one individual made an unsuccessful attempt to penetrate the frontal wall of an autozooid with its proboscis, the proboscis tip rebounding, leaving the frontal wall undamaged.

Two separate lines of study were pursued as a follow up to these observations, the first being a short survey of the *Flustra* zooids, and the second a study of the muscular mechanisms involved in the action of the pycnogonid proboscis during feeding. The fronds of *Flustra foliacea*, dredged from Oxwich Point, bore both autozooids and avicularia, though the latter made up only one-half per cent of the population. The avicularia were commonly seen associated in groups of two or three though no two were adjacent. The dimensions of the autozooids varied according to the position in the frond, those at the sides being narrower. The autozooids of fronds on which pycnogonids had been seen to feed were measured and averages obtained. The operculum is crescent-shaped, the distance across the tips of the crescent being 0·168 mm; the distance between the large spines is 0·255 mm. Both of these compare favourably with the average diameter of the tip of the proboscis of *Achelia echinata*, which is 0·112 mm. Since the pycnogonid touches the larger spines when aligning with the operculum some experiments were carried out on the effect of contact on these spines. When they were touched with a sterile needle it did not cause the zooid to retreat, nor did it prevent a zooid from emerging. When this was repeated using various parts of an *Achelia echinata* the same results were obtained.

At Solva, in Pembrokeshire, there is abundant *Gigartina stellata* growing at the low water mark, and this provides an ideal situation for growths of *Flustrellidra hispida*. Specimens of *Achelia longipes* have been found on this, feeding on the small shoots of red alga which grow on the zooecia. The pycnogonids move over the bryozoan colony, making no effort to feed on the autozooids, even when they emerge in front of them.

The palps are held at the side of the proboscis in the "crooked" position as in *Achelia echinata*. In this species there are five small distal segments which are mobile, *Achelia longipes* having a total of nine palp segments, while *Achelia echinata* has eight. When the palps come into contact with the alga shoots, the proboscis is lowered towards the shoots and pieces are nipped off by the jaws.

The tip of the proboscis is less pointed than that of *Achelia echinata* and so better suited for grazing on the alga shoots. Parts of the mid-gut are reddish in colour, in contrast to that of *Achelia echinata* feeding on *Flustra foliacea*, where it is colourless.

The *Flustra* which is dredged from Oxwich Point has many epizoites growing on it, including *Laomedea angulata*, a hydroid with characteristic tendril-like prolongations of the stem, which is here fed on by *Nymphon rubrum*. Specimens under observation were seen to grasp one of the prolongations of the stem in their chelifores and tear a hole in them. The chelifores were used alternately until a large hole had been torn in the side, at which time the proboscis was placed against the stem and part of the contents removed. After feeding the chelifores were cleaned by the ovigerous legs. The same *Flustra* also has much detritus trapped among the older regions of the colony and this provides a food source for some of the *Pycnogonum littorale* found on the fronds. The proboscis is inserted in the masses of detritus and pieces of it are taken in. *Endeis laevis* is also occasionally found on the *Flustra* though it is more common on the *Antennularia* which grows nearby and it, too, feeds on the detritus, breaking up the masses of detritus with the spines on the tip of the proboscis prior to intake.

A collection of *Bugula turbinata* was made at Tenby, Pembrokeshire, and found to contain many specimens of *Achelia echinata* which were observed feeding on the abundant detritus on the lower regions of the colony. The avicularia in this region were inactive, though those in the distal part of the colony were very active. The detritus was especially abundant around the inactive avicularia, which formed areas where it could aggregate. The *Achelia echinata* walked over the surface of the lower parts of the colony, its palps at the side of the lowered proboscis, in the "crooked" position observed in this and other *Achelia* species when feeding. When it encountered one of the inactive avicularia the pycnogonid paused, with the palps held just above the colony. After remaining in this position for several minutes it pulled back the palps so that the ends were level with the proboscis tip, and then pushed against the avicularium with these three structures. It rapidly pulled back but the avicularium made no movement, and so the process was repeated twice more with similar results. The pycnogonid then began to feed on the detritus around the avicularium. The jaws pulled off small pieces of the detritus, which coloured the pycnogonids gut with orange and green patches. The avicularia of this species are well developed and so experiments were carried out to test their effects on a specimen of *Achelia echinata.* A specimen was placed on the area of the colony which bore active avicularia. It moved over the surface in a manner similar to that observed previously, but each time it came into the vicinity of an avicularium the palps and proboscis tip were withdrawn before actual contact was made. During this time the terminal claw of one of the fourth pair of ambulatory legs became trapped in an avicularium, but it was released after several minutes. The gape of the avicularia is not sufficient to allow them to grasp any other part of the legs. When the proboscis tip and palps were forced into the vicinity of an avicularium they were grasped, but soon released, having suffered no visible damage, though some of the delicate spines on the palp tip may have been broken. Of the species of Bryozoa associated with pycnogonids in this area only *Bugula*, and to a less extent *Flustra*, have noticeable avicularia, and in neither case do they seem to be a defence against pycnogonids. In the case of *Bugula* the pycnogonid does not even feed on the bryozoan and if it keeps to the areas covered with detritus it is not likely to come into contact with avicularia. If the pycnogonid should come into contact with one of the numerous active avicularia then its delicate feeding apparatus is protected by the withdrawal reaction and the rest of the body is not liable to damage by virtue of its size and solid construction.

Flustra foliacea is fed upon by the pycnogonid *Achelia echinata*, but even here the avicularia afford no protection as they are too small and make up a very low proportion (0·5%) of the population. Experiments carried out forcing parts of the pycnogonids into the avicularia showed them incapable of damaging *Achelia echinata*.

ACKNOWLEDGEMENTS

We wish to thank Professor E. W. Knight-Jones, in whose department the work was carried out, Dr Ryland, who collected most of the samples of bryozoa and the N.E.R.C. who financed one of us (DW).

REFERENCES

DOHRN, A. (1881). Die Pantopoden des Golfes von Neapel und der Angrenzenden Meeresabschnitte. *Fauna und Flora des Golfes von Neapel und der Angrenzenden Meeresabschnitte* **3,** 1–252.

FRY, W. (1965). The Feeding Mechanisms and Preferred Foods of Three Species of Pycnogonida "Bulletin of The British Museum (Natural History) Zoology", Vol. 12, (6).

HELFER, H. (1909). Biologisch-faunistiche Beobachtungen an Pantopoden der Nordund Ostsee. *Inaugural Dissertation, Kiel.*

PRELL, H. (1910). Beiträge zur Kenntnis der Lebensweise einiger Pantopoden. *Bergens Mus. Aarb.* **10** (3), 3–30.

19 | The Significance of Areolae in Cheilostome Bryozoa

WILLIAM C. BANTA

Department of Biology, The American University, Washington D.C., U.S.A.

Abstract: A study of 85 cheilostome ascophorans in 18 families supports the hypothesis that areolae (marginal perforations of the frontal wall provided with intrazooidal communication organs) are necessary for support of a hypostegal coelom above the frontal wall. Secondary thickening of the frontal wall and frontal budding appear to be impossible without a hypostega. Only three genera in the sample group definitely lacked a hypostega. The study also supports the hypothesis that areolae are necessary for development and maintenance of adventitious avicularia.

With few exceptions, frontal walls of ascophoran Bryozoa are provided with "pores". Early bryozoan workers realized that there are several different pore types and that they are useful in taxonomy. Unfortunately, the terminology of pores developed before the morphological differences were understood. This resulted in a terminology of pores that is inconsistent, unworkable, and that does not reflect the morphological and physiological differences involved. This is unfortunate because there is evidence that the kinds and distributions of pore types among ascophoran families may be a key to understanding the evolution of the group.

Disregarding pores associated with the operation of the hydrostatic system (Ryland, 1967), there are at least three types of pores found on frontal walls of ascophorans: (1) "pseudopores", uncalcified places in the wall occluded by a "wad" (Tavener-Smith and Williams, 1971) or "skeletal membrane" (Banta, 1971), (2) "windows", uncalcified spots in an otherwise calcified cuticle, and (3) areolae, marginal perforations traversed by cells of communication organs (septulae) (Banta, 1968, 1970, 1971). These pore types are illustrated in Fig. 1. Of these pore types, areolae seem to be of the greatest evolutionary significance because of their association with communication organs. This paper will consider the functional significance of areolae, with emphasis on secondary structures associated with them.

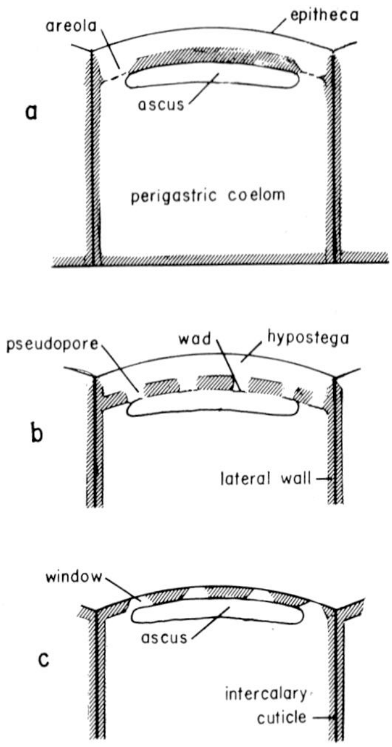

Fig. 1. Schematic drawings of transverse sections of ascophorans showing three different types of pores in the frontal wall. Shaded portions represent calcium carbonate.

Four species have been found to possess "true" (septular) areolae: *Watersipora arcuata*, *Cryptosula pallasiana* (Banta, 1970), *Schizoporella unicornis floridana* (Banta, 1971), and *Parasmittina nitida* (Banta, unpublished). These species are representatives of four widely different families. This suggests that areolae may be widely distributed among a sizable proportion of ascophorans.

Areolae connect adjacent coelomic cavities by communication organs. I have already suggested (Banta, 1971) that areolae are necessary for support of the hypostegal coelom. This is because septulae have been convincingly implicated in the transport of nutrient materials from one zoid to another (Bobin, 1964) and because the hypostega is otherwise isolated from any obvious source of nutrients (Fig. 1). It would seem, therefore, that septulae or some suitable alternative to septulae must be present in all ascophorans which possess a hypostegal coelom. Since there does not seem to be any obvious alternative to septulae, it is perhaps not premature to suggest the hypothesis that all ascophorans with a hypostega must possess areolae.

It is unfortunate that it is sometimes relatively difficult to determine which ascophorans possess areolae. Recognition of septulae requires good sections of material fixed well enough to recognize special cells traversing communication pores. Fixation is frequently a problem because of slow penetration of fixing agents through the body wall. Worse, only a small proportion of the world's museums contain significant quantities of "wet" material; bryozoan workers have relied largely on dried, cleaned skeletons (Hyman, 1959).

There are, however, other criteria by which septulae may be inferred to be present. Paraffin sections of alcohol or formalin-fixed material indicate relationships of cell layers even if the tissues are not fixed well enough to recognize special cells. One may state, for example, that if cell layers of different coelomic compartments appear to concentrate near bases of open pores, that the presence of a communication organ (septula) is suspected. The same inferences may be made from decalcified, stained whole mounts. These relatively direct methods of inferring the presence or absence of septulae are unfortunately impossible in dried or fossil material. Are there ways to infer the presence of septulae without directly observing tissues? Before answering this question, it is necessary to consider the potential functions and morphologies of areolae.

In almost all ascophorans, the ascus underlies the entire frontal wall. In all ascophorans studied, there is no cell layer between the ascus roof and the frontal wall. Areolae, therefore, must pass around the ascus. This accounts for the fact that areolae are necessarily marginal in location. A few species (for example, *Conescharellina*) possess asci located vertically in the cystid (Harmer, 1957, p. 722). In these species, one might expect that a cell layer may eventually be discovered between the ascus and the skeletal wall. Such forms may turn out to be exceptions to the rule that areolae are always marginal, but it is difficult to determine the position of the "frontal" wall in these species, so it is hard to decide if they really are exceptions (see Harmer. 1957, p. 722). For the vast majority of cases, however, we may reasonably presume that areolae are marginal in location.

If areolae are necessary for support of a hypostegal coelom and if areolae are marginal in position, then one would expect species lacking marginal pores to also lack a hypostega. Before this test can be applied, it is necessary to consider the criteria by which a hypostega can be recognized.

The first method is to observe it directly. This can be done in sections or by simple inspection of the colony in frontal view. The hypostega is recognized by the presence of a tenuous cuticular membrane (the epitheca) overlying the frontal wall. This method is useful only with relatively well preserved specimens. Roughly-handled material, especially dry material, may lose the delicate epitheca. In at least one species, *Schizoporella unicornis floridana* Osburn, living autozoids with apparently healthy polypides may occasionally lose the epitheca.

Areolae are presumably sealed off in these cases (S. Cummings, personal communication). Furthermore, the method of direct inspection can not be applied to fossil species, in which cuticular parts are invariably lost (but see Boardman, 1971).

The second method of recognizing the presence of a hypostega stems from the well recognized inferrence that secondary calcification can occur only if a hypostega is present (Harmer, 1902, p. 297; Cheetham, 1971, p. 13 and others). If this inference is correct, one needs only to look for ontogenetic thickening on the outside of the frontal wall to infer the presence of a hypostega. This generalization, however, should be used with care. Apparent "secondary thickenings" may sometimes occur by overgrowths of other organisms, particularly corraline algae. Furthermore, some species with well developed hypostegas may not undergo secondary calcification; one example is *Watersipora arcuata*. Finally, at least one family, Sertellidae, is known to produce kenozoids on the basal (reverse) side of the colony. It is not inconceivable that a similar production of kenozoids may be discovered to occur on the frontal walls of some species. Nevertheless, if used carefully, the observation of secondary calcification of the frontal wall may be taken to mean that a hypostega is present at at least some stage of development. The absence of secondary thickening, however, does not demonstrate the absence of a hypostega.

A third criterion for recognizing the presence of a hypostega is the existence of frontal budding. Harmer (1902, p. 297) was the first to suggest that frontal budding occurs by swelling of the hypostega and reinforcement of the walls of the cavity to form a new zoid above its parent zoid. This theory has been confirmed in one species, *Schizoporella unicornis floridana* Osburn (Banta, 1972). The vertical walls of frontally-budded autozoids are formed by reinforcement of an extension of the intercalary cuticle. It would seem logical to presume, therefore, that if vertical walls of frontally budded zoids originate from vertical walls of underlying zoids, that frontal budding is of the same type as in *Schizoporella*, and that a hypostega is present. This qualification of the rule is necessary because some cheilostomes (e.g., *Antropora tincta* Hastings) form multilaminar colonies by repeated overgrowth of pre-existing primogenial zoid layers. These species may be distinguished from frontally budding species by the presence of a basal wall beneath each zoid; vertical walls arise from this basal wall rather than from vertical walls of underlying zoids. *Hippothoa* may sometimes form multilaminar colonies, but in these cases new zoids are formed as evaginations from normally-placed (primogenial) budding sites. A basal wall is therefore present in each zoid, and vertical walls arise from it. Frontal budding, therefore, implies the presence of a hypostegal coelom above the frontal wall.

With these ideas in mind, I examined 85 species of ascophorans from 18 families (Table 1). I chose these species from my own collections which were conveniently mounted on slides and were reasonably accurately identified. Beyond this, no selection was made. For the most part, specimens were collected in California, Baja California, and Washington State. There seems to be no reason to doubt that the sample is representative of living ascophorans in general.

Table 1. Occurrence of marginal pores, hypostegal coeloms and pore-associated avicularia in a representative sample of ascophoran cheilostomes. Column A: –, marginal pores absent; +, marginal pores present. Column B: –, no evidence for a hypostega; 1, hypostega observed directly; 2, secondary thickening of the frontal wall occurs; 3, frontal budding occurs. Column C: 0, no adventitious avicularia observed; +, adventitious avicularia associated with a marginal pore; –, adventitious avicularia not observed to be associated with a marginal pore. For details, see text

	A Marginal pores	B Hypostega	C Avicularia
FAMILY HIPPOTHOIDAE			
Hippothoa cornuta (Busk, 1854)	–	–	0
Hippothoa distans MacGillivary, 1869	–	–	0
Hippothoa, undescribed species	–	–	0
Trypostega claviculata (Hincks, 1884)	+	2?	0
Trypostega venusta (Norman, 1864)	+	2?	0
FAMILY SCHIZOPORELLIDAE			
Schizoporella unicornis (Johnston, 1847)	+	1, 2	+
Schizoporella unicornis floridana Osburn, 1938	+	1, 2, 3	+
Schizoporella linearis inarmata (Hincks, 1884)	+	2	0
Emballotheca latifrons Osburn, 1952	+	1	+
Emballotheca obscura Osburn, 1952	+	1	+
Arthropoma circinata (MacGillivary, 1862)	+	2	+
Arthropoma cecili (Audouin, 1826)	+	2?	0
Stylopoma informata (Lonsdale, 1845)	+	1, 2	+
Schizomauvella auriculata (Hassall, 1842)	+	2?	+
Dakaria sertata Canu and Bassler, 1930	+	1, 2, 3	+
Dakaria apertura Osburn, 1952	+	1, 3	0
Hippodiplosia insculpta (Hincks, 1882)	+	1	+
Hippodiplosia sp.	+	1	+
FAMILY MICROPORELLIDAE			
Microporella ciliata (Pallas)	+	1, 2	+
Microporella cribosa Osburn, 1952	+	1	+
Microporella vibraculifera (Hincks, 1884)	+	1?	+
Fenestrulina malusi (Audouin, 1826)	+	1	0

(Table 1, continued)

	A Marginal pores	B Hypostega	C Avicularia
FAMILY HIPPOPORINIDAE			
Hippoporina porcellana (Busk, 1860)	+	1	+
?*Aimulosia uvulifera* Osburn, 1914	+	2	+
Hippoporella gorgonensis Hastings, 1930	+	2, 3	+
Hippoporella nitescens (Hincks, 1884)	+	1, 2, 3	+
Hippomonavella longirostrata (Hincks, 1883)	+	1	+
Hippomonavella parvicapitata (Canu and Bassler, 1930)	+	1, 2	–
Hippothyris emplastra Osburn, 1952	+	2?	–
Hippoporidra janthina (Smitt, 1873)	+	2, 3	+
Stephanosella vitrea Osburn, 1952	+	2	+
Stephanosella bolinae Osburn, 1952	+	2	+
Lacerna fistula O'Donoghue, 1923	+	2	–
Lacerna gabriellae Marcus, 1955	+	2	±
Hippopodinella adpressa (Busk, 1864)	+	2, 3	0
Hippalosina rostigera (Smitt, 1873)	+	2	+
Enantiosula manica Canu and Bassler, 1930	+	2	+
FAMILY CHEILOPORINIDAE			
Cheiloporina praelonga (Hincks, 1883)	+	1	0
Anexocona ancorata Osburn, 1950	+	1, 2	0
Cryptosula pallasiana (Moll, 1803)	+	1, 2	–
Veleroa veleronsis Osburn, 1952	+	1, 2	0
FAMILY WATERSIPORIDAE			
Watersipora subtorquata (Busk, 1856)	+	1	0
Watersipora arcuata Banta, 1969	+	1	0
Uscia mexicana Banta, 1969	+	1	0
Pachycleithonia nigra (Canu and Bassler, 1930)	+	1, 2	0
FAMILY SMITTINIDAE			
Smittina spathulifera (Hincks, 1884)	+	1	–
Smittina smittiella Osburn, 1947	+	2	±
Smittina landsborovi (Johnston, 1847)	+	1, 2	–
Smittoidea prolifica Osburn, 1952	+	2	–
Rhamphostomella curvirostrata (O'Donoghue, 1923)	+	1	+
Porella porifera (Hincks, 1884)	+	2	–
Porella patens Osburn	+	2?	–
Mucronella major (Hincks, 1884)	+	2?	0
Parasmittina nitida (Verrill, 1875)	+	1, 2, 3	+
Parasmittina ?*trispinosa* (Johnston, 1825)	+	1, 2, 3	+
Parasmittina californica (Robertson, 1908)	+	1, 2, 3	+
Parasmittina collifera (Robertson, 1908)	+	1, 2, 3	+
Parasmittina crosslandi (Hastings, 1930)	+	1, 2, 3	+

(Table 1, continued)

	A Marginal pores	B Hypostega	C Avicularia
FAMILY SERTELLIDAE			
Rhynchozoon rostratum (Busk, 1856)	+	1, 2, 3	+
Rhynchozoon grandicella (Canu and Bassler, 1930)	+	1, 2, 3	+
Phidolopora labiata (Gabb and Horn, 1962)	+	1, 2	+
FAMILY LAGENOPORIDAE			
Lagenipora punctata (Gabb and Gorn, 1862)	+	1, 2	+
Lagenipora hippocrepis (Busk, 1856)	+	1, 2	+
Lagenipora mexicana (Osburn, 1953)	+	2	+
Lagenipora socialis (Hincks, 1877)	+	3	+
Lagenipora spinulosa (Hincks, 1883)	+	2	+
FAMILY CELLEPORINIDAE			
Celleporina robertsonae Canu and Bassler, 1923	+	2, 3	+
Celleporina ventricosa (Lorenz, 1886)	+	2, 3	+
Celleporina surcularis (Packard, 1863)	+	2, 3	+
Celleporina costazi (Audouin, 1826)	+	2, 3	+
Celleporina procumbens Osburn, 1952	+	2	+
Celleporaria brunnea (Hincks, 1884)	+	3	+
Trematooecia hexagonalis (Canu and Bassler, 1923)	+	2?	+
FAMILY STOMOCHAETOSELLIDAE			
Pachyegis brunnea (Hincks, 1889)	+	1, 2	–
FAMILY UMBONULIDAE			
Umbonula arctica (Sars, 1851)	+	2?	+
FAMILY HIANTOPORIDAE			
Tremogasterina robusta (Hincks, 1884)	+	1	0
FAMILY PETRALIIDAE			
?*Hippopodina californica* Osburn, 1952	+	2	0
Cycloperiella rosacea Osburn, 1947	+	1	0
Coleopora gigantea (Canu and Bassler, 1923)	+	1	0
Robertsonidra ologopus (Robertson, 1908)	+	1	+
FAMILY EURYSTOMELLIDAE			
Eurystomella bilabiata (Hincks, 1884)	–	–	–
FAMILY FIGULARIIDAE			
?*Figularia hilli* Osburn, 1950	–	–	–
FAMILY CREPIDACANTHIDAE			
Crepidacantha poissoni (Audouin, 1826)	+	2	+
Crepidacantha setigera	+ ?	– ?	+ ?
FAMILY MYRIOZOIDAE			
Myriozoum coarcticum (Sars, 1850)	+	1, 2	+

Of the 85 species examined, 79 were found to possess marginal pores, 5 lack them and one is doubtful. 69 species could be demonstrated to possess a hypostegal coelom, 5 species lack a hypostega, and 11 are doubtful. All species possessing a hypostega possess marginal pores (Table 1). The 5 species without marginal pores, *Hippothoa cornuta*, *H. distans*, *H.* sp., *Eurystomella bilabiata*, and ?*Figularia hilli*, are extremely interesting, and deserve closer attention.

Hippothoa is a genus well known to possess a totally imperforate frontal wall. I have examined sections of *H. cornuta* from California and find the frontal wall to be formed by simple reinforcement of the outermost body wall cuticle (Plate I, Fig. 4). Details of this work are being published elsewhere. The other two species of *Hippothoa* were judged also to lack a hypostega because of the absence of evidence for one (see above). A similar situation obtains in *Eurystomella bilabiata*; I find no trace of a membrane overlying the imperforate frontal wall. ?*Figularia hilli* is a species in which marginal pores are often present. Many authors (Osburn, 1950; Bassler, 1953, and others) place *Figularia* among ascophorans, but if we define an ascophoran as an animal with a hydrostatic sac beneath a frontal wall not perforated by acunae, then ?*F. hilli* is an ascophoran (see Banta, 1970). I have examined methacrylate sections of ?*F. hilli* from California and find no evidence of a hypostegal coelom in this animal, although there are lumina of spines within the frontal wall. The pores are windows (Fig. 1c).

To summarize, the available evidence indicates that both marginal pores and hypostegal coeloms occur together, and that if one is lacking, so is the other. This finding supports the hypothesis that areolae are necessary for the possession of a hypostegal coelom, but it is clear that more research is necessary.

The question may now be asked, are areolae associated with structures other than the hypostegal coelom? One possibility is found among adventitious avicularia. Adventitious avicularia are isolated chambers cut off from the body

---

Plate I

Fig. 1. *Eurystomella bilabiata* (Hincks). Note the absence of pores; scale, 500 μ.

Fig. 2. *Hippothoa cornuta* (Busk). Pores of the ovicell are windows. Note the tubular shape of the autozoids and the corrugations, presumably growth lines; scale, 500 μ.

Fig. 3. ?*Figularia hilli* Osburn. The pores are windows. The transverse lines at *z* are interpreted as zones of interrupted growth and not secondary calcification; scale, 500 μ.

Fig. 4. *Hippothoa cornuta* (Busk), longitudinal epoxy section of non-decalcified material taken through the aperture of an autozoid. Note the absence of a hypostegal coelom. Toluidine blue 0, plane polarized light; scale, 100 μ.

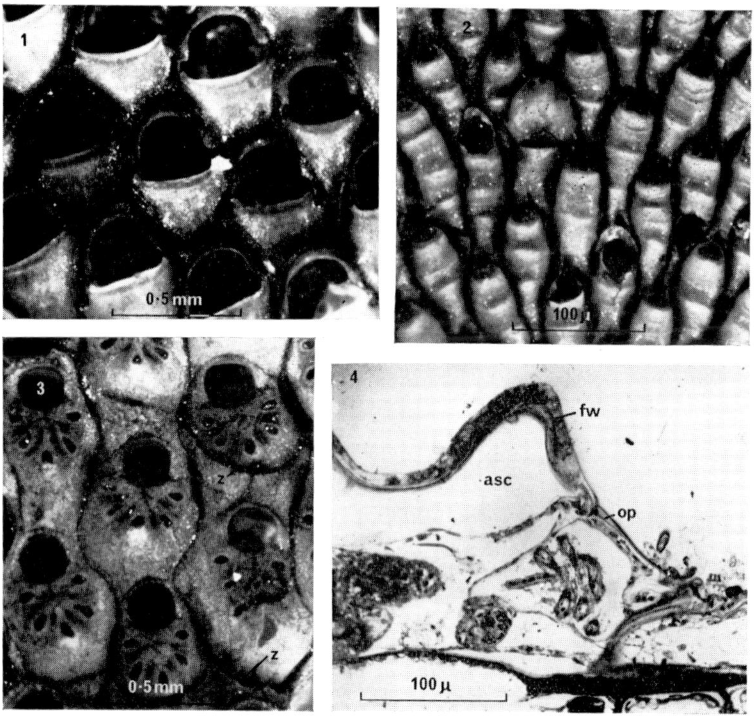
1
0·5 mm
2
100 µ
3
z
z
0·5 mm
4
fw
asc
op
100 µ

cavity of the parent autozoid, except through pores. It seems obvious to speculate that whatever materials are necessary for maintenance of the avicularium must be transmitted through these pores. I postulate, therefore, that all adventitious avicularia possess at least one septular pore; that is, one pore provided with a communication organ. Among ascophorans, the most likely candidates for such septular pores are the areolae.

The same group of ascophorans was examined for adventitious avicularia. Of the 85 species, 64 possess adventitious avicularia. In 48 species, avicularia were seen clearly to be connected to the autozoid through a marginal pore. In 14 cases no connection to marginal pores could be demonstrated; in two cases the connection could be seen only rarely. In most cases failure to demonstrate a connection between avicularia and marginal pores can be explained by the opacity of the skeleton. For example, *Hippomonavella longirostrata* has avicularia placed almost medially behind the aperture. A minute tubule can be seen passing through the transparent skeleton between the avicularium chamber and a marginal pore. If the skeleton were opaque, this minute tubule would be invisible. In most cases where a connection cannot be demonstrated, the opaque skeleton would obscure such a tubule if one were present. It would be tempting to conclude, therefore, that such connections always occur. But in some species, *Hippomonavella parvicapitata* for example, no such connection can be seen, even though the skeleton is transparent enough to allow it. The apparent exceptions to the rule that avicularia connect to marginal pores obviously need closer attention. If the apparent exceptions turn out to be real exceptions, there seem to be three obvious possibilities. (1) Perhaps avicularia can form from septular connections with the coelom of the autozoid without going all the way to the margin. That is, perhaps the ascus is small enough to allow septular connections closer to the center of the zoid that at the very margin. (2) Perhaps septular connections can occur between the avicularium and the hypostegal coelom of the autozoid, rather than with its perigastric coelom. (3) Perhaps the hypothesis of the necessity of septular connections is incorrect, and we need to look to other nutrient sources for the avicularium. It is clear that more research is needed on the functional significance of frontal perforations of ascophoran frontal walls.

ACKNOWLEDGEMENTS

I thank Dr Alan H. Cheetham for helpful discussions and suggestions. I also thank the following persons for technical assistance: Susan Cummings, Penny Pinter, Anna B. Hastings, Phillip Sandberg, Dennis McFaden, Patricia L. Cook and Barbara Gautier.

REFERENCES

Banta, W. C. (1968). The body wall of the encrusting cheilostome *Watersipora nigra* (Canu and Bassler) (Bryozoa) (Preliminary report), *Atti Soc. Ital. Sci. Nat.* **108,** 93–96.

Banta, W. C. (1970). The body wall of cheilostome Bryozoa, III. The frontal wall of *Watersipora arcuata* Banta, with a revision of the Cryptocystidea. *J. Morph.* **131,** 37–56.

Banta, W. C. (1971). The body wall of cheilostome Bryozoa, IV. The frontal wall of *Schizoporella unicornis* (Johnston). *J. Morph.* **132,** 165–184.

Banta, W. C. (1972). The body wall of cheilostome Bryozoa, V. Frontal budding in *Schizoporella unicornis floridana Osburn. Marine Biology*, **14** (1), 63–71.

Bassler, R. S. (1953). Bryozoa. Pt. G. *In* "Treatise on Invertebrate Paleontology", (Moore, R. C., ed.) Univ. Kansas Press.

Boardman, R. S. (1971). Mode of growth and functional morphology of autozooids in some Recent and Paleozoic tubular Bryozoa. *Smithsonian Contrib. Paleontol.* **8,** 51.

Bobin, G. (1964). Cytologie des rosettes de *Bowerbankia imbricata* (Adams) (Bryozoaire cténostome, vésicularine). Hypothèsis sur leur fonctionnement. *Arch. Zool. exp. gén.* **104,** 1–44.

Cheetham, A. H. (1971). Functional morphology and biofacies distribution of cheilostome Bryozoa in the Danian stage (Paleocene) of southern Scandinavia. *Smithsonian Contrib. Paleobiol.* **6,** 87.

Harmer, S. F. (1902). On the morphology of the cheilostomata. *Q. Jl. microsc. Sci.* **46,** 263–350.

Harmer, S. F. (1957). Polyzoa of the Siboga Expedition. Part IV. Cheilostomata Ascophora. II. (Ascophora, except Reteporidae, with additions to part II, Anasca). *Siboga-Expedite Monographie* **28d,** 641–1147.

Hyman, L. H. (1959). "The Invertebrates: Smaller Coelomate Groups." McGraw Hill, New York.

Osburn, R. S. (1950). Bryozoa of the Pacific coast of America. Part II, Cheilostomata Ascophora. *Allan Hancock Pac. Exped.* **14,** 271–611.

Ryland, J. S. (1967). Polyzoa. *Oceanogr. mar. Biol. Ann. Rev.* **5,** 343–369.

Ryland, J. S. (1970). "Bryozoans." 175 pp. Hutchinson Univ. Library, London.

Tavener-Smith, R. and Williams, A. (1970). Structure of the compensation sac in two ascophoran bryozoans. *Proc. R. Soc. Lond.* **175,** 235–254.

20 | Acanthopore Ultrastructure in the Palaeozoic Bryozoan Family Rhabdomesidae

D. BRYAN BLAKE

University of Illinois, Urbana, Illinois, U.S.A.

Abstract: Three morphological types of acanthopores have been found in members of the Palaeozoic bryozoan family Rhabdomesidae. Acanthopores are constructed of differentiated axial structures which are enclosed by laminae arranged in a cone-in-cone pattern and directed toward the zoarial surface. Acanthopore types are distinguished primarily on the basis of their axial structures.

Acanthopore function remains uncertain; however, because acanthopores occur in many genera through the Palaeozoic, they must have been important to the living colony. Studies with the scanning electron microscope suggest acanthopores of various types in diverse taxa of rhabdomesids were integral parts of the skeleton, rather than structures enclosing open passageways. At present acanthopores appear most readily explained as simple protective processes or as supports for soft tissues.

INTRODUCTION

The term "acanthopore" is here applied to elongate cylindrical skeletal structures which are constructed of some form of differentiated axis enclosed by a sheath-like bundle of laminae; these sheath laminae are deflected toward the zoarial surface. Most, if not all, acanthopores in rhabdomesids appear to arise rather abruptly within the skeletal wall, and form prominences at the zoarial surface. They appear to lack connection to the zooecia. Most, if not all, appear to have been closed structures without a continuous space unoccupied by skeletal material.

Although several categories of acanthopores have been recognized, the ranges of variation present within each category suggest that these types may prove to be only convenient end members of a continuum rather than truly distinctive entities. Categories are recognized in part for convenience in discussion.

Acanthopores have been studied by means of both the scanning electron microscope and the light microscope. New major features of acanthopores have not been revealed in scanning electron microscope studies, but the instrument has proved very useful in elucidating detail.

Specimens illustrated with the scanning electron microscope were embedded in Reinhold 37–127 epoxy resin and 37–622 hardener. The epoxy plugs were then cut to expose the desired surfaces; the surfaces were polished, etched in dilute HCl, and metal-coated with gold–palladium. The scanning electron microscope used in the study, a Cambridge Stereoscan Mk IIa, is located in the Central Electron Microscope Laboratory of the University of Illinois. It was acquired with support of the National Science Foundation (GA–1239), the U.S. Public Health Service (FR–07030) and the University of Illinois Research Board. Figured specimens are from collections of the Division of Invertebrate Palaeontology, National Museum of Natural History, Washington, D.C.

A-ACANTHOPORES

Acanthopores of the first category recognized in rhabdomesid Bryozoa, A-acanthopores, correspond most closely to the definition of Bassler (1953) and others: a "Cylindrical tube adjoining zooecial walls and parallel to them in growth, formed of cone-in-cone layers with narrow central tubule which may be crossed by minute diaphragms . . .".

The basic morphology of A-acanthopores, and some of the morphological variation, can be seen in Plate I, Figs 1–5 and Plate II, Figs 5–7. In a specimen of *Acanthoclema* sp. (Plate I, Fig. 1), the axial core (c) ("cylindrical tube" of the Bassler definition) is a continuous structure which is not crossed by apparent

---

Plate I

Figs 1–4. *Acanthoclema* sp., longitudinal sections of portions of A-acanthopores; Hamilton Grp., Dev., New York State; (1) approx. ×500; (2) approx. ×785; (3) USNM 168342, ×400; (4) approx. ×340.

Fig. 5. *Rhombopora crassa* Ulrich, longitudinal section of a portion of an A-acanthopore; Pennsylvanian, Misssouri; approx. ×845.

Figs 6, 8. *Saffordotaxis incrassata* (Ulrich), longitudinal sections of portions of B-acanthopores; New Province Sh., Miss., Kentucky; (6) USNM 168357, ×400; (8) approx. ×550.

Fig. 7. Undescribed Mississippian rhabdomesid, transverse section of B-acanthopores, specimen lost in preparation, Warsaw Fm., near Valmeyer, Ill., approx. ×425.

Figures 3 and 6 are light micrographs, others are scanning electron micrographs. Key to abbreviations: b, boundary between acanthopore core and sheath laminae; c, acanthopore core; d, deflections in sheath laminae; l, laminae which cross acanthopore core; la, sheath laminae around acanthopore core; ms, mural spine; p, area proximal to acanthopore core; s, sheath laminae; sp, spine in core of B-acanthopore; t, core terminus in plane of view; w, wedge-out of zoarial laminae against sheath laminae; z, zooecium; zl, zoarial laminae.

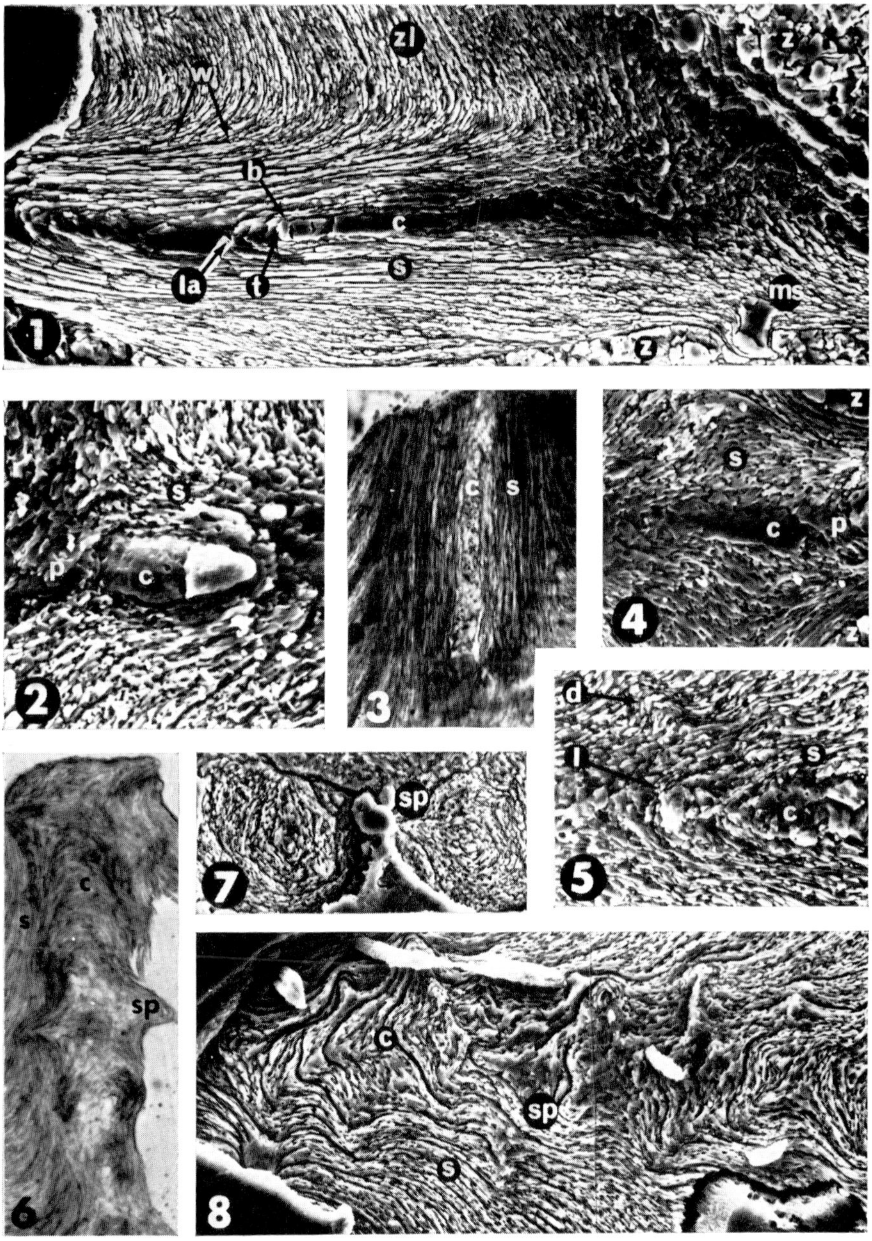
1
zl
z
w
b
c
s
la
t
ms
z
2
s
p
c
3
c
s
4
z
s
c
p
z
5
d
l
s
c
6
c
s
sp
7
sp
8
c
sp
s

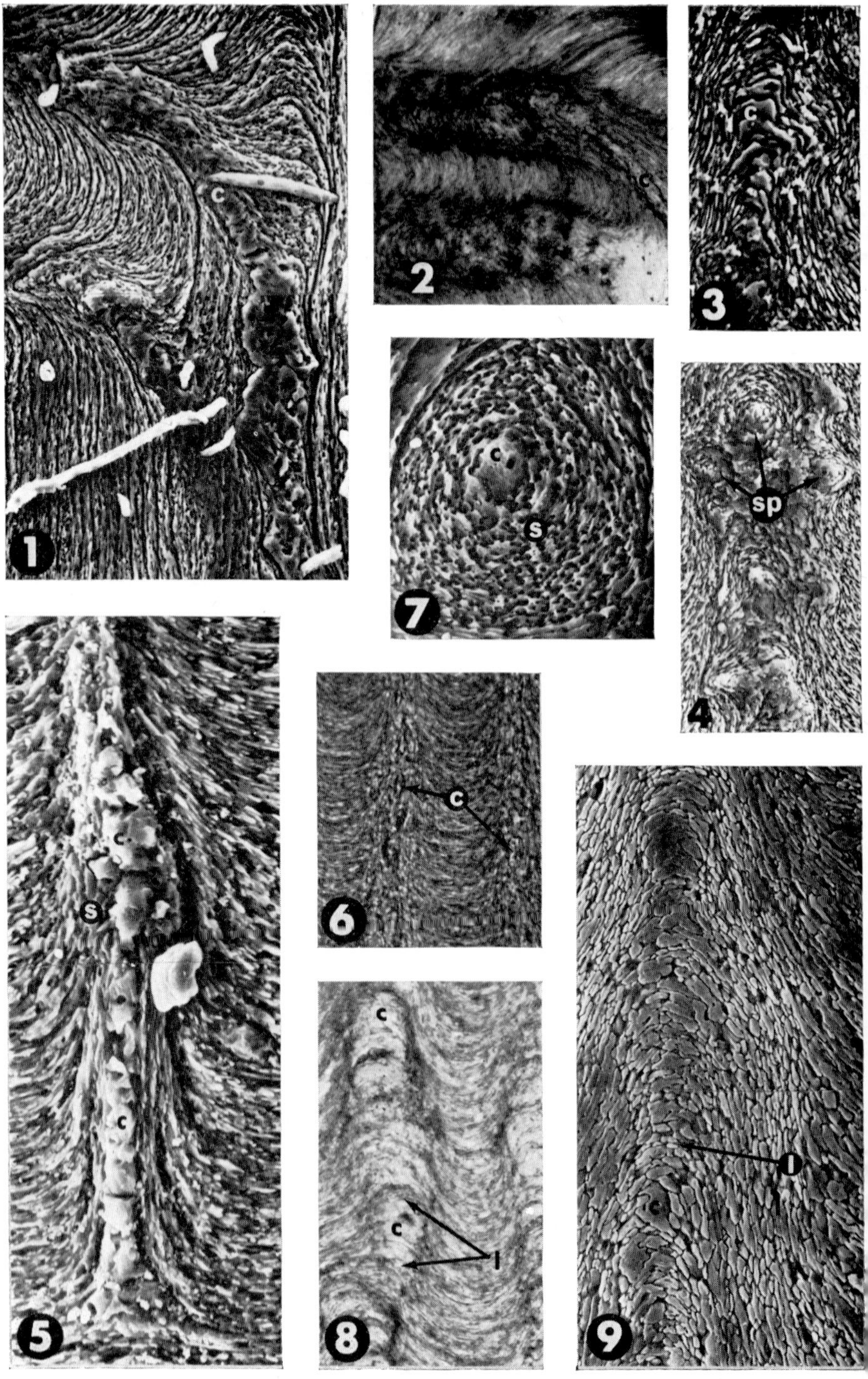
1
c
2
c
3
c
7
c
s
4
sp
5
c
s
c
6
c
8
c
c
l
9
l
c

laminae. The core is enclosed by a sheath of strongly outwardly arcing laminae (s). The boundary (b) between the sheath laminae and the core is well defined, if irregular. The core is oblique to the plane of view and passes above this plane to the left toward the zoarial surface. The terminus (t) of the core in the plane of view is rounded. The core was removed in preparation; the surrounding and underlying laminae (la) form a series of concave layers about the trace of the core.

The sheath laminae can be seen to be continuous with the zoarial laminae (zl). The deposition of zoarial laminae and sheath laminae did not proceed in a completely parallel pattern because some zoarial laminae can be seen to wedge out against the sheath laminae (w). A prominent but overgrown mural spine (ms) is also visible in this micrograph. Light micrographs (Plate I, Fig. 3) of acanthopores of this species also show the unlaminated core (c) and strongly deflected sheath laminae (s). A section perpendicular to the acanthopore axis in *Acanthoclema* sp. (Plate II, Fig. 7) shows the core (c) and concentric sheath laminae (s).

Variation in A-acanthopore morphology is illustrated in Plate 1, Fig. 5 and Plate II, Figs 5 and 6. Acanthopores of *Orthopora* (Plate II, Figs 5 and 6) are quite distinct from those of *Acanthoclema*; the sheath laminae (s) form a tighter bundle and the core (c) appears discontinuous. Core discontinuities in the acid-etched specimen probably are not artifacts of preparation because cores in *Acanthoclema* sp. specimens which were also etched do not show such solution and because cores appear discontinuous in certain unetched specimens of *Orthopora* sp. (Plate II, Fig. 6).

In acanthopores of *Rhombopora crassa* Ulrich (Plate I, Fig. 5), the core (c)

Plate II

Figs 1, 2, 4. *Saffordotaxis incrassata* Ulrich; New Providence Sh., Miss., Kentucky; (1) longitudinal section near proximal end of B-acanthopore, approx. ×405; (2) longitudinal section of B-acanthopores, USNM 168353, ×200; (4) tangential section of a portion of a B-acanthopore, approx. ×550.

Fig. 3. *Rhombopora crassa* Ulrich, longitudinal section of a portion of a B-acanthopore, Pennsylvanian, Missouri; approx. ×890.

Figs 5, 6. *Orthopora* sp., longitudinal sections of portions of A-acanthopores; Keyser Ls., Sil.-Dev., W. Va.; (5) approx. ×800; (6) USNM 168372, ×400.

Fig. 7. *Acanthoclema* sp., section transverse to acanthopore axis, Hamilton Grp., Dev. New York State; approx. ×975.

Fig. 8, 9. *Goldfussitrypa esthonia* (Bassler), longitudinal sections of portion of C-acanthopores; Jewe Beds, Ord., Esthonia; (8) USNM 168363, ×400; (9) USNM 178557, approx. ×960.

appears ill-defined compared to cores in *Acanthoclema* sp. and *Orthopora* sp. Some sheath laminae (s) can be seen to extend across the core in acanthopores of this species (at l). Minor deflections (d) are present in the sheath laminae. Sheath laminae are similar in the three species in that in all they arc strongly toward the zoarial surface.

Where studied, A-acanthopores appear to arise relatively abruptly within the skeleton. In Plate I, Fig. 2, the core of an acanthopore originates in the plane of the photograph because surrounding laminae do not overlap the base of the core. The core then almost immediately passes out of the plane of view. The changes in pattern of the laminae visible about the base of the core are a result of orientation changes in the laminae. This can be illustrated by means of Plate I, Fig. 4 which shows the core base (c) in a section oriented perpendicular to that of Plate I, Fig. 2. Laminae proximal to the core (p) in a plane through the core base (Plate I, Fig. 4) lie essentially perpendicular to the plane of view but parallel to the plane of view in the section of Plate I, Fig. 2. The sheath laminae are arranged in a cone-in-cone pattern about the core. In both orientations they therefore lie parallel to the plane of view along the trace of the core and perpendicular lateral to the core.

B-ACANTHOPORES

B-acanthopores are illustrated by means of light micrographs (Plate I, Fig. 6; Plate II, Fig. 2) and scanning electron micrographs (Plate I, Figs 7, 8; Plate II, Figs 1, 3, 4). In these acanthopores the core is formed of laminae which arc toward the zoarial surface and project laterally in the form of spines (sp). The sheath laminae (s) do not arc as strongly toward the zoarial surface as they do in A-acanthopores.

Breakage can be seen to have taken place along the right boundary between axial laminae and sheath laminae in a light micrograph (Plate I, Fig. 6) suggesting a discontinuity between the two. In the scanning electron micrograph (Plate I, Fig. 8), however, the axis of the acanthopore appears intergrown with the sheath laminae. In certain areas the enclosing laminae apparently can be traced into the axis of the B-acanthopore; in other areas, the core appears to be relatively sharply defined.

In tangential view (Plate II, Fig. 4), the prominent spines (sp) of the core appear as concentric laminae.

B-acanthopores frequently extend from the base of the exozone and may show, near their origin, a continuous single-crystal (or crystalline-aggregate?) core (Plate II, Figs 1, 2) similar to that found in *Acanthoclema*. In both micrographs, the large crystal terminates abruptly toward the zoarial surface.

The concentric appearance of the axial laminae and the deflections of the axial laminae to form the axial spines (sp) are visible in a view transverse to the axis of a B-acanthopore (Plate I, Fig. 7) in a species from an undescribed Mississippian genus.

In some B-acanthopores of *Rhombopora crassa*, well-developed larger crystals (c) occur along the axis and the demarcation between sheath laminae and zoarial laminae appears relatively weak (Plate II, Fig. 3).

C-ACANTHOPORES

C-acanthopores in the Ordovician species *Goldfussitrypa esthonia* Bassler are illustrated by means of a light micrograph (Plate II, Fig. 8) and a scanning electron micrograph (Plate II, Fig. 9). The core of C-acanthopores consists of alternating intervals of clear calcite (c) and laminae (l). These axial laminae are continuous with the sheath laminae. The sheath laminae are directed rather weakly toward the zoarial surface and are continuous with the zoarial laminae. Spines or deflections do not occur in either the cores or sheath laminae. These acanthopores also appear to arise within the skeleton near the base of the exozone.

ACANTHOPORE RELATIONSHIPS AND FUNCTION

Although B- and C-acanthopore morphologies are very different from the form described by the established definition of acanthopores, the A-acanthopore, the B and C types are considered acanthopores for a number of reasons. (1) The lack of reliable demarcation in structure between types, as expressed in a number of features. (a) Sheath laminae are similar in structure and deflection toward the zoarial surface. (b) No clear demarcation exists on the basis of the development of larger crystals (or crystalline aggregates) in the cores. Crystalline cores appear continuous in *Acanthoclema*, intermittent in some *Orthopora* acanthopores and crossed by laminae in *Rhombopora crassa* A-acanthopores. Many longer discrete crystals occur along the axis in *Rhombopora crassa* B-acanthopores and intermittent crystals occur in the axes of B-acanthopores in some species (not illustrated here). Larger crystals (or aggregates) are either lacking or very small along most of the length of acanthopores of *Saffordotaxis*, but are present in the base of some acanthopores of *Saffordotaxis*. (c) There is no demarcation in size among acanthopore categories. (2) In specimens of *Megacanthopora fallacis* Moore, not illustrated here, acanthopores occur which change type along their length (from B- to A-acanthopores). (3) Distribution and surface expression of the various types are closely comparable. Therefore the three categories recognized do appear interrelated and all are considered acanthopores, at least until further information becomes available.

Acanthopore function remains uncertain. Because they occur in most genera of the Rhabdomesidae, their basic morphology maintained through much of the Palaeozoic, acanthopores would seem to have performed an important function. In recent years, a number of workers have gathered evidence which suggests that acanthopores were solid rather than open structures (Tavener-Smith, 1969; Brood, 1970; Armstrong, 1970). Evidence derived from studies with the scanning electron microscope tends to support these authors. The axial crystals of C-acanthopores appear very much intergrown with the zoarial laminae; there is no apparent space for large open volumes. The axis of B-acanthopores consists of laminae with isolated larger crystals sometimes being present. A continuous open space therefore could not have existed except possibly near the base of a few individual B-acanthopores (Plate II, Figs 1, 2). Some A-acanthopore cores were crossed by laminae and therefore must have lacked large open spaces. Only in certain taxa is the core volume continuous and potentially available for occupation by soft tissues. Core diameters of most observed rhabdomesid acanthopores range between 3 and 15 μ; the very small diameter therefore would limit the amount of soft tissue possible. Connections between the acanthopore cores and zooecia do not appear to occur. Acanthopores therefore appear to be integral parts of the skeleton, and not passageways or cavities. Their functions therefore would appear to have been performed either at the surface of the colony or within the skeleton, perhaps in some manner of structural reinforcement. As has been suggested by various authors, acanthopores may have provided protection or support for soft tissues.

In spite of the difficulties encountered in the determination of acanthopore function, the considerable structural variation present among these structures is providing useful taxonomic and evolutionary information. For example, *Saffordotaxis incrassata* and *Rhombopora lepidodendroides* Meek are distinguished (Bassler, 1953) on the basis of "megacanthopore" and "microacanthopore" distribution, yet the microacanthopores of *R. lepidodendroides* and the megacanthopores of *Saffordotaxis incrassata* are B-acanthopores, the megacanthopores of *Rhombopora lepidodendroides* are A-acanthopores. Thus the differences between the two genera are more complex than has been realized. These observations suggest that further studies of skeletal structures with the scanning electron microscope should prove useful in the study of rhabdomesid palaeobiology.

REFERENCES

Armstrong, J. (1969). Zoarial microstructures of two Permian species of the Bryozoan genus *Stenopora*. *Palaeontology* **13**, 581–587.

Bassler, R. S. (1953). Bryozoa. *In* "Treatise on Invertebrate Paleontology." (Moore, R. C., ed.), p. G1–G253. Geol. Soc. Amer. and Kan. Univ. Press.

Brood, K. (1970). On two species of *Saffordotaxis* (Bryozoa) from the Silurian of Gotland, Sweden. *Stock. Contrib. Geol.* **21,** 57–68.

Tavener-Smith, R. (1969). Wall structure and acanthopores in the bryozoan *Leioclema asperum*. *Lethaia* **2,** 89–97.

21 Body Walls and Attachment Organs in Some Recent Cyclostomes and Palaeozoic Trepostomes

RICHARD S. BOARDMAN

National Museum of Natural History, Smithsonian Institution, Washington, D.C., U.S.A.

Abstract: The discovery on a Late Ordovician trepostome of a membrane interpreted to be analogous to the outer membrane of a cyclostome reinforces skeletal evidence of the double-walled organization of trepostomes. The discovery in feeding autozooids of some cyclostomes of perimetrical attachment organs rather than the reported eight radial ligaments, and the possibility of progressive calcification of membranes during later stages of ontogeny of a polypide provide additional bases for morphological comparisons with Palaeozoic trepostomes.

INTRODUCTION

Two recent discoveries of soft part morphology reported here provide advances in continuing efforts to understand the rudimentary biology of fossil and Recent stenolaemate Bryozoa (Boardman, 1971).

BODY WALLS

Laminae generally observable in skeletons of Palaeozoic trepostomes are interpreted as paralleling growth surfaces (Boardman and Cheetham, 1969, p. 210). Three-dimensional configurations of these surfaces suggest the uniformitarian inference that trepostome growth was organized similarly to double-walled growth (Borg, 1926, Fig. 55; 1933, Fig. 26) of some living cyclostomes. Borg's double walls are two separate body walls that together serve to enclose the body cavity of the colony and its member zooids (Fig. 1). The inner body wall includes the interior skeletal walls of the colony that are attached to the basal layer, their depositing epidermis, and peritoneum. The interior skeletal walls are generally compound, that is, grown from two sides, commonly from body cavities of contiguous zooids.

The second or outer body wall includes: (1) the basal layer of the colony which consists of an outermost cuticle and a contiguous exterior skeletal layer,

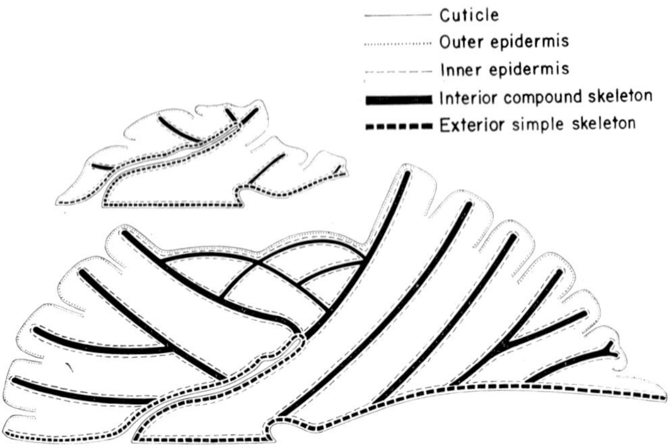

Fig. 1. Idealized diagrams of longitudinal section through center of young lichenoporid colony after Borg (1926, Fig. 55). The central zooid with flattened basal disc is the ancestrula. The exterior simple skeleton forms the basal layer of the colony. The smaller diagram is a hypothetical earlier growth stage with no attempt made to establish actual relative rates of growth of different zooids. It shows early growth of interior compound skeleton from exterior simple skeleton by an infolding of inner epidermis. The fold of exterior skeleton over the basal disc provides a basal skeletal layer for autozooids to left of ancestrula in diagrams. Cystose central region of larger colony depicts brood chambers.

Plate I

Heteroporid cyclostome. Recent specimen from Pacific coast of North America, British Museum (Natural History) 1963.2.6.1.

Fig. 1. Longitudinal hard-soft section, ×100, showing delicate cuticle of outer membranous wall across ends of autozooecia.

Dendroid trepostome. Waynesville Formation, Richmond Group (Upper Ordovician), Hanover, Ohio. USNM 179006.

Fig. 2a. Longitudinal section, ×100, showing brown deposits restricted to dendroid trepostome on right by reddish brown membrane. To left is skeletal basal layer and autozooecia of a monticuliporid overgrowth.

Fig. 2b. Longitudinal section from same zoarium, ×100, showing continuity of membrane across walls between autozooecial cavities.

Fig. 2c. Longitudinal section, ×200, showing brown deposit shaped like a polypide and apparently attached to outer membrane above and mural spine in autozooecial wall on right.

Fig. 2d. Longitudinal section, ×150, showing another polypide-shaped brown deposit apparently fastened to membrane above and mural spine in autozooecial wall on left.

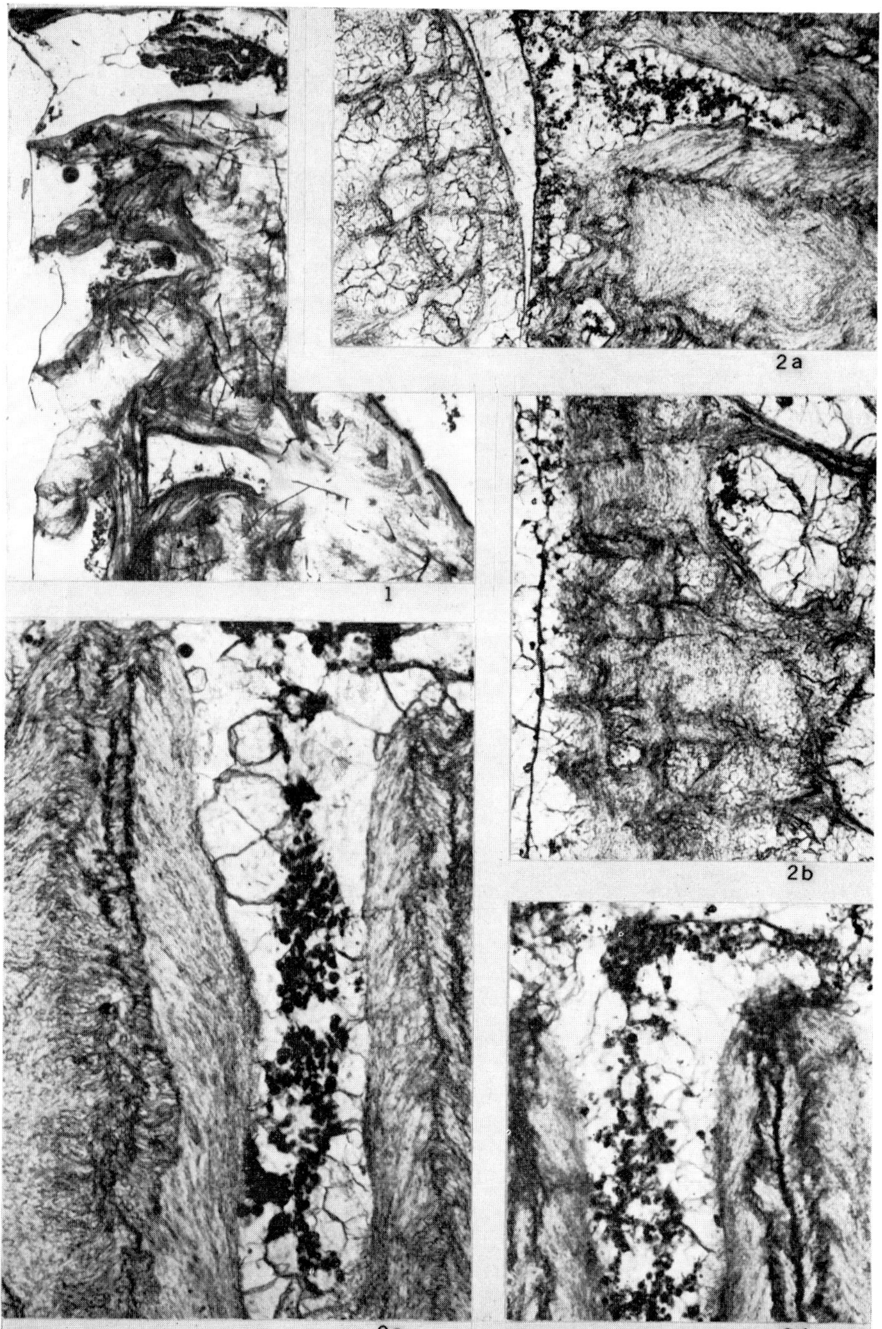
2a
1
2b
2c
2d

and (2) the outer membrane that has an outermost cuticle (Plate I, Fig. 1) and covers the living cyclostome colony everywhere above the basal layer. The exterior skeleton is simple, that is, grown from its inner surface only. The membranous outer wall forms over each zooid a terminal membrane which in the autozooids is connected directly to the lophophore, gut, and membranous sac (Fig. 2). The outer membrane is apparently attached to the skeleton of the colony within the zooids by various ligaments and muscles and around the rim of the basal layer. The sparseness of skeletal attachment of the outer membrane allows for possible interzooidal communication around the ends of zooecia. The exposed position on the outside of the colony commonly results in the

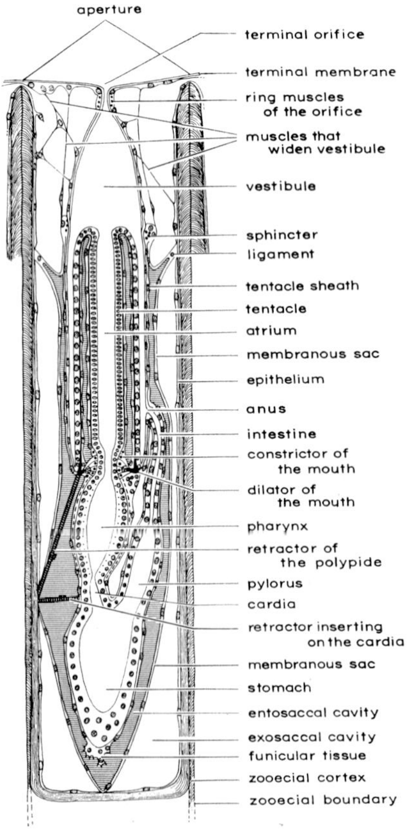

Fig. 2. Model of a trepostome autozooid based on feeding organs of Recent cyclostomes (after Nielsen, 1970, Fig. 13) and autozooecial living chamber of Palaeozoic trepostomes.

removal of this fragile membrane from preserved Recent specimens before sections can be made.

Inferences of a double-walled organization of body walls and mode of growth for fossil forms are necessarily based on studies of their skeletal walls. Such inferences are reasonable, but certainly would be strengthened by finding some direct indication of outer membranous walls in fossils. Recognition of remnants of organic membranes within zooecial cavities of some Palaeozoic Bryozoa (McKinney, 1969; Boardman, 1971) suggests that discoveries of outer membranes are also possible.

A dendroid trepostome of Late Ordovician age is covered by a bright reddish brown membrane inferred to be the cuticle of its outer wall (Plate I, Figs 2a–d). The remarkable preservation of such a fragile and exposed membrane was presumably made possible by the progressive encroachment of an encrusting monticuliporid trepostome during the life of the supporting dendroid colony.

This enveloping membrane appears to be continuous with brown deposits and fragments of membrane within zooecia that have the general shape of a lophophore and gut complex (Plate I, Figs 2c, d). As the zoarium has monomorphic zooecia and some of the deposits extend through most of the length of the living chamber, these deposits are inferred to be indications of the feeding organs.

ATTACHMENT ORGANS

Attachment of the lophophore and membranous sac of feeding autozooids to the enclosing zooecium in living cyclostomes is reportedly accomplished by eight radially arranged ligaments (Borg, 1926, p. 209) placed just inward from the atrial sphincter muscle (Fig. 2). Sections of Recent colonies of the cyclostome genus *Lichenopora* which include both soft and skeletal parts in place (for preparation technique see Nye, Dean, and Hinds, 1972) demonstrate the actual attachment of ligaments to the skeleton of the colony (Plate III, Fig. 1). In longitudinal view (Plate III, Fig. 2) the plane of the section commonly does not pass through ligaments at their attachment points so that the openings of the body cavity between ligaments are seen between the membranous sac and skeleton.

Inspection of hard-soft sections (Plate II, Figs 1a–e) of a specimen of the type species, *N. claviformis* (Waters), of the cyclostome genus *Neofungella* Borg, 1933, reveals that the attachment organ of the feeding autozooids is actually a thick, roughly cone-shaped membrane that is attached to the autozooecial wall by its perimeter. No radially arranged ligaments are involved. This kind of attachment is indicated in longitudinal sections by the extension to enclosing zooecial walls of all attachment membranes within the plane of the section (Plate II, Figs 1c–e). The completeness of the attachment of the membrane around its perimeter can be determined directly in a thicker longitudinal section that includes a significant part of the perimeter (Plate II, Fig. 1c) or a section at right angles to the length of the autozooid in the plane of the perimeter (Plate II, Fig. 1b).

Borg (1933, p. 267) reported ligaments in *N. claviformis.* Inspection of his decalcified sections indicates the difficulty of understanding the exact nature of the attachment if the ligaments were short, or were lacking (Plate II, Figs 2, 3). The exact positions of autozooecial walls are not indicated and the attachment organ, membranous sac, and vestibular membrane have undergone distortion and considerable shrinkage above the ends of the retracted tentacles. The shrinkage apparently occurred when the zooecial walls were dissolved, removing the supports for those organs (compare with the supported specimens in hard-soft sections of the same species, Plate II, Fig. 1e).

Plate II

Neofungella claviformis (Waters, 1904). Recent specimen from Antarctic Ocean, off Colman Island at 100 fathoms, British Museum (Natural History) 1959.8.12.8.; colony sectioned with soft and skeletal parts in place.

Fig. 1a. Tangential section, ×100, showing a feeding autozooid in center of figure flanked by two smaller polymorphs containing radial attachment ligaments.

Fig. 1b. Tangential section, ×150, showing feeding autozooid with perimetrically attached membrane in plane of section in all but N.E. quadrant. Dark mass in center of body cavity is sphincter muscle.

Fig. 1c. Longitudinal thick section, essentially a whole mount, ×200, showing most of the perimeter of the attachment organ above, the top portion of the tentacle sheath in the middle, and the tops of the tentacles at bottom of figure. Membranous sac is suspended vertically from perimeter of attachment organ.

Fig. 1d. Longitudinal section, ×200, section longitudinally offcentered on tentacle sheath showing perimetrical attachment to autozooecial wall.

Fig. 1e. Longitudinal section, ×100, showing position of tentacles in retracted position relative to attachment level above.

Neofungalla claviformis (Waters, 1904). Recent specimens from Antarctic Ocean, Discovery Station 190, at 315 m; stones, mud, rock; Bismark Strait at the Palmer Archipelago, W. of Graham Land. Naturhistoriska Riksmuseet, Stockholm.

Fig. 2. Longitudinal section of a decalcified microtome section of secondary type prepared and studied by Borg, 1933, ×200, showing the attachment organ and thick tentacle sheath, shrunk down so that tentacles are compressed. Notice sizes of structures relative to those in Fig. 1e at ×100.

Fig. 3. Longitudinal section of another Borg type, ×200, showing adjacent polypides with little indication of exact position of decalcified autozooecial walls.

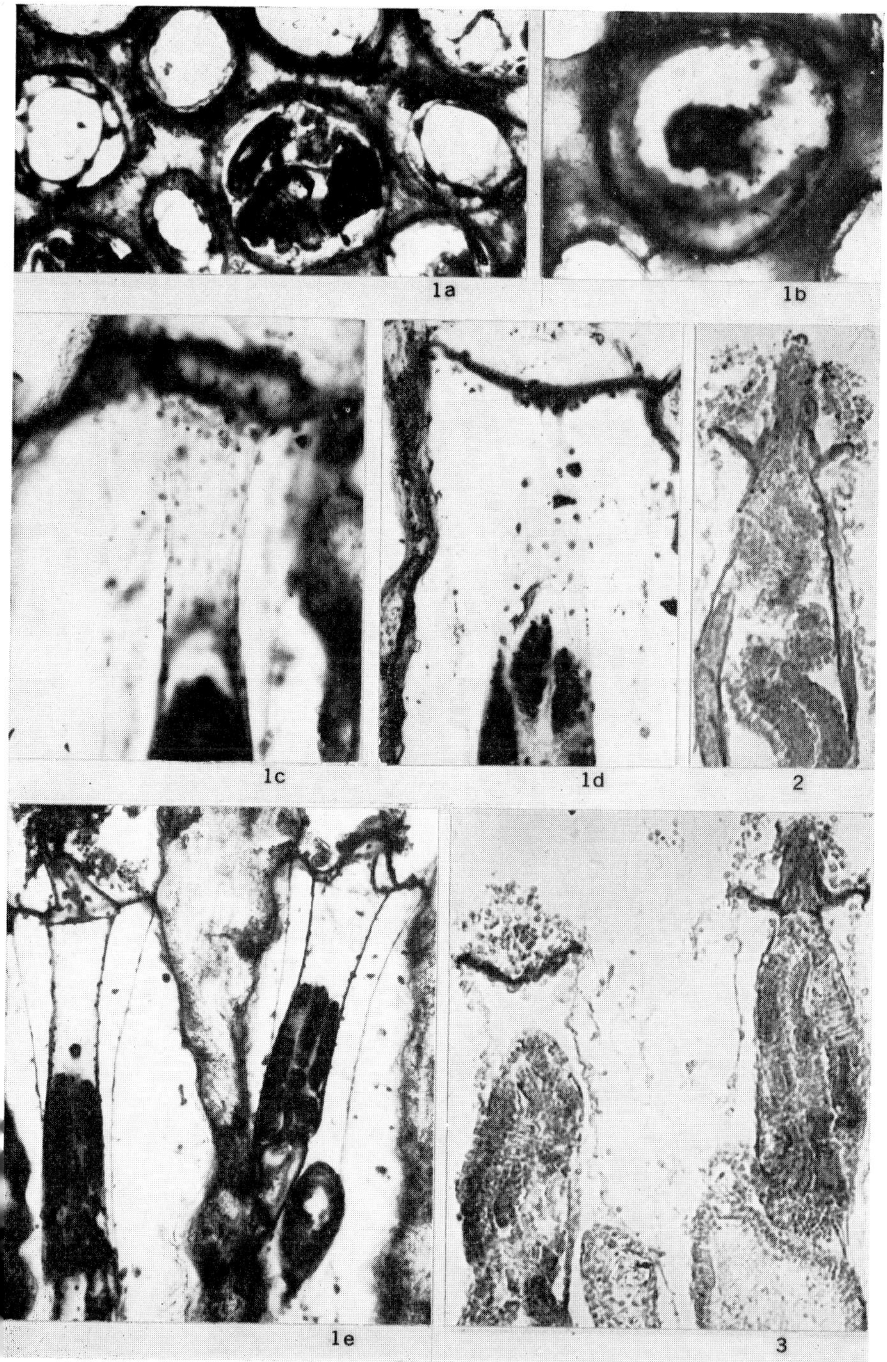
1a
1b
1c
1d
2
1e
3

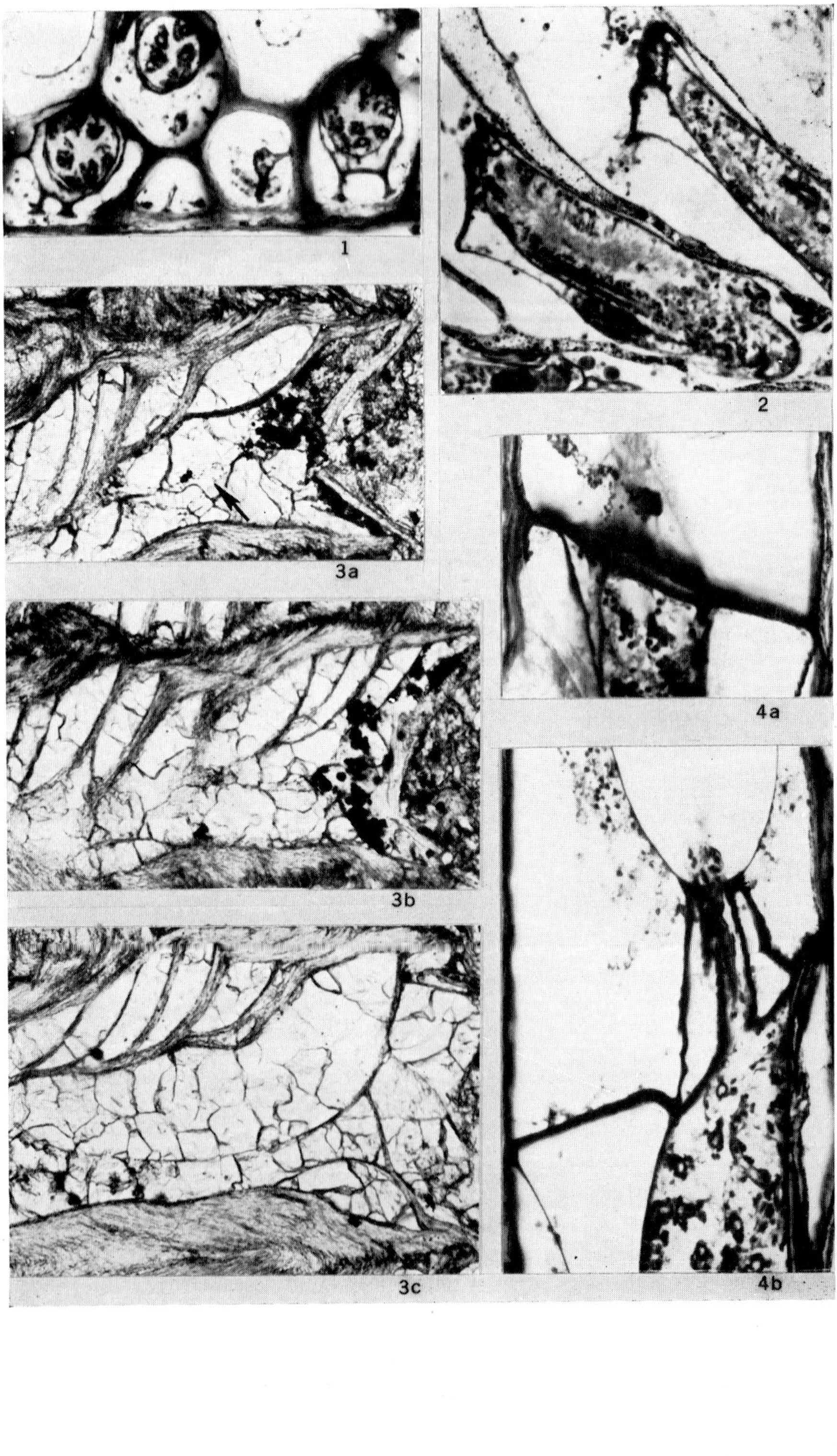
1
2
3a
4a
3b
4b
3c

The functional significance of a perimetrical attachment organ appears to be considerable. The apex of the cone-shaped membrane has a central opening in which the atrial sphincter muscle is placed and through which in turn the tentacles pass when protruded (Plate III, Fig. 4b). This membrane was considered by Borg to be the outermost segment of the membranous sac. The part of the zooidal body cavity outside the membranous sac (exosaccal cavity of Borg) then is divided transversely into two compartments (Fig. 2) by the perimetrically attached membrane, and the two compartments are not confluent. Borg's inferences on tentacle extrusion (Borg, 1926, p. 241) were based on exchange of body fluid from outer to inner parts of the exosaccal cavity through the spaces between radial ligaments. For colonies without those passageways this hypothesis seems unworkable.

Plate III

Lichenoporid cyclostome. Recent specimen from Galapagos Islands, British Museum (Natural History).

Fig. 1. Tangential hard-soft section, ×200, showing tentacles cut transversely, tentacle sheath, and membranous sac attached to autozooecial walls by radially arranged ligaments. It is not clear that there are actually 8 ligaments in any one autozooid.

Fig. 2. Longitudinal hard-soft section, ×200, showing membranous sac pulled out sharply (towards lower left of picture) by radial ligaments not in plane of section.

Peronopora sp. Southgate Formation, Eden Group (Upper Ordovician), railroad cut at Bald Knob, Cincinnati, Ohio. USNM 167682.

Fig. 3a. Longitudinal section, ×150, showing uncalcified, funnel-shaped structure centered longitudinally in living chamber. Faint boundaries of structure extend inward to autozooecial wall below (arrow) and skeletal cystiphragm above. Calcified septum across aperture is from subsequent overgrowth.

Fig. 3b. Longitudinal section from same zoarium, ×150, showing funnel-shaped structure thinly calcified at outer end and attached to autozooecial wall, uncalcified inward to left where it passes out of plane of section.

Fig. 3c. Longitudinal section from same zoarium, ×150, showing calcified funnel-shaped structure capped by subsequent calcified overgrowth. Laminae of funnel difficult to separate from laminae of overgrowth at connection with autozooecial wall. Calcification of funnel thins progressively and nearly disappears to left.

Single-walled cyclostome. Recent specimen from McMurdo Sound, Antarctica. USNM 179007.

Fig. 4a. Longitudinal, hard-soft, thick section, ×150, offcentered on longitudinal axis at peripheral attachment level.

Fig. 4b. Longitudinal thin section from same colony, ×150, showing vestibular membrane above, sphincter, peripheral attachment, tentacle sheath, and membranous sac.

A comparable perimetrical attachment organ has been found in a single-walled cyclostome (Plate III, Figs 4a, b). Preliminary sectioning indicates that other species, both single- and double-walled, also can have perimetrical attachment organs.

Preparation note: To make microtome serial sections of soft parts in original supported position: (1) Impregnate a preserved specimen in an epoxy under vacuum, (2) cut face of desired orientation through the specimen, (3) dissolve the skeleton from face inward for a short interval, (4) refill the resulting voids with the same epoxy under vacuum, and (5) microtome that part of the specimen that has the calcareous skeleton removed. The original position of the skeleton can be determined closely in section by tinting the epoxy used to fill the skeletal voids.

COMPARATIVE INTERPRETATIONS IN TREPOSTOMES

The significance of perimetrically attached membranes to the comparative morphology of Palaeozoic trepostomes also appears to be considerable. Most of the delicate structures of trepostomes that suggest soft parts appear to be perimetrically connected to zooecia. These structures can be grouped empirically into: (1) uncalcified; (2) partly or wholly calcified but not necessarily part of inner body wall; and (3) partly or wholly calcified and continuous with a segment of the zooecial wall, therefore inferred to be a part of the inner body wall. Structures in this third group, generally funnel-cystiphragms (Boardman, 1971, p. 12; Plates 8, 9), are interpreted to have formed an inflexible exoskeletal investment for the zooid (Fig. 3). They are uncommon enough and many are restrictive enough in diameter of living space to suggest smaller polymorphs that alternated in an intrazooidal sequence with feeding forms (Boardman, 1971, p. 26).

Among the specimens having uncalcified structures, the Devonian zooecium inferred to contain membranous remnants of a feeding lophophore and gut (Boardman, 1971, Plate 4, Figs 1a, b) seems to demonstrate at least one level of perimetrical attachment (Fig. 4). The transverse orientation of the membrane at the inner of the two levels of attachment, its appearance on both sides of the body cavity, and its relative position in the autozooecium longitudinally suggest that it could have been the functional equivalent of the perimetrical attachment organ in *Neofungella*. If so, and assuming morphology comparable to Recent cyclostomes, the outer attachment (on the right side of specimen, Fig. 4) would have connected the equivalent of a terminal-vestibular membrane to the skeleton. Borg (1926, Fig. 1) describes musculi extensores vestibuli (muscles that widen vestibule of Fig. 2) in this region in Recent cyclostomes which reportedly

Fig. 3. Partial reconstruction of an autozooid of a monticuliporid (*Prasopora*) showing the skeleton, including overlapping skeletal cystiphragms that encircle the living chamber, a funnel-cystiphragm that forms the neck of a smaller flask-shaped living chamber, and soft parts that are inferred from mode of growth assumptions only.

widen the vestibule when contracted. He describes these muscles as long, narrow, nonstriated muscle fibres (Borg, 1926, p. 230) which attach the terminal-vestibular membrane to the membranous sac or radial ligaments rather than directly to the zooecium. A direct comparison of these delicate fibres with the outer membranous looking attachment in the Devonian specimen seems difficult.

The Late Ordovician zoarium covered by the inferred outer cuticle referred to above contains uncalcified indications of soft parts in the form of brown granular deposits and bits of bounding membrane (Plate I, Figs 2c, d). The shapes of the deposits and their spatial relationships to the autozooecia and inferred outer membrane suggest polypides of feeding autozooids. The best-formed brown deposits in the zoarium suggest attachment to the autozooecial wall at a level of mural spines that is reasonably comparable in position to the attachment levels of the Recent cyclostomes figured. Mural spines are common at this general level throughout the zoarium although not seen in all autozooecia because of their spacing. The use of skeletal spines as ligament

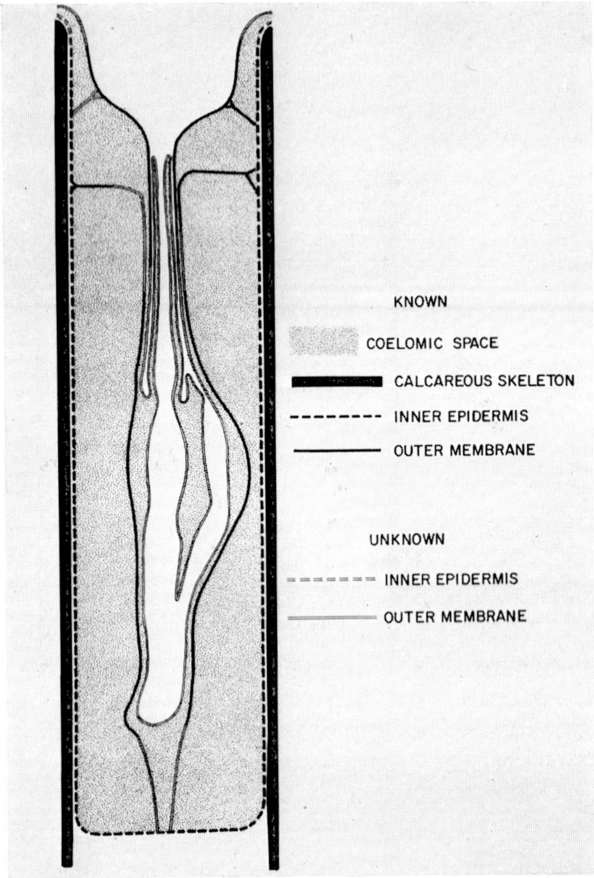

Fig. 4. Reconstruction of the hypothetical feeding organs of *Leptotrypella? praecox* Boardman, from the Lower Devonian of the Ohio Range, Antarctica. The known parts above and below the break midway along the tentacles are based on the suggested relationships of structures seen in two autozooecia (Boardman, 1971, Plate 4, Figs 1a, b, above and Plate 3, Fig. 6b, below). The unknown parts are patterned after generalized cyclostome morphology. The appearance of a transverse membrane attached to both sides of the zooecium at the same general level suggests a perimetrical attachment membrane at the lower of the two attachment levels.

attachment points has not been established in Recent cyclostomes so this interpretation remains speculative. If they had that function (see also halloporid spines, Boardman, 1971, Plate 8, Figs 6a, b) radial ligaments attached to the spines seem to be indicated rather than perimetrical attachments.

The most difficult of the three groups of structures to interpret in Palaeozoic trepostomes are those few that are partly or wholly calcified and on empirical evidence do not necessarily originate as part of the inner body wall. These structures are visibly calcified at their outer ends and the calcification generally thins or disappears inwardly. They appear to have been fastened to zooecial walls or cystiphragms of the inner body wall (Plate III, Fig. 3b), or they are so thinly calcified that the nature of the attachment is not evident in thin section using a light microscope.

It has been generally assumed (Boardman, 1971) that calcification of a structure located in any reasonable position meant that that structure has been calcified by the inner epidermis and was a part of the inner body wall. Part of the argument against calcification of parts of the outer body wall was based on the obvious difficulty of tentacle protrusion by a calcified and stiffened tentacle sheath and vestibular membrane. Certainly, lophophores as we understand them in Recent cyclostomes would be made inoperative by calcification of these outer body wall structures. A generalized reconstruction therefore placed the body cavity and any intrazooecial extension of the outer membranous wall within the total calcified structure of the zooid (Fig. 3) and ruled out the interpretation of funnel-cystiphragms as terminal-vestibular membranes or tentacle sheaths.

The assumption that calcification is produced only by the inner body wall should be questioned. Progressive calcification of parts of the outer body wall with possible loss of function during the later stages in the ontogeny of a single generation of polypides (suggested by John and Dorothy Soule at this symposium) would provide functional terminal-vestibular membranes and tentacle sheaths which would finally begin to calcify and so show varying degrees of calcification after death. Although direct evidence of the exact time of calcification during a regenerative-degenerative cycle seems unobtainable in Palaeozoic Bryozoa, the inward decrease of calcification of funnel-shaped structures suggests, time-progressive calcification. Examples are common throughout the phylum of the progressive covering of autozooidal apertures by skeletal material to make tentacle protrusion unworkable. Progressive calcification of actual terminal-vestibular membranes or tentacle sheaths during the ontogeny of a polypide is apparently unknown in Recent cyclostomes, however, so interpretations made below are necessarily speculative.

The recognition of perimetrical attachment organs in Recent cyclostomes provides additional morphologic comparison with funnel-shaped trepostome structures of possible outer wall origin that also seem to have been attached perimetrically. Inspection of autozooids in Recent cyclostomes indicates two levels of real or apparent perimetrical attachment of membranes to skeletal walls of the living chamber: (1) the real attachment of the tentacle sheath-membranous sac and, (2) the extension or attachment to the zooecial wall of the outermost end of the terminal-vestibular membrane. The configuration of the outermost end of a terminal-vestibular membrane of a cyclostome (Boardman, 1971, Plate 1, Figs 3a, 5a; Plate 2, Figs 1, 4b) most nearly resembles some of the outermost funnel-shaped structures (Fig. 4; Plate I, Figs 2c; Plate III, Figs 3a–c) in trepostomes. The attachment organ of the inferred tentacle sheath-membranous sac in trepostomes might be a transverse membrane just inward from the expansion of the terminal-vestibular membrane (Fig. 4), or the second expansion to the skeletal wall of the living chamber at the bottom of the neck of the flask-shaped structure (Plate I, Figs 2c: Plate III, Figs 3a, c). Alternatively, in some trepostomes the funnel-shaped structure might represent the tentacle sheath attached at its outer end. In Recent cyclostomes studied, the tentacle sheaths and attachment membranes are thicker and more generally preserved than the terminal-vestibular membranes (Plate II, Fig. 2).

Minute, funnel-shaped structures without visible calcification (Plate III, Fig. 3a) can be virtually identical in size and shape to structures in the same zoarium (Plate III, Fig. 3b) in which calcification is discontinuous and just one or two laminae thick. Some of these structures are so small in neck diameter that functional body cavities within them are difficult to visualize and so fragile that a membranous origin seems difficult to rule out (Boardman, 1971, p. 13). All stages of transition to larger and more heavily calcified structures seem to exist. The calcified funnel-shaped structure (Plate III, Fig. 3c) in the same zoarium can be interpreted as part of the inner body wall of a possible intrazooecial polymorph (Boardman, 1971, Fig. 3). If the apparent connection of the structure to the autozooecial wall is actually the connection of the contiguous overgrowing layer that caps the funnel neck, it is possible that the funnel-shaped structure represents the terminal-vestibular membrane that was progressively calcified in later stages of ontogeny of the feeding polypide.

BIOLOGICAL APPROACH TO TAXONOMY

The approach to the selection and treatment of taxonomic characters of trepostomes that is in general use today is almost entirely empirical, based on the skeleton without biologic interpretation. Characters that are used in taxonomy are those that can be seen readily and described without biologic understanding. Broken fragments of zoaria are considered the unit for study; not the colony, its inferred mode of growth, and other biologic features. Taxonomic procedures generally include studies of internal morphology in thin section without three-dimensional conversion, so descriptions are two-dimensional and therefore generally inadequate for biologic understanding or taxonomic use. Taxonomic evaluation of a newly recognized character is commonly made arbitrarily or in proportion to its visual prominence, without inquiring into its possible biologic significance, or occasionally, even its distribution in known taxa.

As a result of this descriptive approach, morphology is ignored that is important biologically but is difficult to understand and describe. The different budding patterns that can produce the same growth habit are a good example. Another is the autozooecial living chamber (Boardman, 1971) which should be described as a morphological unit even though it may be difficult to find intact. What description there is of the living chamber is generally found under the loosely defined concept of aperture which in some usage includes any part of the exozone that falls within the plane of a tangential section. Measurements under the heading of aperture can thus actually be measurements of living chambers or chambers behind basal diaphragms.

Present day approaches to the taxonomy of cyclostomes are even less adequate. The internal morphology of fossil and Recent cyclostomes has been largely ignored so that the same arbitrary approach to taxonomic evaluation is employed on many fewer characters than in trepostomes. The lack of investigation of living cyclostomes is especially unfortunate. Certainly in cheilostomes the soft and skeletal structures involved with tentacle protrusion have been for many years taxonomic characters highly placed in the hierarchy of classification within that order. What will it mean to the classification of cyclostomes when comparable knowledge is finally available for that order? How many different kinds of attachment organs are there in Recent cyclostomes and how many different methods are employed to protrude the tentacles? What does it mean to find smaller polymorphic zooecia nearly empty except for a membranous structure that is attached by radial ligaments (Plate II, Fig. 1a) in a species that has perimetrical attachment organs in the feeding autozooids? The several terms employed to differentiate polymorphs in cyclostomes without knowledge of soft parts or function will provide no answers to biological significance and only limited use in taxonomy.

There is little evidence that Palaeozoic trepostome, cryptostome, or cystoporatid stocks are extinct. The answer to the broad evolutionary patterns of the stenolaemates and the basic information for biologic interpretation of fossils must come from the post-Palaeozoic cyclostomes. Unfortunately, this most critical order is the least known of the stenolaemates. Thousands of Palaeozoic specimens are adequately prepared for detailed study in collections around the world. I am aware of only a few hundred adequately prepared cyclostomes of all ages. The major block to markedly improved understanding of the Palaeozoic tubular Bryozoa is our lack of knowledge of the cyclostomes.

ACKNOWLEDGEMENTS

The preparations were done by A. Janoff of the American University and D. A. Dean. The photography was done by Mr Dean and the reconstructions by L. B. Isham, both of the Department of Paleobiology. The loan of specimens was arranged by Miss P. L. Cook of the British Museum (Natural History) and Valdar Jaanusson of the Naturhistoriska Riksmuseet, Stockholm. The paper was reviewed by A. H. Cheetham.

REFERENCES

BOARDMAN, R. S. (1971). Mode of growth and functional morphology of autozooids in some Recent and Paleozoic Tubular Bryozoa: *Smithsonian Contrib. Paleobiol.* **8,** 51.

BOARDMAN, R. S. and CHEETHAM, A. H. (1969). Skeletal growth, intracolony variation, and evolution in Bryozoa: *J. Paleontol.* **43** (2), 205–233.

BORG, F. (1926). Studies on Recent cyclostomatous Bryozoa: *Zool. Bidrag Uppsala* **10,** 181–507.

BORG, FOLKE (1933). A revision of the Recent Heteroporidae (Bryozoa): *Zool. Bidrag Uppsala* **14,** 254–394.

MCKINNEY, F. K. (1969). Organic structures in a late Mississippian trepostomatous ectoproct (bryozoan): *J. Paleontol.* **43** (2), 285–288.

NIELSEN, CLAUS (1970). On metamorphosis and ancestrula formation in cyclostomatous bryozoans. *Ophelia* **7** (2), 217–256.

NYE, O. B., Jr., DEAN, D. A. and HINDS, R. W. (1972). Improved thin-section techniques for fossil and Recent organisms. *J. Paleontol.* **46** (2), 271–275.

WATERS, A. W. (1904). Bryozoa. *Exp. Antarct. Belge. Res. Voyage Belgica* 1897–1898–1899, 103.

22 | Palaeozoic Cyclostomata (A preliminary report)

KRISTER BROOD

Geologiska Institutet, Stockholms Universitet, Stockholm, Sweden

Abstract: The interzooidal walls of the Palaeozoic Cyclostomata investigated are tripartite structures composed of a middle granular layer flanked on both sides by laminar layers. In species with thin interzooidal walls there is only a double laminar layer. The laminae are oriented either parallel to the growing direction of the wall or in an orally divergent pattern. The exterior walls consist of a single laminar layer. The crystals in the laminar layers consist of long, flat laths, approximately 1 μ thick and up to 50 μ long. They probably grow by edgewise additions on the terminal ends. The skeletal structure agrees with that of the single-walled cyclostomes of Borg (1926). It is probable that the soft tissues were arranged in the same manner. It seems certain that the interzooidal walls were surrounded by a secreting epithelium (the inner epithelium) on both sides, while the exterior body wall consisted of a cuticle, calcareous layer and a secreting epithelium (the outer epithelium) on its inner side. Growing tips and thin sections indicate that the budding of new zooecia took place in a common bud, *sensu* Borg.

The interzooidal walls of the species investigated lack interzooidal pores. In addition, pseudopores are absent in the exterior walls. An exception is the genus *Kukersella*, which exhibits pseudopore-like structures. Calcareous rods resembling the rods in other groups of Palaeozoic tubular Bryozoa have been found in one species (*Clonopora gotlandica*). Terminal diaphragms with aborally flexed laminae and subterminal diaphragms with orally or aborally flexed laminae were found in several species.

In one species, *Sagenella consimilis*, numerous ovicell-like structures were observed which consist of a narrow proximal part, a dilated middle section and a short terminal part. These structures probably represent gonozooecia.

INTRODUCTION

This paper is a preliminary report on investigations of certain Palaeozoic non-cystoporate Cyclostomata. The study is being made on primary types of described species as well as on new material. The specimens are studied chiefly in thin sections and with the aid of a scanning electron microscope. In many cases, the material suffers from recrystallization. Unfortunately, this is true for the primary types of many species. For this reason, scanning studies usually are carried out on well-preserved material collected from the Silurian of Gotland, Sweden. A complete report of these investigations with numerous illustrations will be published in "Stockholm Contributions in Geology" at a later date.

Many of the Palaeozoic forms show interesting similarities in growth-form to post-Palaeozoic genera. Examples of this are *Clonopora*, which is structurally almost identical to *Pustulopora* (*Entalophora*), *Corynotrypa* to *Stomatopora*, *Sagenella* to *Diastopora* and *Mitoclema* to *Spiropora*. It has been suggested by some authors (e.g., Waters, 1884, p. 681) that the Palaeozoic and the post-Palaeozoic forms belong to the same genera and even to the same species. This view, however, is not held by the present writer, who considers the similarity between Palaeozoic and post-Palaeozoic forms to be due to convergent evolution rather than to a close relationship. Numerically the Palaeozoic cyclostomes form a small group with fewer than 100 described species. They are unimportant from the point of view of stratigraphy and palaeoecology. The group, however, is morphologically interesting because it probably represents the most primitive cyclostomes known and perhaps the most primitive Stenolaemata as well. Palaeozoic cyclostomes are reported from the Middle Ordovician to the Permian. Most records are from Silurian and Devonian strata.

WALL STRUCTURE

The interzooidal wall in some forms investigated is a tripartite structure composed of a thin middle granular layer flanked on both sides by a single laminar layer (Plate I, Figs 4, 6) or as in the case of species which have thin interzooidal

---

Plate I

Fig. 1. Interzooidal wall in *Diploclema sparsum* Hall showing laminar structure. GI D 3223. Stereoscan. × 1600.

Fig. 2. *Kukersella boreale* (Bassler) displaying numerous pseudopore-like structures. USNM 57184. × 25.

Fig. 3. Exterior zooecial wall in section *Kukersella boreale* (Bassler) showing pseudopore-like structures and the laminar structure of the zooecial wall. Partly recrystallized specimen. GI D 3103. Stereoscan. × 1500.

Fig. 4. Interzooidal wall in *Clonopora gotlandica* showing the central granular layer and flanking laminar layers. GI D 2924. Stereoscan. × 400.

Fig. 5. Exterior zooecial wall in *Diploclema sparsum* Hall showing laminar structure. GI D 3223. Stereoscan. × 900.

Fig. 6. Contact between granular and laminar layers in interzooidal the wall of *Clonopora gotlandica* GI D 2925. *Stereoscan*. × 700.

Fig. 7. Calcareous rod in the exterior zooecial wall of *Clonopora gotlandica*. GI D 2909. Stereoscan. × 3500.

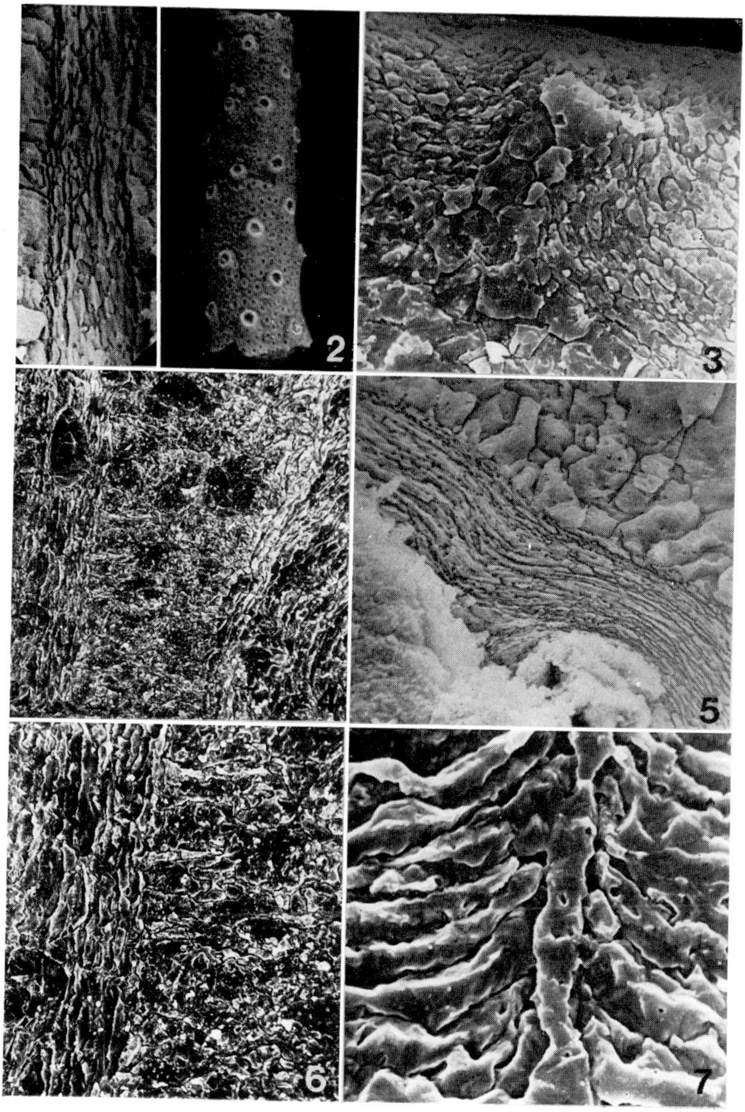
2
3
4
5
6
7

walls (Plate I, Fig. 1), by double-layered laminar structures. The laminar layer is much thicker than the granular one. The laminae of the former are oriented either parallel to the direction of growth or in an orally divergent pattern. The contact between the granular and laminar layers generally is abrupt (Plate I, Fig. 6).

The exterior body wall of the forms investigated is composed of a single laminar layer. The laminae in this layer are arranged almost parallel to the direction of growth (Plate I, Fig. 5). Sometimes the granular layer extends for a short distance into the exterior wall.

The zooecial wall in the forms investigated agrees in structure with that of the single-walled Cyclostomata of Borg (1926). It also seems probable that the soft tissues were arranged in the same manner (Fig. 1). It seems certain that the interzooecial wall was surrounded on both sides by a secreting epithelium (the inner

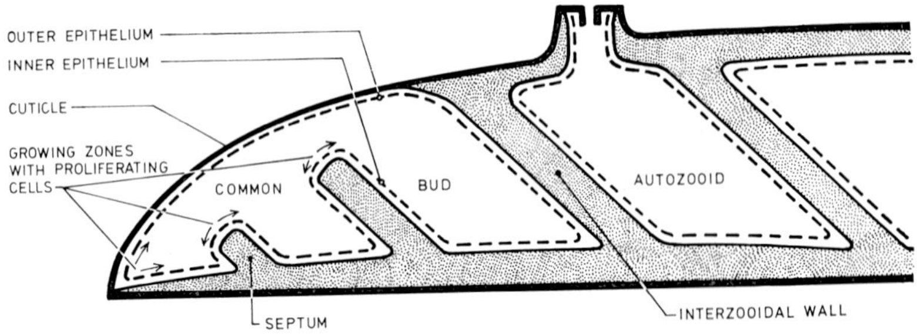

Fig. 1. Diagram of the growing part of an encrusting Palaeozoic Cyclostome showing a common bud with growing zones. The arrows indicate the relative migration of the cells.

epithelium) which lined the zooids. This inner epithelium could have been derived from the basal plate of the ancestrula. The exterior body wall was probably lined on its inner side by a secreting epithelium (the outer epithelium). This is indicated by the one-sided growth from the zooidal side. The outer epithelium could have been derived from the terminal membrane of the ancestrula. The outer surface of the body wall was probably covered by a cuticle, as is the case in post-Palaeozoic single-walled cyclostomes.

Examination of growing tips and thin sections indicates that the budding of new zooecia took place in a common bud *sensu* Borg (1926).

The common bud is considered here to consist of the growing interzooidal walls (septa), the terminal membrane of the growing tip and the enclosed coelomic cavities and soft tissues. The new zooecia probably were formed by the insertion of septa from either the basal lamina (Fig. 2) or from the centre of an erect stem. Later, the septum became the interzooidal walls of the zoarium. The septum will, with continued growth, reach the terminal membrane of the common bud and coalesce with the latter. The secretion of the granular middle layer, when present, was then stopped. The laminar layer in the interzooidal wall continued into the exterior zooecia wall (Plate I, Fig. 5). In some cases an exterior granular layer can be observed externally.

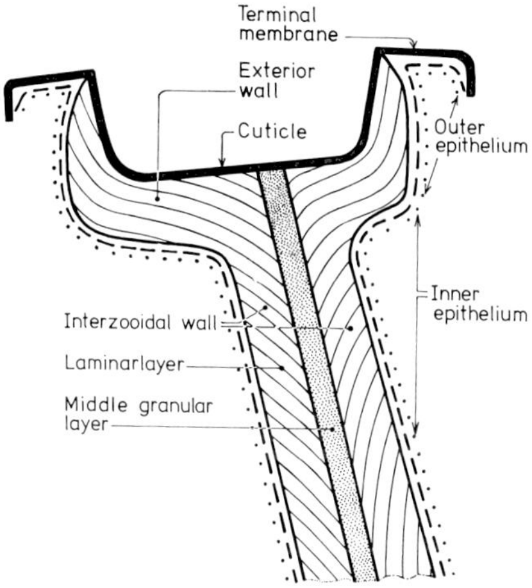

Fig. 2. Diagram of the zooecial wall in *Clonopora gotlandica* a species with a middle granular layer within the zooecial wall.

The crystals which form the laminar layer in the skeleton of the Palaeozoic Cyclostomata investigated consist of thin, elongated laths approximately 0·5–1·5 μ thick (Plate II, Figs 1–4). The crystals are several times longer than they are wide. They can be up to 50 μ long and are arranged in a shingled pattern. The crystals probably grew by edgewise addition of a calcareous substance on the terminal edges of the laths.

PORES

Contrary to the post-Palaeozoic Cyclostomata, pores are generally absent in the skeleton of the Palaeozoic cyclostomes investigated. In no instance have interzooidal pores been found. In one genus, *Kukersella* Toots, pores penetrating the exterior body wall were observed (Plate I, Figs 2, 3). These pores resemble the pseudopores of post-Palaeozoic cyclostomes. It cannot be established, however, whether or not these pores were covered by a cuticle, though this seems probable. In the species *Sagenella consimilis* Lonsdale, the exterior body wall is penetrated by two pores (Plate II, Fig. 7). The function of these pores is unknown.

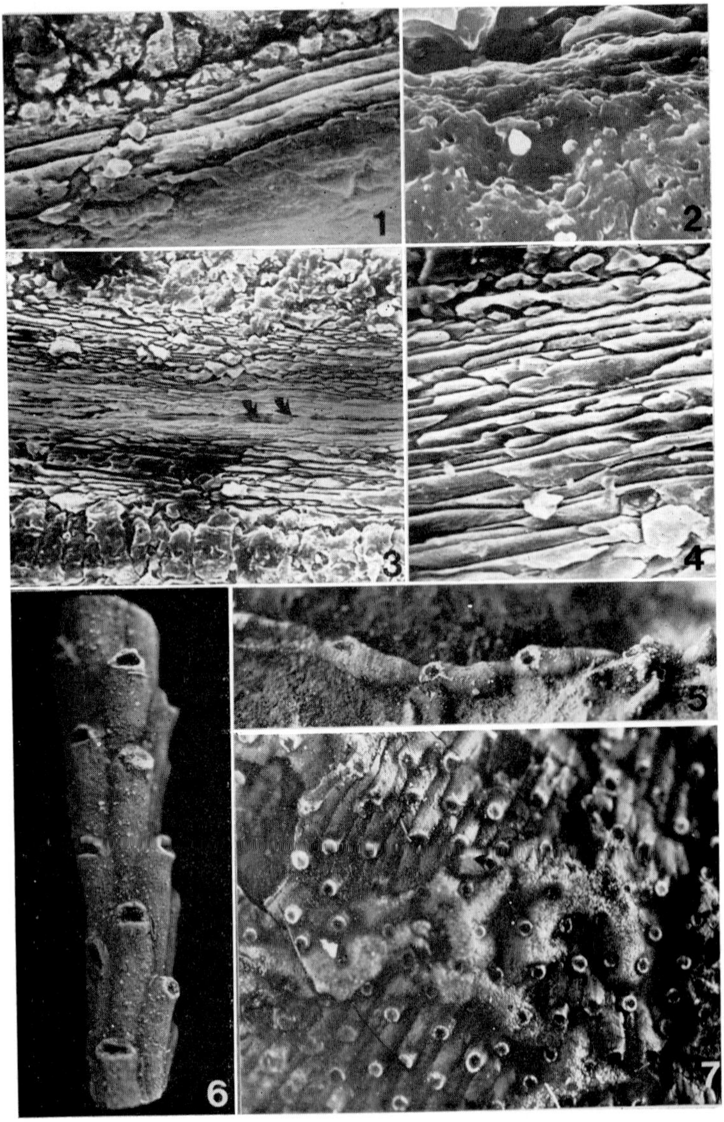
1
2
3
4
5
6
7

The absence of interzooidal pores among the Palaeozoic cyclostomes investigated is interesting from the point of view of interzooidal communications. As this group of Cyclostomata is concluded to be single walled and interzooidal pores are absent, it is concluded that any internal communication between the fullgrown zooids was impossible. Transfer of nutrients between the zooids could not have taken place through the zooidal walls. In addition there was probably no hypostegal coelomic cavity in these forms. Before the zooecia were fullgrown, the zooids could communicate with each other inside the common bud. This was because the septa had not yet coalesced with the terminal membrane of the common bud. Communication between the fullgrown zooids was probably not essential for the well-being of the individual zooids. The autozooid contained all vital organs necessary. Kenozooids were not encountered among the forms investigated.

CALCAREOUS RODS

In one species, *Clonopora gotlandica* calcareous rods are found in the exterior and interzooidal walls. The rods are small solid granular structures (Plate I, Fig. 7), which point towards the interior of the zooecia. The laminae which surround the calcareous rods are flexed upwards in a con-in-cone pattern, indicating that the rods were secreted before the laminae. Their function has not been defined.

Structurally, the rods are similar to the calcareous rods of the fenestellids Tavener-Smith, 1969) and to the acanthopores in Trepostomata as well as certain Cryptostomata.

Plate II

Fig. 1. Section of basal lamina in *Sagenella consimilis* (Lonsdale) showing long crystals cut parallel to the direction of growth. GI D 3609. Stereoscan. × 1500.

Fig. 2. Broken exterior wall of *Sagenella consimilis* (Lonsdale) showing the laminar structure. GI D 3714. Stereoscan. × 600.

Fig. 3. Double basal lamina in bilaminar specimen of *Sagenella consimilis* (Lonsdale). Arrow indicates the position of the hypothetical cuticle. GI D 3617. Stereoscan. × 1500.

Fig. 4. Crystals from the basal lamina in *Sagenella consimilis* (Lonsdale). GI D 3613. Stereoscan. × 3500.

Fig. 5. *Corynotrypa cf. dissimilis* Vine RM 4015. × 20. Lummelunda.

Fig. 6. *Clonopora cf. geniculata* (Hall) USNM 2425. × 20. Tichenor Limestone. Bethlehem. New York.

Fig. 7. *Sagenella consimilis* (Lonsdale) showing several strongly lobated gonozooecia. The arrow indicates an oeciopore. RM 4002. × 20. Djupvik, Gotland.

DIAPHRAGMS

Two types of diaphragms have been identified among the Palaeozoic Cyclostomata investigated; terminal and subterminal diaphragms.

The terminal diaphragms occur as terminal closures in old zooids. They are located near the terminal end of the peristome and are fixed to the zooecial wall by aborally flexed laminae. The diaphragm itself is laminated and nonporous. The aboral flexing of the laminae indicates that the terminal diaphragms were secreted from the aboral side.

The subterminal diaphragms are located below the peristome of the autozooecia. They are fixed to the zooecial wall by laminae which may be flexed either orally, aborally or sometimes in both directions. Subterminal diaphragms are laminated and nonporous. Generally, they are rare among the Palaeozoic forms investigated, but are found in *Kukersella*, where they are numerous, and *Clonopora*. Structurally, the diaphragms found in the Palaeozoic cyclostomes are similar to the corresponding structures in post-Palaeozoic Cyclostomata.

OVICELLAR STRUCTURES

Ovicellar structures have not been reported previously from Palaeozoic non-Cystoporate Cyclostomata.

In two specimens from the Mulde Marl (Silurian, Wenlockian of Västergarn, Gotland) belonging to the species *Sagenella consimilis* more than 70 vesicular structures, considered here to be ovicells, have been found. These structures (Plate II, Fig. 7) consist of a short proximal part, which is structurally not unlike the proximal part of an autozooecium, a strongly dilated middle section and a short terminal part (oeciostome). The terminal part is only slightly modified, being similar in size and shape to the aperture of an autozooecium. The size and shape of the ovicells varies widely within the same colony. They can be observed as small ovid chambers or as large, often highly lobate, sacs.

These cells show a strong similarity to the gonozooecia of the post-Palaeozoic Cyclostomata. The author believes that they served the same purpose. The gonozooecia found in *Sagenella consimilis* differ from the gonozooecia of post-Palaeozoic Cyclostomata in that the former lack interzooidal pores and pseudopores. This fact raises the question of how the embryos were fed inside the gonozooid, since exchange of nutrients must have been impossible through the imperforate walls between the gono- and autozooecia. Possibly the fertile polypide of the gonozooid was fully developed and capable of gathering food, enabling it to feed the growing embryos. However, this assumption cannot be verified in an exclusively fossil group.

To date, ovicell-like structures have been identified only in *Sagenella consimilis*.

For this reason, it is impossible to judge whether or not ovicells were a characteristic of other Palaeozoic Cyclostomata.

DISCUSSION

The Palaeozoic Cyclostomata investigated hitherto demonstrate structural similarities to the post-Palaeozoic cyclostomes in many respects. Palaeozoic Cyclostomata demonstrate a wall structure which agrees well in construction with that of the single walled cyclostomes of Borg. Both groups show a similar budding pattern and the presence of gonozooecia. On the other hand there are important differences such as the absence of interzooidal pores and pseudopores in the Palaeozoic forms. These latter differences are considered by the present author to be of great taxonomic importance and are probably also of importance for discussions considering the phylogenetic relations within the Cyclostomata (cf. Nye, 1968). In post-Palaeozoic cyclostomes the pseudopores and interzooidal pores occur invariably in the forms investigated, whereas (as in other Palaeozoic tubular Bryozoa, except some Cystoporata) they are absent in the Palaeozoic forms examined, except in the genus *Kukersella*, which exhibits pseudopore-like structures. These differences in pore structure means either that the similarities in budding pattern and growth form between Palaeozoic and post-Palaeozoic cyclostomes are a result of convergent development, or that the Palaeozoic genera *Sagenella*, *Corynotrypa*, *Mitoclema* and *Clonopora* have developed *pseudopores* and interzooidal pores independently and thus gave rise to the post-Palaeozoic genera *Diastopora*, *Stomatopora*, *Spiropora* and *Pustulopora*. The first explanation seems the more probable to the present writer.

Because of the absence of pores and interzooidal pores, the Palaeozoic Cyclostomata may be tentatively grouped in a separate suborder, *Paleotubuliporina* subord. n., which can be characterized as follows: Single walled Cyclostomata with a laminated, nonporous skeleton. Laminae in the interzooidal walls are oriented in an orally divergent pattern or are parallel with the growing direction of the wall. A middle granular layer may be present within the interzooidal wall. Gonozooids are present in some forms. Kenozooids are generally lacking. New zooecia are formed in a common bud by budding from a basal lamina or from the centre of an erect stem.

This new suborder differs from Tubuliporina Milne-Edwards in lacking pseudopores, interzooidal pores or kenozooids. Paleotubuliporina differs from the Palaeozoic suborder Hederelloidea Bassler in the presence of a different budding pattern. In the hederelloids the new zooecia are formed from a long central zooecium, which gives rise to all zooecia within the colony. This budding pattern is so different from the Cyclostomata in general that it is impossible, in the writer's opinion, to consider the hederelloids and other Palaeozoic cyclostomes as nearly related.

REFERENCES

BASSLER, R. S. (1911). The early Paleozoic Bryozoa of the Baltic provinces. *U.S. natn. Mus. Bull.* **77,** 1–382.

BASSLER, R. S. (1953). Bryozoa, In "Treatise on Invertebrate Paleontology", (R. C. MOORE, ed.), Lawrence, Kansas.

BORG, G. (1926). Studies on Recent cyclostomatous Bryozoa. *Uppsala Zool. Bidrag* **10,** 182–507.

NYE, O. B. (1968). Aspects of microstructure in post-Paleozoic Cyclostomata. *Atti Soc. It. Sc. Nat. Museo Civ. St. Nat. Milano.* **108,** 111–114.

TAVENER-SMITH, R. (1969). Skeletal structure and growth in the *Fenestellidae* (Bryozoa). Paleontology 12 (2).

TOOTS, H. (1952). Bryozoen des estnischen Kuckersits. *Mitt. Geol. Staatsinst. Hamburg.* **21,** 113–137.

WATERS, A. W. (1884). On fossil cyclostomatous Bryozoa from Australia. *Q. Jl. Geol. Soc.* **XL,** 674–697.

23 | Nature of "Acanthopores" and Related Structures in the Ordovician Bryozoan *Constellaria*

JOHN F. CUTLER

University of Kentucky, Lexington, Kentucky, U.S.A.

Abstract: The "acanthopores" of *Constellaria* are solid rods, termed here "acanthorods", composed of a clear yellowish calcite (yellow tissue) without discernable microstructure. Other structures composed of yellow tissue and their relationship to acanthorods are also considered. The function of acanthorods and other yellow tissue structures is discussed.

Constellaria belongs to the Order Cystoporata (Astrova, 1965). Approximately 15–20 species are known from rocks of the eastern United States and Europe. The genus includes encrusting species as well as ramose and frondose forms. All are characterized by stellate monticules composed of clusters of tightly-packed zooecia; the central portion and the space between stellate "ray" clusters in each monticule is filled with vesicular masses of skeletal tissue (mesopores or cystopores of previous authors). Spaces between monticules are occupied by scattered zooecia which are also separated from each other and from the monticular zooecia by vesicular tissue. *Constellaria* also has "acanthopores" adjacent to zooecia and in the vesicular skeleton between zooecia (especially in the centers of monticules). Acanthopores have generally been held to be minute tubules crossed by diaphragms in which lived "acanthozooids" (Tavener-Smith, 1969, and references therein).

The primary purpose of this paper is to show that these "acanthopores" are in fact solid rods (termed "acanthorods") composed of a clear, yellowish calcite(?) without discernible microstructure (termed "yellow tissue"). (See Cutler, 1968, for details concerning morphology.) Tavener-Smith (1969) has provided a similar, very detailed, picture for the trepostome *Leioclema asperum*.

The secondary purpose of this paper is to describe other structures composed of yellow tissue, to discuss their relationship to acanthorods, and to comment on problems of function for all these structures.

Following are several lines of evidence that support the concept of "rod" (*vs* pore) structure. Electron microscopy, used successfully by Tavener-Smith to show the internal structure of acanthorods, has not yet been tried on *Constellaria*.

(1) Acanthorods associated with zooecia normally have a circular cross section (in tangential view) but these represent only one end member in a series of yellow tissue structures. Rods with a circular cross section grade morphologically into structures with elliptical cross sections, and these into strips of yellow tissue that partly or wholely surround the zooecia. Irregular masses of yellow tissue also exist adjacent to the zooecia and these grade morphologically into normal acanthorods with circular cross section. Acanthorods in the vesicular skeleton (between zooecia) also grade into forms with non-circular cross section: elliptical, elongate strips, or irregular masses, or nearly complete substitution of yellow tissue for the usual laminated material of the vesicular skeleton.

In the stellate monticules, individual "ray" clusters of zooecia often contain a few acanthorods near the center of the cluster. Again, there is a graded series from a few acanthorods to a linear series of acanthorods to merging acanthorods to a "midray partition" of yellow tissue that divides the ray cluster lengthwise into subequal halves. Beyond this, there is an elaboration of shapes of partitions (Cutler, 1968, pp. 58–60, 211–212).

The conclusion must be that, because of this gradational series, the end member called "acanthorods" could not have been pores containing acanthozooids but were instead solid deposits of yellow tissue.

(2) A second argument against pores is the range in size of the acanthorods (those with circular cross section). The observed range is 4 to about 40 μ in diameter; the minimum value seems a bit small to have been a pore housing an acanthozooid.

(3) As seen in longitudinal section, laminae of the vesicular skeleton are often flexed toward the surface of the zoarium ("flexed cystopores" in Cutler, 1968, p. 54). In some cases there are patches of yellow tissue at the point of maximum abaxial flexure; these "incipient acanthorods" grade into normal acanthorods (that is, there is a series with increasing amounts of yellow tissue until a solid rod is formed). This sequence is very similar to that shown by Tavener-Smith (1969, p. 92) for the trepostomes, where the evidence for solid rods is unmistakable.

(4) The larger acanthorods, at least, form short spines on the surface of the zoarium and there is no sign of a central pore. Similarly, on well preserved zoarial surfaces, midray partitions can be seen to form an elevated ridge down the center of the ray cluster (Cutler, 1968, Plate 11, Fig. 1). Since acanthorods grade into partitions and since both are composed of yellow tissue, acanthorods are presumed to have formed as solid structures.

The function of acanthorods is uncertain. They are not always present (especially in certain species) and they vary widely in abundance from place to place within a zoarium. They are usually assumed to have been protective (surface spines) but their occasional absence and frequent rarity means that they were not *required* for protection. The fact that they are sometimes very small (4 μ diameter) and the fact that they grade in shape into irregular masses, midray partitions, etc., argues against the idea of spines primarily for protection.

Yellow tissue also exists in other places in the *Constellaria* skeleton. (1) Midray partitions often extend beyond the ends of the star rays and join with other partitions from rays in adjacent monticules; this produces a "yellow tissue network" (Cutler, 1968, pp. 80–83; Plate 18, Figs 1, 2). Zooecia adjacent to the network are flattened on the side touching the yellow tissue, suggesting that perhaps the network was instrumental in budding the zooecia. Unfortunately, the network is present in only a minority of specimens and so it could not have served an *indispensable* function.

(2) The endozone walls, rather than appearing as normal thin, black lines in longitudinal section, are *sometimes* composed partly or mostly of yellow tissue.

The chemical/mineralogical nature of yellow tissue is unknown, though it is presumed to be calcite. Microprobe analyses are planned for the near future. Under white light, the yellow tissue appears structureless except for occasional black granular inclusions. Under polarized light, yellow tissue appears as a fine to coarse mosaic of randomly oriented crystals. It has the appearance of having recrystallized from some other substance, perhaps aragonite or some organic-rich or chitinous material.

The amount and distribution of yellow tissue varies in several respects. (1) Within a single monticule, only some of the rays may have partitions; the several partitions may contain very different quantities of yellow tissue. (2) From place to place within a single zoarium, there can be wide differences in the amount of yellow tissue no matter in what form it occurs (whether as number and/or size of ray partitions, acanthorods, or irregular masses). (3) From one geographical locality to another there is often an obvious change in the amount of yellow tissue regardless of the species involved. (4) The same can be said for change through time (generally more yellow tissue as one progresses into younger rocks). (5) Lastly, some species typically have more yellow tissue than others. The amount of yellow tissue that one *actually finds* in any given thin section is a function of all 5 of these variables, and it can vary from essentially none to perhaps 30% of all skeletal material.

The role that yellow tissue played in the functioning of *Constellaria* is obscure. Apparently there was some ecological control (variation by locality), but also some genetic control (variation by species and through time-evolution). However, yellow tissue was not a *required* item in the physiology of *Constellaria* because it is entirely absent in some specimens and, as pointed out above, varies between none and much from place to place within the same zoarium. Perhaps yellow tissue was a physiological waste product, the amount deposited as skeletal material being a function of environment plus genetics.

REFERENCES

ASTROVA, G. G. (1965). A new order of Paleozoic Bryozoa. *Int. geol. Rev.* **7,** 1622–1628.

CUTLER, J. F. (1968). Morphology, taxonomy, and evolution of the bryozoan *Constellaria* from the Cincinnati Arch. 297 pp. Unpub. PhD Dissert., Columbia Univ., New York.

TAVENER-SMITH, R. (1969). Wall structure and acanthopores in the bryozoan *Leioclema asperum*. *Lethaia* **2,** 89–97.

24 | Cuticular Pores and Spines, Ion Regulatory Organs in Entoprocta

P. EMSCHERMANN*

Anatomisches Institut der Universität Freiburg, Germany (FRG)

Abstract: Hollow cuticular spines and cuticular pores, which are typical surface differentiations of many Entoprocta, mainly at the stalk-integument of Pedicellinidae and Barentsiidae, are shown to be comparable and by intermediate stages linked structures. Their development is highly dependent on environmental conditions and because of their great variability they are of little value for species characterization. Complexes of specialized epithelial cells underneath the more obvious cuticular structures are interpreted to be ion regulatory organs ("chloride cells"), which are homologous with protonephridia.

Epithelial "gland cells", chimney-like hollow spines and cuticular pores are characteristic features of a number of solitary and colony-building Entoprocta (Plates I and II) (Atkins, 1933; Ehlers, 1890; Harmer, 1885). Particularly in the Pedicellinidae and Barentsiidae these integumental differentiations are often used as criteria in species characterization. Except for a few suggestions, no well founded information on their nature is available in the literature: they have been assumed to be "adhesive glands" (Ehlers, 1890), sensory pits (Hilton, 1923) or gas exchange organs (Maturo and Schopf, 1968). In the present study the relationship between the pores and spines of the colony building Entoprocta, their value as species characteristics, and their possible function are investigated by means of laboratory culture experiments, and with electron microscopica and ultrahistochemical methods in the species *Barentsia gracilis* Sars, *Barentsia discreta* Busk, both having normally cuticular pores only, and *Pedicellina cernua* Pallas, which mostly develops long cuticular spines. The following results have been obtained.

Both structures, pores and spines, are in principle identical integumental differentiations. Their individual appearance is dependent only on the body-area in which they develop, i.e. on the type of cuticle, rigid or flexible, being

* Supported by a grant from the Deutsche Forschungsgemeinschaft. A more detailed account of this subject will be published in *Sarsia*, 1972.

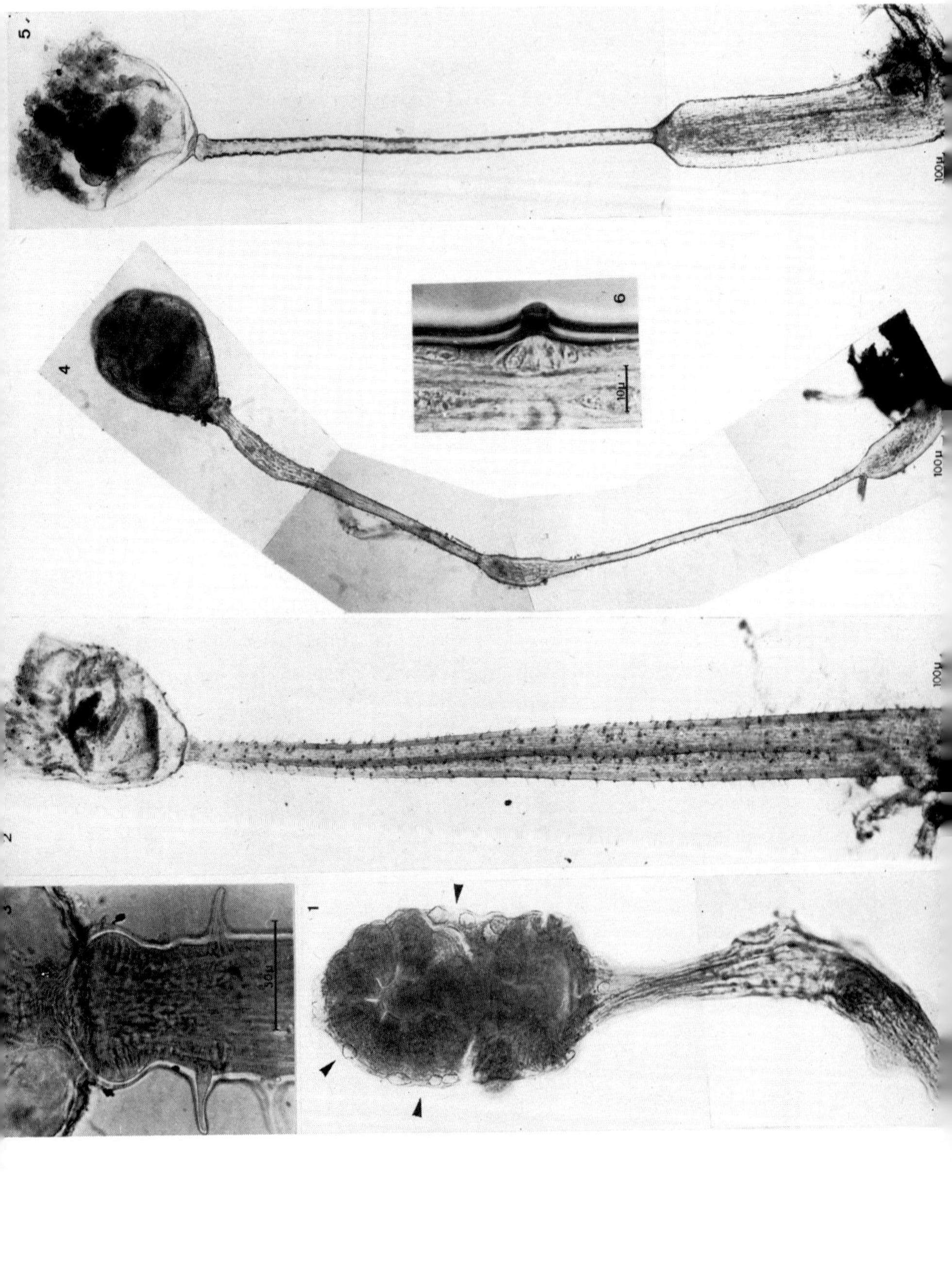
5
100μ
6
10μ
4
100μ
100μ
2
3
30μ
1

secreted in the body region in question. The ability to develop pore-organs seems to be present only in certain species; but in these species pores do not necessarily appear. In zooids from laboratory cultures the pore-organs are fewer in number and smaller than in zooids from their natural marine habitat. The density of pores per unit area of the stalk varies also in marine biotopes during the growth periods of the individual zooids, so that periods in which pores are plentifully developed can precede or follow "pore-free periods".

As observed in cultured animals it is also possible for the pores to be plugged again with a cuticular substance (Plate II, Fig. 1). The differentiation of the pore complexes as well as the frequency with which they occur are induced and governed by external factors as yet unknown. Changes in temperature, salinity and food-composition appear to have no significant influence.

Cuticular differentiations, pore openings and spines, arise from a specially differentiated epithelial cell-complex, which, in the three species investigated, displayed almost the same ultrastructure (Fig. 1) and (Plate III). In the simplest case they are composed each of one central pore cell that is invested by special thickened epithelial cells. The pore- or principal-cells are broad and irregularly pyramid-shaped in outline as shown in Fig. 1. They are characterized by a basal nucleus and an apical pit oriented to the cuticular pore-opening into which numerous microvilli project. On the bottom of the pit protrude some rudimentary, non-motile cilia-stumps, lacking the central microtubule-doublet and sometimes containing an incomplete circle of outer tubules (Plate III, inset).

The inner cell membrane of the principal-cell on the other hand is straight, and without exceptional surface variations. The cytoplasm is more electron dense than that of the usual epithelial cells owing to the rich amount of free ribosomes and polyribosomes. Furthermore large lysosome-like vacuoles with

---

Plate I

The arrangement of epithelial "glandular complexes" in the main types of Entoprocta.

Fig. 1. Loxosomatide (*Loxosomella* (=*Loxocalyx*) *leptoclini* Harmer) with "gland cells" around the margin of the calyx.

Fig. 2. Pedicellinide (*Pedicellina grandis* Ryland) with numerous spine-shaped "glandular complexes" over the whole length of the zooid.

Fig. 3. Part of Fig. 2 in higher magnification showing the subspinal epithelial clusters.

Figs 4–6. Barentsiidae (*Barentsia gracilis* Sars and *Barentsia discreta* Busk) with pore-like "gland-complexes" in the rigid portions of the stalk: Fig. 4. Zooid from laboratory culture with only a few pores; Fig. 5. Zooid from natural habitat with numerous pores; Fig. 6. Pore of Fig. 4 in higher magnification showing the complex of specialized epithelial cells.

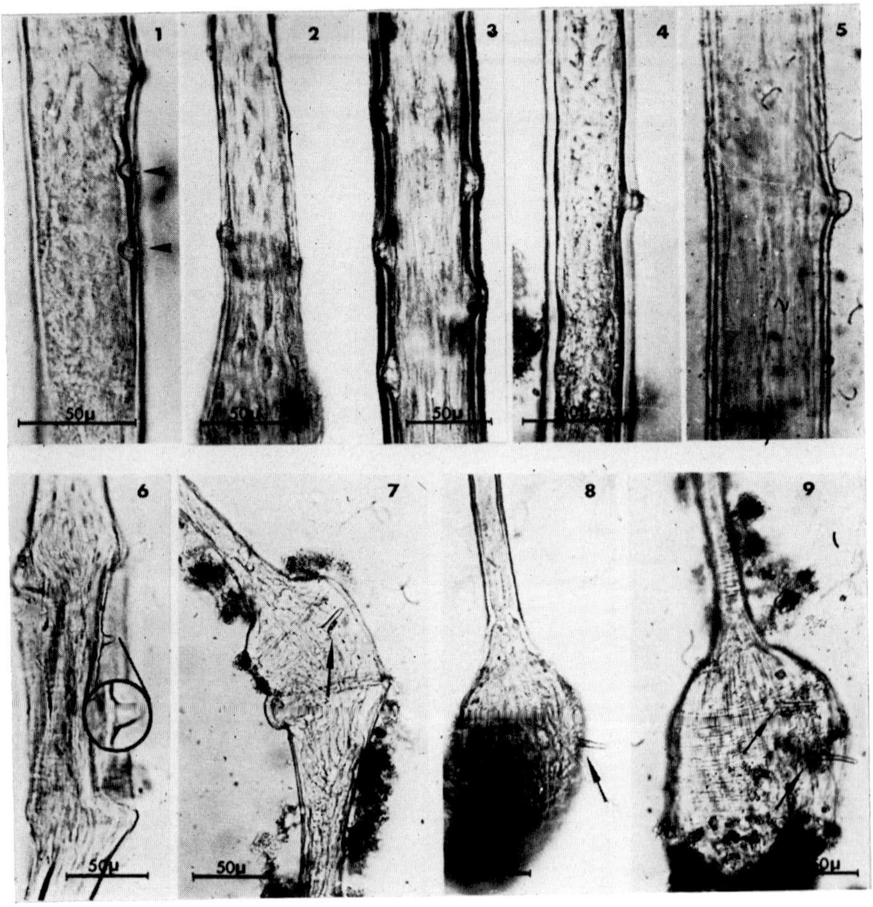
1
2
3
4
5
50µ
50µ
50µ
6
7
8
9
50µ
50µ

heterogeneous contents, extended Golgi fields and a well developed rough endoplasmic reticulum (ER) characterize the appearance of these cells. From the Golgi stacks numerous vesicles are pinched off and spread over the whole cytoplasm. Large elongated mitochondria are arranged mainly around the apical pit. Flattened vesicles and cisternae can be observed lining the lateral and basal cell membranes (Fig. 1) and (Plate III) and are interpreted as parts of a smooth ER; though the inwardly oriented walls of them are partly occupied by ribosomes, the outer leaves, facing the plasma membrane, are smooth in every case. The apical pit is surrounded in a basket-like fashion by a network of microfilamentous bundles.

The lateral cell-membranes interdigitate intimately with neighbouring epithelial cells and are apically affixed to them with tight junctions followed at the inner side by wide septate desmosomes. The pits of the pores are filled with a fibrillar electron dense material, which could be demonstrated to be acid mucopolysaccharides in nature (Behnke and Zelander, 1970; Monis and Zambrano, 1968; Tice and Barrnett, 1965). In the pore-cells, their brush-border and in the ciliary stumps a high sodium concentration was determined by means of precipitation with potassium hexahydroxo-antimonate (Böck, 1970; Komnick, 1962). In the intercellular gaps along the lateral cytomembranes chloride-ions could be precipitated by the silver lactate technique (Böck, 1970; Komnick, 1962), and in the same location a weak activity of alkaline phosphatase could be shown (Daimling, 1964).

As a part of the epithelial layer the pore-cell is separated from the pseudocoel space by a thin filamentous basal lamina. Even in serial sections synaptic connections between the pore-cells and axons of the lateral stalk-nerves of the zooids could in no case be detected.

The whole structure of the pore complexes hints at an ion regulating function like the so-called "chloride cells" described for a great number of aquatic animals,

---

Plate II

Different pore-complexes from *Barentsia gracilis*.

Fig. 1. Zooid from laboratory cultures with pores being closed by cuticular substance during life under culture conditions.

Figs 2 and 3. Different pore frequencies in zooids from culture (Fig. 2) and natural habitat (Fig. 3).

Figs 4–9. Transitions between pores and spines: Fig. 4 shows a normal pore, Fig. 5 shows another one in a thinner, and more flexible cuticularized part of a stalk, and Figs 6–9 show different stages of spine-shaped "pore-glands" in muscular swellings and basal sockets of *Barentsia-gracilis*-zooids.

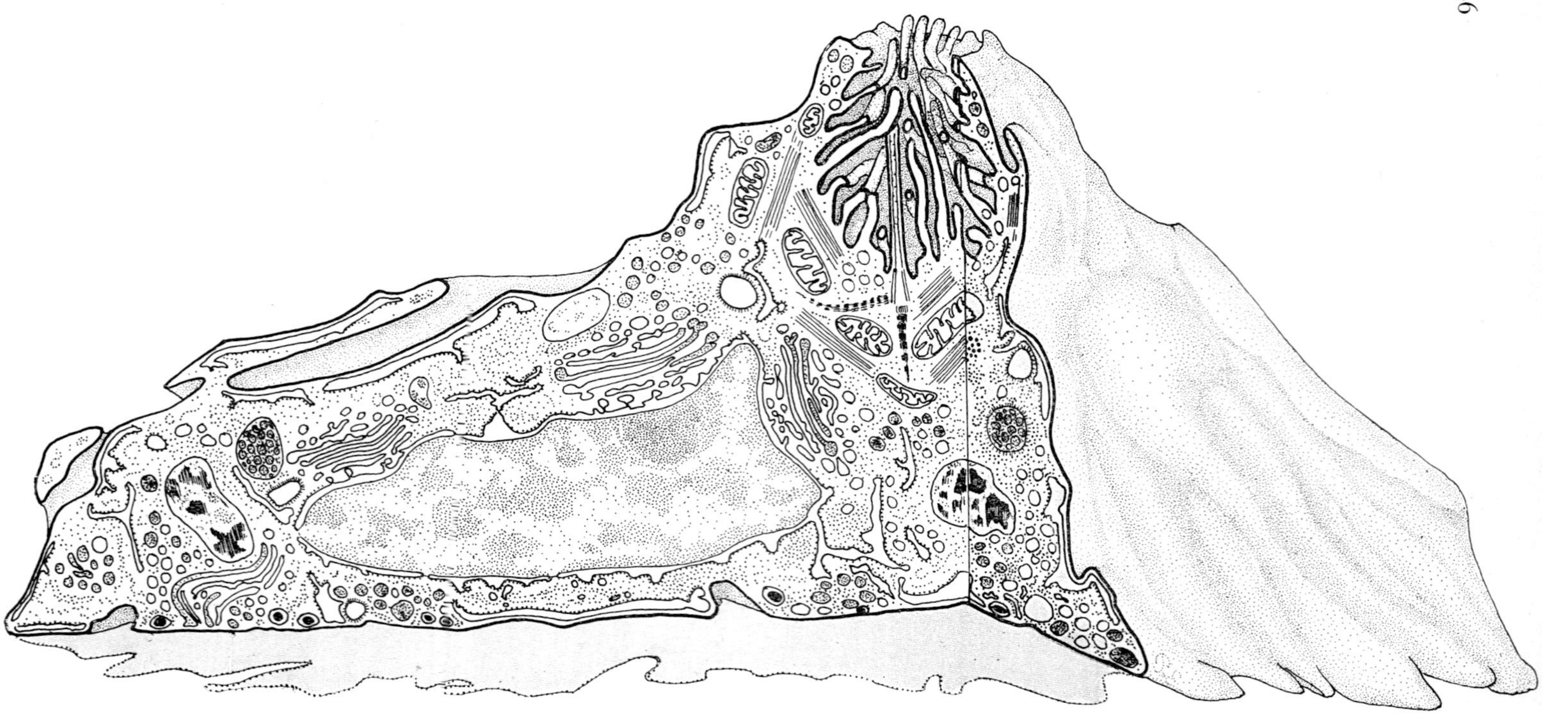

Fig. 1. Diagrammatic three-dimensional reconstruction of a pore cell. On the right hand the outer cell surface with deep infoldings, i.e. interdigitations with neighbouring cells. On the left hand a sector of the cell is cut away the basal area of which conforms with the grey-shadowed field at the bottom.

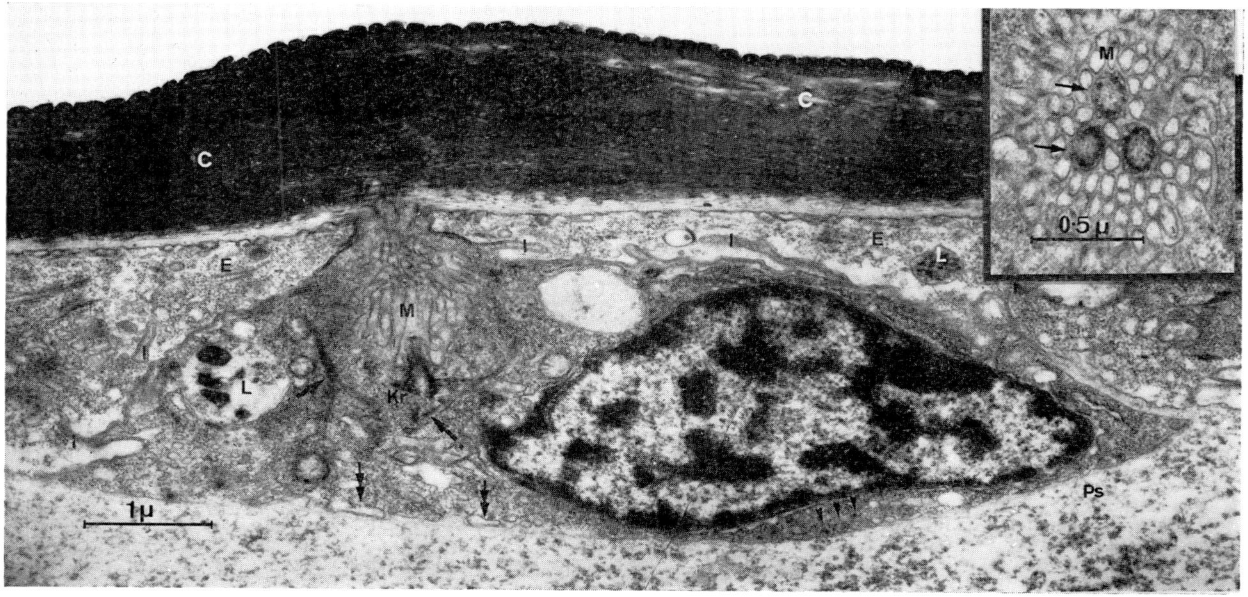

Plate III

Pore cell from *Barentsia gracilis* in low electron microscope magnification; C, cuticle; E, surrounding, specialized epithelial cells; I, sectioned infoldings and interdigitations; Kr, ciliary kinetosome with a short ciliary rootlet; M, microvillous border in the pore-cell's apical pit; L, lysosome-like bodies; Ps, pseudocoel space; —▶▶ submembranous cisternae of smooth ER; —▶ transverse sections of the microfilamentous basket around the apical pit; ▶ Golgivesicles with electron-dense contents in contact with the basal cell membrane. Inset: Part of a transverse section through the apical region and pit of a pore-cell showing cross-sectioned microvilli (M) and ciliary stumps without central tubules and, partly, an incomplete outer circle of peripheral tubules (arrow).

vertebrates and invertebrates. The results of the cytochemical tests support this assumption (Bierther, 1970; Copeland, 1964; Copeland and Fitzjarrel, 1968; Garcia-Romeu and Masoni, 1970; Harb and Copeland, 1969; Kessel and Beams, 1962; Lasker and Threadgold, 1968; Lennep, 1968; Lennep and Lanzing, 1967; Parry *et al.*, 1959; Petřik, 1968; Petřik and Bucher, 1969; Pettengill and Copeland, 1962; Philpott and Copeland, 1963; Schäfer and Schipp; 1968, Skobe, *et al.*, 1970; Threadgold and Houston, 1961; Threadgold, 1964; Wichard and Komnick, 1971). But chloride-cells are never mentioned as having active or rudimentary cilia. This feature on the other hand is reminiscent of protonephridia, which also develop from single specialized epithelial cells, and show in *Urnatella*, a fresh-water living entoproct, for example, a similar distribution pattern along the body, mainly the stalk surface.

As a working hypothesis it is probable that the pore-complex is an ion-regulatory organ, which is homologous to protonephridia. In this case the pore cell must be the point of origin for the evolution (or re-evolution?) of the stalk protonephridia in *Urnatella*, which are further developed in this species (Emschermann, 1965) under the selection pressure of life conditions in fresh water. These considerations also raise the further question of the origin of the pore-organs, for their part. Either one does not impute any phylogenetic relevance to the existence and structure of the pore-organs, i.e. the rudimentary protonephridia, in assuming for the entoproctan epithelia a general ability to develop protonephridia, or, there must have been entoproctan ancestors with numerous protonephridia along all their body surface and the pore-organs may be assumed to be phylogenetic relics of these—a feature which is in general characteristic of fresh-water living animals.

REFERENCES

ATKINS, M. D. (1933). The Loxosomatidae of the Plymouth area. *Q. J. microsc. Sci.* NS **75**, 322–391.

BEHNKE, O. and ZELANDER, T. (1970). Preservation of intercellular substances by the cationic dye alcian-blue in preparative procedures for electron microscopy. *J. Ultrastruct. Res.* **31**, 424–438.

BIERTHER, M. (1970). Die Chloridzellen des Stichlings. *Z. Zellforsch.* **107**, 421–446.

BÖCK, P. (1970). Elektronenmikroskopischer Nachweis von Na^{+}, Ca^{++} und Cl^{-} in der lactierenden Milchdrüse des Meerschweinchens. *Cytobiol.* **2**, 68–82.

COPELAND, E. (1964). A mitochondrial pump in the cells of the anal papillae of mosquitoe larvae. *J. Cell Biol.* **23**, 253–261.

COPELAND, E. and FITZJARREL, A. T. (1968). The salt absorbing cells in the gills of the Blue Crab (*Callinectes sapidus*) with notes on modified mitochondria. *Z. Zellforsch.* **92**, 1–22.

Daimling, O. von (1964). Die Darstellung Phosphat-spaltender Enzyme mittels Schwermetall-Simultanmethoden. *Histochem.* **4,** 48–55.

Ehlers, E. (1890). Zur Kenntnis der Pedicellineen. Abh. kön. Ges. Wiss. Göttingen, *Phys. Kl.* **36,** 1–200.

Emschermann, P. (1965). Das Protonephridiensystem von *Urnatella gracilis* Leidy (Kamptozoa)—Bau, Entwicklung und Funktion. *Z. Morph. Ökol. Tiere* **55,** 859–914.

Garcia-Romeu, F. and Masoni, A. (1970). Sur la mise en évidence des cellules da la branchie des poissons. *Archs Anat. microsc. Morph. exp.* **59,** 289–294.

Harb, J. M. and Copeland, D. E. (1969). Fine structure of the pseudobranch of the flounder *Paralichthys lethostigma*. *Z. Zellforsch.* **101,** 167–174.

Harmer, S. F. (1885). On the structure and development of *Loxosoma*. *Q. J. microsc. Sci.* NS **25,** 261–337.

Hilton, W. A. (1923). A study of the movements of entoproctan Bryozoans. *Trans. Am. microsc. Soc.* **42,** 135–141.

Kessel, R. G. and Beams, H. W. (1962). Electron microscope studies on gill filaments of *Fundulus heteroclitus* from sea water and fresh water with special reference to the ultrastructural organization of the "chloride cell". *J. Ultrastruct. Res.* **7,** 77–87.

Komnick, H. (1962). Elektronenmikroskopische Lokalisation von Na^+ and Cl^- in Zellen and Geweben. *Protoplasma* **55,** 414–418.

Lasker, R. and Threadgold, L. T. (1968). "Chloride cells" in the skin of the larval sardine. *Expl. Cell. Res.* **62,** 582–590.

Lennep, E. W. von (1968). Electron microscopic histochemical studies on salt-secreting glands in Elasmobranchs and marine Catfish. *J. Ultrastruct. Res.* **25,** 94–108.

Lennep, E. W. von and Lanzing, W. J. R. (1967). The ultrastructure of glandular cells in the external dendritic organ of some marine Catfish. *J. Ultrastruct. Res.* **18,** 333–344.

Maturo, F. J. S. and Schopf, Th. J. M. (1968). Ectoproct and Entoproct type material: Reexamination of species from New England and Bermuda named by A. E. Verril, J. W. Dawson, and E. Desor. Postilla, *Peabody Mus. Nat. Hist. Yale Univ.* (*New Haven*) **120,** 1–95.

Monis, B. and Zambrano, D. (1968). Ultrastructure of transitional epithelium of man *Z. Zellforsch.* **87,** 101–117.

Parry, G., Holliday, F. G. T. and Blaxter, J. H. S. (1959). "Chloride-secretory" cells in the gills of teleosts. *Nature, Lond.* **183,** 1248–1249.

Petřik, P. (1968). The demonstration of chloride ions in the "chloride cells" of the gills of eels (*Anguilla anguilla*) adapted to sea water. *Z. Zellforsch.* **92,** 422–427.

Petřik, P. and Bucher, O. (1969). A propos des "chloride cells" dans l'épithélium des lamelles branchiales du poisson rouge. *Z. Zellforsch.* **96,** 66–74.

Pettengill, O. and Copeland, D. E. (1962). Alkaline phosphatase activity in the chloride cell of *Fundulus heteroclitus* and its relation to osmotic work. *J. exp. Zool.* **108,** 235–241.

Philpott, C. W. and Copeland, D. E. (1963). Fine structure of chloride cells from three species of *Fundulus*. *J. Cell Biol.* **18,** 389–404.

Schäfer, A. and Schipp, R. (1968). Vergleichende elektronenmikroskopische Untersuchungen an den Kiemen verschiedener Teleostier zur osmoregulatorischen Funktion der Chloridzellen. *Verh. Anat. Ges.* (*Jena*) **63,** *Vers. Leipzig*, 277–289.

SKOBE, Z., GARANT, PH. R. and ALBRIGHT, J. T. (1970). Ultrastructure of a new cell in the gills of the air-breathing fish *Helostoma temmincki*. *J. Ultrastruct. Res.* **31,** 312–322.

THREADGOLD, L. T. (1964). An electron microscope study of the "chloride cell" of *Salmo salar* L. *Expl. Cell Res.* **34,** 1–23.

THREADGOLD, L. T. and HOUSTON, A. H. (1961). An electron microscope study of the "chloride secretory cell" of *Salmo salar* L., with reference to plasma electrolyte regulation. *Nature Lond.* **190,** 612–614.

TICE, L. W. and BARRNETT, R. J. (1965). Diazophthalocyanins as reagents for fine structural cytochemistry. *J. Cell. Biol.* **25,** 23–41.

WICHARD, W. and KOMNICK, H. (1971). Electron microscopical and histochemical evidence of chloride cells in tracheal gills of mayfly larvae. *Cytobiol.* **3,** 215–228.

25 | Growth in Bryozoans of the Order Fenestrata

T. GARY GAUTIER

Department of Paleobiology, Smithsonian Institution, Washington, D.C., U.S.A.

Abstract: Previous interpretations of fenestellid wall structure, and two fundamental elements of colony formation seen in living cyclostomes are used to interpret variations in skeletal microstructure in some fenestratid bryozoans from the Permian of the Glass Mountains, West Texas. Skeletal variations in these acanthocladiids are explained as representing differences in deposition between areas of wall extension (nonlaminated material) and areas of wall thickening (laminated material). Rods traversing laminated layers could have been grown by special cells.

Studies on Permian fenestratid Bryozoa from the Glass Mountains, West Texas, support statements of Elias and Condra (1957) and Tavener-Smith (1969) that fenestratids grew somewhat like hornerid cyclostomes. Growth in those bryozoans has two basic characteristics: (1) the colony is enveloped by a balloon-like layer of cuticle and associated epidermis which expands from the aperture of the ancestrula and continues to expand in the region of zooidal budding, and (2) coelomic space within the expansion is differentiated into zooidal coelom and hypostegal coelom by growth of interior walls which, by definition, lack intramural extensions of the cuticular colony envelope. Skeletal microstructure is a variable factor in this general pattern of growth, and the characteristic combined development of laminated layers, nonlaminated layers and rods in fenestratids appears to be just one of the many possibilities.

In acanthocladiids from the Permian of the Glass Mountains, as in other fenestratids, rod-bearing laminated skeleton was deposited on external surfaces outside the zooecial cavities. Following the reasoning of Tavener-Smith (1969) for fenestellids, such deposition implies that a mantle consisting of two closely spaced epithelial layers separated by coelom covered those surfaces and provided a channel of communication between tissues and coelom in the zooecial cavities (Fig. 1a). Terminal zooecia of the acanthocladiids often became skeletally complete before the next ones appeared (Fig. 1a). On the outer sides of terminal zooecia, rod-bearing laminated skeleton was deposited (Fig. 1b) which was continuous proximally with similar deposits on the reverse skeleton of the branch. The mantle is inferred, therefore, to have been continuous from the reverse surface over the outer side of terminal zooecial walls to the apertures (Figs. 1a–b).

The outer rod-bearing laminated layer of the zooecial wall is part of a skeletal unit which surrounds most of the zooecial cavity (Fig. 2a). The inner layer of the skeletal unit is nonlaminated (Figs 1b, 2a). Nonlaminated calcite rods extend

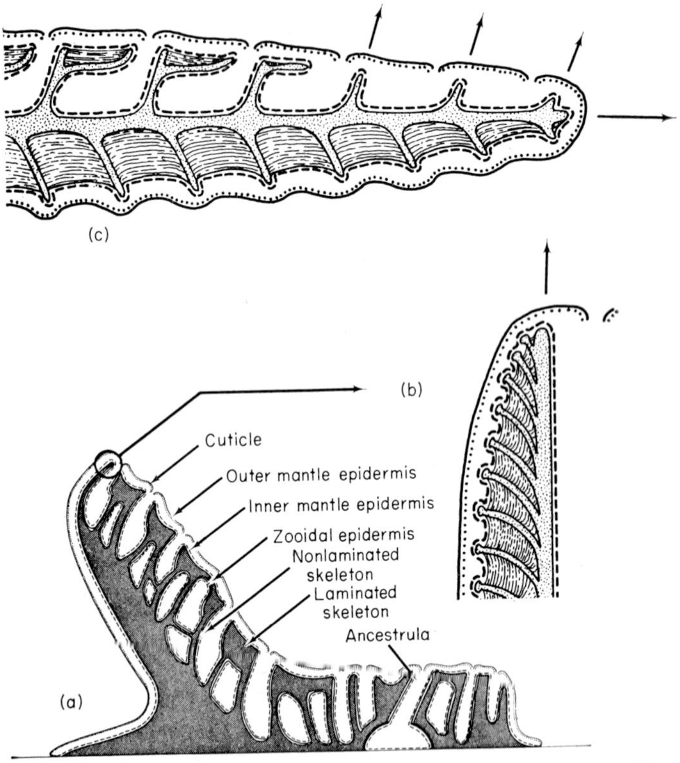

Fig. 1a–c. Arrows denote the direction of extension of various nonlaminated layers which are indicated by random dot patterns in (b) and (c). (a), Diagrammatic longitudinal section through a base of attachment of an acanthocladiid colony. (b), Structure of the zooecial wall segment indicated in (a). Rods protruding at the surface have bulbous tips. The diameter of the rods is generally less than 10 μ and it seems probable that each rod was deposited by a single epithelial cell. Cells were probably proliferated at the wall margin. (c), Diagrammatic longitudinal section through a branch tip of a species of *Fenestella*; partly based on the morphology of silicified branch tips from specimens from the Glass Mountains, West Texas.

from the nonlaminated layer through the laminated layer to the outer surface where they protrude (Fig. 1b), thus confirming that they were present during the life of the animal and not just diagenetic fillings of what were originally tubes.

The skeletal unit was constructed against the upturned reverse surface of the branch tip, beginning as an outgrowth which progressively lengthened (Fig. 2b). Part of the once exterior surface of previously terminal zooecia then became the inner proximal wall of the new zooecial cavity, and rod-bearing laminated mantle deposits thus became part of the thin wall separating zooecial cavities (Fig. 2b). This mode of growth implies that zooidal soft tissues and coelom were derived from the colonial mantle tissues and coelom, not from a parent zooid.

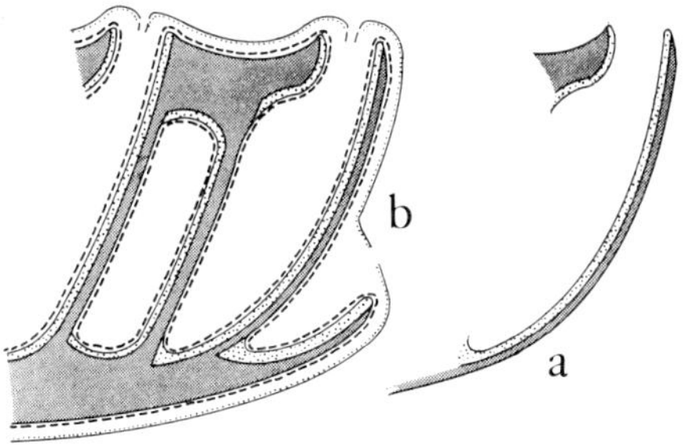

Fig. 2. (a), Detached skeletal unit of an acanthocladiid. (b), Diagrammatic longitudinal section through the branch tip of an acanthocladiid showing the initial outgrowth of the skeletal unit from the reverse surface. Note that the nonlaminated layer of the new skeletal unit abuts onto the outer side of the antecedent skeletal unit. In both figures, nonlaminated skeleton is shown by the random dot pattern and the remaining symbols are as on Fig. 1a.

The outer laminated layer of the skeletal unit is inferred to have been deposited on the side of the wall because laminations are essentially parallel to the plane of the wall (Fig. 1b) and were presumably also parallel to the depositing inner mantle epidermis. Deposition on the margin of the skeletal unit which resulted in its extension must have taken place, therefore, across the end of the nonlaminated layer. This mode of growth is also indicated by the initial low angle at which rods diverge from the nonlaminated layer of some skeletal units (Fig. 1b). Rod axes everywhere in the zoarium extend at a high angle to the surface of deposition. Drawing a line from rod to rod, at a high angle to each rod, produces an approximate reconstruction of growing margin shape showing that the surface of deposition lies across the nonlaminated layer when rods are curved as shown in Fig. 1b.

Nonlaminated layers in these acanthocladiids are inferred, therefore, to represent skeletal secretion in the direction of wall extension, and rod-bearing laminated skeletal layers are inferred to be wall thickening mantle deposits. These ideas and other general features of acanthocladiid growth seem also applicable to fenestellids. The only basic difference between the groups, so far as the skeletal makeup is concerned, is that in fenestellids there are no rod-bearing laminated deposits in walls separating zooecial cavities which implies that no mantle was formed on those walls. Skeleton surrounding zooecial cavities is entirely nonlaminated (not counting zooecial linings, if present) (Fig. 1c).

Ontogenetic gradients observed in some silicified fenestellid branch tips (Fig. 1c) indicate that the nonlaminated walls separating zooecial cavities grew upward from a nonlaminated basal layer (absent in the acanthocladiids studied) which was growing in the direction of branch extension. Rod-bearing laminated skeleton was deposited on the underside or reverse side of that interior basal layer (Fig. 1c). It appears that developing zooidal tissues occupied the whole upper surface of the basal layer at the tip. Lack of obvious external surfaces (outside the zooecial cavities) implies the absence of a mantle in that part of the colony. This in turn explains the lack of rod-bearing laminated mantle deposits in the zooecial walls. Thus, only nonlaminated skeletal deposits representing extension of various wall layers make up the skeleton surrounding zooecial cavities.

These conclusions imply that, as in cyclostomes, fenestratid zoaria grew by progressive secretion of interior walls. No cuticle was found or is inferred to be present in them.

REFERENCES

Elias, M. K. and Condra, G. E. (1957). *Fenestella* from the Permian of West Texas. *Geol. Soc. Am. Mem.* **70,** 158.

Tavener-Smith, R. (1969). Skeletal structure and growth in the Fenestellidae (Bryozoa). *Palaeontology*, **12** (2), 281–309.

26 | A Fine-structure Study of Brown Bodies in the Gymnolaemate *Cryptosula pallasiana* (Moll)

D. P. GORDON

Biology Department, Dalhousie University, Halifax, Nova Scotia, Canada

Abstract: An EM study of brown body formation in *Cryptosula* has shown that this process closely resembles regression of aging cells and tissues of other animals. There is an increase in the number of residual bodies in all tissues, and degradation tends to be self-contained within parent cell membranes. Only in the late stages of lophophore disintegration do cells actually rupture, whereas most gut cells fuse with neighbours or undergo heterophagy, contributing to the final brown body, which comprises huge cells with few or no functional organelles, but with large areas of residue. Glycogen accumulates during regression but becomes depleted as a new distal polypide develops.

INTRODUCTION

Brown bodies have been noted in bryozoans ever since Ellis saw them in 1755, and theories concerning their nature and purpose have been varied. In 1877 Joliet reviewed the ideas up to that time. These were quite diverse; brown bodies had been thought to be ovaries, eggs, incipient larvae, endocystal secretion, the equivalent of phylactolaemate statoblasts, special formations detached from the stomach, or germ capsules from which new polypides arose. Joliet himself, and Nitsche a little earlier in 1871, correctly regarded brown bodies as degenerated polypides. Nowadays we also know that brown body formation is induced by suboptimal water conditions, onset of reproduction (predominantly in those species which brood larvae internally), overgrowth of colonies, and accumulation of inclusions in gut cells, to which natural senescence might be related (Hyman, 1959; Ryland, 1970). Because brown body formation is induced by so many factors its functional significance has not been positively assigned to any of these factors separately, although it is probably related to them collectively. An electron microscopic study of brown body fine structure was undertaken to obtain further information on this noteworthy phenomenon.

MATERIALS AND METHODS

All observations were made on *Cryptosula pallasiana* (Moll), a common marine fouling gymnolaemate. Freshly collected colonies were scraped off rocks or algae and fixed in ice-cold 3% glutaraldehyde in Millonig's buffer (or 0·1 M cacodylate buffer) at pH 7·4 for 3 h. Both buffers gave equally satisfactory results. After washing, specimens were post-fixed in osmium tetroxide (1%) buffered at pH 7·4, and dehydrated in acetone. Polypides and brown bodies were either dissected out of zooecia prior to epon embedment or colonies were decalcified in 10% EDTA (ethylenediaminetetraacetic acid) after glutaraldehyde fixation and subsequently embedded undissected. Thin sections were cut with glass knives or a DuPont diamond knife, mounted on formvar coated 200 mesh copper grids or on uncoated 300 mesh grids, and stained in aqueous 4% uranyl acetate (or saturated in 70% methanol) for 10 min and in Reynold's (1963) lead citrate for 5 min. Micrographs were taken on a Zeiss EM 9 operated at 60 kv. Thick (1–2 μm) sections were cut to supplement electron microscopy, and stained in toluidine blue according to the method of Trump *et al.* (1961) for light microscopic examination. Decalcified colonies were mounted unstained in Permount and photographed using phase contrast.

This paper is primarily concerned with the appearance of the polypide towards the end of regression and the final appearance of the brown body. In order to interpret these stages, however, young and old polypides and polypides beginning to regress were also sectioned, and brief reference will be made to these where pertinent. The fate of the brown body after its formation is not discussed.

RESULTS

1. *General Features of Polypide Regression*

Condensation of the polypide during regression results from cell movement and auto- and heterophagic digestion. It proceeds largely as described for *Alcyonidium polyoum* by Matricon (1960), except that the tentacles do not appear to insert into the rectum via the anus in *Cryptosula*. During regression the muscles of the lophophore retractor maintain much of their structure for a longer time than the intrinsic musculature of the tentacles and gut. The regressing polypide is partly sheathed by mesenchymal cells, by which it is slung from the floor of the ascus and connected to the dorsal body wall. The brown body (or bodies) formed seems to be derived largely from the posterior oesophagus, cardia, central stomach and caecum. The contribution of the tentacles, pylorus and rectum is variable; also some faecal material may be incorporated depending on the state of the gut when regression was initiated. Cells in the mesenchyme and coelom become packed with glycogen as regression proceeds but this falls off in concentration until there is practically no glycogen in the cells surrounding the fully formed brown body.

The intermediate appearance of the regressing polypide (corresponding to stages 2 and 3 of Matricon (1960, pp. 379, 381).

(*a*) *Cells of the lophophore.* Cells at this stage are in an advanced state of regression, with sequestration of organelles for autolysis, and envelopment of sections of the collagenous basal laminae of the tentacles, muscle fibres and ciliary rootlets. Degradation of sequestered material is in evidence. In Plate I, Fig. 1 collagen appears in a finely fibrous subunit form, and the lytic vacuole in Plate I, Fig. 2 can almost be called a primary residual body. Some glycogen is associated with disintegrating muscle and may often be seen enclosed with muscle fibres in autophagosomes.

Myelin figures are common while membranous whorls are less so, occurring mainly at later stages. Endoplasmic reticulum is transformed by vacuolation of cisternae, from which some of the smaller myelin figures may be derived, or reticulum membranes may just disintegrate. In either case ribosomes become liberated to life freely in the cytoplasm.

Mitochondria in the cytoplasm (*ca* 0·3–0·66 μm diam) tend to become vacuolated with loss of cristae. Some of these vacuoles, like those from endoplasmic cisternae, appear to be transformed into incipient myelin figures by "compression" so that opposing vacuole walls come to lie parallel with each other.

Functional Golgi bodies are found in most cells. They are predominantly saucer-shaped and small vesicles 1000 Å diam are pinched off from the periphery of the cisternae, while larger vesicles arise from the concave face. Another type of Golgi body occurs less frequently. It apparently arises only during regression as I have not seen any in the functioning polypide. More deeply cup-shaped, it pinches off coated vesicles 600–1000 Å diam from the periphery (Plate II, Fig. 2). These vesicles in turn appear to be capable of budding off other vesicles, but it is not clear if this budding process also gives rise to the small chains of coated vesicles connected by coated tubules seen in the plate. The most interesting feature of this type of Golgi body is the strange macrovesicles that arise from the concave face. These vesicles contain what look like membranous elements, and these somehow become deposited *inter*cellularly as spheres of material. The purpose of these vesicles and their contents is unknown. Identical macrovesicles have been found in bivalve digestive cells by Owen (1970) but he was not able to ascribe a function either.

Nuclei still have fairly dense contents and one or two nucleoli with granular

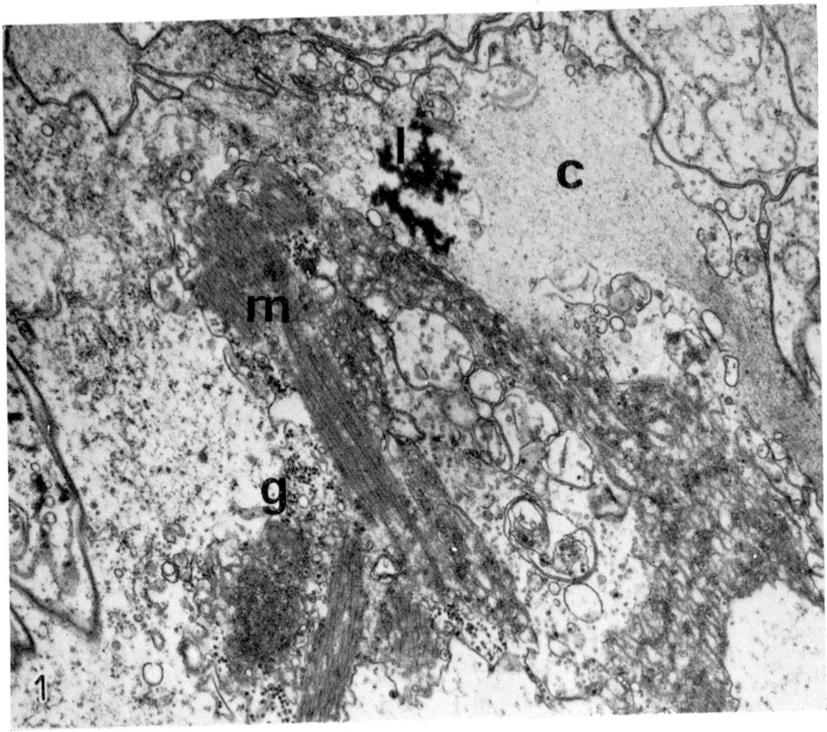
l
c
m
g
1

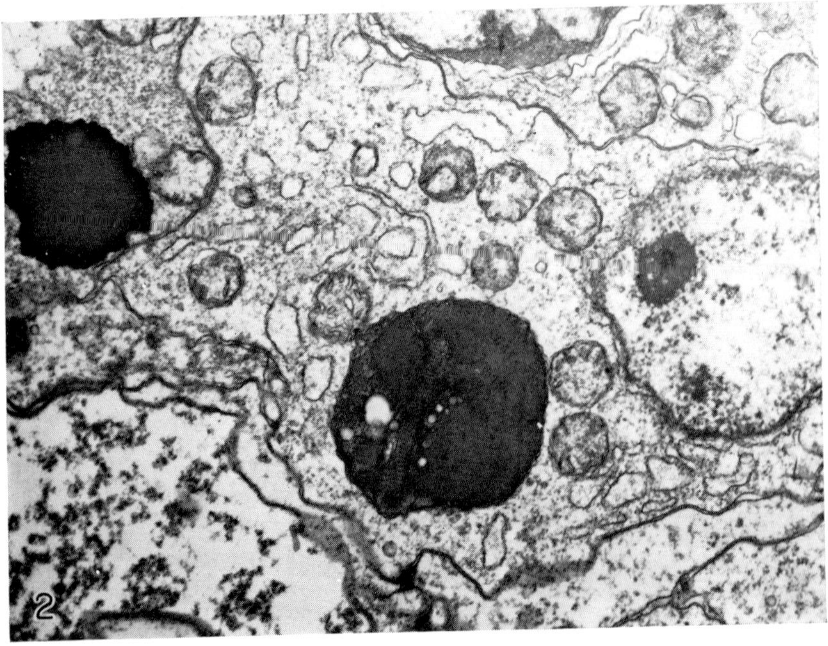
2

nucleonemata. There is no noteworthy peripheral condensation of chromatin, which is sometimes reported to accompany cell degradation (Goldsmith, 1966; David, 1970).

Microtubules are occasional. Nerve axons, recognizable by their neurotubules and neurofilaments, appear to maintain their structural integrity for some time after the beginning of polypide regression.

With further regression of the tentacles, cell membranes rupture, liberating mitochondria, residual bodies and other cytoplasmic material freely into the coelom.

(*b*) *Cells of the gut.* While the gut as a whole, at this stage, is structurally more intact than the tentacles, individual cells may be somewhat further advanced in degradation. This is due to the early accumulation in the gut (virtually from the time a new polypide is fully functional) of the orange-brown inclusions that are commonly seen in the bryozoan gut with the light microscope. These inclusions come to form a considerable part of the brown body (Plate III). Electron microscopic examination of stomach cells from a polypide prior to regression revealed that the inclusions are residual bodies, comprising membrane fragments, amorphous aggregates and osmiophilic granules (Plate IV, Fig. 1). These residual bodies are derived from cellular autophagy rather than the remains of intracellular digestion. Skeletal remains of food organisms do occur, however, and as these may be found intracellularly in degenerating polypides, they are not always eliminated from gut cells prior to the commencement of brown body formation.

Apart from these residual bodies, the remainder of the cytoplasm comprises mitochondria, little granular endoplasmic reticulum, free ribosomes, nuclei and Golgi bodies. The latter are considerably smaller than they originally were in secretory cells of the gut and are like those already described in the regressing lophophore cells. The mitochondria are predominantly spheroidal, with few cristae and partial loss of the outer membrane in some cases. The chromatin of the nuclei is more diffuse than in cells of the tentacles. Many of these organelles

---

Plate I

Cryptosula pallasiana

Fig. 1. Regressing tentacle cells showing fragmentation of muscle m, and areas of collagen c, glycogen g and lipid l. × 18,400.

Fig. 2. Epithelial cells of tentacle showing two autophagic vacuoles, one of which contains a partially digested ciliary rootlet. Note the spheroidal mitochondria with few cristae and the vesicular nature of the endoplasmic cisternae. × 16,700.

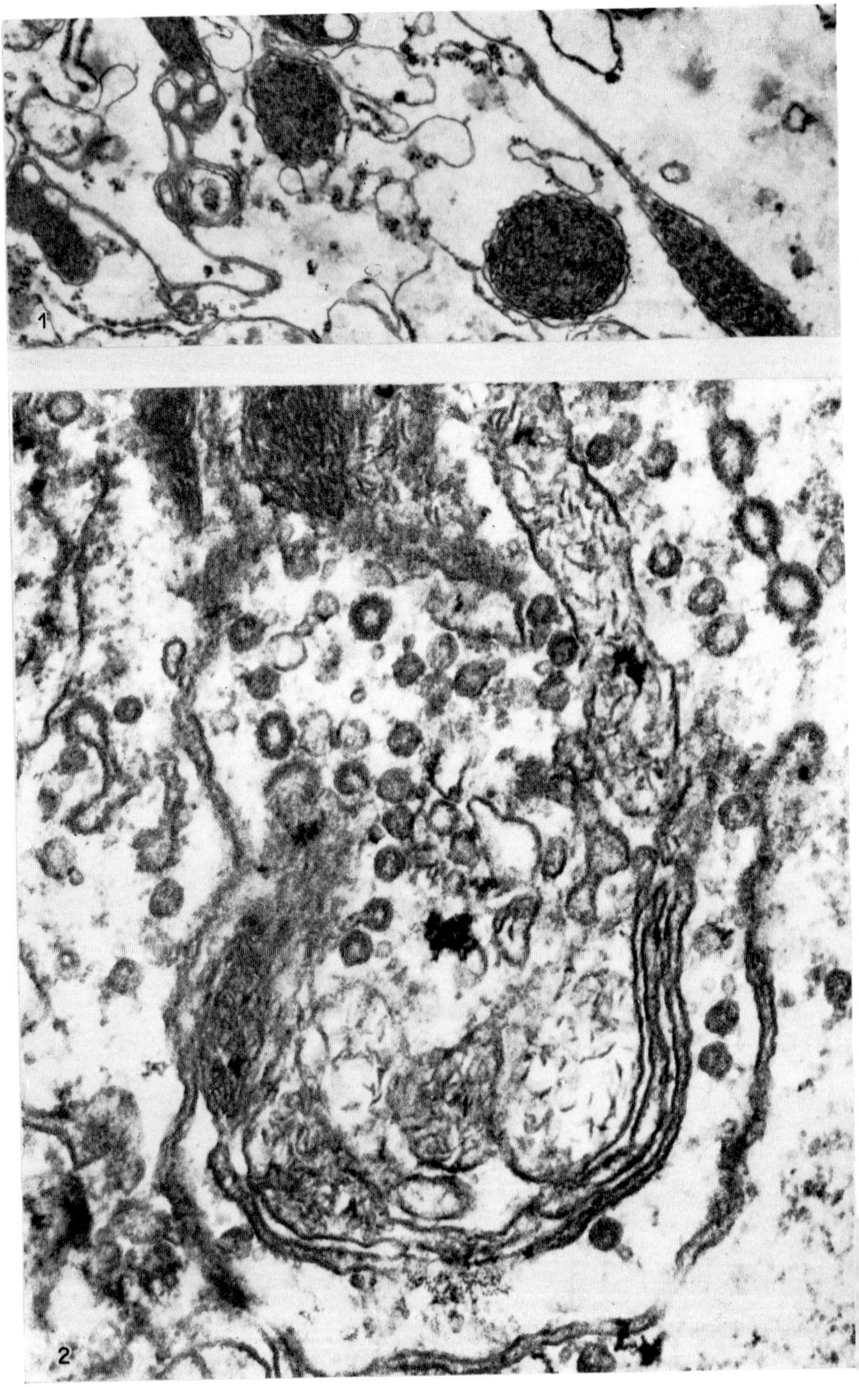
1
2

become incorporated into lysosomes. With heterophagy of cells by neighbours many nuclei are digested leaving apparently only one nucleus per enlarged cell.

2. The Fully Formed Brown Body

This is a mass of residual material, which is contained in greatly enlarged cells up to 18 μm diam. It is not known if these large cells arise entirely because of heterophagy. Fusion of cells with breakdown of membranes at the site of fusion would produce the same result, but I could not always positively distinguish this from heterophagy. Most of the interior of these cells is occupied by a membrane-bounded cluster of residual bodies or one huge residual body. This residual material comprises undigested membrane fragments, myelin figures, amorphous aggregates, osmiophilic granules and occasional small vesicles, all lying in a fine fibrous matrix. A thin band of cytoplasm with a few mitochondria, Golgi bodies, vacuoles and nucleus initially surrounds the residual masses; although this too, disappears by autophagy, vacuolation and membrane fragmentation, the plasma membrane of each of the large cells remains intact. The cell destruction that takes place during brown body formation, is seen in the gut cells to be an orderly process, and while there is much heterophagy and possibly cell fusion, cell death is nonetheless a self-contained phenomenon.

As a larger part of the brown body is derived from gut tissues it is mostly bounded by the fibrous basal lamina of the original gut. While this lamina becomes degraded, it never disintegrates entirely.

Cells surrounding the regressed gut are in a degenerate or ruptured state. Some of these are mesenchymal while most are former lophophore cells. In intact cells the cytoplasm contains residual bodies, myelin figures, whorls, lipid, and bizarre membranous structures quite unlike any in the brown body proper.

DISCUSSION

Qualitative alterations of cell organelles during polypide regression are very similar to changes in cells of organs which have been observed in a state of

---

Plate II

Cryptosula pallasiana

Fig. 1. Remnants of lophophoral tissue showing intercellular deposits of membrane-like elements. × 27,900.

Fig. 2. A deeply cup-shaped Golgi body showing peripheral budding of coated vesicles and the origin of the strange macrovesicles, containing membrane-like elements, from the concave face. × 75,400.

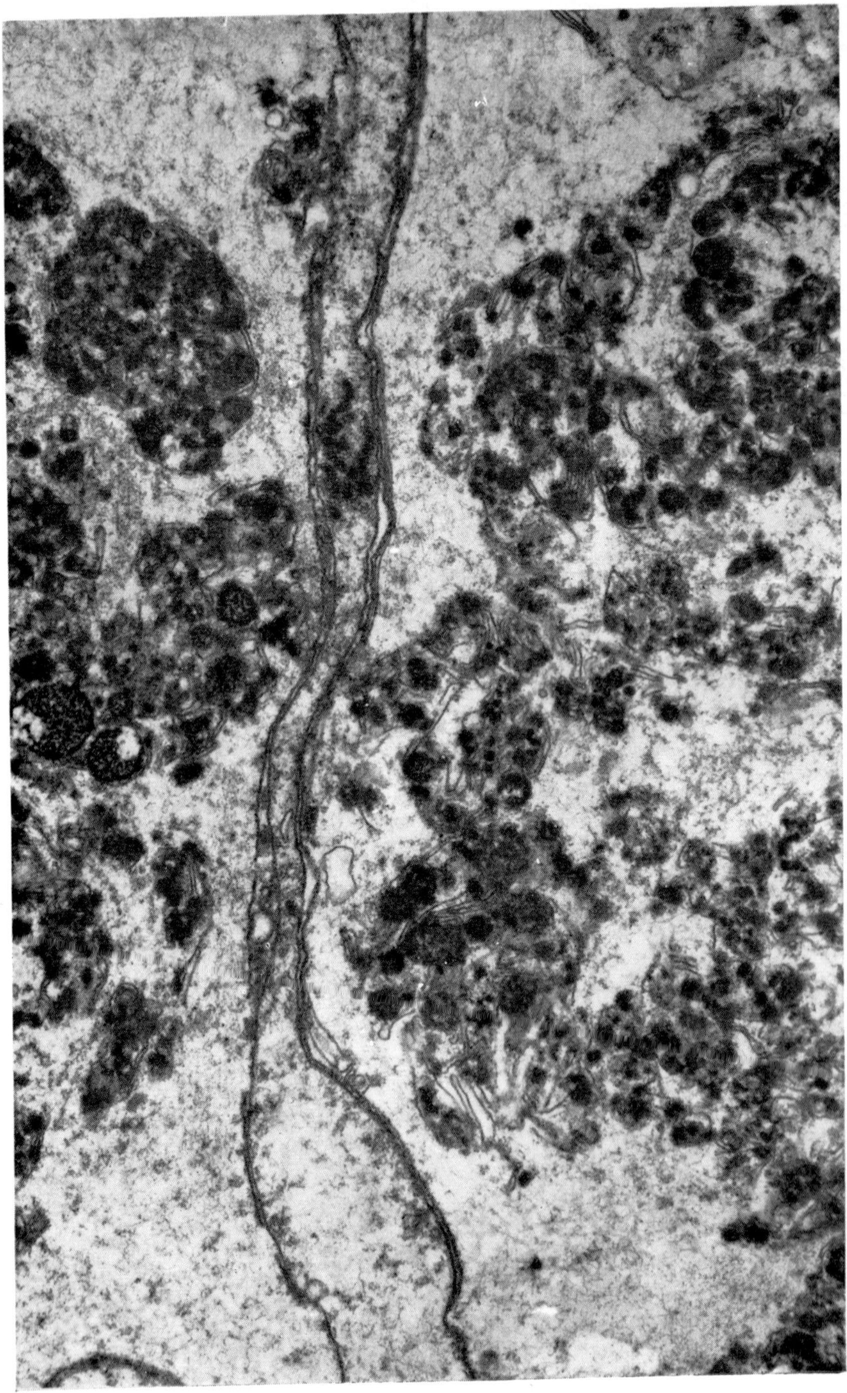

regression in other organisms. These are as follows: progressive accumulation of residual bodies, the tendency of mitochondria to become spheroidal or vacuolated with loss of cristae and vacuolation of endoplasmic cisternae and possible membrane disintegration, with concomitant liberation of ribosomes.

Residual bodies in the fully formed brown body, with osmiophilic granules, aggregates and various membranous structures are practically identical to structures seen in regressing *Campanularia* hydranths (Brock, 1970), cockroach prothoracic gland (Scharrer, 1966), amphibian intestine during metamorphosis (Hourdry, 1971) and amphibian exocrine pancreas during hibernation (Geuze, 1970).

The tendency of mitochondria towards sphericity with cristar loss and sometimes loss of part of the outer membrane is not universal in regressing animal tissues. While a similar tendency was observed in mammalian liver cells adjacent to obstructed blood vessels by David (1970), Bonneville (1963) reported no marked change in mitochondria during regression of gut cells during bullfrog metamorphosis.

Vacuolation of endoplasmic cisternae is common in regressing or malfunctioning animal tissues and has been reported by numerous authors (David, 1970, p. 79). Decrease in numbers of bound ribosomes is reported to occur with increasing age in *Tokophrya* (Rudzinska, 1962) and a rotifer (Herold and Meadow, 1970). However, this is not strictly comparable to the situation in brown body formation, because regression of a polypide to a brown body is a comparatively later stage, and in this respect it would be interesting to compare organelle populations in young and old polypides prior to the commencement of brown body formation.

Cell destruction in degenerating animal tissues is generally self-contained within parent cell membranes or those of neighbouring cells should the former become phagocytosed. This is the case in the degenerating gut cells of *Cryptosula* and in the brown body itself, and (with the persistence of the basal lamina) accounts for the maintenance of the general shape of the gut during regression. The lophophoral tissues, on the other hand, regress in a less organized fashion, in which cells may eventually rupture. The self-contained regression of the gut cells is like that observed in *Rhodnius* by Goldsmith (1966). The

---

Plate III

Cryptosula pallasiana. Portion of brown body showing boundaries between cells containing membranous structures, granules and aggregates in fine, fibrous material. × 27,900.

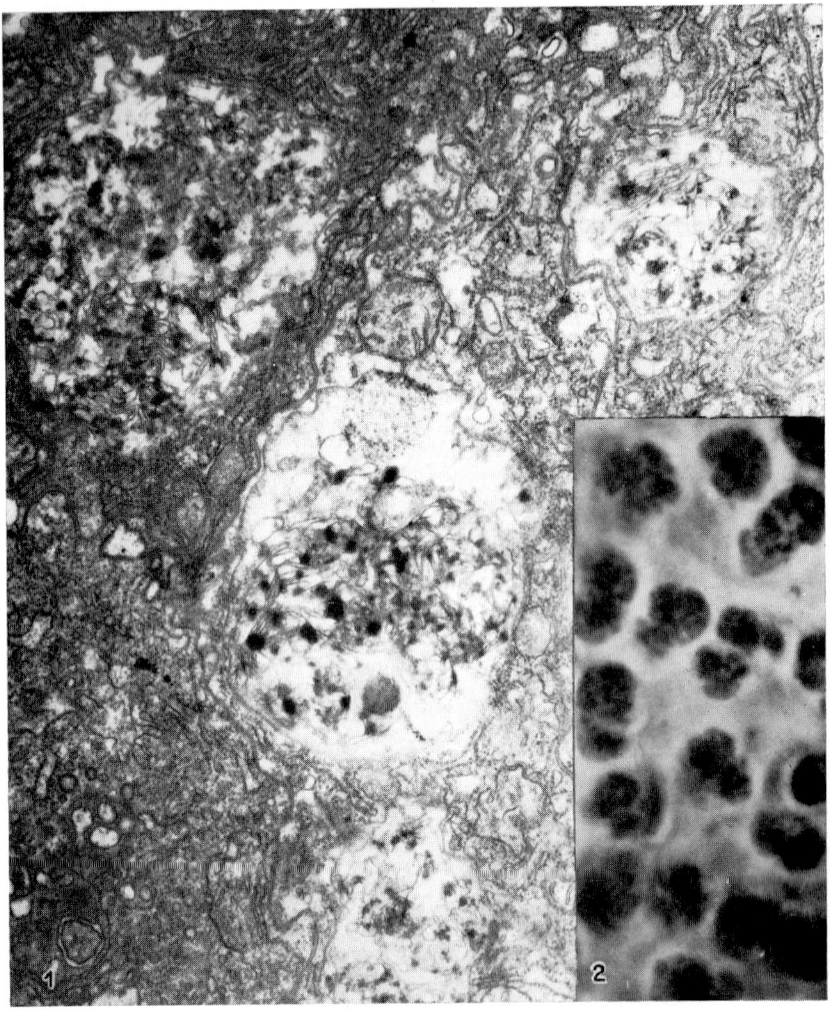

Plate IV

Cryptosula pallasiana

Fig. 1. Cells of central stomach region of a functional polypide showing the fine structure of the orange-brown inclusions. $\times$ 23,000.

Fig. 2. Larger orange-brown inclusions as seen by the light microscope. Phase contrast, $\times$ 1170.

regression of the lophophore is like that observed by Scharrer (1966) who reported complete breakdown of cell membranes in regressing insect prothoracic gland at a time when many organelles had not yet entered into autolysis.

The point to be noted is that while brown body formation (and polypide regeneration) may be a characteristic bryozoan feature, the cellular changes seen during regression are not. Administration of drugs or toxins, or ligation of blood vessels will produce similar regressive responses in cells of vertebrates, and a wide range of external and internal factors will produce these responses in cells of invertebrates, whether or not regression is the end result of ageing. The question to be asked then, is this: is brown body formation in bryozoans a *generalized tissue response* initiated by a range of suboptimal conditions (including accumulation of residual and skeletal material in the gut)? In view of its apparent regularity in the absence of provocative external factors and the fact that residual body formation in gut cells occurs very early in the life of the polypide, the possibility of an internal control of this phenomenon, possibly hormonal, should not be overlooked. In either case, the functional significance of brown body formation would seem to lie in the fact that it is (1) generalized and (2) (because a new polypide arises) rejuvenatory. The extensive glycogen deposits of the mesenchyme that occur with regression and which become finally depleted, should predictably serve as an energy supply for a simultaneously developing polypide.

REFERENCES

BONNEVILLE, M. A. (1963). Fine structural changes in the intestinal epithelium of the bullfrog during metamorphosis. *J. Cell. Biol.* **18,** 579–597.

BROCK, M. A. (1970). Ultrastructural studies on the life cycle of a short-lived metazoan, *Campanularia flexuosa* II Structure of the old adult. *J. Ultrastruct. Res.* **32,** 118–141.

DAVID, H. (1970). Zellschadung und Dysfunktion. *Protoplasmatologia* **10** (1), 1–621.

ELLIS, J. (1755). "An Essay Towards a Natural History of the Corallines." (Not seen.)

GEUZE, J. J. (1970). Light and electron microscopic observations on auto- and heterophagy in the exocrine pancreas of the hibernating frog (*Rana esculenta*). *J. Ultrastruct. Res.* **32,** 391–404.

GOLDSMITH, M. (1966). The anatomy of cell death. *J. Cell. Biol.* **31,** 41A (Abstr.).

HEROLD, R. C. and MEADOW, N. D. (1970). Age-related changes in ultrastructure and histochemistry of rotiferan organs. *J. Ultrastruct. Res.* **33,** 203–218.

HOURDRY, J. (1971). Evolution des processus lytiques dans l'epithelium intestinal de *Discoglossus pictus* Otth (Amphibien Anoure), au cours de sa metamorphose. *J. Microsc.* **10,** 41–58.

HYMAN, L. H. (1959). "The Coelomates: Smaller Coelomate Groups V." 783 pp. McGraw-Hill, New York.

JOLIET, L. (1877). Contributions a l'histoire naturelle des Bryozoaires des côtes de France. *Arch. Zool. exp. Gen.* **6,** 193–304.

MATRICON, I. (1960). Etude histologique d'*Alcyonidium polyoum* (Hassall) caractères structuraux spécifiques d'*Alcyonidium mytili* (Dalyell) (Bryozoaires Cténostomes). *Cah. Biol. Mar.* **1,** 359–395.

NITSCHE, H. (1871). Uber die Anatomie und Entwicklungsgeschichte von *Flustra membranacea*. *Z. wiss. Zool.* **21,** 1–53.

OWEN, G. (1970). The fine structure of the digestive tubules of the marine bivalve *Cardium edule*. *Phil. Trans. R. Soc. Lond.* B **258,** 245–260.

REYNOLDS, E. S. (1963). The use of lead citrate of high pH as an electron-opaque stain in electron microscopy. *J. Cell. Biol.* **17,** 208–212.

RUDZINSKA, M. A. (1962). The use of a protozoan for studies in aging III Similarities between young overfed and old normally fed *Tokophrya infusiorum*. A light and electron microscopic study. *Gerontologia* **6,** 206–226.

RYLAND, J. S. (1970). "Bryozoans." Hutchinson University Library, London.

SCHARRER, B. (1966). Ultrastructural study of the regressing prothoracic glands of blattarian insects. *Z. Zellforsch.* **69,** 1–21.

TRUMP, B. F., SMUCKLER, E. A. and BENDITT, E. P. (1961). A method for staining epoxy sections for light microscopy. *J. Ultrastruct. Res.* **5,** 343–348.

27 | Mode of Growth of the Cupuladriidae (Bryozoa, Cheilostomata)

ECKART HÅKANSSON

Universitetets Institut for Historisk Geologi og Palaeontologi, Østervoldgade, Copenhagen, Denmark

Abstract: A growth model is proposed for the free living *Cupuladria biporosa*, based on structures revealed by sections containing both hard and soft tissue together.

Growth of interior walls which fuse with the frontal cuticle, separates the coelomic space of the settled larva into a triradiate ancestrular triplet, encircled by a narrow colonial hypostegal coelom. With buds developing as swollen bladders in this colonial coelom, the colony is gradually growing by formation of interior walls cutting out portions of the radially expanding colonial hypostegal coelom. The walls separating the autozooidal-vibracular units from the colonial coelom contain a core of primary skeletal material surrounded by secondary skeletal material, in such a way that the primary skeletal material within each zooidal unit is isolated from that of any other unit. The secretion of secondary skeletal material on the basal (extrazooidal) side of the zooidal units continues, perhaps throughout the life of the colony, forming the constantly thickening basal wall.

Preliminary studies on three more species of Cupuladriidae and on other free living anascans indicates that the suggested mode of growth is restricted to and characteristic for the Cupuladriidae.

In order to compare the mode of growth of Cretaceous Lunulites to that of the externally very similar free living Cupuladriidae, thin sections were prepared following methods worked out at National Museum of Natural History, Washington D.C. (described in Nye *et al.*, 1972). As the results obtained for the Cupuladriidae turned out to be rather unexpected, a short preliminary account of this part is given here. Sections of fossil and Recent specimens of four species, *Cupuladria biporosa* Canu and Bassler, *C. canariensis* (Busk), *C. doma* ('d'Orbigny), and *Discoporella umbellata* (Defrance), have been included in the study, with most emphasis laid upon *C. biporosa*, as alcohol-preserved specimens have been available only for this species. Hence the following presentation is primarily concerned with *Cupuladria biporosa*. It is important to note that the sections of alcohol-preserved specimens have been prepared with both skeletal and soft tissue together, and, in order to obtain good contrast between these parts, the sections have been stained using Fleming's triple stain (Nye *et al.*, 1972).

INTERNAL STRUCTURES

The external morphological characters of *Cupuladria biporosa* and the Cupuladriidae as a whole have been discussed by many authors, most notably by Cook (1965a, b). Internal structures, however, have received far less notice.

Orientated thin sections in three directions, longitudinal, transverse, and tangential (see Fig. 1), show that the colony of *Cupuladria biporosa* has a very regular internal pattern resulting from repeated occurrence of a few basic

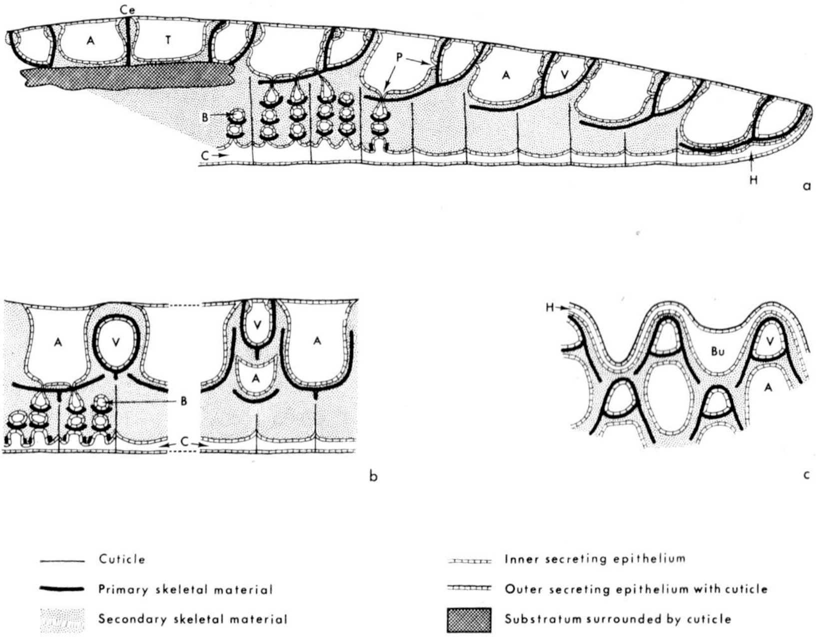

Fig. 1. Idealized diagrammatic sections of *Cupuladria biporosa*, showing the distribution of the major skeletal elements and coelomic spaces. Lateral communication pores with associated primary skeletal material have been omitted. (a), Longitudinal section through the center of the colony. In the distal autozooidal-vibracular units the section is median (showing longitudinal communication pores), and closer to the substratum it is slightly lateral (showing communication pores leading to the basal chambers). (b), Transverse section (with most of the basal chambers omitted). Left and right portion of the figure in slightly different levels within the zooidal units. (c), Tangential section in the growing edge. *Abbreviations:* A, autozooid; B, basal chamber; Bu, young bud in the colonial hypostegal coelom; C, basal cuticular box; Ce, center of the colony; H, colonial hypostegal coelom; P, communication pore; T, autozooid in the triradiate ancestrular triplet; V, vibraculum.

structures. One of the most obvious of the skeletal structures is the autozooecial-vibracular unit, which makes up the whole zoarial frontal. The core in each unit is formed by primary skeletal material, which is in turn surrounded by a layer of secondary skeletal material, typically in such a way that it is separated from the primary material of other units (Plate I, Figs 2, 3 and Plate II, Figs 5, 6). Only in the vibracula is most of the inner surface of the walls formed by primary skeletal material without a secondary cover. In sections the secondary skeletal material is characterized as being composed of long needle-like crystallites arranged in rather large fan-shaped bundles expanding in the direction of growth (transverse fibrous ultrastructure of Sandberg, 1971). The primary skeletal material, on the other hand, has a more granular appearance and is considerably darker. Although the primary and the secondary skeletal material have a quite different appearance, there is no sharp limit between them, and there seems to be a transition from the short randomly arranged crystallites of the primary skeletal material to the gradually longer, well orientated crystallites of the secondary. This is in agreement with the fact that other species of the Cupuladriidae have monomineralic or almost monomineralic skeletons, being composed entirely or chiefly of aragonite (Rucker and Carver, 1969).

The difference in structure of the two types of skeletal material is also reflected in the exterior surface ornamentation, where the primary skeletal material, which is only visible in the extreme margin of the colony, has a more tuberculate appearance (Plate I, Fig. 5).

The deposition of secondary skeletal material on the zooidal side of the primary core is soon interrupted, leaving only a thin layer. On the basal side it continues slowly, perhaps throughout the life of the colony, making up most of the basal wall, which in *Cupuladria biporosa* may reach a considerable thickness. In the basal wall frequent and fairly regularly distributed basal chambers are found, which form distinct rows more or less perpendicular to the surface of the colony (Plate I, Fig. 3). The basal chambers communicate within each row by means of simple pores, as do the first formed in each row with an autozooidal chamber. The under side of each basal chamber is bounded by a small, normally isolated, occurrence of primary skeletal material. The last formed members of the rows of basal chambers, which open into the basal hypostegal coelom, are visible on the basal side of the colony as rounded pores of varying diameter (Plate I, Fig. 5).

The general pattern of the soft parts in the sections, i.e. the distribution of cuticle and coelomic spaces, is, of course, intimately coherent with that of the mineralized skeleton, and accordingly just as regular.

The whole colony is surrounded by a continuous exterior cuticle, which on

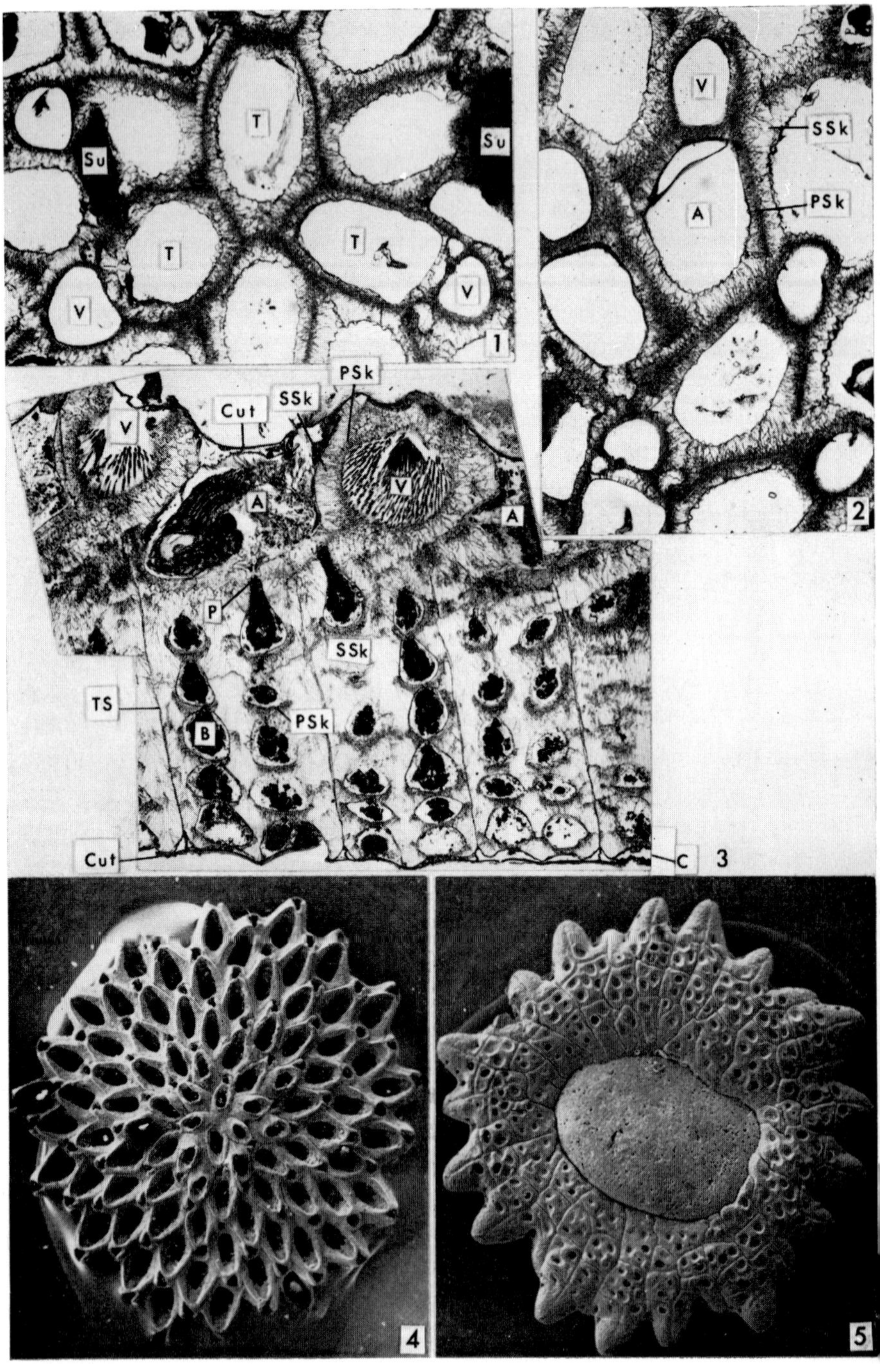
Su
T
Su
T
T
V
V
1
V
SSk
PSk
A
2
PSk
Cut
SSk
V
V
A
A
P
SSk
TS
PSk
B
Cut
C
3
4
5

the frontal side is fused to the mineralized skeleton of the zooidal walls. On the basal side the exterior cuticle is separated from the mineralized skeleton by a coelomic space, but connected to it by means of low, continuous, ribbon-like walls of cuticle joining the basal cuticle and the sectorlines (Plate I, Fig. 3). These connections divide the basal coelom into a system of small, separate, box-like coelomic spaces, which correspond to the pattern formed by the sectorlines on the basal surface of the mineralized skeleton (Plate I, Fig. 5). The internal structures in the sections corresponding to the external system of longitudinal and transverse sectorlines (the former lying in the median plane of the autozooidal-vibracular units) are thin dark lines, which sometimes stain faintly (Plate I, Fig. 3 and Plate II, Figs 5, 6). Furthermore, as they show a very close association with cuticle (see above), this indicates strongly that they are cuticular in origin, and must be considered as almost completely calcified cuticles. These internal sectorlines extend up through the basal wall of the colony reaching the primary skeletal material of the autozooidal-vibracular units, with which they seem closely associated. Consequently, neither the longitudinal nor the transverse sectorlines reach the surface of the colony. Thus, contrary to that which is considered normal among cheilostomes, intercalary cuticles are missing in

---

Plate I

Cupuladria biporosa Canu and Bassler. Recent, Florida. (Sections approximately ×80; SEM micrographs approximately ×17.)

Fig. 1. Tangential section showing the triradiate ancestrular triplet. Dried specimen, USNM 180751.

Fig. 2. Tangential section near the ancestrular triplet, showing the isolated core of primary skeletal material in autozooidal-vibracular units. Same section as Fig. 1.

Fig. 3. Composite photograph of longitudinal section showing a single autozooidal-vibracular unit, with associated rows of basal chambers leading to the basal cuticular boxes. Alcohol-preserved specimen, USNM 180752.

Fig. 4. Frontal view of small specimen showing the radial budding pattern originating from the triradiate ancestrular triplet. Many autozooids, including the autozooids of the ancestrular triplet, are regenerated into vibracula. Dried specimen with all soft parts removed, USNM 180749.

Fig. 5. Basal view of small specimen showing sectorlines and the youngest, open basal chambers. The specimen is growing on a relatively large substratum. Dried specimen with all soft parts removed, USNM 180750. *Abbreviations:* A, autozooid; B, basal chamber; Bu, newly formed bud in the colonial hypostegal coelom; C, basal cuticular box; Cut, cuticle; H, colonial hypostegal coelom; LS, longitudinal sectorline; P, communication pore; PSk, primary skeletal material; SSk, secondary skeletal material; Su, substratum; T, autozooid in the triradiate ancestrular triplet; TS, transverse sectorline; V, vibraculum.

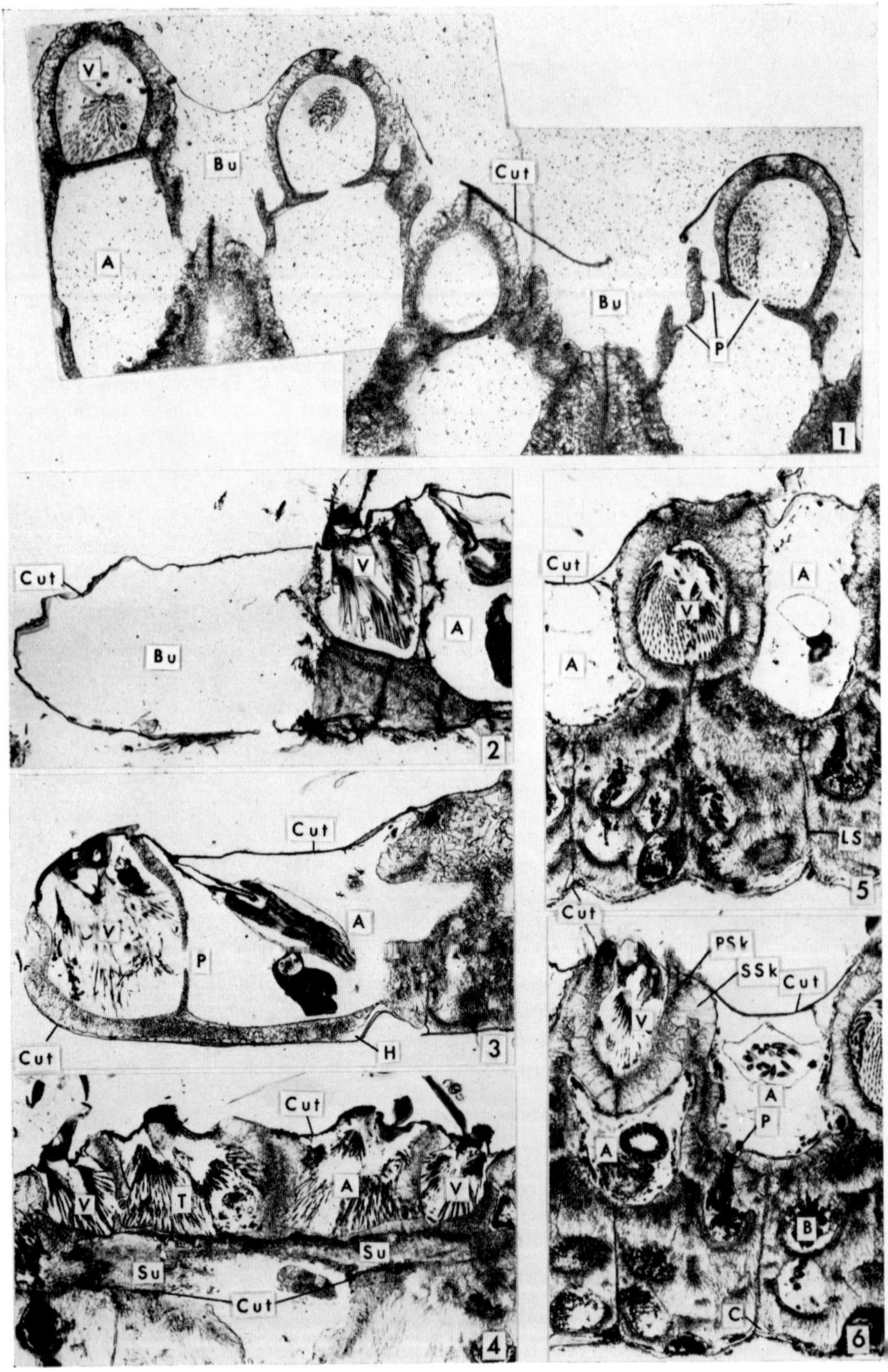
V
Bu
A
Cut
Bu
P
1
Cut
V
A
Bu
2
Cut
V
A
A
LS
5
Cut
A
V
P
Cut
H
3
Cut
PSk
SSk
Cut
V
A
P
A
B
C
6
Cut
V
T
A
V
Su
Su
Cut
4

lateral walls, indicating that these walls must be considered interior and that budding rows do not occur in a traditional sense.

When, as frequently happens in *Cupuladria biporosa*, the substratum becomes completely embedded in the colony by the growth of the basal wall, the exterior cuticle is divided into two isolated parts, an outer part surrounding the colony as a whole, and an inner part surrounding the substratum.

As demonstrated, the mineralized and cuticular walls of the colony form a large number of isolated coelomic spaces, all of which are lined with epithelium. In addition to the autozooidal and vibracular coelomic spaces with well established function, there are some highly differentiated and unusual types, mostly with obscure function. These include the coelomic spaces of the basal cuticular boxes, which probably should be considered extrazooidal. Their function, although uncertain, may provide a means for the continuous secretion of the basal wall. On the other hand, the coelomic spaces of the basal chambers can be considered kenozooids. Their function is unknown, but it is likely that they provide a route for transport of nutrients to the basal cuticular boxes, which would otherwise seem to be without any means of supply. The arrangement of the sectorlines (e.g. longitudinal sectorline lying in the median plane of the autozooidal-vibracular units) provides that each of the basal cuticular boxes

---

Plate II

Cupuladria biporosa Canu and Bassler. Recent, Florida. (All sections approximately × 80.)

Fig. 1. Composite photograph of tangential section in the growing edge, showing young buds developing in the colonial hypostegal coelom. Alcohol-preserved specimen, USNM 180753.

Fig. 2. Longitudinal section showing a single bud. (NB. The mineralized skeleton has cracked during polishing, and parts of the distal and proximal walls of the vibracular chamber have been lost.) Same specimen as Fig. 1.

Fig. 3. Longitudinal section (slightly oblique), showing young zooidal unit, in which the coelomic connection between the autozooid and vibraculum is still open. Same section as Plate I, Fig. 3.

Fig. 4. Longitudinal section through the center of the colony. The substratum is almost completely embedded in the basal wall, and the uncovered part has been severely corroded. A slight corrosion of the basal wall next to the substratum is also seen. Same section as Plate I, Fig. 3.

Fig. 5. Transverse section showing the isolated cores of primary skeletal material in the autozooidal-vibracular units, and their relation to the longitudinal sectorlines. Same specimen as in Plate I, Fig. 3.

Fig. 6. Transverse section in a slightly different level in the zooidal units as Fig. 5. Shows the connection of one basal cuticular box with two autozooidal chambers, through rows of basal chambers. Same section as Fig. 5. *Abbreviations* as in Plate I.

communicates through rows of basal chambers leading to at least two autozooids (Plate II, Fig. 6), thereby securing an uninterrupted supply. Owing to the usually high content of heavily staining globular masses (Plate I, Fig. 3), it may be suggested that the basal chambers also may serve for the storage of nutrients.

The last type of coelomic space found is an extrazooidal hypostegal coelom situated just along the growing edge all around the colony (Plate II, Figs. 1–3). The presence of this often very narrow colonial coelom is due to the lack of lateral cuticles and the consequent lack of normal longitudinal budding rows. The formation of the cuticles connecting the exterior cuticle and the longitudinal sectorlines starts a little below the extreme margin, and that of the cuticles associated with the transverse sectorlines further back, thus forming the proximal wall of this coelomic space.

GROWTH MODEL

Based on the structures of the skeleton of *Cupuladria biporosa*, predominantly as revealed by thin sections and on observations on the growing edge, the following model for the formation of the basal skeletal unit, i.e. the autozooidal vibracular unit, is proposed (Fig. 2).

By localized intussusceptive expansion of the exterior cuticle and the associated

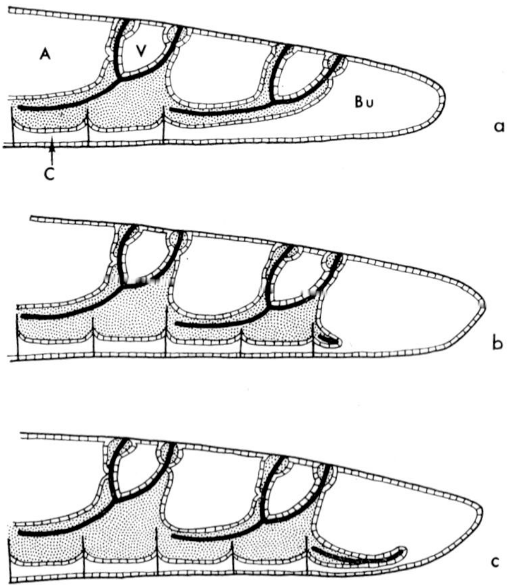

Fig. 2. Diagrammatic series showing the inferred development of an autozooidal-vibracular unit. The complete sequence includes the stage shown in the growing edge of Fig. 1a both at its beginning and at its end. Abbreviations as in Fig. 1.

epithelium which takes place in the angle between the two last formed zooidal units, a swollen coelomic bladder is formed on the growing edge of the colony, not as an individual bud, but as an extension of the colonial hypostegal coelom (Fig. 2a and Plate II, Fig. 1). This in turn is followed by the formation of a distinct fold in the inner secreting epithelium below the proximal and lateral zooidal units (Fig. 2b and Plate II, Fig. 2). By rapid cell proliferation the edge of this fold expands distally in the bud secreting an interior wall composed of primary skeletal material (Fig. 2c). At this point the separation of the two members of the unit, the autozooid and the vibraculum, is initiated by branching of the expanding margin of the fold. It then turns and continues to grow upward to fuse with the frontal part of the exterior cuticle, thus isolating the coelom of the new zooidal unit from the colonial hypostegal coelom (Fig. 1a and Plate II, Fig. 3). Finally the separation of the coelom into the two parts of the unit is completed by the formation of a communication pore closing the gradually decreasing area of coelomic connection between them.

During the continued secretion from the epithelium of the two sides of the fold, the structure of the skeletal material deposited rapidly changes from primary to secondary. While this deposition soon stops on the zooidal side, leaving only a thin layer of secondary skeletal material, it continues on the basal and distal, i.e. the extrazooidal, side of the fold, and eventually a new fold is developed in the distal part of this epithelium, repeating the process in the formation of a new zooidal unit. Along the proximal part of the fold, however, the extrazooidal secretion continues in a more uniform way, forming the basal wall and the basal chambers. Probably the formation of the basal chambers is due to a decreasing secretion of secondary skeletal material from certain parts of the epithelium, resulting in the formation of gradually deeper, round depressions in the basal surface. The depressions are eventually closed by secretion of primary skeletal material from the same epithelium, thus isolating the coelom of the new basal chambers from that of the now established basal cuticular boxes.

It follows from the model just described for the formation of the zooidal units, that the entire colony is formed by growth of interior walls cutting out portions of a constantly expanding colonial hypostegal coelom. And, as the formation of new zooidal units normally takes place simultaneously all the way around the colony, the result is a rhythmic symmetrical expansion of the whole colony in a horizontal plane. The secretion of the basal wall from the lower, extrazooidal side of the units, on the other hand, results in a gradual thickening of the colony, i.e. expansion in a vertical direction.

In addition it follows from the model that the inner secreting epithelium is capable of repeated changes between secretion of primary and secondary skeletal material. There is, however, a transition between the two skeletal types, and buds in an intermediate state (e.g. Fig. 2c) are very seldom seen, indicating that the formation of a skeletal unit is very quick. It therefore seems reasonable to suggest that the difference in structure between what here provisionally is called primary and secondary skeletal material, is due mainly to a difference in the rate of deposition.

This suggestion as to the mode of growth of *Cupuladria biporosa* raises the questions of how the colonial hypostegal coelom is formed and how this particular type of radial budding pattern is established.

Recent observations show that the larvae of at least two species of the Cupuladriidae metamorphose into a triradiate ancestrular triplet from which the radial budding pattern can easily be established (F. J. S. Maturo, unpublished). Unfortunately, no young alcohol-preserved colonies have been available for the present study, so accordingly the following observations are based only on thin sections of the ancestrular area of larger colonies of *Cupuladria biporosa*. These sections show that the walls of the zooidal units in the center of the colony of this species have the same structure as the walls in the rest of the colony, with a core of primary skeletal material surrounded by secondary skeletal material, and without an intercalary cuticle (Plate I, Fig. 1 and Plate II, Fig. 4). This indicates strongly that these walls, like the other walls in the colony, are interior, and formed by secretion from an expanding fold in the secreting epithelium. Further, the sections show that the primary skeletal material, in contrast to what is normally found in the walls of other units, is continuous within the three zooidal units making up the triradiate center of the colony (Plate I, Fig. 1). In view of the general mode of growth of the colony demonstrated above, this indicates that these three zooidal units have been developed simultaneously, thus forming a triradiate ancestrular triplet similar to those observed in other species by F. J. S. Maturo (unpublished).

Consequently the following model for the larval metamorphosis and the early astogeny of *Cupuladria biporosa* is proposed, being in accordance both with Maturo's observations and with the mode of growth suggested above for the later formed, major part of the colony.

The settled larva forms a cuticular bladder lined with epithelium, which encloses an undivided coelomic space. On the fixed lower wall of this bladder a system of continuous folds is developed in the epithelium, which by quick cell proliferation grow upwards, while secreting the primary skeletal material of the interior walls of the ancestrular triplet. When the edge of the expanding fold-system reaches the upper wall of the bladder, the mineralized walls are completed, and fuse with the upper or frontal cuticle, thereby dividing the larval coelom into a number of isolated coelomic spaces. In addition to the coelomic spaces of the three zooidal units of the now completed triradiate ancestrular triplet, these include a narrow extrazooidal hypostegal coelom, which has been isolated merely as a result of the growth of the interior walls, and extends continuously all around the margin of the ancestrular triplet. With this colonial hypostegal coelom established as a direct result of the metamorphosis, the growth of the colony can continue as described earlier in this paper.

Observations on sections of *Cupuladria canariensis*, *C. doma*, and *Discoporella umbellata* in connection with Maturo's observations of the metamorphosis and early astogeny of the two last mentioned species (F. J. S. Maturo, unpublished), show that these species, apart from minor variations, have the same general development and internal structure as *Cupuladria biporosa*. The variations found include lack of one or both types of sectorlines, and lack of basal chambers or a replacement by long tubular pores. The basic structures seem to be common to all species studied. They include the isolated core of primary skeletal material surrounded by secondary skeletal material in each zooidal unit, the lack of intercalary cuticles in lateral walls, and the presence of a colonial hypostegal coelom either restricted along the margin or extending over the whole basal surface. In view of the otherwise established very close relationship between these species (Cook, 1965a, b), it seems justified to suggest that a mode of growth following the same general pattern as described here for *Cupuladria biporosa*, is characteristic for the Family Cupuladriidae as a whole. Investigations just begun on other free living anascans indicate that the mode of growth determined for the Cupuladriidae may be restricted to that family.

ACKNOWLEDGEMENT

The research was completed while the author was a Visiting Research Associate in Department of Paleobiology, Smithsonian Institution, Washington D.C.

REFERENCES

Cook, P. L. (1965a). Notes on the Cupuladriidae (Polyzoa, Anasca). *Bull. Br. Mus. (N.H.) Zool.* **13** (5), 151–187.

Cook, P. L. (1965b). Polyzoa from West Africa. The Cupuladriidae (Cheilostomata, Anasca). *Bull. Br. Mus. (N.H.) Zool.* **13** (6), 189–227.

Nye, O. B., Dean, D. A. and Hinds, R. W. (1972). Improved thin section techniques for fossil and Recent organisms. *J. Paleont.* **46** (2), 271–275.

Rucker, J. B. and Carver, R. E. (1969). A Survey of the Carbonate Mineralogy of Cheilostome Bryozoa. *J. Paleont.* **43** (3), 791–799.

Sandberg, P. A. (1971). Scanning electron microscopy of cheilostome bryozoan skeletons; techniques and preliminary observations. *Micropaleont.* **17** (2), 129–151.

28 | Intrazooecial Structures in Some Tubuliporinid Cyclostome Bryozoa

ROBERT W. HINDS

Department of Geology, Slippery Rock State College, Pennsylvania, U.S.A.

Abstract: Study of thin sections of early Tertiary and Recent cyclostome Bryozoa reveals a variety of intrazooecial structures, some of which have not been noted previously. These structures include diaphragms, hemisepta, spines, and zooecial blades.

INTRODUCTION

Intrazooecial structures in tubuliporinid cyclostome bryozoans are varied in their shape, smaller skeletal structures, and development. Diaphragms, hemisepta, spines and zooecial blades are described. Hemisepta and zooecial blades are newly recognized in cyclostome bryozoans.

DIAPHRAGMS

Calcareous diaphragms in cyclostome taxa have been previously reported by several authors. The discussion by Nye (1968) of diaphragms in post-Paleozoic cyclostomes included three different types of diaphragms: (1) basal; (2) terminal; and, (3) intermediate. Only terminal diaphragms will be discussed herein.

Terminal diaphragms in the cyclostome species studied (Plate I, Figs 1–3) have separable zones of initial calcification and subsequent calcareous laminae (Plate I, Figs 2–3). Calcareous laminae are apposited to the proximal side of the zone of initial calcification so that a secreting epidermis was located on the proximal side of the diaphragm (Plate I, Figs 2–3). In *Diaperoecia californica* (D'Orbigny), diaphragms consist of a thin, more or less hyaline zone of initial calcification which abuts laterally against the subsequent calcareous laminae of the autozooidal outer body-wall (Plate I, Fig. 2). Additional apposition to the diaphragm is in the form of calcareous laminae. In *Erkosonea admota* Canu and Bassler, terminal diaphragms are developed in kenozooecia. The granular zone of initial calcification (Plate I, Fig. 4) is laterally continuous with the same zone in the kenozooecial walls (Plate I, Fig. 3). The terminal diaphragm is usually at

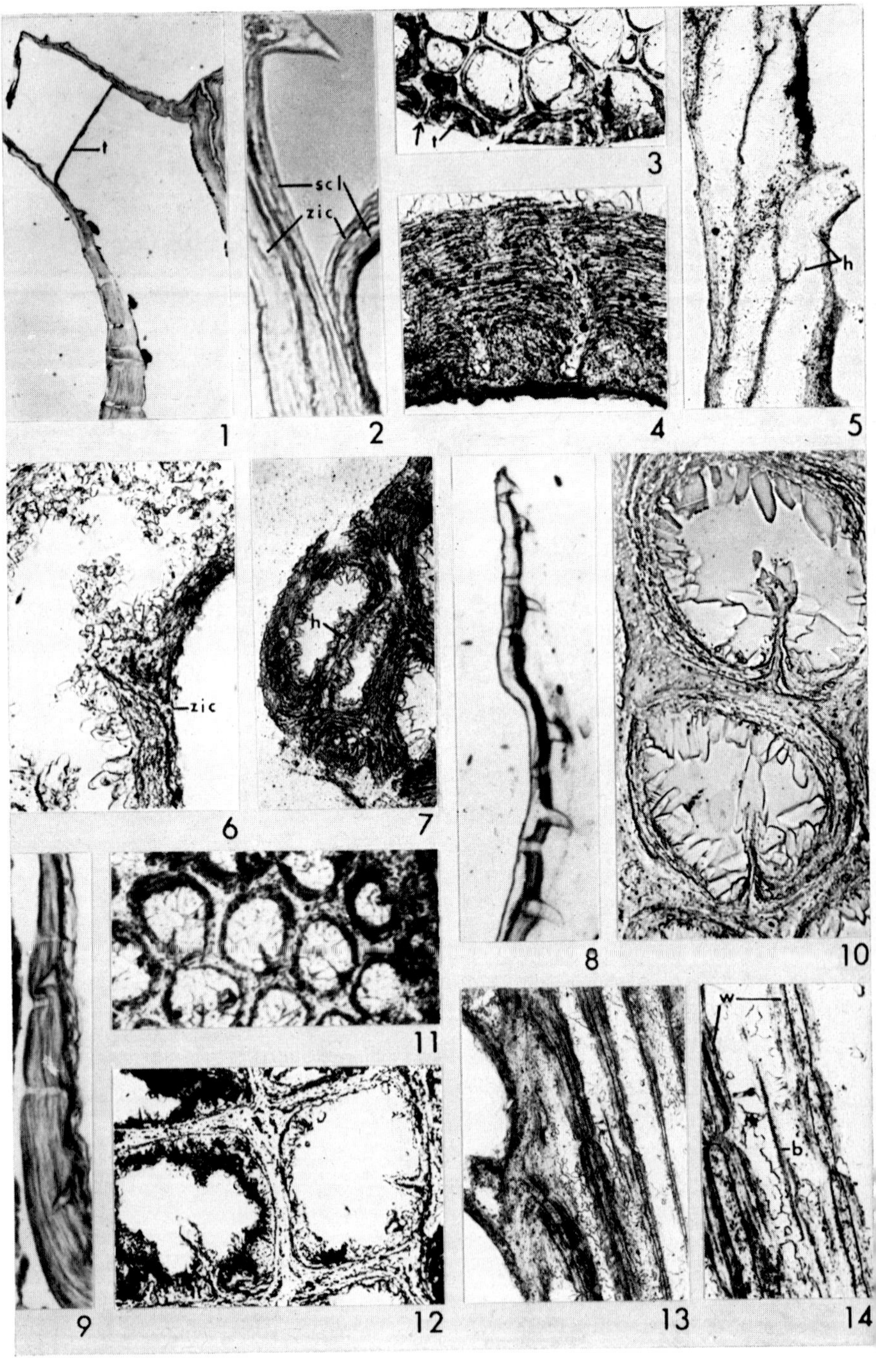
t
scl
zic
h
zic
h
w
b
1
2
3
4
5
6
7
8
9
10
11
12
13
14

Plate I

Fig. 1. Longitudinal view of *Diaperoecia californica* (D'Orbigny) showing a terminal diaphragm (t). ×65, NMNH (National Museum of Natural History, Washington, D.C.) 178583.

Fig. 2. Enlargement of Fig. 1 showing detail of the wall and diaphragm. *Abbreviations:* zic = zone of initial calcification; scl = subsequent calcareous laminae. Also note the calcareous spine near the top of the figure. ×775, NMNH 178583.

Fig. 3. Transverse view of *Erkosonea admota* Canu and Brassler showing: (a) the zone of budding (small tubes); (b) very thick terminal diaphragms, t, along the back surface; and (c) continuity of the zone of initial calcification from the inner body-walls to the terminal diaphragms (see arrow). ×70, NMNH 65357–1.

Fig. 4. Longitudinal view of front wall of *Erkosonea admota.* The zone of initial calcification is finely granular, and subsequent calcification consists of laminae nearly parallel to the wall. Details of the terminal diaphragms in Fig. 3 are comparable to the details of the front wall. ×270, NMNH 65357–1.

Fig. 5. Longitudinal view of *Filisparsa*? sp. showing hemisepta, h. ×65, NMNH 178618.

Fig. 6. Enlargement of Fig. 5 showing bifurcation of the zone of initial calcification (zic) where the hemiseptum forms. ×290, NMNH 178618.

Fig. 7. Transverse view of *Filisparsa*? sp. showing a hemiseptum, h. ×180, NMNH 178619.

Fig. 8. Longitudinal view of distal portion of autozooidal tube of *Diaperoecia californica* showing a hyaline zone of initial calcification and subsequent calcareous laminae. Spines are continuous with the zone of initial calcification and form at the distal end of the autozoidal tube. ×200, NMNH 178583.

Fig. 9. Longitudinal view of a more proximal portion of autozooidal tube of *Diaperoecia californica* than is shown in Fig. 8. The spines are completely covered by subsequent calcareous laminae. ×160, NMNH 178583.

Fig. 10. Transverse view of *Exidmonea magnireversa* (Canu and Bassler) showing zooecial blades. In the lower zooecium, the central portion of the blade has a zone of initial calcification continuous with that of the zooecial wall. In the upper zooecium, the central portion of the blade has a sharp boundary between it and the zone of initial calcification of the zooecial wall. The blade is continuous with the subsequent calcareous laminae of the zooecial wall. ×400, NMNH 65352–1.

Fig. 11. Transverse view of *Exidmonea* sp. A showing blades with thin edges. ×200, NMNH 178599.

Fig. 12. Transverse view of *Exidmonea* sp. B showing blades, one with a dilated edge (in left zooecium), and one with a rounded edge (in right zooecium). ×365, NMNH 178603.

Fig. 13. Longitudinal view of *Exidmonea magnireversa* showing the aspect of the zooecial tubes and blade. ×75, NMNH 65352–1.

Fig. 14. Enlargement of Fig. 13 showing zooecial walls, w, and blade, b. ×180, NMNH 65352–1.

the distal end of the tube, and may be merely a modified type of single, outer body-wall. Structural differences in lateral relationships of the zone of initial calcification in diaphragms and inner body-walls of *Diaperoecia californica* and *Erkosonea admota* possibly reflect variation in the timing of diaphragm development relative to the ontogeny of the individual.

HEMISEPTA

Hemisepta, as per the definition found in the "Treatise on Invertebrate Paleontology" (Bassler, 1953), are found in an Eocene, cyclostome species tentatively assigned to the genus Filisparsa. These hemisepta are curved platforms extending from the proximal and distal zooecial walls into the living chamber (Plate I, Figs 5–7), and form by a bi-furcation of the wall at the zone of initial calcification (Plate I, Fig. 6). The hyaline to granular layer which forms the zone of initial calcification is continuous from the hemisepta to the inner body-wall (Plate I, Figs 6–7). Hemisepta thicken by apposition of calcareous laminae to both sides of the zone of initial calcification.

Living species with hemisepta were not available for study, and function of the hemisepta is unknown. Because of the location of the hemisepta near the apertural ends of the zooecia, it is probable that their function is closely related to the lophophore-gut complex of autozooids. Whatever this relationship might be, it is inferred to be biologically important.

Hemisepta in the questionable *Filisparsa* species and the Cryptostomata are similar in their spatial relationship to each other and to the autozooidal living chamber, and may have had a somewhat similar function. Details of shape, microstructure, and development of the hemisepta in this cyclostome species and the cryptostome species are quite different. Certainly they are not homologous structures. Nevertheless, assuming a similar function, development of hemisepta in this cyclostome species is a remarkable example of convergent homoplasy.

SPINES

Spines in *Diaperoecia californica* are calcareous structures which have the shape of a small, proximally bent cone with the point of the cone protruding into the autozooidal living chamber (Plate I, Figs 1–2, 8–9). Spines appear hyaline when viewed from thin sections through a compound microscope. This appearance is identical to that of the zone of initial calcification, with which the spines are laterally continuous (Plate I, Fig. 8). Spines form essentially at the same time as the other hard parts of the zone of initial calcification near the distal end of the zooecium (Plate I, Fig. 8). Spines can be completely covered by subsequent calcareous laminae during later ontogenetic stages (Plate I, Fig. 9), and they are probably functional only in the distal portion of the autozooid. No soft parts have been seen attached to the spines, and their function is unknown.

In zooecial walls with thick calcareous laminae, spines can be distinguished easily from pseudopores by the direction of inflection of the laminae around the structure. Laminae lap into pseudopores, i.e. bending away from the autozooidal living chamber, whereas, laminae lap onto the spines, i.e. bending in toward the autozooidal living chamber.

ZOOECIAL BLADES

Blade-shaped, calcareous structures that extend from a zooecial wall into the living chamber, and parallel the length of the tube, are herein designated zooecial blades (Fig. 1). The spatial relationship of blade to zooecium is somewhat analogous to the relationship of a single septum to a corallite in the phylum Coelenterata. Discussion of blades is facilitated by adoption of a few descriptive terms (see Fig. 1). Width of a blade is the distance from the base to the edge of the blade. Length is measured in a direction parallel to the length of the zooecium. Thickness is measured at right angles to the width and length.

Zooecial blades were found in Paleocene, Eocene, Oligocene and Recent species belonging to the genus *Exidmonea* Mongereau. Blades have not been seen in other cyclostome genera, but knowledge of internal morphology of most cyclostomes is lacking.

Blades are usually formed of calcareous laminae which are laterally continuous with the subsequent calcareous laminae along the zooecial wall (Plate I, Fig. 10). Laminae of the blades are continuous with the older laminae of the walls, i.e. laminae next to the zone of initial calcification. Only rarely do blades appear to include an infolding of the hyaline zone of initial calcification (Plate I, Fig. 10). If blades increase in width and thickness after their initial stages of formation, they do so at a much slower rate than does the zone of wall laminae. Such thickening has not been observed.

Blade thickness is variable within and between taxa. Thickness of some blades attenuates very gradually away from the zooecial wall, producing a very fine edge (Plate I, Fig. 11). Other blades may have a more blunt, rounded edge (Plate I, Fig. 12), or expanded edges (Plate I, Figs 10, 12). Part of this variation is ontogenetic (as seen in Plate I, Figs 10, 12), and part of it is species dependent. Blade width varies considerably within and between species, up to as much as one-half the total width of the zooidal living chamber (Plate I, Figs 10, 12). Blades are usually found only in the distal third of the zooecium. No more than one blade per zooecium has been seen, and then only along an inner body-wall.

Blades are easily recognized in both transverse (Plate I, Figs 10–12) and longitudinal (Plate I, Figs 13, 14) views.

GENERAL CONCLUSIONS

(1) Intrazooecial structures in cyclostome bryozoans are varied in shape, attitude and growth patterns. Hemisepta, spines, and zooecial blades are interpreted to be closely associated with the lophophore-gut complex of autozooids. Diaphragms are interpreted as more generalized structures in that they are found associated with different polymorphic individuals.

(2) Many intrazooecial structures consist of a hyaline or granular calcareous layer with or without calcareous laminae. Hyaline and granular calcite in the studied forms constitutes a zone of initial calcification, and the laminae are subsequently added. Spines consist exclusively of hyaline calcite which is continuous with the hyaline zone of initial calcification in the zooecial wall. Hemisepta have an inner layer of hyaline to granular calcite surrounded on both sides by calcareous laminae. The hyaline to granular layer in hemisepta is laterally continuous with the zone of initial calcification in the zooecial walls. On their distal side, terminal diaphragms have a thin hyaline layer, but it may

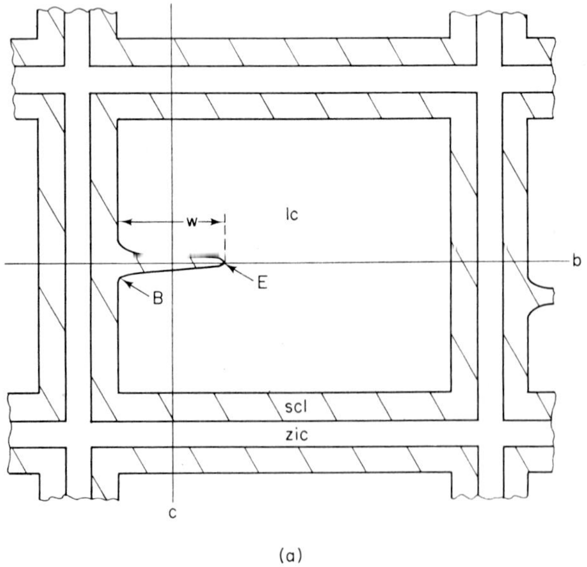

Fig. 1. Diagrammatic sectional views of a portion of a zooecium containing a zooecial blade. *Abbreviations:* Living chamber (lc); subsequent calcareous laminae (scl); zone of initial calcification (zic); base of blade (B); edge of blade (E); length of

or may not be laterally continuous with the hyaline zone of initial calcification of the zooecial walls. Terminal diaphragms thicken by apposition of calcareous laminae to their proximal side. Spines, hemisepta, and diaphragms exhibit slightly varied developmental patterns with respect to the inferred zone of

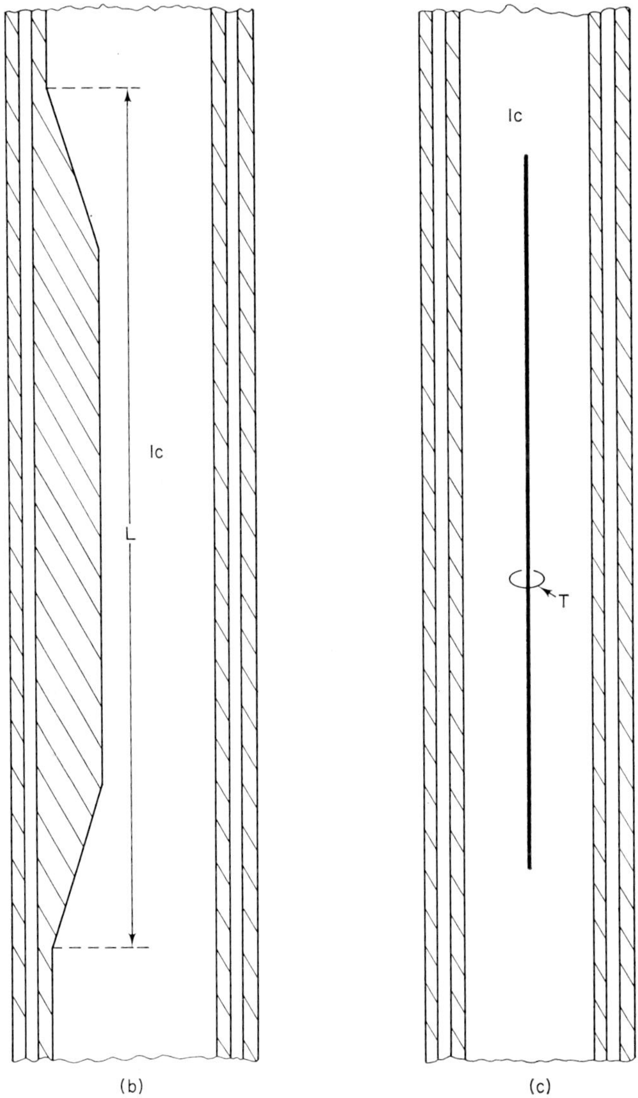

blade (L); thickness of blade (T); width of blade (W). (a), Transverse view, also shows lines of section for views (b) and (c). (b), Sectional view of lateral zooecial walls. Scale = $\frac{1}{2}$ that of (a). (c), Longitudinal view. Scale = $\frac{1}{2}$ that of (a).

initial calcification. Since zooecial walls commonly have a similar hyaline or granular zone of initial calcification, this developmental pattern may reflect a rather basic or common physiological process whose phenetic expression could be anticipated in many, if not most, cyclostome bryozoans.

(3) Zooecial blades are laterally continuous with subsequent calcareous laminae of the zooecial walls. If blades are formed by physiological processes similar to that of other intrazooecial structures [see (2) above], then the blades should have regularly an associated hyaline or granular zone of initial calcification. The nearest hyaline layer is in the zooecial wall adjacent to the blades. An upfold of the hyaline layer from the zooecial wall is only rarely included in the zooecial blades. Therefore, blade formation is interpreted primarily as a localized variation in growth rate and/or growth pattern of the subsequent calcareous laminae of the zooecial wall.

REFERENCES

Bassler, R. S. (1953). Bryozoa, In "Treatise on Invertebrate Paleontology", pt. G. 253 p. Kansas Univ. Press and Geol. Soc. Lawrence, Kansas.

Nye, O. B. (1968). Aspects of Microstructure in Post-Paleozoic Cyclostomata: *Soc. Ital. Sci. nat. Mus. Civ. Storia nat. Milano Atti*, **108,** 111–114.

29 | Different Budding Patterns in the Genus *Stomatopora* (Bryozoa, Cyclostomata)

GISELA ILLIES

Karlsruhe, Germany

Abstract: Within the genus *Stomatopora*, a new budding pattern shown by *S. osterfeldi* n. sp., will be described in addition to the two kinds of supraspecific budding patterns already known. The differences between the three kinds of budding patterns depend on the formation of a varying number of zooecial tubes (1, 2, 4, respectively) in the second generation.

INTRODUCTION

The so-called "uniserial" species, as classified in the genus *Stomatopora*, show different budding patterns. There are two supraspecific budding patterns already known. A further one, shown by a new species, will be described here. The budding pattern may be defined as the pattern produced by the contact of the common walls (septa) of zooecial tubes with the repent basal lamina of the whole colony. Alternatively, budding pattern may be defined as the whole mosaic of zooecia. In the simply structured colonies in which the zooecial tubes occupy the basal and frontal walls of the colony for the whole of their length, structure and budding pattern may be recognized on the surface of the colony by traces of the edges of the common zooecial walls or septa.

Each colony, having one of the budding patterns described in this paper, shows a proximal segment of a younger zooecial tube of the second generation on each side of the ancestrula (Fig. 1a). The two proximal segments of zooecial tubes, growing independently, are separated by the recumbent ancestrula. From this condition the first differences between the three kinds of budding patterns develop.

BUDDING PATTERN I

During the zoarial development, the two proximal segments of zooecial tubes which grow independently from each other on both lateral edges of the branch beside the recumbent ancestrula, meet beneath the rising peristome of the

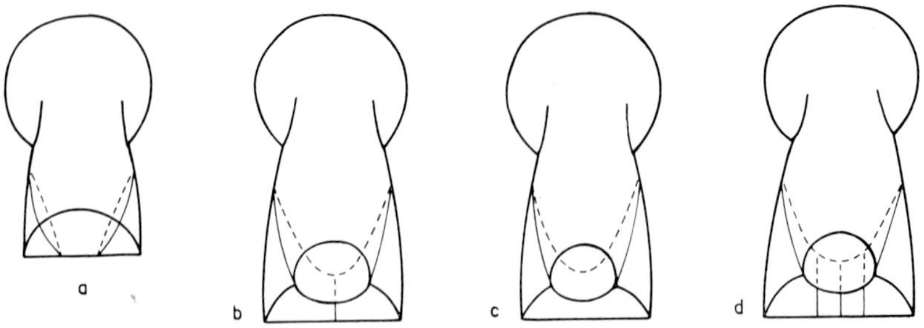

Fig. 1(a–d). Initial stages of three so-called uniserial colonies. (a) This early stage of development is still equivalent in all three species. Along each edge of the branch a proximal segment of a younger zooecial tube grows beside the recumbent ancestrula. (b) In budding pattern I a median vertical septum separates two zooecial tubes under the rising peristome of the ancestrula. (c) In budding pattern II the two proximal segments merge and only one zooecium occupies the branch. (d) In budding pattern III there are three vertical septa separating four zooecial tubes.

ancestrula. At this stage a wall develops between the two proximal segments. This wall was named by Borg (1926, p. 270) the "median vertical septum" (Fig. 1b). The two proximal segments represent the two younger zooecial tubes of the second generation. One of the zooecial tubes develops more strongly and terminates in its peristome, the other zooecial tube remains in an immature condition lengthening its proximal narrow part along the lateral edge of the branch (Fig. 2a). It matures as the next zooecium. Each single zooecium, which is growing towards the middle of the branch, is superseded by a new zooecium which buds as a proximally tapering tube on the same lateral edge of the branch. The zooecial tubes terminate their growth one after the other. Thus the so-called median vertical septum, or to be correct, the uninterrupted chain of median vertical septa, is represented alternately by the left or the right zooecial wall forming a serpentine line (Figs 2a, b). Sometimes two zooecia bud up successively from the same lateral edge of the branch. In this case in a branch the succession of the matured segments of zooecial tubes does not correspond to the sequence of their budding (Fig. 2b). Therefore the succession of mature segments does not correspond exactly to the sequence of generations. On the contrary, the zooecia which bud on the same lateral edge mature strictly in the order of their budding. In this way the colony branches are built-up by two parallel chains of zooecial tubes which are separated by the chain of the so-called median vertical septa (Fig. 2a, b). The succession of the mature segments of tubes, alternately budding on the left and the right edge of the branch, simulate a single

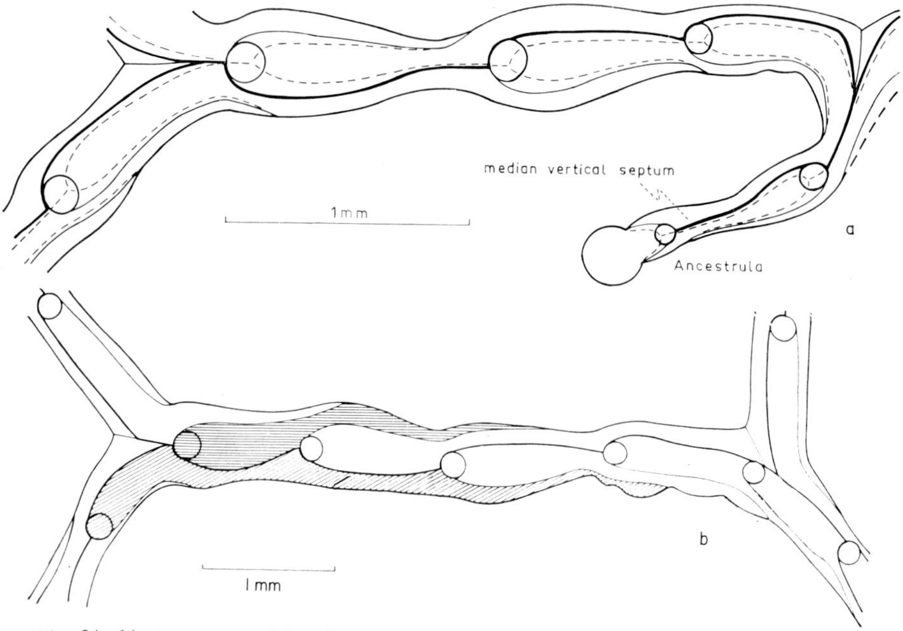

Fig. 2(a, b). *Stomatopora bilix* Illies 1971. Budding pattern I. Broken lines: the pattern caused by the common zooecial walls (septa) on the basal lamina. Weak lines: trace of the same common walls on the colony surface. (a) A young colony. The uninterrupted sequence of the median vertical septa (thicker lines) divides the branch into two chains of zooecia. The sequence of alternating zooecia simulates a single chain. (b) In the colony branches the succession of the mature segments does not correspond to the succession of their budding. On the contrary, within a lateral chain of zooecia the succession of budding and maturing remains strictly the same. (Photographs of these branches cf. Illies 1971, Plate 1, Fig. 2; Plate 2, Fig. 2).

chain of tubes only. Ramification takes place, by widening of the basal lamina and divergence of the two chains of zooecia. The second chain of zooecia, for each new branch arises in the ramification region.

This budding pattern has already been described in an excellent manner by Borg (1926, p. 272, Fig. 29a), from a recent colony, regarded as *Stomatopora granulata* Milne-Edwards 1838. This species Lagaaij (1963) considers a synonym of *S. trahens*, Couch (1841). The same kind of budding pattern was observed on *S. bilix*, Illies (1971) (Middle Liassic) (Fig. 2a, b) and also on *S. macroura*, Illies (1971) (Middle Liassic). The budding patterns of the three different species are modified in a specific manner. Borg was of the opinion that this kind of budding pattern represents a characteristic feature of the genus *Stomatopora*.

BUDDING PATTERN II

In the second kind of supraspecific budding pattern the zooecia grow in a truly uniserial manner. The two proximal segments, of which one is growing on the left and the other on the right side of the recumbent ancestrula, are merged beneath the detaching ancestrula tube (Fig. 1c). Thus the zooecium next to the ancestrula originates from two proximally tapering segments which had grown independently from each other on the two lateral edges of the branch. The zooecium occupies the whole width of the branch and there is no vertical septum. However, ramification takes place while each of the single proximal segments buds separately into zooecial tubes (Fig. 3). A short vertical septum grows between them, separates the branches and terminates where the branch divides.

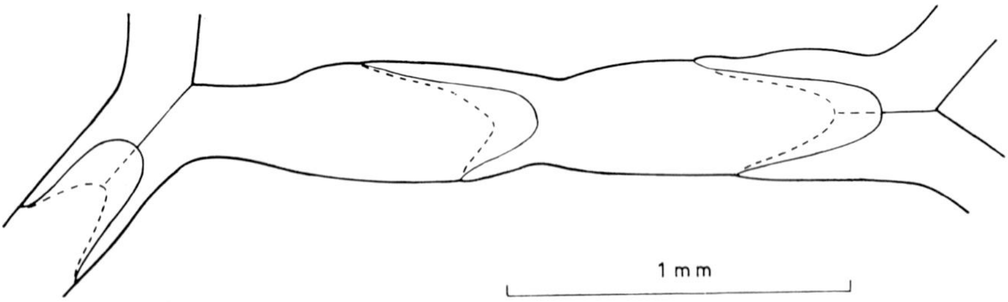

Fig. 3. *Stomatopora bifurca* Illies 1971. Budding pattern II. The colony is uniserially developed. For explanation of broken and weak lines see Fig. 2. (Photograph of this branch cf. Illies 1971, Plate 3, Fig. 4.)

Species having this kind of budding pattern are *Stomatopora dichotoma* (Lamx.) 1821, Illies (1963) (Bajocian), *S. dichotomoides* d'Orb. 1850, Illies (1963) (Bajocian), *S. bajocensis* (d'Orb) 1850, Illies (1963) (Bajocian), and *S. bifurca* Illies 1971 (Middle Liassic) (Fig. 3). Each budding pattern of the four different species is modified in a specific manner. (The budding pattern of the neotype-specimen of *S. dichotoma* (Lamx.) 1821, Walter (1969) from the Bathonian had not been investigated till now.)

BUDDING PATTERN III

A third budding pattern is observed in a new species which incrusts calcareous concretions together with *Stomatopora bilix* and *S. macroura* which follow the first described kind of budding pattern (Fig. 2) and with the colonies of *S. bifurca*, which follows the second kind of budding pattern (Fig. 3). The description with special emphasis on the budding pattern is as follows:

1. Stomatopora osterfeldi **n. sp.** (*Figs 1a, d, 4a, b; Plate I, Figs 1–4*)

(a) *Derivatio nominis.* "Osterfeld" is the name of the locality where the species was found 1961.

(b) *Locus typicus.* Pit of "Am Osterfeld" brick pit near Goslar (Germany) (Voigt 1968, Illies 1971).

(c) *Stratum typicum.* Middle Liassic (delta) (Amaltheenzone, Domerian, Upper Pliensbachian).

(d) *Holotypus.* Zoarium nr. AM8, incrusting concretion Osterfeld nr. 13 (Plate I, Fig. 1) 20 zooecia, generations 1–7, and 12 ramifications.

(e) *Material.* Twenty colonies incrusting eleven calcareous concretions (collection Illies, zoaria nr. AM1–AM20, encrusting concretions "Osterfeld" nr. 3, 4, 8, 12, 13, 16, 19, 20, 22, 39 and 41). Of these colonies, about 180 zooecia

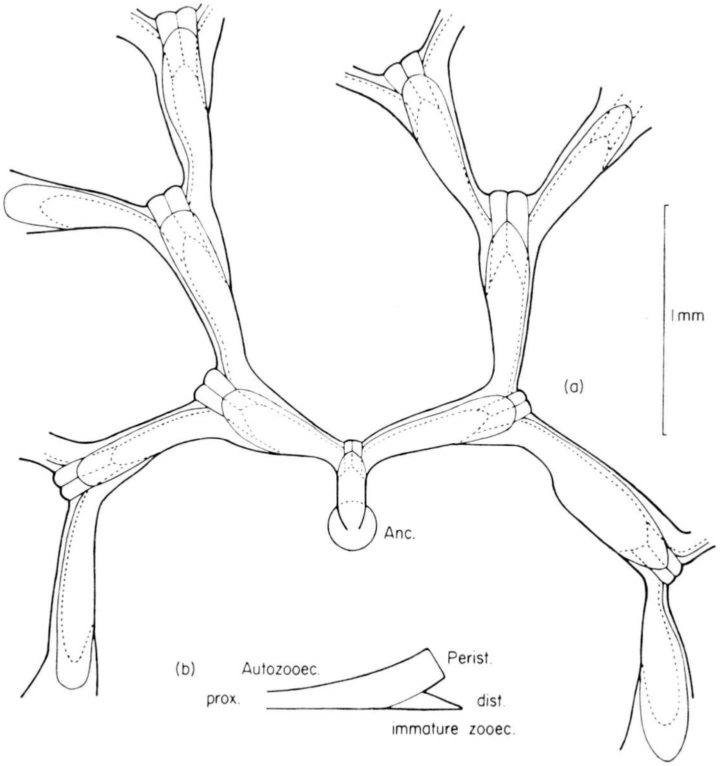

Fig. 4(a–b). *Stomatopora osterfeldi* n. sp., budding pattern III. (cf. Plate I, Fig. 1.) (a) Ramification takes place in each generation. Between the two ramifying regular zooecia there are two aborted zooecia. (b) Typical posture of the peristome over the aborted zooecium. For explanation of broken and weak lines see Fig. 2.

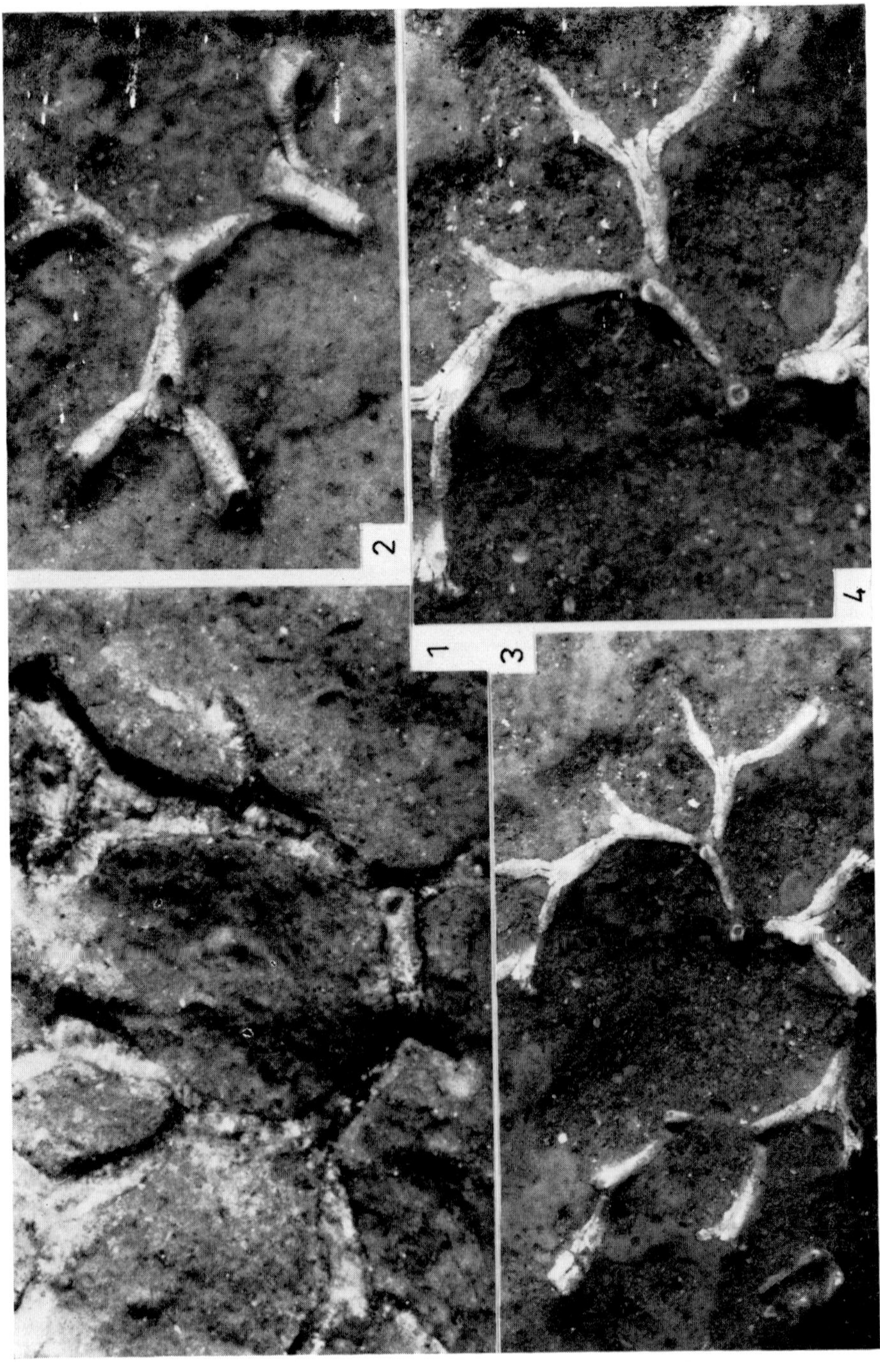
1
2
3
4

with chains of a maximum of 8 generations, 2 ancestrulae and about 60 ramifications are preserved.

(f) *Budding pattern*. Figs 1a, d, 4a, b; Plate I, Figs 1–4). In contrast to the preceding budding patterns, the two proximal segments growing beside the recumbent ancestrula never touch each other. Between them two further zooecial tubes arise beneath the peristome of the ancestrula (Fig. 1d). These two further zooecial tubes cease to grow in the ramification region and remain unfinished. Ramification takes place in each generation and always in the same pattern. Mature zooecia ramify symmetrically. Next to the mature zooecium the proximal segments of two younger normal zooecial tubes grow on both lateral edges of the branch. The younger zooecial tube on that edge of the branch which encloses a branch division began to bud within the preceding ramification between the aborted and the mature zooecium. The proximal segment of this younger normal zooecial tube grows as a narrow tube and remains in an immature stage. The younger zooecial tube on the outer edge of the branch buds distally beside the peristome of the older normal zooecium (Fig. 4a). Each peristome over the aborted zooecia grows in a typical manner upwards at about thirty degrees (Fig. 4b).

If the common walls of the zooecial tubes (septa) are considered in the sense of Borg, this third budding pattern is distinguished by the development of the vertical septa. Borg (1926, p. 270) explains the development of vertical septa by (1) coalescence of two septa, being the normal manner, or (2) by the fission of one septum, which phenomenon Borg describes as a deviation from the normal procedure. In the third budding pattern, the two first septa, which proceed from the lateral edges of the branch in order to complete the ancestral zooecial tube,

---

Plate I

Figs 1–4. *Stomatopora osterfeldi* n. sp. The colonies incrust syngenetic concretions. Middle Liassic (delta) (Amaltheentone, Domerian, Upper Pliensbachian), brickfield "am Osterfeld" near Goslar (Germany). Photos have been taken after the colonies had been moistened.

Fig. 1. Holotype, zoarium nr. AM8 incrusting concretion Osterfeld nr. 13 (collection Illies), with ancestrula. × 25. (Budding pattern see Fig. 4a).

Fig. 2. Paratype, zoarium nr. AM20 incrusting concretion Osterfeld nr. 22 (collection Illies). × 25.

Fig. 3. Paratype, zoarium nr. AM14 incrusting concretion Osterfeld nr. 22 (collection Illies). × 12.

Fig. 4. Enlarged section of the same specimen as in Fig. 3. In some sections the budding pattern is visible. × 20.

bifurcate in this manner before they coalesce. Thus two vertical septa of the second generation are developed and, together with the median vertical septum which originates from coalescence of the two septa, there are now three vertical septa in the second generation. These three vertical septa separate four zooecial tubes, namely the two normal ones and between them the two aborted zooecia. The lateral vertical septa bifurcate again and diverge. Therefore three short vertical septa surrounding the aborted zooecia end within the ramification region. The two lateral extended septa prolong into the new branches, and now separate the mature zooecium from the younger proximal segment in each branch (Fig. 4a). There are thus uninterrupted chains of vertical septa, resulting from fission only (Fig. 4a).

(g) *Additional description of the holotype.* (Zoarium AM8, Plate I, Fig. 1): The symmetrical colony incrusts an area of 2·7 mm². Width of proancestrula 0·25 mm. Length of ancestrula (incl. proanc.) 0·50 mm. Ramification takes place in each generation of the zooecial tubes. Distances of peristomes: generations one to three inclusively 0·73–0·81 mm, gen. four to seven 0·70–0·85 mm. Width of branches, proximally: gen. one to three 0·10–0·13 mm, gen. four to seven 0·13–0·25 mm. Width of branches, dist. gen. one to three 0·13–0·27 mm, gen. four to seven 0·21–0·30 mm. Greatest width of zooecia, gen. one to three 0·12–0·19 mm, gen. four to six 0·16–0·22 mm. Diameter of peristomes, gen. two to six 0·16–0·22 mm. There are two aborted zooecia within each ramification region of the branches. Angles of ramifications: at the first ram. 130°, at the second rams. 100–120°, third rams. 65–90°, fourth rams. 80–100°, fifth ram. 85°. The first space restriction occurs in generations 5 and 6, where some branches already touch each other.

(h) *Observation on paratype.* In zoarium AM14, incrusting concretion Osterfeld nr. 22, one ramification has three (instead of two) aborted zooecia (Plate I, Fig. 3).

(i) *Affinities.* No other species in which the budding pattern has been investigated shows one similar to that described here.

REFERENCES

Borg, F. (1926). Studies on recent Cyclostomatous Bryozoa. *Zool. Bidrag Uppsala* **10,** 181–507.

Illies, G. (1963). Über *Stomatopora dichotoma* (Lamx.) und *St. dichotomoides* (d'Orb.) [Bryoz. Cycl.] aus dem Dogger des Oberrheingebietes. *Oberrhein. geol. Abh., Karlsruhe* **12,** 45–80.

Illies, G. (1971). Drei Arten der Gattung *Stomatopora* [Bryoz. Cycl.] aus dem mittleren Lias bei Goslar und deren verschiedene Knospungsmuster. *Oberrhein. geol. Abh., Karlsruhe* **20,** 125–146.

LAGAAIJ, R. (1963). New Additions to the Bryozoan Fauna of the Gulf of Mexico. *Publ. Inst. Mari. Sci., Texas* **9,** 162–263.

MILNE-EDWARDS, M. H. (1838). Mémoire sur les Crisies, les Hornères et plusieurs autres Polypes vivans ou fossiles dont l'organisation est analogue à celle des Tubulipores. *Ann. Sci. nat. zool.*, Ser. 2, Paris, **9,** 193–238.

VOIGT, E. (1968). Über Hiatus-Konkretionen (dargestellt an Beispielen aus dem Lias). *Geol. Rundsch., Stuttgart* **58** (1), 281–296.

WALTER, B. (1969). Les bryozoaires jurassiques en France. Etude systematique. Rapports avec la stratigraphie et la paléoécologie. *Docum. Lab. Géol. Fac. Sci. Lyon* **35,** 1–328.

30 | The Great Tentacle Sheath Nerve as the Path of an Innervation of the Frontal Wall Structures in the Cheilostome *Electra pilosa* (Linné)

GENEVIÈVE LUTAUD

Laboratoire de Cytologie et Station Biologique de Roscoff, Université de Paris, VI, France

Abstract: The Bielchovsky silver stain applied to two species of *Electra* confirms the results obtained by methylene blue and shows the wide extension of non-motor fibres from the main nerve of the tentacle sheath spreading in the cellular layers of the frontal wall of the zooecium. The ultrastructural characteristics of the nerve are briefly described.

The hypothesis of an innervation of the marginal spines is postulated from the staining of nerve endings of a non-motor branch of the parietal ramification of the nerve at the base of the spines and especially of the long median spine.

The demonstration of nervous extensions in the frontal wall raises the question of whether these or Hiller's parietal network might be homologous to the reticular plexus of the Ctenostomata.

The general pattern of the main nervous pathways in the Cheilostomata Anasca of the family Electridae has already been briefly described from methylene blue vital stains (Lutaud, 1969). From this preliminary account, we may consider three topographical groups of nerves assuming the three possible functions of the activity of entire zooids.

(1) The polypide set with the median dorsal and the twin lateral intestinal nerves proceeding from a small accessory visceral ganglion at the top of the median groove of the oesophagus, and the tentacle motor and sensory nerves arising from the thick and certainly organized nervous belt which prolongs the distal horns of the cerebral ganglion.

(2) The peripheral innervation by the two pairs of classical tentacle sheath nerves arising from the proximal neurons of the ganglion. These meet and unite on their way towards the aperture and are the three-branched motor nerve, the trifid of Bronstein (1937), and a thick direct cord without any autonomous ramification in the first separate part of its course. The two associated nerves account for the

innervation of all the apertural structures and for the command of all the muscles involved in the extrusion and retraction activity of the polypide.
(3) The parietal network discovered by Hiller (1939), using methylene blue stains, at the base of the interzooidal partitions in *Electra pilosa*, which may convey information between adjacent zooids and thus coordinate their responses. However, the nervous nature of this network in partitions is still presumptive and requires confirmation by very careful cytological investigation.

This theoretical diagram has been entirely corroborated by new methylene blue stains and by the complementary use of the Bielchovsky silver stain, according to the Gros-Schultze variant, applied to entire zooids of *Electra pilosa* (Linné) and *Electra posidoniae* Gautier. Silver stained mounts allow a more precise analysis of nerve endings appearing at the end of the vital staining in lethal saturation phases, or hidden by calcified parts of the wall in live animals, and thus reveal the real extension of nerve endings from the ramifications of the main tentacle sheath nerve in the frontal wall of the zooecium.

THE STRUCTURAL FEATURES OF THE MAIN TENTACLE SHEATH NERVE

The roots of the two initially separate tentacle sheath nerves are clearly stained by silver in the neuropile of the heart-shaped cerebral ganglion. They spring out, together with the sheath connectives leading to the partition network, by two side openings of the peritoneal envelope in the median depression and immediately diverge. We know that the first short ramification of the trifid nerve returns to the visceral ganglion, that the second branch turns around the pharynx, spreading between the muscles fibres in the insertion of the retractor muscle of the polypide, and that the third axial branch bends up towards the sheath and displays, at about a quarter of the sheath height, an annular ramification linking the origins of all the longitudinal muscle fibres of the sheath wall. The other nerve runs directly towards the top of the sheath and does not branch before joining the axial branch of the trifid nerve shortly above its annular ramification. Silver stains do not reveal any other ramification from either of the two nerves, before or after their topographical association, in the wall of the sheath itself. The sheath wall does not entirely lack non-motor innervation as

---

Plate I

Cytological characters of the mixed frontal nerve in *Electra pilosa* (transverse microthin section, × 16,000), P: peritoneal layer of the tentacle sheath wall; N: nerve fibre; F pg: peripheral giant fibre; M: longitudinal muscle fibre; Col: collagen; ep. MV: microvill of the epithelial layer; Loph. at: lophophore atrium.

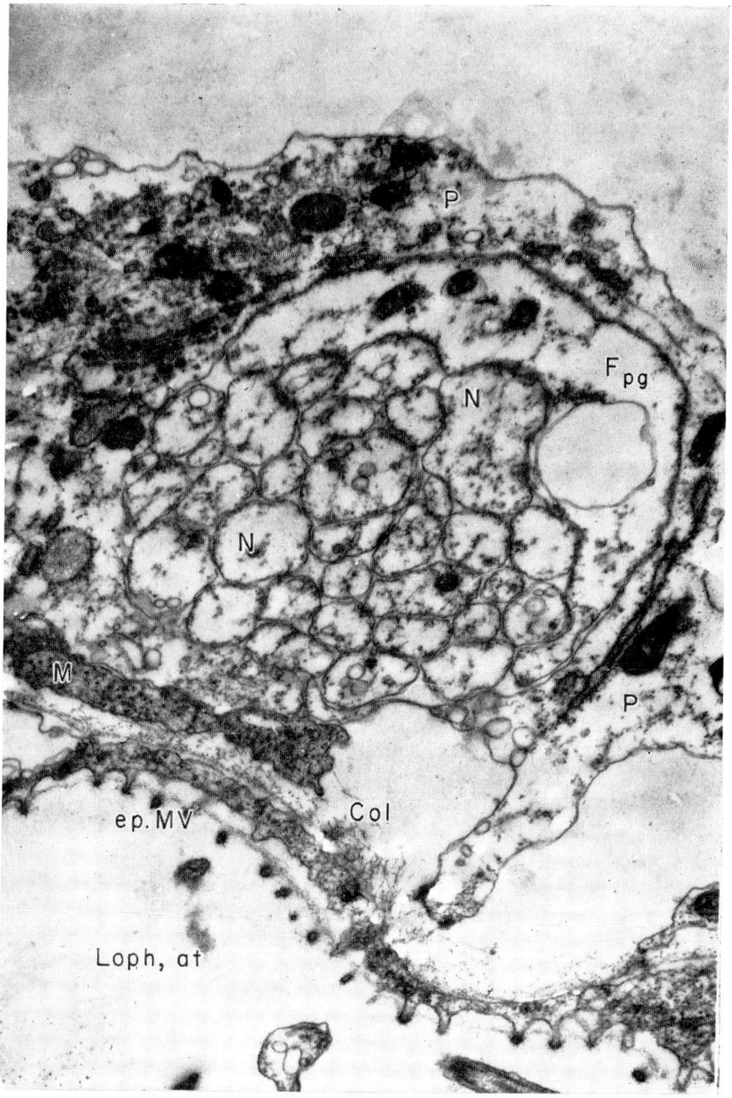
P
F pg
N
N
M
P
ep. MV
Col
Loph, at

methylene blue dyes, at the base of the sheath, a crown of fusiform nodules supported by an ordered fan of thin fibres coming directly from the distal brim of the ganglion.

Investigation at ultrastructural level shows that the association between the two tentacle sheath nerves is more intimate than a simple convergence of pathways. Transverse microthin sections of the sheath show each of the twin compound axes, after the joining of the trifid and the non motor nerves, as a single bundle of fibres running, like the longitudinal muscles in the sheath wall, between the basal collagen of the epithelium lining the lophophore atrium, and the peritoneal layer lining the coelom cavity. The compound nerve is directly imbedded beneath the peritoneal cells without any differentiated glial wrapping, as are also the other nerves of the polypide. However, the nerve fascicle includes two sorts of fibres of different size: about thirty relatively thin fibres, from 0·1 to 0·3 μ in section, with characteristic axial microtubules and long mitochondriae, are partly surrounded by a rather flattened single giant fibre, from 1 to 2 μ in section (Plate I). The interpretation of this peripheral unit as a protective and nutrient cell with the functional meaning of a Schwann cell has not been confirmed as it shows the same clear matrix, microtubules and axial mitochondriae as the minor fibres. We are now trying to determine if this cytological duality exists previously in either of the two constituent nerves and whether it remains after the splitting of motor and non-motor secondary ramifications.

THE MAIN BRANCHES OF THE MIXED FRONTAL NERVE

The symmetrical compound nerves in following their course towards the aperture pass from their dorsal origin in the cerebral ganglion to the side and front parts of the sheath, whereas the connectives joining the partition network in the distal corners of the zooecium remain in dorsal location, and rather abruptly break into several primary branches in the pre-vestibular area of the sheath (Plate II, Fig. 1). The fact that all these branches extend to the apertural structures or into the frontal wall is the reason why the name "frontal nerve" has been chosen in this description; the term "mixed" is used in its complete physiological meaning and accounts at once for the double origin of the compound nerve and for the two sorts of nerve endings it bears either in muscle insertions or between the layers of the vestibule and frontal walls.

1. *The Diaphragmatic Branch*

The first primary branch (b_1d, Fig. 1) separates from the main axis a little below the diaphragmatic parietal muscle. It first passes through the muscle

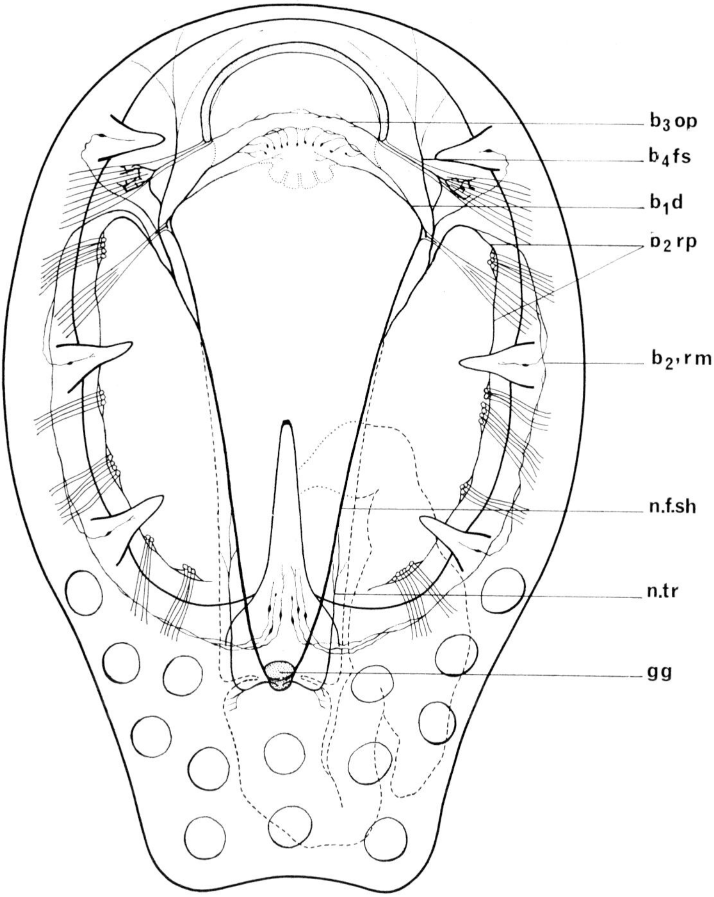

Fig. 1. Interpretative diagram of the ramifications of the great tentacle sheath nerve in *Electra pilosa*. gg: cerebral ganglion; n.tr: trifid nerve; n.f.sh: mixed frontal nerve of the tentacle sheath; b1d: diaphragmatic branch; b2rp: parietal motor recurrent branch; b2, rm: marginal recurrent ramification; b3op: opercular branch; b4fs: residual superficial branch.

insertion on the sheath, where it expands into a characteristic silver stained reticule, and then divides into two secondary ramifications. One ramification reaches the centre of the diaphragm and ends in the neighbourhood of the sphincter muscle, the other splits into numerous and probably single fibres with nodular endings, in regular order, each one penetrating, a fold of the upper vestibular part of the diaphragm (Plate II, Fig. 4). The terminal knobs stained by silver look like fusiform cells when stained by methylene blue, but a more

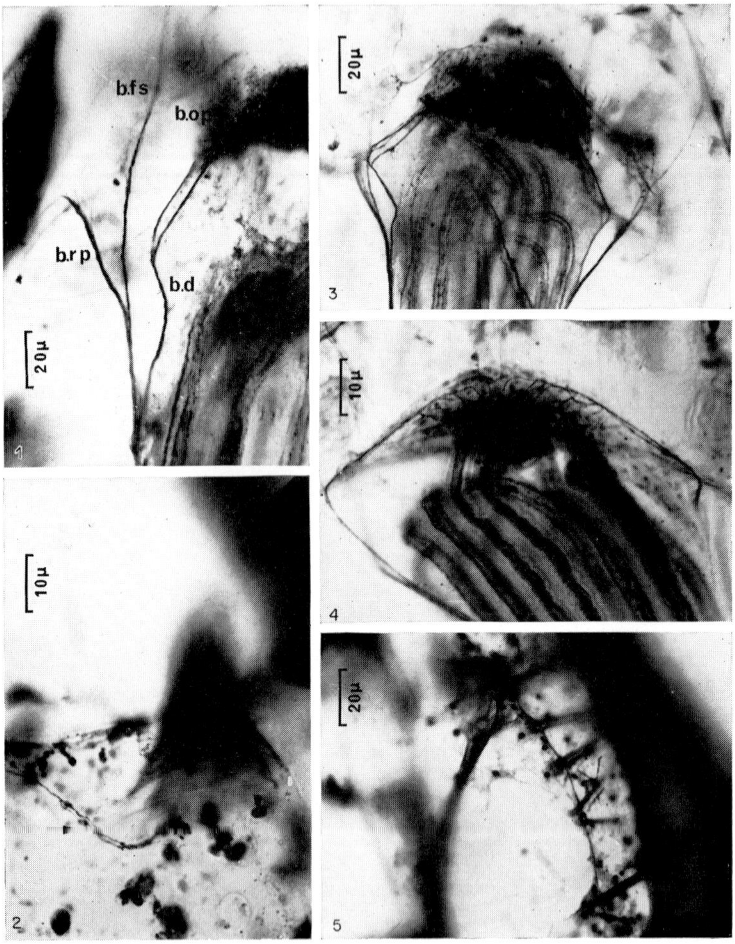
b.f s
b.op
b.r p
b.d
20µ
1
10µ
2
20µ
3
10µ
4
20µ
5

complete cytological investigation is necessary to discriminate between simple terminal processes or true sensory cells.

2. The Opercular Branch

Shortly beneath the aperture, in the vestibule area of the sheath, another primary branch ($b_3$op, Fig. 1) diverges towards the retractor muscles of the operculum and develops its intricate reticule around the opercular end of the muscle fibres. Then, from the muscle insertion, a transverse ramification bearing short lateral expansions without terminal nodules bends back along the hinge of the operculum, and the two symmetrical ramifications, meeting midway, link the opposite muscle insertions like the string of a bow.

3. The Frontal Wall Ramifications

Diverging between the diaphragmatic and the opercular branches, another recurrent branch which is the second in chronological sequence (b_2r.p, Fig. 1) reaches the frontal wall next to the first distal parietal muscle and runs round the border of the frontal membrane through each frontal insertion of all the subsequent parietal muscles (Plate II, Fig. 5). This recurrent motor ramification has been described by Graupner (1930) and by Bronstein (1937) as the "parietal nerve", but, since it proceeds from the main axis, the full designation of parietal recurrent motor ramification has been retained here to keep a terminological unity and to emphasize the interdependence of the different ramifications from the great tentacle sheath nerve.

The axial residual branch ($b_4$fs, Fig. 1), following one of the top ligaments of the sheath, reaches the frontal wall near the proximal margin of the aperture and expands arborescent strands under the epithelial layer of the wall. When stained with methylene blue these superficial fibres appear to be scattered all over the

---

Plate II

The main ramifications of the mixed frontal nerve in *Electra pilosa*.

Fig. 1. The divergence of the primary branches of the mixed frontal nerve at the top of the tentacle sheath (Bielchovsky silver stain).

Fig. 2. Presumed nerve-endings from the recurrent marginal ramification at the base of the median spine (methylene blue stain, decalcified mount).

Fig. 3. Diaphragmatic and opercular ramifications (Bielchovsky silver stain).

Fig. 4. Nerve endings of the diaphragmatic non-motor ramification in the folds of the diaphragm (Bielchovsky silver stain).

Fig. 5. Parietal recurrent motor ramification around the opesia and through the frontal insertions of the parietal muscles (Bielchovsky silver stain).

calcified sides of the frontal wall, but when impregnated with silver they do not show such an extensive spread and only a few constant fibres can be ascertained from the tractus and wrinkles of the decalcified wall. Thus a thin strand usually surrounds the distal margin of the aperture and another reaches the distal septum, but whole mounts are rather dark near the transverse walls and it has not been possible to see if it may extend further into the partition and even, perhaps, join the basal filament of the interzoidal network.

THE RECURRENT MARGINAL RAMIFICATION—THE HYPOTHESIS OF THE INNERVATION OF THE MARGINAL SPINES

In methylene blue stained mounts, at immersion magnification, the recurrent motor branch around the opesia seems to be duplicated by another inconspicuous parallel strand running along the margin of the decalcified gymnocyst. In different zooids of the same colony, some variation is noticed in the relatively early divergence of this non-motor recurrent ramification which may separate from the other recurrent ramification near its origin, but sometimes also arises directly from the main axis a little further on. This marginal ramification (b_2,r.m, Fig. 1) seems to be composed of a few single fibres rather loosely clustered, which separate to enter into the basal enlargement of each marginal spine (Plate II, Fig. 2). The large median spine receives a double supply of these presumed nerve endings with the ultimate fibres of the two opposite marginal ramifications. The fibres do not seem to penetrate far into the tubular body of the spines and bear the same nodular processes as those that have been already mentioned in the diaphragm folds and at the base of the sheath. There is some doubt whether the first distal spines receive fibres from the marginal ramification or whether they are directly derived from the axial branch of the nerve. Silver stained mounts reveal similar nerve fibres, but are difficult to interpret as the marginal ramification runs beneath thickened and naturally brown parts of the ectocyst.

It is necessary to be very careful in advancing any definite interpretation of thin strands detected on decalcified mounts where nerve fibres are certainly injured by such fixatives as the ammonium molybdate used to retain the blue dye, or difficult to discriminate from epithelial tractus. But taking into account the optical continuity of the primary branches of the nerve with the ultimate fibres in spines and the agreement of results using two classical techniques, we may presume a sensory innervation of the marginal spines in the Electridae. These would not only be ornamental or mechanically protective protuberances of the wall, but could also perhaps register shocks or local water current movement in the external environment.

DISCUSSION

Although they do not always derive from a common stem, the two recurrent ramifications, by their parallel paths and by assimilation with the subdivisions of the primary diaphragmatic and opercular branches, could be considered as the secondary motor and sensory ramifications of the same primary mixed recurrent branch. Thus, the main frontal nerve, gathering the axons of a former undoubted motor nerve and of a presumed sensory fascicle, would divide into three main primary branches of mixed functions, each splitting further into motor and non-motor secondary ramifications, the third level of subdivision separating single fibres from the non-motor ramification. The axial residual branch, without any muscle contact, does not fit into this theoretical subdivision and this underlines its residual character; but the superficial strands could just as well be interpreted as derived from the last non-motor ramification across the operculum, or grouped with the spine fibres in a general assortment of parietal nerve endings.

Close study of the topographical territory of the mixed frontal nerve demonstrates that an ordered equipment of nerve endings exists in the cystid wall. The significance of such an extensive parietal spread of the most classical tentacle sheath nerve cannot be disregarded, even if it is restricted to the frontal area in touch with water and related in its distribution to the morphogenetical influence of the sheath in the organization of the aperture and of the opesia. It is to be noted that the presumed sensory fibres expand in hard calcified parts of the wall and that no non-motor ramification could be found below the aperture in the supple and periodically depressed frontal membrane.

The cytological demonstration of terminal sensory cells or processes in the spines, or in the neighbourhood of the aperture, would be conclusive in interpreting the frontal nerve as the agent of a perception function of the cystid wall, and distinguishing between artifact knobs and real differentiated organites is the next aim of this study. But it is difficult to compare the parietal developments of a true axonal nerve and the diffuse reticular plexus that Gewerzhahen (1913) and Marcus (1926; 1934) have stained with methylene blue in the wall of the upright zooids of a few Stolonifera and Phylactolaemata. Although very thin and inconspicuous threads could be revealed, no diffuse superficial network of this sort could be found in *Electra* and it seems plausible that parietal innervation does not occur in this way in the encrusting Anasca or at least in the Electridae. A thorough understanding of the tentacle sheath nerves of the Ctenostomata and of their connection with the parietal plexus, as suggested by Marcus (1926), is necessary before true homologies may be determined. The network discovered by Hiller in *Electra* displays some of the characters of a plexus by its knotty texture, by its presumed nuclear nodes and by its uninterrupted continuity all over the colony. It differs from the reticular plexus of the Ctenostomata by its relatively organized distribution and by its internal location so that it cannot perform an immediate registration of external stimuli as a superficial network is supposed to do. The differentiation of two different paths of parietal innervation in the Electridae may be the consequence of the development of the interzooidal septae and of the opesia in the encrusting colony and a somewhat compensatory extension of the sheath nerve in the free frontal wall.

REFERENCES

BRONSTEIN, G. (1937). Étude du système nerveux de quelques Bryozoaires Gymnolémides. *Trav. St. Biol. Roscoff* **15,** 155–174.

GEWERZHAHEN, A. (1913). Beiträge zur Kenntniss der Bryozoen. *Z. Wiss. Zool.* **107,** 3–345.

GRAUPNER, H. (1930). Zur Kenntniss der feineren Anatomie der Bryozoen. *Z. Wiss. Zool.* **136,** 38–97.

HILLER, S. (1939). The so-called nervous colonial system in Bryozoa. *Nature, Lond.* **143,** 1069–1070.

LUTAUD, G. (1969). Le "plexus" parietal de Hiller et la coloration du système nerveux par le bleu de méthylene chez quelques Bryozoaires Cheilostomes. *Z. Zellforsch. mikrosk. Anat.* **99,** 302–314.

MARCUS, E. (1926). Beobachtungen und Versuche am lebenden Meeresbryozoen. *Zool Jb., abt. Syst. Okol. Geogr.* **52** (1), 1–102.

MARCUS, E. (1934). Über *Lophopus cristallinus* (Pallas). *Zool. Jb., abt. Anat. Ontog.* **58.** 506-606.

31 | New and Little-known Structures of Fenestelloidea

I. P. MOROZOVA

Palaeontological Institute, Academy of Science, Moscow, U.S.S.R.

Abstract: The great taxonomic variety of Recent Bryozoa belonging to the Gymnolaemata is due mainly to different budding patterns and polymorphism. These determine colonial shape and structure, degree of zooidal differentiation, and arrangement of zooecia in the colonies.

Modern methods of investigation applied to the study of fossil Bryozoa give to palaeontologists the possibility of using similar criteria for their classification. These methods also allow changes to be traced in the evolution of different branches in this group in budding patterns and in the morphology of skeletal structures of zooids.

However, among fossil Bryozoa, particularly among those of the Palaeozoic, several taxa were established without due regard to internal colonial structures. As a result, they represent, to a considerable degree, artificial units. Among such taxa first of all the suborder Fenestelloidea (Astrova and Morozova, 1956) of the order Cryptostomata should be mentioned. Elias and Condra (1957) consider this group of Palaeozoic Bryozoa as a separate order Fenestrata. Although these Bryozoa have been studied from many different points of view and many papers devoted to them have been published, they still are combined into a number of heterogeneous families and genera. Although the heterogeneity of this genus is apparent, the successful revision of this taxon is hindered not only by the abundance of species known from different parts of the globe, but, mostly, by inadequate descriptions of the great majority of them. Unfortunately, even in these days, many authors still describe the species of different fenestelloid genera on the basis of external colonial characters alone, without preparing thin sections.

Fenestelloids are extremely widespread in the Palaeozoic of the U.S.S.R.; therefore, this group of Bryozoa has for a long time attracted the attention of

Soviet palaeontologists. The extensive material concentrated in our country allows us to start a revision of fenestelloid genera and families, as well as to determine the precise position of the fenestelloids within the Gymnolaemata.

During the investigation of bryozoan collections available in the Palaeontological Institute special attention has been given to microscopic studies of skeletal structures of zooids composing the colonies of various fenestelloid genera. The material studied has revealed among Bryozoa belonging chiefly to the genus *Fenestella* the existence of several groups of species. Each group develops not only skeletal structures typical for *Fenestella*, but also some that have been virtually unknown previously.

Among these structures are peculiar cavities that have been found in 17 Carboniferous and Permian species of *Fenestella*. These cavities are developed on the reverse of the branches, at the points of their junction with the dissepiments (Plate I, Fig. 1). They are of elongate pyriform shape in thin sections and their enlarged part faces down the direction of growth of the colony. Shulga-Nesterenko (1941), the first to describe these structures in *Fenestella cavifera*, named them "caverns" and regarded them as closed cavities of inorganic origin. Later, having discovered that the caverns are present also in a number of closely related Carboniferous and Permian species, Shulga-Nesterenko (1952) made a suggestion that they might be skeletal remains of some zooids and supposed them to be of supraspecific value. I have confirmed, by means of microscopal studies, the suggestion made by Shulga-Nesterenko and discovered that the caverns have a wall of their own and apertures opening onto the reverse of the branches. Their apertures appear as round openings of much greater size than those of ordinary zooecia. Each of them is provided with a well developed broad caplike lunarium (Plate I, Fig. 2). In the best oriented longitudinal sections hemiseptum-like projections of the "caverns" located near apertures

---

Plate I

Fig. 1. *Fenestella cavifera* Shulga-Nesterenko, 1939. PIN, No. 3229/61. Lower Permian, Middle Urals. Tangential section ×50. Caverns developed on the reverse of the branches, with hemiseptum-like projections located near apertures.

Fig. 2. Same as before. Caverns with well developed cap-like lunaria.

Fig. 3. Same as before. Ex. PIN, No. 3229/63. Tangential section ×70. Caverns with calcareous opercula.

Fig. 4. *Fenestella murengolensis* Shishova. Holotype. PIN, No. 2287/493. Middle Devonian, Mongolia; (a), tangential section ×35; paired structures resembling caverns observed in the upper part of the section; (b), transverse section of a branch with structures resembling caverns.

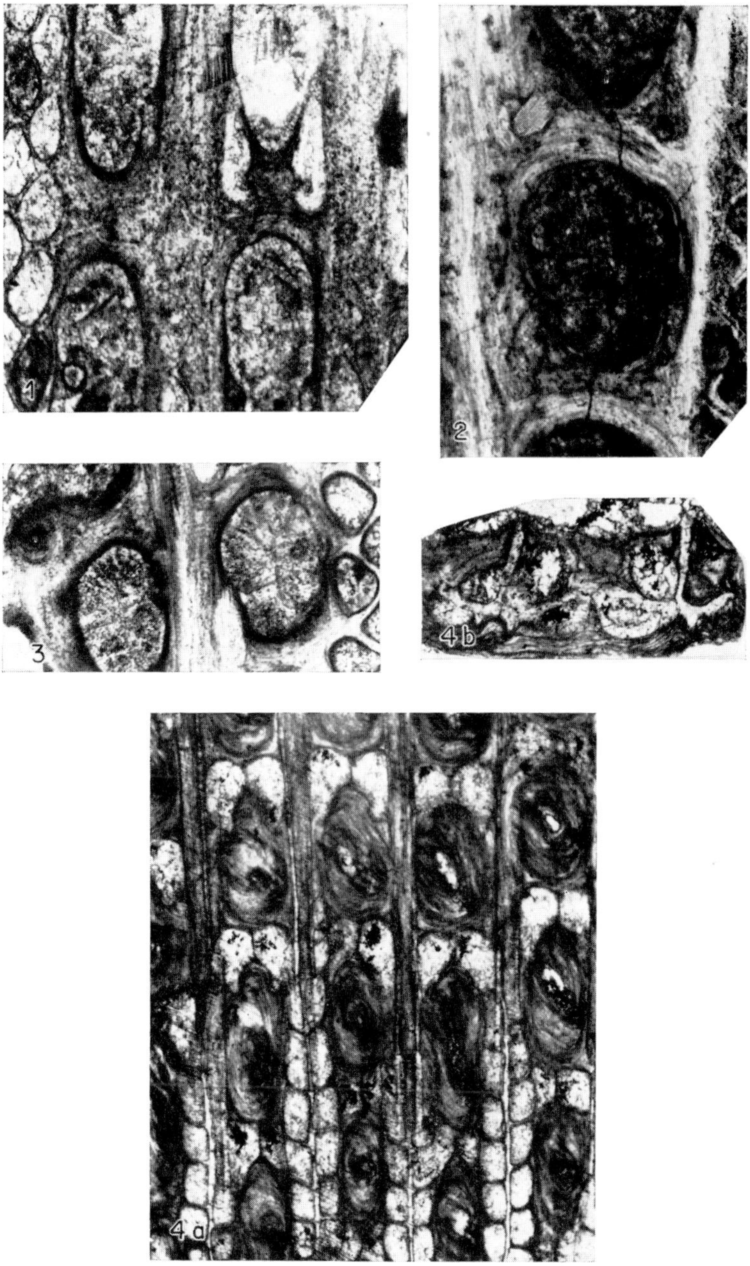
1
2
3
4b
4a

(Plate I, Fig. 1) can be observed. Cross sections of the colonies show that the cavern walls are contiguous with the primary walls of ordinary zooecia. Hence, it may be assumed that either the "caverns" budded immediately after the ordinary zooecia had been formed, or that they arose simultaneously with the latter.

At the colony surface the apertures of the caverns are overgrown in most cases with a thin layer of calcareous matter or, more rarely, are closed by calcareous opercula (Plate I, Fig. 3). The functional significance of "caverns" is unknown. Possibly they fulfilled a defensive role, such as in some Recent cheilostome bryozoans is fulfilled by avicularia developed on the reverse of the colonies. The consistent presence of "caverns" in many widespread species of "*Fenestella*" provides a basis for separating them into an independent genus. It should be noted also that all species with "caverns" share other characters in common, such as similar outline of zooecial section and similar carinae.

Skeletal structures resembling the "caverns" were discovered by Hall about a hundred years ago (Hall, 1883) in a Devonian species named by him *Fenestella labiata*. Hall observed in this species paired cavities of semielliptical section located on the reverse of the broad dissepiments. These structures made Hall doubt that this species belonged to *Fenestella*. In this connection it is interesting to note that a species assigned tentatively to the genus *Fenestella* has recently been described from the Devonian of Mongolia (Shishova, 1970). On the broad dissepiments of this bryozoan, also on the reverse side, there are large paired cavities of pyriform section the enlarged part of which is facing up the direction of growth of the colony (Plate I, Fig. 4). In their shape and arrangement they bear close resemblance to the cavities described by Hall. These two species are similar also in meshwork characteristics and in outline of zooecial section. All these features distinguish them from all known species of *Fenestella*.

Plate II

Fig. 1. *Fenestella labuensis* Morozova, 1970. Holotype. PIN, No. 2830/2. Upper Permian, northeastern USSR. Tangential section ×60. Paired diminuitive zooecia developing between the ordinary ones.

Fig. 2. *Fenestella* sp. PIN, No. 3229/63. Lower Permian, Middle Urals. Reverse of the branches with accessory pores ×60. Some of them closed by centrally perforated calcareous lamina.

Fig. 3. *Polypora* sp. PIN, No. 3230/64. Lower Permian, Middle Urals. Tangential section ×15. Reverse of the branches with accessory pores.

Fig. 4. *Polypora* sp. Tangential section, ×55, showing the apertures of zooecia and accessory pores.

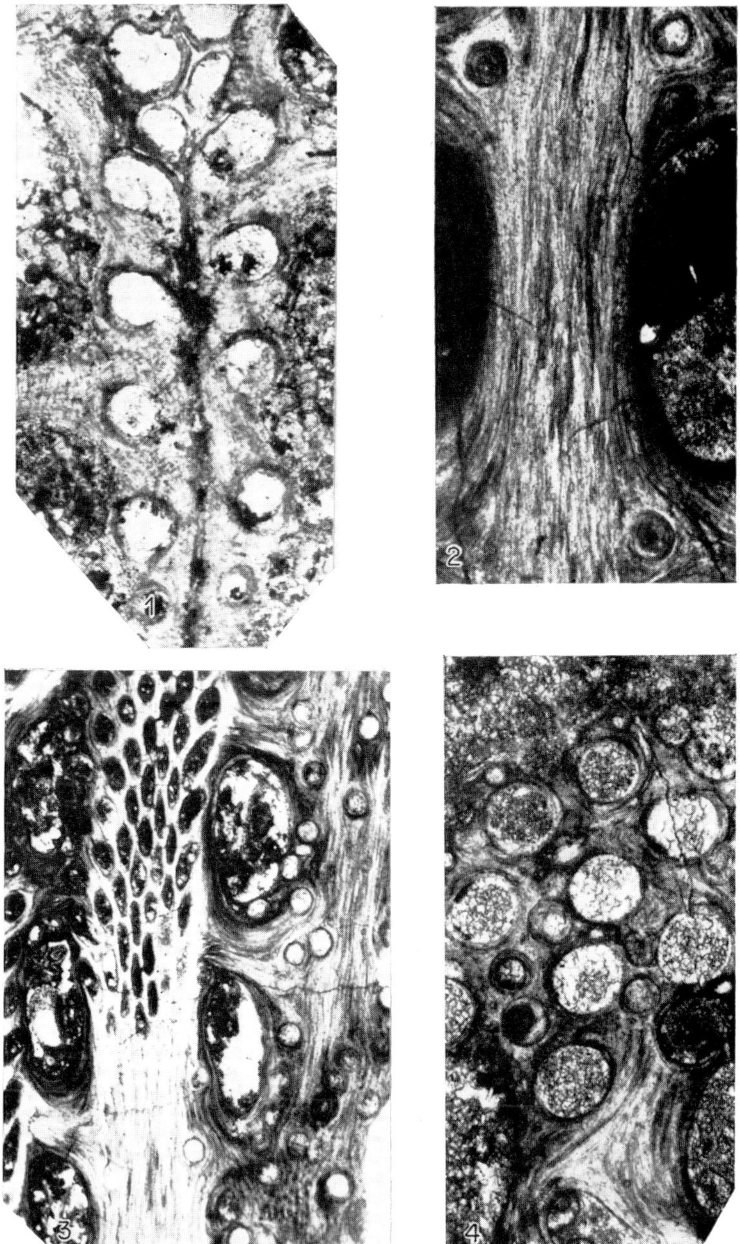
1
2
3
4

There is one more, previously unknown type of skeletal structure, which I have found in the colonies of several species described under the generic name of *Fenestella* from the Permian of the Urals and northeastern U.S.S.R. (Morozova, 1970). This structure consists of small, diminuitive zooecia developing between the ordinary ones (Plate II, Fig. 7). Apertures of diminuitive zooecia open also onto the obverse surface, as do those of normal zooecia, and differ from the latter only by being smaller. The shape of their section differs very little from that of adjacent normal zooecia. These very small zooecia were budded either singly or in pairs, but their number and arrangement in colonies usually varies, in some parts of the colony they are abundant, in others scarce. Their structure and arrangement are reminiscent of those of nannozooids of Recent Bryozoa described by Borg (1926) in the cyclostome genus *Dyplosolen*, and by Marcus (1937) in the cheilostome genus *Tripostega*. Nannozooids in these bryozoans, as well as small zooecia in some "*Fenestella*" species, are dispersed among ordinary zooecia resembling the latter in every detail.

Finally, there are poorly known skeletal structures called "accessory pores". These pores are hemispherical in outline and are developed in many species of different fenestelloid genera in the external walls, both on the obverse and the reverse of the colony. Ulrich (1890) first found and described these structures in a Carboniferous species of the genus *Septopora*. Although this author did not give a detailed description of them, he considered their size, number and arrangement in the colony to be specific features. Later, accessory pores were observed by Nikiforova (1938) and Shulga-Nesterenko (1941) in many Carboniferous and Permian species of various genera (Plate II, Figs 2, 3). First, Shulga-Nesterenko believed them to have been formed by some burrowing organisms, but later, having traced the development of "accessory pores" in a number of species, she made a suggestion that they fulfilled some vital function in the colony.

Investigation of extensive material of the Fenestelloidea has disclosed that the former interpretation of accessory pores as porous structures was erroneous. It seems better to regard them as skeletal remains of regularly developing heterozooids of unknown taxonomic and functional importance. This suggestion is confirmed by the fact that "accessory pores" have a wall of their own and a ring-like thickening in the apertures. In some portions of the colonies "accessory pores" are closed by centrally perforated calcareous laminae (Plate II, Fig. 2).

Arrangement, number and diameter of "accessory pores" may vary in different fenestelloid species (Plate II, Figs 2, 3). In most species "accessory pores" are confined to junction angles of the branches and dissepiments on the reverse of the colony. On the obverse of the colony they are usually observed on the external zooecial walls, one opposite each aperture. In some species of *Polypora* "accessory pores" are unusually abundant and irregularly distributed both on the obverse and the reverse of the colony (Plate II, Fig. 3). Shulga-Nesterenko (1952) and Shishova (1970) traced the evolutionary development of "accessory pores" in a number of species of several fenestelloid genera. They noticed that during evolution in phylogenetically related species "accessory pores" increased in number and in size. The taxonomic and functional significance of "accessory pores" is unknown. Neither is it clear why in some fenestelloid genera (*Fenestella, Polypora*) they are developed only in a comparatively few species, whereas in others (*Septopora, Acanthocladia*) they are present invariably.

In conclusion it may be stated that all these skeletal structures which developed with regularity in a number of fenestelloid species, that have been assigned for the most part to the genus *Fenestella*, demonstrate heterogeneity of this genus, as well as the necessity of revision of all this bryozoan group. This revision can be made on the basis of a thorough microscopic study of skeletal structures of the colony. Also special attention should be paid to polymorphism which needs much more study in this bryozoan group.

REFERENCES

ASTROVA, G. G. and MOROZOVA, I. P. (1956). K sistematike mshanok otriada Cryptostomata. *Doklady Akad. Nauk SSSR* **110** (4), 661–664.

BORG, F. (1926). Studies on recent cyclostomatous Bryozoa. *Zool. Bidrag Uppsala* **X,** 182–507.

ELIAS, M. K. and CONDRA, G. E. (1957). *Fenestella* from the Permian of West Texas. *Geol. Soc. Am.* **70,** 1–178.

HALL, J. (1883). On the mode of growth and relations on the Fenestellidae. *Rept State Geol. Albany* 36–47.

MARCUS, E. (1937). Bryozoen von St. Helena. *Vidensk. Medd. Dansk. Natuh. Foren*, 101.

MOROZOVA, I. P. (1970). K poznaniju pozdnepermskikh mshanok Severovostoka SSSR. *Paleontol. zhurn.* (3), pp. 80–88.

NIKIFOROVA, A. I. (1938). Tipy kamennougolnykh mshanok Evropeiskoi chasti SSSR. Paleontologia SSSR **IV** (5) 1, 1–290.

SHISHOVA, N. A. (1970). Nekotorye novye siluriiskie i devonskie mshanki Mongolii. Zbornik "Novye vidy paleozoisk. mshanok i korral". *Izd-vo* "*Nauka*", pp. 28–31.

SHULGA-NESTERENKO, M. I. (1941). Niznepermskie mshanki Urala. *Paleontologia SSSR, Trudy Paleont. Inst. Akad. Nauk SSSR* **5** (I), 1–276.

SHULGA-NESTERENKO, M. I. (1952). Novye niznepermskie mshanki Priuraliia. *Trudy Paleont. Inst. Akad. Nauk SSSR* **37,** 1–84.

ULRICH, E. O. (1890). Paleozoic Bryozoa. *Geol. Surv. Illinois* **8,** 285–688.

32 | Ultrastructure of the Tentacles of *Flustrellidra hispida* (Fabricius)

L. W. SMITH

Department of Zoology, University College of Swansea, Swansea, Wales

Abstract: The tentacles of *Flustrellidra hispida* have been examined under the electron microscope. They show a microvillous border with a thin layer of mucus between the microvilli, which probably acts as a cushion against mechanical abrasion during retraction of the polypide. The lateral cilia are located in two groups, one group to each of the two lateral cells. The rootlets of these lateral cilia have a curious horizontal branch, the precise function of which is unknown but which might perhaps anchor the cilia during beating or secure them in the cell to resist shearing during retraction of the polypide. The so-called basement membrane in the tentacle has been shown by electron microscopy and histochemistry to be a tube of collagen.

Specimens of *Flustrellidra hispida* (Fabricius) were collected from Mumbles Point, near Swansea, and from Solva and Llangwm, Pembrokeshire, where they occur on the alga *Fucus serratus* L. For light microscopy the entire colony complete with alga was cut into convenient pieces (1·0 cm × 0·5 cm) and was fixed in Bouin's fixative made up with sea water. It was then dehydrated and embedded in paraffin wax (M.P. 55°C) and sectioned. For the electron microscopic study the individual guts were dissected out and fixed in 3% glutaraldehyde using a technique modified from Sabatani *et al.* (1963). The tissue was then washed in cacodylate buffered glucose and post-fixed in osmium-tetroxide, dehydrated in acetone and embedded in Epon 812 (Luft, 1961). Sections were cut on a Huxley microtome, and silver and gold sections were picked up on coated grids and were stained in uranyl acetate and lead citrate (Reynolds, 1963). The grids were then examined on an Akashi Tronscope and an A.E.I. EM6.

In cross section the tentacles are roughly T-shaped, measuring about 35 μm along the long axis by 20 μm across the short, and bear cilia in a frontal and paired lateral tracts. The number of outer cells seen in a section varies, depending on the degree of relaxation of the tentacle (Calvet, 1900), but typically is nine (Fig. 1b). These outer cells are covered with a layer, about 1 μm thick, of simple or (more often) branched microvilli. Between the individual microvilli there is a faintly electron opaque region which is probably a layer of mucus. Using light microscopy of wax sections, the outermost layer of the tentacles stains an intense blue colour with alcian blue (Steedman, 1950), and low iron diamine-alcian blue (Spicer, 1966), indicative of mucopolysaccharides. This same layer also shows metachromasia with toluidine blue, again a positive result for muco-substances. It can therefore safely be assumed that the layer between the micro-villi is mucoid, probably even mucus. The function of the microvillous layer is uncertain. It obviously increases the surface area for gaseous exchange, but possibly it is equally important for the protection of the tentacle during re-traction. The speed of retraction of the tentacles is very great, taking a fraction

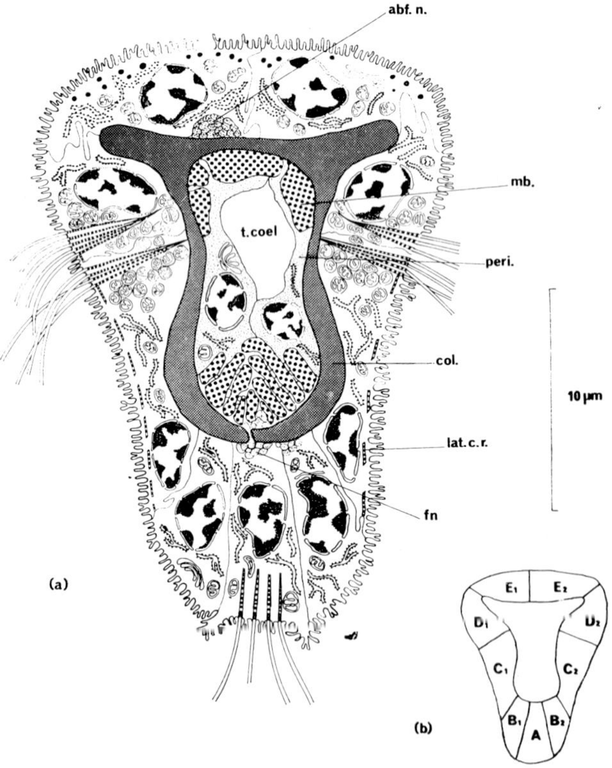

Fig. 1. (a) Generalized view of the tentacle of *Flustrellidra hispida* in cross section showing the major elements and microstructure. (b) Diagram of the cell arrangements in cross section in the tentacle with the numbers referred to in the text. *Abbreviations:* col, collagen; abf. n, abfrontal nerve; fn, frontal nerve; mb, muscle blocks; lat. c. r., lateral ciliary rootlet; peri, peritoneum; t. coel, tentacular coelom.

of a second to pass from the fully expanded state to complete withdrawal into the zooecium. Any small particles (diatoms, sponge spicules etc.) or grains of sand could have quite an abrasive action and tear the cell membrane unless the cell were cushioned by a layer of mucus. Similar functions for microvillous borders on bivalve gills has been put forward by Afzelius (1956, 1960). As well as the bivalve gills, microvilli are known from the outer layers of a great variety of animals such as Pogonophora (Gupta and Little, 1969), Rotifera (Koehler, 1965) and from the Platyhelminthes (Taylor, 1971), and in these animals it may be that the microvillous layer also forms a protective shield.

The lateral cilia are concentrated in a pair of cells on each side of the tentacle (Fig. 1b). The more abfrontal cell, D_1/D_2, is the smaller of the two, with a large nucleus occupying a large proportion of the cell volume. Ciliary rootlets pass through the cell to the basement membrane, converging near the nucleus, or sometimes sweeping around the base of it. The rest of the cell is filled with tightly packed mitochondria, which are slightly larger in the ciliated cells than in the non-ciliated cells, measuring about 1·0–1·5 μm in the former as compared to half that measurement in the latter. Between the mitochondria there are occasional Golgi complexes, but these occur in much higher numbers in the non-ciliated cells, especially the extreme abfrontals (E_1/E_2) or in the non-ciliated region of cells C_1/C_2. The more frontal pair of ciliated cells (C_1/C_2) are much larger than the pair so far considered, occupying almost the whole lateral side of the tentacle section. The ciliated region is restricted to the extreme abfrontal end of the cell, so that its cilia and those of the abfrontal cell (D_1/D_2) appear to form a continuous tract only separated by the cell membrane. The rootlets again pass right through the cell and touch the basement membrane. As in cells D_1/D_2 the mitochondria are tightly packed in the immediate region of the rootlets, and have dimensions as above. In addition to the normal rootlets passing straight down into the cell, there are also lateral rootlets which branch from the main rootlet at the level of the kinetosome (basal body) and travel along the whole length of the cell and join its frontal membrane. In the electronmicrograph profile these rootlets are similar to the normal rootlets, having a banded appearance with an identical periodicity (two major bands separated by 0·13 μm with two thinner bands between them). The lateral rootlets are not oriented parallel to the long axis of the cell but at a slight angle to it (Plate I). The function of these rootlets is not certain. As they appear to be orientated in a direction parallel to the direction of the ciliary beat, they may act as struts anchoring the cilium in the cell during the (violent) power stroke. But against this is the fact that other animals have cilia, and there is no reason to think that the beat in a bryozoan is more powerful than in other animals which do not have these lateral rootlets.

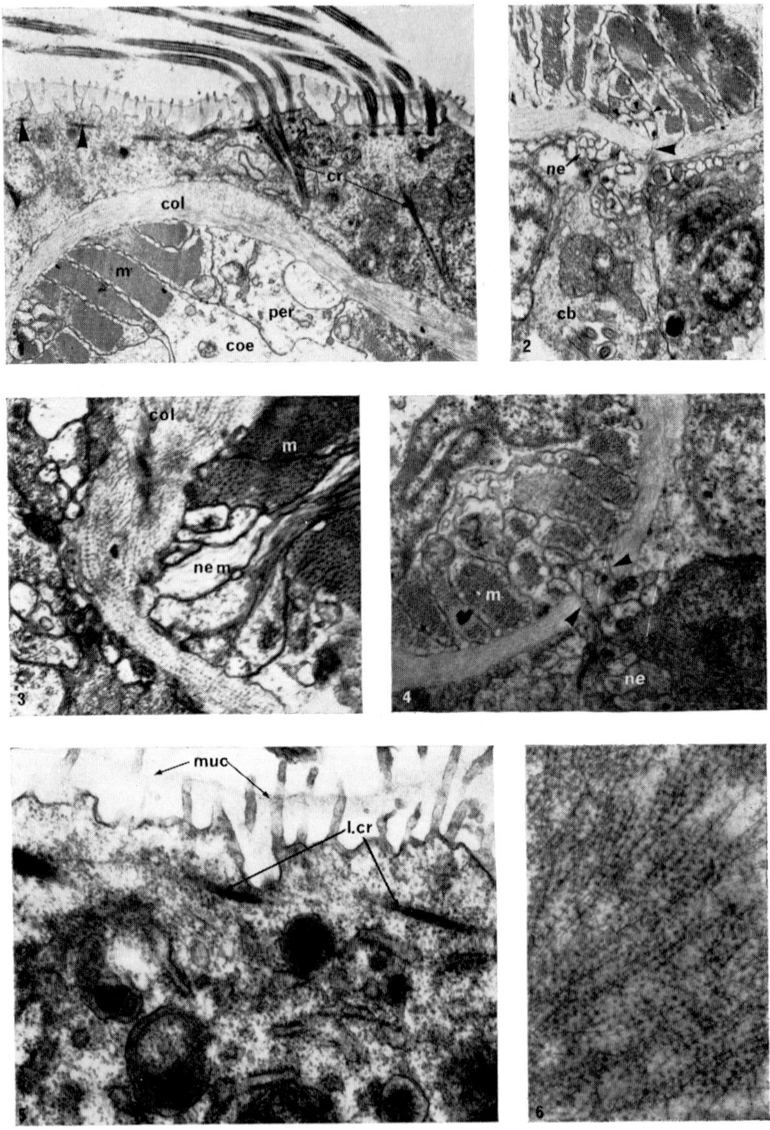
cr
col
m
per
coe
ne
cb
2
col
m
nem
3
m
ne
4
muc
l.cr
6

Branched rootlets have been recorded before, in other animals, for example in *Tetrahymena* (Sleigh, 1962) and *Amphisolops* (Taylor, 1971), but in these, both branches of the bifurcation pass more or less normally down into the substance of the cell, and do not run parallel with the outer cell wall at a mere 0·2 μm below it. It may be significant that the cilia of the abfrontal cells, D_1/D_2, also have lateral rootlets but instead of passing for 10 to 15 μm along the cell, these only extend for about 0·5 μm before reaching the frontal membrane of the cell. Thus lateral rootlets of cilia in both cells run in the same direction, and are joined to the frontal membrane of their respective cells, suggesting that they are subject to common forces acting in this direction. Possibly the lateral rootlets are important not in anchorage during the power stroke of the cilium, but for anchorage during the retraction of the tentacle. When the tentacles are out in the feeding position they form a broad inverted bell, but at full retraction they are in close apposition like a bundle of sticks. As well as being forced together very rapidly, they are also retracted into the zooecium rapidly, and both these forces may act to shear the cilium out of the cell. In order to counteract this tendency the lateral rootlets may have been evolved.

The two abfrontal cells, E_1/E_2, have no ciliation, but have quite large nuclei, in the order of 4·0 μm diameter. There are a number of mitochondria but these are neither so densely packed as in the lateral ciliated cells nor as large, measuring

---

Plate I

Ultrastructure of the tentacles of *Flustrellidra hispida*

Fig. 1. General view of the lateral cells of the tentacle showing cells C and D. Note the convergence of the cilia rootlets towards the cell nuclei, and the lateral ciliary rootlets passing along the outer surface of the cell. This picture also shows the muscle blocks, collagen and the peritoneum and coelom. × 5000.

Fig. 2. Picture of the collagen tube showing the frontal nerve passing through the collagen tube (large arrow) and the nerves innervating the frontal muscle on the other side of the tube (small arrows). × 5000.

Fig. 3. Enlarged view of the nerves innervating the frontal muscle. × 10,000.

Fig. 4. General view of the frontal muscle of the tentacle showing the nerves passing through the collagen tube in two places (arrows). × 7500.

Fig. 5. Higher power view of the lateral cells of the tentacle showing the lateral ciliary rootlets with their typical cross striations, and the microvilli holding the layer of mucus between them. × 20,000.

Fig. 6. High power view of the collagen tube showing the typical collagen "railway-line like" cross striations. × 22,500. *Abbreviations:* coe, coelom; cb, ciliary bases; col, collagen; cr, ciliary rootlets; l. cr, lateral ciliary rootlets; m, muscle; muc, mucus; ne, nerve; nem, nerve innervating muscle; per, peritoneum.

only about 0·4 μm in diameter. There is rough endoplasmic reticulum in these cells, as well as Golgi complexes, and quite often there are large granules (0·1–0·5 μm diam). These may be mucous secretions passing to the outside. The cell membranes both between the cells E_1 and E_2 and between the E cells and the D cells are highly convoluted and bear prominent desmosomes. Often these convolutions pass almost the whole depth of the cell before curving back on themselves several times.

In section the two abfrontal cells are juxtaposed, but not throughout their entire height, for protruding between them towards their bases is a bundle of nerves. Occasionally this bundle of nerve fibrils is found actually under one of the cells, between it and the axial skeletal rod (see later). The number of fibrils in the bundle varies from about ten to twenty depending on the level at which the section is cut. The individual fibrils have a diameter of about 0·5 μm, and the whole bundle of about 2·0 μm. The bundle runs longitudinally along the whole length of the tentacle and may innervate the abfrontal muscle block, but no nerve branches have been seen leading from it.

The frontal surface of the tentacle is made up of a ridge of three cells. In light microscopy these three cells stain deeply with haematoxylin and eosin, while in the electron microscope profiles the cytoplasm is much denser than in the other tentacular cells. These three frontal cells have prominent central nuclei of about 4·5 μm diam. Usually only the median cell is ciliated, although cilia and rootlets are occasionally seen on the two flanking cells. The basal bodies of the frontal cilia stain deeply and are long (0·9 μm) and eventually give rise to the rootlets. These rootlets are not as conspicuous as those of the lateral cilia and do not pass right through the cell, but seem only to penetrate for about 5·0 μm. Also there are none of the lateral rootlets which are such a pronounced feature of the lateral cells. The three cells contain abundant mitochondria as well as a rough endoplasmic recticulum and Golgi complexes and droplets. The latter may be of a secretory nature. At the bases of these cells, and between them and the axial skeletal tube are two bundles of nerve fibrils with dimensions similar to those of the abfrontal nerve fibril bundle.

Immediately internal to the outer cells is a large, thick ring-like structure which has been called the basement membrane (Brien, 1960) or mesogloea (Hyman, 1959). In light microscopic studies this structure was examined using Verhoeff's stain (Verhoeff, 1908), and picro-fuchsin/haematoxylin (Carleton and Leach, 1947) and with Picro Mallory V (Lendrum *et al.*, 1962). The first two stain collagen red, while the last stains it blue. In all these histochemical tests the ring structure of the tentacle stained as for collagen. Under the electron microscope this structure has the appearance of collagen with the familiar parallel fibrils with their "railway line" appearance. The periodicity of the transverse striations on the fibrils in *F. hispida* is 66 μm which is well inside the collagen range. Taken together, the electron microscope appearance and the optical histochemistry demonstrate that this so-called mesogloea is a tube of collagen which probably acts something like a skeleton, and for this reason I have called it the axial skeletal tube. This skeletal tube has a thickness of 0·5–1·5 μm. It forms practically a perfect oval save for a pair of horns at the abfrontal face of the tentacle (Fig. 1a) which protrude about 4 μm between the cells in this region. The orientation of the fibrils of the collagen is usually concentric with the plane of the ring, this being especially true at the lateral and frontal areas of the ring. In the two horns however, the fibrils have a more random appearance with little apparent order. In the horns the fibrils may actually pass longitudinally down the tentacle.

Within the axial skeletal tube are the tentacular muscles, the peritoneal cells and the tentacular extension of the ring coelom. The muscles are in two blocks, the abfrontal being smaller than the frontal. The abfrontal block is composed of about five fascicles arranged roughly in the shape of a horse-shoe. Between the individual muscle bundles are occasional mitochondria and possibly nerves. The larger, frontal, muscle block is made up of ten smaller bundles, between which may be seen a few mitochondria and possibly nerves. Nerve fibrils can most certainly be seen in some sections actually between the muscle block and the axial skeletal tube. A nerve has been seen in one electron-micrograph profile actually passing through the ring of collagen into the area of the muscle block (Plate 1, Figs 2, 4). This has however only been seen once. Both sets of muscles are striated and it is noticeable that there are fewer mitochondria associated with the muscles than there are with the cilia, suggesting that the muscular activity of the tentacles is relatively weak.

Inside the muscles and elsewhere bounding the collagen ring are the peritoneal cells. These are relatively simple cells with large nuclei (diameter about 2·5 μm), a few mitochondria, about 1·0 μm in diameter, and a rough endoplasmic reticulum. Occasionally a Golgi complex can be seen and this may have some function secreting substances into the tentacular coelom to maintain coelomic pressure. Stretching across the coelom between the peritoneal cells are quite often protoplasmic strands of unknown function. It seems probable that the tentacular coelom and the collagenous axial skeletal tube act together to form a firm, but at the same time flexible, support for the tentacles during feeding. They also act together as antagonists to the muscles, giving the latter something to act against during muscular contraction, in a similar principle to the hydrostatic skeleton in the earthworm.

REFERENCES

Afzelius, B. A. (1956). *Expl. Cell Res.* **10,** 257.

Afzelius, B. A. (1960). The function and the formation of the brush border in some types of invertebrate cells. *Prof. Europ. reg. Conf. Electr. Microsc. Delft* **2,** 742–745.

Brien, P. (1960). *In* "Traité de Zoologie" (P-P. Grasse, ed.), Vol. 5 (2), pp. 1054–1335.

Calvet, L. (1900). Contributions a l'histoire naturelle des Bryozaires Ectoproctes marins.

Carleton, H. M. and Leach, E. H. (1947). "Histological Technique." Oxford University Press, London.

Gupta, B. L. and Little, C. (1969). Studies on Pogonophora. II. Ultrastructure of the tentacular crown of *Siphonobrachia. J. mar. biol. Ass. U.K.* **49** (3), 717–741.

Hyman, L. H. (1959). "The Invertebrates," Vol. V, "The Smaller Coelomate Groups". McGraw-Hill, New York.

Koehler, J. K. (1965). A fine structure study of the Rotifer integument. *J. Ultrastruct. Res.* **12,** 113–134.

Lendrum, A. C., Fraser, D. S., Slidders, W. and Henderson, R. (1962). *J. Clin. Path.* **15,** 401.

Luft, J. H. (1961). Improvements in Epoxy resin embedding methods. *J. Biophys. Biochem. Cytol.* **9,** 409.

Reynolds, E. S. (1963). The use of lead citrate at high pH as an electron opaque stain in electron microscopy. *J. Cell. Biol.* **17,** 208.

Sabatani, D. D., Bensch, K. and Barrnett, R. J. (1963). Cytochemistry and electron microscopy: the preservation of cellular ultrastructure and enzymatic activity by aldehyde fixation. *J. Cell. Biol.* **17,** 19.

Sleigh, M. (1962). The biology of cilia and flagella. "International Series of Monographs on Pure and Applied Biology." Vol. 12. Pergamon, Oxford.

Spicer, S. S. (1966). Diamine methods for differentiating mucosubstances histochemically. *J. Histochem. Cytochem.* **13,** 211–234.

Steedman, H. F. (1950). Alcian Blue 8GS; a new stain for mucin. *Q. J. Microsc. Sci.* **91,** 477–479.

Taylor, D. L. (1971). On the symbiosis between *Amphidinium klebsii* (Dinophyceae) and *Amphiscolops langerhansi* (Turbellaria: Acoela). *J. mar. biol. Ass. U.K.* **51** (2), 301–314.

Verhoeff, F. H. (1908). *J. Am. Med. Ass.* **50,** 876–877.

33 | Histological and Histochemical Studies on the Bryozoan-substrate Interface

JOHN D. SOULE

School of Dentistry and Allan Hancock Foundation, University of Southern California, Los Angeles, California

Abstract: Preliminary studies were made on the basal-wall substrate of *Cryptosula pallasiana, Membranipora villosa, M. tuberculata* and *Hippothoa hyalina* utilizing histological staining methods for morphological examination and histochemical procedures for polysaccharides and proteins. A simple acid mucopolysaccharide was revealed as the principal component of the biological adhesive attaching colonies to the substrate.

INTRODUCTION

Studies of biological adhesives are of interest in certain areas of the health professions (i.e. dentistry) because an adhesive is needed which will bind effectively under water, which is of a permanent nature and which is non-toxic to humans. An increased activity in the study of marine bioadhesives has become apparent with the publication of articles by Cook, 1968, 1969, 1970; Lacombe, 1967, 1970; Lacombe and Liguori, 1969, Bocquet-Vedrine, 1970; Walker, 1970, 1971; Liguori, 1970 and Hillman and Nace, 1970. These articles deal with the histology and histochemistry of the cement glands or cement deposits of a variety of barnacles and mussels. The product, commonly referred to as "barnacle cement", is produced by the cement glands of barnacles and is primarily a protein with a small lipid constituent. The bioadhesive of the mussel is also mainly a protein, with very little lipid and no detectable carbohydrate. For a comparison of "barnacle cement" and the mussel bioadhesive, see Cook, 1970, pp. 149–150.

Many species of marine Bryozoa form extensive colonies which become firmly attached to a variety of substrates, following a brief motile larval period. Very little is known about the nature of the bryozoan cementing substance other than the fact that the colonies of many species adhere tenaciously to the substrates and rarely can be removed without damage to the body walls of the colonies.

METHODS AND MATERIALS

A number of bryozoan colonies of *Cryptosula pallasiana*, adherent to shells of the mussel *Mytilis edulis*, and colonies of *Membranipora villosa*, *M. tuberculata* and *Hippothoa hyalina*, adherent to blades of the kelp *Macrocystis* sp., were utilized in this study. Separate specimens of freshly collected material were fixed, in the field, in either 10% formal-sea water, Carnoy's fluid, Rossman's fixative or Bouin's solution. Following fixation for 48 h, specimens were examined under a dissecting microscope and extraneous material carefully removed with a camel hair brush. The material was then decalcified in a 5% trichloracetic acid, then dehydrated, cleared in toluene, infiltrated in paraffin under vacuum and embedded by the usual routine for sectioning at 6 μ.

Sections were mounted in series on numbered slides to permit comparison of closely related sections under each of the various histological–histochemical procedures that were utilized. For purposes of orientation and general histomorphological examination hematoxylin and eosin stain was used. The carbohydrates were ascertained by utilizing the periodic acid-Schiff reaction (PAS), PAS with diastase controls, Alcian blue, colloidal iron, and toluidine blue. To distinguish simple acid mucopolysaccharides and sulfated mucopolysaccharides, methylation was used followed by Alcian blue and toluidine blue, as well as methylation-saponification, followed by Alcian blue and toluidine blue. To detect the presence of proteins the following techniques were used: Biebrich scarlet, bromphenol blue, diazosulfanic acid-Azure A, and the Millon reaction.

RESULTS

Examination of the basal layer of the colonies was particularly rewarding in the areas of junction of the basal wall with the transverse wall. Sections of the basal layer stained with hematoxylin and eosin revealed that the innermost stratum is composed of simple squamous epithelial lining cells. The lining epithelium first formed a very thin exterior cuticular bioadhesive which abutted the substratum, and then formed the organic intercellular ground substance (matrix)

Plate I

Fig. 1. Basal wall-transverse wall union, showing intercellular matrix (m), cuticular layer (c), and epithelial lining cells (e). *Membranipora villosa* on *Macrocystis* sp. The epidermal cells of the kelp form the dark surface beneath the bryozoan colony. PAS reaction. × 950. Transverse walls are formed initially by folding in the basal wall.

Fig. 2. Basal wall-transverse wall union. Note cuticle (c). *Membranipora villosa* on *Macrocystis* sp. Alcian blue stain. × 950.

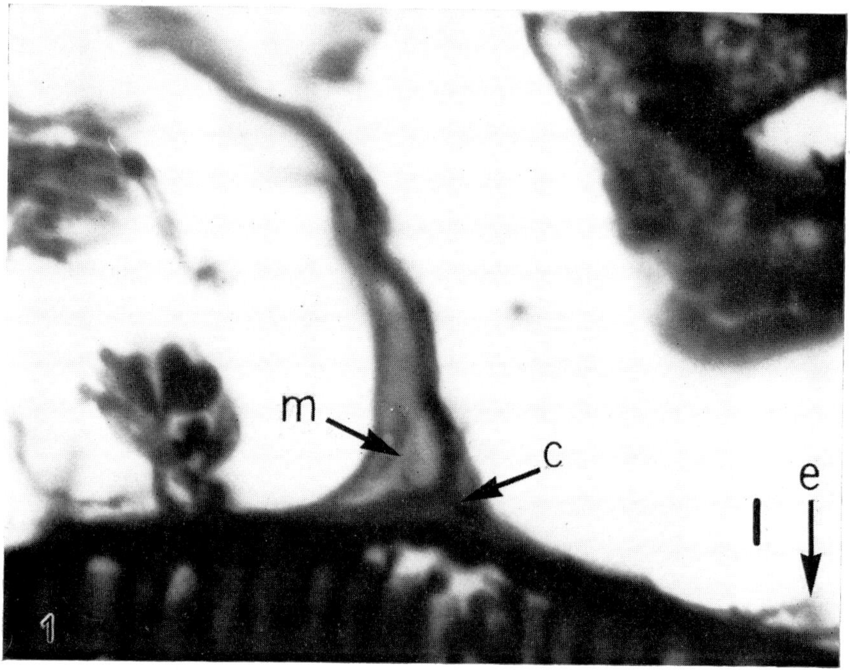
m
c
e
l
1

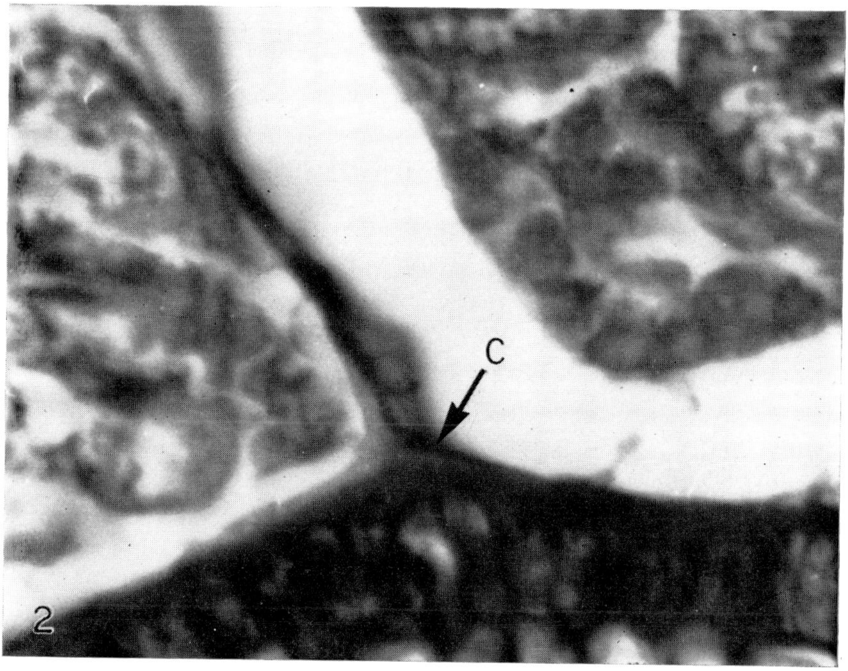
c
2

of the basal wall. The mineral salts which were deposited in the matrix were removed by decalcification during the preparation of the sections. The thin layer of the cuticular adhesive at the bryozoan-substrate interface stains weakly basophilic in the hematoxylin and eosin sections from material fixed in Carnoy's fluid and pale pink in material fixed in formal-sea water.

In sections stained with the periodic acid-Schiff (PAS) reaction for carbohydrates, the intercellular matrix and the basal cuticular layer were negative in material fixed in formal-sea water, but showed a very weak positive reaction in material fixed in Carnoy's fluid. Occasional cells of the simple squamous epithelial lining possess minute PAS positive granules in their cytoplasm, indicating the site of carbohydrate formation.

Material treated with Alcian blue showed a strong positive carbohydrate reaction in the basal cuticle and a weak positive carbohydrate reaction in the intercellular matrix. A similar situation prevailed in sections stained with colloidal iron, in which a strong positive carbohydrate reaction occurred in the cuticular layer and a weak positive reaction was found in the intercellular matrix. Using toluidine blue, a beta (violet) metachromasia was present in the cuticle and in the intercellular matrix.

The reactions for protein ranged from negative for the Millon reaction to the weakest of positive reactions for Biebrich scarlet, bromphenol blue and diazosulfanilic acid-Azure A reaction. Control slides were positive, as were certain bryozoan structures such as the muscle and operculum.

The negative, or at best very weak, PAS reactions appear to eliminate the polysaccharides (glycogen), neutral mucopolysaccharides, mucoproteins, glycoproteins and glycolipids from possible consideration as major constituents of the bioadhesive cuticle. The positive Alcian blue and colloidal iron reactions, and the beta metachromasia of toluidine blue are indicative of the presence of acid mucopolysaccharides, either as a simple acid mucopolysaccharide or a sulfated mucopolysaccharide. The question was resolved by methylating sections for 48 h, then staining some of them with toluidine blue or Alcian blue, while other sections were saponified following methylation and then treated with toluidine blue and Alcian blue. In the methylated sections, the Alcian blue reaction and the beta metachromasia were quenched, while sections that were methylated, saponified and then subjected to toluidine blue and Alcian blue had the metachromasia and the staining reaction restored. This is indicative of the presence of a simple acid mucopolysaccharide.

Histochemists are careful to point out that histochemical methods for carbohydrates are not specific, except to determine the presence of major groups. In addition the picture is complicated in that carbohydrates are most likely to occur in tissues as mixtures. It seems possible that the bioadhesive cuticular structure in the bryozoans is made up of a polysaccharide-protein complex, with a predominance of the carbohydrate, a simple acid mucopolysaccharide. The cuticle is the initial secretory product of the basal lining epithelial cells, being deposited immediately prior to the establishment of the intercellular matrix of the basal body wall. The basal cuticle provides an adhesive interface between the substratum and the mineralized basal body wall. The cuticle might become at least partially mineralized during the deposition of mineral salts in the basal body wall. This would further enhance the fusion of the colony to the substratum. Thus the bryozoan bioadhesive which can be categorized as a "mucin" differs greatly from the "barnacle cement" and the mussel "glue" both of which are predominantly protein.

ACKNOWLEDGEMENT

This investigation was supported by funds provided by General Research Support Grant number 5-S01 RR 05303-10 from the General Research Support Branch, Division of Research Facilities and Resources, National Institutes of Health.

REFERENCES

Bocquet-Vedrine, J. (1970). *Archs Zool. exp. Gen.* **111,** 521.
Cook, M. (1968). *J. Dent. Res.* **47** (supplement), 967, abst.
Cook, M. (1969). *J. Dent. Res.* **48** (supplement), 1124, abst.
Cook, M. (1970). *In* "Adhesion in Biological Systems" (R. S. Manly, ed.), p. 139. Academic Press, New York and London.
Hillman, R. E. and Nace, P. F. (1970). *In* "Adhesion in Biological Systems" (R. S. Manly, ed.), p. 113. Academic Press, New York and London.
Lacombe, D. (1967). *Inst. Pesq. Mar. publ.* 011,1.
Lacombe, D. (1970). *Biol. Bull.* **139,** 164.
Lacombe, D. and Liguori, V. R. (1969). *Biol. Bull.* **137,** 170.
Liguori, V. R. (1970). *In* "Adhesion in Biological Systems" (R. S. Manly, ed.), p. 123. Academic Press, New York and London.
Walker, G. (1970). *Mar. Biol.* **7,** 239.
Walker, G. (1971). *Mar. Biol.* **9,** 205.

34 | Some Aspects of Skeletal Organization in Bryozoa

R. TAVENER-SMITH

Department of Geology, The Queen's University, Belfast, Northern Ireland

Abstract: Ultrastructure in the bryozoan skeleton shows resemblances to that in the Brachiopoda. Within the phylum the wall structure of the Gymnolaemata differs in organization and complexity from that of the Stenolaemata.

A calcareous skeleton represents mineralization on a proteinous organic base: this approach can assist in elucidating the developmental sequence in complex or unorthodox colonies. *Cupuladria* and *Semicoscinium* are considered as examples, and a modification is proposed to the current growth model for fenestrate cryptostomes.

Suggestions are made concerning the structure and significance of acanthopores and mesopores.

A survey of the carbonate skeleton in more than eighty representative genera was recently carried out to ascertain the range of structural and ultrastructural variation (Tavener-Smith and Williams, 1972). As a result, it became evident that the skeletal architecture of Bryozoa is not unique, for in the size, shape and arrangement of crystallites and in other matters there is a correspondence to patterns seen in brachiopod shells. In both phyla the primary (first formed) layer of the skeleton is composed of granular or finely fibrous components with little, if any, proteinous intercrystallite matrix. In the secondary layer a degree of lamination is ubiquitous, and well-shaped tabular, fibrous or lath-like crystallites are intimately associated with an anastomosing web of proteinous organic films. Intraskeletal structures common to Bryozoa and brachiopods include such important features as puncta (pseudopores) and pseudopuncta (acanthopores).

Among carbonate-secreting Bryozoa there are differences between the ultrastructure of the gymnolaematous skeleton and that of the Stenolaemata. The former generally consists of a sequence of one to three distinctive layers with variable components. Finely acicular crystallites lying normal to the depositional surface commonly predominate in the primary skeleton. The secondary layer is usually banded due to the presence of proteinous films separating thin laminae, each composed of closely-packed acicular crystallites. Rarely, tabular crystallites may constitute this part of the skeleton (e.g. *Celleporella*, *Steginoporella*). The tertiary layer may be banded, granular or composed of acicular crystallites and, like the primary one, contains sparse organic constituents.

The fabric of the stenolaemate skeleton (including all Palaeozoic Bryozoa) is consistently double-layered (Plate I, Fig. 1) and of more uniform composition. The primary layer may be either granular or of acicular fibres, but the secondary one is invariably laminar, consisting of spirally-growing tabular plates or lath-like fibres. Study of stenolaematous genera, including *Crisia*, *Hornera*, *Heteropora* and representatives of the Trepostomata, Cryptostomata and Cystoporata suggests that skeletal organization in all these groups has a common basic pattern.

Turning to another aspect of the bryozoan skeleton: the first operation in its formation is the exudation, by cells of the ectodermal epithelium, of a protein-mucopolysaccharide complex which polymerizes to form a peripheral organic sheet, here called the periostracum. This earliest skeletal component is flexible and only later, with the seeding of carbonate crystallites onto the inner side of the periostracum and their subsequent proliferation, does the structure become rigid. It is clear that the morphology of the calcareous skeleton must reflect the initial configuration of the periostracal-epithelial layers. Since these may assume complex shapes it is not surprising that the construction of the mineralized part (all that remains in the fossil) may present some curious and puzzling arrangements.

One may consider as an example *Cupuladria biporosa*, a Recent, free-living cheilostome with a disc-like skeleton of complex structure. On the obverse of

---

Plate I

Fig. 1. *Entalophora*. Part of a transverse section showing inter-zooecial walls with a medial, granular primary and laminar (lath-like fibres) secondary skeletal layers. Scanning electron micrograph. × 1300.

Fig. 2. *Cupuladria*. Outer part of the periostracum (about half the total thickness is seen) showing the triple-unit membrane. Transmission electron micrograph. × 55,000.

Fig. 3. *Semicoscinium*. Transverse section of a branch, with part of the external vesicular investment. S.E.M. × 250.

Fig. 4. *Semicoscinium*. Junction of an external vesicle wall with the branch surface, showing the continuous nature of laminae between them. S.E.M. × 2450

Fig. 5. *Leioclema*. The inner part of an acanthopore showing secondary laminae in contact with the granular axial region. The relationship between the two is gradational. S.E.M. × 6500.

Fig. 6. *Arthrophragma*. A mesopore between two zooecia, showing an initially vesicular structure passing distally into a solid frontal wall. S.E.M. × 120.

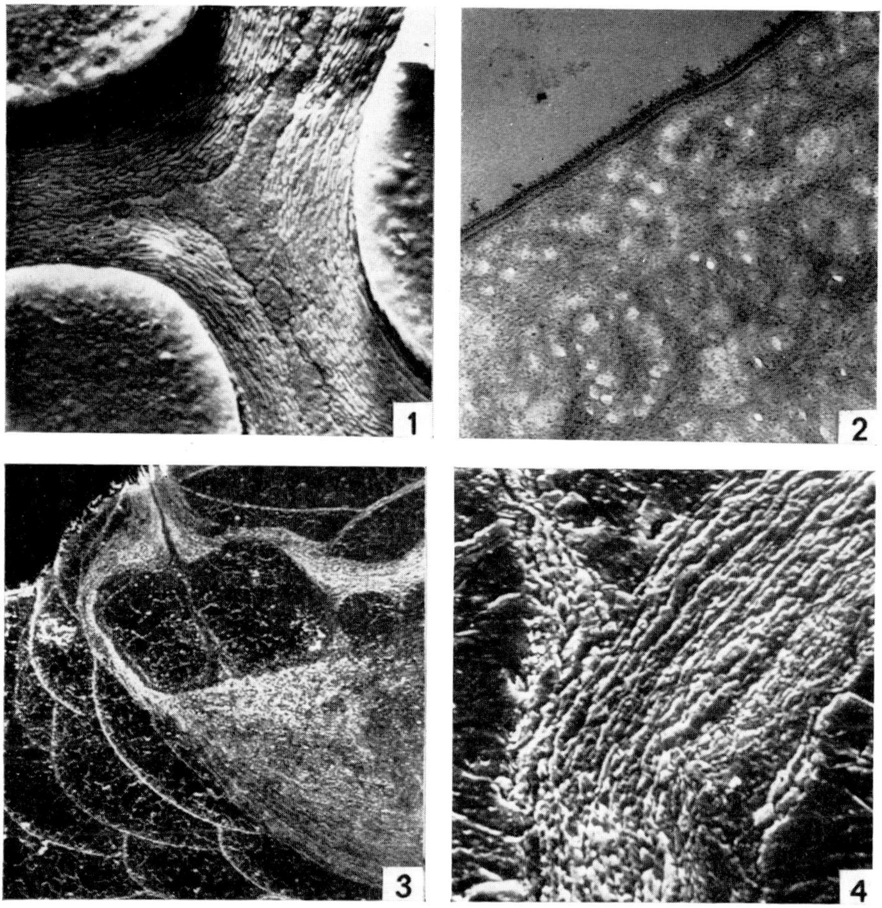

the disc zooecia, alternating with vibracula, are arranged in roughly radiating rows. On the reverse a thick skeletal crust (basal pad) underlies the zooecia and is subdivided by radial and concentric lines into many irregularly quadrate sections. Closer examination shows that the lines are in fact, proteinous sheets embedded in the carbonate skeleton, and the zooecial layer is separated from the basal pad by another such organic sheet (Fig. 1b, c). The quadrate areas generally show on their surfaces three to five variably developed basin-like hollows. Sections show that each of these is the last of a vertical series of spheroidal chambers (kenozooids) within the basal pad (Fig. 1a, b).

In interpreting the construction of *Cupuladria* colonies one recalls that skeletal mineralization is always initiated on a seeding surface of non-cellular proteinous material. Hence an understanding of the origin of organic sheets within the carbonate skeleton is the key to understanding the development of the calcareous shell itself.

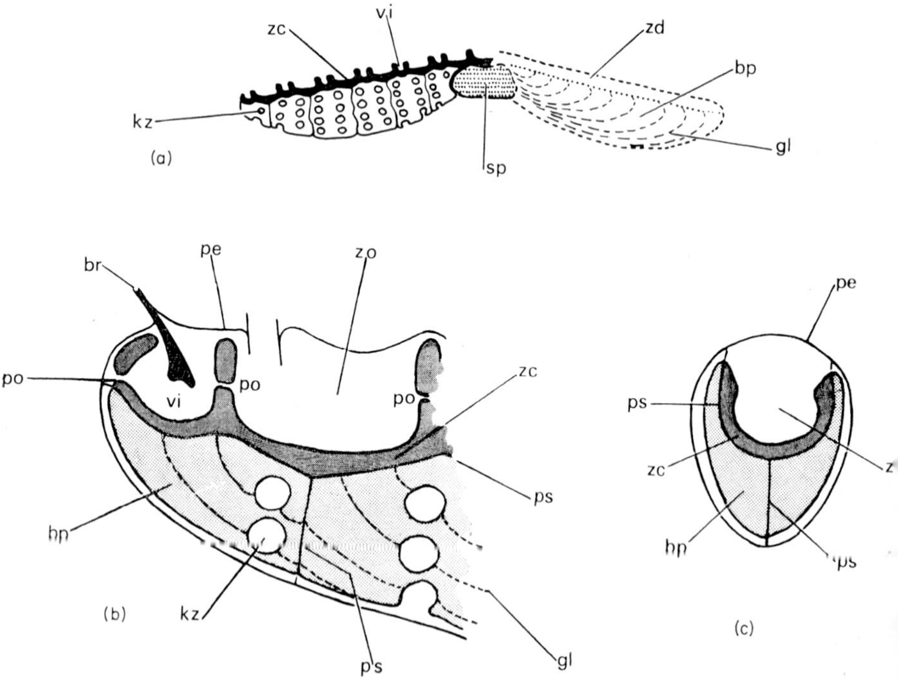

Fig. 1. *Cupuladria*. a: section of a discoid colony; b: longitudinal section of the distal end of a zooidal row; c: transverse section near the distal end of a zooidal row. *Abbreviations:* bp, basal pad; br, bristle; gl, growth line; kz, kenozooid; pe, periostracum; po, pore; ps, proteinous sheet; sp, sedimentary particle; vi, vibraculum; zc, zooecium; zd, zooidal disc.

Cupuladria is a membraniporid, so the first individual of a zooidal row may be considered as a box with an outer investment of periostracum-epithelium. It seems that at an early stage this cover must extrude a flap or fold which extends round the underside of the box (Fig. 2a, b). As new boxes are added the flap is extended lengthways into a flattened sac beneath the developing zooid row (Fig. 2c, d). Three things are now evident. First, the zooids are separated from the basal sac by a thick organic sheet which originated as a doubled periostracal layer. Second, the basal wall of the colony is a coelocyst. Third, formation of the basal pad may now commence by the seeding of crystallites secreted from the hypostegal epithelium onto the lower side of the organic sheet immediately above it (Fig. 3f). Sustained deposition may cause this wall to attain considerable thickness.

In *Cupuladria*, as in *Membranipora*, growth takes place at the distal ends of

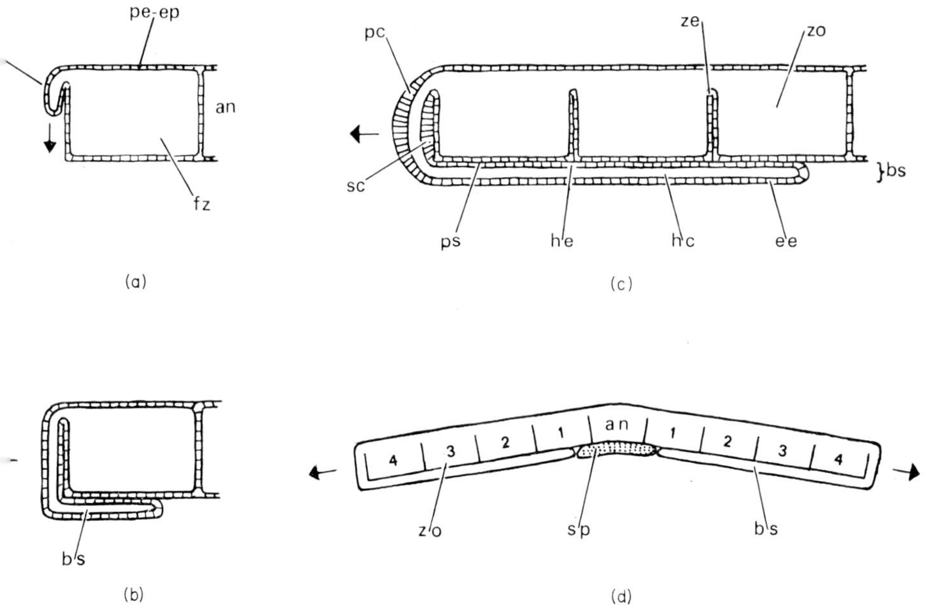

Fig. 2. *Cupuladria.* Stylized and simplified reconstruction of some early developmental stages. a, b: formation of the external flap; c: advancing zooidal row, indicating the positions of growth centres; d: section of a young colony, with zooidal rows radiating from an ancestrula. *Abbreviations:* an, ancestrula; bs, basal sac; ee, eustegal epithelium; fl, flap; fz, first zooid; hc, hypostegal coelom; he, hypostegal epithelium; pc, palisade-cell cap; pe-ep, periostracum and epithelium; ps, proteinous sheet; sc, subsidiary growth centre; sp, sedimentary particle; ze, zooidal epithelium; zo, zooid.

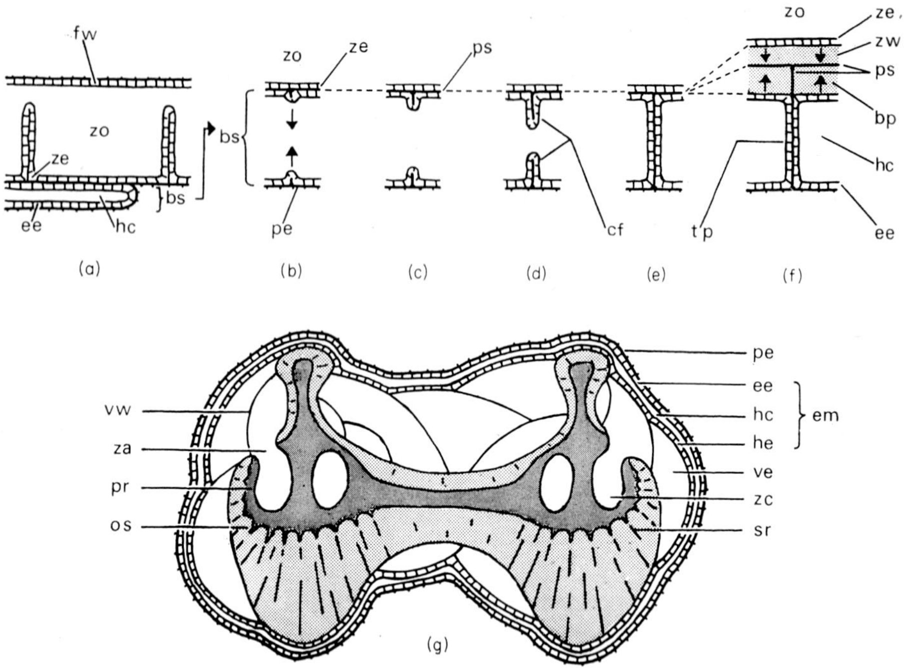

Fig. 3a–f. *Cupuladria.* Stages in the formation of a transverse partition in the basal sac, and the initiation of the basal pad; g: *Semicoscinium.* Reconstructed transverse section through branches, showing position of the external mantle. *Abbreviations* as in previous figures with the following additions: cf, concentric epithelial fold; em, external mantle; fw, frontal wall; os, outer secondary skeleton; pr, primary skeleton; sr, skeletal rod; tp, completed transverse partition; ve, vesicle; vw, vesicle wall; za, zooecial aperture; zc, zooecial chamber; zw, zooidal wall.

zooidal rows (Fig. 2c). Since the skeleton of *Cupuladria* is much the more complex of the two it is not surprising that the organization of secretion from palisade cell caps in that position is also more complex. Exudation from each of these initiates not one, but three proteinous sheets. Only one of these is peripheral (the periostracum), the others being internal. The periostracum is also unique in bearing an outer triple-unit membrane (Plate I, Fig. 2): the other sheets correspond structurally to the inner periostracal layers only. An essentially horizontal sheet spun back from the inner side of the periostracum separates zooidal from colonial skeleton, or basal pad (Fig. 1), and a medial vertical sheet originating beneath each palisade cell cap forms a common lateral wall for adjacent basal sacs (Fig. 1c). This arrangement means that basal sacs do not lie directly beneath zooidal rows, but alternate with them. Calcification on these proteinous sheets results in a massive skeleton of great strength, within which the sheets themselves are immured.

Other organic sheets, forming a roughly concentric pattern on the reverse surface, are initiated within transverse partitions dividing each basal sac into compartments. Partitions of this kind do not originate from the palisade cell cap, but as annular invaginations of the sac epithelium. Each of these closes inward, like an iris diaphragm, until a complete septum containing a proteinous medial sheet is formed (Fig. 3a–e). Once the basal pad is initiated the earlier formed parts of these sheets are progressively enclosed by the thickening aragonitic wall (Fig. 3f).

In these ways a complex of proteinous sheets is initiated within the developing colony, and subsequent calcification leads to the construction of the sophisticated *Cupuladria* skeleton.

Another genus posing a problem in the interpretation of skeletal structure is the Devonian cryptostome *Semicoscinium rhombicum*. Here it is impossible to derive help from an examination of the soft parts, the disposition of which can only be inferred. This genus is said to be distinguished from the much commoner *Fenestella* mainly by the presence of a broad, flattened summit to the median keel, but in the species studied there is another distinctive feature. This is a vesicular superstructure covering branch surfaces and fenestrules on both sides (Plate I, Fig. 3). Careful examination shows that constituent laminae of the vesicle walls are continuous with those of the outer parts of branches (Plate I, Fig. 4). The relationship is identical to that between the laminae of diaphragms or cystiphragms and those of the walls of the trepostome zooecia in which they occur. The vesicles of *Semicoscinium* might, in fact, be considered as a kind of external cystiphragm, and there can be no doubt that they are an integral part of the bryozoan colony.

Considering the structure of the fenestellid skeleton, it is inconceivable that the vesicle walls were formed from within. Carbonate deposition must have taken place from the exterior by secretion from an outer mantle, incorporating epithelial layers and bounded by a periostracum, which invested the colony (Fig. 3g). It can only be concluded that for some unknown reason these tissues, during the latest stages of secretory activity, repeatedly blistered away from the branch surface to form the observed vesicular structure.

By indicating once more the presence of external soft tissues enveloping a fenestellid skeleton, this *Semicoscinium* species confirms the essential feature of the growth model proposed for these bryozoans at Milan (Tavener-Smith, 1968). Nevertheless, that model needs amendment in the light of subsequent studies.

The Milan reconstruction implied that the primary skeletal layer was secreted entirely from the exterior, that is, by the hypostegal epithelium alone, and that an organic sheet probably separated the primary from the inner secondary layer. These conclusions, based on the best evidence available at that time, were supported by the existence of corresponding structures in the living fenestrate Reteporidae and in other cheilostomes, for example *Cupuladria*. From these premises it was logical to suppose that the formation of interzooecial partitions (which contain a medial primary layer) in the Fenestellidae involved an invagination of both organic sheet and hypostegal coelom. More recent work has convinced me that on this point the Milan model is in error, for it is now certain that the primary wall was partly secreted from within, that is, by the zooidal epithelium. Invagination of *this layer alone* is therefore adequate to explain the observed structure of inter-zooecial walls, for the deposition of primary calcite within the fold so formed would be succeeded by that of laminated secondary material on either side. An amended growth model for fenestellid bryozoans is therefore now proposed (Fig. 4) in which inter-zooecial partitions are seen as septa of similar basic construction to those reported by Borg (1926, p. 322) in the Cyclostomata. This model brings the suggested

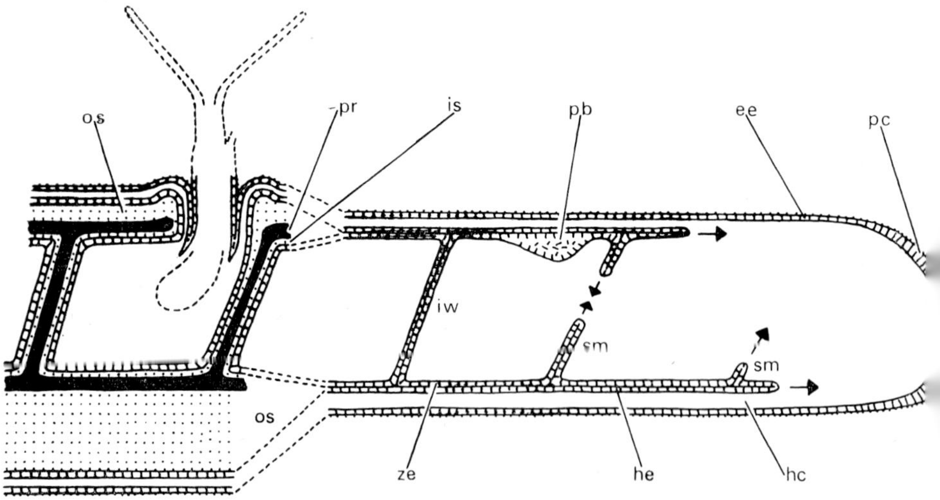

Fig. 4. Longitudinal section showing an amended growth model for *Fenestella*. Inner walls advance within the outer one, as in the living *Heteropora* and other double-walled forms. At the left is a mature part of the branch with completed calcification. *Abbreviations* as for previous figures with the following additions: is, inner secondary skeleton; iw, inter-zooidal wall; pb, polypide bud; sm, septum.

budding pattern of the Fenestellidae into harmony with that of the Stenolaemata in general.

Modern techniques for studying skeletal structure, particularly use of the scanning electron microscope, permit a clearer evaluation of the development and function of kenozooids. Among these acanthopores, best known in the Trepostomata, have attracted some attention in recent years (Tavener-Smith, 1969, p. 87; Brood, 1970, p. 192; Armstrong, 1971, p. 583). Acanthopores are generally characterized by a cone-in-cone structure formed of distally deflected secondary wall laminae; a clear axial "tube"; and a surface projection of some kind. It has commonly been believed that the axial region was originally hollow, and it has been suggested (for example, Hyman, 1959, p. 393) that the hollow accommodated some sort of accessory individual. Neither of these suppositions is true of acanthopores examined by the writer, in which secondary laminae of the cone-in-cone zones pass imperceptibly into granular calcite of the central "tube" (Plate I, Fig. 5). At high magnifications the "tube" is seen to lack a discrete wall, and in places the boundaries of laminae could be traced into, and even across the axial zone, showing the calcite there to be coeval with that of the laminae. Acanthopores with this structure are neither hollow, nor could they ever have housed accessory individuals of a colony. They are integral parts of the skeleton and identical in structure with the pseudopunctae of some brachiopod stocks.

Acanthopores, though characteristic of the Trepostomata, are found as well in all major cryptostome groups (including the Fenestillidae, in which they have been called *skeletal rods*; Tavener-Smith, 1968, p. 87), also in the Cystoporata (called *minutopores*; Astrova, 1964, p. 28) and in certain of the Cyclostomata (Horneridae and Lichenoporidae). All these diverse stocks had at least one feature in common, namely, the possession of coelocystal external walls, so it seems likely that the function of acanthopores was connected in some way with the presence of external soft tissues. In the Brachiopoda it has been suggested (Williams, 1956, p. 252) that the surface projection of granular calcite was a myotest providing an attachment area for muscles or tendons concerned with stabilizing the inner epithelium.

Mesopores are also best known in the Trepostomata, though occurring in many other groups. They are, in general, interstitial tubes with apertures smaller than those of associated autozooids. Little is known concerning their inception, but there is, to the writer's knowledge, no indication that these tubes were budded from autozooids. It seems possible that mesopores originated as residual spaces left between diverging autozooecia in colonies with coelocystal frontal walls (Fig. 5a, b). Such spaces would be lined by hypostegal epithelium in

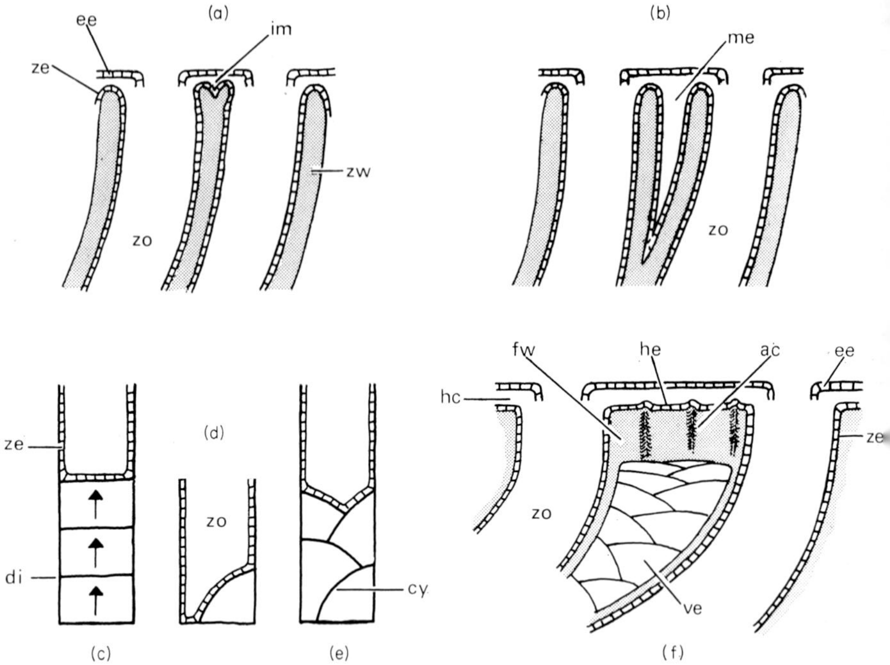

Fig. 5a, b. A suggested method of mesopore inception by zooecial wall divergence in a coelocystal colony; c: formation of diaphragms and cystiphragms in a mesopore; d: vesicular structure in a mesopore space passing distally into a solid frontal wall. *Abbreviations* as for previous figures with the following additions: ac, acanthopore; cy, cystiphragm; di, diaphragm; im, incipient mesopore; me, mesopore.

continuity with that of adjacent zooids, and able to secrete carbonate material for the common walls on either side. As the zooecial tubes extended distally, so would the space between them, thus becoming a mesopore.

As zooids migrated distally in the course of colonial growth, a similar tendency must have affected the soft tissues of mesopores. In many forms this migration occurred in stages, the secretory epithelium becoming detached from the floor of the mesopore, withdrawing a pre-determined distance, then secreting a planar organic sheet on which carbonate deposition recommenced to form a diaphragm (Fig. 5c). In some cases epithelial detachment from the floor was only partial, being followed by blistering into a distally-convex shape and mineralization in that position (Fig. 5d, e). A succession of such partial detachments would have led to the formation of a succession of cystiphragms, and a vesicular structure. Cystopores (in the Cystoporata) and alveoli (in the Lichenoporidae) are names given to spaces between successive cystiphragms. They appear to have originated in the same way.

Mesopores in the bifoliate cryptostome *Arthrophragma* show widely spaced cystiphragms and large vesicles close to the mesotheca, where zooecial tubes were diverging and growth was rapid. As development proceeded the rate of distal extension decreased. There was a corresponding decrease in the magnitude of successive withdrawals by the mesopore epithelium, and vesicles became progressively smaller. Finally, withdrawal became slow and continuous rather than episodic, and sustained carbonate deposition on the last-formed cystiphragm caused the formation of a thick calcitic frontal wall between zooecial apertures (Plate I, Fig. 6; and also Fig. 5f). This demonstration that mesopores may be eliminated during growth, provides an instructive new side-light on the nature of these structures.

ACKNOWLEDGEMENTS

I am indebted to Miss P. L. Cook (B.M.), Dr O. L. Karklins (U.S.N.M.), Dr D. E. Owen (Manchester Museum) and Dr S. Reguant (Barcelona University) for the specimens figured in Plate I.

REFERENCES

Armstrong, J. (1971). Zoarial microstructure of two Permian species of the bryozoan *Stenopora*. *Palaeontology* **13,** 581–587.

Astrova, G. G. (1964). O novem stryade Paleozoyskikh Mshanok (A new Order of palaeozoic bryozoa). *Paleont. Zh.* **2,** 22–31.

Brood, K. (1970). The structure of *Pseudohornera bifida* (Eichwald). *Geol. For. Stockholm Forh.* **92,** 188–199.

Hyman, L. H. (1959). "The Invertebrates," Vol. 5, "Smaller Coelomate Groups". McGraw-Hill, New York.

Tavener-Smith, R. (1968). Skeletal structure and growth in the Fenestellidae (Bryozoa). *Atti. Soc. It. Sci. Nat.* **108,** 85–92.

Tavener-Smith, R. (1969). Wall structure and acanthopores in the bryozoan *Leioclema asperum*. *Lethaia* **2,** 89–97.

Tavener-Smith, R. and Williams, A. (1972). The secretion and structure of the skeleton in living and fossil bryozoa. *Proc. R. Soc. Lond.* B. **177,** 1–65.

Williams, A. (1956). The calcareous shell of the brachiopoda and its importance to their classification. *Biol. Rev.* **31,** 243–287.

35 | Frontal Wall Calcification in a Recent Cribrimorph Genus from Australia

ROBIN E. WASS

Department of Geology and Geophysics,
University of Sydney, Sydney, Australia

Abstract: The frontal wall in a Recent species of the cribrimorph genus *Corbulipora* arises by development of vertical costae from the lateral gymnocyst at a level below the frontal membrane. Further vertical growth with subsequent bifurcation and horizontal growth results in the formation of a frontal wall possessing secondary spines as vertical upgrowths of pelmatidia. Three oral spines on the distal and disto-lateral margins of the orifice grow distally or vertically first and then proximally, fuse and completely obscure the orifice. Further proximal growth of one of the oral spines enables it to coalesce with a secondary spine to form a structure analogous to a tertiary frontal wall. Oral and secondary spines are composed of a central hollow surrounded by a series of round, hollow depressions.

The cribrimorph genus *Corbulipora* was erected by MacGillivray (1895) for specimens which differ "from the other Cribrilinidae in having the zooecia much elongated and readily separable laterally". He described one species, the type species, *C. ornata* from Tertiary strata. Maplestone (1901) described another species, *C. ampulla* also from Tertiary strata. Canu and Bassler (1920) in their monograph on the early Tertiary Bryozoa of North America described *C. collaris*. Brown (1958) records *C. ornata* and *C.* sp. aff. *C. ampulla* from the Tertiary of southwestern Victoria. In addition he includes *Cribrilina elevata* MacGillivray, 1895 and *Cribrilina cornuta* MacGillivray, 1895 in *Corbulipora* and describes a new species of *Corbulipora*, *C. pennata*.

Larwood (1969) summarized previous observations on the structure of the frontal wall but did not mention the work of Canu and Bassler. They stated (1920) that "the frontal is convex; the costules are fine, numerous, ornamented with very small lumen pores; separated by a large, initial slit, followed by two lacunae; the superior costules are more or less united to form a sort of collar".

The species under consideration in this paper is Recent and occurs at the following localities mentioned in Wass *et al.* (1970) off the southern coast of

Australia. They are 90K, 35° 03′ S, 134° 11′ E, 109 fathoms; 123K, 38° 15′ S, 140° 29′ E, 85–120 fathoms; 180K, 40° 48′ S, 148° 43′ E, 65 fathoms.

Terminology used follows that of Larwood (1962), the only difference being that upgrowths from lumen pores, pelma or pelmatidia are termed secondary spines. Observations made are based on distribution of calcification, the only soft part preserved being the frontal membrane; any discussion on soft parts other than the frontal membrane is purely hypothetical.

The shape of zooecia changes throughout the colony; those in the proximal region are very elongate and narrow, those in the centre of the colony become shorter and wider, while those at the distal margins are rounded and oval in outline and much smaller than their proximal equivalents. In zooecia from the proximal and central regions of a colony most costae are normal to the growth direction but in zooecia from the distal region most of the costae are oblique to this direction. Zooecia in the proximal region have up to 20 costae forming the frontal wall; in the central and distal regions of a colony this number is reduced to 13 to 16 (Plate I, Fig. 1). The zooecia have a bizarre appearance because the orifice is obscured below a complex oral shield formed by the three oral spines which first project vertically and then proximally (Plate I, Fig. 2).

The lateral and proximal gymnocyst is smooth beneath the level of costal development. The frontal wall arises by development of vertical hollow costae from the lateral and proximal gymnocyst at a level below the frontal membrane. The costae project outside the inner level of the lateral gymnocyst and are presumably derived by outswellings of the epidermis. Vertical growth takes the costae to a level above the frontal membrane and above the lateral gymnocyst. Bifurcation occurs and the outer element forms a hollow secondary spine directed vertically; the inner element grows in a plane parallel to the frontal membrane and centrally directed; it proceeds rapidly to form a hollow crescent-shaped structure with lateral costae (Fig. 1).

Plate I

All figures are of *Corbulipora* sp., Recent, Bass Strait, Australia.

Fig. 1. Frontal view showing structure of frontal wall, secondary spines and oral shield. ×110.

Fig. 2. Close-up of oral shield showing bifurcations. ×250.

Fig. 3. Frontal view to show elongate Z junction of opposing costae; small holes are "lumen pores", some of which may develop into secondary spines, e.g. inner end, 4th costa, right. ×200.

Fig. 4. Close-up of secondary spine structure on inner ends of two most distal costae. ×350.

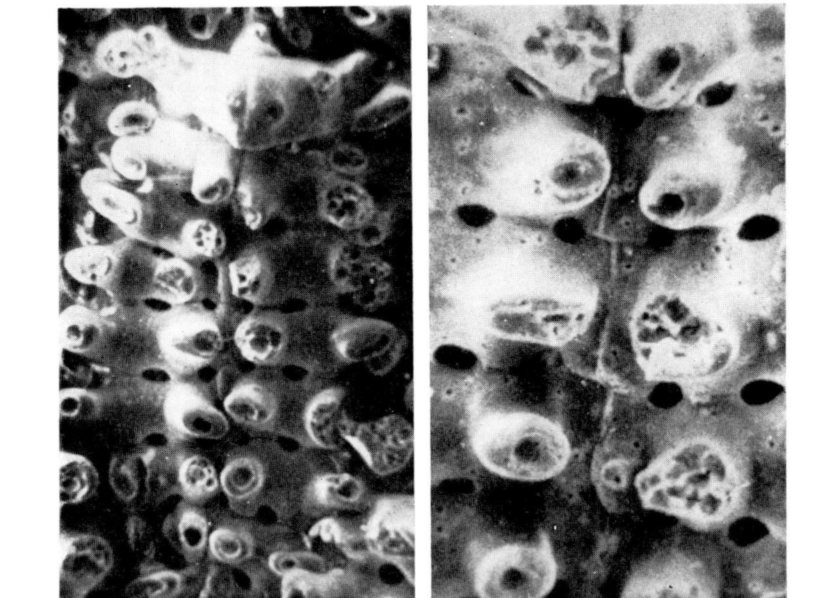
1
3
2
4

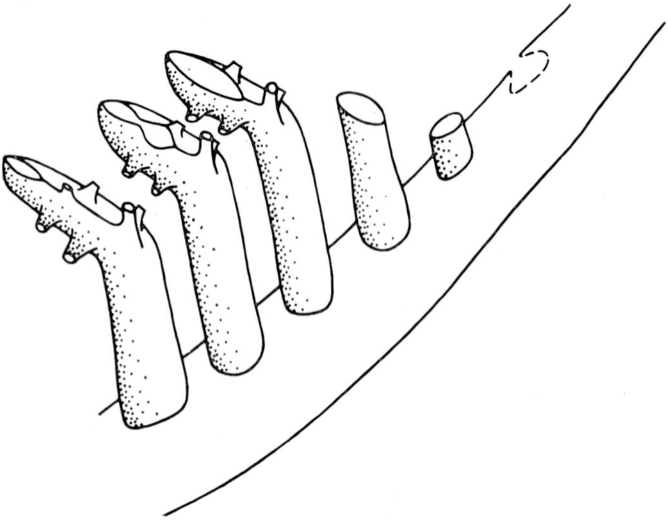

Fig. 1. Diagrammatic reconstruction to show formation of costae from lateral gymnocyst and their development into frontal wall; distal on right; not to scale.

Lateral costae on the proximal side of the costa "unite" with those on the distal side of the previously formed costa. The crescent-shaped structure gradually closes from the point of bifurcation to the central region forming pelmatidia. Vertical upgrowths of pelmatidia form hollow secondary spines. The last formed secondary spines (in the central region) appear to close off before those formed early in costal development. In some zooecia vertical development of the "lumen pores" forms secondary spines. Smaller "lumen pores" are present on the upper surface of the costa and are characteristically present at the inner end. Bifurcation and terminal fusion of the secondary spines may form a suprazooecial structure analogous to a tertiary frontal wall (cf. Larwood, 1962). Also peripheral secondary spines from adjacent zooecia may fuse.

Intercostal spaces are generally oval in outline; lateral costae are generally regularly spaced. There are two or three per costa; the outer or more peripheral is inside the outer line of spines; the second is generally opposite the outer edge of the second line of spines and in some cases a third intercostal space is formed at the extreme end of the costae.

The junction between the two opposing sets of costae is a continuous elongate Z in shape (Plate I, Fig. 3). However, one instance has been observed where two costae are directly opposite and at their ends, pressure ridges are developed in the calcification.

The structure of spines in the secondary frontal wall is unusual. At first they appear to be composed of a thick outer wall with a hollow centre. Removal of a thin portion of the upper level of the spine wall immediately reveals a series of round, hollow depressions surrounding the central tube (Plate I, Fig. 4). These depressions may be separated by thick or thin areas of calcification. This structure of the spine wall has also been observed at the level of the costae and in one instance, a similar structure has been observed in the costae.

The orifice is oblique to the level of the frontal membrane with the distal edge lower than the proximal. Three oral spines are present, one being placed at the distal margin of the orifice and the other two on the disto-lateral margins. The two outer spines first grow vertically and they may bifurcate at a level above the orifice. In any case, once they reach this level growth proceeds in a proximal direction. The outer wall of the spine develops before the base or the inner wall and produces a peristome-like structure above the orifice. Once this is achieved, the base and inner wall develop to form the crescent-shaped structure described previously. Bifurcation from this structure is common. The two spines coalesce either above the proximal margin of the orifice or above the apertural bar. The spine at the centre of the distal margin is last to develop; it follows the same pattern of growth but instead of growing vertically it projects distally before bifurcating and growing proximally. All three spines eventually coalesce and continued proximal growth forms a complete coverage of the orifice. One of the disto-lateral oral spines may continue growing proximally until it coalesces with a spine from a costa. The oral spines may bifurcate while growing proximally.

The formation of the frontal wall by hollow costae which develop at a level below the frontal membrane and the oral shield formed by three oral spines are puzzling developments. Presumed pressure differences associated with depression and elevation of the membrane may allow intrusion of fluids into the coelom through hollow spines and costae. To combat this intrusion, calcification of spinose extremities proceeds but before this can be fully achieved, oral shield development has commenced. From the configuration of the shield and the proximal hinging of the operculum it may be postulated that the tentacles could not be extruded fully.

In one zooecium calcification of the operculum has been completed prior to development of all costae in the frontal wall.

Another interesting feature is the ontogenetic state which reveals that the oral shield has developed in the astogenetically later zooecia at the growing edge before its development is complete in older zooecia.

The pattern of development of the frontal wall has led to a restudy of all species previously included in *Corbulipora*. While this is not complete, some general comments may be made. The frontal wall development in *C. ornata* follows that discussed above. *C. ampulla* appears to comprise two groups. The first has a frontal development similar to *C. ornata*; the second has a frontal wall formed from vertically directed costae but a regular arrangement of costae, intercostal spaces and lateral costal fusions is absent. There is a definite "*Arachnopusia*-type" appearance about the frontal. The frontal development of *C. elevata* is similar to the second group of *C. ampulla*. Frontal wall development in *C. cornuta* is normal for *Corbulipora* as is the development in *C. collaris*.

The genus *Acanthocella* (Canu and Bassler, 1917) has *Cribrilina tubulifera* (Hincks, 1881) from the Recent of Bass Strait, Australia as type species. An examination of the lectotype of this species, BM 99.5.1.726, chosen by Brown (1958) shows that the frontal wall and lateral gymnocyst are of the same construction as in *Corbulipora*. Three oral spines are present but none of the three zooecia on the lectotype possess the oral shield described above.

Morphology considered to be characteristic of *Corbulipora* includes an encrusting or bilamellar zoarium, zooecia rectangular to oval in outline, smooth lateral and proximal gymnocyst, frontal wall developed from vertical or near vertical costae which bifurcate to form a *regular* arrangement of horizontal costae, lateral costal fusions and secondary spines, orifice semicircular, oral spines and avicularia may be present, ovicell entozooidal.

ACKNOWLEDGEMENTS

Most of this work was completed during 1970 while on study leave from the University of Sydney and during tenure of the Eleanor Sophia Wood Travelling Fellowship. Thanks are due to the Trustees of the British Museum (Natural History) for permission to work in the Department of Zoology and more especially to Miss Patricia L. Cook for her patience, understanding and guidance during my introduction to cheilostomatous Bryozoa and for her critical reading of this manuscript.

REFERENCES

BROWN, D. A. (1958). Fossil cheilostomatous Polyzoa from south-west Victoria. *Mem. Geol. Surv. Victoria* **20,** 1–90.

CANU, F. and BASSLER, R. S. (1917). A synopsis of American early Tertiary cheilostome Bryozoa. *Bull. U.S. Nat. Mus.* **96,** 1–87.

CANU, F. and BASSLER, R. S. (1920). North American Early Tertiary Bryozoa. *Bull. U.S. Nat. Mus.* **106,** 1–879.

HINCKS, T. (1881). Contributions towards a general history of the marine Polyzoa, VI–VIII. *Ann. Mag. Nat. Hist.* (5) **VIII,** 1–14, 122–136.

LARWOOD, G. P. (1962). The morphology and systematics of some Cretaceous cribrimorph Polyzoa (Pelmatoporinae). *Bull. Br. Mus. Nat. Hist. (Geol.)* **6** (1), 1–285.

LARWOOD, G. P. (1969). Frontal calcification and its function in some Cretaceous and Recent cribrimorph and other cheilostomatous Bryozoa. *Bull. Br. Mus. Nat. Hist.* (*Zool.*) **18** (5), 171–182.

MACGILLIVRAY, P. H. (1895). A monograph of the Tertiary Polyzoa of Victoria. *Trans. R. Soc. Victoria* **4,** 1–166.

MAPLESTONE, C. M. (1901). Further Descriptions of the Tertiary Polyzoa of Victoria. *Proc. R. Soc. Victoria* **6,** 204–213.

WASS, R. E., CONOLLY, J. R., and MACINTYRE, R. J. (1970). Bryozoan carbonate sand continuous along southern Australia. *Marine Geol.* **9,** 63–73.

36 | Bryozoa Cheilostomata (Anasca) of the Antarctic and Subantarctic

E. I. ANDROSOVA

Zoological Institute, Leningrad, U.S.S.R.

Abstract: During recent years I have studied carefully the Bryozoa of the Antarctic and Subantarctic using rather numerous collections of the Soviet Antarctic expeditions 1955–1958 and the French Antarctic expeditions 1961–1965. The result of these studies are summarized here.

Bryozoa form one of the basic groups of the bottom communities of the Antarctic on stony grounds which predominate there and are used as substrates by Bryozoa colonies. Powerful bottom currents carry plentiful nutrients benefitting bryozoans, sponges and other sessile animals. This is why the Antarctic is often called the kingdom of Bryozoa and Sponges.

Perhaps it would be of interest to the palaeontologists to know that numerous accumulations of dead Bryozoa and Sponges form a continuous layer up to 5 cm thick covering the bottom.

The systematics, vertical distribution and zoogeography of Antarctic Bryozoa are of particular interest.

According to Rogick (1965) 321 species of Bryozoa were known from the Antarctic and Subantarctic, 179 species of these are restricted to the Antarctic. Since 1965 following the investigations of Moyano (1965, 1965–1966, 1969, 1969–1970, 1970), Redier (1965) and Androsova (1968) our knowledge of the Antarctic and Subantarctic Bryozoa has been enriched. Moyano has described, for these regions, 7 new species of Cheilostomata. One new genus and one species were found in the Antarctic for the first time. Redier (1965) notes 6 species (5 of Cheilostomata and 1 of Cyclostomata) for the first time in the Antarctic. Androsova (1968) writing on Cyclostomata and Ctenostomata describes 4 new species and records 5 which are new to these regions. In the Soviet collections Anasca were found in samples from 57 stations, 44 of which were in the Antarctic region from the coast of Princess Ragnhild (20°E) to Rennick Bay (162°E) and Scott Island (180°E). Thirteen other stations are from the

Subantarctic region (Tierro del Fuego, Patagonia, Falkland Islands, Prince Edward Islands, Kerguelen, Heard, Macquarie, Auckland).

In the French collections Bryozoa were found from 43 stations. All of them near Adelie Land.

Thirty-nine species of Cheilostomata Anasca have been determined: *Aetea anguina* (L),*Amastigia antarctica* (Kluge), *A. solida* (Kluge), *Menipea flagellifera* Busk, *Notoplites vanhöffeni* (Kluge), *N. drygalskii* (Kluge), *Farciminellum antarcticum* Hastings, *Himantozoum antarcticum* (Calvet), *Dimetopia cornuta* Busk, *Klugella echinata* (Kluge), *Camptolites angustus* (Kluge), *C. tricornis* (Waters), *C. bicornis* (Busk), *C. retiformis* (Kluge), *Cornucopina polymorpha* (Kluge), *Beania livingstonei* Hastings, *Flustra angusta* Kluge, *F. curva* Kluge, *F. flagellata* Waters, *F. tenuis* Kluge, *F. vulgaris* Kluge, two new species of *Flustra* (*F. arnaudi* and *F. astrovae*), *Callopora gigantea* (Kluge), *Chaperia quadrispinosa* Kluge, *C. furcata* (Busk), *C. galeata* (Busk), *Labioporella adeliensis* Livingstone, *Hemiseptella michaelseni* (Calvet), *Cribrilina projecta* Waters, *Figularia spatulata* (Calvet), *Paracellaria wandeli* Calvet, *Cellaria vitrimuralis* Rogick, *C. moniliorata* Rogick, *Mawsonia extensalata* Rogick, *M. brevimandibulata* Moyano, *Paramawsonia dennanti* (MacGillivray), *Melicerita obliqua* (Thornely), *M. latilaminata* Rogick. They represent 23 genera and 10 families. Three species are new for the Antarctic, 2 being new species. The third species *Chaperia galeata* was previously found in the Subantarctic (Falkland Islands and Kerguelen).

Flustra arnaudi (named in honour of the French hydrobiolgist P. Arnaud) is represented by a big colony of 12–15 cm height, consisting of broad (2·5 cm) leaf-like expansions, which branch many times. The distal ends of the branches are slightly rounded. Colour is from yellowish to dark brown. The colonies are of two layers. Avicularia and ovicells are large, the latter highly convex with a high median ridge. The ovicells occur in zones. The colonies of *F. astrovae* (named in honour of Dr G. G. Astrova) are also big, up to 15 cm height, and consist of leaf-like expansions branching dichotomously. The colour is from yellowish to brown. The colonies consist of two layers. There are no avicularia, ovicells are present and are very big and highly convex, and distributed as in *F. arnaudi*.

There are also 4 species new for the Subantarctic: *Dimetopia cornuta* from Bass Strait and widely distributed in the Antarctic, *Flustra flagellata* and *Labioporella adeliensis* found near Auckland Island. The fourth species, *Paramawsonia dennanti* was previously described from Tertiary deposits in Australia (Victoria) and is found very commonly in the Antarctic. The species occurs also in the Subantarctic collections near Macquarie Island. MacGillivray (1895) attributed it to the genus *Cellaria*. But *C. denanti* is very different from other species of the same genus and is more like *Mawsonia* Livingstone of the same family.

The species of *Cellaria* are chiefly characterized by very narrow jointed branches. The orifice is very broad. Condyles are small. The vicarious avicularia are rather small with triangular mandibles.

Mawsonia consists of species with thick cylindrical stems up to 4 mm in diameter. They are not jointed and rarely branched dichotomously. Orifices are long with large condyles and large vicarious avicularia. The colonies of *Paramawsonia* have very thick jointed stems which do not branch. Orifices are broad, with condyles which are smaller than those of *Mawsonia*. The vicarious avicularia are large with rounded mandibles. *Paramawsonia dennanti* (MacGillivray) has stout cylindrical stems (diameter about 2·5 mm) and lacks joints and branches. Complete colonies were absent in the collections; the maximum length of the stems was 40 mm. The zooecia are hexagonal rather longer than broad, and the orifice rather wider than high. The vicarious avicularia are very large, up to twice the size of the zooecia and have broad rounded mandibles. Ovicells are endotoichal.

If we summarize all the data, we can see that 197 species are now known from the Antarctic alone. One hundred and forty-three species are known from the Antarctic and Subantarctic.

One of the peculiarities of the Antarctic fauna is a very high variability of many species, e.g. *Camptoplites bicornis, C. retiformis, Beania livingstonei.* This peculiarity makes the determination of species limits in the Antarctic Bryozoa most difficult.

A good example of such variability is *Camptolites bicornis* (Busk). The growth and form of zooecia of the main stem of *C. bicornis* varies considerably and differs from the zooecia of the branches, which are also variable. The form and size of the avicularia and the length of their supporting stalks are so variable that several variations have been described. Kluge (1914) described var. *elatior*, var. *compacta* and var. *magna*. Hastings (1943) added to them var. *quadriavicularis*. But the variations overlap and Kluge refers to var. *magna* form α *elongata* and form β *ventricosa*. Other features, the size of the colony, the type of branching, the size and form of the ovicells, vary also.

Among the Antarctic Bryozoa there are species in which parts of the colonies differ highly from one another. *Cornucopina polymorpha* is an example of it. This species has a big bushy highly branched colony. Some branches lack spines and avicularia are rare and very small. Other branches have variable numbers of large, very long and curved spines. Avicularia are also large and tubelike, basally situated but sometimes replaced by small curved frontal avicularia.

Antarctic bryozoan faunas are noted for the large colonies which occur (Rogick, 1965). In *Flustra arnaudi* the colonies reach 15 cm or more. *Flustra astrovae* also has big colonies up to 15 cm high. *Labioporella adeliensis* has large calcareous plates, parts of it reaching 7 cm.

Individual parts of zooecia, such as spines and avicularia, are also often very large. The big stout spines of *Camptolites tricornis* may be observed with the naked eye. The same species has also very large avicularia. *Klugella echinata* is provided with huge and stout spines, which are curved hook-like over the aperture. As the spines and avicularia are used for defence it is possible that their gigantism reflects the struggle for existence in the specific conditions of the Antarctic.

The analysis of vertical distribution shows that most of the bryozoans inhabit depths up to 500–700 m. In other parts of the ocean they are abundant in shallower water at about 200–300 m.

Soviet Antarctic expeditions also explored some abyssal stations, and 5 species are now also noted from the abyssal zone: *Notoplites drygalskii* (5054 m instead of the previous maximum 1080 m), *Himantozoum antarcticum* (2000 m instead of 567 m), *Camptolites tricornis* (2000 m instead of 594 m), *Chaperia galeata* (4700 m instead of 51 m), *Melicerita obliqua* (3560 m instead of 582 m).

The analysis of the distribution of the species studied has shown that the fauna of Antarctic Bryozoa is characterized by a marked endemism of species.

Thirty of the 39 species found, that is 76%, are endemic to the Antarctic. Those endemic to the Subantarctic are only two—*Menipea flagellifera* and *Hemiseptella michaelseni*. Perhaps the quantity of subantarctic endemics would be more, but the collections from this region were rather poor. Three species are common for both regions and four are widely distributed elsewhere.

Of the 23 genera of our collections real endemics of the Antarctic are two: *Mawsonia* and *Paracellaria*; the last has been recently described by Moyano. If we extend the limits to some neighbouring regions (Hastings, 1943) then 9 genera may be recognized—39% (*Amastigia, Farciminellum, Himantozoum, Klugella, Camptolites, Cornucopina, Paracellaria, Mawsonia, Melicerita*).

High numbers of endemic species and genera reflect the ancient origin of the Antarctic fauna.

REFERENCES

ANDROSOVA, E. (1968). Bryozoans of the orders Cyclostomata and Ctenostomata of the Antarctic and Subantarctic. *Explorations of the Fauna of the seas* **VI** (XIV), 35–84.

HASTINGS, A. (1943). Polyzoa (Bryozoa) *I Discovery Reports, Cambridge* **XXII,** 301–510.

KLUGE, H. (1914). Die Bryozoen der Deutschen Sudpolar-Expedition 1901–1903 *Deutsche Sudpolar-Exp. 1901–1903 von Drygalski* **XV**; *Zoologie* **VII,** 599–678.

MACGILLIVRAY, P. (1895). A Monograph of the Tertiary Polyzoa of Victoria. *Trans. R. Soc. Victoria, Melbourne* IV, 1–165.

MOYANO, H. (1965). Bryozoa colectados durante la Expedition Antartica Chilena 1964–1965. *Inst. Centr. Biol. Univ. Concep., Chile* 1–29.

MOYANO, H. (1965–1966). Description de Schizoporella bifrons n. sp. *Apart. del Bol. de la Soc. Biol.* **XL,** 81–89.

MOYANO, H. (1969). Bryozoa celectados por la Expedicion Antarctica Chilena 1964–1965. III. Fam Cellariidae Hincks. *Bol. Soc. Biol. Concep.* **XLI,** 47–77.

MOYANO, H. (1969–1970). Bryozoa colectados por la Expedicion Antartica Chilena 1964–1965. Familia Arachnopusiidae Jullien, 1888. *Bol. Soc. Biol. Concep.* **XLII,** 257–285.

MOYANO, H. (1969–1970). *Flustrapora magellanica* nov. gen. nov. sp. (Bryozoa, Cheilostomata, Anasca). *Bol. Soc. Biol. Concep.* **XLII,** 59–65.

MOYANO, H. (1970). Una Familia, tres Generos y una especie neuvos para la Antartica (Bryozoa, Cheilostomata). *Cah. Biol. Marine.* **XI,** 1–166.

REDIER, L. (1965). Bryozoairos. Expeditions Antarctiques Belges (1959–1960). *Inst. R. Sci. Nat. Belg. Bull.* **41** (40), 1–39.

ROGICK, M. (1965). Bryozoa of the Antarctic. *Monogr. Biol.* **XV,** 401–413.

37 | Devonian Bryozoa from the Southeastern Armorican Massif, Western France

FRANÇOISE BIGEY

Université de Paris, VI, Laboratoire de Paléontologie des Invertébrés, Paris, France

Abstract: An inventory of Devonian Bryozoa from the Ancenis Synclinorium and from the Saint-Julien-de-Vouvantes-Angers Synclinorium is given. Bryozoa are more numerous and more diversified in the Saint-Julien-de-Vouvantes-Angers Synclinorium where biothermal formations are rarer.

In the Palaeozoic beds of the Southeastern Armorican Massif (localities: Fig. 1) Bryozoa are found only in Devonian deposits. Since they are absent from the

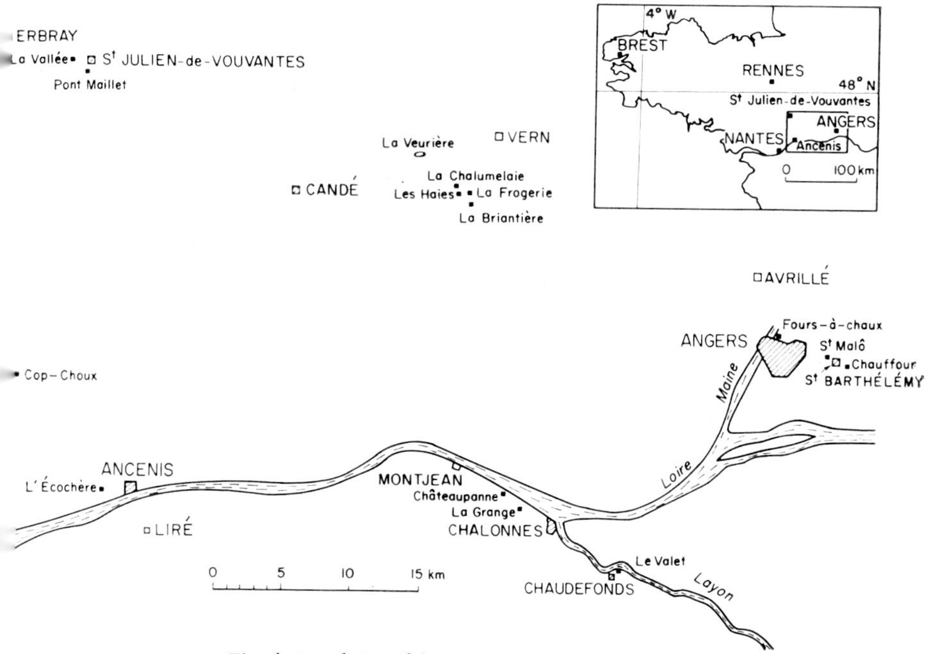

Fig. 1. Localities of the Ancenis-Angers region.

Silurian formations, they cannot provide information on the Siluro-Devonian boundary in this region. The only formations that yield Bryozoa are situated in the Ancenis Synclinorium and in the Saint-Julien-de-Vouvantes-Angers Synclinorium (Fig. 2). Specimens found in collections (Table 1) are kept in the following places: Faculté libre des Sciences de l'Université catholique de l'Ouest in Angers (U.C.O.), Musée paléontologique d'Angers (M.P.A.) and Muséum d'Histoire Naturelle de Nantes (M.H.N.N.). Specimens which it has not been possible to locate in known collections are indicated in Table 2.

ANCENIS SYNCLINORIUM

1. *Châteaupanne-Chaudefonds Syncline*

Discussing the "micrographic characters" of the Chalonnes Limestone (Upper Emsien-Lower Eifelian) Le Maître (1934) indicated fragments of Bryozoa in granular limestones. At the quarry of Châteaupanne, Cavet *et al.* (1970) recorded Bryozoa, Rugosa, tabulate corals and stromatoporoids. These Bryozoa are Trepostomata and are more numerous towards the top of the formation. In the La Grange Limestone, referred to the Chalonnes Limestone formation, the fauna of which was described by Ferronnière (1920–1922), Bryozoa were noticed by Erben *et al.* (1959) and by Cavet *et al.* (1970). The Liré Shales, of uncertain stratigraphic horizon, (Cavet *et al.*, 1971), have yielded numerous Bryozoa (Péneau, 1929). These are poor casts among which fenestellids are recognizable (Bureau Coll., U.C.O.).

2. *Tombeau-Leclerc Horst*

The Valet-en-Chaudefonds Limestone (Upper Emsian-Lower Eifelian) has yielded fenestellids particularly *Fenestella* sp. (Péneau Coll., U.C.O.). I have collected *Fenestella* sp. and *Hemitrypa* sp. from this horizon and locality.

3. *"Culm" Area*

In the L'Ecochère Shales and Limestones (Frasnian) I have found, in a vineyard to the west of the quarry, a few Trepostomata with tabulate corals, Rugosa and brachiopods. Bryozoa were recorded by Milon (1932) in the Cop-Choux Limestone (Frasnian) among organic debris.

SAINT-JULIEN-DE-VOUVANTES-ANGERS SYNCLINORIUM

From the *Platyorthis monnieri* Sandstone (Lower Siegenian of Vern, La Chalumelaie) Pillet (1957) recorded *Fenestella* sp. The same formation in Chauffour (Saint-Barthélémy) yielded fenestellids (M.P.A.).

From the Vern Limestone (*Athyris undata* Shales and Limestones horizon,

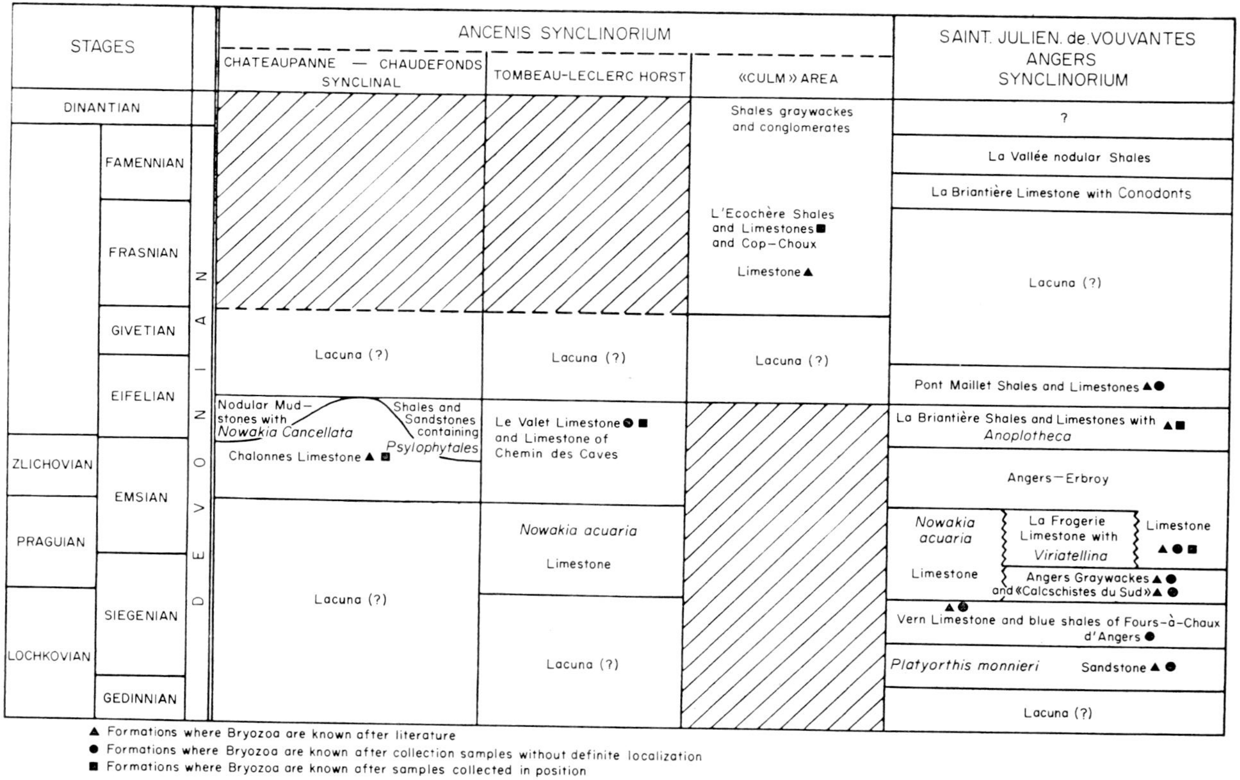

Fig. 2. Occurrence of Bryozoa in the Devonian of the Ancenis-Angers region. (Stratigraphical data after Cavet *et al.*, 1967, 1970; Cavet and Lardeux, 1967; Cavet *et al.*, 1971; and Moreau-Benoit, 1971.)

Table 1. Check-list of Bryozoa located in collections

	Localities	Formations	Age	References	Collections	Proposed determinations
Bryozoa	Liré	Liré Shales	Devonian?	Péneau, 1929	Bureau, U.C.O.	fenestellids
Fenestella sp.	Valet	Le Valet Limestone	U. Emsian-L. Eifelian		Péneau, U.C.O.	*Fenestella* sp.
Fenestella sp.	Chauffour (St-Barthélémy)	*Platyorthis monnieri* Sandstone	L. Siegenian		M.P.A.	fenestellid
Fenestella sp.	Vern	Vern Limestone	M. Siegenian		Couffon, M.P.A.	*Hemitrypa* sp.
Bryozoa	Vern	Vern Limestone	M. Siegenian		M.H.N.N.	fenestellids
Fenestella sp.	Les Haies (Vern)	Vern Limestone	M. Siegenian		Davy, U.C.O.	*Polypora* sp.
Fenestellids	Fours-à-chaux d'Angers	Blue Shales of Fours à-chaux d'Angers	M. Siegenian	Pillet, 1953	Couffon, M.P.A.	fenestellids
Fenestella antiqua Goldfuss	Fours-à-chaux d'Angers	Angers Greywackes	U. Siegenian		M.P.A.	fenestellid
Bryozoa	Chauffour (St-Barthélémy)	Angers Greywackes	U. Siegenian	Pillet, 1948	M.P.A.	fenestellids
Fenestellids	Fours-à-chaux d'Angers	"Calcschistes du Sud"	U. Siegenian	Pillet, 1960	Rondeau, U.C.O.	fenestellids
Retepora sp.	Fours-à-chaux d'Angers	"Calcschistes du Sud"	U. Siegenian		Couffon, M.P.A.	cf. *Polypora* sp.
Monotrypa fibroglobosa (Quenstedt)	St-Malô (St-Barthélémy)	Angers Limestone	Emsian		Couffon, M.P.A.	indeterminate
Monotrypa oehlerti Nicholson	St-Malô (St-Barthélémy)	Angers Limestone	Emsian	Pillet, 1962	Couffon, M.P.A.	indeterminate
Fenestella sp.	St-Malô (St-Barthélémy)	Angers Limestone	Emsian		Couffon, M.P.A.	*Fenestella* sp.
Fenestella cf. *bischofi* Roemer	Erbray	Erbray Limestone	Emsian		Davy, U.C.O.	*Fenestella* cf. *pannosa* Počta
Fenestella cf. *bifurca* Roemer	Erbray	Erbray Limestone	Emsian	Barrois, 1887	Davy, U.C.O.	*Fenestella* cf. *capillosa* Počta
Fenestella cf. *pluma* (Phillips)	Erbray	Erbray Limestone	Emsian		Davy, U.C.O.	*Ptilopora* sp.
Lichenalia cf. *patina* (Roemer)	Erbray	Erbray Limestone	Emsian		Davy, U.C.O.	*Semicoscinium* sp.

Table 2. Check list of Bryozoa which it has not been possible to locate in known collections

	Localities	Formations	Age	References	Remarks
Fenestella sp.	La Chalumelaie (Vern)	*Platyorthis monnieri* Sandstone	L. Siegenian	Pillet, 1957	
Fenestella sp.	Vern	Vern Limestone	M. Siegenian	Couffon, 1925	cf. Table 1: *Hemitrypa* sp.
Fenestella sp.	Fours-à-chaux d'Angers	Angers Greywackes	U. Siegenian	Rondeau, 1890	determined by de Verneuil?
Fenestella sp.	Fours-à-chaux d'Angers	Angers Greywackes	U. Siegenian	Desmazières, 1900	
Fenestella sp.	Fours-à-chaux d'Angers	Angers Greywackes	U. Siegenian	Couffon, 1925	cf. Rondeau, 1890
Retepora sp.	Fours-à-chaux d'Angers	Angers Greywackes	U. Siegenian	Pillet, 1960	
Fenestella sp.	Fours-à-chaux d'Angers	Angers Greywackes	U. Siegenian	Pillet, 1960	cf. Desmazières, 1900
Monotrypa fibroglobosa (Quenstedt)		Angers Limestone	Emsian	D. and P. Oehlert, 1889	cf. Desmazières, 1900
Monotrypa oehlerti Nicholson		Angers Limestone	Emsian	D. and P. Oehlert, 1889	cf. Desmazières, 1900
Fistulipora torrubiae Milne-Edwards and Haime		Angers Limestone	Emsian	D. and P. Oehlert, 1889	cf. Desmazières, 1900
Fenestella sp.		Angers Limestone	Emsian	D. and P. Oehlert, 1889	cf. Desmazières, 1900
Fenestellids	Chauffour (St-Barthélémy)	Angers Limestone	Emsian	Pillet, 1962	cf. Desmazières, 1900
Fenestella sp.	Le Fléchay-en-Avrillé	Le Fléchay-en-Avrillé Limestone	Emsian	Péneau 1950	cf. Desmazières, 1900
Fenestella antiqua Goldfuss	Fours-à-chaux d'Angers	"brown sandstone"	Emsian	Rondeau, 1890	determined by de Verneuil
Ichtyorachis	Fours-à-chaux d'Angers	"brown sandstone"	Emsian	Rondeau, 1890	determined by de Verneuil
Eschara	Fours-à-chaux d'Angers	"brown sandstone"	Emsian	Rondeau, 1890	determined by de Verneuil
Retepara	Fours-à-chaux d'Angers	"brown sandstone"	Emsian	Rondeau, 1890	determined by de Verneuil
Retepora boloniana d'Orbigny	Fours-à-chaux d'Angers	"brown sandstone"	Emsian	Couffon and Dollfus, 1929	
Fenestella nobilis Barrande	La Ferronnière, Erbray	Erbray Limestone	Emsian	de Tromelin and Lebesconte, 1876	cf. Bigey, 1970

referred to the Middle Siegenian) Couffon (1925) recorded *Fenestella* sp. and Pillet (1953) also indicated the occurrence of fenestellids. These were *Hemitrypa* sp. (Couffon Coll., M.P.A.) and other fenestellids (M.H.N.N.). The locality of Les Haies (Vern) has yielded *Polypora* sp. (Davy Coll., U.C.O.).

From the Blue Shales of Fours-à-chaux d'Angers (Middle Siegenian) Pillet (1950) reported Bryozoa and more particularly fenestellids (Pillet, 1953), Couffon Coll., M.P.A.

The Angers Greywackes (Upper Siegenian) contains fenestellids (Pillet, 1953). Several authors record *Fenestella* sp. from the Fours-à-chaux d'Angers quarry (Rondeau, 1890; Desmazières, 1900; Couffon, 1925). *Retepora* sp. and *Fenestella* sp. are also recorded from the same locality by Pillet (1960). I also have found a few casts of fenestellids (M.P.A.). At Chauffour (Saint-Barthélémy) Bryozoa are the most common fossils in the Angers Greywackes (Pillet, 1948), these are fenestellids (M.P.A.).

The "Calcshistes du Sud" of Fours-à-chaux d'Angers, Upper Siegenian according to Pillet (1960), have yielded fenestellids (M.P.A. and U.C.O.) in addition I have recognized cf. *Polypora* sp. (Couffon Coll., M.P.A.).

The Angers-Erbray Limestones (Emsian) contain Bryozoa at several localities. In their study of the Angers Limestone D. and P. Oehlert (1889) listed *Monotrypa fibroglobosa* Quenstedt, *M. oehlerti* Nicholson, *Fistulipora torrubaiae* Milne-Edwards and Haime and *Fenestella* sp. Lists are also given by Bureau (1900) and by Rondeau (1893). Fenestellids were noted by Pillet (1962) at Chauffour quarry and I have observed some of them in thin sections from Chauffour material collected *in situ*. From the limestones of the Saint-Malô (Saint-Barthélémy) quarries Pillet (1962) recorded *Monotrypa fibroglobosa*, *M. oehlerti* and *Fenestella* sp. (Couffon Coll., M.P.A.). Only *Fenestella* sp. may now be determined. One hundred and fifty metres southeast of the Saint-Malô quarries, the top of the Angers Limestone has yielded fenestellids and Trepostomata. The Fléchay-en-Avrillé Limestone, the Emsian age of which has been established by Péneau (1947) yielded a few specimens of *Fenestella* sp. (Péneau, 1950). According to Péneau (1928) the La Veurière Limestones yielded only fenestellids. Lardeux (1969) reported the occurrence of Bryozoa from several parts of the working quarry where I also have collected fenestellids and Trepostomata *in situ*.

At the Fours-à-chaux d'Angers quarry, in a "thin bed of brown sandstone" probably equivalent to the crinoidal limestone of the south of the quarry, the following Bryozoa, determined by de Verneuil, are recorded by Rondeau (1890): *Fenestella antiqua* Goldfuss, *Ichtyorachis*, *Eschara* and *Retepora*. Couffon and Dollfus (1929) recorded *Retepora boloniana* d'Orbigny.

In the Erbray Limestone Tromelin and Lebesconte (1876) recorded *Fenestella nobilis* Barrande at La Ferronnière quarry. Barrois (1887) listed and later (1889) described *Fenestella* cf. *bischofi* Roemer 1855, *Fenestella* cf. *bifurca* Roemer 1850, *Fenestella* cf. *pluma* (Phillips 1836) and *Lichenalia* cf. *patina* (Roemer 1850) (Davy Coll., U.C.O.). I assign these species respectively to *Fenestella* cf. *pannosa* Počta 1894, *Fenestella* cf. *capillosa* Počta 1894, *Ptilopora* sp. and *Semicoscinium* sp. I have found, at La Ferronnière quarry (Bigey, 1970) *Utropora* sp., *Fenestella* cf. *exilis* Počta, *Hemitrypa bohemica* Barrande and *Fenestella* cf. *capillosa* Počta. I have since discovered (Bigey, 1971) *Ptilopora* aff. *bohemica* Prantl at this locality.

The La Briantière Shales and Limestones with *Anoplotheca* (Lower Eifelian) have yielded Bryozoa from several places (Lardeux, 1969), they are fenestellids: *Fenestella* sp. and *Hemitrypa* sp.

Péneau (1928) records fenestellids from the Pont-Maillet Shales and Limestones (Upper Eifelian). I have recognized *Fenestella* sp. in a black shale from this locality (U.C.O.).

ACKNOWLEDGEMENTS

I record my thanks to Madame Baudoin-Bodin, Director of the Muséum d'Histoire Naturelle de Nantes, to Dr M. Gruet Assistant Curator of the Musée paléontologique d'Angers and of the Collections of the Université catholique de l'Ouest, Angers, for permission to work on collections in their care and to borrow specimens.

REFERENCES

Barrois, C. (1887). Note préliminaire sur la Faune d'Erbray (Loire-Inférieure). *Ann. Soc. géol. Nord, Lille* **XIV,** 154–158.

Barrois, C. (1889). Faune du Calcaire d'Erbray. *Mém. Soc. géol. Nord, Lille* **III,** 348 pp.

Bigey, F. (1970). Les Bryozoaires du Calcaire d'Erbray (Loire-Atlantique, Dévonien inférieur): premiers résultats d'une révision. *C. R. somm. Soc. géol. Fr., Paris* (8), 324–325.

Bigey, F. (1971). Présence de *Ptilopora* aff. *bohemica* (Bryozoaire cryptostome) dans le Dévonien du Sud-Est du Massif armoricain. *Bull. Soc. Et. Sc. Anjou, Angers* **VIII** (In press).

Bureau, L. (1900). Notice sur la géologie de la Loire-Inférieure. *In* "La ville de Nantes et la Loire-Inférieure". Vol. III, 99–522, Grimaud, Nantes.

Cavet, P., Cogné, J., Gruet, M. *et al.* (1967). Notice explicative de la feuille d'Ancenis, carte géol. au 80 000ème. *Serv. carte géol., Paris.* 23 pp.

Cavet, P. *et al.* (1970). Notice explicative de la feuille de Chalonnes-sur-Loire, carte géol. au 50 000ème. *Direct. Serv. géol. nat.*, B.R.G.M., Orléans-La Source. 32 pp.

Cavet, P. and Lardeux, H. (1967). Le Dévonien supérieur dans le Sud-Est du Massif armoricain. *Int. Symp. Dev. Syst. Calgary 1967, Calgary* **I,** 143–148.

Cavet, P., Lardeux, H. and Philippot, A. (1971). Ordovicien et Silurien aux environs de Montjean et Chalonnes (Maine-et-Loire, Sud-Est du Massif armoricain). *Coll. Ord.-Silur. Brest 1971, Mém. B.R.G.M., Paris,* No. 73, 199–212.

Couffon, O. (1925). Le Dévonien en Maine-et-Loire. *Bull. Soc. Et. Sc. Angers, Angers* **LIV,** 33–98.

Couffon, O. and Dollfus, G. (1929). A Summary of the Geology of Maine-et-Loire. *Proc. Geol. Assoc., London* **XXXIX** (1928), 369–428.

Desmazières, O. (1900). Compte-rendu de quelques excursions géologiques aux environs d'Angers. *Bull. Soc. Et. Sc. Angers, Angers* **XXIX,** (1899) XLI–XLII.

de Tromelin, G. and Lebesconte, P. (1876). Observations sur les terrains primaires du Nord du département d'Ille-et-Vilaine et de quelques autres parties du massif breton. *Bull. Soc. géol. Fr., Paris* **IV,** 583–623.

Erben, H. K., Lardeux, H., Lys, M., Pillet, J., Serre, B., Ubaghs, G. and Walliser, O. H. (1959). Note préliminaire sur la faune des calcaires de la Grange (Maine-et-Loire). *Bull. Soc. Et. Sc. Angers Angers.* **II** (89), 100–104.

Ferronnière, G. (1920–1922). Le Calcaire de la Grange près Chalonnes (Maine-et-Loire). *Bull. Soc. Sc. nat. Ouest France, Nantes.* **VI,** 121–139.

Lardeux, H. (1969). Le Dévonien du synclinal de Saint-Julien-de-Vouvantes (Sud-Est du Massif armoricain): état des recherches et précisions nouvelles. *Bull. Soc. Et. Sc. Anjou, Angers.* **VII,** 3–40.

Le Maître, D. (1934). Etude sur la faune des calcaires dévoniens du Bassin d'Ancenis. *Mém. Soc. géol. Nord, Lille* **XII,** 267 pp.

Milon, Y. (1932). Étude préliminaire de la Microfaune des Calcaires frasniens de Cop-Choux (Loire-Inférieure). *C.R. somm. Soc. géol. Fr., Paris* **5,** 68–69.

Moreau-Benoit, A. (1971). Recherches de Palynologie et de Planctologie sur le Dévonien et quelques formations siluriennes dans le Sud-est du Massif armoricain. Paris. Thèse ronéot.

Oehlert, D. and Oehlert, P. (1889). Sur le Dévonien des environs d'Angers. *Bull. Soc. géol. Fr., Paris* **XVII,** 742–791.

Péneau, J. (1928). Études stratigraphiques et paléontologiques dans le Sud-Est du Massif Armoricain (synclinal de Saint-Julien-de-Vouvantes). *Bull. Soc. Sc. nat. Ouest Fr.,* Nantes 1–300.

Péneau, J. (1929). Sur les schistes dévoniens de Liré près Ancenis. *Bull. Soc. Sc. nat. Ouest Fr., Nantes* **IX,** XVI–XVII.

Péneau, J. (1947). Révision de la feuille d'Angers au 80 000 e. Terrains primaires. *Bull. Serv. Carte géol. Fr., Paris* **XLVI,** 81–99.

Péneau J. (1950). Contribution à l'étude du Paléozoique de l'Anjou. *Bull. Soc. Et. Sc. Angers, Angers* **LXXIV,** 80ème année, p. 11–18.

Phillips, J. (1836). "Illustrations of the Geology of Yorkshire"; or a description of the strata and organic remains: accompanied by a geological map, sections and diagrams, and figures of the fossils. Part II. "The Mountain limestone district." John Murray, London.

Pillet, J. (1948). La faune des grauwackes de Chauffour près Saint-Barthélémy (Maine-et-Loire). *Bull. Soc. géol. Fr., Paris* **XVIII,** 453–466.

Pillet, J. (1950). Contribution à l'étude des schistes et calcschistes de la grauwacke d'Angers. *Bull. Soc. géol. Fr., Paris* **XX,** 39–48.

Pillet, J. (1953). Contribution à l'étude du Dévonien Armoricain. Les faunes du Siegénien moyen et supérieur en Anjou. *Bull. Soc. Et. Sc. Angers, Angers* **II,** 15–24.

Pillet, J. (1957). Contribution à l'étude du Dévonien Armoricain. II. La faune du Siegénien inférieur en Anjou. *Bull. Soc. Et. Sc. Angers, Angers* 15–25.

Pillet, J. (1960). Sur l'existence de deux niveaux d'âges distincts dans les calcaires des fours à chaux d'Angers. *Bull. Soc. Et. Sc. Angers, Angers* **II,** 95–99.

Pillet, J. (1962). Contribution à l'étude du Dévonien armoricain. III. Préliminaire sur la faune des calcaires d'Angers-Erbray. *Bull. Soc. Et. Sc. Angers, Angers* **III,** 46–61.

Počta, P. (1894). Bryozoaires, Hydrozoaires at partie des Anthozoaires. *In* "Système Silurien du centre de la Bohême" (J. Barrande, ed.), Vol. 8, 229 pp. Prague.

Roemer, F. (1850). Beiträge zur geologischen Kenntniss des nordwestlichen Harzgebirges. *Palaeontographica, Cassel* **3,** I–VIII, 1–52.

Roemer, F. (1855). Beiträge zur geologischen Kenntniss des nordwestlichen Harzgebirges. 3 abt. *Palaeontographica, Cassel* **5,** 1–46.

Rondeau, E. (1890). Étude sur le terrain dévonien aux environs d'Angers. *Mém. Soc. nat. Agr. sc. Arts Angers, Angers* **IV,** 155–191.

Rondeau, E. (1893). Description géologique des environs d'Angers. *Mém. Soc. nat. Agr. Sc. Arts Angers, Angers* **VI,** 209–335.

38 | Study of Cheilostome Polymorphism Using Principal Components Analysis

ALAN H. CHEETHAM

Department of Paleobiology, National Museum of Natural History, Smithsonian Institution, Washington, D.C., U.S.A.

Abstract: The hypothesis that continuous variation in zooid morphology has been parcelled out as polymorphic differences can be tested biometrically in cheilostomes in which groups of zooids are regularly segregated in identifiable regions of colonies. The numerous attributes by which polymorphs differ can be condensed by principal components analysis to a few size and shape factors. Application of this method to evolutionary lineages in *Poricellaria* and *Vincularia* has revealed similar series of gradual changes in both a monomorphic to polymorphic transformation and one in which polymorphism is reduced.

INTRODUCTION

The existence in Cheilostomata of various degrees of differentiation between polymorphic zooids, e.g. the a- and B-zooids of *Steganoporella* as compared to the vicarious avicularia and autozooids of onychocellids (Harmer, 1900), has led to the hypothesis that some if not all polymorphic species have evolved gradually from monomorphic ancestors (e.g. Harmer, 1926). This hypothesis is implicit in many works in which certain polymorphs, such as avicularia and kenozooids, are regarded as modified or reduced autozooids.

There are both biometric and morphologic difficulties in testing this hypothesis on phylogenetic lineages. These problems affect the choice of lineages to be studied, of characters to be measured, and of methods of analysis to be employed. This paper presents one multivariate approach to monomorphic-polymorphic transformations in *Poricellaria* and *Vincularia* that should be applicable to other studies of cheilostome polymorphism.

RELATION OF POLYMORPHISM TO CONTINUOUS MORPHOLOGIC VARIATION

It is taken for granted here, as it has been in other studies (e.g. Mackie, 1963), that morphologic differences between polymorphs correlate with functional specialization of zooids for division of labour, even where the functional basis of a particular polymorphic difference has not been determined. Given this correlation, recognition of evolutionary trends in polymorphism is a biometric problem, namely that of distinguishing discontinuous from continuous morphologic variation among zooids of the same ontogenetic and astogenetic status (Boardman *et al.*, 1970).

Morphologic variation among zooids in an evolutionary lineage can be recognized biometrically across a series of hierarchic levels, the most fundamental of which are the colony and the population. In continuous variation (Fig. 1a) numerous, slightly differing zooid morphologies form a broadly intergrading series within each colony presumably as a result of microenvironmental differences (Boardman *et al.*, 1970). Differences between colonies in cheilostomes, in part the result of genetic differences, are commonly small enough so that overlapping colony ranges of variation produce a continuous population distribution virtually like those of the component colonies. A particular zooid morphology thus cannot be correlated from colony to colony with a given microenvironmental value because of the incompletely separable effects of genetic differences (Boardman *et al.*, 1970).

In discontinuous variation (Fig. 1b, c) the broadly overlapping colony ranges are combined with separation of the zooids in each colony into two or more morphologically distinctive groups. Within each group there is commonly a certain amount of continuous variation which, together with the difference between colonies, may tend to obscure the distinction between groups at the population level. In most polymorphic cheilostomes, however, each polymorph occurs in relatively consistent form from colony to colony so that differences between autozooids and avicularia or between a- and B-zooids in *Steganoporella* are standardized within a species or even from genus to genus.

The standardization or distinctiveness of polymorphs at the population and higher levels has two aspects: (1) the mean difference between the polymorphs and (2) the extent of non-overlap between them. The total distinctiveness, but not necessarily both aspects, is affected by three factors of variation in zooid morphology (Fig. 2a–c): (a) the range of variation within each polymorph, (b) the mean difference between colonies, and (c) the total range of variation among all polymorphs within a colony. Decrease in either of the first two factors or increase in the third results in more distinctive polymorphs.

If polymorphs in cheilostomes evolved gradually, continuous variation in monomorphic ancestors must have been parcelled out to polymorphic descendants through some combination of changes in these three factors (Fig. 2d). Within such lineages, there should have been chronologically intermediate populations having morphologically intermediate stages of differentiation of

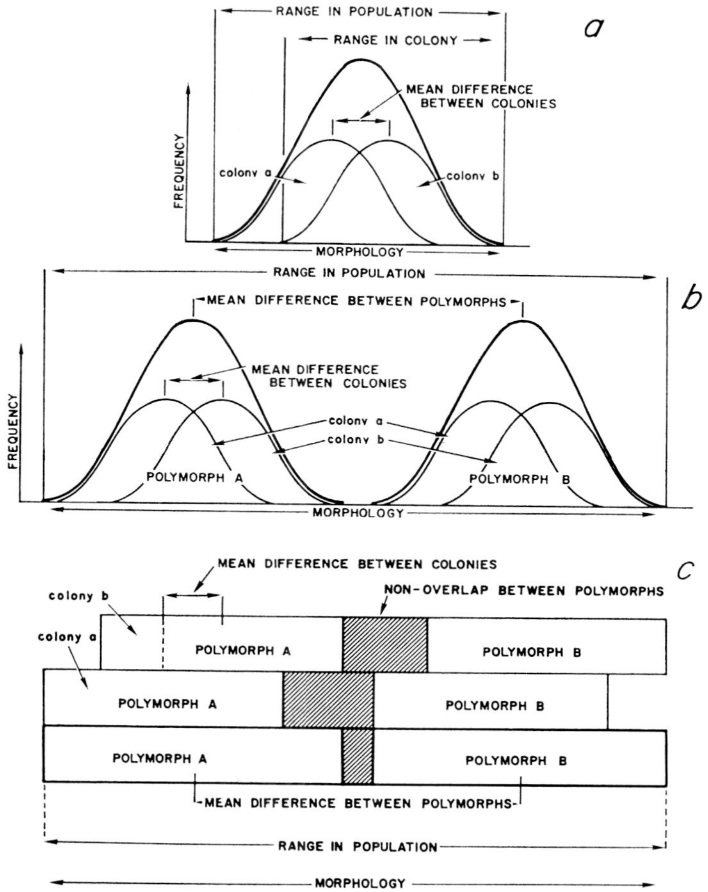

Fig. 1. Theoretical relationships between colony and population variation in zooid morphology. Each population is represented by just two colonies for simplicity of illustration, but more realistic, multiple-colony populations should be the same in principle. Colonies within a population are assumed to have different mean zooid morphologies but the same amount of variation. (a) Continuous variation, as in a monomorphic cheilostome. Population curve (———) is the sum of colony curves (———). (b) Discontinuous variation, as in a polymorphic cheilostome. Variation within each polymorph, A and B, is represented as equal to total variation in monomorphic population shown in a. Polymorphs have same mean difference in each colony and in whole population, even though mean values themselves differ. (c) Discontinuous variation in same population as shown in b, but with variation ranges of population (———) and colonies (———) represented by bars and non-overlap between polymorphs indicated by hatched parts of bars.

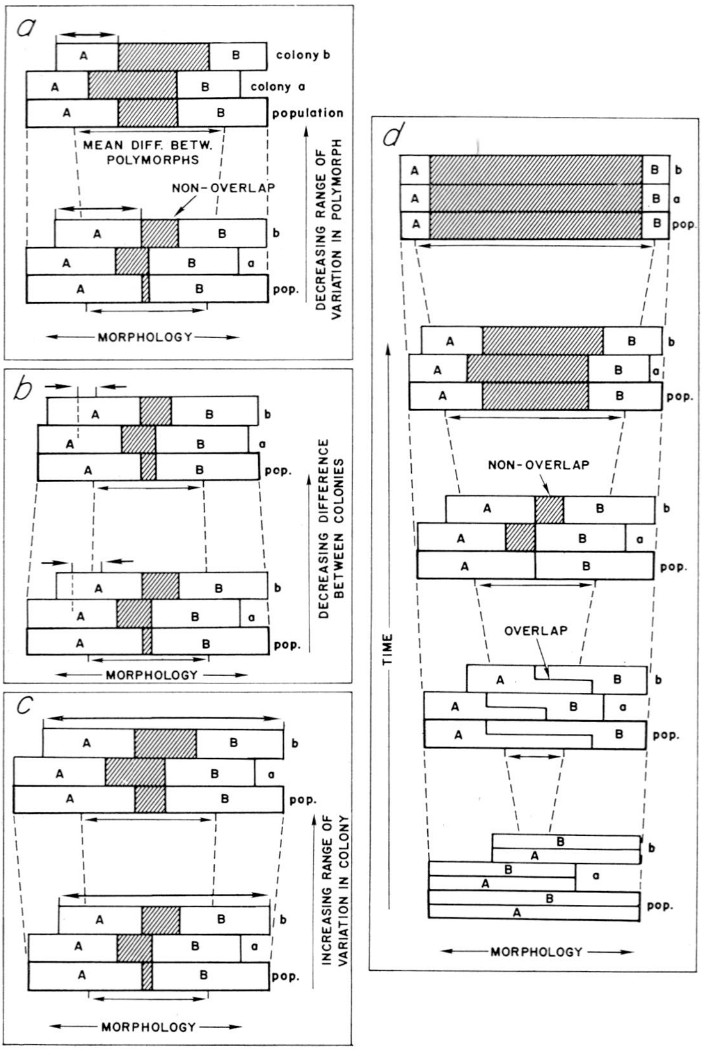

Fig. 2. Theoretical relationships of increasing polymorphism in zooid morphology. Variation ranges of colonies and populations are represented in simplified form as in Fig. 1c. (a–c) Separate effects of three factors increasing distinctiveness of polymorphs A and B. (a) Decrease in *range of variation in polymorphs* (◀——▶) produces significant increases in both mean difference and non-overlap between polymorphs; population variation remains constant. (b) Decrease in *mean difference between colonies* (—▶ ◀—) results in increase in non-overlap without change in mean difference between polymorphs; population range decreases. (c) Increase in *total variation range in colony* (◀——▶) produces significant increase in mean

zooids. The hypothesis can thus be tested by comparing intermediate populations with the theoretical relationships postulated in Fig. 2.

BIOMETRIC APPROACH TO STUDY OF POLYMORPHISM

Polymorphs in cheilostomes commonly differ in numerous morphologic attributes, many of which can be expressed as measurable variates, such as length and width of zooecium. Conventional uni- or bivariate methods of analysis may be not only inefficient for handling the number of variates but also insufficient for portraying the basic ways in which zooid morphologies differ. It seems intuitively reasonable to suppose that two basic properties—size and shape—underlie many of the measurable variates by which polymorphs differ.

A family of multivariate techniques has been used to study morphologic differences between individuals in various animal groups. Of these, principal components analysis has important advantages in providing morphologically interpretable results without requiring unlikely assumptions about the distribution of data (Morrison, 1967). This method extracts uncorrelated linear combinations (eigenvectors) of the measured variates in order of maximum variance (eigenvalue). With morphometric data, the first vector generally expresses size, and among the remaining vectors one or more express shape (Cock, 1966). The multivariate data can then be summarized with respect to size and shape vectors to compare directly zooids of different morphology.

A limitation of this method, as of any morphometric approach, is that all variates must have comparable morphologic significance on every polymorph. This requirement (homeomorphy of Cock, 1966) is easily met for some variates, such as length and width of zooecium. Many polymorphs, however, are heteromorphic with regard to characters of ovicells, polypides or musculature. Such heteromorphic differences can be handled by numerical coding (e.g. presence or absence), but only at the expense of some precision and generality.

Fig. 2. (*contd.*)
difference between polymorphs and less significant increase in non-overlap; population range increases. (d) Hypothetical evolution of polymorphic from monomorphic populations through combined effects of factors a–c. Overlap in zooid morphology is much greater at the population level than in colonies at early stages, so that gaps between polymorph ranges appear first in colonies and then in populations. This difference between colonies and populations is progressively reduced, disappearing entirely if colonies become identical. Population variation range increases only slightly. By varying the relative importance of factors, other patterns can be obtained. Emphasizing a or b decreases the difference in overlap or non-overlap between colonies and population more rapidly, whereas emphasizing c increases the population range more rapidly.

Principal components analysis has been used to study polymorphic differences, especially sexual dimorphism (e.g. Jolicoeur and Mosimann, 1960; Reyment, 1969), in several animal groups. In these studies eigenvectors have been calculated separately for each polymorph and those for different polymorphs compared. In the present study population samples consisting of monomorphic and polymorphic zooecia representing several evolutionary lineages within each genus have been referred to the same set of eigenvectors. Size and shape differences between zooids in a colony, between colonies in a population, and between populations in a lineage can thus be estimated and compared. If, as in the two examples discussed here, the first two eigenvectors account for most of the variation in all measured variates, then these comparisons can be shown in scattergrams similar to those made on two arbitrarily selected variates but having the advantage of presenting a known, major proportion of the total multivariate variation.

SIGNIFICANCE OF POSITIONAL POLYMORPHISM

To detect intermediate stages in zooid differentiation predicted by the hypothesis of gradual evolution of polymorphs, it is necessary that corresponding groups

---

Plate I

Figs. 1, 2. *Vincularia fragilis* Defrance. (All figures × 75).

Fig. 1. Frontal view of part of zoarial fragment in which frontal membranes of zooids, including their opercula, had calcified to form closures. Zooecia are arranged in four longitudinal series: left and right frontal zooecia curve toward each other along midline of zoarial fragment, and left and right lateral zooecia curve toward the frontals and are just visible on the lateral margins of the zoarial fragment in this view. Specimen is syntype of *Heterocella polymorpha* Canu (1907), Lutetian, Chaussy, France.

Fig. 2. View of two left lateral zooecia in another zoarial fragment in which frontal membranes were not calcified, showing morphologic details displayed by typical specimens of this genus (see measurements indicated in Fig. 8 in text). Specimen is Canu's hypotype (1907) Lutetian, Chaussy, France.

Figs 3, 4. *Poricellaria alata* d'Orbigny.

Fig. 3. Part of same zoarial fragment showing details of morphology of left lateral and left frontal zooecia (see measurements indicated in Fig. 3 in text); the distal lateral zooecium is ovicelled. Holotype 9656, d'Orbigny collection, Lutetian, Damery, France. All specimens illustrated are in the collections of the Institut de Paléontologie. Muséum National d'Histoire Naturelle, Paris, France.

Fig. 4. Frontal view of holotype (d'Orbigny 1854), showing frontal zooecia curved toward midline of zoarial fragment, and lateral zooecia curved toward frontals on their outer margins, the same arrangement as in *V. fragilis*.

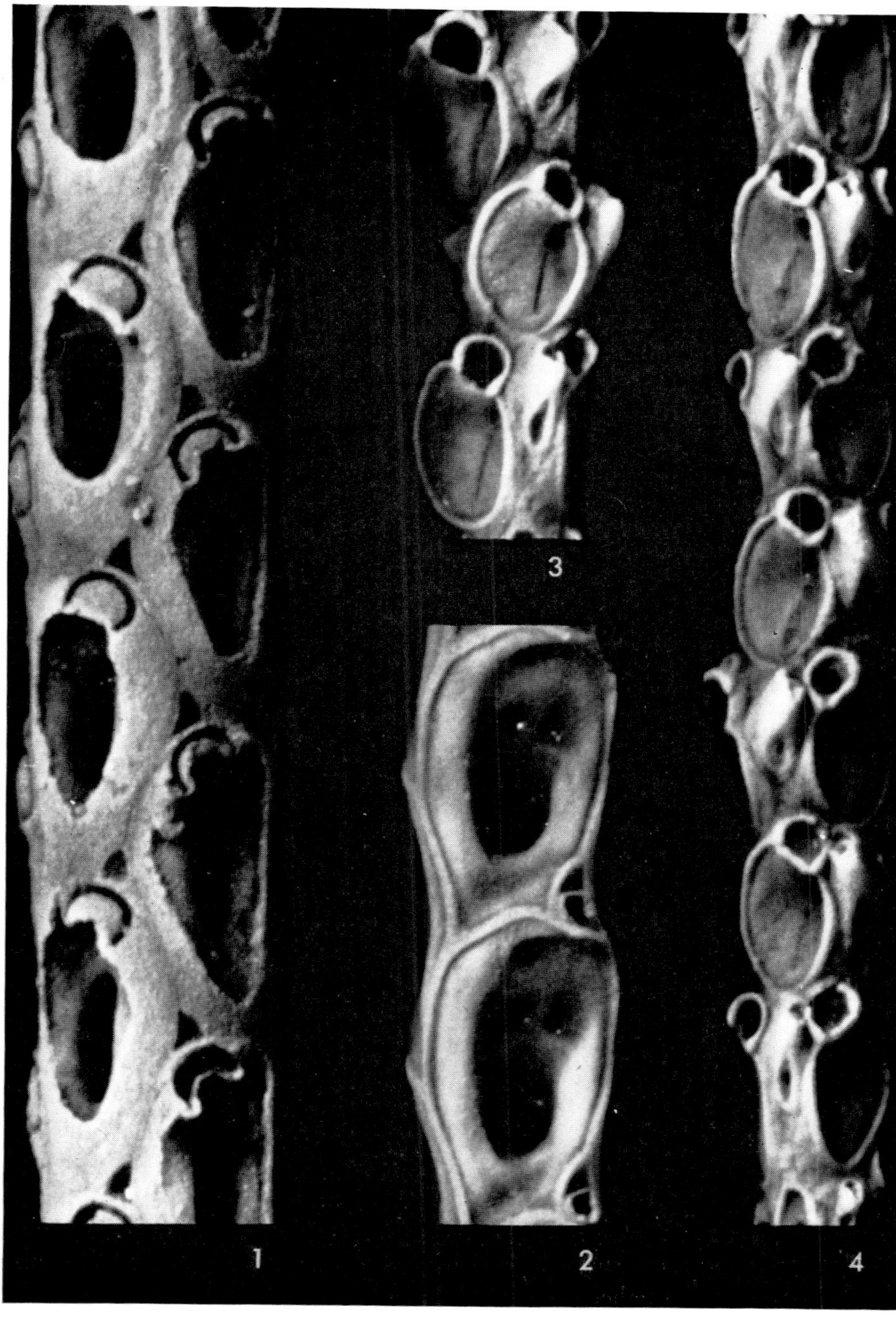
3
1
2
4

Table 1. *Poricellaria* samples and means of zooecial variates.

	Species	Sample(s)	No. Zoarial Fragments
1	*P.* n. sp.	Maastrichtian; Westmoreland Parish, Jamaica (Cheetham, 1968)	12
2	*P. diplodidymioides* (Meunier and Pergens)	Montian; Obourg boring, 21·8 m, Mons, Belgium (from R. Lagaaij)	5
3	*P.* cf. *P. limanowskii* (Canu)	Ypresian; Mont de Magny, near Gisors on road to Paris, France (Loeblich loc. 4703)	4
4	*P. articulata* (Waters)	Bartonian; Brendola and Montecchio Maggiore, Italy	8
5	*P. limanowskii* (Canu)	Lutetian; Orglandes (syntypes), Gourbesville, and Blaye, France	11
6	*P.* n. sp.	Rupelian; Chickasawhay Ls., St Stephens quarry, Alabama, U.S.A.	6
7	*P. complicata* (Reuss)	Rupelian; Gaas (Tartas), France	6
8	*P.* aff. *P. ratoniensis* (Waters)	Burdigalian; Baitoa Fm., Cercado de Mao, Dominican Republic	2
9	*P. ratoniensis* (Waters)	Holocene; Torres Straits; Seria, Brunei; Bombay, India	13
10	*P.* aff. *P. alata* d'Orbigny	"Sparnacian"; Carla-la-Rocquefort, France (Plaziat and Balavoine, 1964)	3
11	*P. alata* d'Orbigny	Ypresian; Mont de Magny (see sample 3)	4
12	*P. alata* d'Orbigny (=*P. negrisi* (Canu))	Lutetian; Damery (holotype), Orglandes, Gourbesville, Blaye, France. Bartonian; Listrac, Ludon, France (see Ducasse and Labracherie, 1967a, b)	19
13	*P. vernoni* Cheetham	Bartonian; Inglis Ls., Levy and Citrus Co.; Williston Ls., Levy Co.; Crystal River Ls., Alachua and Gilchrist Co., Florida, U.S.A.	8
14	*P.* n. sp.	Landenian; Vincentown, New Jersey, U.S.A.	12
15	*P.* cf. *P. alata* d'Orbigny	Landenian; DSDP Leg 12, Hole 117A, Rockall Plateau (see Cheetham and Håkansson)	2
16	*P.* cf. *P. alata* d'Orbigny	"Sparnacian"; Carla-la-Rocquefort (see sample 10)	3

of zooids be identifiable from colony to colony by some other criterion than their discontinuous morphologic variation (Fig. 2d). These groups of zooids must, of course, have the same ontogenetic and astogenetic status within their respective colonies. In diverse bryozoan groups, polymorphism is correlated with position in the budding pattern, with polymorphs regularly segregated as two or more concurrently developed groups of zooids making up separate regions of a zone of astogenetic repetition (Boardman *et al.*, 1970). If budding patterns are consistent enough, such groups of zooids can be compared from colony to colony, whether the zooids are monomorphic or polymorphic.

Means of all variates, except θ and α, are in $dk\mu$

No. Zooecia	Lz	lz	ho	lo	Lcr	lcr	Lg	Lop	θ	α	Lav	Zoarial Position
98	38·30	14·88	5·76	6·90	19·42	10·57	12·29	6·66	39°	2°	5·77	Lateral
	39·87	13·14	5·45	6·55	18·70	10·07	13·66	6·64	25°	10°	6·30	Frontal
20	50·58	18·23	5·48	6·23	22·95	12·94	22·71	5·95	45°	29°	6·48	Lateral
	50·15	17·38	5·42	6·42	19·17	15·09	23·94	5·19	32°	−8°	5·22	Frontal
19	46·58	17·85	5·36	6·21	18·45	10·54	19·04	8·93	49°	16°	7·51	Lateral
	44·86	15·58	5·01	5·57	15·11	10·96	20·21	8·50	27°	12°	6·89	Frontal
64	53·14	18·97	5·40	6·17	25·31	12·54	20·37	9·43	49°	8°	5·77	Lateral
	51·75	16·44	5·40	6·03	23·40	11·98	21·34	9·30	35°	5°	6·13	Frontal
92	40·88	19·55	5·51	6·09	21·11	13·71	13·46	8·28	49°	21°	6·84	Lateral
	40·72	16·72	5·08	5·82	18·27	13·38	14·43	6·72	31°	6°	7·33	Frontal
51	41·58	23·04	5·82	6·83	23·83	15·11	14·89	7·03	59°	71°	8·62	Lateral
	41·32	17·71	5·10	6·30	20·12	13·07	15·19	6·80	38°	57°	9·51	Frontal
43	39·30	22·01	5·95	7·41	24·65	15·73	11·95	6·66	67°	72°	9·23	Lateral
	39·73	17·36	4·65	6·64	19·72	13·68	14·63	6·12	38°	52°	7·92	Frontal
11	36·27	20·40	5·10	8·29	20·61	11·90	11·69	6·67	64°	57°	8·93	Lateral
	37·40	16·32	6·38	7·14	16·49	10·71	15·73	5·44	33°	44°	10·03	Frontal
143	38·00	26·58	6·95	8·11	27·21	17·75	5·94	6·59	66°	77°	9·55	Lateral
	37·06	21·68	6·19	7·57	18·71	15·51	12·64	5·73	36°	51°	8·39	Frontal
20	45·99	22·87	5·61	7·57	21·00	16·41	19·81	10·31	54°	16°	9·44	Lateral
	45·62	20·06	5·19	7·57	21·76	15·47	19·17	9·88	39°	3°	4·86	Frontal
14	59·50	18·13	6·23	7·08	23·66	14·59	23·66	16·01	32°	14°	8·22	Lateral
	58·41	16·68	6·27	6·91	19·98	13·60	25·14	13·39	25°	15°	8·36	Frontal
99	53·36	20·07	6·79	7·54	23·83	16·05	19·47	12·44	25°	11°	8·99	Lateral
	51·86	19·10	6·65	7·23	22·98	15·15	20·78	11·06	21°	2°	9·33	Frontal
42	82·09	30·60	8·05	8·63	38·85	20·97	30·90	3·87	40°	15°	7·96	Lateral
	82·69	27·56	8·20	8·33	34·54	20·70	38·12	3·45	29°	6°	7·73	Frontal
57	57·64	15·79	5·69	7·01	28·26	9·83	21·80	16·39	34°	25°	7·33	Lateral
	59·77	13·39	5·60	6·59	29·93	8·79	22·91	16·07	18°	−4°	6·41	Frontal
8	61·20	19·98	7·65	9·35	26·35	18·28	26·78	7·65	41°	12°	7·65	Lateral
	56·95	20·83	5·95	8·50	23·80	17·85	22·10	4·68	30°	2°	7·23	Frontal
18	68·00	20·78	5·95	9·07	32·68	16·62	28·43	17·96	50°	18°	8·78	Lateral
	65·73	19·08	6·48	9·44	30·03	13·98	26·92	14·81	40°	1°	6·80	Frontal

In many cheilostomes budding patterns are slightly irregular, and consequently polymorphs appear sporadically distributed. A degree of regularity is achieved in genera such as *Thalamoporella* and *Steganoporella*, in which certain polymorphs are associated with bifurcation of lineal series (e.g. Silén, 1938; Powell and Cook, 1966), or *Coscinopleura*, in which they are associated with marginal series in bilaminate colonies (see Voigt, 1956). *Poricellaria* and *Vincularia* exhibit a higher regularity in having a uniform number of lineal series throughout each zone of astogenetic repetition, with each series identifiable in every colony because of the direction of curvature of its zoids (Plate I).

Abundant material of these last two genera has proven to be ideal for this kind of analysis.

EVOLUTION OF POLYMORPHISM IN *Poricellaria*

Principal components analysis of samples of *Poricellaria* ranging in age from late Cretaceous (Maastrichtian) to Holocene (Table 1) has revealed a more complex evolutionary history than was previously suggested by uni- and bivariate analyses of a portion of this material (Cheetham, 1968; Boardman and Cheetham, 1969). Morphometric factors were extracted from a correlation matrix including measurements of both lateral and frontal zooecia in all samples combined. Four factors, contributing more than three-fourths of the total variation (Table 2), can be interpreted as follows: F_1, zooecial size expressing covariation in zooecial length and width, opesial length and width, and cryptocyst length and width; F_2, zooecial shape expressing opposing variation in the angles of zooecial curvature and avicularian orientation on the one hand and zooecial and gymnocyst length on the other; F_3, avicularian length; and F_4, opesiular length. The first two factors, accounting for nearly two-thirds of the total variation, thus describe the form of the zooecia in *Poricellaria*.

Figure 3 shows the general distribution of *Poricellaria* samples, with lateral and frontal zooecia combined, with respect to F_1 and F_2 and the morphologic meaning of these two factors. Variation on the F_1 axis expresses size differences without changes in shape; that on the F_2 axis represents shape changes, more asymmetrical zooecia lying in the positive direction and more elongate ones in the negative direction. Plotted on these axes, samples group in two main morphologic series, each of which represents a chronologic trend. Examined in greater detail (Figs 4, 5), each series is seen to consist of at least two lineages having different rates of evolution but trending in the same morphologic direction. (Lineages within a series follow different trends with respect to F_3 and F_4 which have not been plotted here.) In Lineages A and B there is progressive evolution from small, slightly elongate zooecia to large, highly asymmetrical ones along a distinctly allometric gradient. In Lineages C and D zooecia changed little in shape, remaining elongate, even though they increased in size more rapidly than those in Lineages A and B; this phyletic trend was thus approximately isometric.

Evolutionary trends in the two main series of *Poricellaria* differ not only in whether or not they are allometric, but also in their relative expression in zooecia belonging to lateral and frontal groups. In Lineages C and D (Fig. 4) the two groups evolved at subequal rates so that lateral and frontal zooecia did not differentiate morphologically and populations remained monomorphic. In

Table 2. Correlation between measured variates and significant underlying factors by principal components analysis of *Poricellaria* samples listed in Table 1. Factors were extracted from the correlation matrix based on measurements of frontal and lateral zooecia combined. Data were standardized with pooled mean and standard deviation for each variate. Values of θ and α were transformed to radians $\times 100$. Eigenvectors were calculated to have the length of their corresponding eigenvalue

Factor	Eigen-value	Variance % tot.	Variance Cum.	Lz	lz	ho	lo	Lcr	lcr	Lg	Lop	θ	α	Lav
				Eigenvector (factor loadings)										
F_1	3·664	33·3	33·3	+0·56	+0·85	+0·70	+0·70	+0·76	+0·82	+0·32	0·00	+0·34	+0·26	+0·32
F_2	2·837	25·8	59·1	−0·77	+0·34	−0·11	+0·04	−0·30	+0·18	−0·80	−0·48	+0·66	+0·78	+0·28
F_3	1·058	9·6	68·7	−0·17	−0·15	+0·34	+0·31	−0·26	0·00	−0·19	+0·24	−0·46	−0·07	+0·65
F_4	0·976	8·9	77·6	+0·12	+0·10	−0·07	−0·25	−0·21	+0·19	+0·29	−0·74	−0·23	−0·13	+0·31
Communalities		Factors F_1–F_3		0·933	0·870	0·614	0·583	0·744	0·702	0·776	0·286	0·760	0·686	0·602
Communalities		Factors F_1–F_4		0·946	0·879	0·619	0·646	0·788	0·739	0·861	0·838	0·813	0·703	0·697

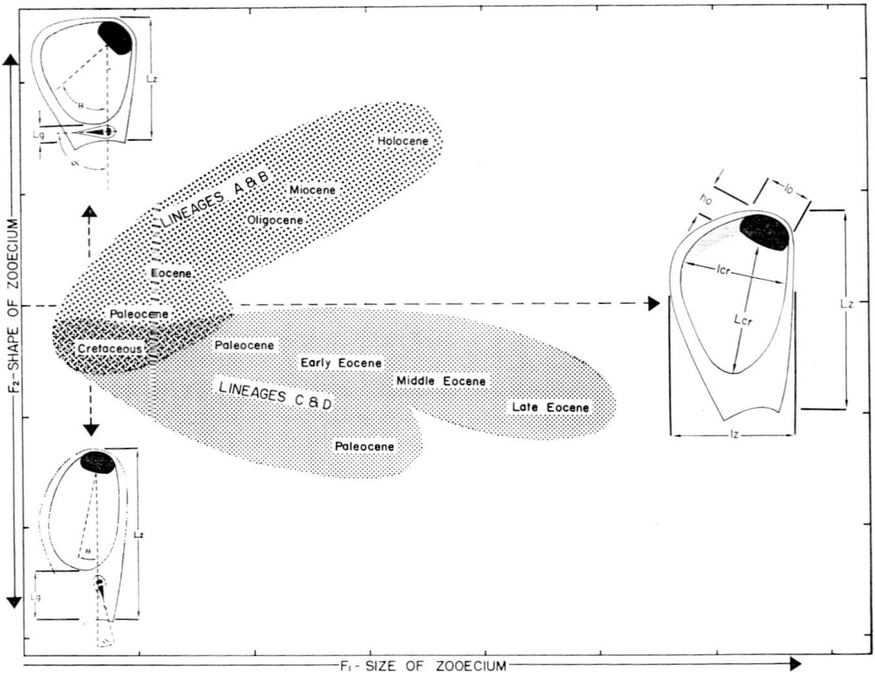

Fig. 3. Generalized evolutionary trends in form of zooecia in *Poricellaria*. F_1 and F_2 axes are first two principal components with equal scales in standardized, dimensionless units (see Table 2). Size and shape of zooecia near end points of axes are represented diagrammatically, with significant variates indicated by conventional symbols (not shown are avicularian length, Lav, and opesiular length, Lop, neither of which is significantly correlated with F_1 or F_2). Actual dimensions of zooecia in each sample are indicated by mean values in Table 1.

Lineages A and B (Fig. 5) lateral zooecia evolved more rapidly along the same allometric trend as the frontal ones so that the two groups of zooecia have become completely non-overlapping in living populations.

Still more detailed examination of samples belonging to Lineages A and B (Figs 6, 7) reveals overall conformity to the gradual evolution of polymorphic from monomorphic populations shown in Fig. 2. The rates of morphologic change may not, however, have been as constant as postulated.

The Cretaceous and Holocene end members of this evolutionary series (Fig. 6) show a high degree of conformity with postulated changes in variation ranges.

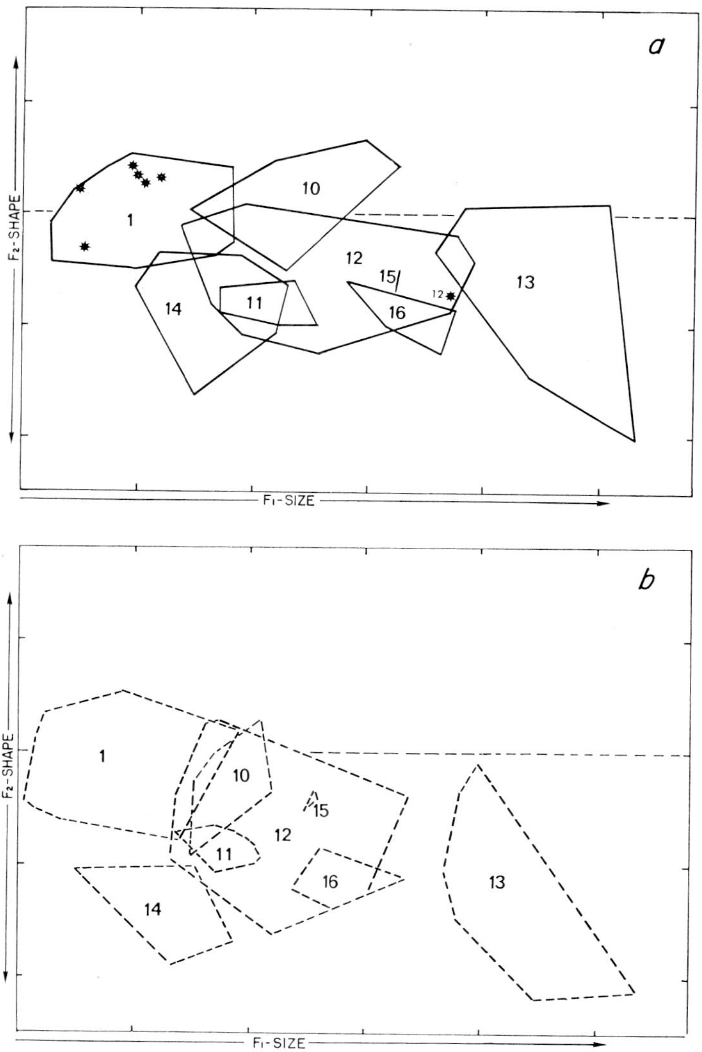

Fig. 4. Scatter polygons of zooecial size (F_1) and shape (F_2) for samples of *Poricellaria* Lineages C and D. Samples are numbered as in Table 1. Scale and origin of axes are the same as in Fig. 3. For zooecia missing one or more measurements because of incomplete preservation or failure to develop an avicularium, sample means were used. Stars indicate ovicelled zooecia. (a) Lateral zooecia; note that ovicelled zooecia fall within range of non-ovicelled ones in same sample. (b) Frontal zooecia.

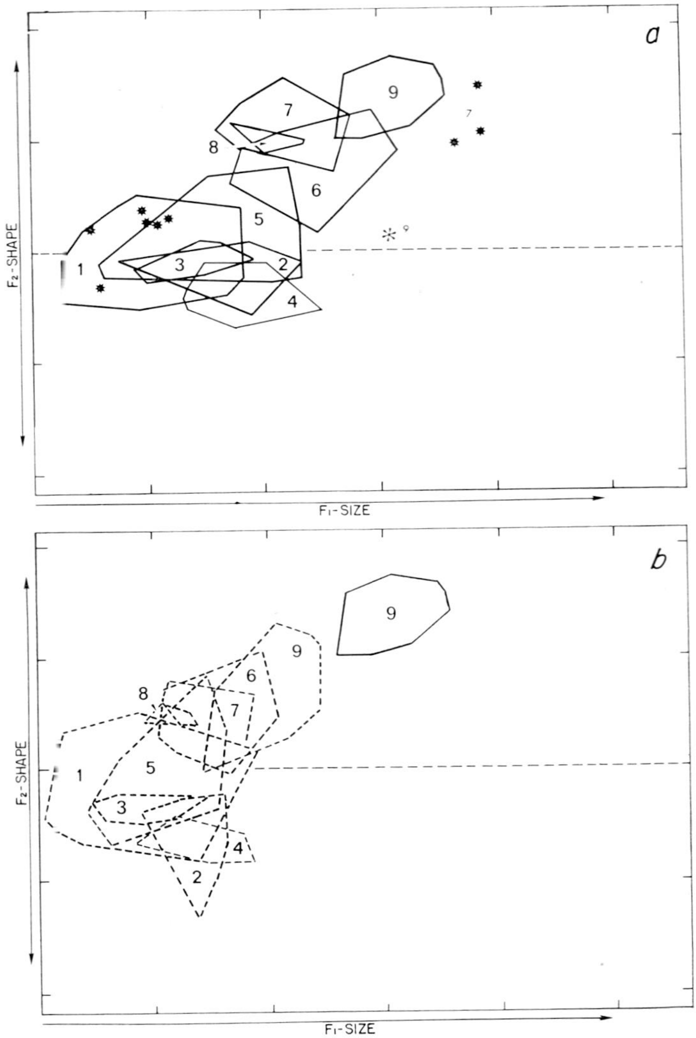

Fig. 5. Scatter polygons of zooecial size (F_1) and shape (F_2) for samples of *Poricellaria* Lineages A and B. Samples are numbered as in Table 1. Scale and origin of axes are the same as in Figs 3 and 4. Sample means substituted for missing data as in Fig. 4. Small stars ovicelled zooecia; * represents aberrant zooecium in sample 9 (see Fig. 6). (a) Lateral zooecia; note that ovicelled zooecia fall within range of non-ovicelled ones except in sample 7. (b) Frontal zooecia, with laterals of sample 9 added to show dimorphism.

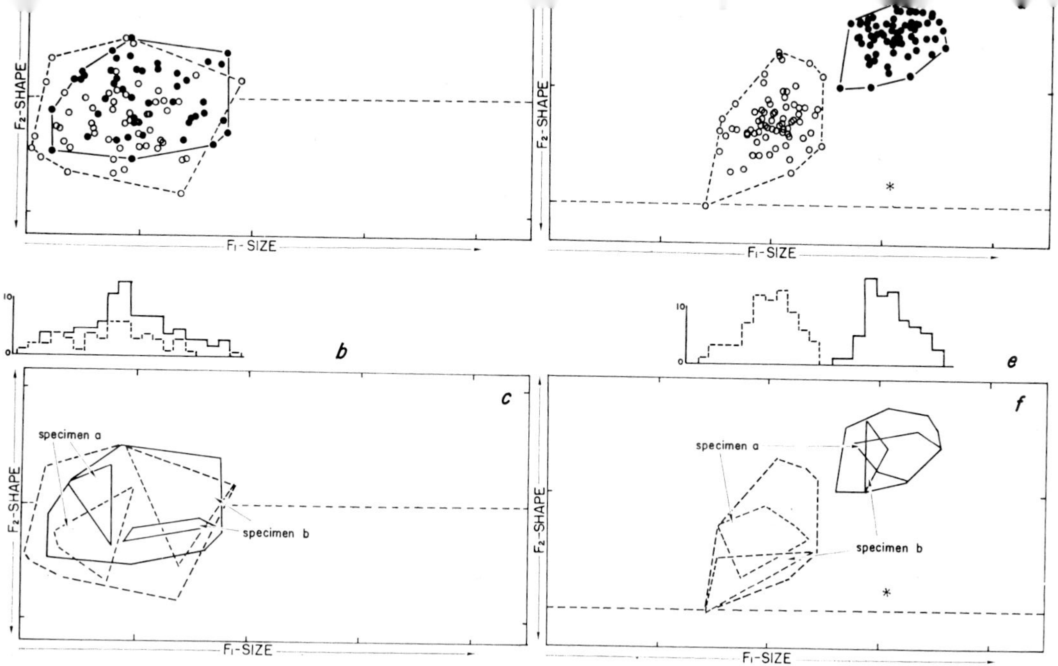

Fig. 6. Variation in zooecial size (F_1) and shape (F_2) in Cretaceous and Holocene representatives of Lineages A and B of *Poricellaria*. Variation ranges of samples and selected zoarial fragments are indicated by areas of scatter polygons. Overlap or non-overlap of lateral zooecia (● ——) and frontal zooecia (○ – – – –) are indicated by scatter polygons and by histograms. Origin and scale of axes are the same as in Figs 3–5. (a–c) *P.* n. sp., Maastrichtian (sample 1, Table 1). (a) Scattergram and scatter polygons for whole sample. (b) Size histogram. (c) Comparison of scatter polygons of whole sample with those of two zoarial fragments. (d–f) *P. ratoniensis* (Waters), Holocene (three pooled samples, 9, Table 1). Lateral zooecium indicated by star occupies position of two zooids in budding pattern. (d) Scattergram and scatter polygons for pooled samples. (e) Size histogram. (f) Comparison of scatter polygons of pooled sample with those of two zoarial fragments: specimen a, Seria, Brunei; specimen b, Bombay, India.

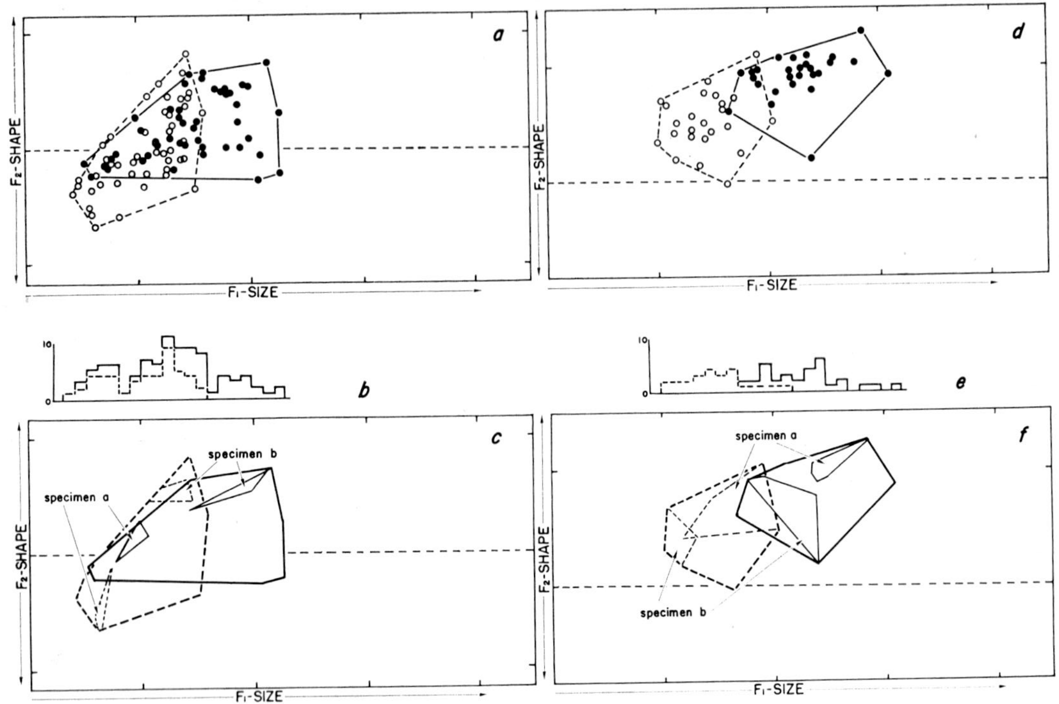

Fig. 7. Variation in zooecial size (F_1) and shape (F_2) in Eocene and Oligocene representatives of Lineages A and B of *Poricellaria*. Variation ranges are indicated as in Fig. 6. Origin and scale of axes are the same as in Figs 3–6. (a–c) *P. limanowskii* (Canu), Lutetian (three pooled samples, 5, Table 1). (a) Scattergram and scatter polygons for pooled samples. (b) Size histogram. (c) Comparison of scatter polygons of pooled samples with those of two zoarial fragments: specimen a, Gourbesville, France; specimen b, Orglandes, France. (d–f) *P.* n. sp., Rupelian (sample 6, Table 1). (d) Scattergram and scatter polygons for whole sample. (e) Size histogram. (f) Comparison of scatter polygons of whole sample with those of two zoarial fragments.

Lateral and frontal zooecia overlap almost completely in the Cretaceous sample and not at all in the Holocene material, even though the total variation range of all zooecia is subequal in the two. Compared to the Cretaceous sample, the Holocene material shows notably greater colony variation ranges, smaller ranges of variation within lateral or frontal sets of zooecia, and smaller differences between colonies.

Samples chronologically intermediate (Fig. 7) between the end members show progressively less overlap between lateral and frontal zooecia and apparently subequal total variation ranges, also in conformity with Fig. 2. The rates of change of within- and between-colony differences, however, appear to have been less nearly constant than postulated. Ranges of variation of laterals and frontals within colonies, even allowing for the small number of zooecia in some of the Eocene zoarial fragments (Fig. 7c), seemingly decreased rapidly at first and then levelled off between Eocene and Oligocene. Total ranges of variation within colonies appear to have remained small and differences between colonies large until the Oligocene, when more rapid changes were initiated. The chief result of these apparently varying rates of change was that lateral and frontal zooecia became non-overlapping polymorphs within colonies even sooner relative to the population level of differentiation than predicted by the sequence in Fig. 2d.

REDUCTION OF POLYMORPHISM IN *Vincularia*

Evolutionary loss of polymorphism is predicted by the theory of division of work in social insects (Wilson, 1968). Comparison of variation ranges in monomorphic and polymorphic species of ants has led to the inference that loss of polymorphism results from a dropping out of one or more polymorphs (Wilson, 1953). Principal components analysis of samples of *Vincularia* ranging in age from early Eocene (Ypresian) to middle Oligocene (Rupelian) (Table 3) has revealed a pattern of gradually decreasing polymorphism through a series of changes in variation that chronologically are the reverse of those shown by Lineages A and B of *Poricellaria*. This pattern contrasts with that in ants.

Four morphometric factors (Table 4), contributing more than three-fourths of the total variation, are similar to those in *Poricellaria*: F_1 zooecial size expressing covariation in zooecial length and width and gymnocyst length; F_2, zooecial shape expressing opposing variation in zooecial length and cryptocyst length on the one hand, and angle of zooecial curvature on the other; F_3, avicularian length with some covariation in zooecial curvature; and F_4, avicularian orientation. Again, the first two factors, accounting for more than half the total variation, describe the form of the zooecia in *Vincularia*.

Table 3. *Vincularia* samples and means of zooecial

	Species	Sample(s)	No. Zoarial Fragments
1	*V.* n. sp.	Ypresian; Mont de Magny, near Gisors on road to Paris, France (Loeblich loc. 4703)	6
2	*V. fragilis* Defrance*	Lutetian; Chaussy, Orglandes, Gourbesville, France	9
3	*V.* aff. *V. fragilis* Defrance	Auversian; Upper Bracklesham Beds, Selsey, Sussex, G.B.; Calcaire de St-Estéphe, Coupe-du Bois-de-Barbe, Villeneuve de Blaye (Gironde), France	8
4	*V.* n. sp.	Bartonian; Williston Ls., Crystal River Ls., Alachua Co.; Inglis Ls., Levy Co., Florida, U.S.A.	6
5	*V.* n. sp.	Rupelian; Gaas (Tartas), France	9
6	*V. vicksburgica* (Canu and Bassler)	Rupelian; Marianna Ls., Salt Mt., Jackson, Alabama (syntypes); St Stephens, Washington Co., Alabama, U.S.A.	9
7	*V. subsymmetrica* (Canu)	Lutetian; Chaussy, France (syntypes)	3
8	*V. grandis* (Canu)	Lutetian; Chaussy, France (holotype)	1

* = *Heterocella monstruosa* Canu; = *H. polymorpha* Canu (part).

The general distribution of *Vincularia* samples is shown in Fig. 8, with larger zooecia lying to the right, more asymmetrical ones above, and more elongate ones below. As in *Poricellaria*, two main morphologic series are apparent, one (Lineage A) lying in a distinctly allometric orientation from small, elongate zooecia at one end to large, highly asymmetrical ones at the other. The chronologic relationships along this series, however, are more complicated than in the corresponding lineages of *Poricellaria*, with an apparent reversal after the middle Eocene.

Within Lineage A of *Vincularia*, frontal zooecia evolved very little (Fig. 9b),

Table 4. Correlation between measured variates and significant underlying factors by principal components analysis of *Vincularia* samples listed in Table 3. All calculations were made as for Table 2

Factor	Eigen-value	Variance % tot.	Variance Cum.	Eigenvector (factor loadings) Lz	lz	Lcr	Lg	θ	α	Lav
F_1	2·099	30·0	30·0	+0·68	+0·82	−0·45	+0·63	+0·25	−0·43	+0·41
F_2	1·527	21·8	51·8	+0·59	+0·11	+0·57	+0·04	−0·70	+0·46	+0·26
F_3	1·199	17·1	68·9	+0·01	−0·12	+0·28	−0·31	+0·54	+0·09	+0·78
F_4	1·041	14·9	83·8	−0·18	−0·08	−0·45	+0·33	+0·01	+0·70	+0·20
Communalities	Factors F_1–F_3			0·811	0·699	0·606	0·495	0·844	0·405	0·844
	Factors F_1–F_4			0·843	0·705	0·808	0·604	0·844	0·895	0·884

variates. Means of all variates, except θ and α, are in $dk\mu$

No. Zooecia	Lz	lz	Lcr	Lg	θ	α	Lav	Zoarial Position
71	49·99	25·11	13·35	1·01	42°	11°	10·87	Lateral
	49·53	20·38	14·78	1·05	16°	11°	9·43	Frontal
135	53·85	35·39	8·03	2·72	46°	−31°	11·34	Lateral
	52·06	28·61	11·75	1·24	27°	−32°	7·65	Frontal
123	49·48	28·67	10·64	1·56	50°	−32°	10·30	Lateral
	49·01	25·00	10·91	1·58	28°	−21°	8·83	Frontal
42	55·05	24·33	7·23	1·56	59°	−34°	9·35	Lateral
	54·88	22·22	9·78	1·43	37°	−33°	9·78	Frontal
59	50·53	27·81	8·23	1·06	40°	−25°	10·46	Lateral
	49·47	21·90	8·91	1·04	25°	−22°	9·35	Frontal
83	58·72	27·40	13·82	2·11	46°	−27°	11·33	Lateral
	58·58	23·61	15·51	3·85	30°	−25°	12·66	Frontal
33	70·65	40·57	13·42	2·24	12°	absent	absent	Lateral
	70·27	37·92	15·25	2·40	7°	−23°	9·35	Frontal
12	97·47	45·19	8·08	4·82	26°	−23°	15·30	Lateral
	97·89	40·80	9·78	3·97	17°	−15°	17·85	Frontal

whereas lateral zooecia rapidly became larger and more asymmetrical and then successively trended back toward smaller and elongate (Fig. 9a). As a result, lateral and frontal zooecia, at first morphologically distinct, became successively less so.

Examined in greater detail (Fig. 10), the evolutionary trend reducing polymorphism in *Vincularia* is seen to have been accompanied by changes that reverse those shown in Fig. 2d. Lateral and frontal zooecia are slightly less distinctive in early Eocene *Vincularia* than in Holocene *Poricellaria*, and they overlap considerably less in Oligocene *Vincularia* than in Cretaceous *Poricellaria*; consequently, the series of changes in *Vincularia* encompasses a smaller part of the sequence in Fig. 2d than do those in *Poricellaria*. However, a general decrease in the range of variation within colonies and an increase in differences between colonies are evident. An increase in variation ranges of lateral and frontal zooecia within colonies is not apparent, perhaps because the trend had not gone far enough by Oligocene time to mirror the early phase of rapid decrease in the ranges of each zooecial group shown by *Poricellaria*. (Miocene specimens of *Vincularia* have been reported by Tewari and Srivastava, 1967, as *Nellia kutchensis*, and by Lagaaij, 1968, but none were available for this analysis.)

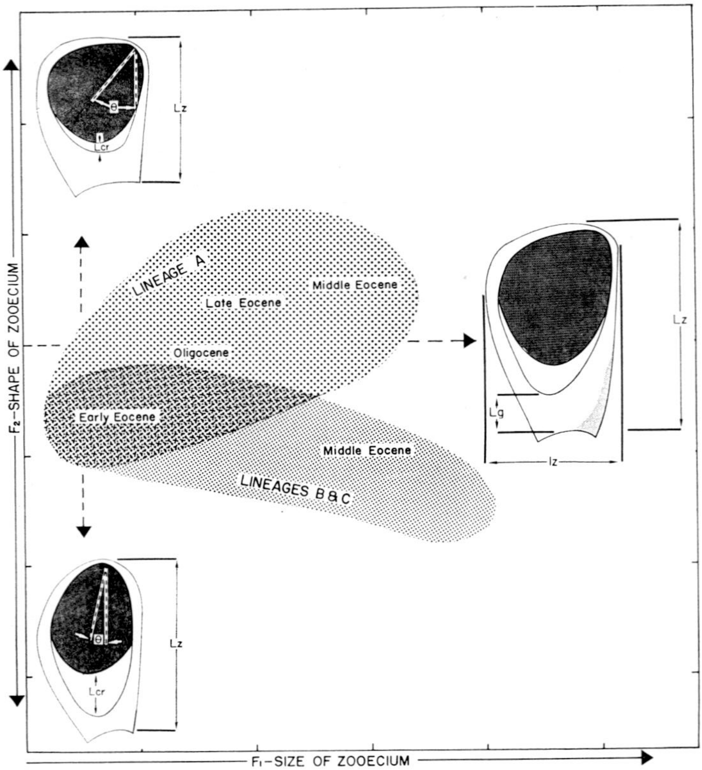

Fig. 8. Generalized evolutionary trends in form of zooecia in *Vincularia*. Representation as in Fig. 3, but from principal components analysis of Table 4. Zooecia near end points of axes are shown diagrammatically, with significant variates indicated by conventional symbols (not shown are avicularian length, Lav, and avicularian orientation, α, neither of which is significantly correlated with F_1 or F_2). Actual dimensions of zooecia in each sample are indicated by mean values in Table 3.

It is interesting to note that it is just with regard to their stages of polymorphism that early Eocene *Vincularia* is like Miocene *Poricellaria* and Oligocene *Vincularia* is like Paleocene *Poricellaria* of Lineage A. With respect to the size and to a lesser extent the shape of their zooecia (cf. Figs 3, 8), coeval populations tend to be more nearly alike.

CONCLUSIONS

Evolutionary trends in morphologic variation in populations of *Poricellaria* and *Vincularia* are consistent with gradual transformations between monomorphic and polymorphic zooecia, whether the transformation was progressive or retro-

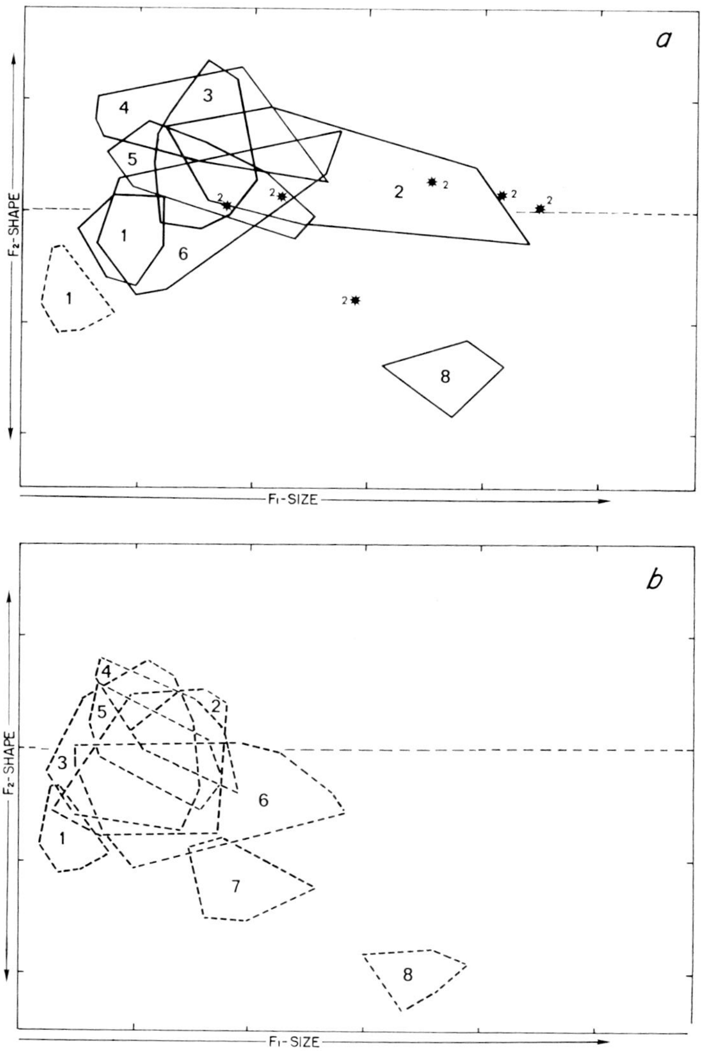

Fig. 9. Scatter polygons of zooecial size (F_1) and shape (F_2) for samples of *Vincularia*. Samples are numbered as in Table 3. Scale and origin of axes are the same as in Fig. 8. Sample means were substituted for missing data. Stars indicate ovicelled zooecia. (a) Lateral zooecia, with frontals of sample 1 added to show dimorphism. Note that ovicelled zooecia fall within or just outside ranges of non-ovicelled ones in same sample. Laterals of sample 7 could not be plotted as they all lack avicularia. (b) Frontal zooecia.

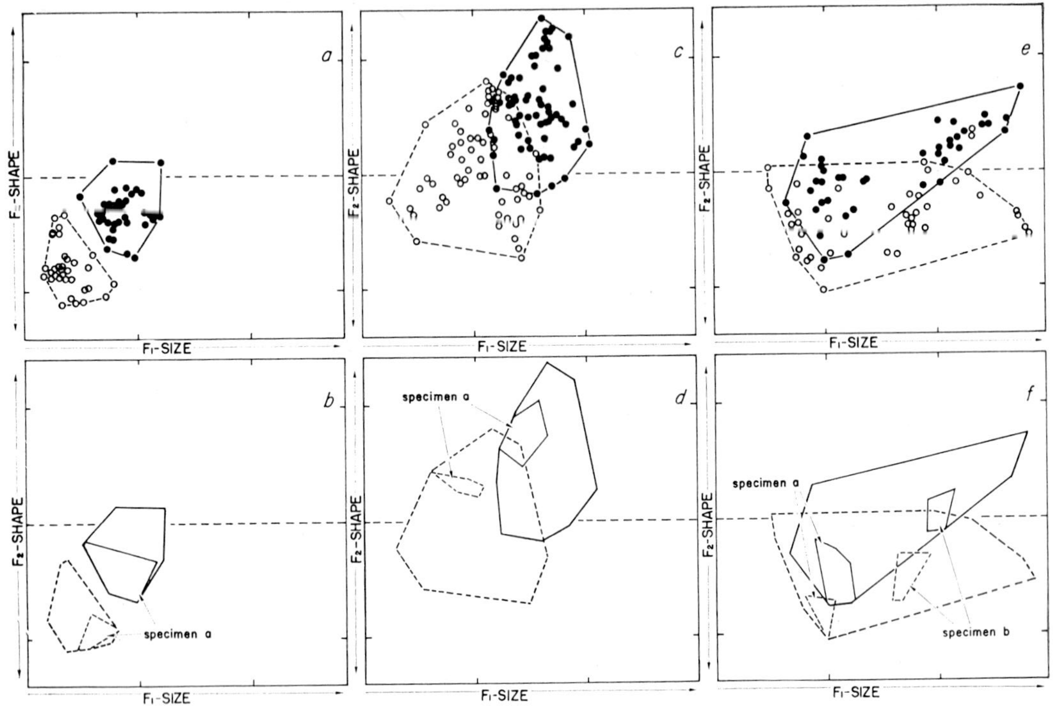

Fig. 10. Variation in zooecial size (F_1) and shape (F_2) in representatives of Lineage A of *Vincularia*. Variation ranges are represented as in Fig. 6. Origin and scale of axes are the same as in Figs 8, 9. (a, b) *V.* n. sp., Ypresian (sample 1, Table 3). (a) Scattergram and scatter polygons for whole sample. (b) Comparison of scatter polygons of whole sample with that of a single zoarial fragment. (c, d) *V.* aff. *V. fragilis* Defrance, Auversian (three pooled samples, 3, Table 3). (c) Scattergram and scatter polygons for pooled samples. (d) Comparison of scatter polygons of pooled samples with that of single zoarial fragment from Selsey, Sussex, G.B. (e, f) *V. vicksburgica* (Canu and Bassler), Rupelian (two pooled samples, 6, Table 3). (e) Scattergram and scatter polygons for pooled samples. (f) Comparison of polygons of pooled samples with those of two zoarial fragments: specimen a, St Stephens, Alabama, U.S.A.; specimen

gressive. In progressive evolution, distinctive polymorphs developed first within colonies, then through whole populations, by early decrease in variation ranges of groups of zooids within colonies and by later increase in total colony variation ranges together with decrease in differences between colonies. In retrogressive evolution, distinctiveness of polymorphs was reduced first at the population level and then within colonies by early decrease in colony variation ranges and increase in differences between colonies; later increase in variation ranges of polymorphs within colonies has not been observed, perhaps because the available material stopped short of attaining fully monomorphic status.

Are the gradual monomorphic-polymorphic transformations in these two genera typical of those in cheilostomes generally, or have some kinds of polymorphs, perhaps some types of heterozooids, appeared or disappeared by quantum evolution? A practical difficulty in answering this question with the method used here is that of heteromorphy, discussed above. If gradualism is the rule, however, there must exist segments of lineages over which polymorphs are sufficiently similar for identification of trends similar to those illustrated here.

Finally, a comment can be made on the possible environmental significance of these transformations in *Poricellaria* and *Vincularia*. The theory of division of work in social insects, a theory which has recently been applied to cheilostome polymorphism by Schopf (1971), predicts that monomorphism should be favoured under fluctuating conditions and polymorphism under stable ones (Wilson, 1968). This suggests that the opposing trends in *Poricellaria* and *Vincularia* were influenced by their adaptation to different environments. Much of the material studied apparently represents sympatric species of the two genera (cf. Tables 1, 3); therefore each genus must have been responding to a different facet of the same general environment.

ACKNOWLEDGEMENTS

Miss JoAnn Sanner, Smithsonian Institution, measured specimens, prepared data for the computer, plotted results, and drafted the figures. Dr H. D. Roth, Smithsonian Institution, wrote the principal components analysis program and provided computations on the Smithsonian CDC6400 computer. Drs R. S. Boardman and H. D. Roth of the Smithsonian and Dr W. C. Banta of American University reviewed the manuscript and provided suggestions for its improvement. Dr E. Buge of the Museum National d'Histoire Naturelle, Miss P. L. Cook of the British Museum (Natural History), Mlle M. Labracherie and Mme J. Prud'homme of the Universite de Bordeaux, Dr R. Lagaaij of Wassenaar, The Netherlands, and Dr P. A. Sandberg of the University of Illinois all provided most valuable assistance in making specimens available for this study. Research was supported by a grant from the Smithsonian Research Foundation. It is a pleasure to acknowledge all of this help.

REFERENCES

Boardman, R. S. and Cheetham, A. H. (1969). Skeletal growth, intracolony variation, and evolution in Bryozoa: a review. *J. Paleont.* **43,** 205–233.

Boardman, R. S., Cheetham, A. H. and Cook, P. L. (1970). Intracolony variation and the genus concept in Bryozoa. *Proceedings of the North American Paleontological Convention*, 294–320.

Canu, F. (1907). Les Bryozoaires des terrains tertiaires des environs de Paris. *Annls Paléont.* **2,** 57–88

Cheetham, A. H. (1968). Evolution of zooecial asymmetry and origin of poricellariid cheilostomes. Proceedings of the First International Conference on Bryozoa. *Atti. Soc. ital. Sci. nat.* **108,** 185–194.

Cheetham, A. H. and Håkansson, E. (In preparation). Preliminary report on Bryozoa from DSDP Leg 12, Site 117, Holes 117 and 117A. *Initial Reports of the Deep Sea Drilling Program.*

Cock, A. G. (1966). Genetical aspects of metrical growth and form in animals. *Q. Rev. Biol.* **41,** 131–190.

d'Orbigny, A. (1854). "*Paléontologie Francaise, Terrains Crétacès*," Vol. 5, "*Bryozoaires*". pp. 985–1192. Masson, Paris.

Ducasse, O. and Labracherie, M. (1967a). Etude paléontologique et stratigraphique des quelques affleurements du Medoc. *Bull. Inst. géol. Bassin d'Aquitaine* **3,** 21–28.

Ducasse, O. and Labracherie, M. (1967b). Observations paléontologiques sur la coupe profonde de Ludon (Gironde). *Bull. Inst. géol. Bassin d'Aquitaine* **3,** 35–58.

Harmer, S. F (1900). A revision of the genus *Steganoporella. Q. J. microsc. Sci.* (n. ser.) **43,** 225–298.

Harmer, S. F. (1926). The Polyzoa of the Siboga Expedition, part II, Cheilostomata Anasca. *Siboga Exped.* **28b,** 181–501.

Jolicoeur, P. and Mosimann, J. E. (1960). Size and shape variation in the painted turtle: a principal component analysis. *Growth* **24,** 339–354.

Lagaaij, R. (1968). Fossil Bryozoa reveal long-distance sand transport along the Dutch coast. *K. Ned. Akad. Wet. Amsterdam, Proc.* (ser. B), **71,** 31–50.

Mackie, G. O. (1963). Siphonophores, bud colonies, and superorganisms. *In* "The Lower Metazoa, Comparative Biology and Phylogeny", (Dougherty, E. C. ed.), pp. 329–337. University of California Press, Berkeley and Los Angeles.

Morrison, D. F. (1967). "Multivariate Statistical Methods." 338 pp. McGraw-Hill, New York.

Plaziat, J. C. and Balavoine, P. (1964). Bryozoaires du "Sparnacien" des Corbieres et du Plantaurel. *Bull. Soc. géol. Fr.* **6,** 2–9.

Powell, N. A. and Cook, P. L. (1966). Conditions inducing polymorphism in *Thalamoporella rozieri* (Audouin) (Polyzoa, Anasca). *Cah. Biol. mar.* **7,** 53–59.

Reyment, R. A. (1969). Some case studies of the statistical analysis of sexual dimorphism. *Bull. geol. Instn Univ. Upps.* (n. ser.) **1,** 97–119.

Schopf, T. J. M. (1971). Ergonomics of polymorphism: latitudinal dependence of distribution of work in colonies in ectoprocts: *Geol. Soc. Am., Abstracts with Programs* **3**(7): 696.

Silén, L. (1938). Zur Kenntnis des Polymorphismus der Bryozoen, Die Avicularien der Cheilostomata Anasca. *Zool. Bidrag. Uppsala* **17,** 149–366.

TEWARI, B. S. and SRIVASTAVA, I. P. (1967). On some fossil Bryozoa from India. *J. geol. soc. India* **8,** 18–28.
VOIGT, E. (1956). Untersuchungen über *Coscinopleura* Marsson (Bryoz. foss.) und verwandte Gattungen. *Mitt. geol. St Inst. Hamb.* **25,** 26–75.
WILSON, E. O. (1953). The origin and evolution of polymorphism in ants. *Q. Rev. Biol.* **28,** 136–156.
WILSON, E. O. (1968). The ergonomics of caste in the social insects. *Am. Nat.* **102,** 41–66.

39 | Variation and Habitat in Some Erect Fossil Bryozoa (Preliminary Report)

FERDINAND D. FLOR

Hamburg, Germany

Abstract: The study of *Spiropora verticillata* (Goldfuss 1826) and *Entalophora proboscidea* (Milne-Edwards, 1838) from several European Upper Cretaceous deposits by biometrical methods made it possible to distinguish near-coastal from off-coastal samples by the construction of variation-polygons as well as by the relation of the extent of variation in stem diameter in relation to the external length of the zooecia.

This paper refers to morphological variation in *Spiropora verticillata* (Goldfuss 1826) in coastal and off-coastal habitat reported by the author to the IBA Congress 1968 in Milan.

More recent investigations have shown further variations in the growth of *S. verticillata* and also in *Entalophora proboscidea* (Milne-Edwards, 1838) and the results of these investigations are presented below.

These species were chosen for investigation because they show a very good display of rheodynamic factors and because the stability of the zoarium is an important factor in resistance to water movement. Naturally such features are found only in erect Bryozoa. In contrast, the encrusting growth forms are less suited for such studies, they have a rheodynamically favourable growth form and the zoarial variants are often caused by the sculpture of the substratum.

Some specimens from various Upper Cretaceous localities in Europe yielded statistical results concerning the diameter of the stems and the outside length of the zooecia, which did not always confirm the conclusion drawn from palaeogeographic and sedimentological research.

From many investigations in Anthozoa, Porifera and other groups I expected to find big variants between coastal biotopes with strong water movement and thin bryozoan material in off-coastal biotope with still water.

The following features were measured (Fig. 1): (a) the diameter of the zoarial stem, (b) the outside length of the zooecia (=whorl distance of *S. verticillata*).

By comparing construction of variation polygons, the extent of variation and

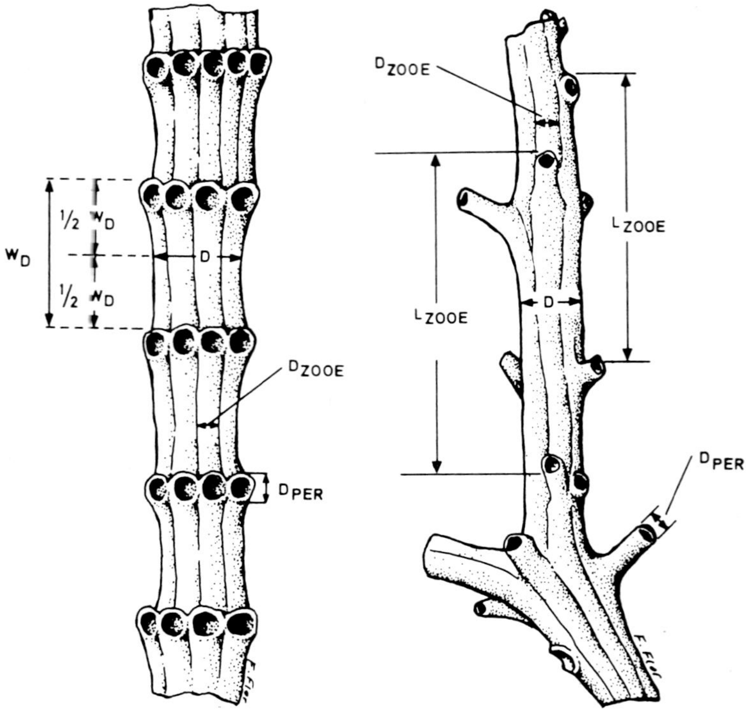

Fig. 1. Sketch of the morphological features measured. Left: *Spiropora verticillata* (Goldfuss, 1826). Right: *Entalophora proboscidea* (Milne-Edwards, 1838). D = diameter of the stem. D_{ZOOE} = diameter of the zooecium. L_{ZOOE} = external length of zooecium. W_D = distance of the aperture-whorls = height of the whorls.

the number of variation classes, it was possible to distinguish unambiguously between populations from off-coastal and near coastal habitats.

It has been known for a long time that off-coastal biotopes are characterized by a relatively homogeneous structure of different ecological factors. In contrast the coastal biotopes are split into well defined sub-biotopes. This fact is expressed in the variation polygons, primarily in the width of variation and in the construction of the variation polygons.

THE CONSTRUCTION OF THE VARIATION POLYGONS

From the compilation of three variation polygons (Fig. 2) for stem diameter of *Entalophora proboscidea* from the off-coastal environment of the Paris Basin of

Meudon, near Paris (Upper Campanian) and from near coastal facies of Hoheneggelsen (Lower Campanian) and Groß-Bülten, near Hannover (Santonian) it may be seen that marked differences between both facies types exist.

Furthermore the variation polygons show morphologically different populations, i.e.: the variants with small stems from the palaeoecological caves from near coastal habitat and the variants with big stems from places with rheodynamic features. In the off-coastal environment variants with thin stems often occur. For this reason no distinct environment can be recognized from few samples with relatively thin stems. However the results clearly show that in general a still water environment is characterized by variants with thin stems only and that variants with thick *and* thin stems indicate strong water movement conditions.

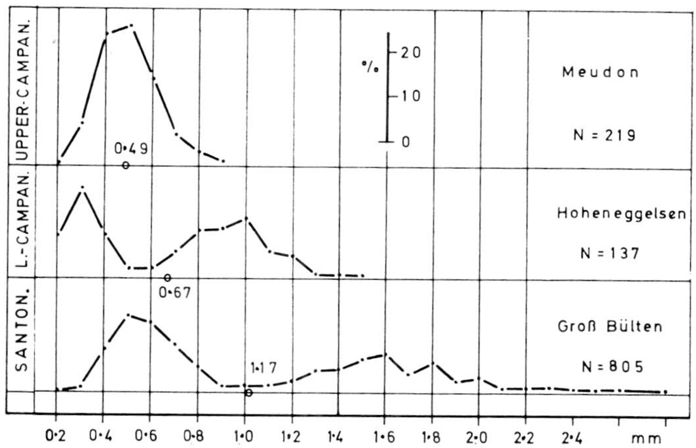

Fig. 2. *Entalophora proboscidea* (Milne-Edwards, 1838). Three samples representing different constructions of the variation-polygons (and variation-statistical mean) from different environments. Meudon (off-coastal environment, Paris basin); Hoheneggelsen and Groβ-Bülten (near coastal environment of the Subherzyn in Northwestern Germany).

THE EXTENT OF VARIATION

There is an interesting relation between the extent of variation in stems diameter and the spacing of the whorls in *S. verticillata* (Figs 3 and 4).

Without a single exception the relatively large populations from a near-coastal habitat showed greater variation in stem diameter than in whorls spacing, and vice versa in respect to off-coastal populations.

Accordingly near-coastal populations are those of

(a) the Tuffeau de Ciply, near Mons (Belgium, Montian)

(b) Voldum, Jutland (Denmark, Danian)
(c) Ilten, near Hannover (GFR, Maastrichtian)
(d) the Curfs quarry from the pietersberg, near Maastricht (Netherlands, Maastrichtian)
(e) Kunrade (Southern-Netherlands, Maastrichtian) and
(f) Groß-Bülten, near Hannover (GFR, Santionian).

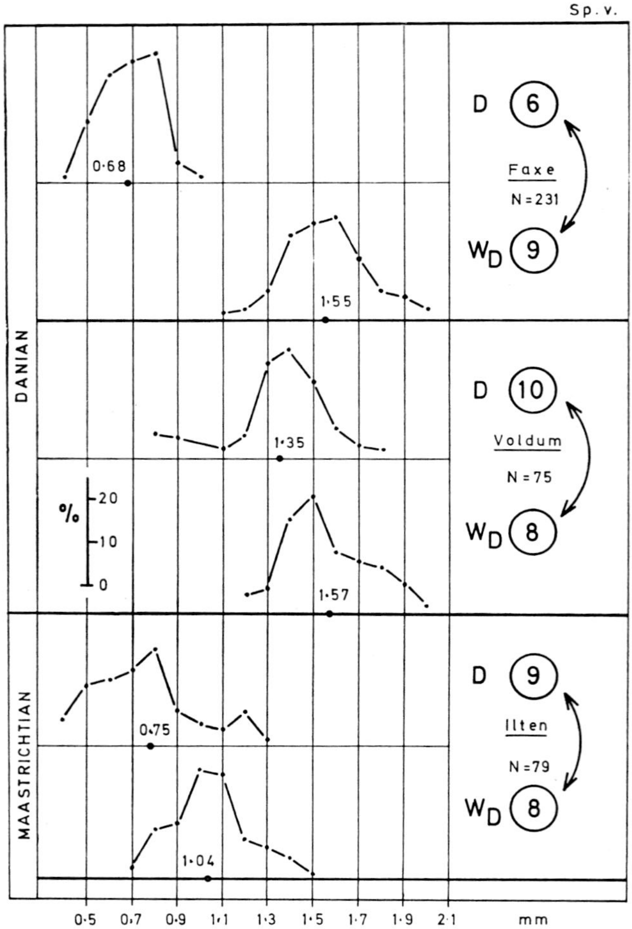

Fig. 3. Some variation-polygons of *Spiropora verticillata* (Goldfuss 1826) which demonstrate the relation between the quantity of variation classes of stem diameter (D) and the height of the whorls (W_D). The members in the circles refer to the variation classification.

In contrast the samples of
(a) Faxe, Sealand (Denmark, Danian)
(b) Lögstedt, Jutland (Denmark, Danian)
(c) Daerstorf, near Hamburg (GFR, erratic boulder) and
(d) Rügen, (Northern GDR, Maastrichtian)
originate from an off-coastal habitat.

The results obtained are illustrated schematically in Fig. 4.

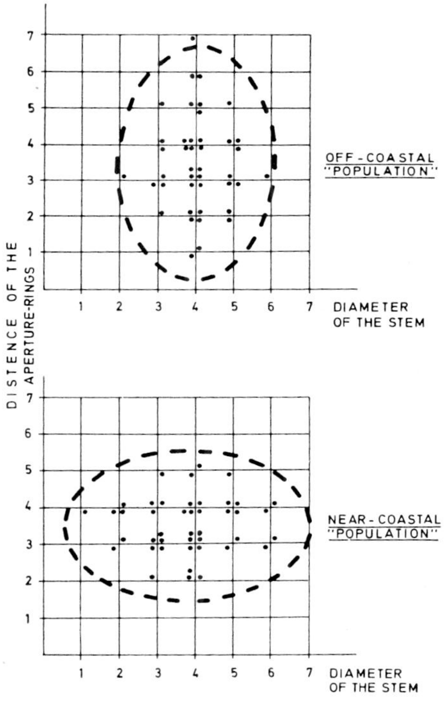

Fig. 4. *Spiropora verticillata* (Goldfuss, 1826). Sketch of the relation between the variation classes of stem diameter and the distance of the aperture-whorls.

It appears that the $D - W_D$ values are controlled by the relative distance from the shoreline (Table 1). Thus the population from the Danian of Lögstedt would be farthest from the coast ($D - W_D = -12$). The Montian habitat of the Canal Albert, near Vroenhoven ($D - W_D = \pm 0$) and the Santonian habitat of Groß-Bülten ($D - W_D = +1$) seem to be relatively near coastal ones.

Table 1. *Spiropora verticillata* (Goldfuss, 1826). Relation between the variation of stem diameter (D) and the distance of the aperture rings (=the height of the whorls=D_W). The figures refer to the quantity of the variation-statistical classes

	D	D_W
MONTIAN		
Tuffeau de Ciply, near Mons	12	8
Canal Albert, near Vroenhoven (Belgium)	8	8
DANIAN		
Faxe (Sealand, Denmark)	7	11
Lögstedt (Jutland, Denmark)	4	16
MAASTRICHTIAN		
Ilten, near Hannover (GFR)	10	5
Curfs quarry, near Berg, Geul-valley, close to Maastricht (Netherlands)	9	7
St Pieter, near Maastricht (Netherlands)	7	5
Kunrade (Southern Limburg, Netherlands)	9	4
Rügen (Northern GDR)	10	18
SANTONIAN		
Groβ-Bülten, near Hannover (GFR)	8	7
ERRATIC BOULDER (DANIAN)		
Daerstorf, near Hamburg (GFR)	8	13

ACKNOWLEDGEMENT

The author is very grateful to Prof. Dr E. Voigt, Geologisch-Paläontologisches Institut der Universität Hamburg, for providing the samples for investigation.

REFERENCES

FLOR, F. D. (1968). The variation of *Spiropora verticillata* (Goldfuss, 1827) from Upper Cretaceous deposits. *Atti. Soc. It. Sc. Nat. Museo Civ. St. Nat. Milano* **108,** 161–164.

VOIGT, E. and FLOR, F. D. (1970). Homöomorphien bei fossilen cyclostomen Bryozoen, dargestellt am Beispiel der Gattung Spiropora Lamouroux, 1821. *Mitt. Geol.-Paläont. Inst. Univ. Hamb.* **39,** 7–96.

VOIGT, E. (1967). Ökologische und stratigraphische Untersuchungen an Bryozoen der oberen Kreide. *Ber. dtsch. Ges. geol. Wiss. A. Geol. Paläont.* **12** (5) 479–491.

40 | Types of Miocene Bryozoa described in Roumania

VIRGIL GHIURCA

Universitates "Babes Bolyai",
Catedra de Geologie. Str. Kogilniceanu, Cluj, Roumania

Abstract: The present paper refers only to the bryozoan types described in the last century from the Roumanian Miocene.

Having obtained a rich collection of Bryozoa from the well known fossiliferous localities of Lăpugiu and Buituri, Reuss (1874) and Manzoni (1877, 1878) described the following new bryozoan types:

(1) *Lepralia intermedia* Reuss 1874, p. 160, Plate 8, Fig. 11. Lăpugiu.
(2) *Lepralia asperrima* Reuss 1874, p. 173, Plate 8, Fig. 9. Lăpugiu.
(3) *Lepralia turgenscens* Reuss 1874, p. 176, Plate 8, Fig. 7. Lăpugiu.
(4) *Lepralia sulcifera* Reuss 1874, p. 176, Plate 8, Fig. 8. Lăpugiu.
(5) *Lepralia anisostoma* Reuss 1874, p. 178, Plate 8, Fig. 12. Lăpugiu.
(6) *Celleporaria crassa* Manzoni 1877, p. 52, Plate 1, Fig. 4. Lăpugiu.
(7) *Hemieschara porosa* Manzoni 1877, p. 55, Plate 2, Fig. 8. Lăpugiu.
(8) *Hemieschara marginata* Manzoni 1877, pp. 55–56, Plate 2, Fig. 9. Lăpugiu.
(9) *Eschara expansa* Manzoni 1877, p. 58, Plate 5, Fig. 17. Buituri.
(10) *Eschara stipitata* Manzoni 1877, p. 60, Plate 12, Fig. 39. Lăpugiu.
(11) *Eschara minax Manzoni* 1877, p. 65, Plate 11, Fig. 37. Lăpugiu.
(12) *Eschara oculata* Manzoni 1877, p. 66, Plate 12, Fig. 40. Lăpugiu.
(13) *Filisparsa elegantissima* Manzoni 1878, p. 9, Plate 8, Fig. 31. Lăpugiu.

Probably all these types from Roumania may be stored in the Naturhistorischen Museum at Vienna, Austria. Unfortunately they were neither revised nor published in a catalogue of types.

If we examine the actual taxonomy of these types (see also Ghiurca, 1961) we can establish the following situation:

Part of the forms were assigned as follows:

(1) *Hippodiplosia asperrima* Reuss 1874, assigned by Canu and Bassler, 1920, p. 394.

(2) *Hippodiplosia sulcifera* Reuss 1874, assigned by Vigneaux, 1949, p. 59, Plate 5, Fig. 6.
(3) *Hippodiplosia oculata* Manzoni 1877, assigned by Canu and Bassler, 1920, p. 394.
(4) *Emballotheca anisostoma* Reuss 1874, assigned by Ardhi, 1964, p. 400.
(5) *Costazzia crassa* Manzoni 1877, assigned by Canu and Bassler, 1924, p. 684.
(6) *Cheiloporina expansa* Manzoni 1877, assigned by Canu and Lecointre, 1930, p. 92.
(7) *Filisparsa elegantissima* Manzoni 1878, assigned by Canu, 1913, p. 127.

The remaining six forms have not so far been assigned to other genera.

The following bryozoan types were recorded at the same time from two or three localities, including some from Roumania, without mentioning the type localities. These are as follows:

(1) *Membranipora semiaperta* Reuss 1874, p. 182, Plate 10, Fig. 2. Rauchstallbrunngraben–Austria and Buituri–Roumania.
(2) *Gemellaria* Sav. Manzoni 1877, p. 73, Plate 17, Fig. 58. Kostel–Austria and Lăpugiu–Roumania.
(3) *Filisparsa seriatopora* Manzoni 1878, p. 9, Plate 8, Fig. 29. Baden, Ehrenhausen–Austria and Lăpugiu–Roumania.
(4) *Eschara neglecta* Manzoni 1877, p. 65, Plate 11, Fig. 38. Grinzig Ehrenhausen–Austria and Lăpugiu–Roumania.
(5) *Cumulipora transilvanica* Manzoni 1877, p. 53, Plate 2, Fig. 7. Niederleis–Austria and Lăpugiu–Roumania.

Of these five species *Eschara neglecta* was referred to *Schizostoma* by Canu and Bassler, 1924, p. 682 and the specimen of *Gemellaria* is probably an undescribed species of *Dittosaria*.

In order to revise the types of bryozoans mentioned by Reuss and Manzoni from Roumania, and to assign them systematically it is necessary to solve the following problems: (a) to collect topotypes; (b) to check the presence and preservation of the appropriate types in the museum in Vienna. (If these types cannot be found it will be necessary to establish neotypes.); (c) to establish precisely the type localities in Roumania; (d) to assign the types systematically; and (e) to found a collection of Roumanian bryozoan types in the museum of the Geological Department of the University of Cluj.

REFERENCES

Ardhi, R. U. (1964). Die Steinbrüche von St. Margarethen (Burgeland) als fossiles Biotop. *Sitzungsber. Österr. Akad. Wissenschaften, Mathem. naturw. Kl.* **173,** 8–10.

Canu, F. (1913). Contributions à l'étude des Bryozoaires fossiles, III *B.S.G. France* (**4**) **XIII,** Paris.

Canu, F. and Bassler, R. A. (1920). North American early tertiary Bryozoa *Smithsonian Inst. Nat. Museum. Bull.* **106,** Washington.

Canu, F. and Bassler, R. A. (1924). Contributions à l'étude des Bryozoaires d'Autriche et de Hongrie. *Bull. Soc. géol. Fr. Paris,* **XXIV**.

Canu, F. and Lecointre, G. (1930). Les Bryozaires Cheilostomes des faluns de Touraine et d'Anjou. *Mém. Soc. géol. Fr. Paris* **VI.**

Ghiurca, V. (1961). Contribuţii la cunoaşterea faunei de Bryozoare din Transilvania. II, Revizuirea taxonomică a Bryozoarelor de la Lăpugiu si Buituri. *Publicate de A. Koch. Stud. Univ. Babes-Bolyai, Geol. Geogr.* **II,** 1, Cluj.

Manzoni, A. (1877). Briozoi fossili del Miocene d'Austria ed Ungheria, II. *Denkschr. Akad. Wissensch.* **XXXVII,** Wien.

Manzoni, A. (1878). Idem. III. *Denkschr. Akad. Wissensch.,* **XXXVIII,** Wien.

Reuss, A. E. (1874). Die fossilen Bryozoen des österreichisch-ungarischen Miocäns, I. *Denkschr. Akad. Wissensch.* **XXXII,** Wien.

Vigneaux, M. (1949). Révision des Bryozoaires Néogènes du bassin d'Aquitaine. *Mém. Soc. géol. Fr. Paris* **60.**

41 | Ordovician Bryozoa of the Altai-Sayan Region*

A. M. JAROSHINSKAYA

Ob-Irtish Expedition. Tomsk, By-street Nachanovicha, U.S.S.R.

Abstract: Ordovician Bryozoa of the western part of the Altai-Sayan mountain region have been fully studied in the areas of Mount Altai, Salair and Gornaya Shoria. There are assemblages of three different ages in Mount Altai. The first—from the Khankharinskaya Formation—is Middle Caradocian, the second—from the Upper Guriyanovskaya Subformation—belongs to the lower part of the Upper Caradocian and the third—from the Cheborskaya Formation—is of Lower Ashgillian age.

The Bryozoa indicate a correlation between the Upper Guriyanovskaya Subformation and the lens-like Gornaya Shoria Formation and the upper part of the Manchurek Formation of western Sayan. The Ashgillian Bryozoa of the Cheborskaya Formation of Mount Altai are like those of the upper part of the Taarlaganskaya Formation and are similar to Bryozoa from the upper part of the Karginskaya Formation of Tuva.

The geographical distribution of Altai-Sayan Bryozoa and the brachiopod/coral fauna of the Siberian region indicate two distinct subprovinces: Eastern Siberia and Altai Sayan.

In the Altai-Sayan mountain region Ordovician deposits occupy a vast area. The deposits of the western part of this region, including Gorni Altai, Gornaya Shoria and Salair, are fully known palaeontologically.

The Ordovician deposits of the western part of the Altai-Sayan region show a typical gradual increase of carbonate material beginning from the lower part of the Ordovician. In the Early Ordovician marine deposits in this region were localized and rocks of this age are mainly terrigenous fragmentals. In the middle Ordovician (Lower Caradoc) there was a wide marine transgression and an accumulation of thick deposits of sandstones, clays and some limestones. In the late Ordovician (Upper Caradoc) with maximum transgression limestones became more abundant in some districts forming thick beds. Irregular development of carbonate deposits in the first part of the late Ordovician is reflected by some peculiarities in the fauna. In the Early Ordovician trilobites were dominant.

* The text was initially translated from Russian by S. R. Raskolenko.

In the first part of the Middle Ordovician the fauna is characterized by trilobites together with occasional brachiopods. In Salair at this time deposition of graptolitic shales increased. Two graptolite zones are to be distinguished: *Didymograptus bifidus* (Llanvirn) and *Glyptograptus teretiusculus* (Llandeilo). In the Lower and Middle Caradoc side by side with the rich fauna of brachiopods and trilobites the Bryozoa appear for the first time. Bryozoa are found in limey-clay seams, they are scarce, and the diversity is low. In the Lower Caradoc of the Anuisko-Chuiski Synclinorium in the deposits of the Bugrishikhinskaya Formation the Bryozoa are rare and they consist of new species of *Diplotrypa*. Bryozoa are more abundant in the Middle Caradoc in the Charishsko-Inskoi Synclinorium. Here in limestones, the bryozoan fauna is almost wholly trepostomatous.

The Bryozoa from the Khankharinskaya Formation are characterized by the dominant genus *Diplotrypa*. The Altai species of this genus have massive and ramose colonies with large zooecial apertures. This is characteristic for the Middle Ordovician *Diplotrypa* species. *D. petropolitana* Nich, is abundant here. This species is widespread, being known in Sweden, and the U.S.S.R. (Baltic, Podolia, Pai-Khoi, Vaigach Island). In Sweden this species is found in the Middle Ordovician and in the Baltic—in the middle and lower part of the Upper Ordovician. Investigation of Altai collections and various published works on *D. petropolitana* (Bassler, 1911; Astrova, 1948, 1965; Modzalevskaya, 1953) as well as a review of the Baltic collection made by Modzalevskaya made it possible for me to determine the great range of variation of species.

Middle Ordovician specimens described from the Azeric, Kukruzec and Iikhviisc horizons have numerous mesopores which almost isolate the round zooecial apertures. Diaphragms are horizontal or oblique. In the Upper Ordovician (Nabalaski horizon of the Baltic) there are specimens with definitely polygonal zooecial apertures, fewer mesopores and different kinds of diaphragms which may be horizontal, oblique and cystlike.

Altai representatives of this species are similar to *D. petropolitana* from the Keilak and Vasalem horizons of the Baltic and to those from the Middle Ordovician deposits of Pai-Khoi. Large encrusting specimens of *D. incrustans* Astrova (first described from the Krivolutski horizon of the Siberian platform) are rather rare. Two new diplotrypid species with phaceloid and ramose zoaria (*D. ramosa* Jaroshinskaya and *D. flexuosa* Jaroshinskaya) may be regarded as endemic local species.

Other genera of Bryozoa such as *Dianulites*, *Homotrypa*, *Monotrypa* and *Fimbriapora* are represented by a few new species. *Fimbriapora* is represented by only one new species. *Batostoma* is represented by a few specimens of *B. inutilis*

Coryell which occurs in the Middle Ordovician of North America (Stones River) and in the U.S.S.R. on Vaigach Island.

In the Upper Ordovician Bryozoa are widespread in the Uimensko-Lebedskoy Synclinorium in the Upper Guriyanovskaya Subformation (Upper Caradoc) and in the Cheborskaya Formation (Ashgill). Deposits of the Upper Guriyanovskaya Subformation are fully developed in the Lebed and Baigol River basins. They are represented by a variety of sediments and particularly by limestones with a rich and varied fauna of brachiopods, bryozoans, corals and stromatoporoids. Remains of Bryozoa occur usually in the clay-limestones and in the limestones.

There are also two main horizons of different lithology with Bryozoa. The first (lower) horizon outcrops along the right bank of the R. Lebed and along the tributaries of the R. Ainka and R. Baigol. It consists of organic limestones with excellently preserved cystoporate Bryozoa as well as corals and stromatoporoids. The Bryozoan genera *Amsassipora*, *Anolotichia*, *Constellaria* and occasionally *Ceramopora* are widespread. *Amsassipora* was found for the first time in the Upper Caradoc deposits of Altai and was represented by three species. At present this genus is found in the Upper Ordovician of Kazakhstan. Representatives of the very rare genus *Anolotichia* known only from the Ordovician and Silurian of North America and the Baltic area of the U.S.S.R. Jaroshinskaya (1960) described one new species *A. commutabilis* represented by a considerable quantity of specimens. *Constellaria* is rare and known only from the Ordovician of North America, England and the Baltic. In our collection it is represented by one new species—*C. floridaeformis* Jaroshinskaya. *Stellipora vesiculosa* Modzalevskaya (1953), typical of the Mangazei and Dolborsk horizons of the Siberian platform, is also found. Organic limestones outcropping on the left bank of the R. Ainka yield representatives of *Nicholsonella*, *Batostoma* and *Homotrypa*.

The second (upper) horizon with Bryozoa is stratigraphically younger and outcrops as a series of clay limestones along the right bank of the R. Lebed, near the villages of Guriyanovka and Suranash. In these deposits trepostome Bryozoa are dominant and are found together with brachiopods. At the base of the horizon is a thin bed of limestone crowded with the remains of *Nicolsonella peculiara* Jaroshinskaya. Higher in the horizon, apart from *Nicholsonella*, are numerous specimens of new species of *Fimbriapora* and *Rhinidictya* which are characteristic of the Middle Ordovician and of the first half of the Upper Ordovician.

Many of the species are allied to those from the Trenton Group of North America; *Anolotichia commutabilis* Jaroshinskaya from the Vazalemsk horizon of the Baltic and *Ceramopora spongiosoformis* Jaroshinskaya (1967) from the Mangazeisk horizon of the Siberian platform may be mentioned. Some of the species are allied to those from the Cincinnati Group of North America: *Constellaria floridaeformis* Jaroshinskaya. In deposits of the Upper Guriyanovskaya Subformation the following species are known from the Mangazeisk and Dolborsk horizons of the Siberian platform: *Fimbriapora multifora* (Nekh.), *Phaenoporella transenna-mesofenestralia* (Schoen.) see Nekhoroshev, (1961) and *Stellipora vesiculosa* Modzalevskaya.

Thus, transitional Middle-Upper Ordovician species predominate and are characteristic of the whole assemblage. The Bryozoa genera from the middle part of the Kabinskaya Formation (Southern Altai) belong to the Upper-Gurjaynovski complex. Neckoroshev (1958) recorded from here *Fistulipora* sp., *Batostoma* ex. gr. *mutabilis*. We can extablish almost the same complex in the "Disk" (lens-like) formation, Amzasskaya Series of Gornaya Shoria. Here, in the area of the R. Amzas and R. Toga grey-black limestones yield Bryozoa which are exclusively cystoporates (*Ceramopora*, *Anolotichia*, *Amsassipora*, *Constellaria*, *Stellipora* and *Fistulipora*). *Anolotichia commutabilis* Jaroshinskaya, *Amsassipora simplex* Jaroshinskaya, *Constellaria floridaeformis* Lar. and *Stellipora vesiculosa* Modzalevskaya are common species of the Upper Guriyanovskaya Subformation and of the "Disk" formation which are thus of the same age.

In the Eastern part of the Altai-Sayan region outcrops along the R. Ak-Sug and R. Manchurek exposed deposits of the Manchurek Series of possibly Middle-Upper Ordovician age. These deposits include sandstones, conglomerates and clay slates and at two places in the upper part of the series poorly preserved Bryozoa occur. Many specimens may be determined only generically, these include *Ceramopora*, *Homotrypa*, *Hallopora* and *Batostoma*. The species *Stellipora mantchurikensis* Modzalevskaya and *Nicholsonella polaris* Modzalevskaya also occur. The genera of Bryozoa represented are the same as for the Upper Gurianovskoi Subformation, but there are no common species. In the western part of the Altai-Sayan region, in the Uimensko-Lebedskoi Synclinorium, the topmost part of the Ordovician is represented by the Cheborskoi Formation. Deposits of this Formation occur in the area of the R. Lebed and of the R. Samish and R. Iogach. The upper part of the Cheborskoi Formation in the districts of the R. Samish and R. Iogach is characterized by carbonate sediments with abundant remains of brachiopods, Bryozoa and corals.

Bryozoa of the Cheborskoi Formation characteristically include very few cystoporates In the fauna the role of trepostomes increases and genera unknown in deposits of the Upper Guriyanovskaya Subformation appear. The cryptostome genus *Ensipora* is widespread and represented by abundant specimens assignable to four species. Three of these species are new, the fourth—*Ensipora erecta* (Nekh.)—is known in the deposits of the Lower Silurian (Llandovery) of the Siberian platform (Nekhoroshev, 1961).

Stictoporella omniba Jaroshinskaya (1967) also occurs. It is allied to *S. lammelata Nekhoroshev* (1961) from the Lower Silurian of the Siberian platform. Representatives of the family Monticuliporidae (*Monticulipora*, *Homotrypa* and *Mesotrypa*) and Trematoporoidae (*Hallopora* and *Diplotrypa*) are widespread.

Most Bryozoa in this complex are new, and only four species belong to previously named taxa: *Stellipora vesiculosa* Modzalevskaya, *Hallopora subnodosa* Ulrich, *Nicholsonella vaupeli* Ulrich and *Batostoma varians* (James). The last three of these are known from deposits of the Richmond Group in North America (Ulrich, 1890; Cumings and Galloway, 1912). The species *Batostoma varians* (James) is known from the Dolbor horizon of the Siberian platform (Astrova, 1955). New species from the Cheborskoi Formation are closely allied to a number of the Upper Ordovician species of North America and of the Baltic and Siberian platform regions of the U.S.S.R. The stratigraphy and fauna of the Cheborskaya Formation indicates that it may correspond to the upper part of the Taarlagansk Formation of the R. Karagema district of the western part of the Northern Chuiskogo mountain range. Nekhoroshev (1958) determined Bryozoa from their including *Amplexopora* sp., *Batostoma* sp., *Hallopora* sp. and *Phaenopora ex. gr. gregaria* Schoen. These are allied to the Bryozoa from the Cheborskaya Formation which in turn show affinity with those from the upper part of the Karginskaya Formation of the southwestern part of Tuva (eastern Altai-Sayan). Here, as well as in the deposits of the Cheborskaya Formation, species of *Monticulipora*, *Homotrypa*, *Amplexopora*, *Hallopora* and *Ensipora* are well developed. There are also common species: *Hallopora subnodosa* Ulrich and *Stellipora vesiculosa* Modzalerskaya.

Characteristic assemblages of Ordovician Bryozoa from different stratigraphic formations of the western part of the Altai-Sayan region are given in the table. Analysis shows that the Ordovician Altai-Sayan Bryozoa, having many endemic species, also have taxa indicating affinity with Bryozoa from other regions of the U.S.S.R. and from other countries. Investigation of the geographical distribution of Altai-Sayan Bryozoa and other widespread faunas (brachiopods and corals) allow the recognition of two subprovinces within the Siberian province: an Eastern-Siberian Subprovince including parts of Siberian platform and Taimir) and an Altai-Sayan Subprovince (including the Altai-Sayan region) Jaroshinskaya (1970).

During the Middle Ordovician in the Altai-Sayan Subprovince the bryozoan assemblage includes forms characteristic of the Baltic and Western-Arctic

Provinces and a connection existed through the Arctic Basin and the geosynclinal Ural Sea. In the Altai-Sayan basin in Lower and Middle Caradoc times there were numerous representatives of *Fibriapora*, *Ensipora* and *Phaenoporella*. Many of them resemble species of Bryozoa from the Baltic and Western-Arctic Provinces. At the beginning of the Late Ordovician the affinity of the Altai-Sayan fauna with that of the Eastern-Siberian Subprovince and of the North American Basin becomes more marked. Different phaenoporines (belonging to *Fimbriapora*, *Ensipora* and *Phaenoporella*) which began to spread widely at the end of the Late Ordovician penetrate the Altai-Sayan Basin. Connection with the Baltic fauna and with Western-Arctic Provinces at this time was slight and thus, in these provinces, there are no representatives of phaenoporines which are the characteristic group among Lower Ordovician Altai-Sayan Bryozoa.

Table 1. Distribution of Bryozoa in the Ordovician of the western part of the Altai-Sayan mountain region

	CARADOC			ASHGILL
	LOWER	MIDDLE	UPPER	
	Bugrishikhinskaya Formation of the Altai Mountains	Khankharinskaya Formation of the Altai Mountains	Upper Gurianovskaya Sub-formation of the Altai; Diskovaya Formation of Gornaya Shoria	Cheborskaya Formation of the Altai Mountain
		I Assemblage	II Assemblage	III Assemblage
Diplotrypa petropolitana Nich	+	+		
D. fruticosa Jar.	+			
Rhinidictya sp. nov.	+			
Diplotrypa incrustans Astr.		+		
Dianulites sp. nov.		+		
D. flexuosa Jar.		+		
D. ramosa Jar.		+		
D. sp. nov.		+		
Monotrypa sp. nov.		+		
Batostoma inutilis Coryell		+		
Homotrypa sp. nov.		+		
Fimbriapora sp. nov.		+		
Ceramopora spongiosoformis Jar.			+	
Anolotichia sommutabilis Jar.			+	
Amsassipora simplex Jar.			+	

Table 1. (*cont'd*)

A. tenuata Jar.	+	
A. sp. nov.	+	
Constellaria floridaeformis Jar.	+	
Stellipora vesiculosa Modz.	+	+
Homotrypa baigoliensis Jar.	+	
Nicholsonella sp. nov.	+	
N. sp. nov.	+	
Trematopora propria Jar.	+	
Batostoma sp. nov.	+	
Fimbriapora sp. nov.	+	
Phaenoporella sp.	+	
Rhinidictya lebedensis Jar.	+	
Rh. altaica Jar.	+	
Pachydictya sp. nov.	+	
Dianulites ramosoformis Jar.		+
Hallopora subnodosa Ulr.		+
H. multitabulata Ulr.		+
Diplotrypa sp. nov.		+
Monticulipora ramosa Jar.		+
Mesotrypa sp.		+
Amplexopora astrovae Jar.		+
Nicholsonella vaupeli Ulr.		+
Trematopora sp. nov.		+
Batostoma varians (James)		+
Stictoporella omniba Jar.		+
Fimbriapora sp. nov.		+

REFERENCES

Astrova, G. G. (1948). Lower Silurian Trepostomata of Pai-Khoi. *Sci. not. Moscow State Ped. Inst.* **LII,** 3–35.

Astrova, G. G. (1955). Ordovician and Gotlandian fauna of Podkamenoi Tunguski. Bryozoa. *Trudi Paleont. Inst. Akad. Nauk SSSR* **LVI,** 128–161.

Astrova, G. G. (1965). Morphology, history of development and Systematics of Ordovician and Silurian Bryozoa. *Trudi Paleont. Inst. Akad. Nauk SSSR* **CVI,** 3–309.

Bassler, R. S. (1911). The Early Paleozoic Bryozoa of the Baltic Province. *Bull. U.S. Nat. Mus.* **77,** 1–382.

Cumings, E. R. and Galloway, J. J. (1912). A note on the Batostomas of the Richmond Series. *Proc. Indiana Acad. Sci. Indianapolis* 147–168.

Modzalevskaya, E. A. (1953). Ordovician Trepostomata from the Baltic and their stratigraphical significance. *Sbornik VNIGRI* **78,** 91–167.

Nekhoroshev, V. P. (1958). Geology of Altai. *Trudi VSEGEI*, pp. 3–255.

Nekhoroshev, V. P. (1961). Ordovician and Silurian Bryozoa of the Siberian Platform. The Order Cryptostomata. *Trudi VSEGEI* **41** (2), 1–246.

Ulrich, E. O. (1890). Paleozoic Bryozoa. *Surv. of Illinois* **VIII,** 285–688.

Jaroshinskaya, A. M. (1960). The type Bryozoa in Biostratigraphy of the Paleozoic of the Altai-Sayan Gornii region. Vol. 1 Lower Paleozoic-*Trudi SNIIGGIMS* **19,** 393–400.

JAROSHINSKAYA, A. M. (1967). Some new Ordovician Bryozoa from Gorni Altai and Salair. Some questions of the geology of Western Siberia. *Sci. not. Tomsk State Uni.* **63,** 183–191.

JAROSHINSKAYA, A. M. (1970). Some peculiarities of the geographical distribution and ecology of Middle and Late Ordovician Bryozoa from the western part of the Altai-Sayan Gorni region. *Bull. MOIP, geol. bran.* **XLV** (5), 99–106.

42 | Restudy of Some Nicholson's Types of Palaeozoic Bryozoans

T. G. PERRY, A. S. HOROWITZ

Department of Geology, Indiana University, Bloomington, Indiana, U.S.A.

and

R. L. ANSTEY

Department of Geology, Michigan State University, East Lansing, Michigan, U.S.A.

Abstract: The H. A. Nicholson Collection of Palaeozoic bryozoans includes representatives of 45 taxa. We outline difficulties involved in studying this collection, comment on procedural methods (qualitative, biostatistical, and numerical-taxonomic) used in their study. Brief remarks are made on species which serve as generic types and one of these, *Monotrypa undulata* (Nicholson), is illustrated.

INTRODUCTORY COMMENTS

During 1967 and 1968 Dr Douglas Nicholson of Edinburgh, grandson of H. A. Nicholson, enhanced the Royal Scottish Museum's collections significantly by donating a large number of type and figured specimens of bryozoans, corals, stromatoporoids and other invertebrate fossils which his grandfather had collected and studied in the closing decades of the last century. H. A. Nicholson's scientific accomplishments have been reviewed in considerable detail by Hinde (1899). The Nicholson Collection includes representatives of 45 bryozoan taxa which are on loan to us through the kindness of Dr Charles D. Waterston, Keeper of the Department of Geology, Royal Scottish Museum.

DIFFICULTIES IN STUDYING THE COLLECTION

Despite their age thin sections in the Nicholson Collection exhibit remarkably good preservation. Studying this collection, however, presents certain difficulties. Many of the sections are too thick to determine adequately the nature of the zooecial walls and of other structural features. Although Nicholson's illustrations generally give an excellent over-all concept of each species, it is virtually impossible to match with certainty the illustrations with any of the thin sections. Thus, for nearly all taxa represented by more than one section, recognition of the section that Nicholson presumably regarded as the holotype is usually not possible. Nevertheless, these sections identified by Nicholson surely must be regarded as the most authentic available representatives of their respective taxa, even if they should not in actuality be primary types.

Most taxa in the collection are not represented by a sufficient number of specimens to permit adequate determination of their intercolonial variation, although for many taxa the thin sections are sufficiently large so that variation of structural features within an individual colony can be determined.

Relating the specimens to a modern stratigraphical framework cannot be accomplished with much certainty.

For some taxa one must assume that the single longitudinal and single tangential section came from the same specimen; for other taxa, however, it is difficult, if not impossible, to demonstrate that a particular tangential and a particular longitudinal section were prepared from the same specimen. For most taxa no material remains from which additional thin sections can be prepared.

METHODS OF STUDY

We aspire to describe each taxon erected by Nicholson or by Nicholson and colleagues; this will involve obtaining additional type material from other repositories, such as the Geological Survey of Canada and the Ohio Geological Survey. In general, our treatment of each species will have a three-fold approach.

Firstly, comments are needed on purely qualitative features to elucidate anatomical characters that cannot be handled adequately by quantitative methods or by numerical taxonomy.

Secondly, a quantitative or biostatistical approach is necessary to obtain measurements of taxonomically significant structural features. Quantitative characters *per se* have not been used to construct a taxonomic system, although measurements of pertinent quantitative characters are useful in the rigorous description of any taxon. As discussed by Anstey and Perry (1970), attempts to develop quantitative classifications of Palaeozoic bryozoans are seriously hampered by lack of detailed knowledge of intrazoarial variability, inadequate information concerning the number of measurements required to characterize each colony for particular characters and insufficient information concerning astogenetic gradients among interacting characters. In studying the Nicholson Collection we try to obtain 20 measurements of taxonomic characters for each section. Further, many of our measurements are correlated with one another in two ways: the unit of measurement is the same for any given taxonomic character throughout the section and measurements of some taxonomic characters are spatially related. For example, the number of zooecia in a one mm^2

area and the number of mesopores in a one mm^2 area are taxonomic characters. Thus, rather than secure 20 measurements of each of the foregoing characters from 20 *different* areas in a section, we obtain wherever possible 20 measurements of these characters from the same areas. This method of obtaining measurements should reflect accurately any correlatable behaviour of taxonomic characters.

Thirdly, utilizing numerical taxonomy in studying Nicholson's types is highly desirable because it offers a convenient method both for adequate description of the specimens and for comparisons of the taxonomic characters represented. Because the Nicholson Collection contains thin sections of almost all of the major groups of Palaeozoic tubular Bryozoa, we have endeavored to include in our list of 150 characters as many of the important structural-morphological features of Palaeozoic bryozoans as possible. The nature of our computer facilities has imposed a limit of 150 characters. Coding each specimen for each 2-state character is highly desirable in order to have both completely consistent descriptions of all specimens and a semi-objective means of comparing all of the specimens with each other, even those from widely separated taxonomic groups. Multi-state (or quantitative) characters have been converted to 2-state characters by standardly used procedures.

Numerical data have been obtained from 53 separate zoaria represented by sections in the Nicholson Collection, and data from additional Nicholson material in other repositories will be incorporated later. The following steps are taken in the numerical analyses:

(a) Q-mode cluster analysis of all specimens using binary similarity coefficients;
(b) R-mode cluster analyses of all taxonomic characters to examine degree of character association;
(c) testing character associations by means of the chi-squared statistic;
(d) elimination of redundant variables and repetition of the Q-mode clustering, thereby isolating the main taxonomic groups; and
(e) use of the previous R-mode analysis to determine the most diagnostic characters subsequently used to construct a diagnostic key of the taxa.

COMMENTS ON GENERIC TYPES

Ascodictyon. The collection includes several specimens of *Ascodictyon stellatum* the type species of *Ascodictyon*; these have been studied only superficially as they are in capsules which must be shattered if the specimens are to be removed. The pyriform vesicles commonly occur in clusters of 3 to 8, are joined by threads or stolons about 0·15 mm in diameter, and the surface of some vesicles have a "roughened up" appearance, suggesting that they may be perforate. This species

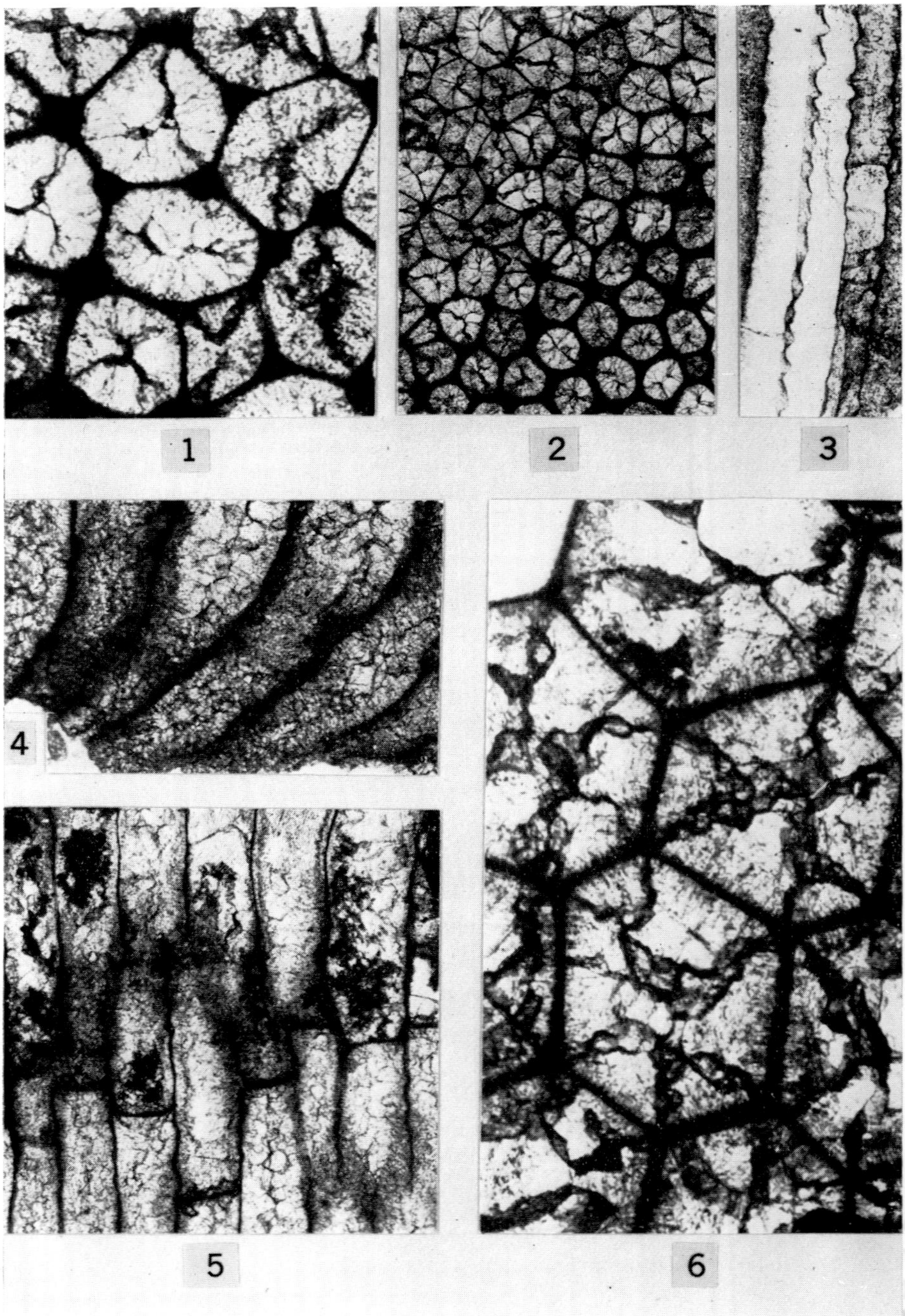
1
2
3
4
5
6

has been referred to the Order Ctenostomata; however, the Ctenostomata commonly are not regarded as having a calcareous skeleton; additionally, these specimens do not have zooecial openings. The exact taxonomic position of *Ascodictyon* is not known; perhaps it belongs among the foraminifera or hydroids.

Carinopora and *Heterodictya*. Nicholson (1874b, c, 1875a, b) established two species, *Carinopora hindei* and *Heterodictya gigantea*, which he regarded as the type species of two new genera. *Carinopora* is considered a synonym of *Semicoscinium* Prout (Nickles and Bassler, 1900; Bassler, 1953), and *Heterodictya* is regarded as a synonym of *Ptilodictya* Lonsdale (Nickles and Bassler, 1900; Bassler, 1953). Incidentally, Nicholson (1874b, c) also erected the new genus *Cryptopora* based on the type species *C. mirabilis* which is not in this collection. This genus, like *Carinopora*, has been regarded as a synonym of *Semicoscinium* (Nickles and Bassler, 1900; Bassler, 1953).

Prasopora. The primary types of *Prasopora grayae* do not materially alter our concept of the genus. Wall structure can be observed only in very local areas because the section is excessively thick and because the walls themselves are very thin. In longitudinal section the wall appears obscurely fibrous but in actuality is probably composed of laminae as in other trepostomatous genera; locally in tangential section, the wall is apparently composed of discontinuous laminae. So-called "brown bodies" are common in longitudinal section and are usually floored by a diaghragm; the "brown bodies" seem different texturally from small bits of sedimentary material locally attached to the zooecia.

Prasoporina. The primary types of *Prasoporina selwyni*, the type species of the

---

Plate I

Figures 1 and 6 are × 90; other illustrations are × 30.

Figs 1–6. *Monotrypa undulata* (Nicholson).

Fig. 1. Tangential section showing prominent acanthopores, Royal Scottish Museum (RSM) 66–518, Hudson River Group, Toronto.

Fig. 2. Tangential section displaying variation in zooecial wall thickness and abundant acanthopores, RSM 66–518.

Fig. 3. Longitudinal section showing prominently crenulated zooecial walls, RSM 66–519, Hudson River Group, Toronto.

Fig. 4. Longitudinal section displaying partly recumbent zooecia at base of zoarium, RSM 66–553, Christiania.

Fig. 5. Longitudinal section very close to zoarial surface and displaying widely separated diaphragms, RSM 66–553.

Fig. 6. Tangential section showing thin-walled angular zooecia and absence of acanthopores, RSM 66–554.

genus, show well the differentiating character of the genus, namely, bulbous or domelike cystiphragms that normally are isolated from one another and commonly, but not invariably, connected to the opposite zooecial wall by one or two diaphragms. The zooecial wall, generally thin throughout large areas of the section, may locally show considerable range in thickness and appears structureless, possibly because of the excessive thickness of the sections. The zooecial apertures may range from nearly circular in outline to essentially angular or regularly polygonal; finally, mesopores vary considerable in their abundance.

Heterotrypa. As argued by Boardman and Utgaard (1966), *H. ulrichi* Nicholson must be regarded as the type species of the genus *Heterotrypa*, one of the most commonly encountered genera of Palaeozoic trepostomes. The thick sections of the primary types clearly show in longitudinal section sturdy amalgamate walls composed of laminae broadly convex outward and flanked by lighter-colored much thinner zooecial linings. Mesopores are so abundant that adjacent zooecia may be only in local contact. Acanthopores are prominent, both in tangential and in longitudinal sections.

Batostoma. Nicholson (1881) based *Batostoma implicatum*, the type species of *Batostoma*, on material from Cincinnatian (Late Ordovician) rocks near Cincinnati, Ohio. Acanthopores in tangential section are abundant, commonly having a transparent lumen, and possessing very thin walls in which definitive microstructure cannot be seen; some are inflecting. A very dark-colored divisional line is commonly observed between zooecial apertures and is flanked on either side by laminated zooecial wall material. Longitudinal and transverse sections of the primary types are very thick, and the former is poorly oriented. These specimens do not alter the concept of the species presented by Boardman (1960) and based on topotypic material.

Monotrypa. For many decades bryozoan palaeontologists have assigned to *Monotrypa* a genus proposed by Nicholson (1879), Palaeozoic specimens normally having angular or subangular thin-walled zooecia but lacking mesopores and acanthopores. We were, therefore, considerably surprised to note that acanthopores, restricted to the corners or angles of the zooecia (Plate I, Figs 1, 2), abound in what presumably are the primary types of *Monotrypa undulata*, the type species of the genus. On the basis of the foregoing types, we cannot detect any appreciable difference between *Monotrypa undulata* and *Amplexopora* Ulrich as the latter was redefined by Boardman (1960), except that conspicuous crenulation (Plate I, Fig. 3) of the zooecial walls (a character long regarded by many workers as of questionable taxonomic value even at the species level) is rarely seen in representatives of *Amplexopora*. Following the

Durham meeting of the International Bryozoology Association, two additional sections (Plate I, Figs 4–6), identified as *Monotrypa undulata* by Nicholson, were obtained from the Royal Scottish Museum. The foregoing specimens have markedly angular to subangular thin-walled zooecia, lack acanthopores, and seem to fit the long-held concept of *Monotrypa*; the trivial name on these sections, however, has been lightly crossed out, indicating that the assignment of the specimens to the species *undulata* was questioned, possibly (but not certainly) by Nicholson. The primary type material on which *Monotrypa undulata* is based came from the Trenton (Ordovician) Limestone in the Belleville–Peterborough area of southern Ontario or the Hudson River Group (Ordovician) of Canada, especially in the Toronto region or, less probably, from the Cincinnati Group of Ohio (Nicholson, 1875b). A problem here exists, namely, the specimens resembling *M. undulata* came from Christiania (now Oslo), presumably from Cambro-Silurian strata in that area, and thus cannot be the primary types; but the specimens (Plate I, Figs 1–3) which, from collecting locality considerations, may be the primary types do not conform to the long-held concept of *Monotrypa* as established by Nicholson and maintained by subsequent workers.

Leptotrypella. The Nicholson Collection includes sections labelled *Monticulipora barrandi* (Nicholson), a taxon Nicholson (1874a) originally described as *Chaetetes barrandi*, from the Hamilton Group (Devonian) of Widder, Ontario; these sections cannot be matched with Nicholson's illustrations (1874a, c, 1879) of the species which Nickles and Bassler (1900) subsequently placed with reservation in *Heterotrypa* Nicholson. Vinassa de Regny (1920) established the genus *Leptotrypella*, nominating *Chaetetes barrandei* Nicholson as the type species We feel the sections under consideration should be assigned to *Atactotoechus* Duncan, having greatest affinity to *A. fruticosus* (Hall).

REFERENCES

ANSTEY, R. L. and PERRY, T. G. (1970). Biometric procedures in taxonomic studies of Paleozoic bryozoans. *J. Paleont.* **44,** 383–398.

BASSLER, R. S. (1953). Bryozoa. *In* "Treatise on Invertebrate Paleontology" (MOORE, R. C., ed.). Univ. Kansas Press, Lawrence, Kansas.

BOARDMAN, R. S. (1960). A revision of the Ordovician bryozoan genera *Batostoma, Anaphragma,* and *Amplexopora. Smithsonian Misc. Collns* **140,** 1–28.

BOARDMAN, R. S. and UTGAARD, J. (1966). A revision of the Ordovician bryozoan genera *Monticulipora, Peronopora, Heterotrypa,* and *Dekayia. J. Paleont.* **40,** 1082–1108.

HINDE, G. J. (1899). Prof. H. A. Nicholson, M.D., D.Sc., F.R.S., F.L.S., F.G.S. *Geol. Mag.* **6,** 138–144.

NICHOLSON, H. A. (1874a). Description of new fossils from the Devonian formation of Canada West. *Geol. Mag.* **1,** 54–60.

NICHOLSON, H. A. (1874b). Descriptions of two new genera and species of Polyzoa from the Devonian rocks. *Ann. Mag. Nat. Hist.* **13,** 77–85.

NICHOLSON, H. A. (1874c). "Report upon the Palaeontology of the Province of Ontario." Hunter, Rose and Co., Toronto.

NICHOLSON, H. A. (1875a). Descriptions of new species and of a new genus of Polyzoa from the Palaeozoic rocks of North America. *Geol. Mag.* **2,** 33–38.

NICHOLSON, H. A. (1875b). On some of the massive forms of *Chaetetes*, from the Lower Silurian. *Geol. Mag.* **2,** 175–177.

NICHOLSON, H. A. (1875c). "Report upon the Palaeontology of the Province of Ontario." Hunter, Rose and Co., Toronto.

NICHOLSON, H. A. (1879). "On the Structure and Affinities of the 'Tabulate Corals' of the Palaeozoic Period." William Blackwood and Sons, Edinburgh and London.

NICHOLSON, H. A. (1881). "On the Structure and Affinities of the Genus *Monticulipora* and its Sub-genera." William Blackwood and Sons, Edinburgh and London.

NICKLES, J. M. and BASSLER, R. S. (1900). A synopsis of American fossil Bryozoa including bibliography and synonymy. *U.S. Geol. Surv. Bull.* **173,** 1–663.

VINASSA DE REGNY, P. (1920). Sulla classificazione dei trepostomidi. *Soc. It. sci. nat. Atti* **59,** 212–231.

43 | The *Hippothoa hyalina* (L.) Complex with a New Species from the Pacific Coast of California

PENNY A. PINTER*

San Francisco State College, Department of Marine Biology, San Francisco, California, U.S.A.

Abstract: In the 19th and early 20th centuries *Hippothoa hyalina* (Hippothoidae) was described under various varietal names on the California coast while many 20th century workers usually described this hippothoid merely as a highly variable species. Statistical analysis and scanning electron micrographs have revealed the presence of three species, *Hippothoa hyalina* (Linnaeus) s.s., *H. cornuta* (Busk) and *H. santacruzana* n. sp., incrusting algae on the California coast. Important species characters are the shape of the autozooid aperture, the morphology of the ovicell and colony formation. These three species occur in distinct but overlapping geographical regions. A fourth, lesser known, species of *Hippothoa*, incrusting shells, is currently under study.

INTRODUCTION

Hippothoa hyalina (Linnaeus) was first described from California by Busk (1854), who besides recognizing *Hippothoa hyalina* s.s., also described variety β, *cornuta*, as having "short tubular processes on each side of the mouth". Hincks (1880) described and figured *H. hyalina* and several varieties from California; his unnamed variety may represent the new species described here. Later workers who followed Busk and Hincks, either added new variety names (Canu and Bassler, 1923; O'Donoghue and O'Donoghue, 1923, 1925, 1926), or described *H. hyalina* as a highly variable species (Robertson, 1908; Osburn, 1952; Soule and Duff, 1957; Soule, 1961; Soule and Soule, 1964).

Examination of specimens has revealed that there are at least three *Hippothoa* species, all of which were previously referred to *Hippothoa hyalina* on the California coast. These three species all commonly incrust algae. A fourth species, which normally incrusts shells, will be considered in a later paper.

In the present paper, a new *Hippothoa* species is described and descriptions of the two other species are given. Patterns of growth in young colonies are discussed. Measurements, in millimetres, of various parameters of the zooids were

* Present address: Department of Paleontology, University of California, Berkeley, California, U.S.A.

taken with the subsequent computation of means and standard errors. These are given with each species. The following abbreviations are used in the text: *ml*, length of male zooid; *mw*, width of male zooid; *fzo*, combined length of ovicell and female zooid; *wo*, width of ovicell; *fzw*, width of female zooid; *sl*, length of autozooid or sterile zooid; *sw*, width of autozooid or sterile zooid; *aw*, width of aperture; *al*, length of aperture.

METHODS

Specimens of *Hippothoa* were collected in central and southern California from red algae which were attached to a substrate. These collections were supplemented by examining west coast *Hippothoa* at the California Academy of Sciences, Allan Hancock Foundation, British Museum (Natural History) and the United States National Museum. The specimens were examined by light microscopy and by scanning electron microscopy and analysed statistically.

1. Hippothoa santacruzana *n. sp.* (*Plate 1, Figs 1, 2*)

Diagnosis. Hippothoa with zooids attenuated proximally; frontal ridged, aperture sinus V to blunted V-shaped; squared corners where the sinus joins the proximal edge of the aperture; female zooid shorter than sterile zooid; ovicell without pores; males narrow and nearly as long as autozooid.

Plate I

Fig. 1. *Hippothoa santacruzana* n. sp. Frontal view of colony showing blunted V-shaped sinus of autozooid (St) and ovicell (Ov) of female zooid. Paratype, Allan Hancock Found. SEM × 105.

Fig. 2. Higher magnification of female zooid of Fig. 1. Denticles (D) are on either side of midline of lower lip aperture; a few pores are on distal portion of ovicell. (Cracks in zooid are result of drying.) SEM × 210.

Fig. 3. *Hippothoa cornuta* (Busk). Female zooid with denticles (D) on the lower lip and ovicell with heavy crenulation (R) on the sides. Ovicell has more than a single row of pores but these are scattered over the surface. SEM × 100.

Fig. 4. *Hippothoa cornuta* (Busk). Frontal view of colony showing truncated aperture of autozooid (St), male and aperture similar to that of the autozooid (Ma) and female zooid with ovicell (Ov). Note row of pores on periphery of the ovicell. SEM × 100.

Fig. 5. *Hippothoa hyalina* (Linnaeus). Showing one of the secondarily developed layers with males, (Ma), females and autozooids (St). Note frontal zooid budding. SEM × 90.

Fig. 6. Higher magnification of Fig. 5. Female zooid with ovicell (Ov) is in the center. Frontal bud (Tu) will develop into a zooid. SEM × 190.

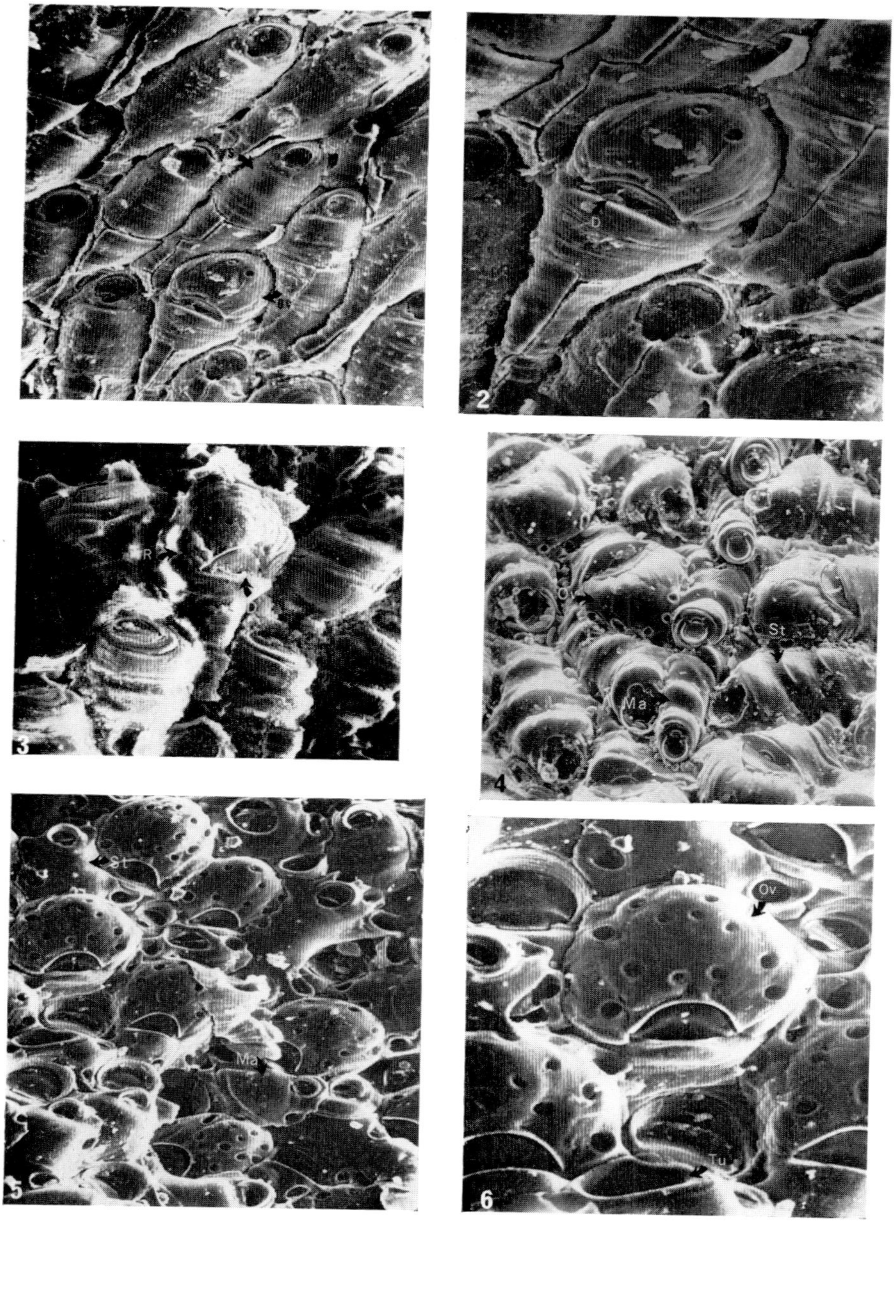
1
D
2
R
3
O
St
Ma
4
St
Ma
5
Ov
Tu
6

Description. The colony is unilaminar and the interzooecial communications are generally poorly developed. The lines of growth are represented by the ribbing on the frontal; an umbo may occasionally be formed by one of the ribs on the midline. Occasionally a single row of these may be formed along the midline.

The colony is hyaline to chalky; the layer of calcium carbonate may be so thin as to crack upon drying (Plate I, Figs 1, 2).

The aperture of the autozooids is oriented at angles less than ninety degrees to the substrate. The aperture is oval with a sinus at the proximal end. The sinus ranges from V to blunted-V shaped within the same colony; the V shape appears on the periphery of the colony. An important diagnostic character is the squared corners where the sinus joins the proximal edge of the aperture.

The female zooid has an ovicell that may or may not have pores. Pores, when present (Plate I, Fig. 2), are few in number and limited to the periphery. A denticle is present on either side of the midline of the lower lip of the aperture and varies from poorly to well developed.

The male zooids, which are sparse in the colony, are narrower, slightly shorter, and have a smaller aperture than the autozooid. The male may be difficult to distinguish from the autozooid since the range in length of the two zooids overlaps.

In comparing this species with *H. hyalina* (Linnaeus) s.s. and with *H. cornuta* (Busk), the following characters are of importance, layers in the colony, shape of autozooid aperture, female zooid and ovicell. *H. hyalina* s.s. is a multilaminar colony; the autozooids are found on the primary, and all three autozooids are found in subsequent layers. The sinus of the aperture is U shaped, the lower lip of the aperture of the female zooid is slightly curved and the ovicell is covered with pores. Both *H. cornuta* and *H. santacruzana* are unilaminar, that is male, female and zutozooids are formed in a single incrusting layer, the lower lip of the aperture of the female zooid has a denticle on either side of the midline. The denticles in *H. cornuta* are generally strongly developed, while those of *H. santacruzana* are poorly developed. The sinus of both of these species is also similar in shape; however, an important diagnostic character of *H. santacruzana* is the junction of the sinus and the aperture (comparison of Plate I, Figs 2 and 4 readily distinguishes the difference in this character). *H. santacruzana* and *H. cornuta* are quite closely related while the differences between *H. cornuta* and *H. hyalina* are the greatest.

H. santacruzana appears to be the unnamed variety which Hincks (1880) figured and briefly described (p. 274; Plate 45, Fig. 3). The Hincks collection has been examined at the British Museum (Natural History) and no specimen resembling either his figure or this species has been found.

Examination of Pleistocene material from Dead Man's Is., San Pedro, described by Canu and Bassler (1923) and deposited in the United States National Museum, has revealed that the material identified as *Hippothoa hyalina* is referable to this species. The specimens and plates are: USNM 68532a, Plate 35, Fig. 6; and possibly USNM 68532b, Plate 35, Fig. 7.

Measurements.

Parameter	N	Mean	Standard error
ml	16	0·28	0·11
mw	16	0·11	0·03
fzo	19	0·37	0·07
wo	19	0·22	0·04
fzw	19	0·16	0·02
sl	26	0·34	0·11
sw	26	0·17	0·02
aw	25	0·09	0·02
al	25	0·09	0·02

Occurrence. Epiphytic on red algae of the sublittoral in the Santa Barbara and Santa Barbara Channel Islands area of California.

Holotype. Allan Hancock Foundation 175, Arroyo Burro Beach, Santa Barbara, California.

Paratypes. Allan Hancock Foundation 175·1, SEM Figs 1, 2. Santa Barbara, California, 80 ft. British Museum (Natural History) 1971.4.4. 1 and 2, San Miguel Island, California, 40 ft.

2. Hippothoa cornuta (*Busk*) (*Plate I, Figs* 3, 4)

Lepralia hyalina var. *cornuta* Busk (1854), p. 84, Plate 95, Figs 3, 4, 5.

Schizoporella hyalina, Robertson (1908), p. 289, Plate 19, Figs 43, 44.

Hippothoa hyalina var. *rugosa* Canu and Bassler (1923), p. 94, Plate 35, Fig. 9 (?).

Hippothoa hyalina Osburn (1952), p. 277, Figs 4, 5. Pinter (1969), p. 207 (*partim*).

Hippothoa cornuta, Levinsen (1909), p. 279.

Description. The colony is hyaline to chalky white, unilaminar, and generally incrusts red algae.

The autozooid is strongly ribbed (Plate I, Fig. 3) and the aperture is oriented at a ninety degree angle to the substrate. The aperture has a sinus at the proximal end, the sides of the sinus are straight, sometimes slanted slightly inward toward the bottom of the sinus. At the junction of the sinus and the proximal end of the aperture, the corners are not squared as in *H. santacruzana* but curve downward and then up (Plate I, Fig. 4). On either side of the aperture, short spine-like processes may be present.

The male zooid (Plate I, Fig. 4) has a smaller aperture of the same shape as that of the autozooid but is also narrower and slightly shorter. There is sufficient variation for the male zooid to be nearly as long as the autozooid; the males may therefore be difficult to distinguish from the autozooids unless widths and aperture sizes are compared.

The female zooid is also shorter than the autozooid, and the ovicell generally has a single row of pores on the periphery (Plate I, Fig. 4). If there is more than a single row of pores (Plate I, Fig. 3), then it is never perforated over the entire surface as in *H. hyalina* s.s. (Plate I, Fig. 6). The sides of the ovicell are crenulated to varying degrees. The two extremes are illustrated in Plate I, Fig. 4 (light crenulation) and Plate I, Fig. 3 (heavy crenulation).

Although Hincks (1880) mentions the var. *cornuta* from California, and California specimens in the British Museum (Natural History) in the Hincks collection are referable to this species, his figure (p. 274, Plate 95, Fig. 3) is referable to another species that is not known from California.

Measurements.

Parameter	N	Mean	Standard error
ml	25	0·29	0·11
mw	25	0·14	0·05
fzo	27	0·42	0·10
wo	27	0·24	0·05
fzw	27	0·17	0·04
sl	38	0·39	0·11
sw	38	0·18	0·04
aw	34	0·10	0 03
al	34	0 09	0·02

Occurrence. Intertidal, San Francisco, California to southern California.

Lectotype. 1854.11.15.198, British Museum (Natural History). Locality, California.

3. Hippothoa hyalina (*Linnaeus*) (*Plate I, Figs 5, 6*)

Cellepora hyalina Linnaeus (1767), p. 1286.

Lepralia hyalina, Busk (1854), p. 84, Fig. 3.

Schizoporella hyalina, Hincks (1880), pp. 271–275, Plate 18, Fig. 9. Levinsen (1894), pp. 66–68, Plate 5, Fig. 47. Robertson (1908), p. 289, Plate 19, Fig. 45.

Hippothoa hyalina, Osburn (1952), p. 277, Plate 30, Fig. 1, 3. Soule and Soule (1964), p. 13 (*partim*). Pinter (1969), p. 274 (*partim*).
Celleporella hyalina, Ryland (1969), p. 223.
Description. The colony is hyaline to chalky white in appearance, multilaminar and often incrusts red algae. The zooids are close together and the interzooecial connections are not readily apparent.

The primary layer is made up entirely of autozooids while the subsequent layers are composed of female, male and autozooids.

The autozooids are not strongly ribbed. The aperture is oriented at angles less than ninety degrees to the substrate and is round with a sinus at the proximal end. The sinus is broadly rounded (Plate I, Fig. 5, 6). An umbo is frequently present proximal to the aperture.

The male zooid (Plate I, Fig. 5) generally has an inflated frontal and the ribbing is weak. The aperture is smaller than that of the autozooid. The male is shorter than that of the other two species.

The female is truncated, and ribbing is poorly developed. The lower lip of the aperture is slightly curved upward, and hints of denticles may be present on either end (Plate I, Fig. 5). The ovicell is covered with pores (Plate I, Fig. 6).
Measurements.

Parameter	N	Mean	Standard error
ml	23	0·23	0·06
mw	23	0·12	0·03
fzo	23	0·33	0·05
wo	23	0·23	0·04
fzw	23	0·14	0·02
sl	27	0·38	0·10
sw	27	0·18	0·04
aw	27	0·12	0·03
al	27	0·11	0·02

Occurrence. Point Barrow, Alaska to southern California and portions of Mexico where cooler waters occur due to upwelling.

GROWTH PATTERN

All of the three species have a similar growth pattern, forming a circular expansion on the substrate. The early budding pattern which is described below is a typical pattern.

The larva metamorphoses into a single ancestrula; from this ancestrula a zooid is budded (Fig. 1, zooid 1). From zooid 1, a distally budded zooid 2 heads a chain of five further zooids which encircles the ancestrula; as can be seen in the diagram, these zooids are budded off at sharp angles. In most cases, a straight line can be drawn through the first, third and seventh zooid. Any of these seven primary zooids can give rise to other buds which radiate out in parallel rows forming the colony. In the two unilaminar species, *Hippothoa cornuta* and *Hippothoa santacruzana*, the males and females are adnate to the substrate. In crowded conditions the interzooecial connections diminish in length in *H. cornuta*. In both species the proximal end of the sexual zooid is adnate to the substrate but the distal end may be attached to the frontal of an autozooid, thus giving a superficial appearance of being multilaminar. *Hippothoa hyalina* s.s. (Plate I, Fig. 6) shows development of the tubules frontally which form zooids of either the male, female or autozooid. *H. cornuta* and *H. santacruzana* have tubules which are formed laterally in the distal half of the zooids; these develop into either male, female or autozooids.

The primary zooids occasionally do not completely encircle the ancestrula. In this instance the colony is fan shaped rather than circular and the ancestrula is located at the apex and not in the centre of the colony.

On round surfaces, such as coralline algae of the genus *Bossiella*, the growth

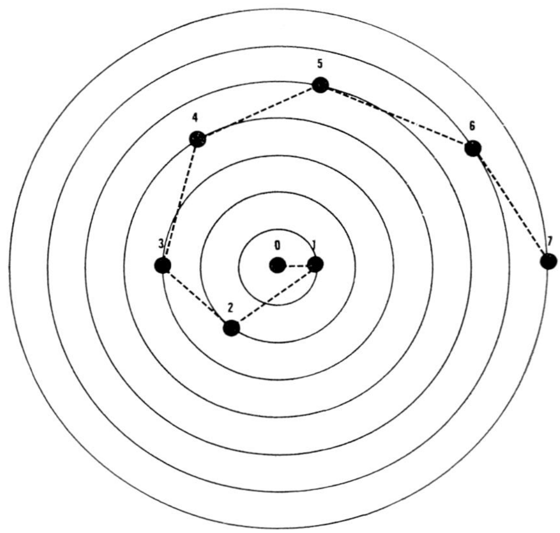

Fig. 1. Diagrammatic development pattern of primary zooids of *Hippothoa hyalina* (Linnaeus), *Hippothoa cornuta* (Busk) and *Hippothoa santacruzana*. Budding angles of zooids approximate those of zoarium. 0, ancestrula; 1, first zooid budded from ancestrula; 2, leading zooid of the chain of those surrounding the ancestrula followed by zooids of generations 3, 4, 5, 6 and 7.

pattern may be modified to suit the surface. The basic pattern of the primary zooids is the same, but zooids on the periphery may curve and bend with the substrate.

Hippothoa cornuta, upon coming into contact with another *Hippothoa* species may take several courses of action. The zooids may curve away from the other species and continue to grow in another direction; the zooids on the periphery may stop development, or the zooids may continue budding and grow over the top of the colony of the other species. Observation of *H. hyalina* and *H. santacruzana* indicates that these two species follow the first and second alternatives, but not the third.

DISCUSSION

In the Hippothoidae, a family of primitive Ascophora (Osburn, 1952), the calcification of the zooecia occurs in successive stages from the proximal to the distal end and is marked by transverse lines. In the species of *Hippothoa* considered here, these lines vary in prominence, *H. hyalina* s.s. generally have the faintest, while *H. cornuta* generally bears the coarsest lines.

The three species produce ovicells at different times of the year, *Hippothoa hyalina* in the spring and early summer and *H. cornuta* in the early autumn. Examination of preserved material of *H. santacruzana* suggests that ovicells are produced in late summer.

The geographical ranges of the species discussed here are distinct but overlapping. *Hippothoa hyalina* has been identified from Point Barrow, Alaska to southern California and is found in Mexico in regions of cold water upwelling. It is a common species from Point Reyes (Marin County, California), north. *Hippothoa cornuta* is common in southern California and is found as far north as the San Francisco region. Where *H. hyalina* and *H. cornuta* are found together, *H. cornuta* is the dominant species. *Hippothoa santacruzana* has only been found in the Santa Barbara and adjacent channel islands in recent material. The species is present in Pleistocene deposits from Dead Man's Island, San Pedro. In the Santa Barbara area, the species is dominated by *H. cornuta* when both are present. In samples from deeper waters, 40 to 80 ft, only *H. santacruzana* has been found; therefore it is a reasonable assumption that *H. cornuta* is limited to shallower waters, while at least in the Santa Barbara region, conditions are conducive for *H. santacruzana* to inhabit deeper water.

ACKNOWLEDGEMENTS

My sincerest thanks to Dr Anna B. Hastings for her comments and willingness to check specimens with those at the British Museum (Natural History); Miss Pat Cook for use of the bryozoan collection during a visit to the British Museum; Drs John and Dorothy

Soule for permission to use the Allan Hancock Foundation collections; Mr D. J. Smith for statistical programming; Mr T. Conway for the scanning electron microscopy; Mrs Jean Firby and Mr James Carlton for reading the manuscript; the California Academy of Sciences for research facilities.

REFERENCES

BUSK, G. (1854). "Catalogue of marine Polyzoa in the British Museum, Part 2, Cheilostomata," pp. i–viii, 55–120. London.

CANU, F. and BASSLER, R. S. (1923). North American Late Tertiary and Quaternary Bryozoa. *Bull. U.S. natn. Mus.* **125,** i–viii, 1–302.

HINCKS, T. (1880). "A history of the British marine Polyzoa," Vol. 1, pp. 1–601. John Van Voorst, London.

LEVINSEN, G. M. R. (1894). *Zool. Dan.* **4** (1), 1–105.

LEVINSEN, G. M. R. (1909). "Morphological and systematic studies on the cheilostomatous Bryozoa." i–vii, 1–431. Copenhagen.

LINNAEUS, C. (1767). "Systema Naturae," Edn. 12, Vol. 1, pp. 1–1327.

O'DONOGHUE, C. H. and O'DONOGHUE, E. O. (1923). A preliminary list of Bryozoa (Polyzoa) from the Vancouver Island region. *Contrib. Can. Biol.* **1,** 143–201.

O'DONOGHUE, C. H. and O'DONOGHUE, E. O. (1925). List of Bryozoa from the vicinity of Puget Sound. *Trans. Puget Sound Biol. Sta.* **5,** 91–108.

O'DONOGHUE, C. H. and O'DONOGHUE, E. O. (1926). A second list of the Bryozoa (Polyzoa) from the Vancouver Island region. *Contrib. Can. Biol. Fish.* **3** (3), 49–131.

OSBURN, R. C. (1952). Bryozoa of the Pacific Coast of America. Part 2, Cheilostomata-Ascophora. *Allan Hancock Found. Publ.* **14,** 270–611.

PINTER, P. (1969). Bryozoan-Algal associations in southern California waters. *Bull. So. California Acad. Sci.* **68,** 199–218.

ROBERTSON, A. (1908). The incrusting cheilostomatous Bryozoa of the west coast of North America. *Univ. Calif. Publ. Zool.* **2,** 235–322.

RYLAND, J. (1969). A nomenclatural index to "A history of the British marine Polyzoa" by T. Hincks (1880). *Bull. Br. Mus. nat. Hist.* (*Zool.*) **17** (6), 14–260.

SOULE, J. (1961). Results of the Puritan-American Museum of Natural History Expedition to western Mexico. 13. Ascophoran Cheilostomata. (Bryozoa) of the Gulf of California. *Am. Mus. Novitates*, 2053, 1–66.

SOULE, J. and DUFF, M. M. (1957). Fossil Bryozoa from the Pleistocene of southern California. *Proc. Calif. Acad. Sci.* S **4** (29), 87–146.

SOULE, D. and SOULE, J. (1964). The Ectoprocta (Bryozoa) of Scammon's Lagoon, Baja California, Mexico. *Am. Mus. Novitates* 2199, 1–56.

NOTE ADDED IN PROOF

In this paper the author uses the term "aperture" to refer to an opening in the calcareous exoskeletons of bryozoans which allows "passage of the lophophore", "orifice would refer to the mouth opening to the digestive tract" (Pinter *in litt.*) (Ed.).

44 | A Jurassic Cheilostome from England

ROBERT A. POHOWSKY*

Department of Geology, University of Cincinnati, Cincinnati, Ohio, U.S.A.

Abstract: Numerous specimens of *Pyriporopsis portlandensis* (gen. et sp. nov.) representing the earliest known occurrence of the Cheilostomata, have been found encrusting bivalves from the Upper Jurassic (Portlandian Stage) of England. Colonies are dominantly uniserial and lack cryptocyst, ovicells, spines and avicularia, but lateral pore chambers are present in most (if not all) zooecia. Palaeoenvironmental factors are discussed, and several aspects of the anatomy of this bryozoan are compared with those in hypothetical ancestral cheilostomes, as well as Lower Cretaceous and Recent species.

INTRODUCTION

The Cheilostomata are the dominant order of bryozoans in modern seas. In taxonomic diversity they surpassed the Cyclostomata as the result of an evolutionary burst in the Upper Cretaceous, and they are common constituents in the fossil record from that time onward. Unimpugnable reports of cheilostomes from older strata, however, are few indeed.

Voigt (1968) listed most or all of the then-published species names attributed to cheilostomes reported from Neocomian to Albian sediments. As he indicated, the only included Neocomian species, *Membranipora neocomensis* d'Orbigny, 1853 (d'Orbigny's published spelling is retained here) is a *nomen nudum*, but if the age is correct, it appears to be the oldest Cretaceous cheilostome reported in the literature to date. The list of Albian species bearing available names currently includes *Pyripora texana* Thomas and Larwood, *Rhammatopora gaultina* (Vine) and *Wilbertopora mutabilis* Cheetham, all listed by Larwood (1967). Subsequent reports of early cheilostomes are those of an unnamed Albian species from Arkansas (Boardman and Cheetham, 1969) and *Membranipora? kiowana* Scott (1970) from the Kiowa Formation in Kansas (Upper Albian—see Scott, 1970). An Aptian specimen from Antarctica was described at the I.B.A. conference (1971) by Dr Brian Taylor of the British Antarctic Survey.

Voigt (1968) persuasively demonstrated that Gregory's (1894) widely pro-

* Present address: Shell International Petroleum, Maatschappij, EP/12, Carel van Bylandtlaan 30, The Hague, Netherlands.

mulgated report of Jurassic Cheilostomata is apocryphal. Oversight of Voigt's iconoclastic paper is undoubtedly responsible for a more recent, textbook reference to Jurassic members of the order (Black, 1970). Putative Palaeozoic cheilostomes have been examined by Voigt (1968) who, at least tentatively, adjudged them to be cryptostomes.

Our knowledge of early Cheilostomata is obviously slight, but existing literature would suggest that the group first appeared in the Lower Cretaceous. However, the bryozoan described below is a Jurassic cheilostome of irrefutably Portlandian age. The author has examined a dozen colonies of this species in the Portland Roach on the Isle of Portland, Dorset.

SYSTEMATICS

Phylum BRYOZOA Ehrenberg, 1831
Class **GYMNOLAEMATA** Allman, 1856
Order **CHEILOSTOMATA** Busk, 1852
Section *MEMBRANIDEA* Silen, 1942
Division ***Malacostega*** Levinsen, 1902
Family Uncertain

The specimens at hand do not fit comfortably in any existing family. Establishment of a new family is avoided in view of taxonomic revision under way for the new *Treatise* volumes on Bryozoa.

1. Pyriporopsis, *new genus*
Etymology. name derived from *Pyripora* d'Orbigny, 1849, and *opsis* (Gr.), "having the appearance of".
Diagnosis. a calcified, encrusting, anascan cheilostome with pore chambers and a poorly to well developed proximal gymnocyst, but without a cryptocyst, spines, avicularia or ovicells. Growth is predominantly uniserial, but some pluriserial growth may occasionally occur. True multiserial growth does not occur. Zooids initially bud median-distally; distal-lateral budding on one or both sides is also frequently observed. Zooecia are longitudinally oval to pyriform, unless deformed by crowding; they undergo a readily observed lengthening of the cauda within zones of astogenetic change. Type species: *Pyriporopsis portlandensis*.

2. Pyriporopsis portlandensis, *new species* (*Fig. 1*; *Plate I, Figs 1–3, 5, 6*)
Diagnosis. as for the genus.
Type specimens and repositories. (Explanation: B.M. = British Museum (Natural History); G.S.M. = Institute of Geological Sciences (London)).

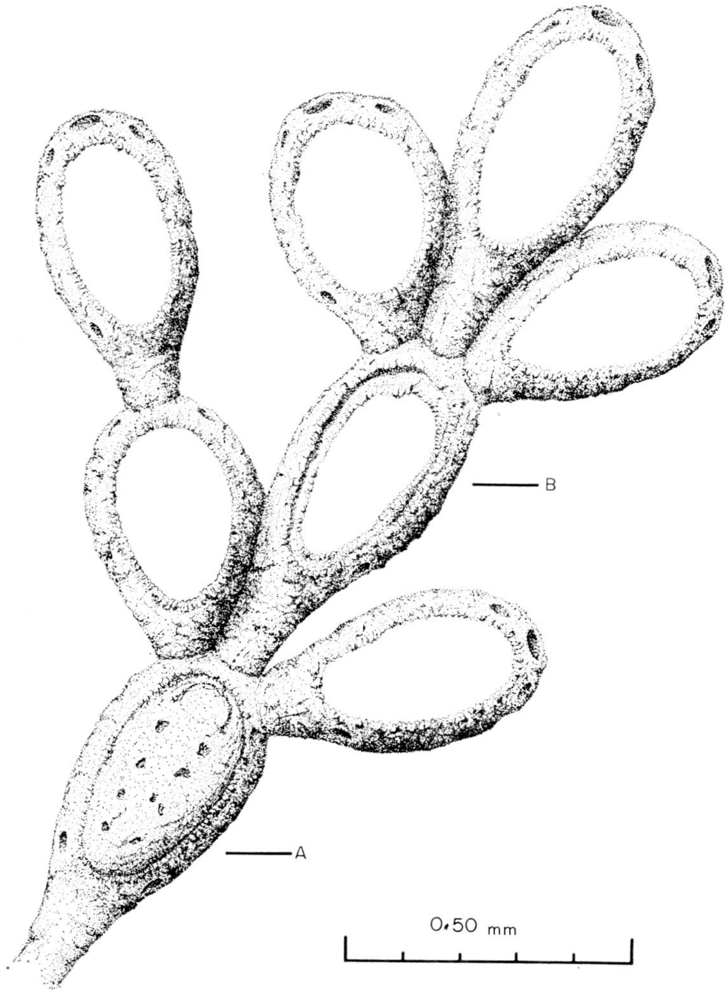

Fig. 1. *Pyriporopsis portlandensis* Pohowsky (gen. et sp. nov.) A=closed zooecium (note scars representing operculum and insertions of transverse parietal muscles); B=zooecium of a regenerated zooid. The illustration shows colony form and zooecial shapes which commonly obtain when local budding is not restricted by the crowding of zooids or the presence of other barriers on the substrate. Although proximal-lateral pore chambers are rare in zooecia having elongated caudae, the position at which they sometimes occur is indicated in the figure. This figure is based on study of the holotype and three of the paratypes.

Holotype. B.M. D.52046, mould of colony encrusting *Laevitrigonia sp.*; Portland Roach (uppermost Portland Stone), Isle of Portland, Dorset.
Paratypes. G.S.M. 48594, encrusting a valve of "*Trigonia*" *wightensis* (Strand) (G.S.M. 48594); top beds of Portland Stone, Swindon, Wilts. Bivalve figured by Cox (1929). B.M. D.52047 and B.M. D.52048, encrusting a valve of *Camptonectes suprajurensis* (Buvignier) (B.M. L.39138); Basal Shell Bed of Portland Stone, west coast of Isle of Portland, Dorset. B.M. D.52049, encrusting a valve of *Camptonectes lamellosus* (J. Sowerby) (B.M. L.24378); Portlandian, Aylesbury, Bucks. B.M. D.52050, mould of colony encrusting *Aptyxiella portlandica* (J. de C. Sowerby) (the "Portland Screw"); Portland Roach (uppermost Portland Stone), 150 m east of West Weare Cliff, in quarry west of Easton, Isle of Portland, Dorset.

British Museum specimens D.52051–60 were also examined by the author.

Description. Colonies of this species are encrusting, but zooids never cover

---

Plate I

Fig. 1. *Pyriporopsis portlandensis*, gen. et sp. nov.; paratype, G.S.M. 48594, ×18·3; arrow indicates zooecium shown in Fig. 5. The colony is very crowded and the size and shape of zooecia vary considerably. There is very little evidence of organized multiserial growth. About 7 or 8 closed zooecia can be seen in the figure.

Fig. 2. *P. portlandensis*; centre: paratype, B.M. D.52047, ×18·3; whitened slightly with ammonium chloride. The group of zooecia on the upper right are part of another paratype, B.M. D.52048, which is growing toward the other paratype. Two zooecia of D.52048 have grown against the central paratype and the cauda of one of these two zooecia therefore has a marked curve.

Fig. 3. *P. portlandensis*; paratype, B.M. D.52049, ×11·4.

Fig. 4. *Conopeum lacinosum* (Shier); B.M. 1966.2.6.5, ×57·4. Recent; Lake Merritt, Oakland, California. Note the low, radially directed ridges along the mural rim and on the adjacent narrow cryptocyst.

Fig. 5. *P. portlandensis*; paratype, G.S.M. 48594, ×57·4; enlarged view of zooid indicated by arrow in Fig. 1. The specimen is here heavily coated with ammonium chloride to emphasize the radially arranged ridges greatly resembling those of *C. lacinosum* (Fig. 4).

Fig. 6. *P. portlandensis*; holotype, B.M. D.52046, ×39·7; specimen showing the underside of a colony preserved as mould. Note the bulbous proximal-lateral and distal-lateral pore chambers and their tubular communications with the interiors of the autozooids. Zooecia with prolonged caudae constitute a zone of astogenetic repetition and generally lack the proximal-lateral pore chambers usually found in zooecia within zones of change. The fine, hexagonal network visible in some areas reflects the microstructure of the bivalve shell (dissolved) on which the colony grew.

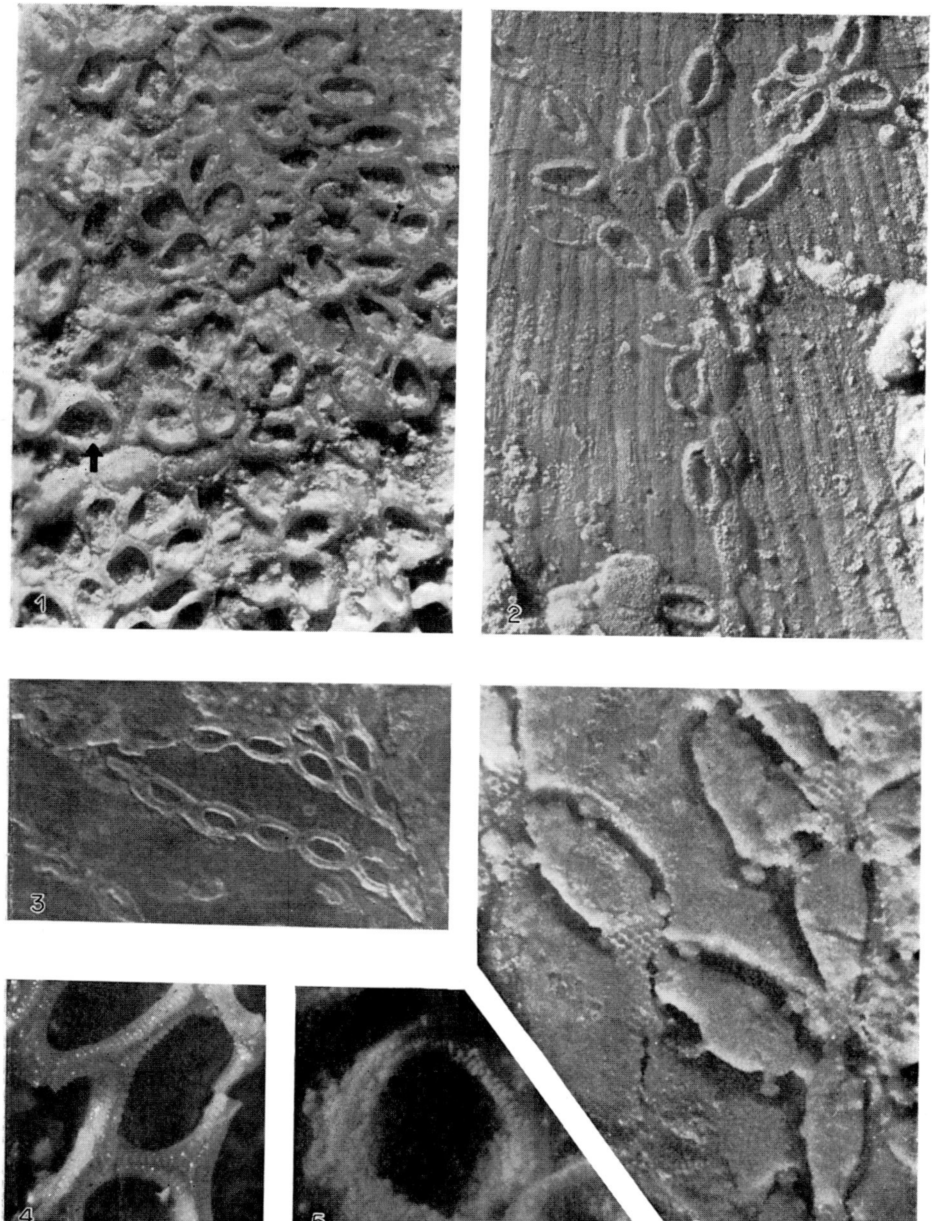
1
2
3
4
5
6

zooecia of the same colony, or of another colony. The zooecia are heavily calcified. A basal wall is definitely present when the substrate is locally very rough, but it seems to be lacking in smooth areas. Lateral and distal walls become thicker toward the base, where a maximum thickness of about 0·06 mm has been observed. The walls are slightly concave internally (convex externally) so that the mural rim (Fig. 2) was both the topmost and innermost calcified zooecial feature in normally functioning zooids (cf. closed zooids, below). No trace of a cryptocyst is in evidence.

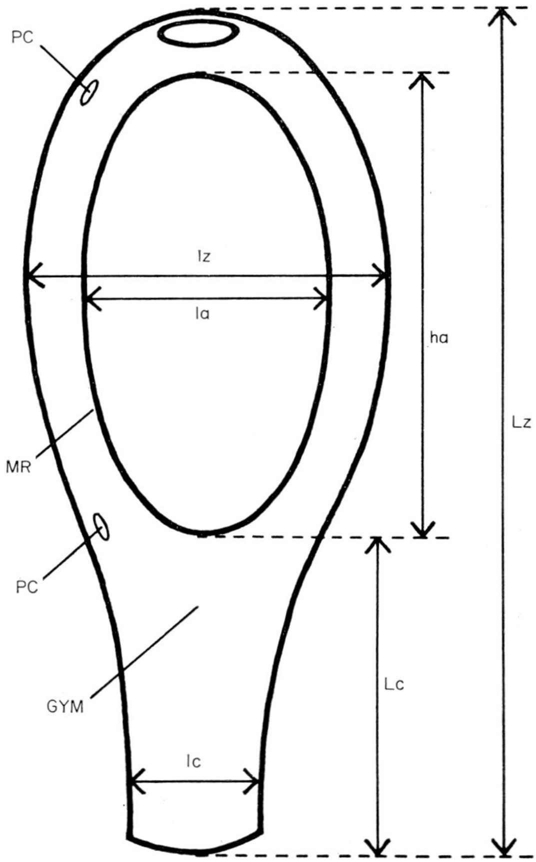

Fig. 2. Zooecial terminology and measured dimensions of *Pyriporopsis portlandensis*. Lz, length of zooecium; lz, width of zooecium; ha, length of aperture; la, width of aperture; Lc, length of cauda; lc, width of cauda; MR, mural rim; GYM, gymnocyst; PC, external opening of pore chamber. (Adapted from Thomas and Larwood, 1960: 374, Fig. 1.)

Except in distorted specimens, the mural rim is longitudinally oval. The well preserved holotype (a mould) and a paratype from Swindown (G.S.M. 48594) show that the narrow mural rim is slightly rounded and bears short, low, radially arranged ridges with a density of approximately 60 per mm (Fig. 1; Plate I, Fig. 5). Very similar ridges with essentially the same spacing are present on the mural rim and adjacent narrow cryptocyst of the living species, *Conopeum lacinosum* (Plate I, Fig. 4).

There is no evidence of spines or spine bases, and there is nothing to suggest that calcified structures of any kind extended over the frontal membrane in normally functioning zooids. However, a number of zooecia in most of the specimens examined are closed by an externally convex calcareous plate (Fig. 1A; Plate I, Figs 1, 2) covering the entire area usually occupied by the frontal membrane. These closed zooids occur most frequently in the older parts of a colony, and there is good evidence suggesting that the plate formed by calcification of the frontal membrane after it bulged outward—presumably when the underlying polypide became moribund. Distal opercular scars can often be detected, and 2–4 bilaterally disposed pairs of uncalcified areas (presumed to represent sites of insertion of transverse parietal muscles) can be seen on most of the closed zooids (see Fig. 1A). The number of pairs of muscle scars seems to bear a direct relationship to zooecial length. Boardman and Cheetham (1969) have reported very similar features in an Albian cheilostome. The type of closure found in this species and in *P. portlandensis* must not be confused with the cryptocystal closure reported in several Recent euryhaline species of *Conopeum* (see Lagaaij, 1963; Cook and Hayward, 1966; Cook, 1968).

The first zooid in a new linear series has a very short proximal gymnocyst (*sensu* Banta, 1970; *non* Bassler, 1953). This feature becomes progressively longer in succeeding zooecia, and (following Lang, 1914a, b; and Thomas and Larwood, 1956, 1960) the resulting, elongated proximal part of the zooecium is here termed a cauda (see Fig. 2). In a new series caudae attain a more-or-less constant length within about four to six generations. Thus, the concept of zones of astogenetic change followed by zones of astogenetic repetition is readily illustrated by *P. portlandensis* (see Boardman, 1968; Boardman *et al.*, 1969). Astogenetic changes and salient mensural features of the zooecia are documented in Table 1. The ancestrula and primary zone of change have not been discerned in available specimens, so the tabulated measurements are presumed to reflect variation within subsequent zones of change, and in primary and/or subsequent zones of repetition.

Budding of new autozooids from a particular zooid always occurs median-distally first. Having produced a median-distal bud, zooids often form additional buds on one or both sides at the distal-lateral position, but within zones of change (as recognized by increments in caudal length), lateral budding of autozooids usually does not occur before the third generation.

Table 1. Zooecial dimensions (mm) of type specimens of *Pyriporopsis portlandensis*

Group		*Lz*	*Lc*	*ha*	*lz*	*la*	*lc*
A.	mean	0·55	0·09	0·40	0·33	0·21	*
	min	0·48	0·08	0·36	0·28	0·14	*
	max	0·68	0·12	0·52	0·44	0·30	*
	n	9	5	6	9	9	0
B.	mean	0·62	0·13	0·43	0·35	0·22	0·17
	min	0·44	0·06	0·34	0·26	0·16	0·08
	max	0·70	0·18	0·50	0·44	0·28	0·22
	n	9	8	7	7	9	5
C.	mean	0·66	0·20	0·38	0·28	0·18	0·11
	min	0·54	0·12	0·28	0·18	0·12	0·08
	max	0·76	0·30	0·48	0·42	0·24	0·16
	n	20	12	12	19	13	16

* Significant data on this parameter cannot be obtained.

Abbreviations (Lc, etc.) are explained in Fig. 2. The data in this table are based on the combined measurements (nearest 0·02 mm) obtained from the holotype and four of the paratype specimens (all illustrated in Plate I). Dimensions of corresponding features on the measured specimens overlap markedly and provide no basis for inference that genetically controlled variation exists within the paradigm. Measurements were made only on those zooecia which showed no substantial evidence of deformation due to crowding by other zooids or unevenness of the substrate. All such zooecia were assigned to one of the following groups: (A) first zooecia in a new series, i.e., zooecia formed through distal-lateral budding; (B) second, third and fourth zooecia in a new series, i.e., in the series produced by median-distal budding of zooecia in group (A); (C) zooecia apparently far from the beginning of a new median-distally budded series, and inferred (initially) to belong to zones of repetition.

Budding of autozooids from median- or proximal-lateral positions has not been observed. The first zooid in a new series is invariably orientated in a generally distal direction, and in most cases it is inclined at 30–45° to the longitudinal axis of the parent zooid (see Fig. 1; Plate 1, Fig. 2). Sometimes, however, it is pressed against the cauda of the median-distally budded "sibling" zooid, and it may diverge at angles up to 60°.

The dominant pattern of growth is clearly uniserial, but there are scattered occurrences of apparent pluriserial growth. Particularly frequent lateral budding often produces closely spaced, misshapen zooecia showing little semblance of orderly packing (Plate I, Fig. 1). True multiserial growth does not occur.

Lateral walls of *P. portlandensis* possess intriguing chambers which are here inferred to be dietellae (pore chambers). The structure of these communication devices is best interpreted from their moulds in the holotype (see Plate I, Fig. 6).

Within zones of change, zooids commonly have four laterally directed chambers arranged in proximal and distal pairs. The exact number and arrangement of these structures may vary slightly, but it is noteworthy that the proximal chambers seem to occur just about as frequently as the distal ones. In contrast, zooids forming zones of repetition usually have dietellae only in the distal-lateral positions. Proximal pore chambers never give rise to autozooids, and no dietellae have been conclusively observed in communication through the lateral walls of other zooecia. Pore chambers are always present between distal-laterally budded autozooids and the parent autozooid. No convincing evidence of dietellae between zooids in distal-median series has been found, though some type of communication device is inferred to have existed at this point.

As shown in Plate I, Fig. 6, the occurrence of a pore chamber is accompanied by an inward deflection of the internal surface of the lateral wall, at least at its point of contact with the basal wall if, in fact, a basal wall is present. The dietellae appear to be shaped like flattened spheres and they are connected with the interior of the zooecium by means of a tube about 0·02–0·04 mm long and 0·04 mm in diameter. The apparent major and minor axes of the bulbous chambers in the holotype are 0·06–0·08 mm and 0·05–0·06 mm long, and the major axis of each chamber lies parallel to the local trend of the zooecial wall. The chambers open directly to the exterior via a small pore that has been observed at anticipated positions on several zooecia of paratype G.S.M. 48594. Simple tubular passages seem to penetrate the lateral walls in some zooecia, but the true nature of these features is obscure.

Regeneration of older zooids was a common phenomenon in this species. The presence of one or more concentric zooecia within the walls of an older zooecium indicates that regeneration has occurred (see Fig. 1B).

Stratigraphic range. Zones of *Kerberites okusensis* and *Titanites titan*, within the Portlandian Stage of the Jurassic System.

Type locality. Isle of Portland, Dorset, England.

DISCUSSION

1. *Stratigraphy and Palaeoenvironment*

Specimens of *Pyriporopsis portlandensis* have been obtained from Portlandian sediments at Portland, Swindon and Aylesbury (Fig. 3). Material from the type area consists of moulds collected from the Portland Roach by the author, and at least two additional specimens (B.M. D.52047 and B.M. D.52048) encrusting a bivalve reputed to have come from the Basal Shell Bed. Inasmuch as the Roach and Basal Shell Bed are, respectively, the uppermost and lowermost beds of the Portland Stone (Arkell, 1933) the present known range of *P. portlandensis*

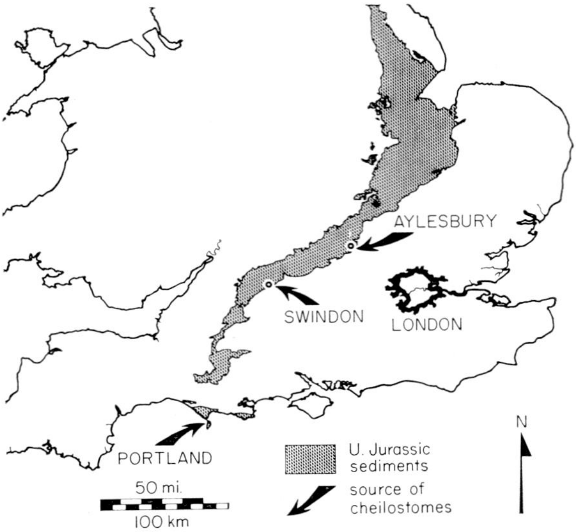

Fig. 3. Locations of Portlandian beds from which *Pyriporopsis portlandensis* has been collected. The stippled area represents the principal distribution of exposed Upper Jurassic sediments in southern England (from the Oxford Clay and Kellaways Beds to the Purbeck Beds). (After the Geological Map of Great Britain: Geological Survey, 1957, 2nd edition.)

appears to be the entire Zones of *Kerberites okusensis* and *Titanites titan*. (The entire Portland Stone may belong to the Zone of *Titanites giganteus*. For a discussion of this zonation, see Arkell *et al.*, 1947; Arkell, 1956; House, 1958.) In any case, the known range of the Cheilostomata is now extended to somewhat beyond the Purbeck Beds, the base of which has been dated at 141 m.y.

The discovery of *P. portlandensis* in the type section of the Portlandian Stage substantially obviates uncertainty about the geologic age of this species. Moreover, in light of their substrates and records of collection, there seems little reason to doubt the essentially equivalent age of the paratypes from Swindon and Aylesbury. The revisional study by Casey and Bristow (1964) did not question the age of the "Portland" beds in these localities.

The highly oölitic Portland Stone has long been viewed as a rather shallow-water marine facies (Woodward, 1895). Imlay (1957) commenting on some marine Jurassic deposits in the Western Interior of the United States corroborates this opinion, stating ". . . two essential conditions for the precipitation of these Jurassic oölites include a large quantity of calcium carbonate and very shallow water." Moreover, whether dominantly fresh-water (Arkell, 1933;

Arkell *et al.*, 1947) or dominantly evaporitic (Brown, 1963), the succeeding basal Purbeck beds in Dorset can only enforce the plausibility of a shallow-water origin for the underlying Portland Stone.

On the basis of palaeomagnetic and palaeoclimatic data, Opdyke (1962) indicated that southern England lay somewhat to the south of 30° N. Lat. during the Jurassic. A generally warmer climate in this Period is certainly suggested by the Oxfordian coral reefs in Yorkshire and the Jurassic temperate flora in east Greenland. Conditions that were at least temperate may also be inferred from the large ammonites, bivalves and gastropods of the Portland Stone. From evidence such as this, Arkell (1956) concluded that "... the Arctic Ocean in the Jurassic had no ice cap and its waters were at least as warm as those of the present temperate zones." It thus seems apparent that the presently known specimens of *P. portlandensis* lived in shallow, temperate or subtropical marine waters.

2. Phylogenetic Considerations

Several modern workers (e.g., Silén, 1944; Banta, 1969; Ryland, 1970) have considered interzooidal communication organs in the lateral walls of cheilostomes (pore plates=septulae; pore chambers=dietellae) to be aborted buds which failed to develop (presumably, though not necessarily, into autozooids) because of spatial restrictions imposed by neighbouring zooids. Silén (1944) regarded the juxtaposition of lateral walls to be a secondarily developed feature in the Cheilostomata. From this premise he concluded that pore plates and pore chambers were also secondary features. If this is so, *P. portlandensis* must have descended from earlier multiserial or strongly pluriserial ancestors, and, except for possible rare communication between lateral walls and the customary participation of distal-lateral dietellae in budding, the pore chambers of *Pyriporopsis* should be regarded as vestigial organs.

The arrangement of four lateral dietellae in many of the zooecia does suggest that they would be appropriately positioned for interzooidal communication in a multiserial colony of closely spaced, hexagonally packed zooids. Furthermore, if astogeny (fundamentally, the ontogeny of a colonial organism) in a general way reflects important preceding developmental states or trends inherent in a particular phylogeny, then the observed astogenetic lengthening of caudae in *Pyriporopsis* also intimates an ancestry of colonies with less strongly caudate (and therefore more closely spaced) zooids. Lack of space prohibits further elaboration on this theory, but it seems that the presence of dietellae in *P. portlandensis* may lend credence to the concepts expounded by Silén in 1944. Alternatively, we could propose that dietellae (or at least the apparent dietellae in *P. portlandensis*) originated in uncrowded, uniserially arranged zooecia.

The phylogenetic status of the Jurassic species may be viewed in the light of interpretive evolutionary studies involving Recent species of *Labiostomella* and *Callopora*. According to theories advanced by Silén (1942) and Voigt (1939), *P. portlandensis* may represent an evolutionary stage more advanced than that of *Callopora*, with regard to the absence of spines. Its lateral budding and encrusting habit place it even further beyond the stage represented by *Labiostomella gisleni* (Silén, 1942; Voigt, 1968). In addition, Silén (1945) expressed the belief that ovicells are a primitive character in the Cheilostomata, and the observation of ovicells in *Wilbertopora mutabilis* led Cheetham (1954) to believe that Silén's theory had been substantiated. However, the complete absence of ovicells in *Pyriporopsis* indicates that palaeontological evaluation of this theory is now largely contingent upon the discovery of additional cheilostomes from the Jurassic.

Until the discovery of the pore chambers, *P. portlandensis* was regarded as a new species of *Pyripora* d'Orbigny. Pore chambers notwithstanding, the new species bears a considerable resemblance to *Pyripora texana* Thomas and Larwood, and it may well be ancestral to this Albian form. If this be true, barring a reversion of some descendants of *P. portlandensis* to multiseriality, we must seek the ancestry of the earliest known multiserial cheilostomes in as yet undiscovered ur assic species.

The question arises whether at least some of the earliest species assigned to *Pyripora* may actually have pore chambers. Unfortunately, the types of *Pyripora texana* are mostly worn, but fracturing or artificial moulding of better topotype material may help to alleviate uncertainty about the generic affinity of this species.

CONCLUDING REMARKS

In retrospect, one of the most striking observations that can be made about the new species is that it was not discovered many years ago. One of the paratype specimens encrusts a bivalve previously illustrated in palaeontological literature, and the holotype, together with six other specimens, was removed from a large block of the unmistakable Portland Roach facies reputed to have been on public exhibition in the British Museum (Natural History). In the old quarries at Portland, a dozen specimens were discovered in the Roach in about an hour, so *Pyriporopsis* could even be considered common at this well-known stratigraphic type locality. Oversight of this species is conceivably due to its confusion with the abundant oöliths in the surrounding matrix. Specimens first came to my attention during an intensive examination of shell surfaces in a search for bryozoan borings.

ACKNOWLEDGEMENTS

This paper is the result of part of the research undertaken during the tenure of a Fulbright-Hays Grant at the Bitish Museum (Natural History). I am indebted to the United States–United Kingdom Educational Commission and the administration of the Museum for the opportunity to undertake this study, and I am grateful to Dr A. H. Cheetham, Dr G. P. Larwood and Prof. Dr E. Voigt for their helpful comments and suggestions. Special thanks are extended to Miss P. L. Cook and Dr N. J. Morris, B.M. (Natural History) for valuable discussions on bryozoan and bivalve specimens, and permission to study the collections under their immediate jurisdiction. Dr R. V. Melville (Institute of Geological Sciences, London) kindly permitted me to borrow a specimen from the Geological Survey collections, and Drs K. E. Caster and W. A. Pryor (University of Cincinnati) provided welcome assistance in proofreading portions of the manuscript.

The opinions expressed herein are not necessarily shared by the persons mentioned above and any misconceptions or inaccuracies contained in the text are those of the author.

REFERENCES

Arkell, W. J. (1933). "The Jurassic System in Great Britain." Oxford University Press, Oxford.

Arkell, W. J. (1956). "Jurassic Geology of the World." Oliver and Boyd, Edinburgh.

Arkell, W. J. *et al.* (1947). The geology of the country around Weymouth, Swange, Corfe and Lulworth. *Mem. Geol. Surv. G. Br.* (2nd Edition.)

Banta, W. C. (1969). The body wall of cheilostome Bryozoa. II. Interzooidal communication organs. *J. Morphol.* **129,** 149–170.

Banta, W. C. (1970). The body wall of cheilostome Bryozoa. III. The frontal wall of *Watersipora arcuata* Banta, with a revision of the Cryptocystidea. *J. Morphol.* **131,** 37–56.

Bassler, R. S. (1953). Bryozoa. *In* "Treatise on Invertebrate Paleontology" (Moore, R. C., ed.), Part G. University of Kansas Press and Geological Society of America, Lawrence, Kansas.

Black, R. M. (1970). "The Elements of Palaeontology." Cambridge University Press, Cambridge.

Boardman, R. S. (1968). Colony development and convergent evolution of budding pattern in "Rhombotrypid" Bryozoa (Annoscia, E., ed.). Proceedings of the First International Conference on Bryozoa. *Atti Soc. It. Sci. Nat. Mus. Civ. Stor. Nat. Milano* **108,** 179–184.

Boardman, R. S. and Cheetham, A. H. (1969). Skeletal growth, intracolony variation, and evolution in Bryozoa: a review. *J. Paleontol.* **43,** 205–233.

Boardman, R. S., Cheetham, A. H. and Cook, P. L. (1969). Intracolony variation and the genus concept in Bryozoa. "Proceedings of the North American Paleontological Convention." Part C, 294–320.

Brown, P. R. (1963). Algal limestones and associated sediments in the basal Purbeck of Dorset. *Geol. Mag.* **100,** 565–573.

Casey, R. and Bristow, C. R. (1964). Notes on some ferruginous strata in Buckinghamshire and Wiltshire. *Geol. Mag.* **101,** 116–128.

Cheetham, A. H. (1954). A new early Cretaceous cheilostome bryozoan from Texas. *J. Paleontol.* **28,** 177–184.

Cook, P. L. (1968). Polyzoa from West Africa. The Malacostega. Part 1. *Bull. Br. Mus. Nat. Hist.* (*Zool.*) **16,** 115–160.

Cook, P. L. and Hayward, P. J. (1966). The development of *Conopeum seurati* (Canu), and some other species of membraniporine Polyzoa. *Cah. biol. mar.* **7,** 437–443.

Cox, L. R. (1929). A synopsis of the Lamellibranchia and Gastropoda of the Portland Beds of England. *Proc. Dorset Nat. Hist. Arch. Soc.* **50,** 131–202.

d'Orbigny, A. (1849). Description de quelques genres noveaux de Mollusques bryozoaires. *Rev. Mag. Zool.* **2,** 499–504.

d'Orbigny, A. (1851–1854). "Paleontologie francaise, Terraine cretace." Vol. 5 (Bryozoaires). Paris.

Geological Society Phanerozoic time scale, 1964. *Q. J. Geol. Soc. Lond.* **120,** 260–262.

Gregory, J. W. (1894). On some Jurassic species of Cheilostomata. *Geol. Mag.* **2,** 61–64.

House, M. R. (1958). On the Portlandian zones of the Vale of Wardour and the use of *Titanites giganteus* as an Upper Jurassic zone fossil. *Proc. Geol. Ass.* **69,** 17–19.

Imlay, R. W. (1957). Paleoecology of Jurassic seas in the Western Interior of the United States. *In* "Treatise on Marine Ecology and Paleoecology" (Ladd, H. S., ed.), Vol. 2 (Paleoecology). *Geol. Soc. Am. Mem.* **67,** 469–504.

Lagaaij, R. (1963). New additions to the bryozoan fauna of the Gulf of Mexico. *Publ. Inst. Mar. Sci., Univ. Texas* **9,** 162–236.

Lang, W. D. (1914a). On *Herpetopora*, a new genus containing three new species of Cretaceous cheilostome Polyzoa. *Geol. Mag.* **51,** 5–8.

Lang, W. D. (1914b). Some new genera and species of Cretaceous cheilostome Polyzoa. *Geol. Mag.* **51,** 436–444.

Larwood, G. P. *et al.* (1967). Bryozoa. *In* "The Fossil Record" (Harland, W. B. *et al.* eds). Geological Society, London.

Opdyke, N. D. (1962). Palaeoclimatology and continental drift. *In* "Continental Drift" (Runcorn, S. K., ed.). Academic Press, New York and London.

Ryland, J. S. (1970). "Bryozoans." Hutchinson and Co., London.

Scott, R. W. (1970). Paleoecology and paleontology of the Lower Cretaceous Kiowa, Formation, Kansas. *Univ. Kansas Paleontol. Contrib.* article 52 (Cretaceous 1).

Silén, L. (1942). Origin and development of the Cheilo-ctenostomatous stem of Bryozoa. *Zool. Bidrag Uppsala* **22,** 1–59.

Silén, L. (1944). On the formation of the interzooidal communications of the Bryozoa. *Zool. Bidrag Uppsala* **22,** 433–488.

Silén, L. (1945). The main features of the development of the ovum, embryo, and ooecium in the ooeciferous Bryozoa Gymnolaemata. *Ark. Zool.* **35A.**

Thomas, H. D. and Larwood, G. P. (1956). Some "uniserial" membraniporine Polyzoan genera and a new American Albian species. *Geol. Mag.* **93,** 369–376.

Thomas, H. D. and Larwood, G. P. (1960). The Cretaceous species of *Pyripora* d'Orbigny and *Rhammatopora* Lang. *Palaeontology* **3,** 370–386.

Voigt, E. (1939). Über die Dornenspezialisation bei cheilostomen Bryozoen und die Nichtumkehrbarkeit der Entwicklung. *Palaeont. Z.* **21,** 87–107.

Voigt, E. (1968). On the Cretaceous age of the so-called Jurassic cheilostomatous Polyzoa (Bryozoa). *Bull. Br. Mus. Nat. Hist. (Geol.)* **17,** 1–45.

Woodward, H. B. (1895). The Middle and Upper Oolitic Rocks of England (Yorkshire Excepted). *In* "The Jurassic Rocks of Britain", Vol. 5. *Memoirs of the Geological Survey of the United Kingdom.*

45 | New Species of *Pyripora* d'Orbigny from the Cretaceous and the Miocene

G. P. LARWOOD

Department of Geology, Science Laboratories,
University of Durham, England

Abstract: *Pyripora magna* sp. nov. from the Upper Chalk (Campanian of England) and *Pyripora minuta* sp. nov. from the Chipola Formation (Miocene of Florida, U.S.A.) are described and compared with other species of *Pyripora*. The Campanian species *P. irregularis* Voigt (1967) and *P. parvicauda* (Voigt), 1930, are discussed and the latter excluded from the genus.

INTRODUCTION

The status of the genus *Pyripora* d'Orbigny was discussed by Thomas and Larwood (1956) and the Cretaceous species of *Pyripora* later revised (Thomas and Larwood, 1960). Voigt (1967) has subsequently described a further Cretaceous species. More recently, three new species of *Pyripora* have been discovered. One of these is described elsewhere by Frey and Larwood (1971) and includes the material described by Shaw (1967) as *Pyripora* sp. The other two species are described here.

SYSTEMATIC PALAEONTOLOGY

Phylum BRYOZOA Ehrenberg, 1831
Class **GYMNOLAEMATA** Allman, 1856
Order **CHEILOSTOMATA** Busk, 1852
Suborder ***Anasca*** Levinsen, 1909
Family Electridae Lagaaij, 1952
Genus *Pyripora* d'Orbigny, 1849

1. Pyripora magna *sp. nov.*
Holotype. Specimen encrusting fragment of echinoid test. Upper Chalk, Campanian, Zone of *Belemnitella mucronata.* Harford Bridges, Norwich, Norfolk. R. M. Brydone Collection. Sedgwick Museum, Cambridge, coll. no. S.M. B.75992.

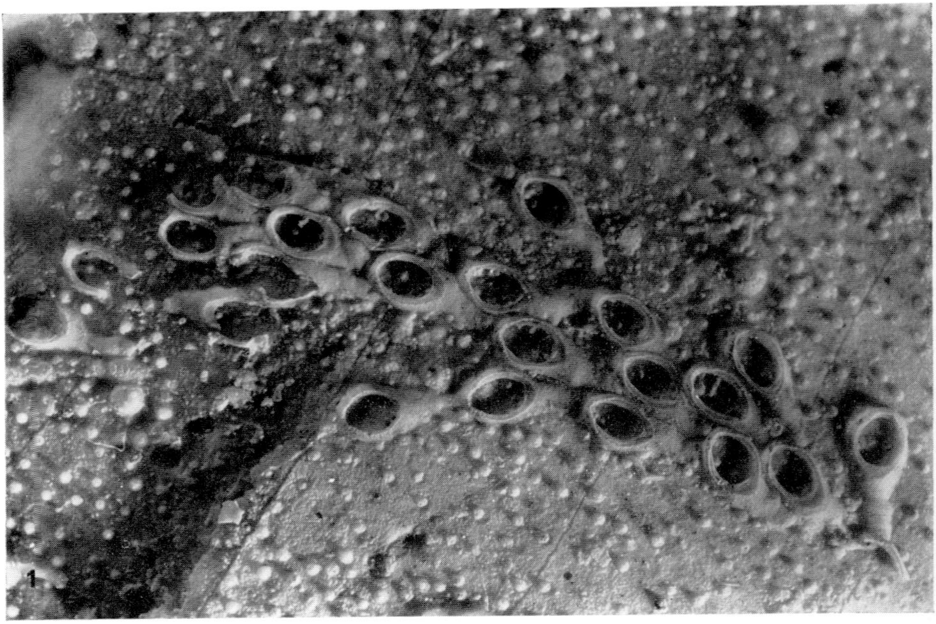

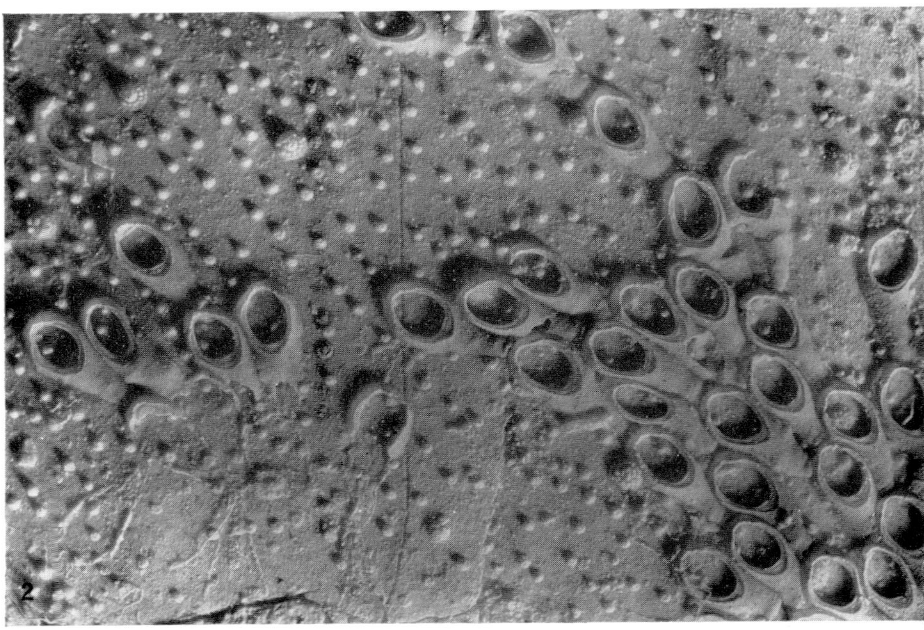

Paratype. Specimen encrusting fragment of echinoid test. Horizon, locality and collection as for the holotype. Sedgwick Museum, Cambridge, coll. no. S.M. B.75993.

Origin of Name. Pyripora magna is so named with reference to the large size of the zooecia.

Diagnosis. Pyripora species with uni- and pluri-serial budding; zooecia large (Lz=1·20 to 1·750 mm, lz=0·525 to 0·825 mm); apertural ratio 1·31 to1·70; caudae relatively broad, nearly as long as the rest of the zooecium.

Description. Zoarium encrusting, commonly with unilateral and, less frequently, bilateral budding. The commonest form of zoarial growth is by the distal-median addition of zooecia producing long uniserial rows. New distal-lateral zooecia may be developed by unilateral budding on either side of a distal-median zooecium. Bilateral budding occurs less frequently with the addition of two distal-lateral zooecia (one on each side of the distal-median zooecium) or, occasionally, by the development of lateral zooecia at right-angles to the parent zooecium. Both uni- and bi-lateral budding frequently give rise to a pluriserial zoarial habit with the zooecia of several rows in contact. The angle varies at which new rows of zooecia diverge.

Caudae relatively wide, straight or only slightly curved, normally just less than half the length of the rest of the zooecium. Distally the caudae expand gently into the smooth, broadly triangular proximal gymnocyst of the zooecium.

Zooecia pyriform, longitudinally oval, the distal half only slightly more inflated than the caudae; lateral and distal walls steep, the distal margin rounded, never pointed; calcified basal wall absent. Aperture longitudinally oval with an apertural ratio of 1·31 to 1·70. Cryptocyst a descending shelf, widest proximally but extending as a narrow shelf to the distal-lateral margins of the aperture forming a longitudinally oval opesia slightly smaller than the aperture. Where growth space is restricted by other zooecia or by irregularities in the substratum the normal pyriform zooecia may be distorted or much reduced in size. There are no examples of regenerated zooecia nor of zooecia with calcified frontal membranes. Kenozooecia, avicularia, spines and ovicells absent.

Plate I

Fig. 1. *Pyripora magna.* Paratype. S.M., B.75993 (× 13). Multi- and uni-serial growth of normal zooecia.

Fig. 2. *Pyripora magna.* Holotype. S.M., B.75992 (× 13). Multi- and uni-serial growth of normal zooecia. Note that the uniserial growth is sustained in areas where there is no restriction of the substrate.

Measurements. (All measurements given in mm.)

Measurement*	Number	Observed range
Lz (length of zooecium including caudae)	65	1·200–1·750
lz (width of zooecium)	66	0·525–0·825
ha (length of aperture)	61	0·625–0·850
la (width of aperture)	64	0·400–0·525
ho (length of opesia)	60	0·550–0·725
lo (width of opesia)	64	0·325–0·475
Lc (length of cauda)	68	0·350–0·900
lc (width of cauda at proximal end)	65	0·175–0·325
Apertural ratio (*ha*/*la*) = 1·31–1·70		

* See Thomas and Larwood, 1960. Data pertaining to the measurements of specimens is deposited in the British Museum (Natural History) London, Department of Zoology, Bryozoa Section, Harmer Library.

Remarks. The large size of the zooecia, their comparatively short and broad caudae and the low apertural ratio together with the extensive areas of pluriserial growth particularly distinguish *Pyripora magna* from other Cretaceous species of the genus. The zoarial habit most closely resembles that of *Pyripora texana* Thomas and Larwood (1956, 1960). Similar pluriserial expansions of zooecia are present in *P. texana* as well as long uniserial rows of zooecia. The pluriserial habit is more firmly established in *P. magna* and the zooecia are considerably larger than those of the American Albian species.

Other Cretaceous species in which the caudae are broad and shorter than the rest of the zooecium lack the distinctive pluriserial areas of zoarial growth. *Pyripora faxensis*, Voigt (1930), has large zooecia but these are more inflated than in *P. magna*, the aperture is relatively wider and the proximal ends of the caudae distinctly narrower.

Stratigraphical Occurrence. Upper Chalk, Campanian, Zone of *Belemnitella mucronata*, "Weybourne Chalk Horizon" of Peake and Hancock (1961).

2. Pyripora irregularis *Voigt (1967) and* P. parvicauda *Voigt (1930)*

Remarks. In a recent paper Voigt (1967) has described a new species of *Pyripora*—*P. irregularis*—from the Campanian of western Kazakstan, U.S.S.R. In discussing *P. irregularis* Voigt has compared it to *P. parvicauda* and figured, for comparison, a specimen (Voigt coll. 4629) which he assigns to the latter species. Thomas and Larwood (1960) accepted the species as correctly belonging to *Pyripora* basing their opinion on Voigt's (1930) description and figures. In his explanation to the

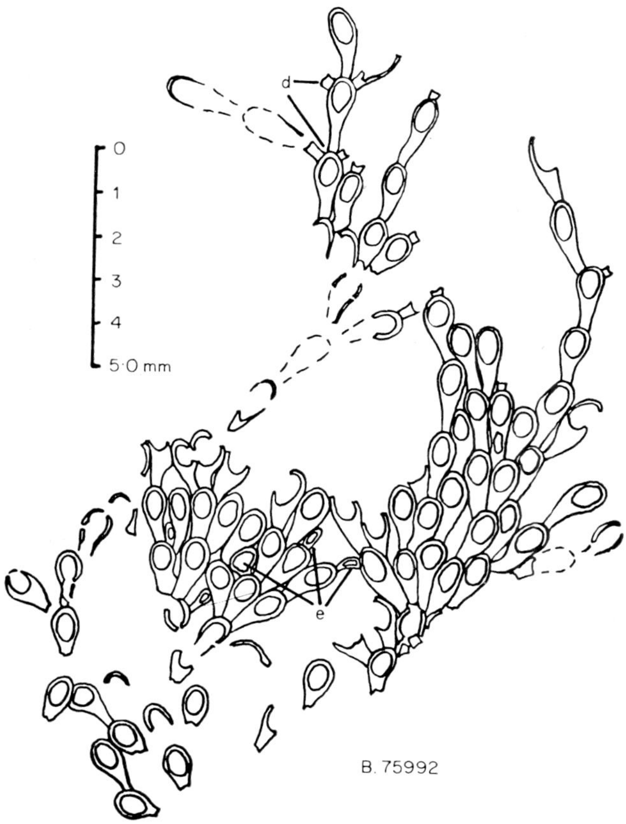

Fig. 1. *Pyripora magna*. Holotype. General distribution of multi- and uni-serial parts of the zoarium. Some zooecia, with limited growth space, are distorted (e). In some uniserial parts of the zoarium distal and distal-lateral budding occurs where there are no space restrictions (d).

illustration Voigt (1967) draws attention to the presence of a well developed hypostomial ovicell. The species, although possessing pyriform zooecia with short narrow caudae is thus excluded from *Pyripora* and may be placed provisionally in *Membranipora* s.l.

Pyripora irregularis Voigt (1967) seems to be most closely related to Cretaceous species of *Pyripora* possessing long caudae, e.g. *P. anglica* (Lang) and *P. laxata* (d'Orbigny), rather than to forms with only slightly developed caudae as Voigt (1967) suggests. The dimensions of the zooecia of *P. irregularis* are encompassed by those of *P. anglica* and *P. laxata*. However, the preservation of the holotype

of *P. irregularis*, the only available specimen, is unfortunately poor (Voigt, 1967) and the presence or absence of a cryptocyst or even of other structures such as spines or ovicells, cannot be determined. The species has been distinguished primarily by the irregularity of its growth pattern but this is not a suitable character on which to establish the species. Until better preserved material is available it is not possible to determine with certainty whether *P. irregularis* should be assigned to one of the caudate Cretaceous species of *Pyripora*. Voigt (1962) has described, as *P. dispersa* (Hagenow), a specimen of *Pyripora laxata* (d'Orbigny) from the Maastrichtian of the Crimea which is geographically, and possibly taxonomically, closest to *P. irregularis*.

3. Pyripora minuta *sp. nov.*

Holotype. Specimen encrusting a small gastropod shell. Miocene, Chipola Formation. Chipola River section, near Clarksville, Florida, U.S.A. Tulane University coll. loc. no. 547.

Origin of Name. P. minuta is named with reference to the very small size of the zooecia.

Diagnosis. Pyripora species with uniserial budding; zooecia very small ($Lz=$ 0·236–0·318 mm, $lz=$0·155–0·200 mm); apertural ratio variable; caudae very short and broad and slightly corrugated transversely.

Description. Zoarium encrusting, with both uni- and bi-lateral budding. Distal-median budding produces uniserial rows of zooecia which frequently give rise to other lateral zooecia. These are budded from the distal-lateral or, more commonly, lateral borders, more nearly at right-angles to the parent zooecium, on either or both sides.

Caudae formed as short and broad proximal extensions of the gymnocyst, often wider than long and occasionally absent.

Zooecia pyriform and sometimes oval or more nearly circular with rather flatly curved lateral and distal walls, the distal margin rounded; calcified basal wall lacking. Aperture large in relation to the whole zooecium with a variable apertural ratio; margin transversely corrugated. Cryptocyst present as a slight descending shelf along the proximal and proximo-lateral borders of the aperture. Opesia oval, only very slightly smaller than the aperture. Regenerated zooecia

---

Plate II

Pyripora minuta. Holotype. Tulane Univ. coll. loc. no. 547. (×53). Normal zooecia following areas between varices of gastropod, some zooecia on upper suture of gastropod with partly calcified frontals.

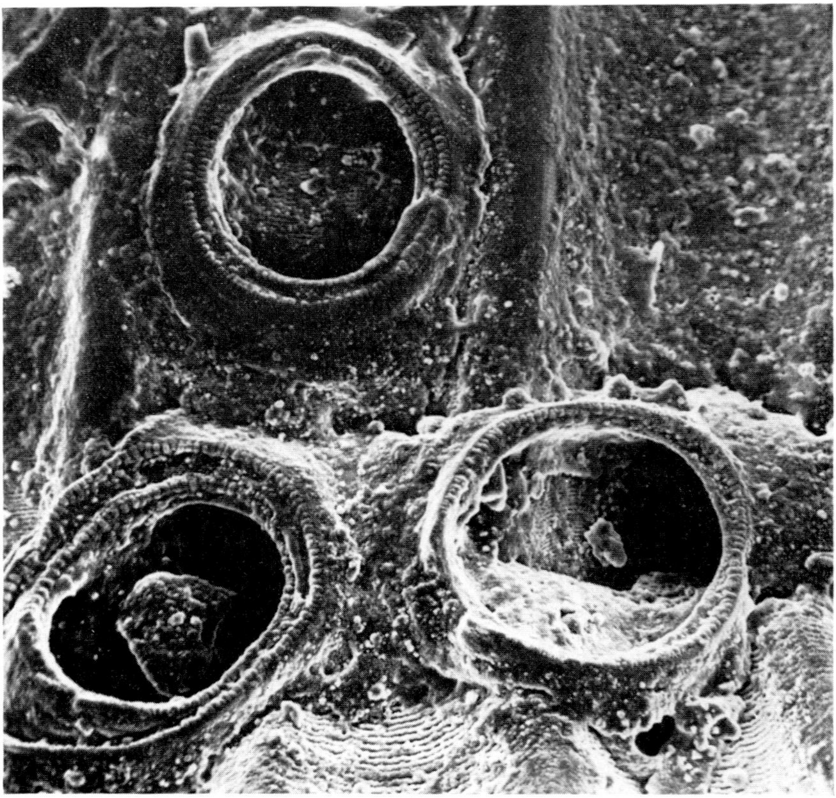

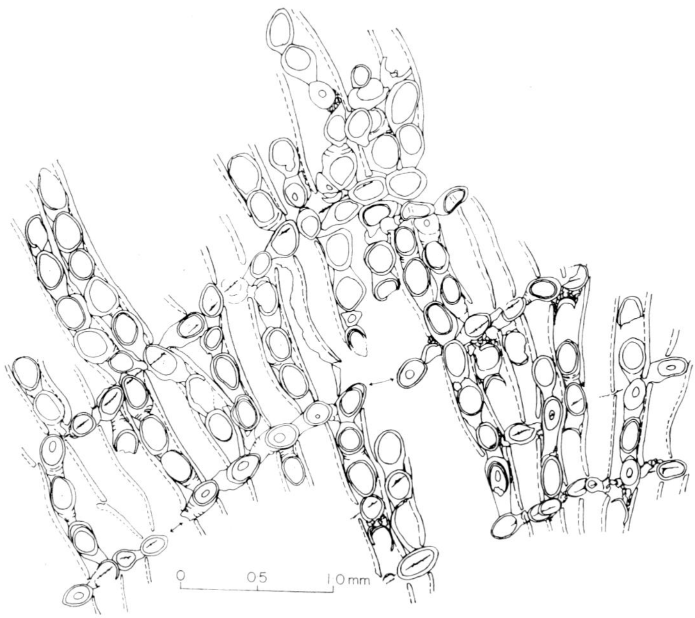

Fig. 2. *Pyripora minuta*. Holotype. General distribution of all zooecia of the holotype as an expanded diagram. Gastropod sutures run from bottom-left to top-right, varices are shown as double lines. Note the sequences of frontally calcified zooëcia following the sutures.

common, frontally calcified zooecia frequent. Calcification of the frontal membrane is usually incomplete leaving an oval or circular uncalcified area in the middle of the frontal wall. Traces of the outline of calcified opercula not seen. Kenozooecia, avicularia, spines and ovicells absent.

Remarks. Pyripora minuta is represented by the holotype only. The zoarium, with well over 100 zooecia, is well preserved but in parts slightly broken. The species is particularly distinct in that the zooecia are of very small size and possess short and broad caudae.

The zoarium of *P. minuta* encrusts a small gastropod with well developed varices. These, and the gastropod shell sutures, closely control the disposition of the rows of zooecia which are approximately at right-angles to one another

---

Plate III

Pyripora minuta. Holotype. (× 220). Normal zooecium (bottom right) and two regenerated zooecia.

following the varices and sutures (Fig. 2). Series of frontally calcified zooecia occur particularly along the gastropod sutures, but isolated examples also occur in some rows of zooecia between the varices. The zooecial apertural borders tend to be raised level with, or just above, the varices. The earlier rows of zooecia apparently were developed along the gastropod sutures and subsequent lateral rows of zooecia occupied the depressions between varices sometimes meeting and producing distortion of the zooecia.

Stratigraphical Occurrence. Miocene, Chipola Formation.

Measurements. (All measurements given in mm.)

Measurement	Number	Observed range
Lz	18	0·236–0·318
lz	18	0·155–0·200
ha	15	0·136–0·227
la	17	0·109–0·164
ho	14	0·127–0·214
lo	14	0·109–0·145
Lc	12	0·055–0·100
lc	12	0·005–0·136
Apertural ratio = 1·25–1·58*		

* Of the 22 zooecia measured, 8 have apertural ratios in the range 1·25 to 1·38 and 5 in the range 1·50 to 1·58, the remaining 9 zooecia were either regenerated or distorted in some way or frontal calcification obscured the borders of the apertures.

ACKNOWLEDGEMENTS

I am indebted to Dr A. W. Medd (Institute of Geological Sciences) for drawing my attention to the specimen of *P. magna* and to Dr R. J. Scolaro (University of Georgia, U.S.A.) for providing the specimen of *P. minuta*. I wish to thank Mr E. L. Håkansson (Institut for historisk Geologi og Palaeontologi, Copenhagen) for photographs of *P. minuta* and for discussion of the material and Mr G. Dresser (Department of Geology, University of Durham) for photographs of *P. magna*.

Part of the work on the material was carried out at the Institut for historisk Geologi og Palaeontologi, Copenhagen, during tenure of a British Council award under its Research Workers Interchange Scheme. Facilities afforded in the Institut and the financial assistance are gratefully acknowledged.

REFERENCES

FREY, R. W. and LARWOOD, G. P. (1971). *Pyripora shawi*; New Bryozoan from the Upper Cretaceous of Kansas (Niobrara Chalk) and Arkansas (Brownstown Marl). *J. Paleontol.* **45,** 969–976.

PEAKE, N. B. and HANCOCK, J. M. (1961). The Upper Cretaceous of Norfolk. *Trans. Nfk. Norwich Nat. Soc.* **19** (6), 293–339.

SHAW, N. G. (1967). Cheilostomata from Gulfian (Upper Cretaceous) rocks of southwestern Arkansas. *J. Paleontol.* **41,** 1393–1492.

THOMAS, H. D. and LARWOOD, G. P. (1956). Some "uniserial" membraniporine polyzoan genera and a new American Albian species. *Geol. Mag.* **93,** 369–376.

THOMAS, H. D. and LARWOOD, G. P. (1960). The Cretaceous Species of *Pyripora* d'Orbigny and *Rhammatopora* Lang. *Palaeontology* **3,** 370–386.

VOIGT, E. (1930). Morphologische und stratigraphische Untersuchungen uber die Bryozoenfauna der oberen Kreide. 1 Teil. Die cheilostomen Bryozen der jüngeren Oberkreide in Nordwestdeutschland, im Baltikum und in Holland. *Leopoldina* **6,** 379–579.

VOIGT, E. (1962). Upper Cretaceous Bryozoa of the European part of the U.S.S.R. and of adjacent regions (In Russian). *Izd. moskov. Univ.* **1,** 1–125.

VOIGT, E. (1967). Oberkreide–Bryozoen aus den asiatischen Gebieten der UdSSR. *Mitt Geol. Staatsinst. Hamburg.* **36,** 5–95.

46 | Calcification in the Colonial Stenolaemate, *Heteropora*

JUNE R. P. ROSS

*Department of Biology, Western Washington State College,
Bellingham, Washington, U.S.A.*

Abstract: Colonies of the stenolaemate ectoproct *Heteropora* from marine waters of western Washington show a sequence of six stages in calcification of the auto- and kenozooecial walls. Colonial and zooecial characters, previously used to differentiate species, are in part phases in the growth of the colony. Species of *Heteropora* show ecotypic variation in colony branching and are shape dependent on depth, substrate, currents and local physical topography.

INTRODUCTION

Ectoproct studies are currently undergoing an interesting change as neontologists and paleontologists are together seeking explanations and relations for the origin and evolutionary development of these lophophorate metazoans. The former trilogy of ectoproct studies isolated chronologically into (1) Paleozoic ectoprocts; (2) Mesozoic, Tertiary and Quaternary ectoprocts; and (3) Recent ectoprocts has now given way to broader studies of the various orders throughout the entire geologic and recent history. The zoologist, Borg (1944, 1965), in a number of studies on both recent and fossil forms provided much of the impetus for reassessing phylogenetic relationships between various orders of ectoprocts.

Borg (1944, 1965) in his examination of both Paleozoic and Recent representatives of the ectoprocts raised many questions about phylogenetic relationships, sometimes in only a few cryptic comments. He again raised the intriguing question which Nicholson (1880) and other paleontologists had raised earlier—namely, the possibility that recent forms of the family Heteroporidae were actually living representatives of the supposedly extinct order Trepostomata. In his analysis of the relations of cyclostome representatives, Borg included, in addition to the Heteroporidae, several other recent families—the Lichenoporidae, Horneridae, Crisinidae, Steghorneridae, Pseudoidmoneidae, Calvetiidae and Discoporellidae, with the fossil trepostomes. However, even after Borg's publications, the order Trepostomata has generally been regarded as an extinct order of Paleozoic bryozoans. For a zoologist, such as Borg, examination of the fossil Trepostomata is greatly handicapped because the order has been so poorly classified that phylogenetic relationships are obscured by taxonomic chaos. Borg's studies (1926, 1933, 1944, 1965) led him to hypothesize a polyphyletic origin for the order Stenolaemata in which he had two sub-orders, the Cyclostomata and the Trepostomata. Borg, in his classification attempted to reflect his phylogenetic interpretations, and posed for later scientists the problems of finding data to support the systematic assignments. For example, what data substantiate phylogenetic relationships between the Mesozoic–Recent stenolaemate families Heteroporidae and Horneridae and the various Paleozoic trepostome families? And what data conclusively establish a phylogenetic relation between recent stenolaematous families and the extinct Paleozoic families Fistuliporidae and Ceramoporidae.

In this paper a few findings are presented on the genus *Heteropora* from my investigations into the relations between the Heteroporidae and extinct trepostome families.

SPECIES OF HETEROPORA

Borg (1933) noted that "like most other Stenolaemata the species of the Heteroporina are not as a rule found in the littoral region proper. They seem to be hard bottom species occurring at moderate or, more often, at rather great depths". Borg (1933) noted that *Densipora corrugata* MacGillivray was an exception because it grew on sea weeds. In the present study *Heteropora* have been found at extreme low tide on the rocky coasts of Oregon and Washington on a hard rocky substrate. Species of the genus appear to occur in the lower part of the littoral zone where environmental conditions are suitable. Colonies of *Heteropora* have also been dredged from depths down to 140 m again from a hard rocky coarse gravelly substrate. Most of the colonies in the collection have been obtained from 3·5 to 35 m where the genus is abundant.

In the marine waters of western Washington and southern British Columbia (Fig. 1), three species of *Heteropora, H. pelliculata* Waters, H. *pacifica* Borg and *H. magna* O'Donoghue and O'Donoghue have been reported in the literature. Borg (1933) provided a key to four species of *Heteropora*, the three species noted above and *H. neozelanica* Busk. He distinguished the species on the form of the colony, characteristics of the distal ends of colonies, and features of the autozooids and kenozooids in the colony.

Examination and measurements on 110 colonies of *Heteropora* from an area

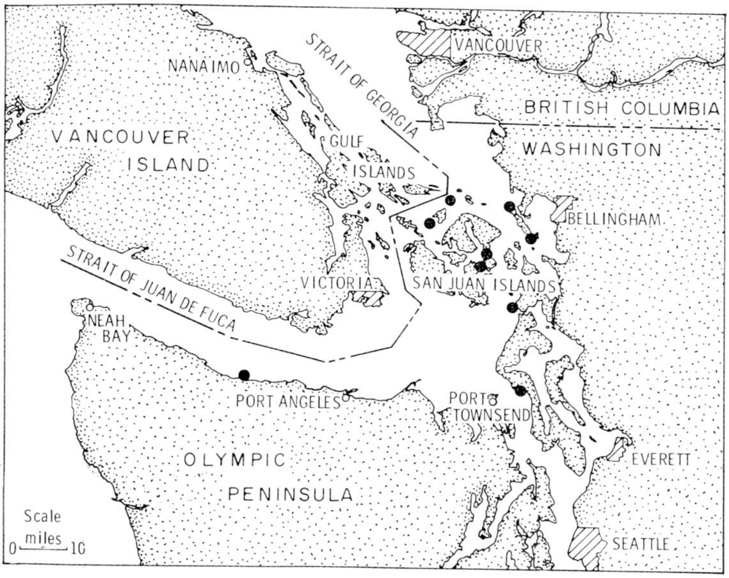

Fig. 1. Location of collections in environs of western Washington.

in the environs of western Washington and Vancouver Island (Fig. 1) presents several difficulties in attempting to identify the different species. Colonies of the three species, *H. pelliculata*, *H. pacifica*, and *H. magna*, occur together at several localities studied here. The major difficulty in using Borg's key arises when key features for different species occur in varying stages of development in a single colony. Further, different combinations of characters used to key out species occur in several colonies so that a single colony could be assigned to two or three species.

COLONY FORM

The colony form including the shape of the colony, the branching distance, and frequency of bifurcation and anastomosing are partly an expression of the age of the colony as well as various environmental factors including depth, substrate, and amount of protection from surge currents afforded by the topography of the immediate physical environment.

Colonies from the intertidal zone do not usually grow as high or as stoutly as colonies from most deeper water habitats. Colonies from the intertidal zone usually have a colony height of 1 cm, colony width of 2 cm and branch diameter of 1·5 to 2·0 mm; rarely a colony reaches colony height of 5 cm, colony width of 6 cm, and branch diameter of 2 to 3 mm. Colonies from a depth of 9 to 12 m may reach a colony height of 1·5 to 3 cm and branch diameter of 1·4 to 5·8 mm. Colonies from a depth of 17 to 25 m may reach a colony height of 3·5 to 4 cm and branch diameter of 1·8 to 7 mm. Colonies from 45 m and 69 to 154 m are low encrustng knob-like colonies without bifurcations.

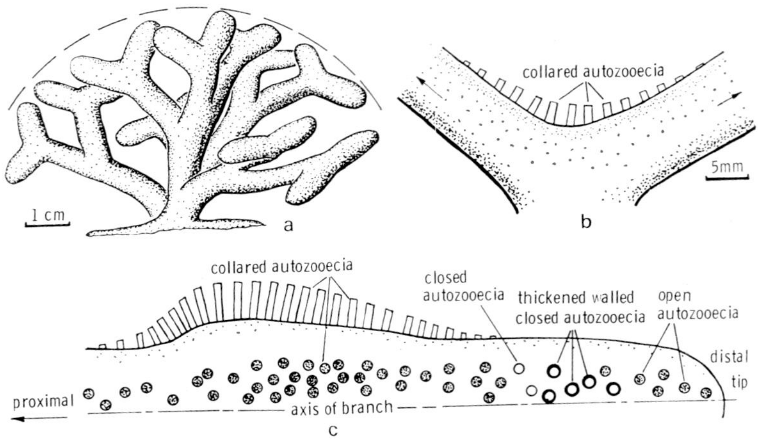

Fig. 2. a, Bifurcating colony of *Heteropora* shows a shape approximating a hemisphere; b, Collared autozooecia on bifurcating branch display taller collars in median region of bifurcation; c, Diagram of distal part of a colony branch shows (1) distal growing tip with open autozooecia, (2) an area immediately proximal to this tip with thick walled closed autozooecia, and (3) an area proximal to (2) with collared open autozooecia; the profile shows the varying length of collared autozooecia on a raised part of the branch and the plan view shows distribution of collared zooecia—the greatest number of collared autozooecia lie at central part of elevation.

The colonies from the intertidal zone grow more erect and generally do not show the hemispherical colony form in which branches bifurcate radially in all directions. Figure 2a shows the distinctive semicircular profile at the distal tips of a hemispherical colony from 17 to 25 m. The more spheroidal colonies are found in deeper water. However, colonies from some localities commonly show a preferred elongation suggesting that the major direction of growth was favored due to current or surge action. It appears that currents, surges, and other dynamics of the water place limitations on the peripheral growth of the colonies.

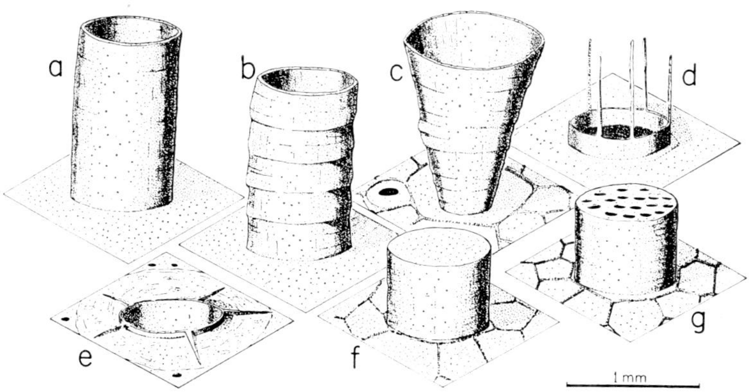

Fig. 3. Various aspects of autozooecia at different stages in zooecial growth. Kenozooecia have not been diagrammed in a, b, and d. a, Long collared open autozooecium (stage 4 of colony calcification); b, Long collared open autozooecium with annulose markings in hyaline calcium carbonate (stage 4 of colony calcification); c, Flared collar on autozooecium (stage 4 of colony calcification); d, Aciculate collared autozooecium (stage 5 of colony calcification); e, Low rim around autozooecium (stage 5), this autozooecium also displays buttressing of autozooecial wall by thickened adjacent kenozooecial walls; f, Short collared open autozooecium (stage 4, short collars precede long collars in Fig. 3a, b); g, Short collared closed autozooecium with perforate cover (modification in stage 4).

DISTAL GROWING TIPS

From an examination of growing tips of numerous colonies, distal branch ends have open autozooecia and kenozooecia. The walls of the autozooecia and kenozooecia are thin and there are no closed spaces between them (Fig. 4a, b; Borg, 1933). In many samples it is difficult to distinguish between autozooecia and kenozooecia as all the zooecia are of uniform size and are angular polygons in cross section. A few samples show autozooecia that are becoming round in cross section. In most samples, where autozooecia and kenozooecia could be differentiated, the kenozooecia were far more abundant than the autozooecia. However, there was one exception to this, where autozooecia and kenozooecia were differentiated, the kenozooecia were very few in number. There are no protrusions or projections of zooecial walls at the distal branch ends.

In profile the distal branch end is usually gently rounded, however, in some samples the growing distal branch end is bulbous and club-shaped (Fig. 4a). Borg (1933) found terminal brood chambers just below the distal ends of bulbous branches of colonies of *Heteropora pelliculata* and used the bulbous character (Borg, 1933) as one of the features for his key to the species of *Hetero-*

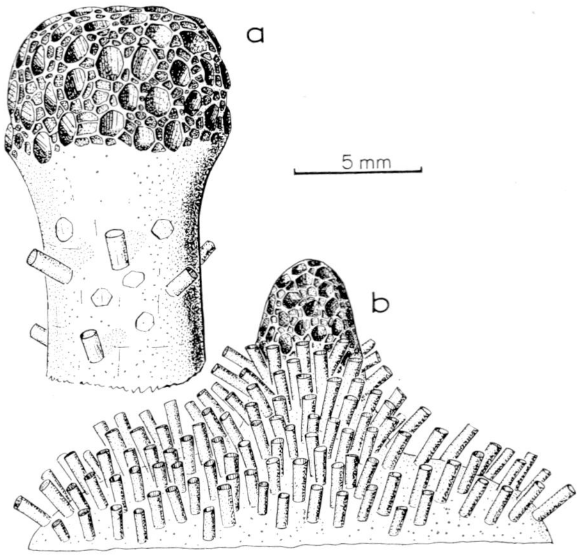

Fig. 4. a, Bulbous growing tip with polygonal autozooecia and kenozooecia; some collared autozooecia in part of branch behind swollen growing tip; b, Knob-like colony with broad encrusting base. Colony has numerous long collared open autozooecia, except at distal growing tip which has polygonal autozooecia and kenozooecia.

pora. However, several bulbous branches in my collection were sectioned and no brood chambers were found. This bulbous feature is not consistently present in all colonies from a single locality and appears to relate to a phase in growth and perhaps the influence of nutrients, including calcium, in the water.

CHARACTERISTICS OF GROWING COLONY

In the majority of colonies examined, autozooecia with long open cylindrical tubes are common and are here termed collared autozooecia (Figs 2b, c; 3a–d, f, g; 4a, b) to clarify discussion of this feature. Autozooecia also at times display short protruding ridges extending above the level of the zoarial surface and they are referred to as low rimmed autozooecia (Fig. 3e).

Growth of a colony, as interpreted by the various phases of calcification which occur in a colony, appears to be separable into six stages. The sequences of stages are:

(1) active zooecial growth of tips of colony with extension of branches and height of colony. This appears to be a continuous phase in the development of the colony;

(2) thickening of zooecial walls (both auto- and keno-zooecial) and the closure of zooecial openings with perforate covers;

(3) resorption of covers over autozooecial openings;

(4) secretion of long tubes (collars) around the autozooecial openings and usually with retention of covers over the kenozooecial openings;

(5) progressive resorption of the long tubes around autozooecia leaving remnants of the long tubes first as aciculate collars and later as low rims around autozooecia. Resorption continues with solution of covers over kenozooecia although sometimes resorption of covers appear to be earlier and to coincide with resorption of collared autozooecial tubes to the level of low rimmed autozooecia;

(6) closure of zooecial openings with perforate covers.

Stages 1 and 2 are continuous processes at the distally growing tips of the colony. Stages 3 through 6 occur in various parts of the colony from the proximal to near the distally growing tips. Distribution of these different stages (3 through 6) show no regular pattern on the surface of the colony, as the various stages seem to occur almost randomly across the colony. In some colonies collared autozooecia are more sparse in the distal parts. Some thickening of zooecial walls may occur during stage 6.

A colony (Fig. 2c) from a depth of 17 to 25 m shows stage 2 in the calcification of the colony with thickened zooecial walls immediately below the distal growing tip of the branch. The growing tip has open autozooecia and kenozooecia whereas proximal to this region some of the autozooecia and kenozooecia are closed by perforate covers. More distally along this same branch long collared open autozooecia and closed kenozooecia show a well advanced phase of stage 4.

A small colony from the lower intertidal zone has closed autozooecia and closed kenozooecia along most of the branches and typifies stage 2. Another colony from this collection has mainly open autozooecia and closed kenozooecia, this appears to be calcification stage 3.

In a collection from a depth of 10 to 12 m, some areas of colonies show closed kenozooecia and autozooecia with initial extensions of collars projecting above the colony surface, indicating the early phase of stage 4. In other parts of the branches the rimmed autozooecia are associated with open kenozooecia or kenozooecia with centrally perforate covers. These parts of the zooecia are undergoing resorption and are part of stage 5. In the perforate kenozooecia resorption appears to be extending radially from the center because the covers are wedge-shaped in thickness, being thinnest at the center.

Another colony from 17 to 25 m has open autozooecia and closed kenozooecia in the upper part of the colony, indicating stage 3, and collared autozooecia (Fig. 3a) mainly in the lower part of the colony, indicating stage 4. A colony collected in the lower intertidal zone at the same time of year as the 17 to 25 m collection has open autozooecia and open kenozooecia and some resorption of the calcareous colony, suggesting the later part of stage 5. Another colony in the same intertidal collection has long collared autozooecia and closed kenozooecia along the branches as far as the distal growing tip, indicating stage 5. These latter zooecial features are similar to Fig. 4b which is a knob-like non-bifurcating colony from 45 m.

In samples collected from 11 to 48 m some colonies are in their early developmental stages and are low mounds or low knobs showing initial bifurcation. These bifurcating colonies have numerous open collared autozooecia especially in the proximal part of the colony. Colonies in this sample commonly have dense opaque calcareous deposits in the most proximal parts of the colony but the autozooecial collars in the same region are thin-walled hyaline calcium carbonate, and in many instances, display annulose markings (Fig. 3b). Calcification in this region of the colony is at stage 5.

In examining the distribution of collared autozooecia, it is apparent that they are not mere accidents of preservation because they are commonly more abundant on the more exposed parts of branches while more protected areas of a colony lack them. In some colonies, as in Fig. 2b from 10 to 12 m, the collared autozooecia are longer at the area of bifurcation and become shorter distally along the branches. In some colonies having aggregations of collared autozooecia on the branches, several of the collars may be flanged (Fig. 3c). In many colonies where the collared tubes of the autozooecia are being resorbed, aciculate collared autozooecia (Fig. 3d) show the remnants of more extensive calcareous tubes. On some branches low autozooecial walls appear to be buttressed by thickened kenozooecial walls (Fig. 3e). These autozooecial walls extend only slightly above the colony surface and seem to represent a phase of wall thickening after resorption (stage 6). Occasionally aggregations of low collared autozooecia are closed by a hyaline perforate or imperforate cover (Figs 3f, g) and it appears these autozooecia represent an early and arrested phase of stage 4. Apparently the autozooecia may be closed and later the covers resorbed so that further extension of the collared autozooecia takes place.

ACKNOWLEDGEMENTS

These studies on *Heteropora* and other stenolaematous ectoprocts have been supported by National Science Foundation research grants NSF GB-4393 and NSF GB-7570, 1965–1971, for which I am most grateful. I greatly appreciate the assistance of Prof. C. A. Ross,

Ms K. McCain and Mr L. Vitt, Western Washington State College, in the collection and preparation of material for this study. Professors R. Fernald, P. Illg and E. Kozloff, Dept. of Zoology, University of Washington, have also been most helpful in the collection of the ectoprocts.

REFERENCES

Borg, F. (1926). Studies on Recent Cyclostomatous Bryozoa. *Zool. Bidr. Uppsala* **10,** 181–507.

Borg, F. (1933). A revision of the Recent Heteroporidae (Bryozoa). *Zool. Bidr. Uppsala* **14,** 254–394.

Borg, F. (1944). "The Stenolaematous Bryozoa," pp. 1–276. Stockholm Naturhist. Riksmuseum.

Borg, F. (1965). A comparative and phyletic study on fossil and recent Bryozoa of the suborders Cyclostomata and Trepostomata. *Ark. Zool.*, ser. 2, **17,** 1–91.

Nicholson, H. A. (1880). On the minute structure of the Recent *Heteropora neozelanica* Busk and on the relations of the genus *Heteropora* to *Monticulipora*. *Ann. Mag. Nat. Hist.* ser. 5, **6,** 414–423.

47 | Morphogenesis of Giant Avicularia and Ovicells in some Pacific Smittinidae

DOROTHY F. SOULE

Allan Hancock Foundation, University of Southern California, Los Angeles, California, U.S.A.

Abstract: Morphogenesis of giant avicularia and ovicells was examined by scanning electron microscopy of 25 new or previously described species of Smittinidae with imperforate frontals. Some giant avicularia form from separate basal wall folds in initial colony layers; others do not. Ovicells have components contributed by the transverse wall, the proximal zooecium, distal zooecium, and adjacent frontals.

INTRODUCTION

The nature of the so-called giant avicularia and the ovicells are significant characters in the systematics of Smittinidae, but lack of information on morphogenesis limits the understanding of the forms which may be found in collected materials. Numerous specimens were collected by D. F. Soule and J. D. Soule from coral reefs and other substrata in the Hawaiian Islands during systematics-ecology field studies. These were compared with selected eastern Pacific specimens by scanning electron microscopy, revealing several basic developmental features previously unreported.

Systematics-ecology studies (accepted for publication elsewhere) show that bryozoans frequently cover the greatest percentage of available substrata among epibionts inhabiting older corals. Thus their modes of development are important to the coral community. Observations on ancestrulae and body wall formation have been published elsewhere (Soule and Soule, 1972).

MORPHOGENESIS OF AVICULARIA

Among previously described genera of the Smittinidae which have a median suboral avicularium (Osburn, 1952), there are few species with frequent occurrences of additional zooecial avicularia, except in *Porella*, and the "giant" median avicularia are usually found only in *Codonellina*. In *Rhamphostomella*, with an asymmetrically placed suboral avicularium, additional frontal and giant avicularia may occur.

In contrast, in the *Parasmittina* species, all of which lack a median suboral avicularium, the avicularia are generally common, occurring around the peristome, or on the frontal in various positions (usually paired), around marginal areolae, on ovicells, and as giant interzooecial structures. Almost any frontal or areolar pore seems to be a potential site for an avicularium. This may be associated with the widespread occurrence of frontal budding in the formation of multilaminar colonies in the genus. The incidence of additional avicularia might also be associated with the high degree of infestation by epizoites such as hydroids, worms, and protozoans. The smittoideid species do not appear to respond similarly.

Several kinds of avicularia are seen in parasmittinids. Frontal zooecial avicularia are those which arise primarily from the few simple tiny pores which lie centrad of the areolae, in the imperforate area of the frontal. Avicularia may be acute or oval, usually small. Oral zooecial avicularia are usually connected to both tiny frontal pores and peristomal pores. Other avicularia may be connected to both frontal pores and areolar pores. Marginal avicularia seem to arise only from areolar pores and may be oriented in any direction on the margins, while others migrate inward on the frontal during development, retaining a tubular connection to areolar pores. These became fairly uniformly positioned and give the appearance of frontal avicularia in mature zooecia.

"Giant" avicularia are characteristically found in a number of parasmittinid species, and are often said to replace one of the smaller, paired avicularia on a zooecium. Many of them have been described as "spathulate", although there are differences as to whether the side of the mandibles are parallel, or whether the tip widens out into a "spoon" shape, and to what extent. Other mandibles are widely triangular or narrowly triangular. There has been disagreement as to whether such avicularia should be considered to be "interzooecial", "vicarious", or "adventitious", and these terms have been used differently by various authors.

Plate I

Fig. 1. Origin of giant avicularium in *Parasmittina* n. sp. "A", cf. *P. parsevalii* (Audouin) of Harmer (1957), photomicrographed at × 174.

Fig. 2. *Parasmittina delicatula* (Busk) cross section of tubular colony with interzooecial channels, × 174.

Fig. 3. *Parasmittina*, n. sp. "B", cf. *P. spathulata* (Smitt), at × 174.

Fig. 4. *P. delicatula*, submandibular area of rostrum with communication pore, × 116.

Fig. 5. *Parasmittina*, n. sp. "C", with lateral wall contributing to rostrum, × 174.

Fig. 6. *P. crosslandi* (Hastings) areolar giant avicularium, × 174.

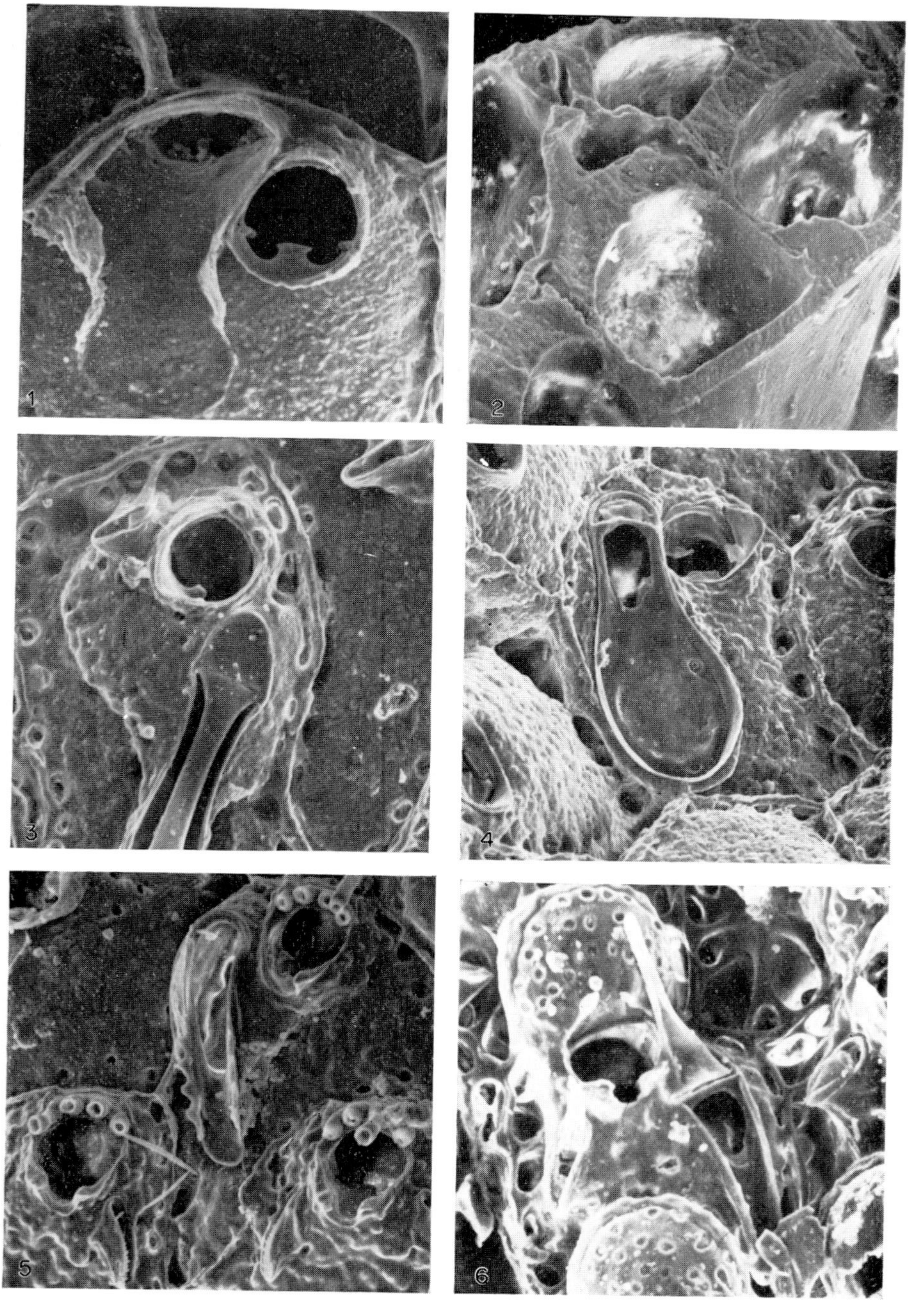
1
2
3
4
5
6

For example, Waters (1889), Hastings (1932), and Harmer (1957) each interpreted the giant avicularium of *P. obstructa* in a different way.

Present studies show that there are several distinct types of giant avicularia, with differences in origin and development. In two Hawaiian species, the giant avicularium was seen to arise simultaneously with the zooecium from separate folds in the basal wall, as if by twinning. Plate I, Fig. 1 shows new species "A", (described elsewhere) similar to the *S. parsevalii* (Audouin) of Harmer (1957); Figs 2 and 4 (Plate I) are of *P. delicatula* (Busk). The avicularium rostrum comes to lie to a greater or lesser extent on the frontal of the twin or "host" zooecium. Communication pores connect the avicularium with the host zooecium along the base of the rostrum, and also occur in the submandibular area of the rostrum (Plate I, Fig. 4). A cross section of the tubular colony of *P. delicatula* (Plate I, Fig. 2) shows the bud channel which produces the giant avicularium as it passes from the initial colony layer to a superficial (younger) layer of the tubular colony. The wall between the avicularian chamber and the host zooecium is thin, but it is double. No superficial calcification would be expected between them, since the exterior epithelium would no longer be in contact there.

In the primary layer of the colony of a new species similar to *P. delicatula*, *Parasmittina*, n. sp. "E", the giant avicularia are located in a more distal and transverse position to the host zooecium, but in older areas of the colony they lie more to the side and parallel (Plate III, Fig. 5); position is thus not necessarily a reliable diagnostic feature. Occasionally, giant avicularia produce smaller daughter avicularia on the submandibular rostral area. The rims of the rostra grow upward just as lateral walls do prior to frontal budding, and they are superficially calcified, as seen at the tip of the rostrum (Plate I, Fig. 4). New species "A" (Plate I, Fig. 1) often has a bilobate or trilobate mandible margin, apparently due to the trimming effect of repeated contact with the irregular rostral margin.

Parasmittina, new species "B" (Plate I, Fig. 3) has another type of giant avicularium in which the developing chamber is confluent with a number of frontal pores, plus one or more areolae on both sides of the zooecium. The chamber arises following primary frontal calcification, and no superficial deposition occurs on the portion of the frontal covered by the avicularium, again suggesting displacement of the secretary epithelium. The rostral chamber shows superficial calcification except in the submandibular area. The giant median avicularium of the codonellinids is formed in a similar manner but the chamber is connected with more distal areolar pores (Plate III, Fig. 6).

A somewhat different mode of development is seen in another new Hawaiian species *Parasmittina*, n. sp. "C", (Plate I, Fig. 5) and in *P. crosslandi* from Panama

(Plate I, Fig. 6). In the former, the heavier exterior wall of the avicularium appears to be an extension of the lateral wall, outside of the areolae. The more central portion of the rostral base rises as a thin partition, extending from the sinus, across the frontal pores, and turning to connect with the lateral wall portion. In *P. crosslandi*, the giant avicularium originates in similar fashion, but the rostral chamber is less inflated. The complexity of the areolar system can be seen well, lying beneath the avicularium and outside of it on the zooecial margin.

In Plate I the avicularia in Figs 1, 2, and 4 all originate from basal wall folds separate from the host zooecia, whereas those in Figs 3, 5, and 6, have not been seen to originate separately from the basal wall. Among the latter figures, the avicularium in Fig. 3 shows a primarily frontal origin, while Figs 5 and 6 incorporate lateral wall contributions as well as frontal and areolar pore connections.

In another new species, *Parasmittina*, n. sp. "I", which was probably included in *S. raigii* (Audouin) by Harmer (1957), two sorts of giant avicularia occur. One is placed along the distolateral shoulder somewhat in the manner of Figs 5 and 6 (Plate I) above; the other appears to originate as a bud between the two layers of the distal wall. In the latter, the rostral base in the submandibular area is divided by a center line which terminates in a tiny pore at the tip, reflecting the two wall layers beneath it (Plate III, Fig. 7). The premandibular area appears to be contributed from the distal portion of the zooecial frontal, behind the spines. In the initial layer of the colony, both kinds of avicularia originate from the basal wall. The shoulder avicularium is formed similarly to that in *P. delicatula*, and the other avicularium is separate, occupying almost the same basal space as a zooecium. Avicularia similar to the last type also form above old zooecia in multilayered colonies, with the apertures beneath clearly visible, especially in colonies invaded by worm tubes and other organisms. Other giant avicularia occur in similar *raigi*form species, oriented variously, and separated from adjacent zooecia by areolae. They may appear to be proximal or distal zooecial avicularia, and may be carried onto ovicells by spreading of the frontals.

Although several modes of giant avicularian formation can be distinguished, it does not seem desirable at this time to give each formation a new name, adding to the already extensive bryozoan terminology, when further work is needed to develop a more complete picture of the various morphogenetic modes involved.

MORPHOGENESIS OF OVICELLS

In the smittinids examined, the ovicells originate somewhat differently from the processes reported by Levinsen (1909), who stated that the distal wall does not take part. He attributed the formation of the two layers of the ovicell to the frontal membrane of the distal zooecium, sometimes with a cryptocyst between them, although it is difficult to decide what he meant in regard to some of the smittinids. Harmer (1903) related the ovicells to spine formation and suggested three possibilities for origin: (1), that the ovicell belongs to the fertile (proximal) zooecium; (2), that it belongs to the distal zooecium; or (3), that it is a modified individual. He relates spine homology as evidence for the first mode of formation by the fertile zooecium. He cites Calvet's evidence of communication pores between ovicells and distal zooecia as relating ovicells to the second mode of formation, and he attributes the third mode to ideas of "Nitsche and others".

There can be little doubt that ovicells relate at least in part to the proximal ("parent") zooecium, since the opercular and peristomal modifications are well known, especially in smittinids. The spines do not seem to be homologous to ovicells since spines are derived by infolding from a single flap extending from the adjacent layer of the distal wall, whereas ovicell formation is more complex. It is true that communication pores are also present around the base of the ovicell, as in separately budded individuals.

In *Parasmittina* observed, the earliest evidence of ovicell formation occurs as a Y-separation in the two layers of the distal wall (Plate II, Fig. 1, new species "D" see arrow). The more distal layer forms a concave calcareous disc, Levinsen's ectooecium, on the distal zooecium, increasing by concentric growth. The adjacent layer of the distal wall turns proximally and a small portion grows up in the spine base, forming an arched support. A second calcareous disc grows from the peristomal rim around the spines and comes to lie atop the first disc (Levinsen's entooecium). A bud apparently arises from between the two distal

---

Plate II

Fig. 1. *Parasmittina*, n. sp. "D", showing outer calcareous disc formed by Y separation of distal wall (arrow) and inner calcareous disc from distal portion of peristome rim. Note fine grained primary frontal in distal zooecia, frontal pores for potential avicularian sites, photomicrographed at × 116.

Fig. 2. Ovicell globe forming around pores, *Parasmittina*, n. sp. "C", × 174.

Fig. 3. Large ovicell with external flaps contributed by encroaching frontals of two distal zooecia and proximal zooecium, in *Parasmittina* n. sp. "E", cf., *P. delicatula*, × 174.

Fig. 4. Ovicell common to two randomly oriented zooecia, in *Parasmittina*, n. sp. "D". Overgrown ovicell in lower left corner, × 116.

Fig. 5. Frontally budded daughter zooecia in *P. marsupialis* (Busk), × 58. Normal site for ovicell in other species is at distal wall opening.

Fig. 6. Earliest evidence of beginning ovicell formation (arrow) in distal wall separation, in *Parasmittina*, n. sp. "C", × 174.

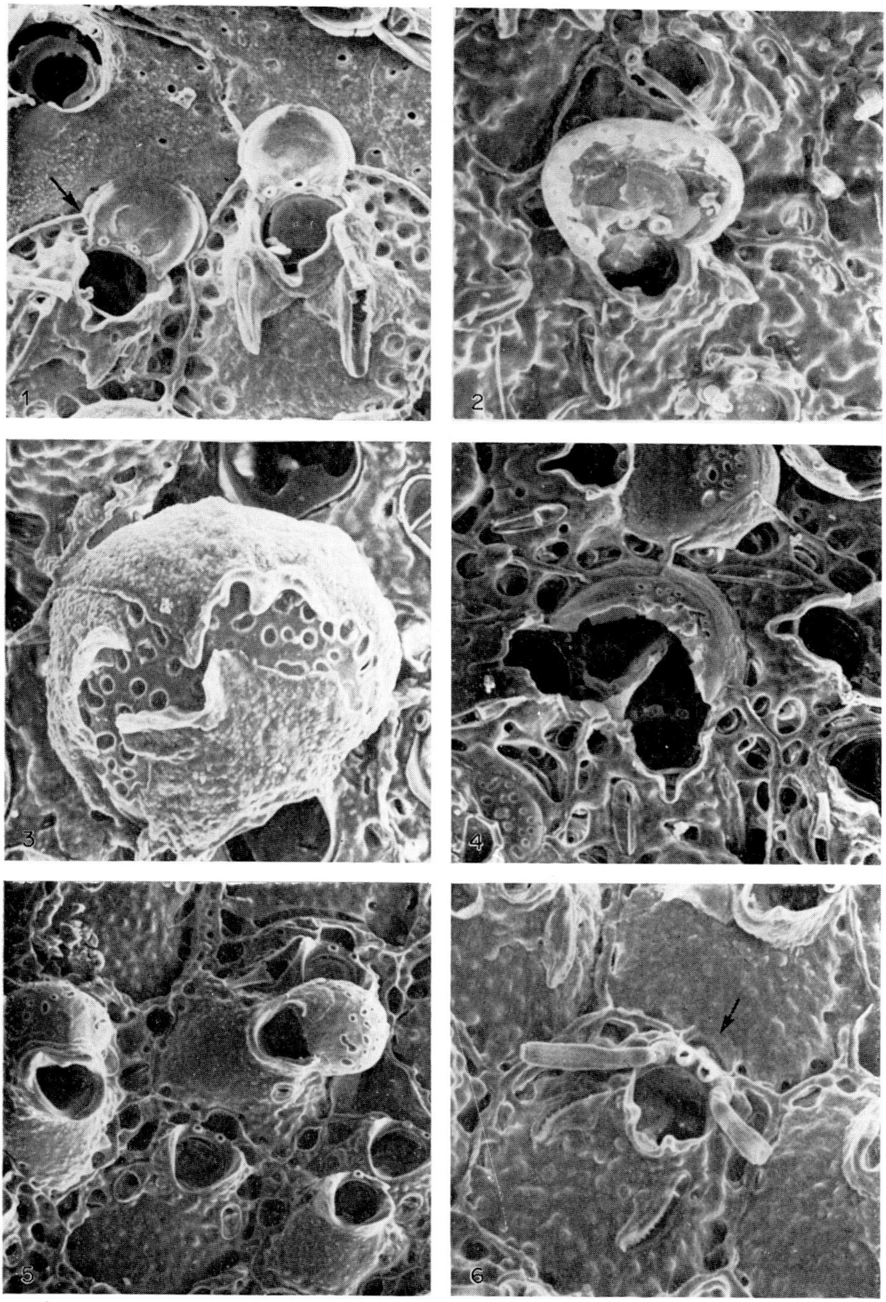
1
2
3
4
5
6

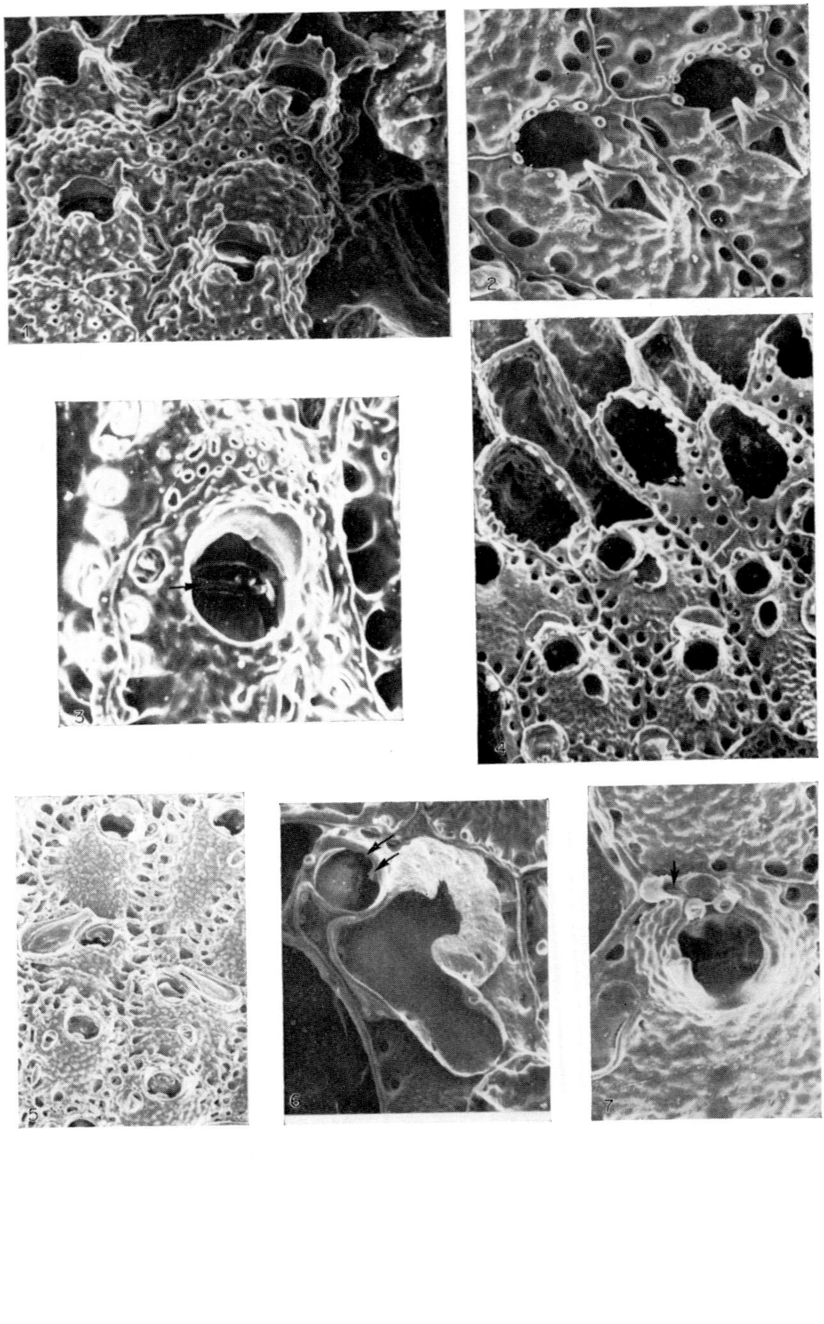

wall layers and forms a pad between the two discs. The edges of the two discs fuse and grow upward and forward to form the ovicell globe.

There are apparently species differences in the various growth rates of structures; in Plate II, Fig. 2, the outer layer forms fingerlike projections around the sites of pores while the interior disc is still small. Another species, a smittoideid (Plate III, Fig. 4), appears to have a calcareous disc which originates from the peristome and an inner membranous disc which comes from around the spine bases, becoming calcified later. A remnant of an inner vesicle or layer is sometimes seen also. In many species, communication pores can be seen connecting either one or both disc layers, to the zooecium beneath. In most species examined, a small, oval operculum was found (Plate III, Fig. 3 arrow), especially in ovicells containing eggs.

The separation in the distal wall where ovicell formation is initiated in most parasmittinids examined, is best seen in *P. marsupialis* (Plate II, Fig. 5), and can be compared with another species, "C" (Plate II, Fig. 6). However, *P. marsupialis* does not produce ovicells in that position. Instead, the "mother zooecium" produces a smaller, frontally budded "daughter zooecium", complete with spines. The daughter rapidly produces an ovicell which is suspended from the distal area of the peristome, sometimes before the daughter zooecial frontal

---

Plate III

Fig. 1. *Smittina*, sp., showing transverse wall separation at sides of peristome, ovicells immersed in distal zooecia, photomicrographed at × 174.

Fig. 2. Smittoideid, n. sp. "F", with transverse wall located similar to *Smittina*, sp., × 174. Note *Thalamoporella* spicules in Figs 1, 2.

Fig. 3. Ovicell operculum (arrow) opened into peristome. Zooecium operculum, displaced beneath, usually closes both openings in codonellinid, n. sp. "G", which lacks median denticle.

Fig. 4. Forming walls and frontals, ovicell calcareous outer disc and uncalcified inner disc during development. Not communication pores linked to areolae beneath calcareous disc, in smittoideid, n. sp. "H".

Fig. 5. Varied orientation of giant avicularia in *Parasmittina*, n. sp. "E", cf. *P. deliculata*, × 58.

Fig. 6. Giant frontal avicularium connected to distolateral areolae in codonellinid, n. sp. "G", × 116. Lower primary apertural rim ends at tiny cardelles midlaterally, upper primary rim ends at larger cardelles in normal proximolateral position instead of merging into median denticle as in parasmittinids.

Fig. 7. Beginning of distal giant avicularium in *Parasmittina*, n. sp. "I", cf. *P. raigii* (Audouin) of Harmer. Note rostrum base separation along transverse wall similar to ovicell formation. Giant lateral areolar avicularium is directed proximally, × 174.

is completely closed. The globe may drop down onto the parent frontal, but does not usually occlude the parent aperture, which seems to remain functional. This species also has peculiar vertical wall structure, in which the layers do not fuse completely, leaving small vertical channels open to the basal surface. The tops of the separate lateral wall layers overgrow above the level of the frontal and fuse in places, forming irregular tubules on the margins of the zooecia. These may be associated with the production of frontally budded ovicells.

Following the normal completion of ovicells, frontal from adjacent zooecia builds up superficial calcification (Harmer's ectooecium). The patterns are fairly constant for different species; some are only ringed distally by encroaching frontal and never covered. Others may become almost completely submerged except for a small crescent of pores, while others have a flap, or several flaps, covering the distal part (Plate II, Fig. 3). The number of flaps varies with the number of adjacent zooecia whose frontals can come in contact with the ovicell, rather than having a certain number of flaps for a given species.

After completion of the ovicell, the peristome usually fills in the distal gap where the spines were, and often forms a distinctive pattern, perhaps after the ova have been shed. The peristome may become tubular, or form a brim on the ovicell globe. Encroaching frontal may carry adjacent avicularia onto the ovicell as well. Much material must be examined before it can be concluded that as a species character, the peristome never fills in, or that the frontal never encroaches, since growth stages differ.

Some influence apparently operates between zooecia, for ovicells merge in orderly fashion (Plate II, Fig. 4) where frontal budding produces random orientation. Similarly, where two zooecia bud distal to one ovicelled zooecium a two chambered ovicell is formed with one aperture. Ovicells examined appear to be produced by all three of Harmer's methods mentioned above. The layers of the transverse wall contribute the "ectooecium", a buttress, and possibly part of the "entooecium", with a tissue pad between the layers. The proximal zooecium forms the entooecium, plus an internal vesicle, the operculum, and the ovum. The distal zooecium provides support and is connected to the ovicell by communication pores; along with adjacent zooecia it provides the superficial coverings or flaps derived from expanding frontals. Although the ovicell belongs primarily to the zooecium, it should probably be considered a function of the colony, even though it is not communal as it is in cyclostomes.

Insufficient material was available to trace development closely in smittoideids and *Smittina*, but body wall formation appears to be more variable, and apertures are set more terminally than in parasmittinids. Further investigation by scanning microscopy and in living material is needed to clarify the various processes.

ACKNOWLEDGEMENTS

Scanning microscopy was done through the kindness of Dr Henry Lee, South El Monte, California. Systematics-ecology studies were supported in part by NSF Grants GB5208 and 7723.

REFERENCES

HARMER, S. F. (1903). On the morphology of the Cheilostomata. *Qu. J. Micros Sci.* **46,** 263–350.

HARMER, S. F. (1957). Polyzoa of the Siboga Expedition, Pt. IV. Cheilostomata Ascophora. II. "Siboga Expeditie," 28 d., 507 pp. E. J. Brill, Leiden.

HASTINGS, A. B. (1932). The Polyzoa with a note on an associated hydroid. Great Barrier Reef Expedition 1928–29. *Br. Mus. nat. Hist.* **4** (12), 349–458.

LEVINSEN, G. M. R. (1909). Morphological and systematic studies on the cheilostomatous Bryozoa. *Nat. Forfat. Forlag.*, 431pp. Copenhagen.

OSBURN, R. C. (1952). Bryozoa of the Pacific Coast of America. Univ. Southern California. *Allan Hancock Pacific Exped.* **14** (2), 271–612.

SOULE, D. F. and SOULE, J. D. (1972). Ancestrulae and body wall morphogenesis of some Hawaiian and eastern Pacific Smittinidae (Bryozoa). *Trans. Am. Microsc. Soc.* **91** (3), 251–260.

SOULE, D. F. and SOULE, J. D. (1973). Morphology and speciation of Hawaiian and eastern Pacific Smittinidae (Bryozoa, Ectoprocta). *Bull. Am. Mus. nat. Hist.* (accepted for publication).

WATERS, A. W. (1889). Bryozoa of New South Wales. *Ann. Mag. nat. Hist.* ser. 6 **4,** 1–24.

48 On the Morphology and Systematics of Some Late Cretaceous Cyclostomatous Bryozoa

L. A. VISKOVA

Palaeontological Institute, Academy of Science, U.S.S.R., Moscow, U.S.S.R.

Abstract: Study of some Late Cretaceous Cyclostomata (*Tubuloporina*, *Cancellata*) by thin sections has revealed the fact that polymorphism, distinctive interarrangement of zooids, similar or different in their structure and functions, as well as microstructure of skeletal elements are of a great importance for the systematics of the group in question. The idea of the special importance of ovicells in the systematics of cyclostomatous Bryozoa has been questioned: in most of the generally accepted families including forms with sharply different zoarial structures but with similar gonozooecia (the genera *Petalopora*, *Crisina*, *Sulcocava* and many others in the *Petaloporidae*). Genetic unity of each family belonging to this order should be established on the grounds of similarities in budding and growth patterns of zooids which determine morphological constancy of zoaria. This view-point allows us to establish the family *Siphoniotyphlidae* and revise the family *Petaloporidae*.

Cyclostomatous Bryozoa are widespread in the Upper Cretaceous deposits of the U.S.S.R. They are notable for good preservation, large numbers and diversity of composition. Late Cretaceous Cyclostomata are, in fact, represented in U.S.S.R. territory by almost all kinds of taxa referred to in the "Treatise". The relationships of existing cyclostome taxa are uncertain. According to Bassler (1953) Cyclostome families include forms with similar gonozooecia. But, as a rule, many genera in most of the families have a completely different zoarial structure and have practically nothing in common but the similarity in the ovicell structure (for example, *Petalopora*, *Crisina*, *Sulcocava* in the Petaloporidae). On the other hand, not infrequently, the zoaria differ only in the shape and size of their ovicells, e.g. *Cavarinella*, *Leioscoecia*, *Tretocycloecia* are included in different families, the Petaloporidae, Leiosoeciidae and Tretocycloeciidae respectively. If the gonozooecium is absent, the researcher can either classify the material under study with any of the above families or with the family

Diastoporidae comprising the forms with no gonozooecia. The fact of their absence, however, does not necessarily mean that they have not been developed and may be accounted for by fragmentary material.

A formal consideration of one criterion only, without taking into account other morphological characteristics of the cyclostomatous zoaria, does not allow an adequately accurate systematization of the genus and results in the formation of heterogeneous families.

The present study of Petaloporidae included an analysis of the complex of their characteristics and leads to the conclusion that the family consists of three groups, which can be classified as separate families. The groups differ in their style of budding, in the peculiar mutual positions of some zooecia, and in special developments of polymorphism and in the microstructure of skeletal elements.

The first group comprises the Petaloporidae proper. The zoaria are branched, as a result of axial budding, the zooecia protruding in all directions. On the marginal parts of zoarium, between zooecia, mesopores are to be found (Plate I, Fig. 1). This group may include the following genera: *Petalopora*, *Cavarinella*, *Multicavea*, *Multizonopora*, *Sparsicavea* and others.

The second group is the Crisinidae. Zoaria branching formed by dorsal budding, zooecia protruding to the frontal only. On the dorsal side tergopores, or pore-ducts can be detected, the mesopores being either on the frontal or absent (Plate I, Fig. 3). Crisinidae may include the following genera: *Crisina*, *Reteporidea Filicrisina*, *Hemicellaria*, *Laterocavea*, *Multicrisina* and others.

The third group is the Sulcocavidae. Zoaria branching, formed by axial budding, zooecia protruding onto two opposite sides. The narrow lateral areas are smooth or with longitudinal ribs. Between zooecia peculiar tear-shaped pores, known as composite pores (Viskova, 1969), can be developed. They look like large curved cavities, which, obviously, owe their formation to the fusion of pore-ducts extending from the pores in the walls of two adjacent zooecia.

Plate I

Fig. 1. *Petalopora costata* d'Orbigny (after Bassler, 1953). Showing axial budding of zooecia with intervening mesopores; a, c, d, × 20; b, × 8.

Fig. 2. *Sulcocava cristata* d'Orbigny, PIN N2052/751, Volsk; Lower Maastrichtian; a, frontal view, × 10; b, lateral view, × 10; c, longitudinal section, × 32, showing axial budding of zooecia and composite pores (c.p.); d, transverse section, × 32.

Fig. 3. *Crisina macropora* Marsson, PIN N2052/1751, Belogorsk; Upper Maastrichtian; a, frontal view, × 10; b, dorsal view, × 10; c, longitudinal section, × 32, showing dorsal budding of zooecia and tergopores (t.); d, transverse section, × 32.

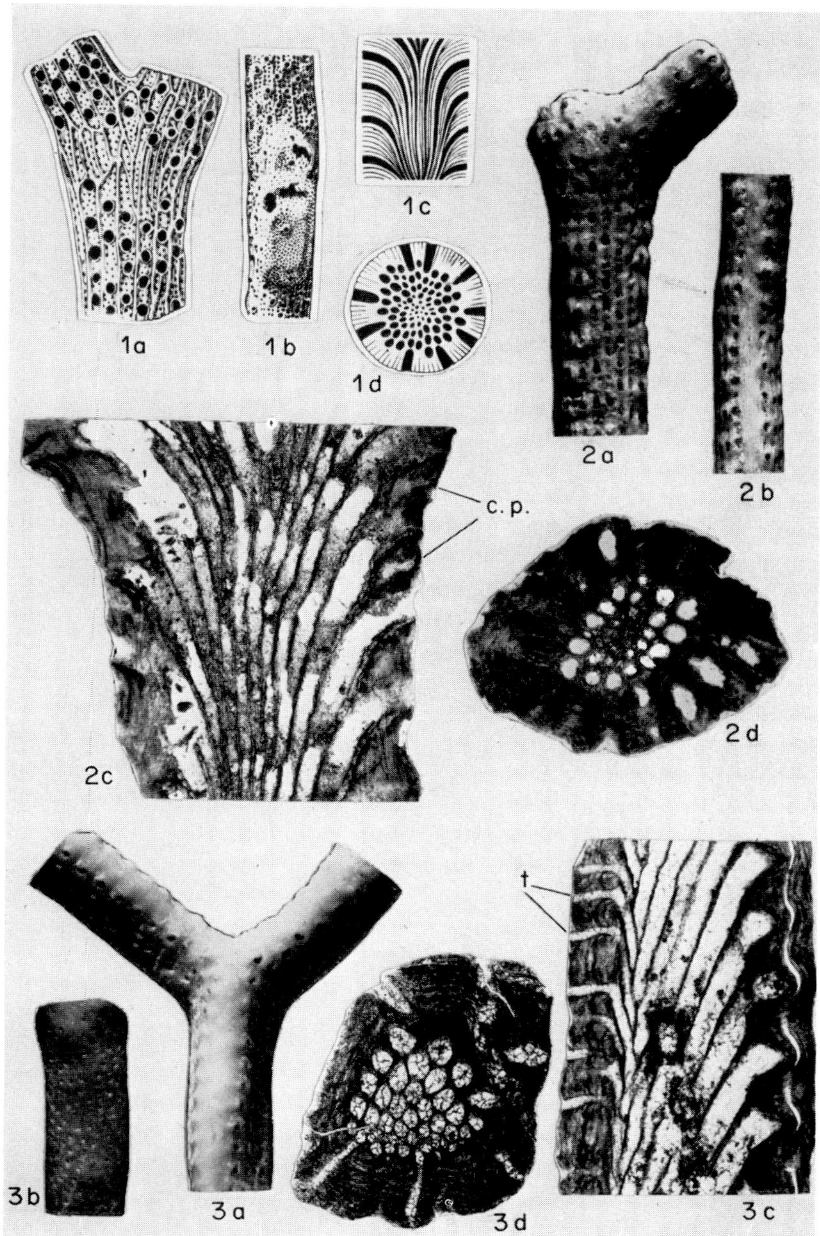
1a
1b
1c
1d
2a
2b
c. p.
2c
2d
t
3b
3a
3d
3c

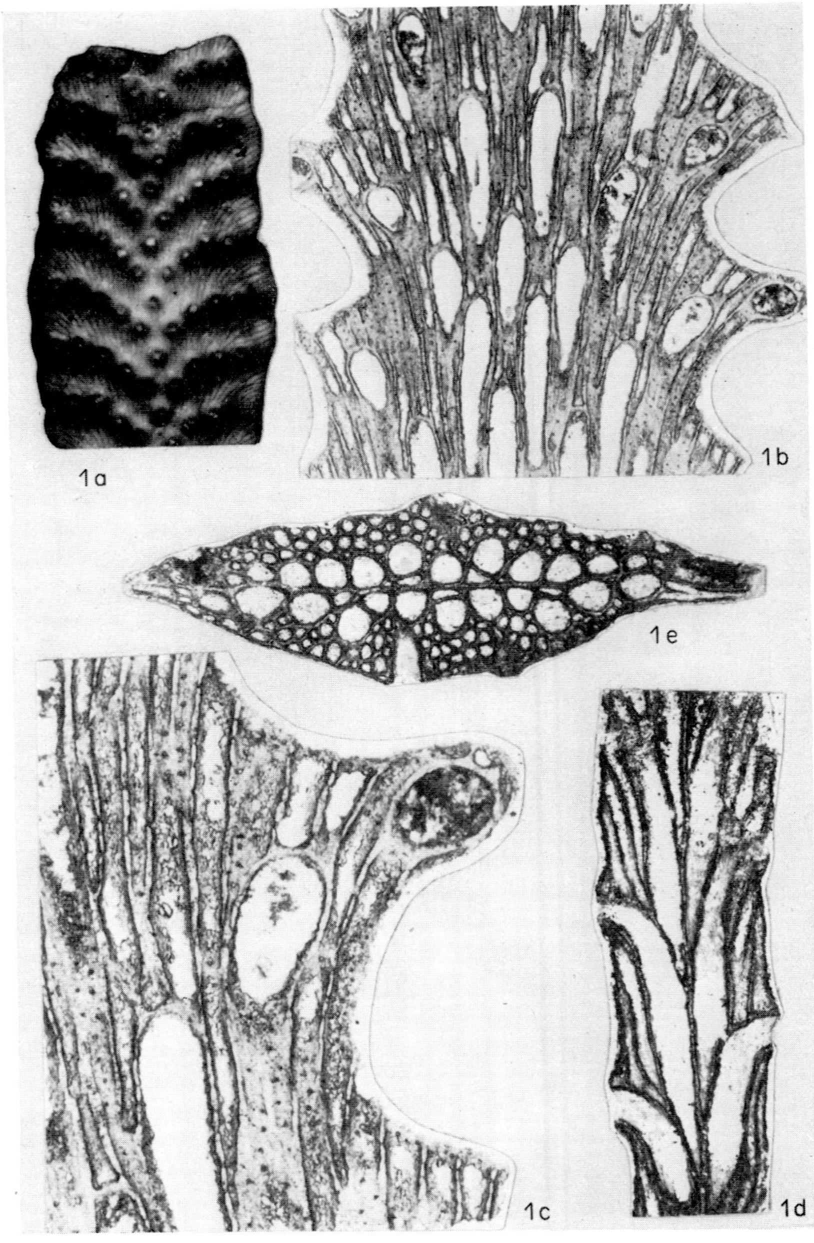
1a
1b
1e
1c
1d

With the help of these canals the cavities of zooecia were connected with the composite pores, which, in turn, were connected with the environment by oval or tear-shaped openings (Plate I, Fig. 2). This family includes two genera: *Sulcocava* and *Laterocava*.

The differences in the structure of zoaria of these three groups are beyond doubt. Of great significance are not only the different modes of budding and the resulting peculiarities of the mutual position of zooecia and heterozooecia but the degree and character of polymorphism is also very important. Petaloporidae have mesopores only which ordinarily appear at later stages of zoarial development. In most members of Crisinidae, apart from mesopores developed only on the frontal, tergopores are also present forming the dorsal side of the zoarium. The tergopores were budding beginning from the base of zooecia and were developing simultaneously with the latter. It should be emphasized that mesopores and tergopores are zooecial features. They are skeletal elements of heterozooids resulting from the division of functions in the zoarium. The pore-ducts and composite pores are structures resulting from the considerable thickening of the outer wall of zooecia and a necessity to retain gas exchange processes.

The structure of gonozooecia would seem to have specific importance rather than family significance.

As mentioned, the members of the family Diastoporidae are characterized by one negative feature, the absence of ovicells. Voigt (1967) discovered gonozooecia in two genera of this family—*Siphoniotyphlus* and *Clinopora* and separated them into an independent family Siphoniotyphlidae. Apart from the peculiarities of gonozooecia Voigt (1967) stated in the diagnosis of this family, that kenozooecia form a peculiar striped or netted structure on the zoarial surface. Unfortunately, the structure of kenozooecia, the character of their development, their connection with zooecia and their role in zoarial life have not been determined. The specimens of Siphoniotyphlidae available to me have no gonozooecia, but nevertheless it is quite obvious that they should be classified with this family (Plate II, Fig. 1). Substantial evidence obtained from the study of Siphoniotyphlidae in thin section revealed a peculiar structure in their bifoliate zoaria which were budded with the formation of a median lamina. The structure

---

Plate II

Fig. 1. *Siphoniotyphlus marginatus* Viskova, sp. nov. holotype PIN N2922/3, Gubkin; Coniacian; a, frontal view, × 10; b, tangential section, × 32, c, tangential section, × 72, showing kenozooecia and interzooecial pores; d, longitudinal section, × 32; e, transverse section, × 32.

of kenozooecia is especially remarkable. They differ from all the types of hetero-zooecia known among the fossil Bryozoa. The kenozooecia gemmate from the distal end of the zooecia going vertically upward parallel to the walls of the adjacent zooecia. These tubular formations have no openings. They either lean to the lower wall of the over-lying zooecium at its distal end, or turn away to the margin to become flat at the very edge of a zoarium, their walls forming a thin edge or border. Lying between individual zooecia, or their rows, keno-zooecia are likely to have strengthened the zoarium and connected individual zooids and their rows. This connection was accomplished by means of numerous pores penetrating the walls of zooecia and kenozooecia. It was through these pores that the kenozooids having no polypides are likely to have received, from other zooids, nutritients for their development. The structure of *Siphoniotyphlus* and *Clinipora* is peculiar enough for them to be correctly classified with the Siphoniotyphidae even when no gonozooecia are present. If present, the structure of gonozooecia can characterize peculiarities of species.

It should be emphasized in conclusion that regularities of budding and growth of zooecia with similar or different structures and functions determining the morphological constancy of zoaria should be used for establishing the genetic relationship of each family included in the order Cyclostomata. The peculiarities of structure and position of gonozooecia should be considered when establishing the species and in some cases the genus too.

REFERENCES

Bassler, R. S. (1953). Bryozoa. "Treatise on Invertebrate Paleontology," Part G, Kansas, pp. 1–253.

Viskova, L. A. (1969). On Some morphological Peculiarities of the Genus *Sulcocava* and its distribution. *Bull. UOTP, Geol.*, **44** (3), Moscow, 150–151.

Voigt, E. (1967). Oberkreide–Bryozoen aus den asiatischen Gebieten der UdSSR. *Hamburg* **36,** pp. 5–95.

49 | The Freshwater Ectoprocta: A Zoogeographical Discussion

JOHN H. BUSHNELL

Division of Environmental Biology, Department of Biology, University of Colorado, Boulder, Colorado, U.S.A.

Abstract: Zoogeographical information on both Gymnolaemata and Phylactolaemata, is presented (49 species). Forty per cent of all species are endemic to one faunal region, the largest number occurring in regions which embrace the equator. Successful dispersal and colonization of new habitats are related to statoblast design and production potentials, correlations of peak statoblast production periods with bird migrations and wide environmental adaptability (including epibenthic in *Fredericella* and *Paludicella*), and colonial fragmentation (*Fredericella*). *Plumatella repens* and *Fredericella sultana* are given special attention.

Both the Phylactolaemata, and Gymnolaemata, occur in freshwater, but the Phylactolaemata predominate and are restricted to freshwater. The remaining species are ctenostomes in the class Gymnolaemata. Irrespective of the need for more distributional data and better agreement on systematics, zoogeographic patterns are emerging, and by contemplating available information on the biology and ecology of the better studied species it is possible to draw some conclusions about dispersal, e.g. why certain forms have had more success than others.

George James Allman, regius professor of Natural History at the University of Edinburgh, in his splendidly illustrated Ray Society Monograph of 1856 contributed the first major review of the freshwater ectoprocts. He makes these remarks about distribution:

> "The number of forms of fresh-water Polyzoa which, after careful comparison, I have deemed it right to retain as distinct species, amount to 21 in all. Of these, sixteen are British; and of the remaining five, one has only been described as occurring in the fresh waters of Belgium, while four are confined to the United States of America.
>
> The most northern point at which we have any positive information of the discovery of a species is probably the neighbourhood of Stockholm, while the neighbourhood of Nice would seem as far as is yet known to be the most southern limit of

the group in the Eastern Hemisphere, and Philadelphia to be the most southern limit at which we have any record of a fresh water Polyzoan having been discovered in the Western Hemisphere. Vlademir, in Central Russia, is the most eastern point, while, Philadelphia must for the present be viewed as the extreme limit towards the west.

The species, so far as we are as yet acquainted with their distribution, are thus confined to the North Temperate Zone, not one having been recorded as occurring south of the Mediterranean in the Old World, or of Phila. in the New."

This interesting, if narrowly anglocentric, distributional summary of little more than 100 years ago has suffered considerable alterations. Freshwater species now number 45–50. The synonymy of a few of the 49 species which are recognized in this paper probably will be substantiated in time. While the number of world species has increased from 21 to approximately 49, the number of British species has decreased from 16 to 9 with taxonomic revisions. Aside from Allman (1856) and Kraepelin (1887, 1892) extensive regional information on ectoproct distribution published after 1900 is given in the following selected references: Abricossoff (1924, 1927a, b) Russia; Annandale (1907, 1908, 1910, 1922) India, Ceylon; Borg (1936a, b, 1940, 1941) Africa, Sweden; Bushnell (1965a, b, c, 1968, 1971) Michigan, Colorado, Mexico; Davenport (1904) U.S.A.; Lacourt (1968) Europe; Marcus (1926, 1941, 1942) Brazil; Rogick (1934a, b, 1935, 1937, 1940a, 1941, 1943) Ohio, New York, Canada; Toriumi (1941a, b, 1942a, b) Japan, Korea, Taiwan; Wiebach (1964, 1967, 1970a, b) Africa and S. America. The valuable monograph of Lacourt (1968) provides distributional information only on the Phylactolaemata (his maps omit specific collection sites). Lacourt had not had access to several recent papers considerably extending world ectoproct ranges, or to certain papers describing new species or confirming questionable forms, e.g. Bushnell (1965c), Wiebach (1970a, b), Grancarova (1971). Lacourt (1968) proposes extensive systematic revision of the Phylactolaemata. The present paper accepts certain of the revisions but considers others premature until we have better information on perplexing aspects of growth and form, physiological ecology, and sexuality.

The northern limit for freshwater species is 75° N latitude, in Novaya Zemlya and in lakes near Spitzbergen and the southern limit is 55° S latitude (Tierra del Fuego). The greatest altitude is Lake Titicaca, 4151 m (Marcus, 1946). The greatest water depth is 170 m for *Fredericella sultana* in Lake Lucerne (Zschokke 1906) The distribution of ectoprocts within major faunal regions is given in Table 1. The number of species shared by major faunal areas are shown in Table 2. Distribution maps for the Phylactolaemata are given in Figs 1–5. A Distribution map for the Gymnolaemata is given in Fig. 6. The symbols do not necessarily represent single records, only locales from which one or more collections were made. Forty per cent of the species are endemic to a single region, while only four species are completely cosmopolitan. More endemics are known from the Neotropical, Ethiopian and Oriental regions, the three faunal regions which are equatorial. Except for *Hislopia lacustris*, none of the gymnolaematous Hislopiidae are known from outside a narrowly circumscribed locale within one faunal region, e.g. *Echinella placoides* from Lake Baikal and rivers north of Lake Baikal (Wiebach, 1966). Among the Gymnolaemata (Fig. 6) only *Paludicella articulata* is widely distributed, except in the Ethiopian region. However, the occurrence records for this species are clustered largely in eastern North America and Europe. It can be suggested that the species originated in the eastern Nearctic or western Palearctic with a subsequent north Atlantic dissemination. There are decidely a greater number of gymnolaemates in the Neotropical and Oriental regions than in any of the other regions (Table 1). Perhaps certain species are relict species.

Among the gelatinous Lophopodidae most species occur in Africa and south-east Asia (Fig. 5). The genus *Lophopodella* may have originated in the Ethiopian or possibly Oriental region with *L. carteri* being either an earlier or more vagile species than the others. *Lophopus crystallinus* is limited to the eastern Nearctic and western Palearctic (principally). There is no overlap between the ranges of the two species of *Pectinatella*. The three species of Lophopodidae known from the Western Hemisphere are limited to eastern North America. Two of these species, *L. crystallinus* (no recent records) and *L. carteri* are rare. This suggests dispersal to North America from Europe via the North Atlantic (Birds? Man? Scandic bridge?), and, presuming continental drift, evolution of the family subsequent to separation of the continents. *P. magnifica*, in all likelihood, had a later, Nearctic origin.

Cristatella mucedo (Fig. 5) is the only species, except *Plumatella fungosa* (Fig. 2), known with a completely circumpolar holarctic occurrence pattern. Of the two species of *Fredericella* (Fig. 4), *F. sultana* is cosmopolitan, and *F. australiensis*, while rather widely distributed, is rarely common. The latter species is seemingly often associated with more arid regions (Bushnell, 1971).

Among the Plumatellidae many species are endemic to one or two regions. The distributional dominance of the cosmopolitan *Plumatella repens* over the cosmopolitan *Plumatella emarginata* in the Nearctic and western Palearctic is reversed in the eastern Palearctic, Oriental and Australian regions where *P. emarginata* predominates (Fig. 1). This is made even clearer by examinations of Table 3, which summarizes collecting records from Japan and Michigan (U.S.A.) during two extended collection periods. *P. repens* was the dominant species in collections from Michigan and *P. emarginata* was the dominant species in collections from Japan. *P. emarginata* occurred with 30% the frequency of *P. repens* in Michigan, and *P. repens* occurred with 41% the frequency of *P. emarginata* in Japan. The world distribution patterns for these ectoprocts may reflect different centers of origins for the two species. Table 3 provides quantitative information on frequency of occurrence for all species of ectoprocts found in Japan and Michigan.

Table 1. Distribution of freshwater Ectoprocta within major faunal regions

				PALEARCTIC*					
	COSMOPOLITAN	NEARCTIC	NEOTROPICAL	European and Mediterranean Subregions	Siberian and Manchurian Subregions	ETHIOPIAN	ORIENTAL	AUSTRALIAN	ENDEMIC to one region
PHYLACTOLAEMATA:									
FREDERICELLIDAE									
Fredericella sultana	×	×	×	×	×	×	×	×	
F. australiensis		×	×	×	×			×	
PLUMATELLIDAE									
Plumatella repens	×	×	×	×	×	×	×	×	
P. emarginata	×	×	×	×	×	×	×	×	
P. fruticosa		×	×	×	×				
P. casmiana		×	×	×	×	×	×		
P. longigemmis			×			×	×		
P. marcusi			×						×
P. vorstmani					×		×		
P. siolii			×						×
P. javanica		×	×	×		×	×	×	
P. ruandensis						×			×
P. marlieri						×			×
P. fungosa		×		×	×				
Hyalinella punctata	×	×	×	×	×	×	×	×	
H. carvalhoi			×						×
H. vaihiriae		×						×	
H. lendenfeldi			×					×	
H. minuta					×		×		
H. orbisperma		×							×
H. africana						×			×
H. indica						×	×		
Afrindella philippinensis						×	×		
A. tanganyikae					×	×	×		
Stephanella hina					×				×
Stolella indica		×				×	×		
S. evelinae		×	×						

Table 1 (*Continued*)

	COSMOPOLITAN	NEARCTIC	NEOTROPICAL	European and Mediterranean Subregions	Siberian and Manchurian Subregions	ETHIOPIAN	ORIENTAL	AUSTRALIAN	ENDEMIC to one region
S. agilis			×						×
Gelatinella toanensis					×		×		
Internectella bulgarica				×					×
LOPHOPODIDAE									
Lophopus crystallinus		×		×	×				
Lophopodella carteri		×			×	×	×	×	
L. capensis						×			×
L. thomasi						×			×
L. stuhlmanni						×	×		
L. pectinatelliformis							×		×
Pectinatella magnifica		×		×					
P. gelatinosa					×		×		
CRISTATELLIDAE									
Cristatella mucedo		×		×	×				
GYMNOLAEMATA:									
PALUDICELLIDAE									
Paludicella articulata		×	×	×	×		×	×	
P. pentagonalis			×				×		
Pottsiella erecta		×	×						
HISLOPIIDAE									
Hislopia lacustris			×				×		
H. corderoi			×						×
H. malayensis							×		×
H. moniliformis							×		×
H. cambodgiensis							×		×
Echinella placoides					×				×
Arachnoidia ray-lankesteri						×			×
Total	4	19	20	14	19	19	24	10	19
Total Endemic, by region		1	6	1	2	6	4	0	

* Total species in Palearctic is 24

Plumatella casmiana has been better known in the eastern Palearctic and Oriental regions. Only recently was it reported from Europe (Wiebach, 1963; Vigano, 1968). Until recently it was considered rare in the U.S.A. However, a form originally reported as the "Pony Creek" form of *P. emarginata* from Michigan (Bushnell, 1965b) is now known to be *P. casmiana*. Most recently *P. casmiana* has been reported from several locales in Colorado (U.S.A.) by

Table 2. Number of species of Ectoprocta shared by major faunal regions

Regions and numbers of species reported	NEARCTIC	NEOTROPICAL	PALEARCTIC* European and Mediterranean Subregions	PALEARCTIC* Siberian and Manchurian Subregions	ETHIOPIAN	ORIENTAL	AUSTRALIAN
NEARCTIC 19		11	14	12	8	9	9
NEOTROPICAL 20	11		9	8	7	10	8
PALEARCTIC European and Mediterranean Subregions 14	14	9		11	7	8	7
Siberian and Manchurian Subregions 20	12	8	11		7	12	7
ETHIOPIAN 19	8	7	6	7		13	6
ORIENTAL 24	9	10	7	12	13		7
AUSTRALIAN 10	9	8	7	7	6	7	

* Total species in Palearctic is 24

Bushnell (1968) and Bushnell and Wood (1971). If, as continuing collections suggest, the species is most common in western North America, then there is reason to believe *P. casmiana* has had a trans-Pacific dissemination.

Plumatella javanica has a wide distribution, but it is far more frequent and luxuriant in tropical and subtropical regions. *Plumatella fruticosa* (Fig. 2) is decidedly more common in Europe, but recent studies (Bushnell, 1965b; 1968) indicate it is more common in North America than previously believed, particularly in the montane lakes and glacial moraine ponds of the Rocky Mountains.

The only species of *Hyalinella* with a cosmopolitan distribution (Fig. 3) is *H. punctata*, but with occurrences greatest in Europe, and secondly eastern North America. The other seven species of the genus are narrowly restricted in range, five of these to a single faunal region. *H. orbisperma* (formerly genus *Plumatella*; Kellicott, 1882) is known only from Michigan, and recently received fuller

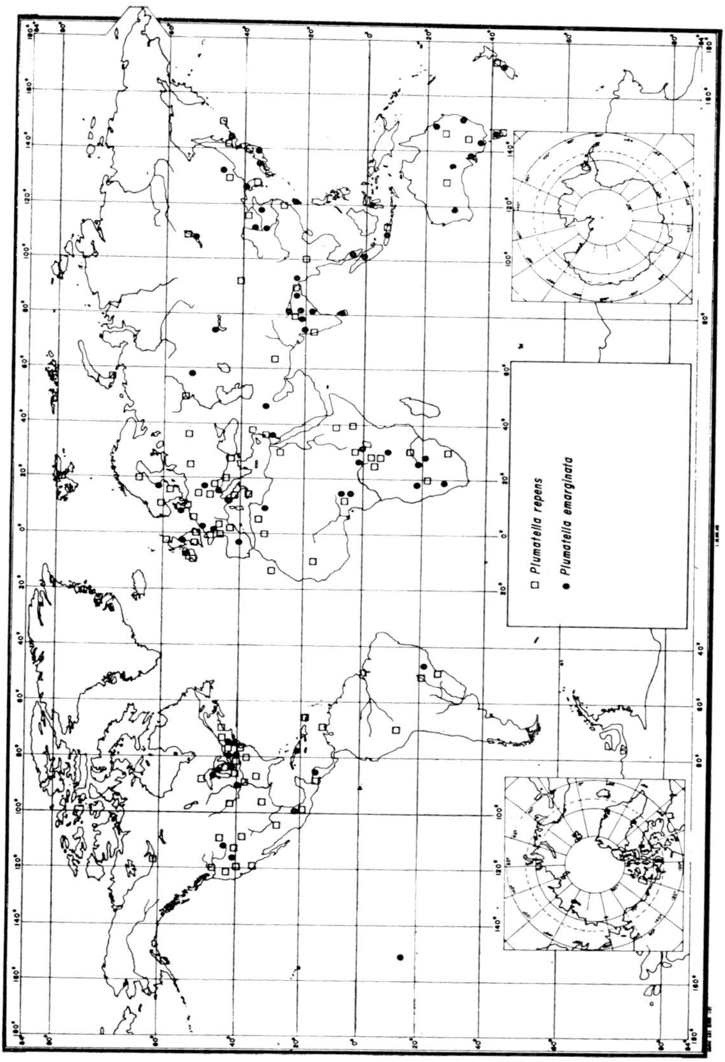

Fig. 1

Figs 1–5. Distribution maps for species of Phylactolaemata

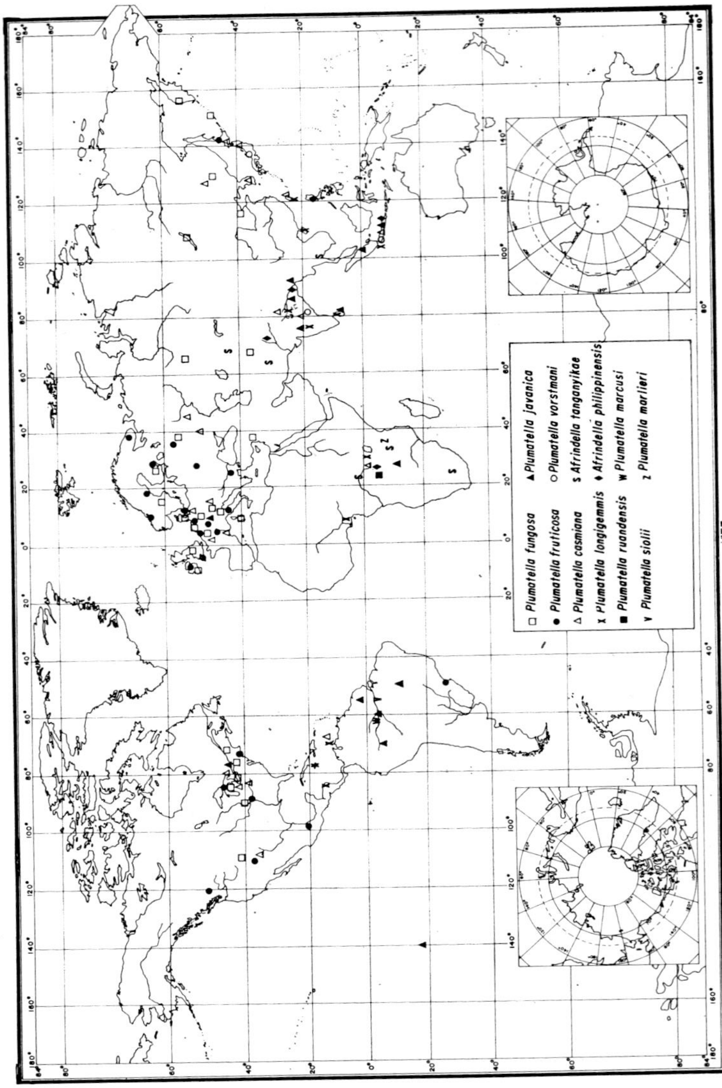
□ Plumatella fungosa
● Plumatella fruticosa
△ Plumatella casmiana
x Plumatella longigemmis
■ Plumatella ruandensis
v Plumatella siolii
▲ Plumatella javanica
○ Plumatella vorstmani
s Afrindella tanganyikae
♦ Afrindella philippinensis
w Plumatella marcusi
z Plumatella marlieri

Fig. 2

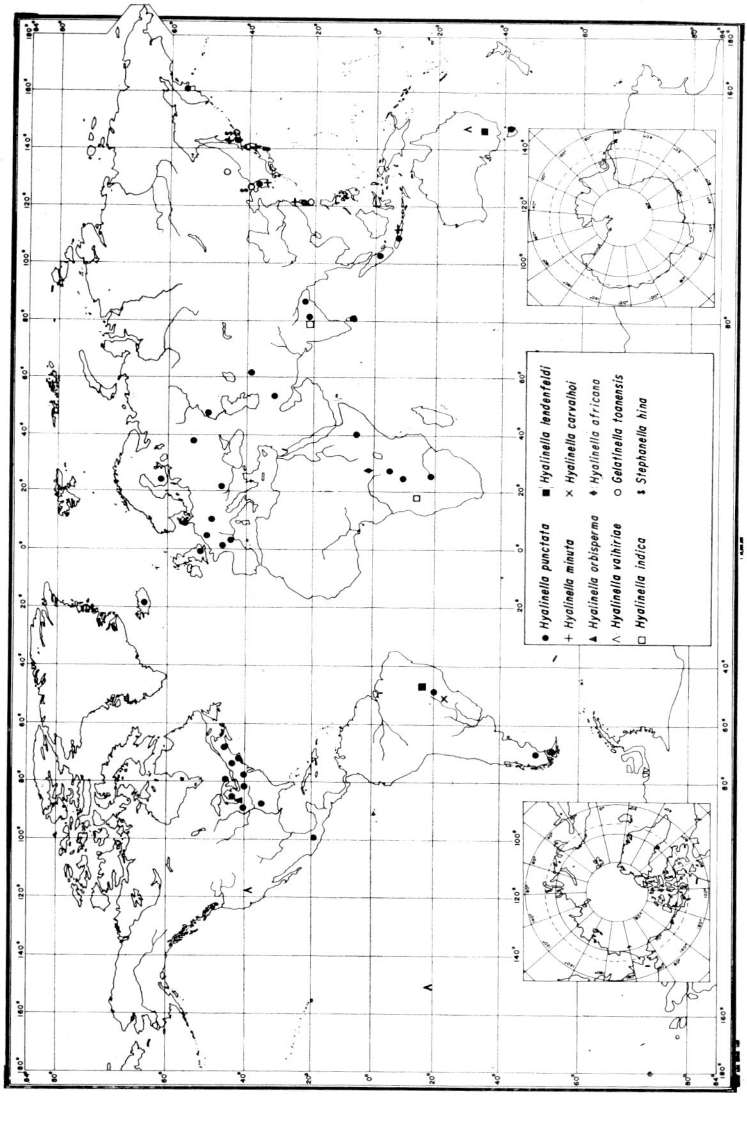
Hyalinella punctata
Hyalinella minuta
Hyalinella orbisperma
Hyalinella vaihiriae
Hyalinella indica
Hyalinella lendenfeldi
Hyalinella carvalhoi
Hyalinella africana
Gelatinella toanensis
Stephanella hina

Fig. 3

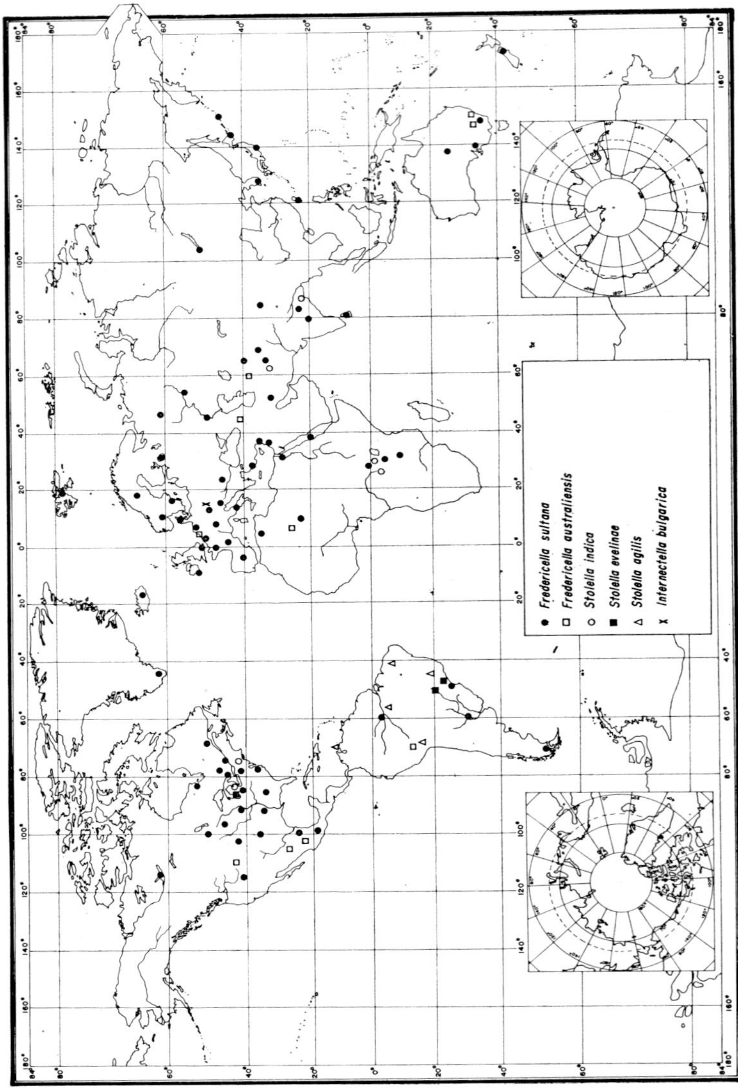
● Fredericella sultana
□ Fredericella australiensis
○ Stolella indica
■ Stolella evelinae
△ Stolella agilis
x Internectella bulgarica

Fig. 4

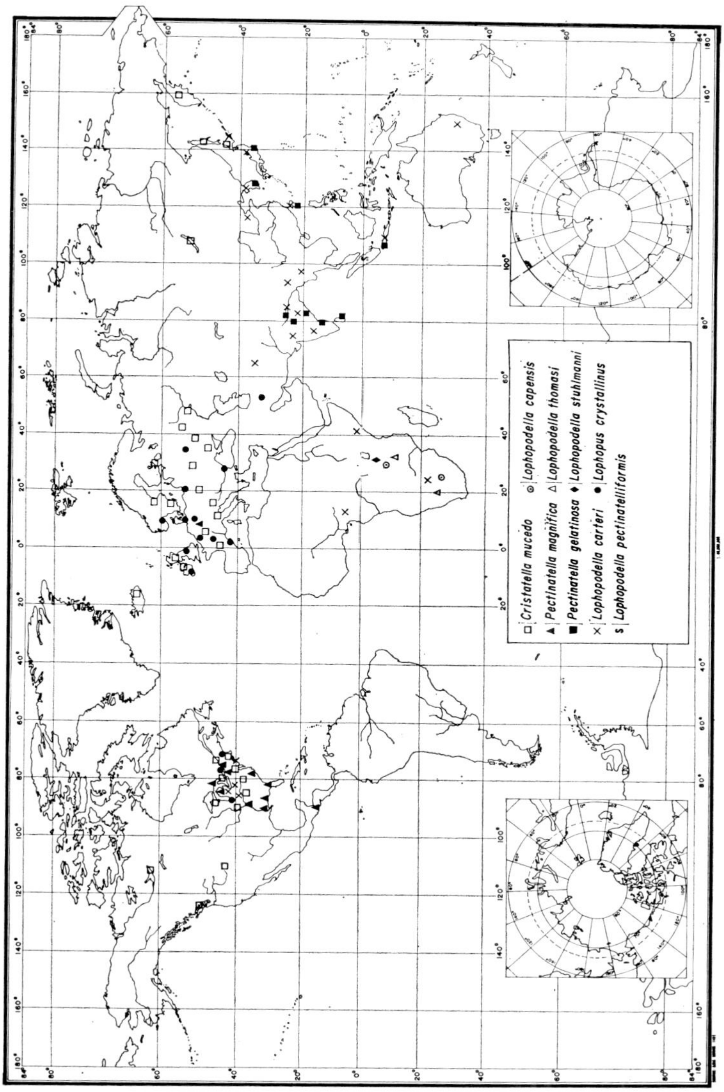
Cristatella mucedo
Pectinatella magnifica
Pectinatella gelatinosa
Lophopodella carteri
Lophopodella pectinatelliformis
Lophopodella capensis
Lophopodella thomasi
Lophopodella stuhlmanni
Lophopus crystallinus

Fig. 5

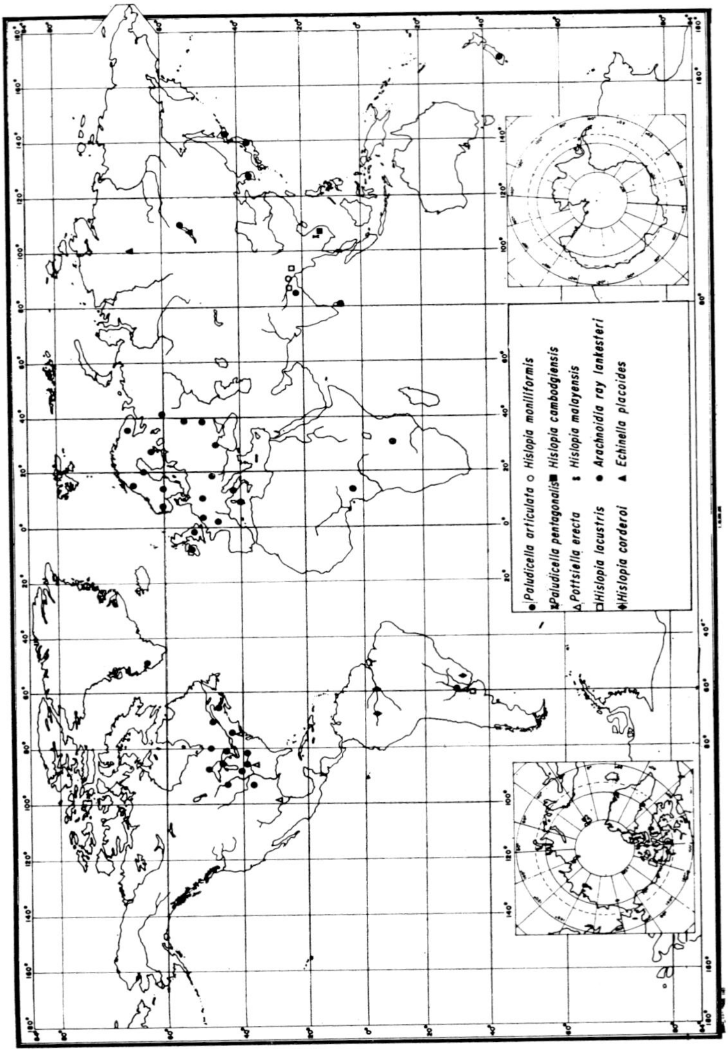

Fig. 6. Distribution map for species of freshwater Gymnolaemata

Table 3. Species of Ectoprocta collected in Japan and Michigan during two studies. The number of locations at which a species occurred is given in the left column. The frequency of occurrence of each species is expressed as the percentage of the number of locations at which it occurred in comparison to the dominant species. The dominant species in both Michigan and Japan are given a value of 100%. Percentages are to the nearest whole number.

Species	U.S.A.—Michigan 1957–1961 123 locations†		Japan 1951–1955 373 locations‡	
	No. of coll.	% Dom. spec.	No. of coll.	% Dom. spec.
* *Plumatella repens*	83	100	45	41
* *P. emarginata*	24	30	109	100
* *P. fruticosa*	10	12	~8	7
* *P. casmiana*	6	7	26	24
P. vorstmani	—	—	27	25
* *Cristatello mucedo*	22	26	9	8
* *Lophopodella carteri*	1	1	19	17
Pectinatella magnifica	6	7	—	—
P. gelatinosa	—	—	11	10
Gelatinella toanensis	—	—	12	11
Stephanella hina	—	—	17	15
* *Hyalinella punctata*	5	6	15	13
H. minuta	—	—	32	29
H. orbisperma	3	3	—	—
Stolella indica	1	1	—	—
S. evelinae	1	1	—	—
* *Fredericella sultana*	69	83	39	36
* *Paludicella articulata*	44	53	7	6
Total species	13		14	

* Species common to both areas.
† Bushnell publications 1965.
‡ Toriumi publications 1951–1955.

descriptive treatment (Bushnell, 1965c) and the new generic designation (species status strongly confirmed, upon examination of specimens, by the late Dr M. Rogick and Dr F. Wiebach). Other species with a particularly narrow range are *Stephanella hina* and *Gelatinella toanensis*, reported only from the curve of Southeast Asia, and *Internectella bulgarica* from Bulgaria.

Statoblasts are a principle means of dispersal for phylactolaematous ectoprocts. De Gerne (see Allman, 1856) was the first to report the finding of a *Plumatella* statoblast in mud on a duck's foot. Distribution occurs within connected waterways via currents, and within the gut of cold-blooded animals. Brown (1933) investigated the germination potential of *Pectinatella magnifica* statoblasts which had been passed through the digestive tract of several animals and recorded a maximum of 32%, 69%, and 73% germination for the turtle, frog, and salamander respectively, but only up to 4% germination for *Pectinatella* statoblasts fed to ducks. Osburn (1921) found statoblasts in fish. The author dissected viable statoblasts from the stomachs of Blue Gills and observed these same fish nibbling at Cristatella colonies. Statoblasts are doubtless carried overland for short distances between one pond, lake, or river and another in the gut of vertebrates and perhaps flying insects. Statoblasts could be transported within the gut of birds for greater distances. Brown found statoblasts remained in the gut of a duck from 14 to 26 h. Ridley (1930) gives a maximum of seven hours for the passage of food through the gut of ducks, geese or swans. Overland the flight speed of birds is 80 km/h. Mallards and pintails have been checked by radar at speeds as low as 50 km in 7 h, but presuming that metabolism is increased during flight and that birds fly more slowly over water they may not travel farther than 150 km in 7 h. Birds could therefore carry statoblasts, hibernacula or seeds in this manner for 150 to 600 km.

Two species of *Stolella* and *Lophopodella carteri* were all found in one major water fowl sanctuary in Michigan (Bushnell, 1965a, 1966). None of these three species were found again in any of the collections throughout the state. Thousands of ducks and geese and a number of shore birds come into this lake during migrations. The Blue Wing Teal winters as far south as Colombia, and the Green Wing Teal winters as far south as Brazil. Statoblasts, like algae, seeds, and resistant eggs, could be carried externally for great distances in the feathers or in mud on the feet or bill.

Many statoblasts, particularly soon after release, have adherent qualities. The author has information with respect to their adherence in fur. A number of Meadow Voles (*Microtus pennsylvanicus*) were captured in a damp Michigan grassland in which were located a chain of small ponds. Three statoblasts were washed from the fur of 27 voles, evidence enough to confirm transportability. Thirty voles were permitted to walk in shallow basins of water in which approximately 1000 *P. repens* statoblasts had been placed. By brushing animals at different times following immersion large numbers of statoblasts (up to 70) were retrieved after 15 min and 2 h. Up to five statoblasts were washed from the fur of animals which had been permitted to move about in a grassy pen for six hours after immersion. One or two statoblasts were washed from the fur of some animals after 8 h. None were obtained thereafter. Statoblasts are carried possibly even longer between the toes. Some *Cristatella* statoblasts were recovered ten hours after immersion. It may be that the peripheral hooks afford *Cristatella* statoblasts a better purchase in fur, feathers or mud. The longest dispersals may occur in dried mud on the feet or bill of birds. Shore birds are likely animals for such dispersal. Many, e.g. plovers and sandpipers, are also the most notorious wanderers. The Upland Plover ranging between the Yukon and the Pampas of South America has wandered westward as far as Australia (Elliot, 1895; Bent, 1962). The occurrence of *Hyalinella vaihiriae* in Australia, Tahiti and Utah could be attributable to the odyssey of an Upland Plover. The Spotted Sandpiper is another wandering shore bird. The Arctic Tern is known for the extent of its migration (Austin, 1931; 1961) up and down both sides of the Atlantic, regularly crossing Davis Strait and other marine barriers of the north Atlantic. Though essentially an ocean bird, these terns are seen frequently inland on the shores of fresh water lakes. Dr A. Löve (personal communication) has seen this bird often along freshwater lakes near Reykjavik, Iceland. The Arctic Tern is a potential carrier between Nearctic and Palearctic locations.

Floatoblasts and even shore-marooned fragments of colonies containing sessoblasts or hibernacula are subject to wind dispersal. If as Dahl (1958) states, seeds must have essentially the same dimensions as fine loess or sand for effective wind dispersal, then diameters of greater than 0·2 mm become limiting. Löve (1963), in discussing plant dispersal, refers to experimental data indicating that the limiting factors for the effective dispersal of seeds is a density of not more than 2·6 g/cm^3 and a diameter of not more than 0·6 mm. On the basis of these criteria the large (1 mm) statoblast of the Lophopodidae and *Cristatella* would be least effectively distributed by wind, while the statoblasts of the Plumatellidae, particularly *Plumatella repens*, would be most effectively distributed by this means. Sufficient turbulence could carry statoblasts into the colder and faster upper air streams, where statoblast tolerance to cold and dessication become of extreme importance. Tolerance to dessication is also important for long distance external transport by birds. In general, statoblasts of *Plumatella repens* tolerate sub-zero temperatures very well (Brown, 1933). Freezing of *Pectinatella magnifica* statoblasts at $-12°C$ affected the per cent germination only after 175 days. Brown concluded that dessication and low humidity kill statoblasts in a few days and he found a decrease in the rate of germination of *Pectinatella* statoblasts after 10 days drying. Rogick (1940b) reported a high germination for *Fredericella sultana* sessoblasts which had been dried for two years, with no germination for *Plumatella repens* statoblasts kept dry for two or more years. The highest viability she found was 70% germination after 4¼ years for *Lophopodella carteri*. Oda (1959) found two statoblasts of *L. carteri* viable after 2296 days. A long term viability obviously permits statoblasts much more time to find a suitable habitat for germination, thus facilitating distributional success.

Other factors peculiar to the biology of certain ectoproct species may help to explain the relative distributional success of the freshwater Ectoprocta. Statoblast productivity may be one of the most important. Statoblast productivity for

P. repens (Brown, 1933) has been given as 513 cm² with *Pectinatella* only approximately 20% of this. I have found polypides in luxuriant colonies of *P. repens* with 15 to 20 statoblasts. Recently I found up to 27 statoblasts in polypides of *P. repens* from Colorado. Maximal productivity in other species is less than half this number. High statoblast productivity, and a nearly optimal size for wind dispersal, have doubtless aided the distribution of *P. repens.* The statoblasts of *P. magnifica* require a considerably longer time interval prior to germination (Brown, 1933). The longer period would appear to allow greater time for dispersal to occur. Brown (1933), however, suggests that the hooks on *Pectinatella* statoblasts restrict the distribution of the species. This seems unlikely in the case of bird and mammal dispersals. Also statoblasts are eaten by insect larvae (Bushnell, 1966) and possibly snails, so the period of exposure to potential statoblast predation is long for *P. magnifica.*

Assuming transport by birds, the time of statoblast production becomes important. *Pectinatella* and *Lophopodella* seemingly have only one production peak, in the summer. By contrast the spring and late summer and early autumn production peaks of *P. repens* (Brown, 1933; Bushnell, 1966) coincide with bird migration times. Also, *P. repens* has an ectocyst that disintegrates more easily than certain other species, thereby releasing statoblasts earlier.

The distributional success of the cosmopolitan and common *F. sultana* may be attributed to other factors. This form produces only a small number of piptoblasts. An obvious aid to successful dispersal is the broad adaptability of this species (Bushnell, 1966) to a wide variety of ecological situations. Adaptability to turbulent water is shared only with *Paludicella articulata, Plumatella emarginata,* and possibly *Plumatella fruticosa.* The gelatinous *Lophopodella, Pectinatella* and *Cristatella* are least adapted structurally for survival in turbulent waters. *F. sultana* is the only species known to thrive all year under the ice (regularly observed in Colorado). It sometimes grows in vertical clusters attached to bottom sand or debris. *Paludicella articulata* is also encountered frequently on benthic debris (Bushnell, 1966). Benthic deposits are particularly subject to attachment and transport on the feet of wading animals. We should expect a species having only non-floating piptoblasts or hibernacula (Fredericellidae and the Gymnolaemata) to have greater adaptability to benthic and epibenthic conditions than forms with floatoblasts. The former depend largely on ciliated larvae to reach near surface locations, and sexual maturity for most species has been only rarely observed.

The two species of *Fredericella* are the principal forms having asexual structures (piptoblasts) which are neither cemented to the substrate nor capable of floating. Unless caught by an intermediate substrate, or retained within the mother zooecium until germination, they sink to the bottom mud or sand. Preliminary data (unpublished) indicate that *Fredericella* statoblasts have significantly greater germination potential than *Plumatella* statoblasts after extended wet storage in clay-loam soil.

An interesting and possibly important mechanism for the dispersal of *F. sultana* was first observed by the author while growing colonies in a slowly flowing artificial stream in Pennsylvania. Fragments of the colony would frequently break off and later become attached and continue growing downstream. Observations of a similar nature were later made on epibenthic colonies in the slowly flowing outlet of a lake. Recently Dr T. Wood (personal communication) has repeatedly observed this same phenomenon in both field and laboratory.

A number of factors, both biological and physical, interact to facilitate or impede the dispersal and effect the distribution of each species. Only selected species, for which more knowledge is available, have been referred to in discussing mechanisms of dispersal and adaptability among freshwater ectoprocts.

REFERENCES

Abricossoff, G. G. (1924). To the knowledge of the fauna of the Bryozoa of the Lake Baikal. *Russ. Hydrobiol. Z. Saratov* **3,** 260–266.

Abricossoff, G. G. (1927a). To the knowledge of the fauna of the Bryozoa of the Caucasus. *Russ. Hydrobiol. Z. Saratov* **6,** 84–92.

Abricossoff, G. G. (1927b). Uber die Susswasser-Bryozoen der USSR. *C.R. Acad. Sci. Russ.* **19,** 304–312.

Allman, G. J. (1856). A monograph of the fresh-water Polyzoa, including all the known species, both British and foreign. *Ray Soc.* **28,** 1–119.

Annandale, N. (1907). The Polyzoa occurring in Indian fresh- and brackish-pools. Notes on the fresh-water fauna of India. No. 12. *J. Asiatic Soc. Bengal.* **3,** 83–93.

Annandale, N. (1908). Three Indian Phylactolaemata. *Rec. Ind. Mus.* **2,** 169–174.

Annandale, N. (1909, 1910). Materials for a revision of the phylactolaematous Polyzoa of India. *Rec. Ind. Mus.* **5,** 37–57.

Annandale, N. (1922). Polyzoa in the Colombo waterworks. *Spolia Zeylanica* **12,** 207–209.

Austin, O. L. (1931). Birds of Newfoundland and Labrador. *Mem. Nutt. Orn. Club, Cambridge, Mass.* 229 pp.

Austin, O. L. (1961). "Birds of the World." A survey of the twenty-seven orders and one hundred and fifty-five families. 316 pp. Golden Press, New York.

Bent, A. C. (1962). "Life histories of North American Shore Birds." Dover Publ., New York (orig. publ. 1927. pt. 1 Bull. 142, Smithsonian Inst., 1929. pt. 2 Bull. 146, Smithsonian Inst.).

Borg, F. (1936a). Uber die Susswasser-Bryozoen Afrikas. *Senckenbergiana* **18,** 20–36.

Borg, F. (1936b). Sur quelques Bryozoaires d'eau douce Nord-Africains. *Bull. Soc. Hist. Nat. Alger.* **27,** 271–283.

BORG, F. (1940). On a collection of freshwater Bryozoa from Belgian Congo. *Rev. Zool. Bot. Afr.* **34,** 33–46.

BORG, F. (1941). Uber die Susswasserbryozoen Schwedens. *In* Festskrift tillagnad Prof. Sven Ekman. *Zool. Bidr. Uppsala* **20,** 479–494.

BROWN, C. J. D. (1933). A limnological study of certain fresh-water Polyzoa with special reference to their statoblasts. *Trans. Am. Microsc. Soc.* **52,** 271–314.

BUSHNELL, J. H. (1965a). On the taxonomy and distribution of fresh-water Ectoprocta in Michigan. Part 1. *Trans. Am. Microsc. Soc.* **84,** 231–244.

BUSHNELL, J. H. (1965b). On the taxonomy and distribution of fresh-water Ectoprocta in Michigan. Part II. *Trans. Am. Microsc. Soc.* **84,** 339–358.

BUSHNELL, J. H. (1965c). On the taxonomy and distribution of fresh-water Ectoprocta in Michigan. Part III. *Trans. Am. Microsc. Soc.* **84,** 529–548.

BUSHNELL, J. H. (1966). Environmental relations of Michigan Ectoprocta, and dynamics of natural populations of *Plumatella repens. Ecol. Monogr.* **36** (2), 95–123.

BUSHNELL, J. H. (1968). Aspects of ecology, zoogeography and architecture of freshwater Ectoprocta (Bryozoa). *Atti Soc. Ital. Sci. Nat. Mus. Civ. St. Nat. Milano* **108,** 129–151.

BUSHNELL, J. H. (1971). Porifera and Ectoprocta in Mexico: Architecture and Environment of *Carterius latitentus* (Spongillidae) and *Fredericella australiensis* (*Fredericellidae*). *Southwest. Nat.* **15,** 331–346.

BUSHNELL, J. H. and WOOD, T. (1971). Honeycomb colonies of *Plumatella casmiana* OKA (Ectoprocta: Phylactolaemata) *Trans. Am. Microsc. Soc.* **90,** 229–231.

DAHL, E. (1958). Amfiatlantiske Planter, Problems of Amphiatlantic Plant Distribution. *Blyttia* **16,** 93–121.

DAVENPORT, C. B. (1904). Report on the fresh-water Bryozoa of the United States. *Proc. U.S. nat. Mus.* **27,** 211–221.

ELLIOT, D. G. (1895). "North American Shore Birds." 268 pp. F. P. Harper, Publ. New York.

GRANCAROVA, T. (1971). *Internectella bulgarica* nov. gen., nov. sp., a new Bryozoa species (Bryozoa Phylactolaemata). *Comptes Rendus Acad. Bulg. Sci.* **24,** 361–364.

KELLICOTT, D. S. (1882). Polyzoa. Observations on species detected near Buffalo, N.Y. *Proc. Am. Soc. Microsc.* **4,** 217–229.

KRAEPELIN, K. (1887, 1892). Die deutschen Susswasser-Bryozoen. Eine Monographie. *Abh. Naturwiss. Verein Hamburg* **10** (1887), 1–168, **12** (1892), 1–67.

LACOURT, A. W. (1968). A monograph of the freshwater Bryozoa-Phylactolamata. Uitgeg. *Rijksmus. Nat. Hist. Leiden*, (93), 1–159.

LÖVE, D. (1963). Dispersal and survival of plants. *In* "North Atlantic Biota and their history," pp. 189–205. Pergamon Press, Oxford.

MARCUS, E. (1926). Beobachtungen und Versuche and lebenden Susswasserbryozoen. *Zool. Jahrb. Syst.* **52,** 279–350.

MARCUS, E. (1941). Sobre Bryozoa do Brazil. *Bol. Fac. Fil. Cienc. Letr. Sao Paulo Zool.* **5,** 3–208.

MARCUS, E. (1942). Sobre Bryozoa do Brazil II. *Bol. Fac. Fil. Cienc. Letr. Sao Paulo Zool.* **6,** 57–96.

MARCUS, E. (1946). New Brazilian form of *Fredericella sultana. Comm. Zool. Mus. Hist. Natur. Montevideo* **2,** (31), 1–10.

ODA, S. (1959). Germination of the statoblasts in freshwater Bryozoa. *Sci. Rep. Tokyo Kyoiku Daig.* **9** (B) (135), 90–131.

OSBURN, R. C. (1921). Bryozoa as food for other animals. *Science, N.Y.* **53,** (1376), 451–453.

RIDLEY, H. N. (1930). "The Dispersal of Plants Throughout the World." L. Reeve and Company Ltd., Ashford, England.

ROGICK, M. D. (1934a). Additions to North American freshwater Bryozoa. *Ohio J. Sci.* **34,** 316–317.

ROGICK, M. D. (1934b). Studies on freshwater Bryozoa. I. The occurrence of *Lophopodella carteri* Hyatt 1866 in North America. *Trans. Am. Microsc. Soc.* **53,** 416–424.

ROGICK, M. D. (1935). Studies on freshwater Bryozoa. II. The Bryozoa of Lake Erie. *Trans. Am. Microsc. Soc.* **54,** 245–263.

ROGICK, M. D. (1937). Studies on freshwater Bryozoa. V. Some additions to Canadian fauna. *Ohio J. Sci.* **37,** 99–104.

ROGICK, M. D. 1940a. Studies on freshwater Bryozoa, IX. Additions to New York. Bryozoa. *Trans. Am. Microsc. Soc.* **59,** 187–204.

ROGICK, M. D. (1940b). Studies on freshwater Bryozoa XI. The viability of dried statoblasts of several species. *Growth* **4,** 315–322.

ROGICK, M. D. (1941). Studies on freshwater Bryozoa. X. The occurrence of *Plumatella casmiana* in North America. *Trans. Am. Microsc. Soc.* **60,** 211–220.

ROGICK, M. D. (1943). Studies on freshwater Bryozoa. XIV. The occurrence of *Stolella indica* in North America. *Ann. N.Y. Acad. Sci.* **45,** 163–178.

TORIUMI, M. (1941a). Studies on freshwater Bryozoa of Japan, I. *Sci. Rep. Tohoku Imp. Univ.* **16,** 193–215.

TORIUMI, M. (1941b). Studies on freshwater Bryozoa of Japan, II. Freshwater Bryozoa of Tyosen (Korea). *Sci. Rep. Tohoku Imp. Univ.* **16,** 413–425.

TORIUMI, M. (1942a). Studies on freshwater Bryozoa of Japan, III. Freshwater Bryozoa of Hokkaido. *Sci. Rep. Tohoku Imp. Univ.* **17,** 197–205.

TORIUMI, M. (1942b). Studies on freshwater Bryozoa of Japan, IV. Freshwater Bryozoa of Taiwan (Formosa). *Sci. Rep. Tohoku Imp. Univ.* **17,** 207–214.

VIGANO, A. (1968). Note su *Plumatella casmiana* OKA (Bryozoa). *Riv. Idrobiol.* **7,** 421–468.

WIEBACH, F. (1963). Studien uber *Plumatella casmiana* OKA (Bryozoa). *Vie Milieu* **14,** 579–596.

WIEBACH, F. (1964). Untersuchungen an Susswasser-Bryozoa aus Zentral-afrika. *Ann. Mus. Afr. Centrale, Sci. Zool.* **129,** 1–42.

WIEBACH, F. (1966). Ein Bryozoen mit Kaumogen aus dem Baikalsee (*Echinella placoides* Korotnev, Bryozoa Ctenostomata). *Zool. Anz.* **176** (2), 132–142.

WIEBACH, F. (1967). Amazonische Moostiere (Bryozoa). *Amazoniana* **1,** 173–187.

WIEBACH, F. (1970a). Amazonische Moostiere (Bryozoa) II. *Amazoniana* **11** (3), 353–362.

WIEBACH, F. (1970b). Susswasser-Bryozoen aus Brasilien und Zentral-afrika. *Rev. Zool. Bot. Afr.* **81,** (1–2), 62–81.

ZSCHOKKE, F. (1906). Ubersicht Uber Die Tiefenfauna des Vierwaldstattersees. *Arch. Hydrobiol.* **2,** 1–8.

50 | The Occurrence of Double Polypides in Freshwater Bryozoa

S. ODA

Biological Laboratory, St. Paul's University, Tokyo, Japan

and

R. M. NAKAMURA

Department of Tuberculosis, National Institute of Health, Tokyo, Japan

Abstract: Rare examples of twins or double polypides have been reported in freshwater Bryozoa. In our research we have observed many double polypides among the primary polypides deriving from the statoblasts of a known stock of *Lophopodella carteri*.

Abnormal polypides may be classified into the following groups: (1) Polypides with an X-shaped lophophore, (2) Polypides with an E-shaped lophophore, (3) Polypides with separated lophophores, and (4) Polypides with a half lophophore.

The individual zooids of a colony in freshwater Bryozoa do not specialize and are similar in shape. Generally, in a normal zooid, the polypide arises from the cystid. However, rare examples of twins or double polypides have been reported in freshwater Bryozoa (Rogick, 1945; Oda, 1954b; Wiebach, 1959). Such abnormal polypides cannot be found unless zooids or colonies are reared under laboratory conditions and a large number of polypides are carefully observed.

In our research during the past several years we have had the opportunity to observe many double polypides among the primary polypides deriving from the statoblasts of a known stock of *Lophopodella carteri*.

MATERIAL AND METHODS

Important characteristics of Bryozoa (Ectoprocta) are that the mouth opens in the center of the lophophore and the anus is situated outside the lophophore. In the freshwater Phylactolaemata the mouth is partially overhung by the epistome, and the lophophore is horseshoe-shaped or U-shaped. A single zooid consists of two parts, cystid and polypide. The polypide is tubular and is topped by a horseshoe-shaped lophophore bearing numerous ciliated tentacles. The body wall of living materials is very thin and transparent. Hence, the digestive tracts, such as the oesophagus, stomach, and rectum can be observed through the body wall. Accordingly, the general structure of a zooid can easily be understood in the primary zooids deriving from the germinated statoblasts, even if they have two valves of a statoblast and the parts of the body wall are dotted with whitish yolk. In a zooid the side on which the anus is situated is called the anal side, and the opposite side is called the oral side.

We collected colonies of *Lophopodella carteri* from the Furutone Pond near the Tone River every summer and autumn. Numerous statoblasts, taken from these colonies and dried, were kept in a small glass tube under laboratory conditions. For germination a set of 100 statoblasts was immersed in water regulated at the proper temperature. If the water was regulated to 25°C, statoblasts germinated easily about five days after immersion. Ordinarily one young polypide appears between two valves of a statoblast. In the summer of 1962 we accidentally found many double polypides among the primary polypides, when the statoblasts of a known stock, collected in August of 1961, were immersed in the water.

OBSERVATIONS

The abnormal polypides which were observed in 1962 will be classified into four groups according to the shape of the lophophore and the state of the digestive tracts.

1. *Polypides with an X-shaped Lophophore*

Abnormal polypides with an X-shaped lophophore were observed (Fig. 1c–g), although the normal lophophore exhibited the U-shape. In such polypides usually one mouth opened in the center of the lophophore (c, d). But in some cases the mouth was overhung by two epistomes facing each other (d). From these facts the X-shaped lophophore might be caused by fusing of two U-shaped lophophores on the oral side. Moreover, in another example, two mouths opened inside the lophophore and the oesophagus branched (e). In contrast examples having two rectums situated at the upper part of the stomach were also found (f). In such cases two anuses opened on each side and the buds developed on two sides of the cystid. Moreover, a polypide in which two rectums were situated in the upper part also possessed a branched stomach (g).

2. *Polypides with an E-shaped Lophophore*

Some polypides had an irregular lophophore, in which the middle part of the U-shaped lophophore protruded inside its arc (j–l). Such a lophophore is distinguishable from the X-shaped one, and is termed an E-shaped lophophore.

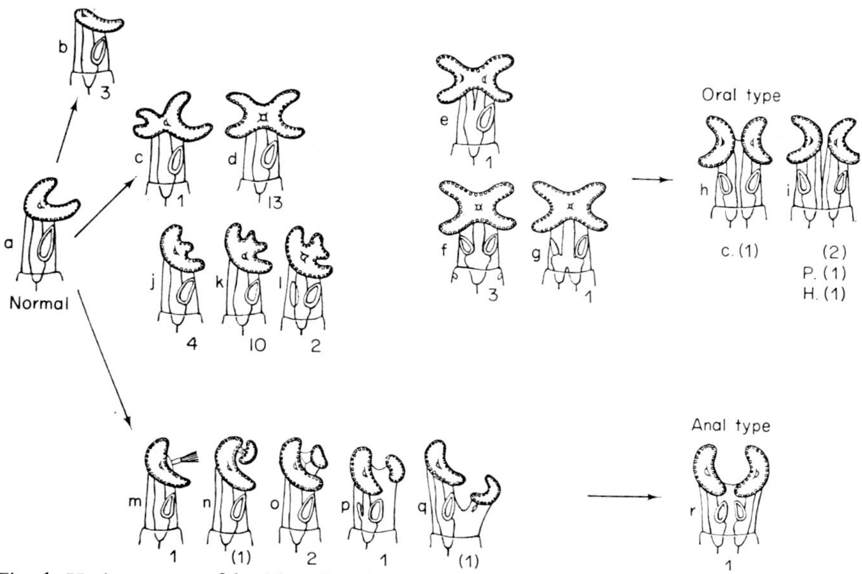

Fig. 1. Various types of double polypides in freshwater Bryozoa.
In this illustration, tentacles are not shown. Small circles exhibit the situation of tentacles on the lophophore. The figures show the number of abnormal polypides which were observed, and the parentheses show examples which were recorded in previous papers (Rogick, 1945; Oda, 1954b).
C: *Cristatella mucedo*, H: *Hyalinella punctata*, P: *Pectinatella gelatinosa*,
Others: *Lophopodella carteri*.

In some cases the middle arm of the lophophore markedly protruded inside its arc and it slightly branched at the distal part. In other examples the mouth was overhung by two epistomes and these examples also had two rectums (l). The process-like arm in this type might be caused by fusing of the adjoining arms of two U-shaped lophophores. Accordingly, it seems that the E-shaped lophophore has the possibility of becoming the X-shaped one, if the middle arm of the lophophore greatly develops and completely separates. Therefore, it will be considered as another type of duplication making a mirror image on the oral side.

3. *Polypides with Separated Lophophores*

A polypide with two separated lophophores of nearly the same size was found (r). In this polypide, having a common tentacular sheath, two digestive tracts were completely separated and made a mirror image on the anal side with each other. Moreover, some polypides with two separated lophophores of different sizes were also observed (o, p). These polypides will be regarded as the middle type with respect to the typical twin polypides making a mirror image on the anal side. In another example there was a club-like process with several tentacles attached to the lophophore of the autosite (m). The "parasite" of this polypide, that is, the club-like process with tentacles, might be caused by the extreme degeneration of the second lophophore. A certain polypide with two lophophores of different sizes had the process-like rectum at the upper part of the stomach in addition to the normal size rectum (p).

4. Polypides with a Half Lophophore

Some polypides bearing only one arm of the lophophore were also found (b). Such polypides will be regarded as having a defect in one arm of the U-shaped lophophore. But the defect of the arm is not limited to a particular side.

In addition to those mentioned above, some polypides without the lophophore-arms were found in germination tests in 1970. They were topped only by several tentacles arranged in a circle.

Forty double polypides were observed, but only 3 polypides with a half lophophore. These numbers corresponded to about 3–4% of the total numbers of zooids which germinated from the statoblasts of a known stock.

DISCUSSION

The double polypides so far found in freshwater Bryozoa are broadly divided into two types, oral and anal. The complete twin polypides of the oral type make a mirror image on the oral side with each other, and in the anal type the polypides make a mirror image on the anal side. The polypides with an X-shaped lophophore exhibit the middle stage with respect to the oral type of duplication because it seems they may be caused by the fusion of two U-shaped lophophores on the oral side of a polypide (Fig. 1c, d). Polypides of the oral type are of two kinds, distal and proximal. In the distal kind the separation of the digestive tract proceeds from the mouth to the oesophagus, that is, at first the mouth is overhung by two epistomes, and successively two mouths open and the oesophagus branches (e). In the proximal kind it proceeds from the rectum to the stomach, that is, in such a polypide two rectums are situated at the upper part of the stomach and two anuses open in each position (f). In the more separated polypide the stomach also branches (g). In extreme examples the digestive tracts may be completely separated producing twin polypides making a mirror image on the oral side.

Typical examples of the oral type in the duplication of polypides have already been recorded as twin polypides with a common tentacular sheath in *Cristatella* by Oda (1954b) (see Fig. 1h) and as "separate" twin polypides arising from the same cystid in *Hyalinella* by Rogick (1945) and in *Lophopodella* and *Pectinatella* by Oda (1954b) (see Fig. 1i).

The E-shaped lophophore is considered to be a modification of the oral type of duplication, because such a lophophore may be caused by fusing of adjoining arms at one side of two U-shaped arms (j, k). In the polypides with well developed E-shaped lophophores, in which the middle arm protrudes from the arc of the U-shaped lophophore branches at its distal part, the mouth is overhung by two epistomes, and two rectums are situated at the upper part of the stomach (l). From the above facts it seems that these polypides were at the middle stage with respect to the typical oral type of duplication of polypides.

There is an example of the typical anal type (r) in contrast with the typical oral type shown in *Cristatella* (h). In such twin polypides with a common tentacular sheath, both the two lophophores and the two digestive tracts made a mirror image on the anal side. The two lophophores in the twin polypides described by Oda (1954b) (see Fig. 1q) distinctly made a mirror image on the anal side, although those lophophores were of different sizes.

In another example, reported in 1954 (see Fig. 1n) the smaller lophophore attached to the arm of the autosite. In the "parasite" of this polypide the digestive tracts were only slightly developed, a sign of great degeneration. Thus this polypide can also be regarded as the anal type of duplication, because these lophophores faced each other. The polypide bearing the U-shaped lophophore and the additional club-like process with several tentacles (m) is also considered to be an extreme example of the anal type of duplication. The polypide with two lophophores (p) is regarded as the anal type of duplication because of the relation between the two lophophores of different sizes, although this polypide had an additional process-like rectum.

Almost all double polypides were found in the primary polypides which emerged from the germinated statoblasts. Polypides which emerged asexually by budding from such double polypides were normal in shape.

In earlier work, we have repeatedly attempted, but without success, to prove experimentally the regulation of buds in *Lophopodella carteri*. The various stages of buds in a colony and in a germinating statoblast were exposed to different doses of X-rays, but no double polypides appeared (Oda, 1954a; 1959).

REFERENCES

Oda, S. (1954a). Radiosensitivity in *Lophopodella carteri,* a freshwater bryozoan. *Sci. Rep. Tokyo Bunrika Daigaku, Sec.* B, **7,** 211–217.

Oda, S. (1954b). On the double monsters of polypides in freshwater Bryozoa. *Collecting and Breeding* **16,** 15–18. (In Japanese.)

Oda, S. (1959). Germination of the statoblasts in freshwater Bryozoa. *Sci. Rep. Tokyo Kyoiku Daigaku*, Sec. B, **9,** 90–132.

Rogick, M. D. (1945). Studies on freshwater Bryozoa, XV. *Hyalinella punctata* growth data. *Ohio J. Sci.* **45,** 55–79.

Wiebach, F. (1959). "Bryozoa." (Brohmer, P., Ehrmann, P., Ulmer, G., eds) *Tierwelt* Mitteleur. **1** (8), 1–56.

51 | Studies on Freshwater Bryozoa—III. The Bryozoa of the Narmada River System

K. S. RAO

Department of Zoology, Government College,
Mhow (Madhyapradesh) India

Abstract: The Bryozoa of the Narmada River System were studied systematically. The material was collected from 250 stations and about ten freshwater Bryozoan species were recorded. The study adds two species, *Plumatella casmiana* and *Hyalinella minuta*, to the Indian Bryozoan fauna. Two populations were noticed in *Plumatella casmiana* and these are briefly discussed. All the species recorded are new to the Narmada River System and their distribution is extended to central and western India. *P. emarginata* and *Fredericella sultana* were established as the most common species in the region. The abundance of *Lophopodella carteri* in Madhyapradesh is reported. The sessoblast of *Stolella indica* is recorded and figured for the first time, and the branching pattern of *S. indica* and *S. himalayana* commented upon. *Hislopia moniliformis* seems to be a constant species in the region, but the species recorded are not generally uniformly represented in the region.

INTRODUCTION

Annandale pioneered work on Indian Bryozoa, and in his publications from 1906 to 1928 he described many freshwater bryozoan species from India and neighbouring countries. The Indian Bryozoa were later totally neglected and but for the works of Rao and Kulshrestha (1962–1964), and Rao and Ghosh (1962–1964) no references of value could be traced on the subject. The present paper is a report based on a study of a part of the collections made from the Narmada River System and a few important freshwater bodies in the Madhyapradesh region. The collecting localities and the extent of collection areas are indicated in the map (Fig. 1).

The freshwater Bryozoan species recorded by Annandale from India are: 1. *Fredericella sultana;* 2. *Plumatella emarginata;* 3. *P. fruticosa;* 4. *P. javanica;* 5. *P. tanganyikae;* 6. *Stolella indica;* 7. *S. himalayana;* 8. *Hyalinella punctata;* 9. *Lophopodella carteri;* 10. *Lophopus* spp.; 11. *Lophopodella himalayana;* 12. *Pectinatella burmanica;* 13. *Hislopia lacustris;* 14. *H. moniliformis.*

From the present collections there are definite records of the following ten

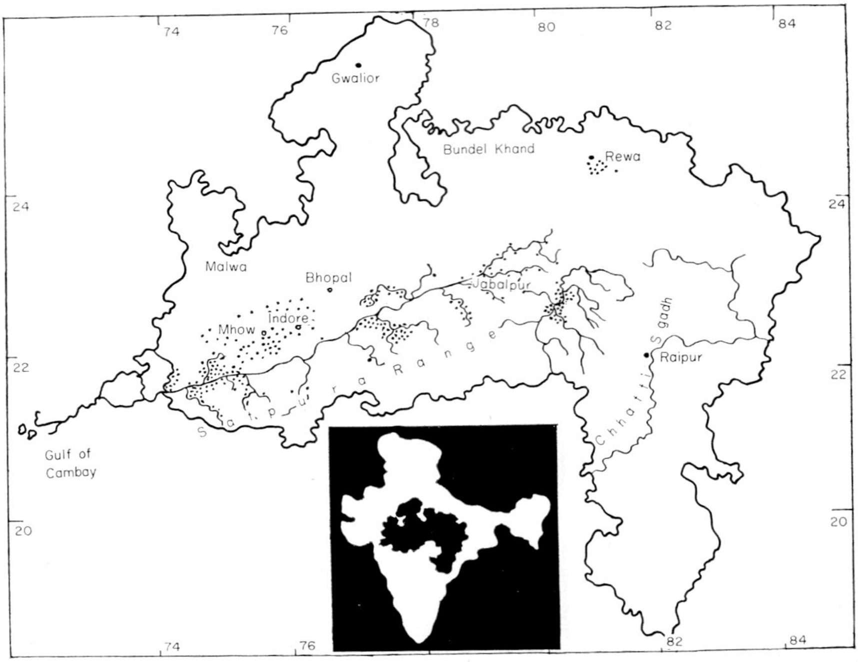

Fig. 1. Map of Madhyapradesh, India showing the Narmada River System and other collecting localities. (The black dots indicate the approximate positions of the collecting localities.)

species: 1. *Fredericella sultana;* 2. *Plumatella emarginata;* 3. *P. fruticosa;* 4. *P. javanica;* 5. *P. casmiana;* 6. *Stolella indica;* 7. *Hyalinella minuta;* 8. *Lophopodella carteri;* 9. *Hislopia lacustris;* 10. *Hislopia moniliformis.*

The Narmada Bryozoan species thus add *Plumatella casmiana* and *Hyalinella minuta* to the Indian freshwater bryozoan fauna.

COLLECTION DATA

Collections were made mainly from the Narmada River System. Pond collections from nearby localities were also included. The material was collected from 250 localities. For each collecting station the details were noted separately in the station register. *Lophopodella carteri* was collected in abundance on submerged rocks only in the Kaveri branch of the Narmada River System. In the same way *Hyalinella minuta* was collected only from Nageshwar Ghat of the Narmada River System and from no other station. It was observed in general that each species is not uniformly distributed but has its own particular grounds from where it can be recorded in abundance. Strictly speaking *Plumatella casmiana* was not recorded from the Narmada River System but from the nearby Patalpani Falls and Govindhgadh Lake areas. Many samples of doubtful material from the Narmada River System are not included in the paper.

DISCUSSION

Fredericella sultana. The *Fredericella* material was recorded in abundance. The entire material was assigned to the species *F. sultana* (Blumenbach). Variations in the size and structure of the paiptoblasts were noticed. They are mostly oval with a smooth or mamillated margin and a simple or ornamented surface. The number of tentacles in each polypide varies from 22 to 28. The detailed variation is given in the following table.

	St. 5	St. 6	St. 18	St. 26	St. 38	St. 45	St. 47	St. 80
Tentacle number	22	28	25	20	24	22	20	27
Nature of paiptoblasts	Simple	marginally ornamented	Simple	Hexagonal markings on surface	Simple	Simple	marginally ornamented	marginal processes
Width (mm)	0·20	0·25	0·18	0·20	0·20	0·22	0·26	0·20
Length (mm)	0·35	0·40	0·33	0·38	0·38	0·37	0·42	0·40

Different varieties of *Fredericella* were reported from Japan (Toriumi, 1956) from Germany (Kraepelin, 1892, 1914) and from North America (Rogick and Brown, 1942). Annandale (1910) reported this species from Eastern India, which seems to have larger paiptoblasts than the present material from central and western India. The zooecia of the samples collected from submerged rocks are less recumbent and radiating than those observed on twigs or leaves. The width of the zooecial tip is highly variable in the material collected from different localities. It is generally almost colourless but in a few samples appears brownish and uniform in colour with the rest of the zooecial tube. The kidney shaped paiptoblasts observed by Annandale (1910) could not be observed in the present collection. They range from oval to elongated with a surface which is smooth, ornamented or with marginal processes (Fig. 3a, b, c).

Plumatella emarginata. *P. emarginata* was recorded in rich quantities from the Narmada material. The extensive variation of the species is well known and needs no special mention. The extensive overlap of this species with other species of *Plumatella* suggests the need for its reconsideration.

Plumatella fruticosa. This is sparingly represented in the present collections. The long elliptical statoblasts are larger in size than those described by Annandale (1910). The material from Jabbalpore is on a flat encrusting type, which is sparingly branched, while it is usually shrubby as reported from other stations.

Plumatella javanica. This species is represented only sparingly in the present collections. The dichotmous branching observed by Annandale (1910) in this species is not found in the present material. The zooecia are recumbent and the statoblasts are small and typically slightly drawnout at the sides. Annandale (1910) reported the occurrence of this species in Java, Penang and India. Wiebach (1967) considered it in detail in his report on Amazon Bryozoa. The various aspects of the species are discussed at length by Toriumi (1952). The Indian localities of this species are Bengal, Calcutta, Berhampore, Murshidabad, R. Jharai, Siripur, Madras and Travancore. The species is reported from Madhyapradesh for the first time.

Plumatella casmiana. The material of *P. casmiana* was recorded from only two localities, Govindhgadh (St. No. 41) and the Patalpani Falls area (St. No. 102). The former material resembles *P. emarginata* very closely but for the presence of two types of floatoblasts characteristic of *P. casmiana.* This material was found profusely grown, covering a considerable area with interwoven branches. The zooecia are repent, cylindrical and have a thick ectocyst. The latter is dark brown or brownish black in colour with colourless tips. They are clearly emarginate at the tip with a distinct furrow and well developed keel. The polyp is elongate and the number of tentacles varies from 38 to 44. The velum, at the base of the tentacles, extends upwards for a considerable distance without being festooned. The Patalpani Falls material is light brown and appears fungoid with recumbent zooecia. The tip of the zooecium is slightly emarginated having a faintly visible furrow and keel. The material from both the localities was recorded as overgrowths on small pieces of submerged sticks.

Available reports record this species from Japan (Toriumi, 1942a), North America and Canada (Rogick, 1941, 1943; Rogick and, Schalie 1950) and Bushnell and Wood, 1971) and from West Germany (Wiebach, 1963). The present record shows its occurrence in Indian waters. The material recorded from both the localities in India appears to suggest the occurrence of different populations in this species. The variation range is high as the colonies vary from thick, interlaced brownish black masses to light brown recumbent, fungoid growths (Patalpani material) or to luxuriant honey-comb groups reported from Colorado (Bushnell and Wood, 1971). The statoblasts of the Indian material, however, conform closely with the description given by Oka (1907), Toriumi (1942a, b) and Rogick (1941, 1943). (Fig. 3d, e.)

Stolella indica. This species is of common occurrence in the Narmada River System, especially in the localities of the Indore and Mhow regions. The branching and the mode of formation of the zooecia are typical of the species as is particularly the shape of the floatoblasts. The material recorded from the

	Patalpani Falls material	Govindhgadh material
Length of the floatoblast	0·34 mm	0·36 to 0·40 mm
Length of the floatoblast (Thin type)	0·32 mm	0·33 to 0·36 mm
Width of the capsule (Thin type)	0·26 mm	0·22 to 0·28 mm
Width of the capsule	0·10 mm	0·10 to 0·15 mm
Length of the capsule	0·25 mm	0·23 mm
Length of the capsule (Thin type)	0·28 mm	0·25 mm
Length of the sessoblast	0·41 mm	0·44 mm
Width of the sessoblast	0·35 mm	0·39 mm

Patalpani and Gambhir localities resembled *S. himalayana* Annandale when they are luxuriantly grown, but their younger colonies resembled *S. indica* (Fig. 2c, d). The author prefers to discuss these two species in detail elsewhere. The contention of Annandale (1910) that this species is strictly confined to the Indo-gangetic plains cannot be maintained any more as it is already reported from the Vindhyan region (Rao and Kulshrestha, 1962–1964). The present record extends its distribution from eastern to southwest India.

A single sessoblast is reported and figured here for the first time (Fig. 3g) from the Gambhir River. The formation of the sessoblasts seems to be rare but their presence may call for a reconsideration of the evolutionary status of the genus as discussed by Toriumi (1956) and Abricossoff (1959).

Hyalinella minuta. This is recorded from a single locality, the Nageshwar ghat (St. No. 97) near Badawaha in the Narmada River System. The material was observed as an encrustation on small pieces of sticks. The colony was entirely repent with no erect branches, and is dense in growth (Fig. 2b). It has a continuous ectocyst of the older and newly formed zooecia. The zooecia are small and flat with a swollen soft and colourless ectocyst, with small foreign particles attached to them. Its tip is slightly upright with an obscure emargination and keel. The septa are not visible. The floatoblasts differ slightly from those of *H. punctata* and the sessoblasts are not recorded. The floatoblasts agree with those of the Japanese material (Fig. 3).

Kraepelin (1887) described the species as *H. punctata* var *densa* and figured it. Vortsman described *H. punctata densa*, but Toriumi (1955) observed it to be different from that reported by Kraepelin and established the species from Japanese waters. The present material resembles the latter very closely.

Lophopodella carteri. This seems to be abundant in the Narmada River System and the surrounding freshwater bodies. The Indian *Lophopodella* was recorded from clear water ponds in stone quarries. The species is completely unrepresented in Bengal. In Madhyapradesh it is extremely common even in muddy ponds. Extreme variations were observed in its spinoblasts. The extensive variation of these was described by many workers and needs no special mention. The species was reported to occur in Bombay, West Himalayas (Bhimtal 1500′) and

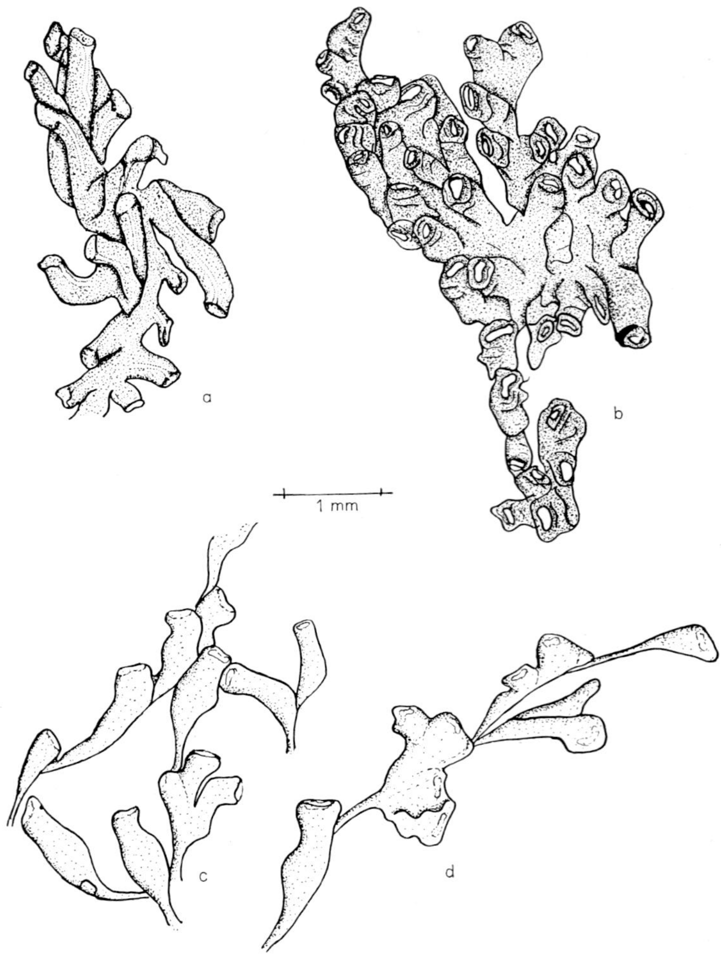

Fig. 2. a. Part of a colony of *Plumatella casmianai*; b, Part of a colony of *Hyalinella minuta*; c, Thickly grown colony of *Stolella indica*; d, Young colony of *S. indica*.

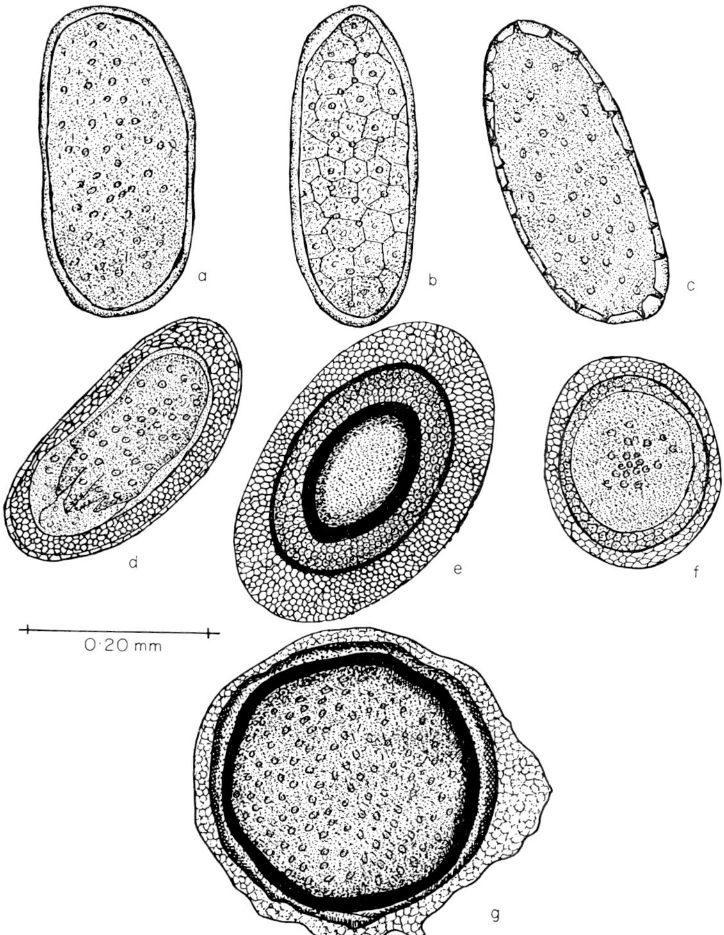

Fig. 3a, b, c. Paiptoblasts of *Fredericella sultana*. d, e, The two types of statoblasts of *Plumatella casmiana*. f, Floatoblast of *Hyalinella minuta*. g, Sessoblast of *Stolella indica*.

Madras. The present record of the species from the Narmada River System extends its geographical distribution to Madhyapradesh.

Hislopia lacustris. This species occurs throughout south Asia. In the Narmada River System *H. lacustris* is very common. The material from different localities exhibits many variations and needs more detailed study. Wiebach (1967) gave a detailed account of the genus from the German material. Annandale (1911) reported its abundant occurrence from Calcutta. The present material agrees with the typical description of the species in general.

Hislopia moniliformis. Annandale (1911) reported this species from Cambodia, Siam and India. The present author recorded it from the Vindhyan region (1962–1964) and now from the Narmada River System in India. The zooecia of this species are typically arranged in linear series. Sometimes the ancestrula develops growth in different directions, but always forms rows of zooids in a single line. Its variations were considered in detail by Rao and Ghosh (1962–1964) in the report on Vindhyan Bryozoa. The same variations could be observed in the present material also. Further records of the species may prove its constancy and discount the contention that it might be a variant of *H. lacustris*.

ACKNOWLEDGEMENTS

The author is grateful to Dr Wiebach, Dr Toriumi, Dr Bushnell and Dr Anna B. Hastings for providing important references and useful suggestions. My especial thanks are due to Dr Raviprakash, Principal, Holkar Science College, Indore, for giving facilities and constant encouragement throughout the course of the study.

REFERENCES

ABRICOSSOFF, G. G. (1959). The generic classification of the freshwater Polyzoa. The phylactolaemata. *Dokl. Acad. Nauk. S.S.S.R.* (Translation in English by Amer. Inst. Biol. Sci. Washington) Nos. 1–6.

ANNANDALE, N. (1910). Materials for a revision of the phylactolaematous Polyzoa of India. *Rec. Ind. Mus.* **5,** 37–57.

ANNANDALE, N. (1911). "The Fauna of British India." Freshwater Sponges, Hydroids, and Polyzoa, pp. 161–238, 240, 245–247.

BUSHNELL, J. H. and WOOD, T. S. (1971). Honey comb colonies of *Plumatella casmiana* Oka (Ectoprocta: Phylactolaemata). *Trans. Am. Microsc. Soc.* **90** (2), 229–231.

KRAEPELIN, K. (1887). Die Deutschen susswasser Bryozoen. I. *Anatomisch-Systemetischer Teil. Abh. Naturw. Ver. Hamburg* **10.**

KRAEPELIN, K. (1892). Die Deutschen susswasser Bryozoen. II. Entwicklungs geschichtlicher. Teil. *Abh. Naturw. Ver. Hamburg* **12.**

KRAEPELIN, K. (1914). Bryozoa. *In* "Beitrage zur kenntnis der land und susswasser fauna Deutsch-Sudwest Afrikas." Hamburg.

OKA, A. (1907). Zur Kenntnis der susswasser Bryozoen fauna von Japan. *Annotnes. Zool. Jap.* **6,** 117–123.

RAO, K. S. and KULSHRESTHA, K. S. (1962–1964). Studies on freshwater Bryozoa. I. The Bryozoa of the Vindhyan region. *J. Univ. Saugor.* **11–13,** Part. II, Sec. B, 50–64.

RAO, K. S. and GHOSH, N. (1962–1964). On the extension of the geographical distribution of *Hislopia lacustris* sub. sp. *moniliformis* Annandale (Ectoprocta: Gymnolaemata) along with a note on its morphological peculiarities. *J. Univ. Saugor.* **11–13,** Part II, Sec. B, 47–49.

ROGICK, M. D. (1941). Studies on freshwater Bryozoa. X. The occurrence of *Plumatella casmiana* in North America. *Trans. Am. Microsc. Soc.* **60** (2), 211–220.

ROGICK, M. D. (1943). Studies on freshwater Bryozoa. XIII. Additional *Plumatella casmiana* data. *Trans. Am. Microsc. Soc.* **62** (3), 265–270.

ROGICK, M. D. and BROWN, C. J. D. (1942). Studies on freshwater Bryozoa. XII. A collection from various sources. *Ann. N. Y. Acad. Sci.* **43** (3), 123–144.

ROGICK, M. D. and SCHALIE, H. VAN DER (1950). Studies on freshwater Bryozoa XVII. Michigan Bryozoa. *Ohio. J. Sci.* No. 1 **3,** 136–146.

TORIUMI, M. (1942a). Studies on freshwater Bryozoa of Japan III. Freshwater Bryozoa of Hokkaido. *Sci. Rept. Tohoku. Univ. Sandai. Japan.* Ser. 4. *Biol.* **17** (2), 197–205.

TORIUMI, M. (1942b). Studies on freshwater Bryozoa. IV. Freshwater Bryozoa of Taiwan (Formosa). *Sci. Repts Tohoku. Univ. Sendai. Japan* Ser. 4. *Biol.* **17** (2), 207–214.

TORIUMI, M. (1952). Taxonomical study on freshwater Bryozoa. IV. On *Plumatella javanica* Kraepelin. reported by Vortsman in 1928. *Sci. Repts Tohoku. Univ.* Ser. 4. *Biol.* **29** (3), 264–269.

TORIUMI, M. (1955). Taxonomical study on freshwater Bryozoa. XII. On *Plumatella repens* var. *minuta. Sci. Repts Tohoku. Uni.* Ser. 4. *Biol.* **XXI** (2) 57–88, 137–144.

TORIUMI, M. (1956). Taxonomical studies on freshwater Bryozoa. XVII. General consideration and intraspecific relations of described species and phylogenetic considerations. *Sci. Repts Tohoku. Univ.* Ser. 4 **XXII** (2), 57–88.

WIEBACH, F. (1963). Das Moostierchen *Plumatella casmiana. Mikrokosomos* **52**. *Jahrgang* (3), 76–79.

WIEBACH, F. (1967). Amazonische Moostiere (Bryozoa). *Amazoniana* **1, 2,** 173–187

52 | Preliminary Notes on a Revision of the Genus *Hyalinella*

F. WIEBACH
Plön, West Germany

Abstract: A discussion of *Plumatella punctata* Hancock = *Hyalinella punctata* (Hancock) and of topotype material collected by Dr A. B. Hastings leads to the result that Hancock's species and the topotype samples rather represent a *Plumatella*, probably a phenotype of *P. repens*, than a valid species of the genus *Hyalinella*. This statement should be re-examined by rearing experiments with statoblasts from respective samples.

Hyalinella, established by Jullien on a species (*Plumatella vesicularis* Leidy) which probably is neither a valid *Plumatella* species nor a *Hyalinella*, but rather a phenotype of *P. repens*, should, however, be regarded as a valid genus.

Some *Hyalinella* problems are discussed (the nature of the ectocyst, the sessoblast problem, the ash-blue swim-ring of *Hyalinella punctata* floatoblasts). An attempt is made to establish a diagnosis of the genus *Hyalinella*. Eighteen *Hyalinellae* mentioned in literature are discussed, 8 of which should, with certain reservations, be regarded as valid species.

Bromley Lough (1056 by 440 yards, i.e. 950 by 400 m) and Crag Lough (880 by 176 yards, i.e. 800 by 150 m) (Dr A. B. Hastings, *in litt*) are lakes situated at a distance of about 30 miles west of Newcastle, Northumberland, England. Hancock (1850) found several species in these lakes one of which he described as new and named *Plumatella punctata*. Hancock illustrated his description by three figures: "(1) Side view of exserted tentacular apparatus much enlarged, (2) *Plumatella punctata* five or six times magnified, exhibiting polypes exserted, (3) Three cells of ditto more highly magnified and more produced than usual, with the polypes exserted." There is no doubt that Hancock had only collected very young colonies. It is a fact that young *Plumatella* colonies can often hardly be distinguished from one another, and this particularly applies to *P. repens* and *P. punctata*. The "eggs" he found were sessoblasts only.

Allman (1856) adopted *Plumatella punctata* Hancock, quoting Hancock's description, and reproducing two of Hancock's three figures. He remarked "it comes very near to *P. repens*, closely resembling the young state of this species" and I consider that Allman's reservation was justified.

In 1937 Dr A. B. Hastings visited the lakes, and has kindly put at my disposal her unpublished notes. Remarking on the differences between *P. punctata* and *P. repens* found in Bromley Lough she concluded "It is clear that the differences are not due to age, nor, as all the colonies were developed from fixed statoblasts, to differences in origin, and as both kinds could be found on a single stone, their environment may be supposed to be the same." Although two apparently different species were sometimes found on a single stone, there are two reasons that make me hesitate in believing in one being *P. punctata* and the other *P. repens*. Dr Hastings was also kind enough to let me have specimens collected by her and supposed to be *P. punctata* for close examination; there is no doubt about the presence of septa in some of them, which means they cannot be *Hyalinella*. In Hancock's and Dr Hasting's specimens no floatoblasts were traceable. Sessoblasts alone (except a few *Plumatellas*, such as *P. fruticosa* and *P. tanganyikae*) are not generally sufficient to determine a species of *Plumatella*. I am inclined to determine the *P. punctata* material from Bromley Lough as a phenotype variety of *P. repens*, and in my opinion both kinds of *Plumatella* mentioned above, although they were found on a single stone, are phenotypes of *P. repens*. I have found two apparently different *Plumatella* colonies under one single leaf of *Nuphar luteum* and taken one of them for *P. repens* and the other for *P. fungosa*. I am now sure that both of them were *P. repens* colonies. I have therefore come to the conclusion that the specimens defined by Hancock as *P. punctata* and the topotype specimens collected by Dr Hastings do not belong to *P.* (*Hyalinella*) *punctata*, but rather to *P. repens*.

P. repens and *P. fungosa* have hitherto been considered as two valid species. Toriumi found that their wide ranges of variation overlapped and intermediate forms occurred.

According to information received from Dr Hastings there is no known type-material of Hancock's in the Hancock Museum at Newcastle-upon-Tyne or in the collection of the British Museum. We may, however, assume that the specimens collected by Dr Hastings are of the same kind as those on which Hancock established his *P. punctata*. For a definite re-examination as many specimens as possible should be collected from Bromley Lough and colonies reared from statoblasts. I consider, however, that the colonies reared will turn out to be *P. repens*.

Jullien (1885) defining the genus *Hyalinella*, said "This genus differs from *Plumatella* only by its ectocyst which is gelatinous instead of being coriaceous. This ectocyst is not always colourless, it can be brown at some spots, but mostly it remains transparent". I consider this definition to be incomplete. Toriumi (1956) remarked "The thickness of the ectocyst in *Hyalinella* has hithero been believed by many workers to be distinctly thicker than that of *Plumatellas*, but it is very difficult to distinguish *Hyalinella* from *Plumatella* upon such a feature".

Another problem in *Hyalinella* is the sessoblast. In describing *P. punctata* Kraepelin (1887) remarked "Statoblasts always provided with a swim-ring", which means that he had not found sessoblasts. In another paper (1914) he described samples of *P. punctata* collected in East Africa. In these specimens he found sessoblasts "which thoroughly correspond to the sessoblasts of other *Plumatella* species". I am sure that this material is identical with No. B 96 in the collection of the Hamburg Zoological Institute. It was originally named "*Hyalinella punctata*", and later erroneously renamed, "*Hyalinella indica*". After carefully examining the sample I have come to the conclusion that it represents neither of the two, but rather *P. osburni*.

Grančarova (1968) dealt with Bulgarian *Hyalinella punctata*, stating that sessoblasts were present in her material, but I could not confirm her finding, in specimens which she kindly loaned.

Another *Hyalinella punctata* problem is that of the ash-blue swim-ring of the floatoblast in fresh field specimens which have not been treated with chemicals or alcohol. This colour has been mentioned by Kraepelin, Wesenberg-Lund, Borg and by E. Rüsche (unpublished).

Dr. A. Viganò was kind enough to send me some samples which he had collected in a small Italian lake (Laghetto del Nivolet). The small colonies, growing on stones, reminded me at once of Dr Hastings' *P. punctata* samples, and I had no doubt that they represented a variation of *P. repens*. After dissecting three floatoblasts from one single cystid, I found that one of them (the biggest) looked like a *punctata* floatoblast, another could also be taken for a *punctata* floatoblast or an intermediate form, the third one was clearly a *repens* floatoblast.

When Jullien (1885) erected the genus *Hyalinella*, he established it on *Plumatella vesicularis* Leidy, of which he had seen no specimen himself; as a synonym of *Hyalinella vesicularis* (Leidy) he mentioned *Plumatella vitrea Hyatt*. Concerning *Hyalinella punctata* he only mentioned that Allman had stated some similarity between this form and *Hyalinella vesicularis*. Dr Toriumi (*in litt.*) informs me that there are no specimens of Leidy's *P. vesicularis* extant, and topotype specimens should be collected. As to *Plumatella vitrea*, I have not been able to ascertain if any material has been preserved; if any samples are available, they should be closely examined. Hancock's *P. punctata* was considered by Jullien to be a phenotype of *P. repens*, and he was probably right.

Although the genus *Hyalinella* has probably not been established on a valid

Hyalinella species it is to be regarded as a valid genus. Jullien's definition was enlarged for *P. punctata* by Kraepelin, who considered the following features as specific for this species: a constant transparence of the cuticula, the polypides rising from numeroys almost vertical, "volcano-like" orifices (blunt cones), colony entirely attached to the substratum, no erect branches, floatoblasts only, colonial tube common to all polypides (no septa present). Dr Toriumi gave a diagnosis of *Hyalinella punctata* in 1955.

The characters of the genus would appear to be: Colonies either dendritic or compact (massive), sometimes spindle-like, mostly entirely attached to substratum, no erect branches, in very rare cases like garlands hanging in the water. Ectocysts constantly transparent and gelatinous, never chitinized; they can agglutinate into a compact mass. Colonial tube common to all polypides (no septa present). Polypides rising from almost vertical blunt cones (often very flat). Short pseudostolons possible. Floatoblasts agree in structure with those of *Plumatella*. Sessoblasts may (seldom) occur.

The following list is, to the best of my knowledge, an enumeration of all *Hyalinellas* mentioned hitherto in literature, including erroneous denominations:

1. *Hyalinella punctata* (Hancock) is not likely to represent the type genus; this species will require a full-scale re-examination, especially with regard to the two so called seasonal forms *densa* and *prostrata*; a renaming of this species may prove necessary.

2. *H. africana* Wiebach 1964; considered by Lacourt (1968) a synonym of *Stolella evelinae* Marcus see Wiebach (1970). It might be a synonym of *H. longigemmis*; further examination of this species and of *H. africana* is required. The type specimen is in the Musée Royal de l'Afrique Centrale (Tervuren, Belgium). This species and the Bulgarian *Hyalinella punctata* (see above) are the only species of *Hyalinella* in which sessoblasts have been found.

3. *H. agilis* (Marcus) is a name given to *Stolella agilis* Marcus by Lacourt (1955, 1968).

4. *H. auricomis* Annandale, established by Annandale for *Plumatella auricomis*, and described by him again (1915b) and by Abricossoff (1927) and Hastings (1938) who remarked "If *Hyalinella*, type *H. punctata* (Hancock), is to be maintained as a genus or subgenus distinct from *Plumatella*, it is clear that *P. auricomis* should be placed in it". Marcus (1941) remarked "In *P. auricomis* Annandale . . . the ectocyst is hard and hyaline as in *Stolella*, but it differs from the latter by being thick and getting shrivelled in alcohol". Hozawa and Toriumi (1941) mention that the present species is *P. auricomis* Annandale and remarked "The ectocyst is greatly thickened and is either colourless or more or less a pale yellowish-brown hue. It is hyaline and stiff, differing from the ecto-

cyst of *Hyalinella punctata* which is soft and contractile". Their Figs 4b, c also point to a *Plumatella*, and Plate 14, Fig. 6, leaves a doubt as to whether it represents a *Hyalinella* or a *Plumatella*. Lacourt (1948) mentioned *P. auricomis* under this name, making it, however, a synonym of *P. pseudostolonata* Borg and of *P. tanganyikae*, the first of which is a *Stolella*, whilst the second is a valid species of its own. Lacourt (1968) considered *P. auricomis* a doubtful species. A specimen which I received for inspection from the British Museum (No. 14.6.10.1.) collected by Behning in the Volga River, and determined by Annandale, seemed to me to be rather a *Plumatella*, probably a phenotype of the polytypic *P. repens*.

5. *H. bigemmis*, established by Annandale (1919, as *P.* (*Hyalinella*) *bigemmis*) was considered by Toriumi (1955) a synonym of *P. repens*. I accept his statement "All the features of *P. bigemmis* agree with these of *P. repens*", all the more so as Annandale's figures plead more for *Plumatella* than for *Hyalinella*, and his description does not allow a clear conclusion in favour of *Hyalinella*.

6. *H. carvalhoi* Marcus (1942). The author stated that this species belonged to the *Hyalinella* forms with a moderately thick ectocyst in which he included *H. vaihiriae* and *H. auricomis*. Lacourt (1968) placed this species (*H. carvalhoi*) among the *Plumatellas*, considering it a synonym of *P. repens var. osburni* Rogick and Brown. Comparison of the cross-sections of the floatoblasts of these two species makes this synonymy impossible. *H. carvalhoi* requires re-examination; it may turn out to be a *Plumatella*.

7. *H. iheringi*. Established by Meissner (1893) as *Lophopus iheringi;* and regarded by Kraepelin (1892) as identical with *Lophopus lendenfeldi*. Kraepelin (1914) discussed the problem of this supposed identity. Annandale (1915a) mentioned *Australella iheringi* (Meissner), and in a systematic survey (1919) he referred the two species *lendenfeldi* and *iheringi* to *Hyalinella*. Hastings (1929) remarked "In its statoblasts *P. inheringi* is nearer to *Pectinatella* than to either of the other two genera" (viz *Lophopus* and *Hyalinella*). I think the assignment to *Pectinatella* by Hastings, though she puts a quotation mark after *Pectinatella*, should be regarded as definite. It is strange that this species, found more than 75 years ago in Brazil (Rio Grande do Sul), should not have been found again. The type specimen is available in the Institut für Spezielle Zoologie in Berlin (East), as No. 454, named *Lophopus iheringi*.

8. *H. indica* (Annandale) was established as *Australella indica* (1915a), but later (1919) referred to as *Plumatella indica*. Specimens are available in the British Museum. This is no doubt a valid species, well characterized by its specific floatoblast, which was figured by Annandale (1915a). Annandale (1919) himself dropped the genus *Australella* in favour of *Hyalinella*. Lacourt (1968)

mentioned "*Plumatella emarginata*", determined by Borg, as a synonym of *H. indica*. He referred to the sample SMF 795 in the Senckenberg Museum, Frankfort, now labelled SMF 836, which certainly is neither *H. indica* nor *P. emarginata*, but *P. osburni*.

9. *H. lendenfeldi* (Ridley), found in the Paramatta River, New South Wales, was described as a new species, *Lophopus lendenfeldi*, in 1890. The description by the author is not very clear, but his figures indicate that it is unlikely to be assigned to *Lophopus*. Dr Hastings, who re-examined a slide of statoblasts of this species stated (British Museum 1886.9.22.9.) "Ridley's figure gives a very true representation of the appearance of the statoblasts, except that the capsule should be very slightly wider in proportion to its length". I had the opportunity of examining one tube and four slides of the present species (British Museum 1886.9.22.10.); floatoblasts were absent. In contrast to Ridley (1890) who stated 'Epistome not observed" (which would be a very extraordinary thing in Phylactolaematous Bryozoa), I could clearly recognize this organ. I think Rousselet (1904) was right in assigning this species to *Hyalinella*, and it should be considered as a valid species.

10. *H. longigemmis* (Annandale), established (1915a) as *Plumatella punctata* var. *longigemmis*, was illustrated by two figures (colony and floatoblasts), which are not quite satisfactory. The description reads: "It agrees with *P. punctata* in every respect except that the gemmules are uniformly more elongate and have relatively smaller capsules than is usually the case. They have a swim-ring slightly curved in cross-section, in this respect somewhat resembling those of *Australella indica*, though the feature is less marked. . . ." This last feature is important. It also appears in *H. africana*, in which this feature is clearly different from (1) *H. punctata*, where the suture-line is horizontal, and (2) from *H. indica*, where it is strongly curved. A re-examination should be made of both *H. africana* and *H. longigemmis*. A sample declared to represent *H. longigemmis* (B.M. 1920.12.16 (stained), Lale at Colombo Municipal Water Supply, sent by Dr L. F. Hirst), and two slides labelled *P. longigemmis* did not allow a reliable examination; there were to be seen numerous unspecific floatoblasts which I should rather have taken for those of *P. javanica* or *P. repens*, the ectocyst was found to be chitinized, septa were present—so that there was no doubt that the material represented a *Plumatella* and not *H. longigemmis* Annandale.

11. *H. minuta* (Toriumi) originally described (1955a) as *P. repens* var. *minuta*, proves by the figures given to be a valid species. It occasionally forms pseudostolons as sometimes have been observed in other species of *Hyalinellae*. I do not think Lacourt 1968 is right in putting this species in the synonymy of *H. punctata*.

12. *H. orbisperma* (Kellicott), assigned to *Plumatella*, was re-examined and discussed by Rogick and van der Schalie (1950) under the old denomination. They pointed to the similarity between the floatoblasts of this species and those of *Stephanella hina* Oka and concluded that further investigation would be necessary. This was undertaken by Bushnell (1963) who stated "The taxonomic position of *P. orbisperma* has been uncertain for the past 80 years. Recently collected and studied specimens of this form recommend a reassignment of the *orbisperma*-form from the genus *Plumatella* to the genus *Hyalinella* . . .". Bushnell (1965) discussed the matter in detail, particularly pointing to the similarity between the floatoblasts of *H. orbisperma* and *S. hina*, adding "If sessoblasts should be found in *H. orbisperma* colonies, then a comparison with *S. hina* would again be demanded". After examining some *H. orbisperma* specimens which Dr Bushnell kindly sent to me, I have come to the conclusion that this is a valid species.

13. *H. osburni* (Rogick and Brown, 1942) was considered to be a synonym of *P. repens* var. *osburni* by Lacourt (1955) who later (1968) placed it in the synonymy of *P. longigemmis*. I consider that Toriumi (1952a) was right in raising the variety *osburni* to specific rank as *Plumatella osburni*.

14. *H. toanensis* Hozawa and Toriumi (1941) was placed in the genus *Gelatinella* by Toriumi (1955c). Lacourt (1968) certainly does not do justice to the problems of this species in referring it to *Plumatella*. I do not accept his reasons. I consider that *toanensis* is a valid species, which belongs to either *Hyalinella* or *Gelatinella* and I prefer to refer it to *Gelatinella*.

15. *H. vaihiriae* Hastings (1929). Dr Hastings (*in litt.*) writes "Over the years, and after examining the topotype of the type-species, I have begun to doubt whether *Hyalinella* should be maintained as a distinct genus". Dr Toriumi (*in litt.*) writes "*Plumatella vaihiriae* sent from Dr Hastings falls within the variation range of *P. repens* in all features". After examining samples put at my disposal (*Hyalinella vaihiriae* Hastings. Lake Vaihiria, Tahiti. Crossland Coll. 78A. B.M. 1928.12.18.1.) I have come to the same conclusion. I do not think Lacourt (1968) is right in considering this species as a valid one.

16. *H. vesicularis* (Leidy) upon which Jullien 1885 established the genus *Hyalinella*, should be considered a *nomen dubium* until further notice; it may be a *Plumatella*, possibly a phenotype of *P. repens*. Kraepelin regarded it as a synonym of *P. punctata*, and it was originally referred to *Plumatella*.

17. *H. vitrea* (Hyatt) originally referred to *Plumatella*, should be considered a *nomen dubium*. As with *P. vesicularis*, there are probably no preserved specimens available. *H. vitrea* may be a phenotype of *P. repens*, and Kraepelin regarded it as a synonym of *P. punctata*.

18. *H. vorstmani* (Toriumi, 1952b) was referred to *Hyalinella* by Lacourt (1968). After examining samples of the present species put at my disposal by Dr Toriumi, I agree with him in regarding *vorstmani* not as a *Hyalinella*, but as a *Plumatella*. I leave open to reconsideration whether *P. vorstmani* is a valid species or only a variation of the polytypic *P. javanica*, especially as Toriumi (1952b) says: "Therefore specific difference between the two species (viz. *P. javanica* and *P. vorstmani*) is recognised in the shape and in the dimensional characters of the floatoblast."

ACKNOWLEDGEMENTS

Thanks are due to Dr A. B. Hastings who kindly put at my disposal her 1939 manuscript on "Some British Freshwater Polyzoa" and her Bromley Lough topotype material of *P. punctata*, to Miss P. L. Cook (British Museum) who kindly sent me various samples, to Mrs T. Grančarova, Sofia, for sending me Bulgarian *Hyalinella punctata* material, to Dr John H. Bushnell for letting me have *Hyalinella orbisperma* samples, to Dr Toriumi for valuable and important letters on *Plumatella* and *Hyalinella* problems, and to Dr A. Viganò who sent me samples helping to elucidate *Hyalinella punctata* problems.

REFERENCES

Abricossoff, G. G. (1927). Über die Süsswasserbryozoen der USSR. *C.R. Acad. Sci. U.S.S.R.* **19**, 307–312.

Allman, G. J. (1856). "A Monograph of the Freshwater Polyzoa," including all the known species, both British and Foreign. Ray Society, London.

Annandale, N. (1915a). The Genus *Australella* and some allied Species of Phylacolaematous Polyzoa. *J. Proc. As. Soc. Bengal* **XI,** 163–169.

Annandale, N. (1915b). Report on a Collection of Phylactolaematous Polyzoa from the Volga Region. *Trav. Soc. Nat. Saratov* **V,** 73–80.

Annandale, N. (1919). Sponges, Hydrozoa and Polyzoa of Seistan. *Rec. Ind. Mus.* **XVIII,** 91–97.

Bushnell, J. H. (1963). The reassignment of *Plumatella orbisperma* and ecological information on the freshwater Ectoprocta. *American Zoologist* **3.**

Bushnell, J. (1965). On the taxonomy and distribution of freshwater Ectoprocta in Michigan. III. *Trans. Am. Microsc. Soc.* **84** (4).

Grančarova, T. (1968). Neue Bryozoen in der bulgarischen Fauna (bulgar. mit deutscher Zusammenfassung). *Bull. Inst. Zool. Mus., Acad. Bulg. Sci.* **XXVIII.**

Hancock, A. (1850). On the anatomy of freshwater Bryozoa, with description of three new Species. *Ann. Mag. Nat. Hist.* **V,** 173–204.

Hastings, A. B. (1929). Notes on some little-known Phylactolaematous Polyzoa and description of a new species from Tahiti. *Ann. Mag. Nat. Hist.* (10) **III,** 300–310.

Hastings, A. B. (1938). Polyzoa from Lake Huleh. *Ann. Mag. Nat. Hist.* (11) **II,** 529–535.

Hozawa, S. and Toriumi, M. (1941). Some Freshwater Bryozoa found in Manchoukuo. *Sci. Rep. Tôhoku Univ.* 4th Ser. *Biol.* **XVI**/3, 233–241.

Jullien, J. (1885). Monographie des Bryozoaires d'eau douce. *Bull. Soc. Zool. Fr.* **10,** Meulan.

KRAEPELIN, K. (1887). Die Deutschen Süsswasserbryozoen, I. *Abh. Naturw. Ver. Hamburg* **10**.

KRAEPELIN, K. (1892). Die Deutschen Süsswasserbryozoen, II. *Abh. Naturw. Ver. Hamburg* **12**.

KRAEPELIN, K. (1914). Bryozoa. *In* Michaelson: "Land- u. Süsswasserfauna Deutsch-Südwestafrikas."

LACOURT, A. W. (1948). On two Freshwater Bryozoa (Phylactolaemata) from Belgian Congo. *Rev. Zool. Bot. Afr.* **XL,** 4.

LACOURT, A. W. (1955). Freshwater Bryozoa (Phylactolaemata) from Curaçao, Aruba and Bonaire (Studies on the Fauna of Curaçao and other Caribbean Islands, Nr. 27).

LACOURT, A. W. (1968). A Monograph of the Freshwater Bryozoa (Phylactolaemata). *Zool. Verh.* **93,** Leiden.

MARCUS, E. (1941). Sôbre Bryozoa do Brasil. *Bol. Fac. Fil. Ciên. Letr. Univ. São Paulo* **22,** Zool. Nr. 5.

MARCUS, E. (1942). Sôbre Bryozoa do Brasil, II. *Bol. Fac. Fil. Ciên. Letr. Univ. São Paulo* **25,** Nr. 6.

MEISSNER, M. (1893). Eine anscheinend neue Süsswasser-Bryozoer (*Lophopus Iheringi* n. sp.) aus Brasilien *Sitzber-Ges. nat. Freunde Berlin*, 260–263.

RIDLEY, S. O. (1890). On the Characters of the Genus *Lophopus*, with Description of a new Species from Australia. *Linn. Soc. J. Zool.* **XX,** 61–64.

ROGICK, M. D. and BROWN, C. J. D. (1942). Studies XII: A collection from various sources (*Ann. N.Y. Acad. Sci.* **43,** 123–144.

ROGICK, M. D. and SCHALIE, H. VAN DER (1950). Studies XVII: Michigan Bryozoa *Ohio J. Sci.* **50,** 136–146.

ROUSSELET, CH. F. (1904). On a new Freshwater Polyzoon from Rhodesia. *J. Quekett Microsc. Club* (2) **IX**/54, 45–56.

TORIUMI, M. (1952a). Taxonomical studies on freshwater Bryozoa III. *Plumatella osburni* (Rogick and Brown). *Sci. Rep. Tôhoku Univ.* (4) **19** (3), 260–263.

TORIUMI, M. (1952b). *Op. cit.* IV. On *Plumatella javanica* Kraepelin reported by Vorstman in 1928. *Ibid.* (4) **19** (3), 264–269.

TORIUMI, M. (1955). Tax. Study IX: *Plumatella repens* (L.). *Sci. Rep. Tôhoku Univ.*, 4th Ser., *Biol.* **XXI** (1).

TORIUMI, M. (1955a). Tax. Study XII: On *Plumatella repens* var. *minuta* Toriumi, 1941. *Sci. Rep. Tôhoku Univ.* **XXI** (2).

TORIUMI, M. (1955b). Tax. Study XIII: *Hyalinella punctata* (Hancock). *Sci. Rep. Tôhoku Univ.* **XXI** (3–4).

TORIUMI, M. (1955c). Tax Study **XIV**: Reconsideration on *Hyalinella toanensis* Hozawa and Toriumi. *Sci. Rep. Tôhoku Univ.* **XXI** (3–4).

TORIUMI, M. (1956). Tax. Study XVII: General Consideration: Interspecific Relation of described Species and phylogenic Consideration. *Sci. Rep. Tôhoku Univ.* **XXII** (2).

TORIUMI, M. (1970–1971). Additional Observations on *Plumatella repens* (L.), I–V. *Bull. Mar. Biol. Stat. Asamushi, Tôhoku Univ.* **XIV** (1–2).

WIEBACH, F. (1964). Untersuchungen an Süsswasser-Bryozoen aus Zentralafrika. *Mus. R. Afr. Centr. Tervuren, Ann.* (8), 129.

WIEBACH, F. (1970). Süsswasser-Bryozoen aus Brasilien und Zentralafrika. *Rev. Zool. Bot. Afr.* **LXXXI** (1–2).

53 An Improved Classification, Based upon Numerical-Taxonomic Analyses, for the Higher Taxa of Entoproct and Ectoproct Bryozoans

ROGER J. CUFFEY

Department of Geosciences (Deike Building), Pennsylvania State University, University Park, Pennsylvania, U.S.A.

Abstract: Use of computer-based numerical-taxonomic techniques, in combination with sound conventional-taxonomic principles, to explore the taxonomic implications of the total morphological patterns displayed by the major bryozoan groups provides the basis for a revised classification of bryozoan higher taxa. This revision, presented fully herein to furnish a suitable taxonomic framework until future work elucidates evolutionary relationships among bryozoans, represents a significant improvement over current classificatory arrangements.

INTRODUCTION

In addition to being one of the most challenging tasks facing systematists, formulating higher taxa for fossil and living organisms is an important practical matter, because such taxa serve as the framework within which new understandings of an animal group are developed. Due to the recent flourishing of bryozoan studies, the present moment is a highly appropriate time to consider the state of higher taxa among bryozoans.

PRESENT STATUS OF BRYOZOAN CLASSIFICATION

Current higher-level classifications of bryozoans are deficient for three underlying reasons. First, such classifications do not reflect the evolutionary relationships among various bryozoan groups, because these relationships are so poorly known at present (Ryland, 1970). Second, such classifications have been based upon few of the morphological characters available for study, instead of the total morphological patterns (composed of all examinable characters considered collectively) exhibited by bryozoans. Bryozoologists now recognize such situations as undesirable (Boardman, *et al.*, 1970). Moreover, zoologists have relied predominantly on soft parts, while paleontologists have emphasized skeletal characters (Cuffey, 1971b). Third, bryozoan classifications have been formulated entirely by the relatively subjective methods of conventional taxonomy and thus show considerable variation in the hierarchical structures used by different bryozoologists in the past (Cuffey, 1971b).

Due to these three reasons, problems arise when considering bryozoan higher taxa. Because conventional taxonomic procedures have not produced identically repeatable results in practice, a universally acceptable hierarchical structure describing classificatory relationships among bryozoans eludes us. Defining these taxa has sometimes proved difficult (Hyman, 1959; Weller, 1969). Also uncertain is whether bryozoan taxa classified at the same taxonomic rank are in fact comparable groupings. Piecemeal changes in the ranks of some taxa have left gaps in the classificatory structure which have not yet been formally filled by the monotypic taxa required. In general, therefore, higher-level classifications of bryozoans leave much to be desired.

PROCEDURE FOR FORMULATING REVISED CLASSIFICATION

Application of computer-based numerical-taxonomic techniques, in combination with sound conventional-taxonomic principles, can yield a revision of bryozoan higher-level classification which represents a significant improvement over present arrangements. Proposing this revision for practical taxonomic use is the principal contribution of this paper; detailed discussion of the methods involved in formulating this revision will be presented elsewhere (Cuffey, in preparation).

Use of computerized numerical techniques to analyse data on as many morphological characters as possible constitutes a first step (Table 1), the formulation of a phenetic, polythetic classification based on overall resemblances. Enough is summarized about bryozoan morphology (Hyman, 1951, 1959; Buge, 1952, 1960; Bassler, 1953; Brien, 1959, 1960; Astrova *et al.*, 1960; Ryland, 1970; Cuffey, 1971a) that doing this is feasible now. A second step, at present impractical because of incomplete knowledge concerning bryozoan evolutionary history, is modification of that phenetic classification to make it consistent with phylogenetic history.

The procedures followed in formulating the revised classification presented herein are briefly summarized in Fig. 1. Perusal of the summary literature cited above suggests that bryozoologists have tended to recognize fifteen major bryozoan groups (Table 2), treated formally at approximately subordinal rank. Little general data exists for taxa ranked below these major groups, while considerable disagreement reigns concerning union of these major groups into

Table 1. Historical stages in the development of taxonomy

Sequential historical stage	Basis for classifying	Comprehensiveness of classifications	Techniques for making classificatory comparisons
4 (early 1970s)	Phenetic and distributional observations, modified by phylogenetic interpretations (Rowell, 1970, p. 268–269, 292; Mayr, 1969; Cheetham, 1968)	Polythetic (total morphological pattern)	Explicit or objective (computerized numerical comparisons, modified by implications of distributional data)
3 (early 1960s)	Phenetic observations alone (Sokal and Sneath, 1963)	Polythetic (total morphological pattern)	Explicit or objective (computerized numerical comparisons)
2 (late 1800s)	Phenetic observations, modified by phylogenetic inferences or speculations	Monothetic (a few key or diagnostic characters)	Intuitive or subjective (conventional taxonomy)
1 (late 1700s)	Phenetic observations alone	Monothetic (a few key or diagnostic characters)	Intuitive or subjective (conventional taxonomy)

higher-rank taxa. Consequently, these fifteen major groups constitute a practical starting point in formulating revisions of bryozoan higher taxa (Fig. 1a, b). Placement of a few, poorly known, peculiar forms is uncertain, but does not affect this overall picture. The cluster-analysis program (Fig. 1f; Bonham-Carter, 1967; Cuffey, 1970) run on Penn State's IBM 360/67 computer, generated 16 different numerical dendrograms (Fig. 1g); because no method yet exists by which one particular dendrogram can be selected as "best" for taxonomic purposes (Minkoff, 1965; Sokal and Michener, 1967), another approach was employed to obtain practical taxonomic results (Fig. 1h, i, j). The taxonomic scale used (Figs 1j and 2) was calibrated by selecting as phylum the level at which all ectoprocts cluster into one taxon (Cuffey, 1967; Cuffey, 1969), and as order a level which permits retention at that rank of several widely utilized orders (to minimize changes from current usage). Finally,

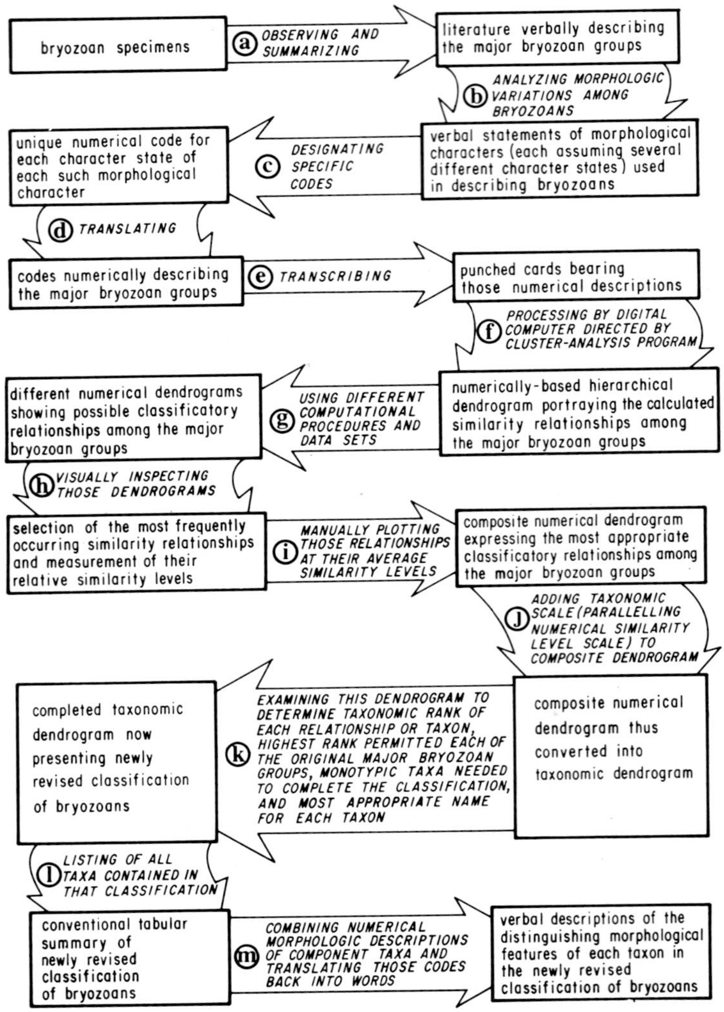

Fig. 1. Procedures followed in formulating the improved bryozoan higher-level classification presented herein.

Table 2. Major bryozoan groups on which classification is based

Abbreviation	Major bryozoan group	Generic examples	Position and name within revised classification herein
SOL EN	solitary entoprocts	*Loxosoma, Loxocalyx*	Suborder Loxosomatina
COL EN	colonial entoprocts	*Pedicellina, Barentsia*	Suborder Pedicellinina
CHT PH	chitinous phylactolaemates	*Plumatella, Fredericella*	Infraorder Plumatellita
GEL PH	gelatinous phylactolaemates	*Pectinatella, Cristatella*	Infraorder Pectinatellita
STL CT	stoloniferous ctenostomes	*Avenella, Vesicularia*	Infraorder Stoloniferita
CRN CT	carnose ctenostomes	*Alcyonidium, Paludicella*	Infraorder Carnosita
ANA CH	anascan cheilostomes	*Membranipora, Electra*	Suborder Anascina
CRB CH	cribrimorph cheilostomes	*Cribrilina, Rhabdopora*	Suborder Cribrimorphina
ASC CH	ascophoran cheilostomes	*Schizoporella, Mucronella*	Suborder Ascophorina
FEN CR	fenestrate cryptostomes	*Fenestella, Polypora*	Suborder Fenestrina
BIF CR	bifoliate cryptostomes	*Ptilodictya, Sulcoretepora*	Infraorder Ptilodictyita
RMB CR	rhomboporoid cryptostomes	*Rhabdomeson, Rhombopora*	Infraorder Rhabdomesita
TREP	trepostomes	*Tabulipora, Prasopora*	Suborder Trepostomina
STN CY	stony cyclostomes	*Fistulipora, Coeloclema*	Suborder Cystoporina
DEL CY	delicate cyclostomes	*Crisulipora, Tubulipora*	Infraclass Cyclostomata

currently used names are retained (Fig. 1k) whenever possible without doing serious violence to the taxonomic concepts which they at present denote (Cuffey, 1969).

PROPOSED IMPROVED CLASSIFICATION

Employment of the procedures outlined above leads to the improved, revised, higher-level classification of bryozoans proposed here (Fig. 2; Table 3). This revised classification is a significant improvement over currently employed ones, for several reasons.

First, this classification is based upon the total morphological patterns exhibited by the bryozoan groups, rather than a few selected diagnostic or key characters. Second, it has been formulated by relatively objective numerical-taxonomic techniques, so that much of the subjectively caused instability inherent in conventional schemes is removed. Third, the ranks of various taxa are consistent and comparable throughout the bryozoans. Fourth, all monotypic taxa necessary to complete the formal classification are included. Fifth, each taxon included is described (or redescribed) by giving all the studied characteristics common to all its component subtaxa, in order to facilitate practical usage of this classification; colony forms seen are separately summarized (Fig. 3). Sixth, the highest rank permitted each of the fifteen major groups from which this classification was formulated is indicated (Table 2); future revisionary work at lower taxonomic levels can thus be fitted into this classification without upsetting desirable higher-level features like the internal consistency of

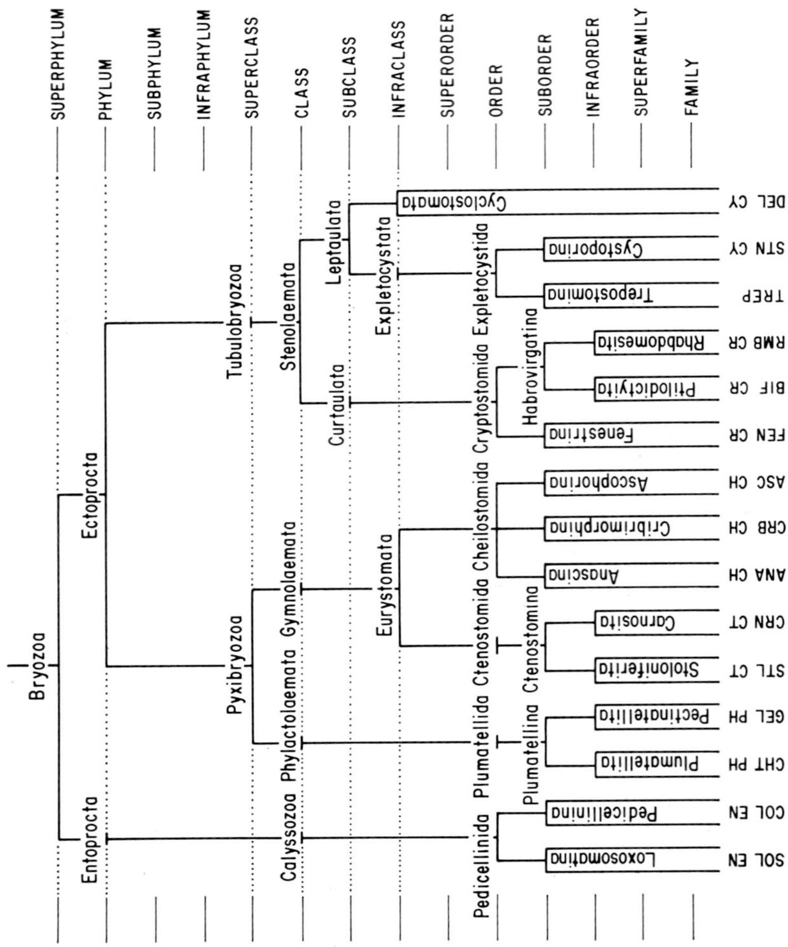

Fig. 2. Improved bryozoan classification presented in this paper, in taxonomic-dendrogram form.

Table 3. Improved bryozoan classification proposed in this paper

Superphylum Bryozoa Ehrenberg, 1831
Phylum Entoprocta Nitsche, 1869
Class Calyssozoa Clark, 1921
Order Pedicellinida Boettger, 1952
Suborder Loxosomatina Cuffey, *nov.*
Suborder Pedicellinina Boettger, 1952
Phylum Ectoprocta Nitsche, 1869
Superclass Pyxibryozoa Cuffey, *nov.*
Class Phylactolaemata Allman, 1856
Order Plumatellida Pennak, 1953
Suborder Plumatellina Pennak, 1953
Infraorder Plumatellita Pennak, 1953
Infraorder Pectinatellita Cuffey, *nov.*
Class Gymnolaemata Allman, 1856
Infraclass Eurystomata Marcus, 1938
Order Ctenostomida Busk, 1852
Suborder Ctenostomina Busk, 1852
Infraorder Stoloniferita Ehlers, 1876
Infraorder Carnosita Gray, 1841
Order Cheilostomida Busk, 1852
Suborder Anascina Levinsen, 1909
Suborder Cribrimorphina Lang, 1916
Suborder Ascophorina Levinsen, 1909
Superclass Tubulobryozoa Cuffey, *nov.*
Class Stenolaemata Borg, 1926
Subclass Curtaulata Cuffey, *nov.*
Order Cryptostomida Vine, 1883
Suborder Fenestrina Elias and Condra, 1957
Suborder Habrovirgatina Cuffey, *nov.*
Infraorder Ptilodictyita Astrova and Morozova, 1956
Infraorder Rhabdomesita Astrova and Morozova, 1956
Subclass Leptaulata Cuffey, *nov.*
Infraclass Expletocystata Cuffey, *nov.*
Order Expletocystida Cuffey, *nov.*
Suborder Trepostomina Ulrich, 1882
Suborder Cystoporina Astrova, 1964
Infraclass Cyclostomata Busk, 1852

taxonomic ranks. Finally, for each taxon, its known stratigraphic and ecologic ranges are given (Fig. 4), its only synonyms listed are names which have been used to denote essentially the same combination of bryozoans, and the ending of its name has been made uniform (Table 4); also, the derivation of newly proposed names is given (Table 5).

Because this revised classification is a decided improvement over current usage, I propose it here as a suitable framework for providing taxonomic

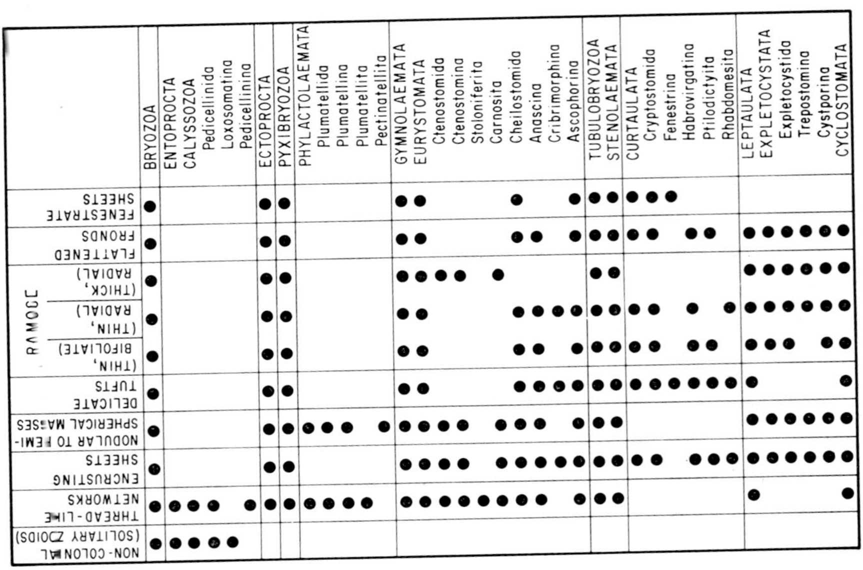

Fig. 3. Distribution of colony forms among the taxa included in the revised classification presented herein.

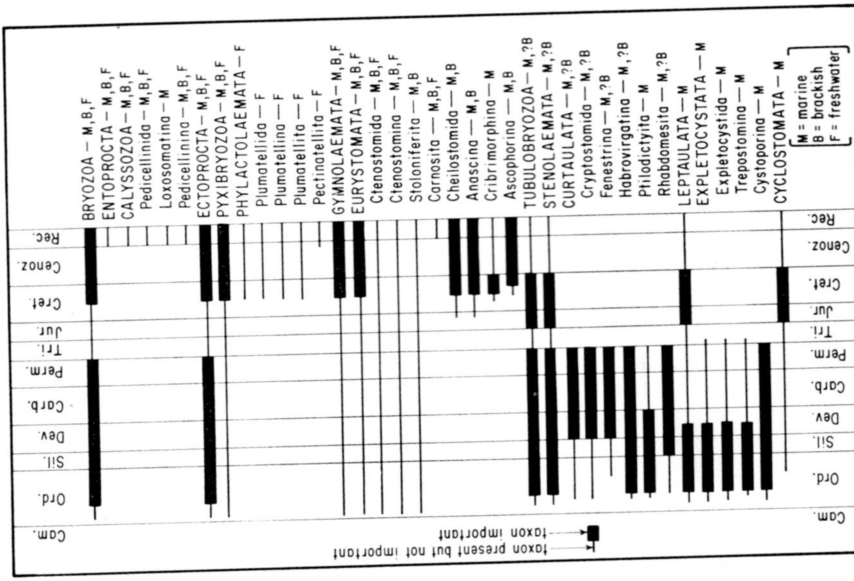

Fig. 4. Stratigraphic and ecologic ranges of the taxa included in the revised classification presented herein.

Table 4. Uniform endings adopted for taxonomic names employed in revised bryozoan classification

Taxonomic rank	Ending of name
Phylum- and class-group taxa	-a (often -ata)
Superorders	-oida
Orders	-ida
Suborders	-ina
Infraorders	-ita

Table 5. Etymological derivation (Jones, 1956) of taxonomic names newly proposed herein

Name—derivation
Curtaulata—*curtus* (shortened), *aulos* (tube).
Expletocystata, Expletocystida—*expletus* (complete), *cystos* (wall).
Habrovirgatina—*habros* (dainty), *virgatus* (twig-like).
Leptaulata—*lepto* (slender), *aulos* (tube).
Loxosomatina—from the genus *Loxosoma*.
Pectinatellita—from the genus *Pectinatella*.
Pyxibryozoa—*pyxis* (box; "box-like bryozoans").
Tubulobryozoa—*tubulus* (tube; "tubular bryozoans").

perspectives for bryozoan-oriented studies, until future advances develop much more understanding of bryozoan relationships.

Superphylum BRYOZOA Ehrenberg, 1831

Polyzoa *sensu lato* Thompson, 1830; Bryozoa *sensu lato* Ehrenberg, 1831; Holobranchia *fide* Blackwelder, 1963.

Characteristics summarized elsewhere (Cuffey, 1971a).

Phylum ENTOPROCTA Nitsche, 1869

Endoprocta *auct.*

Same characteristics as only included order.

Class **CALYSSOZOA** Clark, 1921

Kamptozoa Cori, 1927.

Same characteristics as only included order.

Order **PEDICELLINIDA** Boettger, 1952

Zooid epifaunal; consisting of basal stalk, cup-like calyx (periodically shed and regenerated), and tentacles (8–30; arranged in circle or oval; retracted by individually rolling inward); pseudocoelomate; usually hermaphroditic, sometimes dioecious. Body wall composed of outer corneous cuticle, cuboidal-cell epidermis, and inner longitudinal-and-diagonal muscle layer. Digestive tract with ciliated epithelium; rectum distinct; anus inside tentacular circlet, often on anal cone. Protonephridia (2) with duct; also gonoduct. Cleavage spiral-determinant. Larvae sheltered under tentacular umbrella, rounded, briefly free-swimming, partially degenerating at metamorphosis. No hardened zooecium, ascus, peritoneum, direct interzooidal visceral-cavity connections, lophophore-retractor or funiculus muscles, epistome, digestive-tract-wall muscles, ovicells, polyembryony, brown bodies, or resistant resting bodies.

Suborder **Loxosomatina** Cuffey, nov.

Zooids solitary (completely isolated), non-polymorphic. Asexual budding from body wall of zooid stalk.

Suborder **Pedicellinina** Boettger, 1952

(as Pedicellinida Boettger, 1952).

Colonies not regionated; usually hermaphroditic, rarely dioecious. Autozooids relatively isolated, arising from stolons. Asexual budding from body wall of stolon or zooid stalk.

Phylum ECTOPROCTA Nitsche, 1869

Polyzoa *sensu stricto* Thompson, 1830; Bryozoa *sensu stricto* Ehrenberg, 1831.

Zooid consisting of soft-bodied trunk, anchored within hardened zooecium (most commonly with round or polygonal aperture flush with colony surface), and tentacular crown (retractable as a mass into zooecium). Living representatives (Pyxibryozoa, Cyclostomata) eucoelomate, with peritoneum (thin, flat cells), well-developed lophophore-retractor and funiculus muscles, digestive-tract-wall muscles, anus outside tentacular circlet, digestive tract with non-ciliated epithelium, and holoblastic cleavage, but no rectum, anal cone, excretory organs, or gonoduct.

Superclass **PYXIBRYOZOA** Cuffey, nov.

"Box-like bryozoans" Boardman and Cheetham, 1970.

Colony not regionated. Zooecium short, wide, straight, vase- or box-like. Spermatozoa developing in large masses. No large intrazooecial calcareous partitions (like diaphragms); not polyembryonic.

Class **PHYLACTOLAEMATA** Allman, 1856

Lophopoda Dumortier, 1835; Hippocrepia *fide* Blackwelder, 1963.

Same characteristics as only included suborder.

Order **PLUMATELLIDA** Pennak, 1953

(as Plumatellina Pennak, 1953).

Same characteristics as only included suborder.

Suborder **PLUMATELLINA** Pennak, 1953

Colony epifaunal, hermaphroditic, non-polymorphic. Zooid hermaphroditic, with 20–106 tentacles, epistome, direct interzooidal visceral-cavity connections. Body wall including epidermis (thick, tall cells) underlain by longitudinal-and-circular muscle layer; never calcified. Zooecium simple; aperture same diameter as zooecium. Larvae sheltered within parent zooid, rounded, briefly free-swimming, not degenerating at metamorphosis. Asexual budding from invagination of zooid body wall; polypide inception preceding cystid differentiation; statoblasts produced. No ovicells, ascus, or brown bodies.

Infraorder PLUMATELLITA Pennak, 1953

(as Plumatellina Pennak, 1953).

Zooids relatively isolated, arising from earlier ones. Zooecial wall usually thin and membranous, sometimes thickened and chitinous (very rarely thinly gelatinous). Lophophore kidney-shaped to markedly U-shaped.

Infraorder PECTINATELLITA Cuffey, nov.

Zooids tightly packed, embedded side-by-side over surface of gelatinous colony, with thick gelatinous (zooecial) wall separating adjacent ones. Lophophore markedly U-shaped.

Class **GYMNOLAEMATA** Allman, 1856

Infundibulata Gervais, 1837; Stelmatopoda Van der Hoeven, 1846–1855.

Same characteristics as only included infraclass.

Infraclass **EURYSTOMATA** Marcus, 1938

Cheiloctenostomata Silén, 1942.

Colony hermaphroditic. Zooid usually hermaphroditic, sometimes dioecious; with 8–34 tentacles; lophophore circular; periodically forming brown bodies. Body wall including outer cuticle; also epidermis (thin, flat cells). Larvae degenerating completely at metamorphosis. Asexual budding producing new zooids in branching series (often contiguous) behind growing points; cystid differentiation preceding polypide inception. No epistome, body-wall muscle layer, or direct interzooidal visceral-cavity connections.

Order **CTENOSTOMIDA** Busk, 1852

(as Ctenostomata Busk, 1852).

Same characteristics as only included suborder.

Suborder **CTENOSTOMINA** Busk, 1852

(as Ctenostomata Busk, 1852).

Zooecium simple; wall usually chitinous and thin to somewhat thickened, sometimes weakly calcified (where in contact with substrate); aperture same diameter as zooecium, closed by pleated collar. Larvae usually sheltered within parent zooid, rounded, briefly free-swimming; but sometimes shed immediately, flattened, extendedly free-swimming. No ascus or ovicells.

Infraorder STOLONIFERITA Ehlers, 1876

(as Stolonifera Ehlers, 1876).

Colony epifaunal or infaunal (boring). Zooids relatively isolated, arising from stolons. Asexual budding from body wall of stolon only. No resistant resting bodies.

Infraorder CARNOSITA Gray, 1841

(as Carnosa Gray, 1841).

Colony epifaunal; sometimes including rhizoids or kenozooids. Autozooids arising from previous autozooids; usually relatively isolated, sometimes loosely bundled; proximal parts prostrate, distal parts erect. Body wall occasionally including an outermost thick gelatinous layer. Asexual budding from partitioned-off portion of visceral cavity. Freshwater representatives producing hibernacula.

Order **CHEILOSTOMIDA** Busk, 1852

(as Cheilostomata Busk, 1852); Chilostomata *auct.*

Colony epifaunal; often with avicularia or vibracula. Zooids usually loosely bundled. Zooecium complex; wall usually thickly calcified (solid or with "pseudopores"); frontal wall often highly ornamented; aperture smaller than transverse zooecial diameter, with operculum (often complicated in outline), sometimes also with peristome or apertural spines. Larvae usually sheltered within skeletal-walled ovicell (developed as attachment on autozooid), usually rounded and briefly free-swimming. Asexual budding from partitioned-off portion of visceral cavity. No resistant resting bodies.

Suborder **ANASCINA** Levinsen, 1909

(as Anasca Levinsen, 1909).

Colony often with rhizoids or kenozooids. Zooids occasionally relatively isolated. Zooecial wall sometimes only weakly calcified, or even uncalcified (chitinous). Frontal wall membranous, open to surrounding water; no ascus. Larvae sometimes sheltered within parent zooid (rather than ovicell); occasionally shed immediately, flattened, and extendedly free-swimming.

Suborder **CRIBRIMORPHINA** Lang, 1916

(as Cribrimorpha Lang, 1916); Acanthostega Levinsen, 1902.

Frontal wall membranous, overarched by calcareous spines partly protecting it from surrounding water; no ascus.

Suborder **ASCOPHORINA** Levinsen, 1909

(as Ascophora Levinsen, 1909).

Colony often with rhizoids and kenozooids. Frontal wall calcareous, solid; with ascus. Larvae sometimes sheltered within parent zooid (rather than ovicell).

Superclass **TUBULOBRYOZOA** Cuffey, nov.

"Tubular bryozoans" Boardman and Cheetham, 1969.

Same characteristics as only included class.

Class **STENOLAEMATA** Borg, 1926

Colony epifaunal, usually regionated. Zooecial wall calcareous; aperture same diameter as zooecium; zooecium often with complete diaphragms. Asexual budding from partitioned-off portion of visceral cavity; budding occurring in a common colonial growing zone. No ascus.

Subclass CURTAULATA Cuffey, nov.

Same characteristics as only included order.

Order **CRYPTOSTOMIDA** Vine, 1883

(as Cryptostomata Vine, 1883).

Endozone and exozone quite distinct; transition between them abrupt, often marked by 1–2 hemisepta per zooecium; endozonal zooecial walls thin, exozonal walls very thick. Zooecium intermediate in complexity, medium-length to moderately short, averaging medium-width (but variable), straight to markedly curved, vase-like to short-prismatic; walls solid, laminated; usually with acanthopores; aperture often with peristome. No ovicells.

Suborder **FENESTRINA** Elias and Condra, 1957

(as Fenestrata Elias and Condra, 1957).

Colony sometimes with rhizoids. Zooecia loosely bundled. "Endozonal" portion of zooecium relatively short.

Suborder **HABROVIRGATINA** Cuffey, nov.

Colony often with mesopores. Zooecia tightly packed.

Infraorder PTILODICTYITA Astrova and Morozova, 1956

(as Ptilodictyoidea Astrova and Morozova, 1956).

Colony occasionally with monticules or maculae. Endozonal portion of zooecium relatively short; aperture sometimes with lunarium.

Infraorder RHABDOMESITA Astrova and Morozova, 1956

(as Rhabdomesoidea Astrova and Morozova, 1956); Rhabdomesonata Shishova, 1967.

Endozonal portion of zooecium relatively medium-length.

Subclass LEPTAULATA Cuffey, nov.

Colony often with mesopores. Zooecia usually tightly packed; long, slender, proximally tapering, tubular to elongate-prismatic, straight to markedly curved; endozonal portions relatively long.

Infraclass EXPLETOCYSTATA Cuffey, nov.

Same characteristics as only included order.

Order **EXPLETOCYSTIDA** Cuffey, nov.

Colony often with monticules or maculae. Endozone and exozone quite distinct. Zooecia often with hemiphragms. No ovicells.

Suborder **TREPOSTOMINA** Ulrich, 1882

(as Trepostomata Ulrich, 1882); Tripostomata *auct.*; Monticuliporoidea *fide* Blackwelder, 1963.

Endozone-exozone transition moderately gradual; endozonal zooecial walls thin, exozonal walls thick. Zooecium of intermediate complexity; often with perforated diaphragms, heterophragms, or cystiphragms; frequently with acanthopores; walls solid, markedly laminated.

Suborder **CYSTOPORINA** Astrova, 1964

(as Cystoporata Astrova, 1964).

Colonies with cystopores (vesicles). Endozone–exozone transition very gradual, zooecial walls thin throughout both. Zooecia sometimes loosely bundled. Zooecium simple; aperture often with lunarium; wall finely granular, often solid but sometimes with mural pores.

Infraclass CYCLOSTOMATA Busk, 1852

Stenostomata Marcus, 1938

Colonies and zooids usually hermaphroditic, rarely dioecious. Colony sometimes including rhizoids, nanozooids, or kenozooids. Endozone and exozone not to slightly distinct, transition between them very gradual, zooecial walls thin throughout both. Zooids with 8–34 tentacles; lophophore circular; sometimes only loosely bundled or even relatively isolated; periodically forming brown bodies. Body wall including outer cuticle, zooecial wall (heavily calcified, but perforated by pseudopores and sometimes communication pores), and epidermis (thin, flat cells). Zooecium simple; occasionally with diaphragms (complete or perforated); aperture sometimes with peristome. Spermatozoa developing in small tetrads. Polyembryonic; larvae sheltered within skeletal-walled ovicell (developed from specialized gonozooid), rounded, briefly free-swimming, degenerating completely at metamorphosis. Cystid differentiation following polypide inception. No epistome, body-wall muscle layer, direct interzooidal visceral-cavity connections, or resistant resting bodies.

ACKNOWLEDGEMENTS

T. G. Perry, A. S. Horowitz, R. S. Boardman, A. H. Cheetham, and F. K. McKinney all provided significant stimulus and interest. The Kansas Geological Survey made the cluster-analysis program available to me. G. B. Newton, A. B. Thorn, D. L. Bernitt, and J. L. Politzer helped with computer programming and processing, which was performed with facilities and funds provided by Pennsylvania State University. I gratefully acknowledge all this help.

REFERENCES

Astrova, G. G., *et al.* (1960). Bryozoa—Mshanki. "Osnovy Paleontologii" **7,** 13–112. Akademii Nauk S.S.S.R., Moskva.

Bassler, R. S. (1953). Bryozoa. "Treatise on Invertebrate Paleontology" **G,** 1–253. University of Kansas Press, Lawrence, Kansas, U.S.A.

Boardman, R. S., Cheetham, A. H. and Cook, P. L. (1970). Intracolony variation and the genus concept in Bryozoa. *N. Am. Paleont. Conv.* (*Chicago, 1969*) *Proc.* **C,** 294–320.

Bonham-Carter, G. F. (1967). FORTRAN IV program for Q-mode cluster analysis of nonquantitative data using IBM 7090/7094 computers. *Kan. Geol. Surv. Comp. Contrib.* **17,** 1–28.

Brien, P. (1959). Classe des endoproctes ou kamptozoaires. *In* "Traité de Zoologie" Vol. 5 (1), pp. 927–1007. Masson, Paris.

Brien, P. (1960). Classe des bryozoaires. *In* "Traité de Zoologie" Vol. 5 (2), pp. 1053–1335. Masson, Paris.

Buge, E. (1952). Classe des bryozoaires. *In* "Traité de Paléontologie" Vol. 1, pp. 688–749. Masson, Paris.

Buge, E. (1960). Bryozoaires fossiles. *In* "Traité de Zoologie" Vol. 5 (2), pp. 1336–1379. Masson, Paris.

Cheetham, A. H. (1968). Morphology and systematics of the bryozoan genus *Metrarabdotos*. *Smithson. Misc. Coll.* **153** (1), 1–121.

Cuffey, R. J. (1967). Bryozoan *Tabulipora carbonaria* in Wreford Megacyclothem (Lower Permian) of Kansas. *Univ. Kan. Paleont. Contrib. Bryozoa* **1,** 1–96.

Cuffey, R. J. (1969). Bryozoa versus Ectoprocta—The necessity for precision. *Syst. Zool.* **18,** 250–251.

Cuffey, R. J. (1970). "Paleontological Users' Manual for Cluster-Analysis Programs", 139 pp. Pennsylvania State University (syllabus), University Park.

Cuffey, R. J. (1971a). Bryozoa, Cheilostomata, Cryptostomata, Ctenostomata, Cyclostomata (Bryozoa), Ectoprocta, Entoprocta, Gymnolaemata, Phylactolaemata, Stenolaemata, Trepostomata. *In* "McGraw-Hill Encyc. Sci. Tech.", 3rd Edn; **2,** 386–394; **3,** 21, 630–631, 658–659, 695; **4,** 453; **5,** 7–8; **6,** 344; **10,** 220–221; **13,** 133; **14,** 89–90.

Cuffey, R. J. (1971b). History and comparison of conventionally and numerically based classifications of the higher taxa of bryozoans (abs.). *Classif. Soc. 2nd Ann. Mtg. Abstr. of Pap.* 6.

Hyman, L. H. (1951). The pseudocoelomate Bilateria—Phylum Entoprocta. *In* "The Invertebrates", Vol. 3, pp. 521–554; Vol. 5, p. 747. McGraw-Hill, New York.

Hyman, L. H. (1959). The lophophorate coelomates—Phylum Ectoprocta. "The Invertebrates", Vol. 5, pp. 275–515. McGraw-Hill, New York.

Jones, D. J. (1956). "Introduction to Microfossils", 406 pp. Harper, New York.

Mayr, E. (1969). "Principles of Systematic Zoology", 428 pp. McGraw-Hill, New York.

Minkoff, E. C. (1965). The effects on classification of slight alterations in numerical technique. *Syst. Zool.* **14,** 196–213.

Rowell, A. J. (1970). The contribution of numerical taxonomy to the genus concept. *N. Am. Paleont. Conv. (Chicago, 1969) Proc.* **C,** 264–293.

Ryland, J. S. (1970). "Bryozoans", 175 pp. Hutchinson Univ. Libr., London.

Sokal, R. R. and Michener, C. D. (1967). The effects of different numerical techniques on the phenetic classification of bees of the *Hoplitis* complex (Megachilidae). *Proc. Linn. Soc. Lond.* **178,** 59–74.

Sokal, R. R. and Sneath, P. H. A. (1963). "Principles of Numerical Taxonomy", 359 pp. Freeman, San Francisco.

Weller, J. M. (1969). "The Course of Evolution", 696 pp. McGraw-Hill, New York.

54 The Importance of Different Growth Directions in the Phylactolaemata and Gymnolaemata for Reconstructing the Phylogeny of the Bryozoa

D. JEBRAM

Zoologisches Institut der Technischen Universität, Braunschweig, Germany

Abstract: The Bryozoa Phylactolaemata bud in an oral direction, the Gymnolaemata bud in an anal direction. The difference between these directions is bridged by the introduction of a change in the colony arrangement. The anal budding direction seems to have developed by budding of cystids on the anal side of the zooids upwards from the encrusting basal colony. The "missing links" between modern Gymnolaemata and ancestral Phylactolaemata may have been serially arranged, erect growing forms, which are not preserved as fossils, because they are not calcified. Models for this intermediate stage are shown by the extinct Fenestellidae, and by recent *Labiostomella gisleni* (Protocheilostomata) and *Benedenipora catenata* (Protoctenostomata), which seem to be most closely related to the supposed extinct basic group of the Gymnolaemata. The Stenolaemata would then be a very early side branch from the basic group. The spines and other adventive heterozooids on the oral side of the zooids in the Ctenostomata and Cheilostomata and the regenerative polypide buds of the oral cystid wall in most Gymnolaemata may be derived as vestiges of the original oral budding in the ancestors, the Phylactolaemata. A contrary concept with derivation of the Phylactolaemata from the Gymnolaemata is considered improbable.

INTRODUCTION

Observations on the ontogeny and growth potential in cultivated colonies of *Farrella repens*, *Bowerbankia* and other species have led to an improved understanding of morphology and of the true nature of the stolons in the Ctenostomata (Jebram, 1970, and in press). Such understanding aids systematic revision of the Ctenostomata provided the colony-form is known for ancestors of the recent Ctenostomata. This is either a plane encrusting colony as in the Alcyonidioidea, or a serial arrangement as in the "Paludicellea" (*sensu* Silén, 1942a) and in the stolonate groups.

The two subphyla of the Bryozoa grow in different directions. Colonies of the Gymnolaemata grow in the anal direction, building first cystid buds on the anal side of established zooids. Later, in the new cystid bud, a new polypide bud is developed on its oral side (Silén, 1942a, b). In the colonies of the Phylactolaemata at first new polypide buds are built on the oral sides of established zooids; each polypide bud influences the building of its own new cystid space.

Kraepelin (1892) and Herwig (1913) have supposed the existence of "anal" polypide buds in the phylactolaematous genus *Plumatella*. New polypide buds in the Phylactolaemata can only grow in special growing areas which are always situated at the oral side of the zooids. The polypide bud under discussion is not anally related to the other slightly more advanced polypide bud, but is related orally to the descending fullgrown zooid. The small distance between the polypide buds depends on the limited growing areas available for the propagation of new polypide buds.

TERMINAL DOMINANCE AS DEVELOPMENTAL POTENTIALITY IN THE PHYLOGENY OF THE BRYOZOA

1. *The Development of the Anal Growth Direction*

In the Bryozoa the polypide and the cystid are parts of a single, polar-differentiated body. Changes in the polarity arrangement of an organism during phylogeny may be traced by interpreting anatomical or ontogenetic derivations. No facts are known which may account for the rotation of about 180° of the polypide within the cystid. We should, therefore, consider whether or not a change of the colony arrangement could have caused the difference in the growth directions. In the Phylactolaemata the existence of a protocoel in the epistome and of a U-shaped lophophore show that the Phylactolaemata are closely related to the origin of the entire Tentaculata. No calcified Phylactolaemata are known. The earliest fossil Bryozoa are of Upper Cambrian age and are of calcified Stenolaemata, and, therefore of the Gymnolaemata (Allman, *non* Borg), which grow in the anal direction (Borg, 1926, confirmed by the figures of Nielsen, 1970). For this reason the change in colony construction should have been developed in the Precambrian period (*comp.* Buge, 1960) in uncalcified forms which can hardly be preserved as fossils.

The change in colony construction can, therefore, be determined only theoretically. The colony-forms of the Stenolaemata may be traced back to a serial arrangement. A serial colony-form exists in some Ctenostomata (*Benedenipora, Hislopia, Paludicella, Victorella*, a.o.) and also in the early developmental stages of some compactly encrusting Cheilostomata (some *Electra* and *Cono-*

peum). The more primitive Cheilostomata (*Aetea, Scruparia, Pyripora,* and some *Electra*) build serially arranged colonies. *Labiostomella gisleni* Silén relates most closely to the origin of the Cheilostomata, and, therefore, Silén has introduced a new group "Protocheilostomata". The colony of *Labiostomella* is built out of serially arranged zooids in branches growing upwards from the ancestrula. This colony-form resembles the form of the extinct Fenestellidae (Tavener-Smith, 1969). I therefore conclude that the intermediate group at the origin of the Gymnolaemata had serially arranged zooids in erect growing branches.

The central phylogeny of the Bryozoa can be reconstructed and described in the following way (Fig. 1).

The supposed earliest Tentaculata were single and hemi-sessile animals (A), which had potential for regenerative growth (as with recent Phoronidea). Regeneration of the forebody happened by an infold of the epidermis. When the tentacles were built in inner buds, a tentacle sheath must have been developed also, and in this way the possibility of retracting and invaginating the tentacles was developed. Regeneration of the forebody, the polypide, led to the possibility of asexual propagation (B). By incomplete separation colonies of animals were developed (C). At this early stage orifices were directed towards the outside of the colony and encrusting colonies grew in an oral direction (cf. recent Phylactolaemata). Then the potential for building regenerative buds on the oral side of each zooid led to intercalary growth (D). By interaction of environmental selection and genetic variation the potentiality for continuous growth upwards from the substratum ("terminal dominance" Jebram, 1970) was developed. Erect growing series of zooids were developed in the Plumatellidae and in related forms (E). Nevertheless this group represents a phylogenetic dead-end because it continued the oral direction of budding only. Another group was more successful in developing a new type of serial erect zooidal growth. First a new cystid space was built on the anal side of established zooids in which a new polypide was then developed on the oral side; later the new zooid on the anal side becomes separated from its descending zooid by a septum (F).

During the intermediate stages (F) both types of budding might have been realized. Later the erect series growing in an anal direction was dominant. Competition for space on the substratum caused a restriction of the orally growing basal colony leading to a general reduction of oral budding. Gradually the orally growing basal colony became restricted to the size of a single animal. Ryland and Stebbing (1971) have recently pointed out that "During metamorphosis there is a reversal of internal polarity: the distal end of the polypide and the blastogenic face of the ancestrula correspond to the posterior of the larva". So the settling larva represents the remnant of an orally growing ancestral colony-form, the early Phylactolaemata. The polarity change during the metamorphosis of the gymnolaematous larvae is a clear ontogenetic indicator of the change of growth direction during the phylogenetic development from the Phylactolaemata to the Gymnolaemata. The first polypide in the ancestrula after the metamorphosis in recent Gymnolaemata represents the "first true gymnolaematous animal" in each colony. In recent Phylactolaemata, however, the ontogenetic stage of a true larva is suppressed, and a "primary colony" with one, two, or three polypides develops directly within the egg.

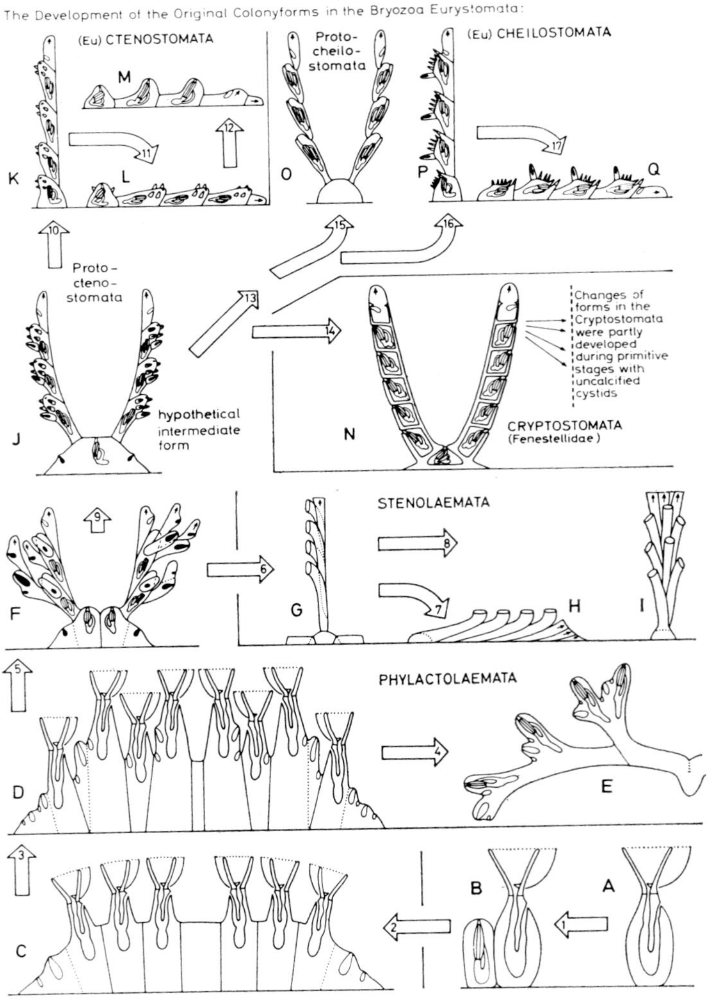

Fig. 1.

Anally directed budding in erect zooidal series led to a restriction of the body size. In this way the epistome was eliminated, the U-shaped lophophore reduced, and the tentacle number decreased to 8, which is common in many species of the Stenolaemata and in many serially arranged Ctenostomata, and which perhaps was also characteristic of the Cryptostomata (Tavener-Smith, 1971). (Later with an increase of body size and/or advances in anatomy the tentacles became more numerous again in many Gymnolaemata.)

Accordingly we may suppose that the transformation of the colony arrangement has happened once, or twice separately for Stenolaemata and Eurystomata, we may suggest that the Gymnolaemata represent a truly monophyletic group or only a phylogenetic stage of colony development. It seems improbable that an anal direction of growth could be developed twice in independent ways. Therefore, I suggest, the Gymnolaemata (Allman, 1856) is a natural monophylectic group which includes the Stenolaemata.

If the anally directed growth was developed by erect growing zooidal series, species with related colony-forms should presumably be close to the origin of the different orders of the Gymnolaemata. In the Stenolaemata, as an early side branch of the Gymnolaemata, we can find secondarily erect growth forms, but also colonies growing straight upwards from the ancestrula, like *Crisia*. The erect colony of *Crisia* with serially arranged zooids resembles a form which perhaps the ancestors of the entire Stenolaemata had, but *Crisia* itself is found only in comparatively young geological deposits. I suppose the intermediate forms in the Stenolaemata had uncalcified walls like the suggested central group for the entire Gymnolaemata. In this view *Crisia* represents the oldest stenolaematous colony-form with a late calcification of the walls. This theory could explain why different groups of the Stenolaemata appear as isolated units during geological history (Ryland, 1970).

Among the Cryptostomata, representing the next branch of the Gymnolaemata, the Fenestellidae represent the most primitive colony-form. Since the earliest occurrence of Fenestellidae is in deposits later than those containing examples of other cryptostomatous families it may be assumed that, as in the Stenolaemata, the early primitive stages in the Cryptostomata were not calcified or only poorly so. The development of different new colony forms in the Cryptostomata could have occurred in this less calcified and unpreserved basic group.

No fossil examples of the poorly calcified erect *Labiostomella* (Protocheilostomata) are known. In the Cheilostomata, *Actea* and *Scruparia* are the most primitive forms. They are poorly calcified and are therefore seldom found fossil. Their zooecia are serially arranged and, as in the Ctenostomata, possess collars but they are also operculate. The earliest occurrence of Cheilostomata is of

Upper Jurassic age and of a pyriporimorph species (Pohowsky, 1973, p. 447) in which the zooecia are again serially arranged, but more strongly calcified. Recent *Pyripora*-species have no collar, so this genus may be less primitive.

The ctenostomatous *Benedenipora catenata* has a colony of erect serially arranged growing zooids. This form seems to be closely related to the origin of the Ctenostomata (K). The encrusting, serially arranged Hislopiidae, Paludicellidae, Victorellidae, and related groups (L, M) have been developed later in the Ctenostomata, when the erect zooidal series has returned to the substratum. More details of the phylogeny of the Ctenostomata are described by Jebram (1970, and in press).

We find, in the phylogeny of the entire Bryozoa, that in each group some forms advanced their colony-form and anatomy and developed calcified body walls, while other forms preserved a less advanced colony-form and developed calcification of the body wall later than the advanced forms, but developed few special anatomical or ontogenetic details and/or changed their ecology. For this reason we find in the geological record fossils of calcified forms appearing as isolated anatomical groups. The central phylogeny of the entire Bryozoa developed by successive changes in the colony arrangement ("*Aromorphosen*" Sewertzoff, 1931), while at different stages in the development of colony structure the forms became differentiated and different ecological niches were conquered in radiative developments ("*Idioadaptationen*" Sewertzoff). The "mising links" carrying on the central phylogeny were uncalcified or only poorly so. The Protoctenostomata (Jebram, 1970, and in press) and the Euctenostomata seem to represent this central group, from which only few fossils could be preserved under special conditions (*comp.* Voigt, 1966).

From the central phylogeny envisaged it seems probable, in contrast to the opinions of Silén (1942a, 1944) that most recent Ctenostomata form a single monophyletic group as well as the larger part of the Cheilostomata. Thus the Eurystomata comprise: Protoctenostomata (incl. *Benedenipora*), Euctenostomata, Cryptostomata, Protocheilostomata, Eucheilostomata, and perhaps Inovicellata as a separate group. The Eurystomata and the Stenolaemata together from the Gymnolaemata.

Similarities of wall structures have been proposed recently as demonstrating a close relationship between the Cryptostomata and the Stenolaemata. However, several authors have shown that similar wall structures may be developed in different groups and different wall structures in related groups. If a true close relationship between the Cryptostomata and the Stenolaemata does exist, the central phylogeny of the Gymnolaemata s.l. may still have been carried on by a group with uncalcified, erect colonies of serially arranged zooids.

2. The Development of Polypide Replacement

Most species of the Gymnolaemata have the potential to substitute for damaged or physiologically exhausted polypides new regenerative polypide buds. Such polypide replacement in the same cystid space is not known in the Phylactolaemata, where damaged animals may be replaced only as a whole by entire new zooids, because each polypide bud influences the growth of an own new cystid space (total regeneration).

Harmer (1891, 1892) and Marcus (1926) have stressed the importance of polypide replacement as a mode of excretion. While the Phylactolaemata have a coelomic pore which may perhaps be used for excretory processes, the Gymnolaemata have no real nephridia or other excretory organs. For this reason replacement of old polypides charged with excretory products by new vital polypides is a significant development in the Gymnolaemata. In undisturbed growing colonies of several species of Ctenostomata and Cheilostomata I have observed a continuous replacement of polypides. Zonation of the colonies into areas with active and with reduced polypides, in many species, shows that polypide replacement is a normal physiological event in the life of these zooids.

Borg (1926, 1947) has denied that polypide replacement in the Gymnolaemata has been developed primarily as a mode of excretion. He has supposed that polypide replacement relates to the way in which gymnolaematous Bryozoa release ova through a special intertentacular tube or intertentacular pore. Borg suggested that the formation of an intertantacular organ is indispensably combined with polypide replacement. I have observed, in some cases, in *Conopeum seurati* (Cheilostomata) the formation of an intertentacular tube within the individual life of the polypides. Although newly evaginating polypides of the first generation never possess an intertentacular tube I was able sometimes to find in older polypides, in the same zone of the first generation small incomplete or nearly fully developed intertentacular tubes. Thus polypide replacement is not necessary for growth of the intertentacular organ even when, in normal cases, the development of an intertentacular organ happens during a polypide change. On the other hand, in some stolonate species polypide replacement never occurs, but ova are released through an intertentacular organ, and the excretory process is not carried out by polypide replacement.

It may be concluded that polypide replacement has been developed independently of modes of excretion and of releasing the ova.

My theory of change of the growth direction may explain how polypide replacement could be developed in the Gymnolaemata. The increasing "terminal dominance", resulting from upward growth, reduced budding of new zooids on the oral side, but this mode of budding was not completely suppressed. In anally growing zooidal series polypides developed on the oral sides of cystids, not only one polypide in each cystid, but also, in the course of time, subsequent polypides were developed in the same cystid. The subsequent polypides lost the potentiality to influence the growth of their own cystid spaces, so the new polypides remained in the old cystid. Perhaps, under the influence of "terminal dominance", new polypides moved towards the terminal direction and became connected with the old aperture. Present day observable positions of polypide buds in the Gymnolaemata, first proximally and later moving distally to the definitive aperture may partly demonstrate a repetition or recapitulation of phylogeny in the ontogeny of the individuals (Braem, 1951; Jebram, 1970, and in press).

3. Spines and Other Adventive Zooids

Recently some authors have denied that spines of *Flustrellidra* (Silén, 1947) and of *Electra* (Bobin, 1968) are really homologous to zooids, because these spines have a different anatomy and a different ontogeny compared with those of *Callopora*. The spines of *Callopora* (Cheilostomata) have been shown to be heterozooids by comparison with the position of the zooid series on the ancestrula of *Labiostomella* (Silén, 1942a). The anatomy and ontogeny of the spines of *Membranipora* (Cheilostomata) also differ from those of *Callopora*, but comparison with reticulumzooids of *Conopeum reticulum* (Cheilostomata) shows the spines of *Membranipora* also to be heterozooids (Jebram, 1968). The spines of *Flustrellidra* (Ctenostomata) and of *Electra* (Cheilostomata) are closely related to those of *Membranipora*. The spines of some species of *Electra* represent a reduced and simplified form and those of some species of *Flustrellidra* and *Electra* are developed in a special way compared with *Membranipora*-type spines (cf. "Anaboli" Sewertzoff, 1931; "Hypermorphosis" de Beer, 1958). Apart from changes in anatomy, the mode of ontogeny may also be altered during phylogeny and for this reason differences between the spines of *Electra* and *Callopora* may not necessarily indicate a non-zooidal nature for the spines of *Electra*. In most species of Ctenostomata and Cheilostomata spines are homologous to real zooids, as with the spines in the stenolaematous *Crisia*.

Recognition of spines in the Ctenostomata and Cheilostomata as real adventive heterozooids on the oral side of descending autozooids has great importance for our understanding of the phylogeny of the Bryozoa Eurystomata.

The ancestors of the Ctenostomata and Cheilostomata had spines, not only on the ancestrula but also on succeeding zooids. *Benedenipora catenata* (Protoctenostomata) and *Labiostomella gisleni* (Protocheilstomata) lack spines and are representatives of the phylogenetic stage with an ancestral colony-form of serially arranged zooids in primary erect branches. I suggest, that in a kind of "*Spezialisationskreuzung*"*, in *B. catenata* and *L. gisleni* the plesiomorph colony-form has been maintained, while the species have totally suppressed the formation of spines and adapted their biology to life in deep water regions.

Spines have been reduced independently in several lineages of the Eurystomata, so that we often find no spines in advanced forms. In a few cases, on the contrary, spines seem to have been secondarily increased in number (e.g. *Alcyonidium parasiticum*, *Flustrellidra hispida*). The potential to form adventive heterozooids on the oral side has enabled different structures to develop from them. In the Cheilostomata especially many different structures (e.g. scutes, avicularia, vibracularia, frontal zooid wall) were developed from this origin leading to a multiplicity of forms.

All these adventive heterozooids on the oral side may also be derived as vestiges of the original type of budding in the oral direction in the Phylactolaemata.

ORIGIN AND DIRECTION OF DEVELOPMENT IN PHYLOGENY

Investigations of phylogeny often pose the problem of deciding between two apparently possible but contrary ways of development. In which direction along a derived lineage of morphological differentiation has phylogeny proceeded? In the Bryozoa we need to decide whether the Gymnolaemata have originated from the Phylactolaemata or vice versa.

We find no definite "apomorph" anatomical detail which could support derivation of the Phylactolaemata from the Gymnolaemata. In the Phylactolaemata we find an attribute, which is certainly not primitive: the statoblasts. Even the statoblasts are only adaptive features by which this bryozoan group was able to overcome ecological difficulties in fresh water environments and so survive until recent times. In any case there was time enough to develop the statoblasts which cannot be used for a derivation of a gymnolaematous ancestry.

In contrast the derivation of the Gymnolaemata from the Phylactolaemata accords well with the given facts and explains several problems. If the Phylactolaemata should have developed their oral growth direction from the anal growth direction of the Gymnolaemata two new problems appear. Why have the Phylactolaemata never developed heterozooids, and why should the Phylacto-

* "*Spezialisationskreuzung*": In many cases species with some plesiomorph anatomical structures have developed additionally some apomorph structures or attributes, especially as adaptations to ecology and environment.

laemata have reduced the potential for polypide replacement without any discernable necessity? Polypide replacement and formation of heterozooids are features which have facilitated the success of the gymnolaematous groups.

Borg (1926) has suggested, that the "Gymnolaemata" *sensu* Borg, the Stenolaemata, and the Phylactolaemata might have originated together from an extinct group named "Protobryozoa". The "Protobryozoa" should have grown in both directions, oral and anal, and should have had erect colony branches. At first sight this idea seems to be closely related to my concept, but firstly no zooid arrangement in an encrusting colony is possible when the zooids can grow in both directions. Only in erect growing branches may both growth directions be realized, as supposed for the first stage of the intermediate basic group of the Gymnolaemata in my theory. A construction of the colony-form for the basic group of the entire Bryozoa from erect branches only, growing directly upwards from the ancestrula, is quite improbable. When the anally growing Gymnolaemata have successfully developed calcified walls, polypide replacement and heterozooids, why have the orally growing Phylactolaemata not done so? If the Phylactolaemata developed parallel to the Gymnolaemata, from a common ancestral group, there exists no reason why the Phylactolaemata should have developed none of these useful features.*

I conclude that the Gymnolaemata have originated from the Phylactolaemata and that the epistome and the U-shaped lophophore have been reduced during this development. When the basic group of the Bryozoa possess a protosoma, the epistome, like the Phoronidea and Brachiopoda, this most resembles the protosoma in the Pterobranchia and the Echinodermata. The position of tentacles on the second metamere in most basic groups of coelomate animals could be used as a basis for deciding homology of similar formations in similar structural arrangement. This leads to the conclusion that the concept of the group of the "Archicoelomata" (Ulrich, 1949; Remane, 1949, 1967; Siewing, 1967) in which the ancestors of all coelomate animals had a trimerous or oligomerous organization and tentacles on the second metamere currently is the most acceptable.

* Bushnell (in his paper at the I.B.A. Conference Durham 1971) has recently found replacement of polypides in the same cystid in the phylactolaematous *Plumatella casmiana*. This single example is not representative for the whole Phylactolaemata and cannot indicate an original polypide replacement in this group. In my opinion the polypide replacement in *P. casmiana* has been developed secondarily and parallel to the Gymnolaemata. It may be significant that no other phylactolaematous species have polypide replacement.

ACKNOWLEDGEMENTS

I wish to thank very much Prof. Dr A. Remane for supervision at the start of my work on ecology, morphology, and phylogeny of the Bryozoa. The figure is prepared, after my sketches, by Mrs G. Kleber, Kiel. I thank Dr J. S. Ryland and Dr G. Larwood for their assistance in translating the manuscript.

REFERENCES

Allman, G. J. (1856). A Monograph of the Fresh-Water Polyzoa including all the Known Species, Both British and Foreign. *Ray Soc. Lond.*, 1–121.

Bobin, G. (1968). Morphogenese du termen et des epines dans les zoecies d'*Electra verticillata* (Ellis et Solander) (Bryozoaire Chilostome, Anasca). *Cah. Biol. mar.* **9,** 53–68.

Borg, F. (1926). Studies on recent Cyclostomatous Bryozoa. *Zool. Bidr. Uppsala* **10,** 181–507.

Borg, F. (1947). Zur Kenntnis der Ökologie und des Lebenzyklus von *Electra crustulenta* (Bryozoa chilostomata) nebst Bemerkungen über den sog. braunen Körper bei den Bryozoen. *Zool. Bidr. Uppsala* **25,** 344–377.

Braem, F. (1951). Über *Victorella* und einige ihrer nächsten Verwandten, sowie über die Bryozoenfauna des Ryck bei Greifswald. *Zoologica* **102,** 1–59.

Buge, E. (1960). Bryozoaires Fossiles. *In* "Traité de Zoologie," Vol. 5, pp. 1336–1379. Masson, Paris.

de Beer, G. (1958). "Embryos and Ancestors." 3rd Edn., pp. 1–197. Oxford University Press.

Harmer, S. F. (1891). On the regeneration of lost parts in Polyzoa. *Rep. 60th (Leeds) Meetings Br. Ass. Adv. Sci. 1890*, pp. 862–863. London.

Harmer, S. F. (1892). On the Nature of the Excretory Processes in Marine Polyzoa. *Q. Jl microsc. Sci.* N.S. **33,** 123–167.

Herwig, E. (1913). Beiträge zur Knospung bei den Bryozoen. (Inaug.-Diss. Marburg) *Arch. Naturgesch.* Berlin **79A**(12), 1–24.

Jebram, D. (1968). A Cultivation Method for Saltwater Bryozoa and an Example for Experimental Biology. *Atti Soc. it. al. Sci. nat.* **108,** 119–128.

Jebram, D. (1970). Stolonen-Entwicklung und natürliches System bei den Bryozoa Ctenostomata. Inaugural-Dissertation Kiel 1970, 1–106.

Jebram, D. (in press). Stolonen-Entwicklung und Systematik bei den Bryozoa Ctenostomata. *Z. zool. Syst. Evolutionsforsch.*

Kraepelin, K. (1892). Die deutschen Süßwasserbryozoen. II. Entwicklungsgeschichtlicher Teil. *Abh. Wiss. Ver. Hamburg* **12,** 1-67, Taf. I–V.

Marcus, E. (1926). Beobachtungen und Versuche an lebenden Meeresbryozoen. *Zool. Jb. (Abt. Syst. Geograph. Biol. Tiere)*, **52**(1), 1–102, Taf. 1–2.

Nielsen, C. (1970). On the metamorphosis and ancestrula formation in cyclostomatous bryozoans. *Ophelia* **7**(2), 217–256.

Pohowsky, R. A. (1973). "Living and Fossil Bryozoa,'' p. 447. Academic Press, London.

Remane, A. (1949). Die Entstehung der Metamerie der Wirbellosen. *Verh. dtsch. Zool. Ges. Mainz 1949*, 16–23.

Remane, A. (1967). Die Geschichte der Tiere. *In* Heberer, "Die Evolution der Organismen," 3. Auflg., pp. 589–677, Gustav Fischer, Stuttgart.

Ryland, J. S. (1970). "Bryozoans," pp. 1–175. Hutchinson University Library, London.

Ryland, J. S. and Stebbing, A. R. D. (1971). Settlement and orientated growth in epiphytic and epizoic bryozoans. *In* "Fourth European Marine Biology Symposium" Crisp, D. J., ed., pp. 105–123. Cambridge University Press, 1971.

Sewertzoff, A. N. (1931). "Morphologische Gesetzmäßigkeiten der Evolution." Gustav Fischer, Jena.

Siewing, R. (1967). Diskussionsbeitrag zur Phylogenie der Coelomaten. *Zool. Anz.* **179,** 132–176.

Silén, L. (1942a). Origin and Development of the Cheilo-Ctenostomatous Stem of Bryozoa. *Zool. Bidr. Uppsala* **22,** 1–59.

Silén, L. (1942b). On the Formation of the Interzoidal Communications of the Bryozoa. *Zool. Bidr. Uppsala* **22,** 433–488.

Silén, L. (1944). The Anatomy of *Labiostomella gisleni* Silén (Bryozoa Protocheilostomata) with Special Regard to the Embryo Chambers of the Different Groups of Bryozoa and to the Origin and Development of the Bryozoan Zoarium. *Kgl. Svenska Vetensk. Hand.*, Stockholm, Ser. 3, **21** (6), 1–111.

Silén, L. (1947). On the Spines of *Flustrella* (Bryozoa). *Zool. Bidr. Uppsala* **25,** 134–140.

Tavener-Smith, R. (1969). Skeletal structure and growth in the Fenestellidae (Bryozoa) *Palaeontology* **12**(2), 281–309.

Tavener-Smith, R. (1971). *Polypora stenostoma*: a Carboniferous bryozoan with cheilostomatous features. *Palaeontology* **14**(1), 178–187.

Ulrich, W. (1949). Über die systematische Stellung einer neuen Tierklasse (Pogonofora K. E. Johansson), den Begriff der Archicoelomaten und die Einteilung der Bilaterien. *Sitzber. dtsch. Akad. Wiss. Berlin, Math. nat. Kl.*, 1949.

Voigt, E. (1966). Die Erhaltung vergänglicher Organismen durch Abformung infolge Inkrustation durch sessile Tiere. *N. Jb. Geol. Paläont. Abh.* **125,** 401–422.

55 | Offspring Variation from Known Maternal Stocks of *Parasmittina nitida* (Verrill)

FRANK J. S. MATURO, JR.
Zoology Department, University of Florida, Gainesville, Florida, U.S.A.

Abstract: Two morphotypes of *Parasmittina nitida* (Verrill) were studied to determine whether they represent genotypes within common or separate gene pools. Offspring from isolated colonies of each morphotype were raised under similar conditions. Each differentiated daughter colony exhibited the morphotype of its maternal parent (the paternal parent was unknown). The evidence strongly suggests that the two morphotypes represent separate species.

INTRODUCTION

Intercolony variation in Bryozoa has posed difficult problems for the systematist trying to characterize certain species. In general, three approaches have been taken. Where one conspicuous feature or a particular grouping of traits appeared repeatedly in samples of a "species" from broad geographical regions or within strata, the investigator has considered such specimens as (1) belonging to a single, broadly variable species without subspecific divisions; (2) constituting named "varieties" or "forms" of a single, variable species; or (3) belonging to separate species identified by the trait or traits. The subspecies category has been avoided by most bryozoologists probably because many such assemblages of traits have sympatric distributions and may even be found in the same microhabitat. Which interpretation is used seems dependent on the previous experience of the investigator.

The genera *Parasmittina*, *Microporella*, and *Schizoporella* include species which are good examples of this problem. In *Parasmittina trispinosa* (of authors) alone there have been at least 14 "varieties" designated (Canu and Bassler, 1929); Osburn (1940) recognized 4 from one area, Puerto Rico. Recently, *Parasmittina nitida* (Verrill) has come under scrutiny (Maturo and Schopf, 1968). It was recognized that Verrill's type material contained two categories of specimens. Examination of the U.S. National Museum collections and those of Maturo (1968) emphasized the widespread and sympatric distribution of the two groups. We decided to designate these categories as two "morphotypes", A and B (see below).

What is the significance of these morphotypes? Are these phenotypes representing genotypes of the same gene pool? Conversely, do these phenotypes represent genotypes of different gene pools, i.e. separate species? Ideally, one would approach the problem by performing simple single or two character crosses. Unfortunately, we cannot make matings with our present state of knowledge of bryozoan reproductive processes. We can make certain assumptions, however:

(a) cross fertilization occurs between members of different colonies;
(b) some colonies should be heterozygous for the traits examined;
(c) progeny of two heterozygous parental colonies should include both morphotypes in some identifiable ratio;
(d) if interbreeding can occur between the two morphotypes, it will occur since both morphotypes live together in the same habitat and have overlapping reproductive periods.

The present study was undertaken to examine variation in offspring from isolated parental stocks of *Parasmittina nitida* in an attempt to determine whether or not the morphotypes represent Mendelian trait combinations of a common gene pool.

DESCRIPTION OF THE MORPHOTYPES

The following descriptions of the morphotypes of *Parasmittina nitida* are excerpted from Maturo and Schopf (1968).

The most salient features of morphotype A are: (1) avicularium acute, single, directed proximally, placed on the frontal to one side of the mid-line, partly to completely *proximal* to the orifice (Plate I, Fig. 1); (2) frontal area of ovicell

---

Plate I

Fig. 1. *Parasmittina nitida*, morphotype A, showing proximally directed, acute avicularium placed proximal to the orifice. × 133.

Fig. 2. The same, showing characteristics of ovicell and avicularium. × 133.

Fig. 3. The same, showing detail of frontal area of ovicell. Orificial border is along lower right. × 350.

Fig. 4. *Parasmittina nitida*, morphotype B, showing characteristic types and position of avicularia. × 134.

Fig. 5. The same, showing ovicell with its associated avicularium and typical pattern of secondary calcification. × 146.

Fig. 6. The same, showing detail of ovicell frontal area. Orificial border is toward lower left. × 350.

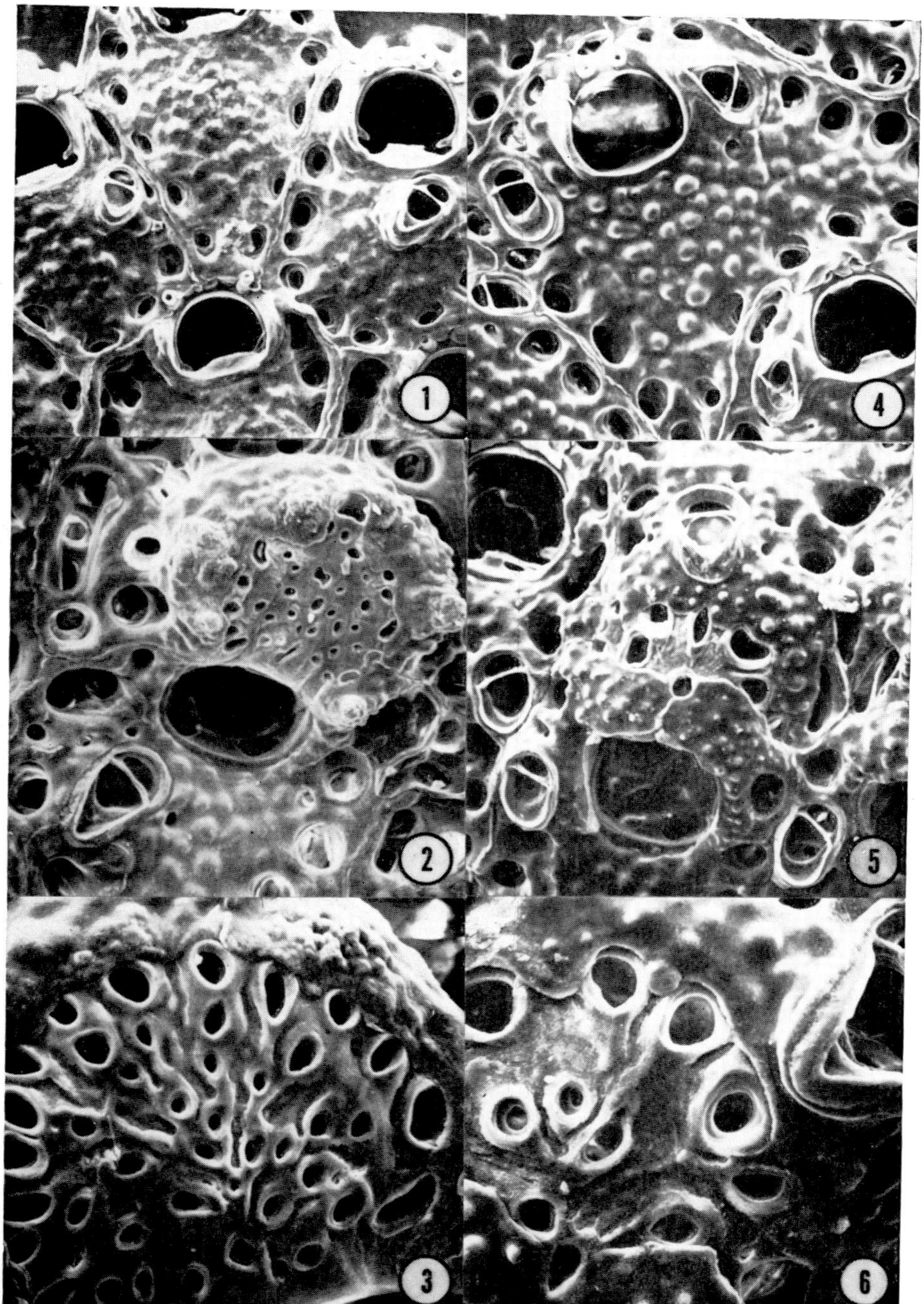
1
4
2
5
3
6

evenly perforated by many (usually more than 10) small, round pores; its orificial margin lacking a beak or border formed by the orificial collar; no avicularium associated with ovicell (Plate I, Figs 2 and 3).

Morphotype B is distinguished by (1) avicularia most frequently of two types (a) and (b), but a third type (c) is also found. (a) A small, narrow, acute, single form placed *lateral* to the orifice, its rostrum elevated on the side of an orificial lappet, its orientation disto-medial. This type is absent in some colonies. Rarely it has a reverse (proximal) orientation, but in this instance, its position relative to the orifice and its narrowness distinguish it from those of morphotype A. (b) An oval avicularium, often widened distally, with variable orientation, is found on the side of the orifice opposite to that of the acute avicularium; additional avicularia of this type may be scattered around the frontal margin of the zooids. The ovoid avicularium type is nearly always present (Plate I, Fig. 4). (c) Giant avicularia of various shapes and with enlarged chambers are present on a few zooids. (2) Ovicell perforated by fewer (up to about 10), larger, and more irregularly shaped pores than those of morphotype A (Plate I, Fig. 5). Orificial collar continued around proximal border of ovicell to form a beak or raised border. An acute avicularium with its rostrum raised on the sides of the ovicell is present in most ovicells (Plate I, Figs 5 and 6).

METHODS

Shell litter containing colonies of *Parasmittina nitida* were dredged from Bogue Sound near the southwestern edge of the Port Terminal basin, Morehead City, N.C. Collections were made as needed from mid-July to mid-August, 1970.

Colonies containing embryos in ovicells were removed from the shells with tile cutters and placed in separate plastic petri dishes with seawater. The colonies were assigned numbers and identified as to morphotype. The dishes were either put in a dark cabinet and exposed later to light or were subjected to normal night-day sequence near a north window in the laboratory. Each dish was examined after several days for larvae and ancestrulae. Each ancestrula was marked on the dish bottom by scratching a circle around it with a dissecting needle. The number of ancestrulae was recorded and the dish placed upside down on a rack in a water table with running seawater. If the parent colony still had embryos in ovicells, it was placed in another petri dish and the procedure repeated. Ultimately, all parental colonies were preserved in 70% alcohol and saved for comparison with their progeny.

Offspring colonies were allowed to develop from 53 to 94 days on the water table. The experiment was terminated on 10 October 1970. Each petri dish was examined under a Wild M5 dissecting microscope and the marked colonies counted and assigned to appropriate morphotypes. Assignment was possible only in those colonies with zones of astogenetic repetion or completed zones of astogenetic change (see Boardman, *et al.*, 1969). Only those colonies possessing avicularia were considered differentiated. The number of missing colonies was noted.

RESULTS

Some colonies produced large numbers of offspring; notably, morphotype A colonies no. 41, 56, 62, and 65. These produced 237, 127, 107, and 106 ancestrulae, respectively (Table 1).

In general growth of the progeny was poor. Many ancestrulae failed to bud new zooids; others developed only a few zooids, all of which were in stages of astogenetic change. Occasional colonies developed extensively, some having a radius of as many as 11 zooids.

Twenty-nine maternal colonies of morphotype A produced 1377 offspring. Of these, 275 differentiated sufficiently to assign to a morphotype. All belonged to morphotype A (Table 1).

Twenty-one maternal colonies of morphotype B produced 676 offspring. One hundred sixty-three of the progeny differentiated; all were of morphotype B (Table 2).

DISCUSSION

The maternal parent of each offspring examined in this study is obviously known, but the paternal parent is unknown. It is possible, but not established for *Parasmittina nitida*, that eggs can be fertilized by sperm from zooids from the same colony. It is assumed here that sperm from other colonies can and do fertilize the eggs of the maternal zooids. As long as the latter premise is possible, the former is inconsequential to the interpretation of the results; a mechanism is provided for new combinations in genotypes and the occurrence of heterozygosity.

The constancy of the characters ascribed to the two morphotypes seems well established. The shapes, positions, and orientations of the avicularia are so characteristic for each morphotype that the features of the ovicell, etc., are not necessary to distinguish between them (none of the offspring in this study had ovicells). Undoubtedly then these are genetically controlled features subject to minimal modification by the environment.

If the condition of the avicularia is definitive for each morphotype and without intergrades, then the differences could be due to alleles in small linkage

Table 1. Differentiation of offspring from colonies of morphotype A of *Parasmittina nitida* (Verrill)

Culture no.	Total no. offspring counted	No. differentiated	Type A offspring	Type B offspring	Missing
13	18	6	6	0	5
23	60	15	15	0	2
25	19	8	8	0	3
29	15	10	10	0	1
32	81	17	17	0	2
34	40	7	7	0	0
35	49	22	22	0	3
36	25	4	4	0	4
37	36	1	1	0	0
38	25	9	9	0	0
39	45	15	15	0	1
40	29	4	4	0	
41	237	10	10	0	0
42	17	4	4	0	0
43	34	8	8	0	0
48	17	0	0	0	
50	41	8	8	0	2
53	38	9	9	0	1
54	46	11	11	0	1
56	127	3	3	0	0
57	19	3	3	0	2
59	22	4	4	0	0
60	10	5	5	0	0
61	23	13	13	0	4
62	107	22	22	0	3
63	26	11	11	0	0
65	106	25	25	0	0
66	46	20	20	0	1
67	19	1	1	0	2
Totals	1377	275	275	0	37

groups. Therefore, progeny from heterozygous matings, whether by selfing in a heterozygote colony or by intercolony cross fertilization, would be expected to consist of both morphotypes in an identifiable ratio. Since both parental morphotypes were collected at random from the same habitat (often from the same piece of shell substrate), cross fertilizations could have occurred. Progeny of some colonies used in this study should have included both morphotypes.

Table 2. Differentiation of offspring from colonies of morphotype B of *Parasmittina nitida* (Verrill)

Culture no.	Total no. offspring counted	No. differentiated	Type A offspring	Type B offspring	Missing
20	69	17	0	17	0
21	65	22	0	22	0
22	27	9	0	9	2
24	18	5	0	5	0
26	59	24	0	24	3
27	34	18	0	18	0
28	30	4	0	4	0
30	4	3	0	3	0
31	4	1	0	1	0
33	16	8	0	8	1
44	26	5	0	5	4
45	38	2	0	2	0
46	13	0	0	0	0
47	70	3	0	3	0
49	41	0	0	0	3
51	15	2	0	2	1
52	15	2	0	2	4
55	44	15	0	15	0
58	27	2	0	2	0
64	32	13	0	13	2
68	29	8	0	8	0
Totals	676	163	0	163	20

Such was not the case; all offspring possessing avicularia resembled the maternal parent. The most logical interpretation is that the two morphotypes are not parts of the same gene pool and therefore represent separate species.

Since only the maternal parent is known and all F_1 are like that parent in phenotype, it is possible that a cytoplasmic factor(s) forms the basis of variation. Only reciprocal crosses could prove it; i.e. sperm A × ova B and sperm B × ova A. This type of inheritance is so rare that I have made the assumption that it does apply here.

On the whole, only 20–25% of the offspring colonies differentiated in the time period of the experiment (53–94 days). Mortality of ancestrulae was high but many colonies alive at the conclusion of the study had arrested growth and had not developed more than a few circles of zooids. The culture conditions were not controlled. The colonies were subjected to natural fluctuations of salinity and temperature of the running seawater of the laboratory system and had to obtain their own food. It is possible that the salinity range of water from Beaufort Harbor is sufficiently different from that of Bogue Sound to produce the poor development of colonies. The salinity in Bogue Sound is known to be slightly higher and less variable than that at Beaufort Harbor (Maturo, 1959).

Another factor may be the speed of the water current in the water table where the culturing was done. It most likely was not as fast as that normal to a tidal cycle in Bogue Sound. This perhaps might affect development by not providing adequate food supply and gas exchange.

I am assuming that mortality in acestrulae was not the death of morphotypes B having morphotype A maternal parent and vice versa. Were this the case it would further support the thesis of genetic isolation of the morphotypes.

The experimental design employed in this study offers a simple method for analysing genetic variation in bryozoan colonies. Using this and more sophisticated plans, it is hoped that bryozoologists may soon identify conspicuous morphological traits that are inherited according to simple Mendelian principles.

ACKNOWLEDGEMENTS

This study was completed during tenure of a National Science Foundation Postdoctoral Fellowship at the Duke University Marine Laboratory. I wish to thank Dr John D. Costlow, Jr., Director, for the many courtesies extended during my stay. Able and appreciated assistance in this work was provided by Mr James Vansant and Sr Luiz M. Braga; their support was provided by the University of Florida Division of Biological Sciences and a UNESCO Institute of International Education Fellowship, respectively. My colleague Dr F. Clifford Johnson critically evaluated the manuscript and provided helpful discussion.

The photomicrographs were made by Miss Edythe Humphries who also prepared the specimens for scanning electron microscopy.

REFERENCES

Boardman, R. S., Cheetham, A. H. and Cook, P. L. (1969). Intracolony variation and the genus concept in Bryozoa. *Proc. N. Am. paleont. Conv*. 294–320.

Canu, F. and Bassler, R. S. (1929). Bryozoa of the Philippine region. *Bull U.S. natn. Mus.* **100,** 1–685.

Maturo, F. J. S. Jr. (1959). Seasonal distribution and settling rates of estuarine Bryozoa. *Ecology* **40** (1), 116–127.

Maturo, F. J. S. Jr. (1968). The distributional pattern of the Bryozoa of the east coast of the United States exclusive of New England. *Atti. Soc. ital. Sci. nat.* **108,** 261–284.

Maturo, F. J. S. Jr. and Schopf, T. J. M. (1968). Ectoproct and entoproct type material: re-examination of species from New England and Bermuda named by A. E. Verrill, J. W. Dawson, and E. Desor. *Postilla* **120,** 1–95.

Osburn, R. C. (1940). Bryozoa of Porto Rico with a résumé to the West Indian bryozoan fauna. *Scient. Surv. P. Rico* **16** (3), 321–486.

56 | Population Genetics of Ectoprocts: Status as of January, 1972

THOMAS J. M. SCHOPF

Department of the Geophysical Sciences,
and Committee on Evolutionary Biology,
University of Chicago, Chicago, Illinois 60637, U.S.A.

Abstract: The use of proteins as markers for specific genes has revealed that local populations of ectoprocts (1) routinely outbreed, (2) have as much genetic variability as any other organism, and (3) can be significantly genetically differentiated over distances as short as 15 km, provided that the few species so far examined are typical for the phylum.

INTRODUCTION

The purpose of this brief article is to summarize what is currently known of the population genetics of the Phylum Ectoprocta. This summary follows the introduction of new techniques with new results obtained in the past three years. It was prepared for those with little formal training in genetics.

Prior to 1966, summaries of reproduction in ectoprocts stressed the large number of hermaphroditic species, and the probable existence of wide-spread self-fertilization (Hyman, 1959; several general invertebrate zoology text books; specialist literature reviewed by Silén, 1966). Because sexual reproduction is very difficult to observe in the tiny individual ectoproct zooids, cross-fertilization was not even shown to be feasible until special studies were undertaken for that purpose (Silén, 1966). The need for direct observation to understand the details of the sexual process also excluded from consideration the vast majority of ectoprocts because they are well calcified. The possibility for out-breeding of course does not demonstrate the extent to which it may occur. As Silén fully realized, fertilization by sperm from another zooid in the same colony, which was a distinct possibility in his observations, is only another type of self-fertilization since all zooids in a colony are budded asexually from the same initial individual.

Very few determinations of chromosome number exist for ectoprocts. In the "Atlas of Chromosome Number of Animals," *Membranipora pilosa* is reported to have a haploid chromosome number determined from a primary oocyte of 11 (Makino, 1951). *Plumatella fungosa* has a haploid chromosome number determined from a primary spermatocyte of 6–7. Polyploidy, which is so common in plants, would be worth checking for in ectoprocts, but this has not been done in these sessile, hermaphroditic animals.

This summary is concerned with 3 aspects of ectoproct population biology: I. Outbreeding; II. Genetic Variability; and III. Genetic Differentiation.

OUTBREEDING

About a century ago, Gregor Mendel discovered that heredity results in the transmission of particulate characters. These particulate characters (genes) are responsible for single proteins, or the polypeptide module from which complicated proteins are constructed. In some cases, the phenotypic expression of a gene (for example, eye color in some species) is also a particulate character. Alternative forms of a gene at a locus on a chromosome are known as alleles. If a gene produces only one form of a product, then there is said to be one allele at that locus; if a gene produces two more products, then there are said to be two or more alleles at that locus. A monomorphic locus has one allele; a polymorphic locus has two or more alleles.

In a simple, but common situation with two alleles at a locus, some individuals in a diploid population will be homozygous and bear only one allele, others will be homozygous and will bear the other allele, and some will be heterozygous and bear both alleles. In shorthand:

p = frequency of A, where A is a particular allele

q = frequency of a, where a is a particular allele

A and a are at the same locus; the genotypic classes in a diploid organism would be AA, Aa, and aa.

If p and q are known, and if the population is freely interbreeding with insignificant selection, immigration and mutation, then the percentage of the population of a given genotype (AA, Aa, aa) is predictable for the following reason: the discrete alleles will reassemble at fertilization in combinations proportional to their frequency in the parent population. Clearly the higher that p is, for example, the more AA homozygotes are to be expected. In the general case, p and q will occur according to the expression $p^2 + 2pq + q^2$; this is known as the Hardy-Weinberg equilibrium. If the genotypic classes in a local population match the frequencies predicted by this binomial distribution, then the population is assumed to have mixed its gametes during mating in a fashion characterized as "freely interbreeding". A statistically significant deviation from the predicted occurrence of genotypes can and does occur in nature. For example, significant selection may result in a deficiency of heterozygotes.

I will first give a hypothetical example to show that extensive inbreeding could be detected. Let us assume that we have recorded in the local population, 20 individuals homozygous for one allele (A), 20 individuals homozygous for another allele (a), and 10 individuals heterozygous for the two alleles; ($n = 50$). In this diploid, the frequency of allele A is $40/100 + 10/100 + 0/100 = 50/100 = 0{\cdot}50$. The frequency of allele a is $0/100 + 10/100 + 40/100 = 0{\cdot}50$. What would be the expected distribution of genotypic classes based on these observed allele frequencies? Following the Hardy-Weinberg distribution, the expected distribution of genotypes would be $(0{\cdot}5)^2 + 2(0{\cdot}5 \times 0{\cdot}5) + (0{\cdot}5)^2$; or 0·25, 0·50, and 0·25, respectively, for homozygous class, heterozygous class, homozygous class. However, the observed distribution of genotypic classes in 0·40, 0·20 and 0·40 (i.e. 20/50, 10/50, 20/50). In this hypothetical natural population, there is a strong deficiency of heterozygotes (and an excess of homozygotes) compared to what would be expected if we had been sampling a random breeding population.

Now I will present data from an actual case for the cheilostome ectoproct *Schizoporella errata* (*S. unicornis* of authors; material from Woods Hole, Massachusetts). In the sampled local population, 25 individuals were homozygous for one allele, 3 were homozygous for another allele, and 22 were heterozygous for both alleles ($n = 50$). The frequency of allele A is $50/100 + 22/100 + 0/100 = 72/100 = 0{\cdot}72$. The frequency of allele a is $6/100 + 22/100 + 0/100 = 28/100 = 0/28$. Following the Hardy-Weinberg distribution, the expected distribution of genotypes would be $(0{\cdot}72)^2 + 2(0{\cdot}72 \times 0{\cdot}28) + 0{\cdot}28)^2$; or, 0·52, 0·40, and 0·08. The observed distribution of genotypic classes was 0·50, 0·44, and 0·06; (i.e. 25/50, 22/50, 3/50). This is direct evidence that the alleles of this particular local population are segregating in an approximately random fashion at each generation's mating.

What are these particular, Mendelian characters which were utilized in the study of *S. errata*? No external morphological characters have ever been found to be under the control of single genes in ectoprocts. However, since genes make proteins, these substances (especially the enzymes) can and do serve equally well. Indeed, since proteins are the direct expression of gene activity, the marker genes revealed by examining proteins are very close to a direct reading of the genome.

The protein markers are revealed by electrophoresis. This is a technique which separates proteins of slight charge difference. A single amino-acid substitution in one protein with respect to another may be sufficient to cause a charge difference. Such charge differences can occur in the proteins made by two alleles of the same genetic locus.

The electrophoresis procedure works as follows. From a few to several hundred zooids of a single colony (a single genetic unit) are ground up, centrifuged, and the supernatant drawn off. An aliquot of the supernatant is placed in a slot on a polyacrylamide or starch gel. This procedure is repeated for many colonies (often 10–25) so that on a single gel many genetic individuals will be electrophoresed at the same time, each in its own slot. By analysing many individuals simultaneously, the potential effect of different conditions from gel to gel is equalized; this then permits the easy comparison of different individuals. During electrophoresis the proteins migrate in the electrical field established through the gel. After 2–6 h the gel is taken out of the electrophoresis apparatus and is stained for specific proteins. For ectoprocts, some 30 different protein stains have been tried; almost all of these are for specific enzymes. When the staining procedure works well then one or more bands appears in the column on the gel for each genetic individual.

If a single band appears in each column at the same distance from each origin for each colony, then this implies an identical charge for the protein stained in each sample. Generally speaking, this is sufficient evidence to suggest an identical form of that particular protein in each individual sampled. Within a single ectoproct species, this implies the same allele in all colonies sampled. Such a result indicates a monomorphic gene locus; all individuals samples are homozygous for the allele.

Another pattern is that in which a single band appears in several columns but at 2 different distances from the origin depending upon which specimen is being examined. In addition, in this situation, other individuals commonly have a double band each part of which is identical in mobility to a single band of the remaining individuals. A protein with this pattern is interpreted to result from a 2 allele locus. Some individuals are homozygous for one allele, other individuals are homozygous for the other allele (hence the single band of different mobilities in different individuals) and some individuals are heterozygous (hence the double band). The genotypes would be *AA*, *aa*, and *Aa*.

In this way, from the phenotypic pattern of the enzyme as revealed on the gel, the genotype of each individual is determined. Gene frequencies are then computed from the genotypes as in the examples previously given.

Not too uncommonly, significant band variability may exist from one individual to the next, but the resolution is not good enough to permit genotypes to be unambiguously scored. Such a situation is most commonly improved by changing the buffer, pH, gel permeability, etc., and takes experimentation in procedures. This accounts for the ability to score some loci as provisionally polymorphic but yet for the allele frequencies to remain undetermined.

At least one polymorphic locus has been identified in each of the 4 ectoproct species in which they have been sought: *Schizoporella errata*, *Bugula stolonifera*, *Bugula turrita*, and *Bugula simplex*. However, only in *S. errata* and *B. stolonifera* have allele frequencies been deciphered.

With respect to outbreeding, in 16 populations of *Schizoporella errata* the observed genotype frequencies and the predicted genotype frequencies computed from the observed gene frequencies are not statistically different (Table 1). Further, allele frequencies differ widely in the observed populations. There is no reason to believe that this locus would yield results compatible with Hardy-Weinberg equilibria in many populations, and with widely different allele frequencies, unless it was reflecting independent assortment of alleles at each mating in each local population; i.e. the species was routinely outbreeding.

In *Bugula stolonifera*, 4 esterase loci are polymorphic and resolvable into alleles that can be scored (Gooch and Schopf, 1970). A local population in Eel Pond, Woods Hole, Mass., was sampled, and 3 of the 4 polymorphic loci showed close agreement in genotype frequency with the genotype frequency predicted from the measured allele frequencies. The fourth locus is a curious and really more interesting case. The genotype of an individual indicated by that locus can be predicted from a knowledge of one of the other polymorphic loci as though they were in linkage. This is an interesting genetic problem and could be further pursued. The chief interpretation for our purposes, however, is that *B. stolonifera* is routinely outbreeding.

Although the procedures outlined above are essentially a product of the 1960s, they have been very widely used, and a literature of many hundreds of papers now exists (see the journals *Genetics* and *Biochemical Genetics* for a sampling). In every case where it has been desirable to test electrophoretic data by breeding tests (and there are many instances in the literature on *Drosophila*, etc.), the electrophoretic interpretation has been supported.

GENETIC VARIABILITY

Three biochemical measures have been used as an index of genetic variability. There are (1) the proportion of gene loci which have 2 or more alleles in frequencies of 5% or more (i.e. are polymorphic); (2) the number of alleles at a polymorphic locus; and (3) the genome size as measured by the weight of DNA per cell, or the haploid number of nucleotide pairs. Information in ectoprocts is available for the proportion of loci polymorphic, and the number of alleles per polymorphic locus.

Thirty-six loci have now (1973) been identified in *Schizoporella errata*, and 3 (and

Table 1. Observed and expected genotype frequencies for the ectoproct *Schizoporella errata* at the biallelic "leucine" amino peptidase locus

Locality	Genotype* frequency: *Schizoporella errata*																				
	1969							1970							1971						
	n	observed			expected			*n*	observed			expected			*n*	observed			expected		
Shark Shoal Jetty								20	20	0	0	20	0	0							
Duke Marine Laboratory								15	15	0	0	15	0	0							
Indian River Inlet								13	8	5	0	8	4	0							
Cuttyhunk								30	6	9	15	4	14	13	31	6	12	13	5	15	12
Robinsons Hole								36	4	14	18	3	15	17							
Vineyard Sound	33	9	14	10	8	16	9														
Sheep Pen Harbor								29	10	16	3	11	14	4							
MBL Dock	50	25	22	3	26	20	4	45	14	27	4	17	21	7	58	22	23	13	19	28	10
Green Pond	43	25	15	3	25	16	3	47	28	15	4	27	17	3							
Quissett Harbor															48	20	20	8	19	22	7
Monument Beach															48	30	17	1	31	15	2
Cape Cod Canal								15	10	4	1	10	5	1							

* Expected genotype frequencies are computed from observed gene frequencies. The close similarity of expected and observed genotype frequencies supports the view that alleles at this biallelic locus are segregated as in a randomly breeding population. *n* = number of colonies examined. Number of each genotype class are given in the sequence of low mobility homozygotes, heterozygotes, and high mobility homozygotes; mobility is with respect to the migration on polyacrylamide gel. Due to arithmetic rounding of fractions, the *n* for computed genotypes may differ by 1 from the *n* observed.

possibly 7) of these are polymorphic. Thirteen loci have been identified in *Bugula turrita* and two (and possibly three) of these are polymorphic. These figures of 11 to 23% polymorphism are well within the range found in other animals.

Where particular alleles have been able to be assigned without question, from 2 to 5 alleles per locus occur. This figure is again comparable to most other organisms so far examined.

Thus these marine ectoprocts are not unusual in their extent of genetic variability in comparison to other animals so far sampled.

GENETIC DIFFERENTIATION

In *Schizoporella errata* the extent of genetic differentiation has been investigated on scales of less than a square meter, tens of square meters, and along linear transects of a few tens of kilometers to more than 1000 km.

On a very local scale, genetic differentiation has been examined with respect to deviations from Hardy-Weinberg frequencies, and an excess or deficiency of one of the genotypic classes, as a function of depth (0–3 m), exposure to waves (different sides of exposed pilings, and exposed versus unexposed pilings), and different sides of a floating dock. Although 4 of 17 tested distributions were significant at the 0·05 level, none were significant at the 0·01 level. None of the significant deviations have been found at more than one locality. Thus there appears to be a rather uniform distribution of genotypes over a scale of less than a square meter to 10 m^2.

Much more important has been the observed changes in allele frequencies along coastal transects. Allele frequencies may change from 0·3 to 0·7 over a distance as short as 15 km for a species (*S. errata*) whose probable distance of larval travel is less than 1 km. Continuing strong selection is not required to maintain such a cline. Since the local cline reverses itself regionally parallel to a regional reversal in water temperature, selection is attributed to the effect of temperature *per se* on enzyme activity, or to a factor which is temperature dependent. Such a factor might be the occurrence of phytoplankton of different species at different temperatures. Selection at the enzyme level would, in that case, be related to food resources.

The use of new techniques in genetics permits us to learn many new things about populations of ectoprocts. In the next few years, we hope that these studies will be extended so that we gain a knowledge of the genetic differentiation of a few species with different types of life history. These species could serve as analogues for extinct forms in the paleontologist's attempt to interpret the static patterns of fossil remains in terms of the dynamic processes of speciation.

ACKNOWLEDGEMENTS

This report summarizes data obtained during the summers of 1969, 1970 and 1971 at the Marine Biological Laboratory, Woods Hole, Massachusetts. During this time I have had the fortune to work with Dr James L. Gooch, Juniata College, and he shares equal responsibility for the data thus far obtained (see publications cited below). I also benefited greatly from discussions with Dr Edward Berger, State University of New York, Albany, who worked on ectoprocts in our laboratory during 1971. I acknowledge with sincere thanks the support provided by NSF grant GB-7325, the Block Fund of the University of Chicago, and more recently by NSF grant GB-30870.

REFERENCES

GOOCH, J. L. and SCHOPF, T. J. M. (1969). Genetic studies on marine species of the Phylum Ectoprocta. *Biol. Bull. mar. biol. Lab., Woods Hole* **137,** 400–401.

GOOCH, J. L. and SCHOPF, T. J. M. (1970). Population genetics of marine species of the Phylum Ectoprocta. *Biol. Bull. mar. biol. Lab., Woods Hole* **138,** 138–156.

GOOCH, J. L. and SCHOPF, T. J. M. (1971). Genetic variation in the marine ectoproct *Schizoporella errata. Biol. Bull. mar. biol. Lab., Woods Hole* **141,** 235–246.

HYMAN, L. H. (1959). "The Invertebrates: Smaller Coelomate Groups," 783 pp. McGraw-Hill Book Company, Inc., New York.

MAKINO, S. (1951). "An Atlas of the Chromosome Number in Animals," 290 pp. Iowa State College Press, Ames, Iowa.

SCHOPF, T. J. M. and GOOCH, J. L. (1970). A natural experiment in the population genetics of a marine ectoproct suggesting changes in gene frequencies with changes in temperature. *Biol. Bull. mar. biol. Lab., Woods Hole* **139,** 435.

SCHOPF, T. J. M. and GOOCH, J. L. (1971). Gene frequencies in a marine ectoproct: a cline in natural populations related to sea temperature. *Evolution* **25,** 286–289.

SILÉN, L. (1966). On the fertilization problem in the Gymnolaematous Bryozoa. *Ophelia* **3,** 113–140.

57 | Morphological and Developmental Affinities of the Lophophorates

RUSSEL L. ZIMMER

Santa Catalina Marine Biological Laboratory,
University of Southern California,
Avalon, California, U.S.A.

Abstract: Nine developmental and morphological criteria used in the characterization of the two invertebrate superphyla (Protostomia and Deuterostomia) are compared with the features of the three lophophorate phyla (Brachiopoda, Bryozoa and Phoronida). It is demonstrated that the lophophorates share only one unambiguous character with the Protostomia, but eighteen with Deuterostomia. Of the characters held in common with the deuterostomes, regionation of the larval and adult bodies into an oligomerous pattern in lophophorates is the most decisive argument for allying the three lophophorate phyla not only to each other, but also to the deuterostome branch of the animal kingdom.

For many years, developmental patterns have been used as the basis for the positioning of phyla within evolutionary trees. Our broadest division of the invertebrates, that into protostomes and deuterostomes, is based on developmental criteria either exclusively or in large part depending on the number of features considered. The criteria most frequently used are all developmental (Table 1, 1–6), but certain adult features (Table 1, 7–9) are useful in diagnosis also.

It is not always possible to utilize all nine criteria in evaluating the position of a phylum or group of phyla, but it is valuable to explore as many as possible if we are to evaluate the true affinities of the group. Therefore, the main body of this paper consists of a confrontation of the three lophophorate phyla with the above criteria.

As there is general agreement that the Bryozoa, Brachiopoda and Phoronida constitute a closely related group, the lophophorates, it is not necessary to review all their similarities. It should be pointed out that their approximation is based on the congruities in adult morphologies and not on developmental features, among which there are surprisingly few identities. The fundamental pattern of all lophophorates is the regionation of the body into (1) a trunk, (2) a distinctive circumoral lophophore and (3), in all but the gymnolaemate

Table 1. Criteria used to distinguish Protostomia from Deuterostomia

Character	Expression in Protostomia	Expression in Deuterostomia
1. Origin of adult mouth	From larval blastopore	Independent of larval blastopore
2. Cleavage pattern	Spiral	Radial
3. Mosaicness of development	Typically highly mosaic	Usually regulative (non-mosaic)
4. Origin of adult mesoderm	Endomesoderm from 4d cell	Endomesoderm from archenteric pouches
5. Method of coelomogenesis	Schizocoely	Typically by archenteric pouching
6. Type of larva	"Trochophore-type"	"Dipleurula-type"
7. Position of adult nervous tissue	Usually subepidermal	Usually intraepidermal
8. Degree of cephalization	Strong cephalization	Cephalization frequently very limited
9. Body regionation of the adult (and larva)	Various patterns, but true segmentation in higher forms	Typically trimerous with proto-, meso-, and meta-somes

bryozoans, a preoral epistome. The significance of this minute preoral fold is highly controversial, but I shall attempt to demonstrate that it is of profound significance.

Because the alliance of these three phyla into a common group is based primarily on configuration of the adult body, it would seem that body regionation should be a most important factor to consider in broader phylogenetic considerations. Inexplicably, however, the most persistent comparison made is that of whether the lophophorates have larvae that are similar to the trochophore larvae of protostomes. This is surprising not only for the reason just advanced, but also because the question should be: "Do larvae of the lophophorates have more in common with the protostomous or deuterostomous larvae?" not "Do they share features with the typical protostomous larva?" Because a one-sided question has usually been asked, it is not surprising that a one-sided response has been given. The end result has been that the lophophorates' affinities with the deuterostomes have been ignored and they have been erroneously linked with the Protostomia.

Of the three lophophorate phyla, the Phoronida are widely accepted as being the least modified from the ancestral pattern and so our consideration will begin with them. Much of the information presented on phoronids will be unpublished personal observations based largely on the development of *Phoronis vancouverensis* (Zimmer, 1964, 1967).

In all phoronids which have been studied, the adult mouth is derived from the persistent anterior portion of the blastopore. The cleavage is best described as biradial as the 16-cell stage features two tiers of eight cells each and has only two planes of symmetry through the animal-vegetal axis. Rather interestingly, there are occasional spiral arrays in the early cleavage stages of phoronids as noted by many workers. I consider such arrays to be fortuitous and the probable consequence of mutual compression of the blastomeres, distorting the mechanically unbalanced orthogonal arrangement of cells into the more stable and highly packed spiral pattern. Such compression could be generated if the eggs which are contained in a tough hyaline membrane were exposed to a hypotonic medium. The presence of typical radial arrays and the occurrence of 8-celled tiers during later cleavages dictate that cleavage in phoronids is radial not spiral.

In *Phoronis vancouverensis* there is full regulation of development after separation of blastomeres at both the 2- and 4-cell stages.

The origin of mesoderm in phoronids has been reported in a bewildering variety of ways, suggesting that there may be species differences (and probably differences between investigators). In *Phoronis vancouverensis* the adult mesoderm is produced by the budding of mesodermal cells from the archenteric wall in two phases. The earlier proliferation is produced from the future buccal region and by ameboid wandering the component cells subsequently line the blastocoelic space of the developing preoral hood, forming the preoral hood coelom. The existence of this protocoel has previously been denied, but in the six species representing both genera I have examined it is always present at least transitorally as a typical coelom. As that portion of the hood containing the ganglion and more ventral component of the preoral hood coelom is retained through metamorphosis as the adult epistome, one should consider the hood and the epistome as a true protosome. Initially the protocoel is discrete, but after metamorphosis it apparently becomes confluent with the lophophoral coelom which arises very late in larval life; the origin of the lophophoral coelom remains enigmatic.

The second proliferation of mesodermal cells gives rise to the trunk coelom. About eight primordial cells, thought to wander out from the archenteric wall at the level of the future cardiac stomach, proliferate to form a solid mass which by schizocoely forms the trunk cavity.

It is apparent that these localized proliferations of multiple mesodermal cells from the archenteron are not typical archenteric diverticulae, but similar proliferations in enteropneusts are accepted as modified diverticulae. Clearly the mesoderm formation in phoronids is totally dissimilar from that in typical protostomes, but is compatible with that in some deuterostomes.

Comparisons of the actinotroch larvae of phoronids with the trochophore of protostomes *and* the dipleurula (or other larva) of deuterostomes have not been made in many years, but one-sided comparisons of the actinotroch with the trochophore are common and consistently result in the conclusion that the two are related. Of recent authors, Hyman (1959) has made the most extensive comparison of the actinotroch and trochophore. Citing several similarities between these two larvae (but ignoring the similarities of actinotrochs with deuterostome larvae), she concluded: "that the actinotroch is an altered trochophore does not appear open to doubt". Certainly the derivation of the adult phoronid mouth from the blastopore is a protostomian feature, but can we conclude that the trochophore affinities of the actinotroch are irrefutable on the basis of unilateral comparisons and then attach great phylogenetic significance to the decision reached? It would seem that any decisions on larval types require the characterization of the typical protostomous and deuterostomous larvae and that such characterizations will utilize those embryological features already considered of superphyletic importance. It is then redundant to include a comparison of larval types in our confrontation.

Considering the adult characteristics, phoronids have an intraepidermal nervous system and lack cephalization. With reference to the controversial subject of body regionation, we have already indicated that both the larval and adult phoronid are trimerous. Some scholars find abhorrent the concept of trimery (or oligomery), but it is a fact that a tripartite regionation has been utilized in virtually every proposed archetype of an adult or larval deuterostome, and that such regionation is not known in protostomes. It is also a fact that both larval and adult phoronids share this regionation with typical deuterostomes and that the mesosome is tentaculated in pterobranchs and echinoderms as it is in all lophophorates. That this remarkable identity could be due to convergence seems most improbable.

Within the largest and most diverse lophophorates phylum, the Bryozoa, it is generally conceded that the Eurystomata have the most basic development although even this is highly aberrant. For simplicity and with little risk of deception, only the development of this order is considered in the following paragraphs.

Whether an open blastopore ever exists in any of these bryozoans has been questioned, but there can be no doubt that the adult mouth of bryozoans is not related to the site at which gastrulation occurred. The cleavage is of a biradial nature as in phoronids, but is much more uniform in pattern. I have separated the blastomeres of the 2-cell stage of *Membranipora isabelleana* and full regulation followed; I know of no other reports on developmental potencies.

Neither the origin of the mesoderm nor its differentiation into the coeloms has been adequately explained for any bryozoan. In a recent study, Woollacott and Zimmer (1971) concluded that the lophophoral coelom and the splanchnic portions of the visceral coelom are derived from an embryonic blastemal tissue (the internal or mesodermal thickening of many authors) which is probably of ectomesodermal rather than entomesodermal origin, whereas the somatic lining of the visceral cavity is from pigmented mesodermal cells derived from the mesendodermal mass. While there are other interpretations of the origin and differentiation of the mesoderm, it is difficult to derive any of them from the patterns typical for either protostomes or deuterostomes.

Several authors, including Hyman (1959), Nichols (1962), Rudwick (1970) and Ryland (1970) have indicated the bryozoan larva is a modified trochophore, but the latter three scholars did not justify their statement. Hyman's judgement was based on the argument that the actinotroch is a trochophore and therefore the bryozoan larva must also be one although only the corona and apical plate of the bryozoan could be compared with features of trochophores. The corona does have morphological similarities with the prototroch of trochophores, but Correa's (1948) assertion that the cell lineages of the two structures are identical is erroneous. Apical plates are extremely abundant in deuterostomous larvae and cannot be considered characteristic of protostomes.

Considering adult features, the bryozoans lack cephalization, but the nervous tissue is subepidermal. The marine bryozoans possess as adults only two body cavities, the lophophore and trunk, but in freshwater forms the trimerous pattern typical of deuterostomes is complete. With the possible exception of *Flustrellidra* (Pace, 1906), eurystome larvae have no differentiation of the mesoderm into coeloms and there is no regionation of the larval body.

The remaining lophophorate phylum, the Brachiopoda, has cleavage that is consistently recognized to be radial (or a biradial modification). Long (1964) and Zimmer (unpublished study) have independently demonstrated that development is regulative after separation of blastomeres at the 2-cell stage. In the articulates, the origin of the mesoderm and its differentiation into coelomic components appears clearly related to the archenteric pouching of deuterostomes, but in the one inarticulate reported on, these features are rather modified (but still unlike the protostomous patterns). The division of the adult body into a preoral epistome, circumoral lophophore and postoral trunk is reflective of deuterostomes. In inarticulates the epistome has coelomic channels which are confluent with the lophophoral coelom and articulates have a solid epistome.

It may be well to digress here for a moment and point out that the comparison of the various epistomes of lophophorates with the protosome of an idealized deuterostome usually ends with the criticism that the lophophorate structure either lacks a peritoneally-lined space or that this space is confluent with the lophophoral coelom. Yet, there is never question of the echinoderm protocoel which in the adult is invaded by the tissue of the axial gland and which in the larva and adult is characterized by its connection with the hydrocoel. Phylogenetic considerations should not be plagued by double standards such as this. To return to the main theme, the body regions of the adult inarticulates are fully anticipated in the larvae. In the articulates, the three lobes of the larva do not correspond to the three regions of the adult, although the meso- and metacoels of the adult are already developed.

The nervous system is largely intraepidermal, although controversy on this point exists in the early literature; no brachiopod exhibits even moderate cephalization.

Two features of the development have been cited as indicating protostomous affinities of the brachiopods. Of these, the claim that the articulate larva is a modified trochophore—although an apical plate is the only shared feature—is unworthy of further discussion. The second feature relates to the potential accord between the adult mouth and the closed larval blastopore. Earlier workers suggested that there was a correspondence in positions, but among recent workers, Percival (1944) could not relate the mouth to the site of the closed blastopore and Franzen (1969) and Long (1964) have concluded that the two are not spatially related.

In summary of this extended confrontation, derivation of the adult mouth from the blastopore in phoronids constitutes the *single* irrefutable feature which any lophophorate shares with protostomes, although the subepidermal position of the nervous system in bryozoans, the presence of solenocytic protonephridia in larval phoronids, and the prototroch-like corona of the bryozoan larvae have parallels with protostomes rather than deuterostomes. In marked contrast there are at least eighteen significant similarities of lophophorates with deuterostomes as revealed in this paper. Of these similarities, I consider the most important to be the possession by the adult and some larval lophophorates of a trimerous body plan that is diagnostic of deuterostomes. The conclusion that lophophorates are allied with the deuterostomes and not the protostomes appears beyond doubt.

REFERENCES

CORREA, D. (1948). A embryologia de *Bugula flabellata. Zoologia* (*São Paulo*) **13,** 7–71.

FRANZÉN, Å. (1969). On larval development and metamorphosis in *Terebratulina*, Brachiopoda. *Zool. Bidrag Uppsala* **38,** 155–174.

HYMAN, L. H. (1959). "The Invertebrates. V. Smaller Coelomate Groups," vii + 783 pp. McGraw-Hill, New York.

LONG, J. (1964). "The Embryology of Three Species Representing Three Superfamilies of Articulate Brachiopods," 239 pp. Ph.D. Dissertation, Univ. Washington, Seattle.

NICHOLS, D. (1962). "Echinoderms," 200 pp. Hutchinson and Co., London.

PACE, R. M. (1906). On the early stages in the development of *Flustrella hispida* Fabricius. *Q. Jl microsc. Sci.* **50,** 435–478.

PERCIVAL, E. (1944). A contribution to the life history of the brachiopod, *Terebratalia inconspicua* Sowerby. *Trans. Roy. Soc. N.Z.* **741,** 1–23.

RUDWICK, M. J. S. (1970). "Living and Fossil Brachiopods," 199 pp. Hutchinson and Co., London.

RYLAND, J. S. (1970). "Bryozoans," 175 pp. Hutchinson and Co., London.

WOOLLACOTT, R. M. and ZIMMER, R. L. (1971). Attachment and metamorphosis of the cheilo-ctenostome bryozoan *Bugula neritina* (Linné). *J. Morph.* **134,** 351–382.

ZIMMER, R. L. (1964). "Reproductive Biology and Development of Phoronida," x + 416 pp. Ph.D. Dissertation, Univ. Washington, Seattle.

ZIMMER, R. L. (1967). Unpublished studies on the development of *Phoronis vancouverensis.*

58 | Information Services Desired by Bryozoologists

F. KENNETH McKINNEY

Department of Geography–Geology, Appalachian State University, Boone, North Carolina, U.S.A.

and

ROGER J. CUFFEY

Department of Geosciences, Pennsylvania State University, University Park, Pennsylvania, U.S.A.

Abstract: Surveying bryozoologists' preferences for information services indicates that most want a variety of such services, but that no one particular type of service is overwhelmingly preferred. A catalogue of new-species descriptions reproduced from the bryozoan literature (without taxonomic modifications or updatings) seems to be the most practical augmentation of the information services presently available, because nothing like it exists at present, and because other types of information services considered are either economically unfeasible or already available to some degree. An informative directory of bryozoologists also seems highly desirable.

INTRODUCTION AND BACKGROUND

Work on bryozoans has greatly increased in the past few decades, and the volume of literature concerning these interesting animals has grown correspondingly (Schopf, 1967). While bryozoologists seem not yet overwhelmed by their expanding literature, current trends and others' experiences suggest that this situation may soon change. Other fields, having grown to the point of inability to cope with their increasing literature, have tried—with widely varying degrees of success—to develop information storage and retrieval techniques to supplement traditional methods (Becker and Hayes, 1963; Committee on Geoscience Information, 1971; Schneider, 1971). We therefore should consider how such techniques might be profitably applied to the bryozoan literature, while there is still opportunity for them to develop optimally.

Bryozoologists usually have ready access to only part of the widely scattered bryozoan literature, and are often dependent upon the serial holdings of nearby libraries and upon the number of reprint mailing lists in which they are included. Standard bibliographic services include most major works on bryozoans (both entoprocts and ectoprocts); however, some references still escape detection. Moreover, such bibliographies contain limited information about each publication; only one (Zoological Society of London, 1864ff) lists the taxa referred to in the publications included.

Bryozoologists' information needs have never been specified precisely. However, some have developed bryozoan-oriented information systems to aid their research efforts. These range from standard taxonomic bibliographies and very large reprint collections, to standard card files, semi-automated edge-notched-card systems, and automated or computerized systems.

POTENTIALLY AVAILABLE INFORMATION SERVICES

Various kinds of information services have been developed in recent years. Biological and paleontological information-retrieval systems are usually built to handle only one of three types of materials:

(1) Documents (choices a and f, below).
(2) Taxa, especially species (choices b, c, d, e, and g, below).
(3) Individual specimens (i.e. curating large collections rather than handling the general literature; therefore omitted below).

While design of information systems capable of handling the first two or all three types of materials simultaneously is technically feasible, the great cost of building and operating such systems has thus far been prohibitive.

With the foregoing in mind, we can consider several, different, potentially available types of information services which could help bryozoologists cope with their literature. These, together with references to examples of each type, are listed below (Table 1).

BRYOZOOLOGISTS' PREFERENCES AND THEIR IMPLICATIONS

In order to determine what types of information services bryozoologists desire, a check-list questionnaire listing the seven types described above was sent to bryozoan oriented scientists throughout the world. Table 1 presents the responses obtained from the 78 questionnaires returned; many written comments were also received.

Inspection of these responses suggests that many bryozoologists would like access to several information services, but that no one type of service is strongly preferred over all the others.

A straightforward bibliography of bryozoan papers (a, in Table 1), was slightly preferred over the other possible information services. Standard geo-

Table 1. Types of potentially available information services and frequency with which each type is desired by bryozoologists returning questionnaires

Type of information service	% of returns in which type was checked as desirable
(a) Bibliography of bryozoan-related papers, alone (Toomey, 1954), or with indexes (U.S. Geological Survey, 1923ff), or also with summaries (Biosciences Information Service, 1927ff).	65
(b) Bibliographic-reference listing of bryozoan species, with topically arranged indexes (Camp *et al.*, 1968).	32
(c) Bibliographic-reference listing of bryozoan species, with topically arranged indexes and also short abstracts or summaries or annotations (Zoological Society of London, 1864ff).	59
(d) Catalogue of new-species descriptions reproduced from the bryozoan literature, without any attempt to modify or update them taxonomically (Ellis and Messina, 1940ff, 1952ff).	44
(e) Catalogue of new-species descriptions reproduced from the bryozoan literature, extensively modified or updated taxonomically (Traverse *et al.*, 1957ff).	49
(f) Computer-based information-retrieval system capable of rapid response to queries concerning documents within the bryozoan literature (Blackwell *et al.*, 1969; Hopps *et al.*, 1968, 1969).	51
(g) Computer-based information-retrieval system capable of rapid response to queries for specific information concerning particular bryozoan species (Wilson *et al.*, 1969; Hopps *et al.*, 1968, 1969).	57

logical and biological bibliographies already partly satisfy this need, although a single comprehensive bibliography devoted to all aspects of bryozoans (or compiled order by order) would be welcomed. A bibliographic listing of references to bryozoan taxa, especially species (b), attracted somewhat less interest. But, this type of listing supplemented by appropriate summary comments (c) did evoke considerable interest, perhaps in part because this latter already exists in the bryozoan section of the *Zoological Record* (Zoological Society of London, 1864ff). Principal disadvantages of existing representatives of these three types (a, b, and c) are the sizeable time lag between the appearance of the papers referenced and the publication of the bibliographic work and also the fragmented or serial nature of the bibliographies. However, without a tremendous increase in the volume and speed requirements of systematic bryozoology, the increase in cost and effort which could significantly reduce this time lag seems prohibitive.

Nothing like a catalogue of new-species descriptions exists at present for bryozoans. However, a significant proportion of respondents believed such catalogues would be desirable, particularly for systematists; moreover, other palaeontologists underscore their value (Easton, 1971, pp. 778–779). This type of information service could be provided economically if simple automatic data-processing equipment and inexpensive but adequate reproduction techniques were used, because the bryozoan literature has not yet swollen inordinately. Responses were evenly divided between those favoring catalogues without taxonomic updating (d) versus catalogues with updating (e). However, many favoring the former (d) strongly objected to the cataloguers' making taxonomic modifications of original species descriptions and generic assignments, because of the possibilities for misjudgements by the cataloguers and because of the probability that what is updated for publication in the catalogue would itself eventually become outdated and require further modification. Moreover, a catalogue involving updating (e) would require a considerably larger professional-level staff to make the necessary taxonomic judgements, whereas a catalogue not including modifications (d) could be produced by a predominantly clerical-level staff.

Computer based information retrieval systems, whether document-oriented (f) or species oriented (g), provide a desired capability for rapidly answering specific queries. Experience with similar systems in the field of medical ecology (Hopps *et al.*, 1968, 1969) and remarks by several bryozoologists suggest that, while technically feasible, such systems are probably at present economically impractical for bryozoologists. Building a computerized data file up to the size where it can answer a reasonable number of queries concerning bryozoans would require very many man-hours of preparatory work. Programming the query system is likely to be quite time-consuming, if not difficult. In practice, computerized systems are used primarily by those situated very near the organizations housing them; mailed queries and responses, especially, would travel slowly enough to discourage most distant users. Finally, the amount of bryozoan work requiring very rapid responses is quite small. Therefore, although a computer-based bryozoan-oriented information system is technically feasible, building and maintaining it would not be feasible at present.

In addition to asking bryozoologists their preference for various types of information services, our questionnaire also asked whether or not the respondent would find useful a directory of active bryozoan-oriented scientists, with their systematic, stratigraphic, geographic, and methodologic interests indicated.

Most respondents (73%) answered affirmatively. Such a directory could most appropriately be compiled under the auspices of the International Bryozoology Association.

CONCLUSIONS AND RECOMMENDATIONS

A survey of information-service preferences of many bryozoologists indicates that most want a variety of such services, but that there is no clearly evident strong preference for any one particular type.

The most practical augmentation of the information services available would be a catalogue of new-species descriptions reproduced from the bryozoan literature, without taxonomic modifications or updatings. Such a catalogue is not at present available in any guise, and would be useful to a significant portion of those who study bryozoans.

Other types of information services considered, while potentially available and useful, have various disadvantages compared with that recommended above. The inherent subjectivity and potential cost involved in a catalogue of new-species descriptions updated taxonomically make it less desirable and practical than that recommended above. Computerized information systems, while technically feasible, are prohibitively expensive. Bibliographic references of various kinds, although both their scope and timeliness could be greatly improved, are already available to some extent.

Finally, a directory of currently active bryozoologists (and their various interests) would be useful to many scientists. Consequently, we recommend that the International Bryozoology Association consider expanding its membership list into such a directory.

REFERENCES

Becker, J. and Hayes, R. M. (1963). "Information Storage and Retrieval: Tools, Elements, Theories," 448 pp. Wiley, New York.

Biosciences Information Service (1927ff). "Biological Abstracts." Vol. 1ff. Biosci. Info. Serv., Philadelphia.

Blackwell, P. W., Morrison, J. L. and Smith, W. E. Jr. (1969). GIPSY computer retrieval of geologic literature. *Okla. Geol. Notes* **29,** 6–14.

Camp, C. L. *et al.* (1968). Bibliography of fossil vertebrates, 1959–1963. *Geol. Soc. Am. Mem.* **117,** 644 pp.

Committee on Geoscience Information (1971). Geoscience information. *Geotimes* **16** (3), 19.

Easton, W. H. (1971). Paleontologists and paleontology—an appraisal and a proposal. *J. Paleont.* **45,** 773–780.

Ellis, B. F. and Messina, A. R. (1940ff). "Catalogue of Foraminifera." Vol. 1ff. Am. Mus. Nat. Hist., New York.

Ellis, B. F. and Messina, A. R. (1952ff). "Catalogue of Ostracoda." Vol. 1ff. Am. Mus. Nat. Hist., New York.

Hopps, H. C. *et al.* (1968). "The Mapping of Disease (*MOD*) Project—The geographic distribution of infectious diseases," 435 pp. Armed Forces Institute of Pathology, Washington.

Hopps, H. C. *et al.* (1969). "Computerized mapping of disease and environmental data," 418 pp. Armed Forces Institute of Pathology, Washington.

Schneider, J. H. (1971). Selective dissemination and indexing of scientific information. *Science* **173,** 300–308.

Schopf, T. J. M. (1967). The literature of the phylum Ectoprocta 1555–1963. *Syst. Zool.* **16,** 318–327.

Toomey, D. F. (1954). A bibliography of the family Fusulinidae. *J. Paleont.* **28,** 465–484.

Traverse, A. *et al.* (1957ff). "Catalogue of Fossil Spores and Pollen." Vol. 1ff. Penn. State Univ., Univ. Park.

U.S. Geological Survey (1923ff). "Bibliography of North American Geology." U.S. Geol. Surv., Washington. (Many v.)

Wilson, L. R., Morrison, J. L. and Reid, W. E. (1969). Computer research in palynology demonstrated by use of Oklahoma University General Information Processing System (GIPSY)—Permian palynology of North America and some associated problems. *Univ. Okla. Info. Sci. Ser. Mon.* **2,** 143 pp.

Zoological Society of London (1864ff). "The Zoological Record." Vol. 1ff. Zool. Soc. Lond., London.

Author Index

A

Abbott, M. B., 42, *51*
Abricossoff, G. G., 504, *519*, 533, *536*, 542, *546*
Afzelius, B. A., 337, *342*
Albright, J. T., 268, *270*
Allen, H. A., 82, *94*
Allen, J. R. L., 143, 145, 148, *150*
Allman, G. J., 503, 504, 515, *519*, 539, *546*, 566, 569, *575*
Androsova, E., 369, *372*
Androsova, E. I., *105*
Annandale, N., 504, *519*, 529, 531, 532, 533, 535, 536, *536*, 542, 543, 544, *546*,
Annoscia, E., *63*
Anstey, R. L., 430, *435*
Ardhi, R. U., 418, *418*
Arkell, W. J., 455, 456, 457, *459*
Armstrong, J., 7, *10*, 228, *228*, 357, *359*
Astrova, G. G., 5, 7, *10*, 257, *260*, 327, *333*, 357, *359*, 422, 425, *427*, 550, *563*
Atkins, M. D., 261, *268*
Aulenbach, D. B., 117, *127*
Austin, O. L., 517, *519*

B

Balavoine, P., 392, *408*
Banta, W. C., 209, 210, 212, 216, *219*, 453, 457, *459*
Barnes, H., 159, *163*
Barrnett, R. J., 265, *270*, 335, *342*
Barrois, C., 378, 381, *381*
Bassindale, R., 27, *35*
Bassler, R. A., 417, 418, *419*
Bassler, R. S., 216, *219*, 222, 228, *228*, 248, *256*, 302, *306*, 361, 366, *366*, 422, *427*, 433, 435, *435*, *436*, 437, 441, *446*, 453, *459*, 497. 498, *502*, 550, *563*, 577, *584*
Bayne, B. L., 30, *34*
Beams, H. W., 268, *269*
Becker, J., 601, *605*
Beer, G. de, 572, *575*
Behnke, O., 265, *268*
Bellan, G., 54, 59, 62, *63*
Benditt, E. P., 276, *286*
Bensch, K., 335, *342*
Benson-Evans, K., 23, 27, *36*
Bent, A. C., 517, *519*
Berthois, L., 143, *150*
Bierther, M., 268, *268*
Bigey, F., 379, 381, *381*
Black, R. M., 448, *459*
Blackwell, P. W., 603, *605*
Blaxter, J. H. S., 268, *269*
Boardman, R. S., 212, *219*, 231, 235, 240, 242, 243, 244, 245, *246*, 386, 392, 394, *408*, 434, *435*, 447, 453, *459*, 549, *563*, 581, *584*
Bobin, G., 210, *219*, 572, *575*
Bock, K. J., 22, *34*
Bock, P., 265, *268*
Bocquet-Vedrine, J., 343, *347*
Bonham-Carter, G. F., 551, *564*
Bonneville, M. A., 283, *285*
Borg, F., 4, *10*, *105*, 308, 309, 313, *314*, 332, *333*, 475, 476, 479, *483*, 504, *519*, *520*, 566, 571, 574, *575*
Borg, Folke, 231, 232, 235, 236, 239, 240, 241, *246*
Borg, G., 247, 250, *256*
Braem, F., 572, *575*
Braiko, V. D., 4, *10*
Bregant, D., 53, *63*
Brehmer, M. L., 125, *127*

Brien, P., 340, *342*, 550, *564*
Bristow, C. R., 456, *459*
Brock, M. A., 283, *285*
Bronstein, G., 317, 323, *326*
Brood, K., 228, *228*, 357, *359*
Brouwer, J., 139, *150*
Brown, C. J. D., 515, 517, 518, *520*, 531, *537*, 545, *547*
Brown, D. A., 361, 366, *366*
Brown, P. R., 457, *459*
Bucher, O., 268, *269*
Buchsbaum, R., 80, 81, *93*
Buge, E., *105*, 550, *564*, 566, *575*
Bulman, O. M. B., 11, 16, *20*
Bureau, L., 380, *381*
Bushnell, J. H., 504, 505, 507, 508, 515, 516, 518, *520*, 532, *536*, 545, *546*
Busk, G., *105*, 437, 442, *446*
Buzas, M. A., 47, *51*

C

Calder, D. R., 125, *127*
Calvet, L., *63*, *105*, 335, *342*
Camp, C. L., 603, *605*
Cano, M., *63*
Canu, F., 361, 366, *366*, 390, *408*, 417, 418, *419*, 437, 441, *446*, 577, *584*
Carleton, H. M., 340, *342*
Carver, R. E., 289, *298*
Casey, R., 456, *459*
Cavet, P., 376, 377, *381*
Cheetham, A. H., 67, *71*, 97, *105*, 212, *219*, 231, *246*, 386, 392, 394, *408*, 447, 453, 458, *459*, *460*, 549, 551, *563*, *564*, 581, *584*
Chupakhin, M. S., 81, *94*
Clark, P. J., 170, *172*
Clayton, R. N., 79, 81, 92, *93*, *94*
Cock, A. G., 389, *408*
Cogné, J., 377, *381*
Colman, J., 21, *34*
Condra, G. E., 271, *274*, 327, *333*
Conolly, J. R., 361, *367*
Cook, M., 343, *347*
Cook, P. L., 28, *34*, 65, 70, *71*, 162, *163*, *288*, 297, *297*, 386, 392, 393, *408*, 453, *459*, *460*, 549, *563*, 581, *584*
Copeland, D. E., 268, *269*
Copeland, E., 268, *268*
Correa, D., 597, *599*
Cory, R. L., 125, *127*
Couffon, O., 379, 380, *382*
Cox, L. R., 450, *460*
Craig, H., 79, 85, *93*
Crisp, D. J., 28, *34*
Croasdale, H., 22, 30, *35*, 42, *51*
Crosnier, A., 143, *150*
Cuffey, R. J., 550, 551, 553, 557, *564*
Cummings, E. R., 425, *427*
Cutler, J. F., 257, 258, 259, *260*

D

Dahl, E., 517, *520*
Daimling, O. von, 265, *269*
Dansgaard, W., 86, *93*
Davenport, C. B., 504, *520*
David, H., 279, 283, *285*
David, L., *105*
Dean, D. A., 235, *246*, 287, *298*
Degens, E. T., 79, 81, 92, *93*
Delhanty, J. E., 23, 27, *36*
Desmazières, O., 379, 380, *382*
Dill, R. F., 150, *150*
Dohrn, A., 200, *207*
Dollfus, G., 379, 380, *382*
Domke, W., 192, *197*
d'Orbigny, A., 390, *408*, 447, 448, *460*
Downing, F. S., 159, *164*
Ducasse, O., 392, *408*
Duff, M. M., 437, *446*
Dunaeva, N. N., 4, *10*
Dybowski, W., 2, *10*

E

Easton, W. H., 604, *605*
Ebling, F. J., 26, 27, 28, 29, *35*, *36*
Eggleston, D., 30, *34*, 108, *113*
Ehlers, E., 261, *269*

Elias, M. K., 271, *274*, 327, *333*
Elliot, D. G., 517, *520*
Elliott, J. M., 166, 167, *172*
Ellis, B. F., 603, *605*, *606*
Ellis, J., 275, *285*
Emiliani, C., 80, 81, 86, *93*
Emschermann, P., 268, *269*
Epstein, S., 79, 80, 81, 82, 86, 87, *93*, *94*
Erben, H. K., 376, *382*
Evans, E. C., 170, *172*

F

Ferronnière, G., 376, *382*
Fierro, G., 53, 62, *63*, *64*
Fisk, D., 23, 27, *36*
Fitzjarrel, A. T., 268, *268*
Flor, F. D., 411, 412, 413, 414, 415, 416, *416*
Florkin, M., 11, *20*
Franzen, A., 598, *599*
Fraser, D. S., 340, *342*
Fraser, J. H., 162, *163*
Frey, R. W., 463, *472*
Friedman, G. M., 47, *51*
Fry, W., 200, *207*

G

Galloway, J. J., 425, *427*
Garant, Ph. R., 268, *270*
Garcia-Romeu, F., 268, *269*
Gautier, Y. V., 54, *64*, 137, *138*
Geuze, J. J., 283, *285*
Gewerzhahen, A., 325, *326*
Ghelardi, R. J., 29, *34*
Ghiurca, V., *419*
Ghosh, N., 529, 536, *537*
Gibson, T. G., 47, *51*
Giresse, P., 143, *150*
Goldfuss, G. A., 185, *197*
Greig-Smith, P., 166, 167, 168, *172*
Goldsmith, M., 279, 283, *285*
Gooch, J. L., 589, 590, *592*
Grancarova, T., 504, *520*, 541, *546*
Graupner, H., 323, *326*
Gregory, J. W., 447, *460*
Gross, M. G., 80, *93*
Gruet, M., 377, *381*
Gupta, B. L., 337, *342*

H

Håkansson, E., 392, *408*
Hagerman, L., 22, 28, 30, *35*
Hall, J., 330, *333*
Hancock, A., 539, *546*
Hancock, J. M., 466, *472*
Harb, J. M., 268, *269*
Harmelin, J. G., *105*
Harmer, S. F., *105*, 211, 212, *219*, 261, *269*, 385, *408*, 486, 488, 489, 490, *495*, 571, *575*
Hartley, H. O., 167, *172*
Hastings, A., 371, 372, *372*
Hastings, A. B., 488, *495*, 542, 543, 545, *546*
Hayes, R. M., 601, *605*
Hayward, P. J., 166, 169, *172*, 453, *460*
Helfer, H., 200, *207*
Heller, C., *105*
Henderson, R., 340, *342*
Hennig, W., 573
Herold, R. C., 283, *285*
Herwig, 566, *575*
Hiller, S., 318, *326*
Hillman, R. E., 343, *347*
Hilton, W. A., 261, *269*
Hincks, T., 41, *51*, *105*, 173, 175, *183*, 366, *366*, 437, 440, 442, *446*
Hinde, G. J., 429, *435*
Hinds, R. W., 235, *246*, 287, *298*
Hodgson, W. A., 80, *93*
Holliday, F. G. T., 268, *269*
Hopps, H. C., 603, 604, *606*
Houbolt, J. J. H. C., 139, *150*
Hourdry, J., 283, *285*
House, M. R., 456, *460*
Houston, A. H., 268, *270*
Hozawa, S., 542, 545, *547*
Hutchins, L. W., 42, *51*
Hyde, P. J. W., 14, *20*

Hyman, H. L., 340, *342*
Hyman, L. H., 211, *219*, 275, *285*, 357, *359*, 550, *564*, 585, 592, 596, 597, *599*

I

Iizima, H., 158, 162, *163*
Illies, G., 309, 310, 311, *314*
Imlay, R. W., 456, *460*

J

Jaroshinskaya, A. M., 423, 424, 425, *427*, *428*
Jebram, D., 565, 567, 570, 572, *575*
Jenkins, M., 23, 27, *36*
Johnston, G., 95, 105, *105*
Jolicoeur, P., 390, *408*
Joliet, L., 21, *35*, 275, *286*
Jones, D. J., 557, *564*
Jullien, J., 540, 541, 545, *547*

K

Kaplovsky, A. J., 117, *127*
Kawahara, T., 158, 162, *163*
Keith, M. L., 80, 92, *93*
Kellicott, D. S., 508, *520*
Kessel, R. G., 268, *269*
Kitching, J. A., 26, 27, 28, 29, *35*, *36*
Klingebiel, A., 130, *138*
Kluge, G. A., 2, *10*, 105, *106*
Kluge, H., 371, *372*
Knight-Jones, E. W., 31, *35*, 108, 112, *113*, 166, *172*, 182, *183*
Koehler, J. K., 337, *342*
Komnick, H., 265, 268, *269*, *270*
Kozlowski, R., 11, 15, *20*
Kraepelin, K., 504, *520*, 531, 533, *536*, 541, 543, *547*, 566, *576*
Krinsley, D. H., 14, *20*
Kulshrestha, K. S., 529, 533, *536*

L

Labracherie, M., 129, 132, *138*, 392, *408*
Lacombe, D., 343, *347*
Lacourt, A. W., 504, *520*, 542, 543, 544, 545, 546, *547*
Lagaaij, R., 54, *64*, 309, *315*, 392, 403, *408*, 453, *460*
Lang, W. D., 453, *460*
Lanzing, W. J. R., 268, *269*
Lardeux, H., 376, 377, 380, 381, *381*, *382*
Larwood, G. P., 361, 362, 364, *366*, 447, 452, 453, *460*, 463, 466, *472*, *473*
Lasker, R., 268, *269*
Leach, E. H., 340, *342*
Lebesconte, P., 379, 380, *382*
Le Calvez, Y., 143, *150*
Lecoimtre, G., 418, *419*
Le Maître, D., 376, *382*
Lendrum, A. C., 340, *342*
Lennep, E. W. von, 268, *269*
Levinsen, G. M. R., 441, 442, *446*, 489, *495*
Lewis, J. R., 26, *35*
Liguori, V. R., 343, *347*
Lilly, S. J., 27, 29, *35*, *36*
Linnaeus, C., 442, *446*
Little, C., 337, *342*
Lloyd, R. M., 80, 81, 86, 92, *93*
Long, J., 597, 598, *599*
Loosanoff, V. L., 28, *35*
Löve, D., 517, *520*
Lowenstam, H. A., 79, 80, 81, 87, *93*, *94*
Luft, J. H., 335, *342*
Lutaud, G., 317, *326*
Lynch, W. F., 160, *163*
Lys, M., 376, *382*

M

McCrea, J. M., 82, *94*
McDougall, K. D., 127, *127*, 158, *163*
MacGillivray, P., 370, *372*
MacGillivray, P. H., 361, *367*
MacIntyre, R. J., 361, *367*
Mackie, G. O., 385, *408*
McKinney, C. R., 79, 80, 81, 82, 87, *94*
McKinney, F. K., 235, *246*
Makino, S., 586, *592*

Manzoni, A., 417, 418, *419*
Maplestone, C. M., 361, *367*
Marcus, E., 176, *183*, 325, *326*, 332, *333*, 504, *520*, 542, 543, *547*, 571, *576*
Masoni, A., 268, *269*
Masse, J. P., 145, *151*
Matricon, I., 276, 277, *286*
Maturo, F. J. S., 28, *35*, 97, *106*, 261, *269*, 296, 297,
Maturo, F. J. S., Jr., 42, *51*, 125, 126, *127*, 577, 578, 584, *584*
Mawatari, S., 157, 158, *163*
Mayeda, T., 79, 86, *93*, *94*
Mayr, E., 551, *564*
Meadow, N. D., 283, *285*
Meissner, M., 543, *547*
Menon, N. R., 153, 154, 155, 160, 161, 162, *163*
Messina, A. R., 603, *605*, *606*
Michener, C. D., 551, *564*
Miller, M. C., 32, *35*
Milne, A., 162, *163*, 166, *172*
Milne-Edwards, M. H., 309, *315*
Milon, Y., 376, *382*
Minkoff, E. C., 551, *564*
Miyaku, S., 158, *163*
Modzalevskaya, E. A., 422, 423, *427*
Molinier, R., 54, 59, 62, *63*
Mongereau, N., *105*, *106*
Monis, B., 265, *269*
Moore, P. G., 24, 26, 27, *35*
Moreau-Benoit, A., 377, *382*
Morgan, G., 23, 27, *36*
Morozova, I. P., 4, *10*, 327, 330, 332, *333*
Morrison, D. F., 389, *408*
Morrison, J. L., 603, *605*, *606*
Mosimann, J. E., 390, *408*
Moyano, H., 369, *373*
Moyano, H. I., 182, *183*
Moyse, J., 31, *35*, 108, 112, *113*, 166, *172*, 182, *183*

N

Nace, P. F., 343, *347*
Naidin, D. P., 81, *94*
Nair, N. B., 153, 154, 157, 158, 161, 162, *163*, *164*
Nair, N. U., 153, *164*
Nekhoroshev, V. P., 424, 425, *427*
Neviani, A., *106*
Nichols, D., 597, *599*
Nicholson, H. A., 2, *10*, 433, 434, 435, *435*, *436*, 475, *483*
Nickles, J. M., 433, 435, *436*
Nielsen, C., 566, *576*
Nielsen, Claus, 234, *246*
Nier, A. O., 82, *94*
Nikiforova, A. I., 332, *333*
Nitsche, H., 275, *286*
Norman, A. M., *106*
Nye, O. B., 255, *256*, 287, *298*, 299, *306*
Nye, O. B., Jr., 235, *246*

O

Oda, S., 517, *521*, 523, 525, 526, 527, *527*
O'Donoghue, C. H., 437, *446*
O'Donoghue, E. O., 437, *446*
Oehlert, D., 379, 380, *382*
Oehlert, P., 379, 380, *382*
Oka, A., 532, *536*
Opdyke, N. D., 457, *460*
Osburn, R. C., 41, 42, *51*, 105, *106*, 216, *219*, 437, 441, 443, 445, *446*, 485, *495*, 516, *521*, *576*, 577, *584*
Owen, G., 108, *113*, 277, *286*

P

Pace, R. M., 597, *599*
Parker, P. L., 85, *94*
Parker, R. H., 92, *93*
Parry, G., 268, *269*
Peake, N. B., 466, *472*
Pearson, E. S., 167, *172*
Péneau, J., 376, 378, 379, 380, 381, *382*
Percival, E., 598, *599*
Pérès, J. M., 59, *64*, 96, *106*
Perry, T. G., 430, *435*

Petřik, P., 268, *269*
Pettengill, O., 268, *269*
Philippot, A., 376, 377, *381*
Phillips, J., 381, *382*
Philpott, C. W., 268, *269*
Picard, J., 54, 59, 62, *63*, *64*, 96, *106*
Pielou, E. C., 168, 170, *172*
Pillay, K. K., 161, *164*
Pillet, J., 376, 378, 379, 380, *382*, *383*
Pinter, P., 441, 443, *446*
Plaziat, J. C., 392, *408*
Počta, P., 381, *383*
Podestà, G., 54, *64*
Pohawsky, R. A., 570, *575*
Powell, H. T., 159, *163*
Powell, N. A., 41, *51*, 393, *408*
Prell, H., 200, *207*
Prenant, M., 21, *35*
Prud'homme, J., 129, *138*,
Pyefinch, K. A., 159, *164*

R

Rao, K. S., 529, 533, 536, *536*, *537*
Raup, D. M., 80, *94*
Redier, L., 369, *373*
Reid, W. E., 603, *606*
Remane, A., 574, *575*
Reuss, A. E., 417, 418, *419*
Reyment, R. A., 390, *408*
Reynolds, E. S., 276, *286*, 335, *342*
Rickards, R. B., 14, *20*
Ridley, H. N., 516, *521*
Ridley, S. O., 544, *547*
Riley, G. A., 48, *51*
Robertson, A., 437, 441, 442, *446*
Roemer, F., 381, *383*
Rogick, M., 369, 371, *373*
Rogick, M. D., 22, 30, *35*, 42, *51*, 504, 517, *521*, 523, 525, 526, *528*, 531, 532, *537*, 545, *547*
Rondeau, E., 379, 380, *383*
Roper, R. E., 23, 33, *35*
Ross, J. P., 2, *10*
Round, F. E., 28, *35*
Rousselet, Ch. F., 544, *547*
Rowell, A. J., 551, *564*
Rubinson, M., 79, *94*
Rucker, J. B., 289, *298*
Rudwick, M. J. S., 597, *599*
Rudzinska, M. A., 283, *286*
Ryland, J., 441, *446*
Ryland, J. S., 22, 28, 29, 32, *35*, 127, *127*, 159, 160, 161 *164*, 166, 168, 169, *172*, 173, *183*, 209, *219*, 275, *286*, 457, *460*, 549, 550, *564*, 567, 569, *575*, *576*, 597, *599*

S

Sabatani, D. D., 335, *342*
Sandberg, P. A., 289, *298*
Sanders, H. L., 50, *51*
Sanders, J. E., 47, *51*
Scarratt, D. J., 22, 29, *36*
Schäfer, A., 268, *269*
Scharrer, B., 283, 285, *286*
Scheer, B. T., 126, *127*
Schipp, R., 268, *269*
Schneider, J. H., 601, *606*
Schopf, T. J. M., 97, *106*, 261, *269*, 407, *408*, 577, 578, *584*, 589, 590, *592*, 601, *606*
Schopf, T. J. M., 190, 196, *197*
Scott, R. W., 447, *460*
Seed, R., 30, *36*
Seiglie, G. A., 145, *151*
Serre, B., 376, *382*
Sewertzoff, A. N., 570, 572, *576*
Shaw, N. G., 463, *473*
Shishova, N. A., 330, 333, *333*
Shulga-Nesterenko, M. I., 328, 332, 333, *333*
Siewing, R., 574, *576*
Silén, L., 165, *172*, 393, *408*, 457, 458, *460*, 565, 566, 570, 572, *576*, 585, *592*
Skellam, J. G., 170, *172*
Skobe, Z., 268, *270*
Sleigh, M., 339, *342*
Slidders, W., 340, *342*
Sloane, J. F., 27, 28, 29, *35*, *36*
Smith, S. T., 23, 27, *36*

Smith, W. E., Jr., 603, *605*
Smitt, F. A., 105, *106*
Smuckler, A. E., 276, *286*
Sneath, P. H. A., 551, *564*
Sokal, R. R., 551, *564*
Soule, D., 437, 443, *446*, 485, *495*
Soule, J., 437, 443, *446*, 485, *495*
Spicer, S. S., 336, *342*
Srivastava, I. P., 403, *409*
Stach, L. W., *64*, 193, *197*
Stebbing, A. R. D., 30, 31, *36*, 166, 170, *172*, 173, 181, *183*, 567, *576*
Steedman, H. F., 108, *113*, 336, *342*
Stickney, A. P., 42, *51*
Stocchino, C., *63*
Stringer, L. D., 42, *51*
Sudbury, M., 16, *20*
Swennen, C., 32, *36*

T

Tarutani, T., 79, *94*
Tavener-Smith, R., 7, *10*, 209, *219*, 228, *229*, 253, *256*, 257, 258, *260*, 271, *274*, 349, 355, 357, *359*, 567, 569, *576*
Taylor, D. L., 337, 339, *342*
Teissier, G., 21, *35*
Tewari, B. S., 403, *409*
Theis, R. V., 81, *94*
Thomas, H. D., 452, 453, *460*, 463, 466, *473*
Thompson, T. E., 23, 27, 32, *36*
Thorson, G., 127, *128*
Thorsteinsson, R., 19, *20*
Tice, L. W., 265, *270*
Tokuoka, Y., 158, *164*
Tommers, F. D., 28, *35*
Toomey, D. F., 603, *606*
Toots, H., 251, *256*
Toriumi, M., 504, 515, *521*, 531, 532, 533, *537*, 540, 541, 542, 543, 544, 545, 546, *547*
Tracey, J. L., Jr., 80, *93*
Traverse, A., 603, *606*
Threadgold, L. T., 268, *269*, *270*
Tromelin, G. de, 379, 380, *382*
Trotti, L., 53, *64*
Trump, B. F., 276, *286*

U

Ubaghs, G., 376, *382*
Ubaghs, J. G., 185, 186, *197*
Ulrich, E. O., 2, *10*, 332, *333*, 425, *427*
Ulrich, W., 574, *576*
Urbanek, A., 16, *20*
Urey, H. C., 79, 80, 81, 82, 87, *93*, *94*
Ursin, E., 26, *36*
Utgaard, J., 434, *435*

V

van der Schalie, H., 532, *537*, 545, *547*
Veillon, M., 130, *138*
Verhoeff, F. H., 340, *342*
Verrill, A. E., 48, *51*
Vigano, A., 507, *521*
Vigneaux, M., 418, *419*
Vinassa de Regny, P., 435, *436*
Viscova, L. A., 498, *502*
Voigt, E., 185, 186, 190, 192, *197*, 311, *315*, 393, *409*, 411, 412, 413, *416*, 447, 448, 458, *460*, *461*, 463, 466, 467, 469, *473*, 501, *502*, 570, *576*
Von Hagenow, F., 185, *197*

W

Wade, A. E., 23, 27, *36*
Walker, G., 343, *347*
Walliser, O. H., 376, *382*
Walter, B., 310, *315*
Wass, R. E., 361, *367*
Waters, A. W., 4, *10*, *106*, 236, *246*, 248, *256*, 488, *495*
Weber, J. N., 80, *93*, *94*
Weller, J. M., 550, *564*
Wichard, W., 268, *270*
Wiebach, F., 504, 505, 507, *521*, 523, *528*, 532, 535, *537*, 542, *547*
Williams, A., 209, *219*, 349, 357, *359*
Williams, G. B., 28, *34*
Wilson, E. O., 401, 407, *409*

Wilson, L. R., 603, *606*
Wisely, B., 108, 112, *113*, 168, 169, *172*
Wood, T., 508, *520*
Wood, T. S., 532, *536*
Woodinard, H. B., 456, *461*
Woollacott, R. M., 597, *599*

Z

Zambrano, D., 265, *269*
Zelander, T., 265, *268*
Zimmer, R. L., 595, 597, *599*
Zschokke, F., 504, *521*

Subject Index

A

Abiesgraptus, 19–20
Abiotic environment, 48
Aborted zooecia, 4–9, 457
Acanthocella, 366
Acanthocladia, 333
Acanthocladids, 271–274
Acanthoclema, 222–227
Acanthodoris, 32
Acanthopores, 2–9, 221–228, 253, 257–259, 349–358, 433–435
Acanthorods, 257–259
Acanthostega, 561
Acanthozoids, 257, 258
Accessory pores, 332, 333
Achelia, 200–206
Actinia, 200
Actinopora, 189–194
Actinotroch larvae, 596
aculeata, *Ophiopholis*, 26
adeliensis, *Labioporella*, 370–372
Adeonella, 89
Adeoniform, 55–60, 131–134
Adhesive glands, 261
admota, *Erkosonea*, 299–302
adpressa, *Hippopodinella*, 214
Aechmella, 187
Aetea, 43, 182, 370, 567–569
Aeverrilla, 155–159
africana, *Hippopetraliella*, 143
africana, *Hyalinella*, 506, 542–544
Afrindella, 506
agilis, *Hyalinella*, 542
agilis, *Stolella*, 507, 542
Aimulosia, 214
Aisenvergia, 73–75
alata, *Membranoptera*, 30
alata, *Poricellaria*, 390–393
Alcyonaria, 187
Alcyonidioidea, 565
Alcyonidium, 24–32, 107–113, 169–170, 177–182, 276, 553, 573
Alderina, 155–159
Algae, 21–34, 80, 96, 107, 190, 344, 441–443
Alleles, 581, 586–591
altaica, *Rhinidictya*, 427
alula, *Onychocella*, 66
Alveoli, 358
Amastigia, 370, 372
americana, *Hippoporina*, 42, 47
Ammotheidae, 200
Amphiblestrum, 40–47
Amphisolops, 339
Amphistegina, 141–150
Amplexopora, 425, 427, 434
Amplexoporoidea, 5–9
ampulla, *Corbulipora*, 361, 366
Amsassipora, 423–427
Anal budding, 565, 573
Anasca, 297, 317, 325, 369, 561
Anascina, 553–561
Ancestrula, 65–71, 232, 250, 271, 272, 288–297, 307–314, 353, 443, 444, 453, 567, 580–583
ancorata, *Anexocona*, 214
Anexocona, 214
anglica, *Pyripora*, 467
anguina, *Aetea*, 43, 370
angulata, *Electra*, 158
angulata, *Laomedea*, 205
angusta, *Flustra*, 370
angustus, *Camptoplites*, 370
Anisograptidae, 17
anisostoma, *Emballotheca*, 418
anisostoma, *Lepralia*, 417
Anisotrypa, 74, 75

Anolotichia, 423–426
anomalopora, *Ditaxia*, 189–194
Anomia, 31
Anoplodactylus, 200
antarctica, *Amastigia*, 370
antarcticum, *Farciminellum*, 370
antarcticum, *Himantozoum*, 370, 372
Antennularia, 205
Anthozoa, 411
antigua, *Fenestella*, 378–380
antigua, *Floridina*, 141
Anus, 234
apertura, *Dakaria*, 213
Apertural spines, 15
Aptyxiella, 450
arabiensis, *Alderina*, 155–159
Arachnoidia, 507
Arachnopusia, 366
Araxoporidae, 8, 9
arctica, *Umbonula*, 215
arcuata, *Watersipora*, 210–214
arenicola, *Ascorhynchus*, 200
Areolae, 209–219
argentus, *Monograptus*, 15
arnaudi, *Flustra*, n. sp., 370, 371
Arthrophragma, 350, 359
Arthropoma, 213
Arthyris, 376
articulata, *Microporina*, 83, 84
articulata, *Paludicella*, 505–518
articulata, *Poricellaria*, 392
Ascidians, 26, 182
Ascodictyon, 431, 433
Ascophora, 74, 210–218, 561
Ascophorina, 553–561
Ascorhynchus, 200
Ascus, 210–218, 276
asperrima, *Hippodiplosia*, 417
asperrima, *Lepralia*, 417
asperum, *Leioclema*, 257
Associations, 199–207
Astogeny, 12–19, 66–71, 232, 250, 271–274, 287–297, 307–314, 352–355, 365–366, 392, 444, 465–471, 479–482, 488, 497, 565–567
astrova, *Amplexopora*, 427
astrovae, *Flustra*, n. sp., 370, 372
Atactotoechidae, 9
Atactotoechus, 435
atlantica, *Idmonea*, 95–106
Atrium, 234, 239
Attachment, 125–127, 186, 193
Attachment organs (internal), 231–235
auricomis, *Hyalinella*, 542, 543
auricomis, *Plumatella*, 542, 543
auriculata, *Schizomavella*, 143, 213
aurita, *Callopora*, 25, 26, 42–47
Australella, 543, 544
australiensis, *Fredericella*, 505, 506
Austrodecus, 200
Autosyndrome, 182
Autothecae, 11–16
Avenella, 553
Avicularia, 1, 68, 202–206, 213–218, 330, 366, 370–372, 285–405, 448, 485, 486, 580, 581
avicularis, *Cellepora*, 43

B

Bacterial inhibitors, 29
baigoliensis, *Homotrypa*, 427
bajocensis, *Stomatopora*, 310
Barentsia, 261–266, 553
Barentsidae, 261
barrandi, *Chaetetes*, 435
barrandi, *Monticulipora*, 435
Basal diaphragm, 299
Batopora, 55
Batostoma, 422–427, 434
Beania, 370–371
Beisselina, 193–194
Beisselinopsis, 193
Belemnitella, 466
Benedenipora, 565–573
bengalensis, *Electra*, 155–162
beralicus, *Stenodiscus*, 75
Berenicea, 27, 173–175, 187
Bicellaria, 43
bicornis, *Camptoplites*, 370–371
bicornis elongata, *Camptoplites*, 371
bicornis ventricosa, *Camptoplites*, 371

bicornis compacta, *Camptoplites*, 371
bicornis elatior, *Camptoplites*, 371
bicornis magna, *Camptoplites*, 371
bicornis quadriavicularis, *Camptoplites*, 371
bicuspidata, *Obelia*, 126
bifidus, *Didymograptus*, 422
bifurca, *Fenestella*, 378, 381
bifurca, *Stomatopora*, 310
bigemmis, *Hyalinella*, 543
bilabiata, *Eurystomella*, 215–216
bilix, *Stomatopora*, 309–310
Bioadhesives, 343–344
biporosa, *Cupuladria*, 287–297, 350
bischofi, *Fenestella*, 378
Bithecae, 11–16
Body walls, 231–245
bolinae, *Stephanosella*, 214
bohemica, *Ptilopora*, 381
bohomica, *Hemitrypa*, 381
boloniana, *Retepora*, 379–380
boreala, *Kukersella*, 248
borealis, *Spirorbis*, 108, 112, 168
boryi, *Caberea*, 54–58, 141
bosqueti, *Thalassocharis*, 190–192
Bossiella, 444
Botryllus, 182
Bowerbankia, 42, 118–126, 155–159, 200, 565
Brachiograptus, 18
Brachiopoda, 80, 574, 593
brevimandibulata, *Mawsonia*, 370
brongniartii, *Chorizopora*, 42, 47
Brooding, 4, 65–69, 232
Brown bodies, 275–286, 433
brunnae, *Celleporaria*, 215
brunnae, *Pachyegis*, 215
Bryozoa as food, 199
Budding, 2–7, 12–17, 212, 245, 250, 255, 293, 296, 307–315, 330, 332, 357, 366, 392, 438, 443–444, 448–458
Bugala, 43, 118–127, 200, 205–206, 155–161, 589–591
bulgarica, *Internectella*, 507, 515
burmanica, *Pectinatella*, 529
buskii, *Steganoporella*, 65

C

Caberea, 54–58, 141
Calcareous rods, 253
Calcification, 5, 8, 65–71, 87, 192, 211, 216, 239–244, 251–253, 257–259, 271–274, 288–297, 299–306, 330–332, 349–358, 361–372, 390, 452–453, 475–482, 570
californica, *Diaperoecia*, 299–302
californica, *Hippopodina*, 215
californica, *Parasmittina*, 214
Callopora, 24–28, 42, 47, 189, 370, 458, 572
Calpensia, 55
Calvetiidae, 475
Calyssozoa, 554–557
cambodgiensis, *Hislopia*, 507
Campanularia, 283
Camptonectes, 450
Camptoplites, 370–372
canalifera, *Vincularia*, 193
canariensis, *Cupuladria*, 287, 297
Cancellata, 479
cancellata, *Crisina*, 194
capensis, *Lophopodella*, 507
Capillaries, 7
capillosa, *Fenestella*, 378, 381
Carbon ratios, 89–92
Cardia, 234
carinata, *Crisisina*, 193, 194
Carinopora, 433
Carnosita, 553–556, 560
carteri, *Lophopodella*, 505–507, 515–517, 523–527, 529–530, 534
carvalhoi, *Hyalinella*, 506, 543
casina, *Venus*, 59, 62
casmiana, *Plumatella*, 506–508, 515, 529–532, 574
Catalogue systems, 603
catenata, *Benedenipora*, 565, 570–573
Catenicelliform, 54–60
catenularia, *Pyripora*, 47
Cavaria, 192
Cavarinella, 194, 497, 498
Caverns, 328–330

cavifera, Fenestella, 328
cavolini, Eunicella, 96
cecili, Arthropoma, 213
Ceidomonea, 193
Cellaria, 54–58, 141, 370, 371
Cellariiform, 54–63, 130–136
Cellarinella, 200
Cellepora, 25–27, 43
Celleporaria, 26, 417
Celleporella, 24–32, 350
Celleporiform, 56–60, 193
Celleporina, 83, 84, 88, 215
Celleporinidae, 215
cellulariform, 43
cellulosa, Sertella, 54–58
Cementing substances, 343
Cephalization, 594
Cephalodiscus, 13
Ceramopora, 423–426
Ceramoporidae, 476
Ceriopora, 193
cernua, Pedicellina, 261
Chaetetes, 435
Chaperia, 370–372
Cheiloctenostomata, 559
Cheiloporina, 214, 418
Cheiloporinidae, 214
Cheilostomata, 369, 447, 448, 457, 463, 560, 567, 572
Cheilostomida, 554–556, 560
Chorizopora, 42, 47
Chromosomes, 585
cilia, 65, 66, 263–266, 335–341
ciliata, Bicellaria, 43
ciliata, Microporella, 25, 26, 42, 118–121, 125, 126, 213
Cirripedes, 31, 108, 166, 343
circinata, Arthropoma, 213
Cladia, 16–20
Classification, 549
clathrata, Retecava, 194
Clausotrypa, 74, 75
clavata, Paramuricea, 96
claviculata, Trypostega, 213
claviformis, Neofungella, 235, 236
Cleidochasma, 143
Clinopora, 501, 502
Clonograptus, 18
Clonopora, 248–255
Clytia, 29
coarctum, Myriazoum, 87, 92, 215
coccineus, Escharoides, 24–28
cochinensis, Schizoporella, 155–157
Codonellina, 485
Coeloclema, 553
Coelocochlea, 194
Coelocyst, 353
Coelom, 7, 69, 70
Coenoecium, 13
Coleopora, 215
collaris, Corbulipora, 366
collifera, Parasmittina, 214
Colony form, 54–63, 99–102, 129–138, 187, 411–416, 477–482
Colony overgrowth, 173–183
commissuralis, Obelia, 126
Communication organs (see septulae), 209–211, 218, 251–255, 288, 289, 294, 295, 455–457
communis, Heterocrisina, 194
commutabilis, Anolotichia, 423–426
Competition, 31, 75, 101, 108, 166, 173–183, 567
complicata, Poricellaria, 392
Composite pores, 498
compressa, Plagioecia, 194
compressa, Porella, 89
Computer based systems, 603
concentrica, Nikiforopora, 74, 75
Conescharellina, 89, 211
Conescharelliniform, 131
Conopeum, 118–126, 155–157, 162, 450, 453, 566, 571, 572
consimilis, Sagenella, 251–254
Constellaria, 257–260, 423, 424, 427
Copidozoum, 141
cornuta, Hippothoa, 438–445
Corallium, 96
Corbulipora, 361, 365, 366
corderoi, Hislopia, 507
Cornucopina, 370–372
cornuta, Cribrilina, 361

cornuta, Dimetopia, 370
cornuta, Hippothoa, 213, 216
coronata, Doto, 32
Coronotrypa, 248–253
Cortical layer, 13, 14
Coscinopleura, 194, 393
Costae, 362–366
costata, Petalopora, 498
costazi, Celleporina, 215
Costazzia, 418
Costules, 361
crassa, Celleporaria, 417
crassa, Costazzia, 418
crassa, Heteropora, 193
crassa, Rhombopora, 222, 227
Crassimarginatella, 56
Crassostrea, 28
craticula, Callopora, 42, 47
Crepidacantha, 215
Crepidacanthidae, 215
cribosa, Microporella, 213
Cribrilaria, 55, 143
Cribilina, 370, 42, 47, 54, 55, 361, 366, 553
Cribrilinidae, 361
Cribrimorpha, 187, 190, 561
Cribrimorphina, 553–556, 561
Crisia, 25, 26, 54–58, 89, 141, 165, 170, 171, 182, 200, 350, 569, 572
Crisina, 193, 194, 497, 498
Crisinidae, 475, 498
cristata, Sulcocava, 498
Cristatella, 505–507, 515–518, 525–527, 553
Cristatellidae, 507
Crisisina, 193, 194
Crisulipora, 553
crosslandi, Parasmittina, 214, 486–489
Crustopora, 74
Crustoporidae, 5, 9
crustulenta, Electra, 155–162
Cryptocyst, 65, 448–453
Cryptostomata, 75, 245, 253, 302, 327, 350, 561, 569, 570
Cryptostomida, 554–556, 561
Cryptosula, 160, 210, 214, 275, 344
crystallinus, Lophopus, 505, 507
Ctenostomata, 2, 42, 317, 326, 365, 433, 565, 569–572
Ctenostomida, 554–556
Ctenostomina, 554–556, 560
cucullata, Bugula, 155–157, 160, 161
cucullata, Watersipora, 158, 162
Cucullograptus, 16, 17
Cumulipora, 418
Cupuladria, 89, 287–297, 350–356
Cupuladriidae, 287–298
Curtaulata, 554–557, 561
curva, Flustra, 370
curvirostrata, Rhamphstomella, 214
cuvieri, Stamenocella, 193
Cuticle, 231–235, 250–253, 261–263, 271–274, 288–291, 294–297, 346, 347
Cycloperiella, 215
Cyclostomata, 2, 5, 42, 56, 90, 95–105, 187, 193, 231, 247–256, 303, 369, 447, 476, 497, 501, 553–556
cyclostoma, Onychocella, 189, 196
Cyclotrypa, 75
cylindrica, Aisenvergia, 73, 75
Cyrtograptus, 19, 20
Cysticella, 88, 89
Cystiphragms, 239–243, 355, 358, 359, 434
Cystoporata, 75, 245, 350, 357, 358, 423, 424, 255
Cystopores (see mesopores)
Cystoporina, 553–556

D

Dakaria, 213
Decapods, 187
Dekayia, 2, 9
Delessaria, 30
delicatula, Parasmittina, 486–493
Dendrograptus, 14
Dendroid graptolites, 12–17
Dendroidea, 16, 17
dennanti, Paramawsonia, 370, 371
Densipora, 476
Denticles, 438, 440

denticulata, Crisia, 25, 26, 165
Depth variation, 50
Deuterostomia, 593, 594
Dianulites, 422–427
Diaperoecia, 299–302
Diaphragms, 5–8, 254, 279, 299–306, 320–325, 355, 358, 433, 434
Diastopora, 25, 26, 54, 55, 248, 255
Diastoporidae, 498, 501
Dicalycal theca, 17, 18
dichotoma, Grammascosoecia, 194
dichotoma, Obelia, 126
dichotoma, Onychocella, 194
dichotoma, Stomatopora, 310
dichotomoides, Stomatopora, 310
Dictyonema, 14
Didymograptus, 422
Dietellae, 448–458
digitata, Laminaria, 168
Dimetopia, 370
Dimorphocella, 55
Dimorphostylus, 193
Dipleura larvae, 594, 596
Diploclema, 248
diplodidymioides, Poricellaria, 392
Diploporaria, 75
Diplosolen, 55, 332
Diplotrypa, 2, 422–427
Discoporella, 297
Discoporellidae, 476
discreta, Barentsia, 261, 263
Discritella, 7, 74, 75
dispersa, Pyripora, 469
Displaced fauna, 139
Disporella, 173, 175
distans, Hippothoa, 213, 216
Dissepiments, 13
dissimilis, Coronotrypa, 253
disticha, Actinopora, 189–194
distincta, Volnovachia, 73, 75
Distribution, 22–25, 115–127, 153–163, 165–171, 375–381, 503–514
Ditaxia, 189–194
Ditrupa, 190
Dittosaria, 418
divaricata, Hippothoa, 43
Diversity, 37–51
Diversograptus, 19, 20
DNA, 589
doma, Cupuladria, 287, 297
donaica, Fenestella, 74, 75
donbassica, Nikiforovella, 74
dorsata, Idmonea, 189, 193
Doto, 32
Double polypides, 523
Drosophila, 589
drygalskii, Notoplites, 370, 372
duboisii, Stylopoma, 65–71
Dynamena, 200
Dyscritellidae, 5–9

E

eburnea, Crisia, 170, 171, 182
echinata, Achelia, 200–206
echinata, Klugella, 370, 372
Echinella, 505, 507
Echinodermata, 31, 189, 574, 598
Echinus, 31
Ecological studies, 37–208, 475–483, 503–537
Ectoprocta, 554–558, 585
edulis, Mytilus, 30, 344
Electra, 24–31, 42, 137, 155–162, 167, 168, 175–181, 200, 317–326, 553, 566, 572
Electridae, 317–326, 463
Electrophoresis, 587, 588
elegantissima, Filisparsa, 417, 418
elegantula, Hallopora, 2
elevata, Cribrilina, 361, 366
emarginata, Plumatella, 505–519, 529, 530, 544
Emballotheca, 213, 418
Embryo extrusion, 65–69
Embryology, 595
Embryos, 1–4, 112, 254
emplastra, Hippothyris, 214
Enantiosula, 214
Endeis, 200, 205
Endozone, 5, 7, 259
Ensipora, 424–426

Entalophora, 54–58, 137, 193, 248, 350, 411–413
Entoprocta, 261–270, 554–557
Entosaccal cavity, 234
Epidermis, 231, 271
Epiphyte zonation, 29
Epistome, 594
Epitheca, 211
equina, *Actinia*, 200
erecta, *Ensipora*, 424
erecta, *Pottsiella*, 507
Eridopora, 75
Eridotrypellidae, 9
Erkosonea, 299–302
errata, *Schizoporella*, 587–591
Eschara, 379, 380, 417, 418
Escharifora, 194
Eschariform, 132, 193
Escharina, 25–27, 143
Escharoides, 24–28
esculentus, *Echinus*, 31
esperi, *Virgocella*, 193
etaformis, *Brachiograptus*, 18
esthonia, *Goldfussitrypa*, 225, 227
Eucheilostomata, 570
Euctenostomata, 570
Euginoma, 139, 141
Eunicella, 96
Eurystomata, 554–559, 570–573, 596
Eurystomella, 215, 216
Eurystomellidae, 215
Euteleia, 143–150
evelinae, *Euteleia*, 143, 145
evelinae, *Stolella*, 506, 515, 542
Exidmonea, 301, 303
Exilapores, 4–9
exilis, *Fenestella*, 381
Exosaccal cavity, 234, 239
Exozone, 2–7
expansa, *Cheiloporina*, 418
expansa Eschara, 417
Expletocystata n. infracl., 554–562
Expletocystida n. ord., 554–562
Expressivity, 16
extensalata, *Mawsonia*, 370
Extrusion of embry 65–69

F

fallacis, *Megacanthopora*, 76, 227
fallax, *Petalopora*, 194
Farciminellum, 370, 372
Farella, 565
Fauna lists, 23–34, 40, 41, 54–58, 73–77, 146–147, 155, 194, 375, 379
Faunal relationships, 30
faxensis, *Pyripora*, 466
Feeding, 1, 466
femoratum, *Phoxichilidium*, 200
Fenestella, 74, 75, 327–332, 355, 356, 376–381, 553
Fenestellidae, 357, 567, 569
Fenestellids, 253, 274, 376–381
Fenestelloidea, 327–333
Fenestrata, 271, 327, 562
Fenestrina, 553–562
Fenestrulina, 213
Fertilization, 165, 578, 581, 585
fibroglobosa, *Monotrypa*, 378–380
Figularia, 215, 216, 370
Figulariidae, 215
Filicrisia, 54–57
Filicrisina, 498
Filifascigera, 187
Filisparsa, 193, 301, 302, 417, 418
filix, *Truncatulipora*, 194
filosa, *Stolonicella*, 192
Fimbriopora, 422–427
Fine structure, 275
fistula, *Lacerna*, 214
Fistulipora, 75, 379, 380, 424, 553
Fistuliporidae, 476
flabellata, *Obelia*, 29
flagellata, *Flustra*, 370
flagellifera, *Menipea*, 370, 372
flemingii, *Amphiblestrum*, 40–47
flexuosa, *Diplotrypa*, 422, 426
Floats, 14
floridaeformis, *Constellaria*, 423–427
Floridina, 141
Flustra, 30, 31, 83–85, 170, 171, 200–206, 370–372

Flustrellidra, 28, 177–179, 200, 205, 335–342, 572, 573, 597
foetida, Jullienella, 65
foliacea, Flustra, 30, 31, 83–85, 170, 171, 200–206
Foraminifera, 65, 80, 139
Fouling, 115, 153–163
foveolata, Cellarinella, 200
fragilis Ophiothrix, 26
fragilis, Vincularia, 390–406
Fredericella, 503–519, 529–535, 553
Fredericellidae, 506, 518
Frontal budding, 212, 486
Frontal membrane, 68, 453
Frontal wall, 68–69, 211–218, 318–325, 361–366
frutex, Fenestella, 75
fruticosa, Diplotrypa, 426
fruticosa, Plumatella, 506–518, 529–531, 540
fruticosus, Atactotoechus, 435
Fucus, 22–30, 107, 112, 167, 169, 176–181, 335
fulva, Bugula, 125, 126
fungosa, Plumatella, 505, 506, 586
furcata, Chaperia, 370
Fusellar layer, 13–15
Fusicellaria, 193

G

gabriellae, Lacerna, 214
galeata, Chaperia, 370, 372
Gametogenesis, 47
Gas exchange organs, 261
gaultina, Rhammatopora, 447
Gelatinella, 507, 515, 545
gelatinosa, Pectinatella, 507, 515, 525
Gemellaria, 418
geniculata, Clonopora, 253
geniculata, Obelia, 29
Genotypes, 578, 587, 589
gibbera, Adeonella, 89
gibbosa, Amphistegina, 141–145
gigantea, Callopora, 370
gigantea, Coleopora, 215
gigantea, Heterodictya, 433
giganteus, Titanites, 456
Gigartina, 205
gisleni, Labiostomella, 458, 565–573
glaciale, Austrodecus, 200
Globorotalia, 139
Goldfussitrypa, 225, 227
Goniocladia, 75
Goniograptus, 18
Gonoecia, 66
Gonozooecia, 1–4, 98–104, 254, 255, 497
gorgonensis, Hippoporella, 214
gotlandica, Clonopora, 248, 251
gracile, Nymphon, 200
gracilis, Barentsia, 261–266
gracilis, Bowerbankia, 42, 118–126, 155–157
Grammascosoecia, 194
grandicella, Rhynchozoon, 215
grandiporosa, Conescharellina, 89
grandis, Pedicellina, 263
grandis, Vincularia, 402
granulata, Stomatopora, 307
Graptolithina, 11
Graptoloid graptolites, 12
Graptoloidea, 12–16
Graptozooid, 13
grayae, Prasopora, 433
gregaria, Phaenopora, 425
Growth, 16, 71, 100, 110–113, 129, 173–176, 187, 271, 303–306, 359
Growth rates, 92
gurievensis, Leptotrypella, 7
Gymnocaulus, 12, 13
Gymnocyst, 362, 366, 394, 401, 448–453
Gymnolaemata, 2, 275, 328, 349, 503–518, 554–559, 565–573

H

Habrovirgatina, n. subord., 554–562
Halimeda, 96
Hallopora, 2, 424–427
Halloporidae, 8
Halloporoidea, 5, 8

Hammia, 190
Hederelloidea, 255, 256
Hemicellaria, 498
Hemichordata, 11
Hemieschara, 417
Hemipneustes, 190
Hemisepta, 302–306, 328
Hemiseptella, 370, 372
Hemitrypa, 376–381
Heterocella, 390, 402
Heterocrisina, 194
Heterodictya, 433
Heteropora, 193, 350, 356, 475–479
Heteroporid, 232
Heteroporidae, 475, 476
Heterotrypa, 9, 434, 435
Heterotrypidae, 8, 9
Heterozooids, 1–9, 573
hexagonalis, Trematoecia, 215
Hexites, 75
Hiantoporidae, 215
hilli, Figularia, 215, 216
himalayana, Lophopodella, 529
himalayana, Stolella, 529, 533
Himantozoum, 370, 372
hina, Stephanella, 506, 515, 545
hindei, Carinopora, 433
Hippaliosina, 214
Hippocrepia, 558
hippocrepia, Phoronis, 335
hippocrepis, Lagenipora, 215
Hippodiplosia, 213, 417, 418
Hippomonavella, 214, 218
Hippopetraliella, 143
Hippopodina, 215
Hippopodinella, 55, 143, 214
Hippoporella, 214
Hippoporidra, 214
Hippoporina, 42, 47, 55, 214
Hippoporinidae, 214
Hippothoa, 42, 43, 168, 212–216, 344, 437–446
Hippothoidae, 213, 445
Hippothyris, 214
hirsutum, Alcyonidium, 28, 107–113, 169, 170, 177–182
Hislopia, 505, 507, 529–536, 566
Hislopiidae, 505, 507
hispida, Disporella, 173, 175
hispida, Flustrellidra, 28, 177–179, 200, 205, 335–342, 573
hofkeri, Fusicellaria, 193
Holoporella, 58, 88
Homosyndrome, 175–183
Homotrypa, 422–427
Homotrypella, 424
Hormones, 15
Hornera, 350
Hornerid, 271
Horneridae, 357, 475, 476
hyadesi, Membranipora, 182
hyalina, Celleporella, 24–32
hyalina, Hippothoa, 42, 43, 168, 344, 437–446
hyalina cornuta, Hippothoa, 437
Hyalinella, 506–517, 525, 529–535, 539–546
Hydractinia, 182
Hydroids, 28–32, 125, 126, 283
Hydrostatic skeleton, 341
hyperboarea, Laminaria, 23–30
Hypostegal coelom, 70, 210–216, 271, 288–297, 356

I

Ichtyorachis, 379, 380
Idmidronea, 54–58, 190, 194
Idmonea, 89, 95–106, 189, 193
iheringi, Australella, 543
iheringi, Hyalinella, 543
iheringi, Lophopus, 543
imbricata, Bowerbankia, 157, 159, 200
implicatum, Batostoma, 434
Incipient ovicells, 47
incrassata, Saffordotaxis, 222–225
incrustans, Diplotrypa, 422, 426
indica, Australella, 543, 544
indica, Hyalinella, 506, 543, 544
indica, Plumatella, 543
indica, Stolella, 506, 515, 529–535
Inferred environments, 53–63

informata, *Stylopoma*, 213
Information services, 602
Infundibulata, 559
Initial bud, 14–16
Inovicellata, 570
insculpta, *Hippodiplosia*, 213
Intercalary bud, 68
Intercalary cuticle, 212
intermedia, *Lepralia*, 417
Intermediate diaphragm, 299
Internectella, 507, 515
Interspecific competition, 31
Interzooidal communication, 177, 181, 234, 250–253
inutilis, *Batostoma*, 422, 426
Inversaria, 193
Ion regulating organs, 261
irregularis, *Pyripora*, 463–469
isabelleana, *Membranipora*, 597
Isolate theca, 17
Isotope ratios, 79

J

janthina, *Hippoporidra*, 214
javanica, *Plumatella*, 506, 508, 529–532, 544, 546
johnstoni, *Clytia*, 29
Jorunna, 32
Jullienella, 65, 66

K

karakubensis, *Fenestella*, 75
Kenozooids, 2–8, 67, 212, 293, 385, 479–482
Kerberites, 455, 456
kiaeri, *Kiaerograptus*, 13
Kiaerograptus, 13
kiowana, *Membranipora*, 447
kirchenpaueri, *Hippopodinella*, 55
Klugella, 370, 372
Koremagraptus, 14
Kukersella, 248–255
kutchensis, *Nellia*, 403

L

labiata, *Fenestella*, 330
Labioporella, 370, 372
Labiostomella, 458, 565–573
labuensis, *Fenestella*, 330
Lacerna, 214
lacinosum, *Conopeum*, 450, 453
lacustris, *Hislopia*, 505, 507, 529–535
laevis, *Endeis*, 205
Laevitrigonia, 450
lafontii, *Savignyella*, 155–159
Lagenipora, 215
Lageniporidae, 215
lamarcki, *Onychocella*, 194
lamellosus, *Camptonectes*, 450
Laminaria, 22–30, 166, 168
lammelata, *Stictoporella*, 425
lamourouxii, *Coscinopleura*, 194
landsborovi, *Smittina*, 214
Laomedea, 205
Larvae, 66, 69, 108, 112, 160–162, 166–170, 182, 186, 189, 275, 296, 297, 343, 443
Larval dispersal, 47, 50, 127, 166
Larval settlement, 66–69, 108–113, 115, 159–162, 166–171, 182, 186, 189
Larval types, 594, 596
lata, *Hippopodinella*, 143
Laterocavea, 498, 501
latifrons, *Emballotheca*, 213
latilaminata, *Melicerita*, 370
laxata, *Pyripora*, 467, 469
lebedensis, *Rhinidictya*, 427
Leioclema, 257, 350
Leiosoecia, 497
Leiosoeciidae, 497
lendenfeldi, *Hyalinella*, 506, 544
lendenfeldi, *Lophopus*, 543, 544
lepidodendroides, *Rhombopora*, 228
Lepralia, 417
Leptaulata n. subcl., 554–562
leptoclini, *Loxosomella*, 263
Leptotrypella, 7, 242, 434
lesueuri, *Thyracella*, 194
levinseni, *Smittipora*, 65, 66

Lichenalia, 378, 381
lichenoides, *Reteporidea*, 194
Lichenoporid, 232, 239
Lichenopora, 54, 55, 187–194, 235
Lichenoporidae, 357, 358, 475
Ligament attachments, 235–242
limanowskii, *Poricellaria*, 392, 400
linearis inarmata, *Schizomavella*, 155–157
linearis inarmata, *Schizoporella*, 213
lineata, *Callopora*, 24–28
Linograptus, 19, 20
Lithothamnia, 185, 196
Lithothamnion, 27
littorale, *Pycnogonum*, 201, 205
littoralis, *Umbonula*, 25–28
livingstonei, *Beania*, 370, 371
Lobograptus, 16, 17
longigemmis, *Hyalinella*, 542–545
longigemmis, *Plumatella*. 506
longipes, *Achelia*, 200, 205
longirostrata, *Hippomonavella*, 214, 218
longispinus, *Rastrites*, 17
longissima, *Obelia*, 126
Lopholepis, 194
Lophophorates, 593
Lophophore, 65–70, 234–243, 276–285, 304, 320, 523–527, 566–569, 594
Lophophore retractor, 276
Lophopoda, 558
Lophopodella, 505–518, 523–527, 529–534
Lophopodidae, 503–517
Lophopus, 505, 507, 529, 543, 544
Loxocalyx, 263, 553
Loxosoma, 553
Loxosomatina n. subord., 553–558
Loxosomella, 263
Lumen pores, 362, 364
Lunulites, 287
Lunulitiform, 131, 190
luteum, *Nuphar*, 540

M

macilenta, *Idmidronea*, 190, 194
Macroacanthopore, 7
Macrocystis, 29, 344
macropora, *Crisina*, 498
macroura, *Stomatopora*, 309, 310
Maculae, 2–9
maculata, *Dekayia*, 2
maculata, *Reticrisina*, 189, 193
maculosa, *Scrupocellaria*, 141
maderensis, *Scrupocellaria*, 141
madreporacca, *Mecynoecia*, 194
magna, *Heteropora*, 476, 477
magna, *Pyripora*, n. sp., 463–467
magnifica, *Pectinatella*, 505–518
magnireversa, *Exidmonea*, 301
major, *Mucronella*, 214
malayensis, *Hislopia*, 507
Male zooids, 11
malusi, *Fenestrulina*, 213
manica, *Enantiosula*, 214
mantchurikensis, *Stellipora*, 424
marcusi, *Plumatella*, 506
Margaretta, 143
marginata, *Hemieschara*, 417
marginatus, *Siphoniotyphlus*, 501
marlieri, *Plumatella*, 506
marsupialis, *Parasmittiana*, 493
martis, *Polypora*, 74, 75
Mathematical analysis, 98–102, 109–110, 166–170, 431
Mawsonia, 370–372
Mecynoecia, 194
Meganthopora, 75, 76, 227, 228
Meganthopore, 228
Melicerita, 370, 372
Melobesia, 59
membranacea, *Membranipora*, 29, 31, 175–183
Membranipora, 29, 31, 42, 56, 70, 141–150, 155–159, 175–183, 187, 344, 353, 418, 447, 553, 572, 585, 597
Membraniporella, 24–27
Membraniporiform, 54–63, 132, 187–193
Membranoptera, 30
menardi, *Globorotalia*, 139
Menipea, 370, 372
Mesoderm, 2, 595
Mesopores, 257, 357–359, 431, 434, 501
Mesotrypa, 425, 427

Mesotrypidae, 8
Mesozooecia, 2–9
Metamorphosis, 66–69, 297
Metasicula, 15, 16
Metrarabdotos, 65–70, 143
mexicana, *Lagenipora*, 215
mexicana, *Uscia*, 214
michaelseni, *Hemiseptella*, 370, 372
Microacanthopores, 7, 228
Micropora, 55, 88
Microporella, 25, 26, 42, 55, 118–126, 213, 577
Microporellidae, 213
Microporina, 83, 84
microvilli, 261–266, 318, 336–339
miliaris, *Psammechinus*, 31
minax, *Exchara*, 417
minuta, *Hyalinella*, 506, 515, 529–535, 544
minuta, *Pyripora*, n. sp., 463–471
Minutopores, 357
mirabilis, *Cryptopora*, 433
Mitoclema, 248, 255
Mollia, 141
Mollusca, 28–32, 59, 62, 80, 113, 189, 344, 449
moniliformis, *Hislopia*, 507, 529, 536
moniliorata, *Cellaria*, 370
Monograptus, 12–20
monostachys, *Electra*, 42, 162
Monotrypa, 378–380, 422, 426, 433–435
monstriosa, *Heterocella*, 402
Monticules, 4, 257–259
Monticulipora, 425, 427, 435
Monticuliporid, 232–241
Monticuliporidae, 8, 425
Monticuliporoidea, 562
Morphotypes, 578–581
mosae, *Ditrupa*, 190
mucedo, *Cristatella*, 505–515, 525
mucronata, *Belemnitella*, 466
Mucronella, 214, 553
mulleri, *Escharifora*, 194
Multicavea, 498
Multicrisina, 498
multifora, *Fimbriapora*, 424
Multilamellar growth, 187, 212
Multiramous dichograptids, 17
Multisonopora, 498
multitabulata, *Hallopora*, 427
murengolensis, *Fenestella*, 328
muricata, *Onchidoris*, 32
Muscle scars, 449–453
mutabilis, *Batostoma*, 424
mutabilis, *Wilbertopora*, 447, 458
Myriozoidae, 215
Myriozoum, 87–92, 215
Mytilis, 30, 32, 344

N

Nannozooids, 5, 332
Negative "skewness", 54, 59
neglecta, *Eschara*, 418
negrisi, *Poricellaria*, 392
Nellia, 141, 403
Nematopora, 75
neocomensis, *Membranipora*, 447
Neofungella, 235, 240
neozelanica, *Heteropora*, 476
neritina, *Bugula*, 127, 155–161
Nervous system, 317–323, 336, 339, 340
Nicholsonella, 423, 424, 427
nigra, *Pachycleithonia*, 214
Nikiforopora, 74, 75
Nikiforovella, 74
Nipponostenopora, 76
nitescens, *Hippoporella*, 214
nitida, *Membraniporella*, 24–27
nitida, *Parasmittina*, 42, 210, 214, 577, 578, 581–583
nobilis, *Fenestella*, 379–381
Nolella, 155–159
Notoplites, 370–372
Numerical analysis, 431, 549
Nuphar, 540
nutans, *Pedicellina*, 335
Nutrient transport, 210
Nymphon, 200, 205
Nymphonidae, 200
nysti, *Onychocella*, 194

O

Obelia, 29, 126
obliqua, *Melicerita*, 370–372
obscura, *Emballotheca*, 213
obstructa, *Parasmittina*, 488
oculata, *Eschara*, 417
oculata, *Hippodiplosia*, 418
oculata, *Nellia*, 141
oehlerti, *Monotrypa*, 378–380
okusensis, *Kerberites*, 455, 456
oligopus, *Robertsonidra*, 215
omniba, *Stictoporella*, 425–427
Onchidoris, 32
Ontogeny, 66–71, 365
Onychocella, 54, 66, 70, 187–196
Onychocellaria, 193
Onychocellids, 385
operculata, *Sertularia*, 28
Operculum, 365, 390, 449
Ophiopholis, 26
Ophiothrix, 26
Ophioroidea, 26
Oral budding, 565, 573
Oral shield, 362–366
orbisperma, *Hyalinella*, 506–508, 515, 545, 546
ornata, *Corbulipora*, 361, 365
Orthograptus, 15
Orthopora, 225–227
osburni, *Hyalinella*, 545
osburni, *Plumatella*, 544, 545
Osculipora, 194
osterfeldi, *Stomatopora*, n. sp., 311, 313
Overgrowth, 173–183
Ovicells, 47, 254, 255, 448, 485, 490–494, 497, 501
owenii, *Cupuladria*, 141
Oxygen ratios, 85–89

P

Pachycleithonia, 214
Pachydictya, 427
Pachyegis, 215
pacifica, *Heteropora*, 476, 477
Palaeoecology, 53, 79, 129, 148–150, 185, 411, 455, 456
Palaeosalinity, 79, 117
Palaeotemperature, 79–87
Paleotubuliporina, n. subord., 255
pallasiana, *Cryptosula*, 160, 210, 214, 275, 344
palmata, *Rhodymenia*, 30
Paludicella, 503–518, 553, 566
Paludicellea, 565
Paludicellidae, 507
pannosa, *Fenestella*, 378, 381
papuensis, *Nolella*, 155–159
Paracellaria, 370, 372
parallela, *Dyscritella*, 7
Paramawsonia, 370, 371
Paramuricea, 96
parasiticum, *Alcyonidium*, 573
Parasmittina, 42, 175, 183, 210, 214, 486–493, 577–583
Parietal network, 317
parsevalii, *Parasmittina*, 486, 488
parvicapitata, *Hippomonavella*, 214, 218
parvicauda, *Pyripora*, 463, 466
Pasythea, 143–150
patellaria, *Mollia*, 141
patens, *Porella*, 214
patina, *Berenicea*, 173, 175
patina, *Diastopora*, 25, 26
patina, *Lichenalia*, 378, 381
pavida, *Victorella*, 155–162
Pectinatella, 505–518, 525, 526, 529, 543, 553
pectinatelliformis, *Lophopodella*, 507
Pectinatellita, n. infraord., 555–557
Pectocaulus, 12, 13
peculiara, *Nicholsonella*, 423
Pedicellina, 261, 263, 335, 553
Pedicellinida, 554–557
Pedicellinidae, 261
Pedicellinina, 553–558
pelliculata, *Heteropora*, 476–479
Pelma, 362
Pelmatidium, 362, 364
Penetrance, 16
pennata, *Corbulipora*, 361

Penniretepora, 75
pentagonalis, *Paludicella*, 507
pentagonalis, *Rhombotrypella*, 7
Perenopora, 239
Periderm, 14, 15
Periostracum, 350–353
Peritoneum, 231
Permopora, 7
pesanseris, *Escharina*, 143
Petalopora, 194, 497, 498
Petaloporidae, 497, 498
petiolatus, *Anoplodactylus*, 200
petiolata, *Udotea*, 96
Petraliidae, 215
petropolitana, *Diplotrypa*, 422, 426
Phaenopora, 425
Phaenoporella, 424–427
Phaenoporinidae, 426
Pharynx, 234
Phenotypes, 578
Phidolopora, 215
philippinensis, *Afrindella*, 506
Phoronida, 593
Phoronidea, 567, 574
Phoronis, 335, 595
Phototropism, 66, 69, 127, 160, 189
Phoxichilidiidae, 200
Phoxichilidium, 200
Phoxichilus, 200
Phylactolaemata, 275, 325, 503, 506, 523, 529–536, 554–558, 565–573
Phylogeny, 549, 565
pilosa, *Acanthodoris*, 32
pilosa, *Electra*, 24–31, 167, 168, 175–181, 200, 317
pilosa, *Membranipora*, 585
piriformis, *Onychocella*, 196
placoides, *Echinella*, 505, 507
Plagioecia, 194
Platyhelminthes, 337
pluma, *Fenestella*, 378, 381
Plumatella, 503–519, 529–532, 539–546, 553, 566, 574, 586
Plumatellida, 554–559
Plumatellidae, 506, 517
Plumatellina, 554–559
Plumatellita, 553–559
Pogonophora, 337
poissoni, *Crepidacantha*, 215
polaris, *Nicholsonella*, 424
Polycera, 32
polymorpha, *Cornucopina*, 370, 371
polymorpha, *Heterocella*, 390, 402
Polymorphism, 1–11, 95, 100, 327–333, 385–408
polyoum, *Alcyonidium*, 29, 32, 177–181, 276
Polypides, 1, 2, 211, 234, 244, 275–279, 317–320, 356, 389, 453, 523–527, 566
Polyploidy, 586
Polypora, 74, 75, 330, 333, 378, 380, 553
polyschides, *Saccorhiza*, 29
Population analysis, 37–43, 111
Population genetics, 585
porcellana, *Hippoporina*, 214
porcellanum, *Cleidochasma*, 143
Pore cells, 261, 265
Pore chambers, 448–458
Pore organs, 261
Pore plates, 183, 210
Porella, 42, 43, 89, 214, 485
Pores. 209–219, 251–255, 261, 303, 330, 332, 349, 352, 438–442
Poricellaria, 385–407
Porifera, 26, 32, 95, 411
porifera, *Porella*, 214
Porina, 193, 194
porosa, *Escharina*, 143
porosa, *Hemieschara*, 417
porosa, *Hippoporina*, 42
portlandica, *Aptyxiella*, 450
portlandensis, *Pyriporopsis*, n. sp., 448–456
posidoniae, *Electra*, 318
posthumus, *Linograptus*, 19
Pottsiella, 507
praecox, *Leptotrypella*, 242
praelonga, *Cheiloporina*, 214
Prasopora, 241, 433, 533
Prasoporina, 433
Predators, 31, 133, 199
Preoral lobe, 13
Principle cells, 261, 265

Principle components analysis, 385–407
proboscidea, Entalophora, 137, 411–413
Proboscina, 187
procera, Ceidmonea, 193
procera, Vincularia, 193
procumbens, Celleporina, 215
progenitor, Cucullograptus, 17
projecta, Cribrilina, 370
prolifica, Smittoidea, 214
propinqua, Onychocella, 194
propria, Trematopora, 427
Prosicula, 15
Prostomia, 593, 594
Prothecal segment, 13
Protocheilostomata, 570, 572
Protoctenostomata, 570, 572
Protonephridia, 268, 598
Protrusion of lophophore, 66
Psammechinus, 31
Pseudoidmoneidae, 475
pseudopetropolitina, Diplotrypa, 2
Pseudopores, 70, 209
Pseudopuncta, 349
Pterobranchia, 11, 13, 574
Ptilodictya, 433, 553
Ptilodictyita, 553–562
Ptilodictyoidea, 562
Ptilopora, 378, 381
pumicosa, Cellepora, 25–27
pumila, Dynamena, 200
punctata, Cribrilina, 42, 47
punctata, Hyalinella, 506, 508, 525, 529, 539–544
punctata var. *densa, Hyalinella*, 533
punctata, Lagenipora, 215
punctata, Plumatella, 539–545
punctata var. *longigemmis, Plumatella*, 544
Pustulopora, 190, 194, 248, 255
pustulosa, Cavaria, 192
pustulosa, Pustulopora, 190, 194
Pycnodonte, 190
Pycnogonida, 199–205
Pycnogonium, 201, 205
pygmaeus, Anoplodactylus, 200
Pylorus, 234
pyrifera, Macrocystis, 29
Pyriform vesicles, 431
Pyripora, 47, 189, 192, 447–458, 463–471, 567, 590
Pyriporopsis, n. gen., 448–458
Pyxibryozoa, n. supercl., 554–558

Q

Quadricellaria, 192
quadrilineata, Polycera, 32
quadrispinosa, Chaperia, 370

R

radians, Lopholepis, 194
radiata, Cribrilaria, 55, 143
radiata, Cribrilina, 54, 55
radiatus, Hempneustes, 190
Radicells, 193
raigii, Parasmittina, 489, 493
ramosa, Cavarinella, 194
ramosa, Diplotrypa, 422, 426
ramosa, Monticulipora, 427
ramosoformis, Dianulites, 427
ramulosus, Clonograptus, 18
Rastrites, 16, 17
rationensis, Poricellaria, 392, 399
ray-lankesteri, Arachnoidia, 507
rectangulata, Rhombotrypella, 75
reduplicata, Porella, 42, 43
repens, Farrella, 565
repens, Plumatella, 503–506, 515–518, 539, 540, 543–545
repens minuta, Plumatella, 544
repens osburni, Plumatella, 545
repens, Truncatulipora, 194
Reproduction, 275
Reptadeonella, 54, 143
reptans, Scrupocellaria, 28, 165, 168, 200
Resorption foramen, 15, 16
Retecava, 194
Retepora, 378–380
Reteporidae, 194, 356, 498
Reteporiform, 54–63, 131, 193
Retecrisina, 189, 193
Reticular plexus, 317–326

reticulata, Lichenopora, 194
reticulum, Conopeum, 155–157, 162, 572
retiformis, Camptoplites, 370, 371
Retractor muscle, 234
Rhabdomesidae, 221–228
Rhabdomesita, 553–556, 562
Rhabdomesoidea, 562
Rhabdomeson, 74, 553
Rhabdomesonata, 74, 562
Rhabdopleura, 12, 13
Rhabdopora, 553
Rhabdosome, 13–20
Rhammatopora, 447
Rhamphostomella, 57, 214, 485
Rhinidictya, 423–427
Rhodnius, 283
Rhodymenia, 30
rhombea, Onychocellaria, 193
rhombium, Semicoscinium, 355
Rhombopora, 222–228, 553
Rhombotrypella, 7, 74, 75
Rhynchozoon, 215
rigidus, Clonograptus, 18
rigidus, Cyrtograptus, 19
robertsonae, Celleporina, 215
Robertsonidra, 215
robusta, Tremogasterina, 215
Rods, 272–274
rosacea, Cycloperiella, 215
rostratum, Rhynchozoon, 215
rostrigera, Hippaliosina, 214
Rotifera, 337
ruandensis, Plumatella, 506
rubrum, Corallium, 96
rubrum, Nymphon, 200, 205
rudis, Schizomavella, 55
runcinatus, Diversograptus, 19
rylandi, Bugula, 125, 126

S

saccata, Cysticella, 88, 89
Saccorhiza, 29
Saffordotaxis, 222–228
Sagenella, 248–255
salicornioides normani, Cellaria, 141
Salinity, 50, 79, 117–125, 137, 154–162, 263, 583
sanguinea, Delesseria, 30
santacruzana, Hippothoa, n. sp., 438–445
savartii, Membranipora, 155–159
Savignyella, 155–159
scanicus scanicus, Cucullograptus, 17
scanicus amphirostris, Cucullograptus, 17
scanicus parascanicus, Cucullograptus, 17
schindewolfi, Stolonicella, 189, 192
Schizomavella, 55, 155–157, 213
Schizoporella, 25, 26, 42, 55, 58, 143, 155, 156, 210–213, 553, 577, 587–89
Schizoporellidae, 213
Schizostoma, 418, 482
scindulata, Stamenocella, 193
Sclerotized stolons, 13
Scruparia, 567, 569
Scrupocellaria, 28, 54–58, 141, 165, 168, 200
scruposa, Scrupocellaria, 28
Seasonal distribution, 42–46, 120–125
Seasonal settlement, 115
Seasonal skeleton deposition, 92
Secondary spine, 362–366
Sediments, 27–30, 53–63
selwyni, Prasoporina, 433
semiaperta, Membranipora, 418
Semicoscinium, 350–355, 378, 381
Sensory pits, 261
Septa, 307–314, 326, 355, 356
Septopora, 75, 332, 333
Septulae, 210, 211, 457
Serialaria, 200
seriatopora, Filisparsa, 418
serratus, Fucus, 22–30, 107, 112, 167, 169, 176–181, 335
sertata, Dakaria, 213
Sertella, 54–58
Sertellidae, 212, 215
Sertularia, 28
setifera, Taeniocellaria, 192
setigera, Aeverrilla, 155–159
setigera, Crepidacantha, 215
Settlement, 65, 115, 159, 160, 190–192
seurati, Conopeum, 571

sexangularis, Onychocella, 194
Sexual maturation, 100
Sexual precocity, 100
Sexual reproduction, 4, 9, 15, 37–50
Shore zonation, 28
Sicula, 14–20
simplex, Amsassipora, 424, 426
simplex, Bugula, 589
simplex, Cucullograptus, 17
sioli, Plumatella, 506
Siphoniotyphlidae, 497, 501
Siphoniotyphlus, 501, 502
Skeletal composition, 79
Skeletal membrane, 209
Skeletal rods, 357
Skeletal tissues, 271–274, 349
smittiella, Smittina, 214
Smittina, 214
Smittinidae, 214, 485
Smittipora, 65–70
Smittoidea, 214
socialis, Lagenipora, 215
solida, Amastigia, 370
Sparsicavea, 498
sparsum Diploclema, 248
Spat, 30
spathulata, Parasmittina, 486
spathulifera, Smittina, 214
Spatial competition, 75, 166, 173, 567
Spatial distribution patterns, 165
spatulata, Figularia, 370
Species lists, 392, 426, 427
Spermatozoa, 165
Sphincter ligament, 234–239
Spines, 179, 226, 261, 302–306, 324, 442, 448–453, 490, 493, 572, 573
spinifera, Escharina, 25–27
spinosus, Endeis, 200
spinulosa, Lagenipora, 215
Spiropora, 194, 248, 255, 411–416
Spirorbis, 108, 112, 168–170, 182
spirorbis, Spirorbis, 169
spongioseformis, Ceramopora, 424, 426
Stamenocella, 193
Statistical analysis, 97–103, 110, 112, 431
Statoblasts, 515–519, 524, 531–535, 539
Steganoporella, 65, 70, 141, 385–393
Steghorneridae, 475
Steginoporella, 350
stellata, Gigartina, 205
stellata, Triporula, 143
stellatum, Ascodictyon, 431
Stellipora, 423–427
Stelmatopoda, 559
Stenodiscus, 74, 75
Stenolaemata, 248, 349–357, 475, 476, 554–561, 565–570
Stenophragmidium, 74, 75
Stenoporidae, 7, 9
Stephanella, 506, 515, 545
Stephanosella, 214
Stictoporella, 425, 427
stigmatophora, Onychocella, 194
Stipe, 18–20
stipitata, Eschara, 417
Stolella, 506–516, 529–535, 542
Stolon system, 11–16
Stolonate, 43
Stolonicella, 189, 192
Stolonifera, 325
stolonifera, Bugula, 125, 126, 589
Stoloniferita, 553–560
Stolothecae, 11–14
Stomach, 234
Stomachaetosellidae, 215
Stomatopora, 55, 187–190, 248, 255, 307–314
Storage areas, 294
Storage and retrieval techniques, 601
Stratigraphical indicators, 73–77
Streblotrypa, 74
striata, Beisselina, 194
stuhlmanni, Lophopodella, 507
Stylopoma, 65–71, 213
subgracile, Myriozoum, 88
subnodosa, Hallopora, 425, 427
subovoidea, Watersipora, 155–161
Substrate interface, 343
Substrata, 20, 28–34, 66, 190–194
subsymmetrica, Vincularia, 402
subtorquata, Watersipora, 214
sulcifera, Hippodiplosia, 418

sulcifera, *Lepralia*, 417
Sulcocava, 497, 501
Sulcocavidae, 498
Sulcoretepora, 75, 553
sultana, *Fredericella*, 503–519, 529–535
Sun exposure, 115–127
suprajurensis, *Camptonectes*, 450
surcularis, *Celleporina*, 83–88, 215

T

Tabulata, 75
Tabulipora, 74, 75, 553
Taenicellaria, 192
tanganyikae, *Afrindella*, 506
tanganyikae, *Plumatella*, 529, 540, 543
Temperature, 47, 79, 117–125, 154, 162, 263, 457
tenellus callavei, *Clonograptus*, 18
tenellus tenellus, *Clonograptus*, 18
Tentacle extrusion, 239–245, 337–339, 365
Tentacle sheath, 66, 234–244, 317–326
Tentacles, 234–243, 341–355, 365, 567
Tentaculata, 566, 567
tenuata, *Amsassipora*, 427
tenuirostre, *Copidozoum*, 141
tenuis, *Flustra*, 370
tenuis, *Membranipora*, 42
tenuissimum, *Conopeum*, 118–126
teretiusculus, *Glyptograptus*, 422
Tergopores, 501
Terminal diaphragm, 299
Terminal dominance, 567–572
Terminal zooid, 12
Tessaradoma, 55–57
Tetrahymena, 339
tetrastichus, *Dimorphostylus*, 193
texana, *Pyripora*, 447, 458, 466
Thalamoporella, 141, 393, 493
Thalassinoides, 187
Thalassocharis, 190, 192
thomasi, *Lophopodella*, 507
thureaui, *Goniograptus*, 18
Thyracella, 194
Tidal effect, 117, 126
tincta, *Antropora*, 212
titan, *Titanites*, 455, 456
Titanites, 455, 456
toanensis, *Gelatinella*, 507, 515
toanensis, *Hyalinella*, 545
Tokophrya, 283
tomentosa, *Jorunna*, 32
torquata, *Coelocochlea*, 194
torrubliae, *Fistulipora*, 379, 380
transenna-mesofenestralia, *Phaenoporella*, 424
transilvanica, *Cumulipora*, 418
Transport and dispersal, 515–519
Trematoecia, 215
Trematopora, 427
Trematoporidae, 8, 9, 425
Tremogasterina, 215
Trepostomata, 1–9, 73–75, 231, 253, 258, 350–357, 376, 380, 423, 475, 476, 562, 553–562
Tretocycloecia, 497
Tretocycloeciidae, 497
triangularis extremus, *Monograptus*, 17
triangularis seperatus, *Monograptus*, 17
triangularis triangularis, *Monograptus*, 17
tricornis, *Camptoplites*, 370, 372
Trigonia, 450
Trinchesia, 32
Triphyllozoon, 89, 90
Triporula, 143
trispinosa, *Parasmittina*, 175, 183, 214, 577
Trochophore larvae, 594, 596
truncata, *Osculipora*, 194
Truncatulipora, 194
Trypostega, 143, 213, 332
tuberculata, *Membranipora*, 141, 145, 344
tubiporacea, *Inversaria*, 193
Tubucellaria, 54, 58
tubulifera, *Cribrilina*, 366
tubulifera, *Filisparsa*, 193
Tubulipora, 553
Tubuliporina, 255, 497
Tubuliporinid, 299
Tubulobryozoa, n. supercl., 554–561
tulipifera, *Pasythea*, 143, 145
tuna, *Halimeda*, 96
Turbidity, 21, 26, 115–127, 162

turbinata, *Bugula*, 200, 205
Turbulence, 27, 75, 115–127, 132, 162, 166, 176, 186–196, 324, 369, 411, 413
turgenscens, *Lepralia*, 417
turrita, *Bugula*, 43, 589, 591
Type specimens, 417, 429

U

Udotea, 96
ulrichi, *Heterotrypa*, 434
Ultrastructure, 14, 221, 320, 335
umbellata, *Discoporella*, 297
Umbonula, 25–28, 215
Umbonulidae, 215
undata, *Arthyris*, 376
undulata, *Monotrypa*, 433–435
unguiculatum cookae, *Metrarabdotos*, 65–68, 143
Uniavicularium, 143
unicornis, *Schizoporella*, 25, 26, 42, 158, 587
unicornis floridana, *Schizoporella*, 210–213
Uniserial stipe, 17
Urnatella, 268
Uscia, 214
Utropora, 381
uva, *Valkeria*, 25, 26
uvulifera, *Aimulosia*, 214

V

vaihiriae, *Hyalinella*, 506, 517, 543, 545
Valkeria, 25, 26
vancouverensis, *Phoronis*, 595
vanhoffeni, *Notoplites*, 370
varians, *Batostoma*, 425, 427
Variation, 75, 95, 371, 411
vaupeli, *Nicholsonella*, 425, 427
Veleroa, 88, 214
veleronis, *Veleroa*, 314
ventricosa, *Celleporina*, 215
ventricosa, *Escharella*, 55
Venus, 59, 62
venusta, *Trypostega*, 143, 213
vermiformis, *Euginoma*, 139, 141
vernoni, *Poricellaria*, 392
verrilli, *Hippoporina*, 42, 47
Verruca, 31
Vertical distribution, 120, 121, 132, 142, 143, 153–163, 445, 476
verticillata, *Spiropora*, 194, 411–416
Vesicularia, 553
vesicularis, *Hyalinella*, 545
vesicularis, *Plumatella*, 539–545
vesiculosa, *Stellipora*, 423–427
Vestibule, 234–243
vibraculifera, *Microporella*, 213
Vibraculum 1, 288–295, 352
vicksburgica, *Vincularia*, 402, 406
Victorella, 155–162, 566
villosa, *Membranipora*, 344
Vincularia, 193, 385–407
Vinculariiform, 54–63, 131–137, 193
violacea, *Reptadeonella*, 54
Virgella spine, 15
virginica, *Crassostrea*, 28
Virgocella, 193
viridis, *Trinchesia*, 32
vitrea, *Hyalinella*, 545
vitrea, *Plumatella*, 541, 545
vitrea, *Stephanosella*, 214
vitrimuralis, *Cellaria*, 370
Volnovachia, 73, 75
vorstmani, *Hyalinella*, 546
vorstmani, *Plumatella*, 506, 515, 546
vulgaris, *Flustra*, 370

W

Wad, 209
Walkeria, 26
Wall structure, 5–8, 248–254
wandeli, *Paracellaria*, 370
Water currents, 21–28, 107, 478
Watersipora, 155–162, 210–214
Watersiporidae, 214
Web structures, 14
wightensis, *Trigonia*, 450

Wilbertopora, 447, 458

X

X-rays, 527

Z

Zonation, 28, 29
Zoogeography, 42, 371, 503